Real Life Applications of Soft Computing

OTHER AUERBACH PUBLICATIONS

Cloud Computing: Technologies and Strategies of the Ubiquitous Data Center
Curtis Franklin Jr. and Brian J.S. Chee
ISBN 978-1-4398-0612-8

Cognitive Radio Networks: Architectures, Protocols, and Standards
Edited by Yan Zhang, Jun Zheng, and Hsiao-Hwa Chen
ISBN 978-1-4200-7775-9

Essential Project Management Skills
Kerry Wills
ISBN 978-1-4398-2716-1

Handbook of Public Information Systems, Third Edition
Edited by Christopher M Shea and G. David Garson
ISBN: 9781439807569

GIS in Hospital and Healthcare Emergency Management
Ric Skinner
ISBN 978-1-4398-2129-9

Healthcare Informatics: Improving Efficiency and Productivity
Edited by Stephan P. Kudyba
ISBN 978-1-4398-0978-5

HSDPA/HSUPA Handbook
Edited by Borko Furht and Syed A. Ahson
ISBN 978-1-4200-7863-3

Information Security Management: Concepts and Practice
Bel G. Raggad
ISBN 978-1-4200-7854-1

Information Security Risk Analysis, Third Edition
Thomas R. Peltier
ISBN 978-1-4398-3956-0

ITIL Release Management: A Hands-on Guide
Dave Howard
ISBN 978-1-4398-1558-8

Mobile Device Security: A Comprehensive Guide to Securing Your Information in a Moving World
Stephen Fried
ISBN 978-1-4398-2016-2

Orthogonal Frequency Division Multiple Access Fundamentals and Applications
Edited by Tao Jiang, Lingyang Song, and Yan Zhang
ISBN 978-1-4200-8824-3

Overlay Networks: Toward Information Networking
Sasu Tarkoma
ISBN 978-1-4398-1371-3

Process Improvement and CMMI® for Systems and Software
Ron S. Kenett and Emanuel Baker
ISBN 978-1-4200-6050-8

Project Management Tools and Techniques for Success
Christine B. Tayntor
ISBN 978-1-4398-1630-1

Real Life Applications of Soft Computing
Anupam Shukla, Ritu Tiwari, and Rahul Kala
ISBN 978-1-4398-2287-6

The Project Manager's Communication Toolkit
Shankar Jha
ISBN 978-1-4398-0995-2

Transmission Techniques for Emergent Multicast and Broadcast Systems
Mario Marques da Silva, Americo Correia, Rui Dinis, Nuno Suoto, and Joao Carlos Silva
ISBN 978-1-4398-1593-9

Underwater Acoustic Sensor Networks
Edited by Yang Xiao
ISBN 978-1-4200-6711-8

Wireless Sensor Networks: Principles and Practice
Fei Hu and Xiaojun Cao
ISBN 978-1-4200-9215-8

Real Life Applications of Soft Computing

Anupam Shukla
Ritu Tiwari
Rahul Kala

CRC Press is an imprint of the
Taylor & Francis Group, an **informa** business
AN AUERBACH BOOK

CRC Press
Taylor & Francis Group
6000 Broken Sound Parkway NW, Suite 300
Boca Raton, FL 33487-2742

First issued in paperback 2019

ISBN-13: 978-1-4398-2287-6 (hbk)
ISBN-13: 978-0-367-38401-2 (pbk)

Library of Congress Cataloging-in-Publication Data

Shukla, Anupam, 1965-
Real life applications of soft computing / Anupam Shukla, Ritu Tiwari, Rahul Kala.
p. cm.
Includes bibliographical references and index.
ISBN 978-1-4398-2287-6 (hardcover : alk. paper)
1. Soft computing. I. Tiwari, Ritu, 1977- II. Kala, Rahul. III. Title.

QA76.9.S63S56 2010
006.3--dc22 2010005445

Visit the Taylor & Francis Web site at
http://www.taylorandfrancis.com

and the CRC Press Web site at
http://www.crcpress.com

To my mother, Kanta Devi Shukla

Anupam Shukla

To my mother, Parwati Tiwari

Ritu Tiwari

To all young researchers round the world...

Rahul Kala

Contents

SECTION I Soft-Computing Concepts

SECTION II Soft Computing in Biosystems

SECTION IV Soft Computing *Implementation Issues*

APPENDICES

REFERENCES

Foreword

Soft Computing today may be portrayed as a massively dynamic concept whose pace of development presents a promising picture of the future world. The mammoth of literature that has been compiled in various forms not only showcases the current development, but also speaks about the massive scope of work that is to come. The constant and exponential growth of the field has been a source of great encouragement and inspiration for all of us to contribute tirelessly to make an intelligent tomorrow where anything that can be perceived is available in reality. At one end, we find very positive steps being taken by numerous researchers worldwide who are making significant contributions towards the theoretical domains and addressing various problems that erstwhile restricted the performance of the systems in general. The increasing use of evolutionary concepts, modularity in design, hardware integration, massively hybrid systems, self-adaptive, and dynamic and online systems are a part of the complete landscape that is immense, but beautiful. Not only does it attract people to explore it in depth, but also presents its boundless and never-ending characteristics. At the other end we find a lot of people contributing towards the application domain where they use soft computing as an effective problem solving tool that ultimately results in effective systems of practical use. The bulk of research over the domains of robotics, wireless networks, optical character recognition, etc. is very encouraging. Each of these fields today stands as a well-established discipline with a long history, issues, concepts and future, mostly involving a multi-disciplinary treatment which makes the whole task an even bigger challenge.

We have made a significant movement toward making a computationally more intelligent world where intelligence lies in all machines that surround us. We have many more challenges to conquer and many more issues to be uncovered. The past decade especially has not only seen positive developments in research labs worldwide, but also witnessed a lot of practical use. Now intelligent systems using soft computing principles are being packaged as the hardware or software in various applications of commercial use. The R&D labs of all major corporates worldwide are using a variety of soft computing tools to provide effective solutions to the ever-increasing human demands. At this juncture, there is a need to educate people about the principles, concepts, applications, issues and solutions to the use of soft computing. I am sure that this book will be great contribution toward the same and would further carve a deep mark in the soft computing literature for reference of enthusiasts for a very long time.

Real Life Applications of Soft Computing is a one-stop book for all knowledge related to soft computing, computational intelligence, machine intelligence, and other related areas. The book is of a special interest to me due to the manner in which it has been conceptualized and ultimately penned down. The authors cover all the theoretical foundations with a practical approach and later present a variety of applications that further help in understanding soft computing systems. The scientific approach with discussion of commonly faced problems would certainly help readers build systems rather than just reading for pleasure. From our teaching experience, we also realize that it is always good to have a practical component attached to the courses whose performance ultimately reflects the overall understanding of the students. Such courses without practical components are like machines without electricity. Making students perform the practical is a much more difficult task due to the various issues that are not commonly found in what they read. Ample coverage of applications, issues and perspectives, makes the book rich and diverse. This book is a significant movement in this direction.

This book will deeply impact the intellect of those who own it, or can access it in their libraries. It is the kind of work that will be consulted often, providing enduring value to those who read it.

I hope that each section, chapter and unit of this book turns out to be thought provoking for the readers and helps them understand, appreciate, implement and contribute to the domain. I herby congratulate the authors for the book and wish the readers a happy reading and lifelong learning.

Prof. Amit Konar
Professor, Electronics and Telecommunications
Jadavpur University
Jadavpur, India

Preface

Soft computing has been a major field of development in the previous couple of decades. The field attracts the attention of a large number of people. The highly multidisciplinary nature of the field further results in a large number of students studying soft computing. The various developments in application areas of soft computing in the previous few years have shown a very large potential for the development of highly-scalable systems using the soft computing tools and techniques.

The present literature beautifully covers the theoretic concepts, but the practical applications might still be very difficult for students to comprehend and implement. Many students find it difficult to visualize the application of soft computing from a theoretical text alone. This book not only aids them in understanding these systems, but also exposes them to the current developments over various fields. This empowers them to pursue research in these areas and contribute towards the research and development at large. The basic purpose of this book is to expose the real life applications and systems of soft computing to the readers. This will enable them to get an in-depth, practical exposure to the methodology of the application of the soft computing approaches in various situations. This book also discusses the various issues and difficulties involved in the process.

The book can also be used as a standard text or reference book for courses related to soft computing, pattern matching, artificial neural networks, etc. The book may be rated as an "elementary" to "advanced" book that introduces the concepts of soft computing to a good extent and then advances to their real life practical examples. Every care has been given to nicely explain all the theoretical concepts, algorithms, applications, tools, and techniques. We have tried to ensure that readers using this as the first text of soft computing are also able to understand each and every line of book.

The book incorporates some of the recent developments over various areas. Many other models have been proposed in the recent years that still exist in theoretical studies and elementary experimentations. These models have not been discussed in the book. Also, due to limited space we have not been able to cover many new exciting areas where soft computing finds applications, such as weather monitoring and music compositions. We would be looking forward to include these in the future editions.

SALIENT FEATURES

- In-depth practical exposure to real life systems
- Discussion on various problems and issues in application of soft computing tools and techniques
- Updated information on developments in various real life application fields
- Multidimensional coverage of the concepts
- Detailed description of the problem and solution for easy reproduction
- Scope for future work and active areas of research presented at various places

WHO SHOULD READ THIS BOOK

Students of Universities as a Text or Reference for Soft-Computing Courses:

This is meant to be a good text or reference book for all the courses related to soft computing, pattern matching, machine learning, etc.

Young Researchers for BSEE/MSEE/PhD Thesis

The book will give them a good insight into the theoretical and application areas and would be of a great help to them to themselves develop the systems.

Researchers

A good coverage of latest technologies would give all researchers a good insight into the developments of the discussed domains. Further, a practical approach would help them easily experiment and reproduce the results.

ORIGIN OF THE BOOK

The book is a result of the research done over the time by the authors. Much of the work presented is done by the authors themselves. Databases to the various problems have either been self-generated by the authors or used from the public database repositories. At some places the authors have incorporated a comprehensive discussion of the problems and the solutions present in the literature. This work was largely possible as a result of the efforts put in by various students of Indian Institute of Information Technology and Management Gwalior in forms of projects: BTech dissertations, MTech dissertations, and PhD thesis.

Many topics are the result of the current research being pursued by the authors. This makes the book state of the art with respect to coverage. Newer advances into the application areas of soft computing have been incorporated. Many of the systems presented are still in the experimental phase and have not yet been implemented by the industry. Many others are being deployed at the industry.

The book has been made possible as a result of the course of artificial intelligence and soft computing being taken by the authors at Indian Institute of Information Technology and Management Gwalior besides other universities. The feedback provided by the students towards the course and their motivation to learn beyond the curriculum have helped a lot in improving the quality of the book.

MATLAB® is a registered trademark of The Math Works, Inc. For product information, please contact:

The Math Works, Inc.
3 Apple Hill Drive
Natick, MA
Tel: 508-647-7000
Fax: 508-647-7001
E-mail: info@mathworks.com
Web: http://www.mathworks.com

Acknowledgments

This book is our first step into the world of authored books. It all started as a dream to use our past research as a base to create a big landmark into the research horizon of the world. It is natural that the dream required a lot of encouragement, guidance, help, and support that was provided by numerous people in various phases of the book. Before the book unveils its multidisciplinary contents, it would be wise to acknowledge the people who made this possible.

The authors wish to express their sincere gratitude to Dr. M.N. Buch, chairman, Board of Governors, ABV-IIITM Gwalior whose visionary ideas made it possible for us to think big and formulate the book. The authors further wish to thank Prof. S.G. Deshmukh, director, ABV-IIITM Gwalior for extending all sorts of help and support during the course of writing of this book. The book would not have been initiated without his encouragement and support.

The authors would like to thank Prof. Amit Konar, Jadavpur University, for providing the inspiration for this book, as well as the foreword. His precious words have indeed added a greater charm to the book.

The authors also express their sincere regards to the anonymous reviewers of the various journals and conferences for their valuable suggestions over numerous pieces of work that oriented the authors and guided them in various ways.

The authors further express their thanks to Dr. A.S. Zadgaonkar, vice chancellor–Dr. C.V. Raman Technical University Bilaspur, for the guidance and motivation that he has bestowed during the entire course of work with the authors. The spirit to learn, comprehend, develop, and dream is an inspiration from him that made this book possible. The strong foundation to research inculcated by him largely led to the immense amount of work that has been presented in the book.

The authors express their thanks to Sourabh Runta, director technical, Rungta College of Enineering and Technology, Bhilai for his assistance in developing various codes for speech synthesis and speaker recognition, which forms an integral part of the book.

The book would have never been possible without the pioneering work carried out by the co-researchers associated with the authors. The authors also wish to thank the student fraternity of the Institute for taking deep interest into the courses titled "Artificial Intelligence" and "Soft Computing." Their interest to learn beyond the classroom program has been the key motivation for the book. The authors are highly indebted to all the students who undertook challenging problems as a part of their PhD, MTech or BTech thesis and came up with excellent solutions and experimental results. The authors wish to thank Dhirender Sharma, Chandra Prakash, Rishi Kumar Anand, Sourabh Sharma, Piyush Sharma, Hemant Kumar Meena, Prateek Agrawal, Mamta Agrawal, Rahul Dubey, Akhil Kumar Meshram, Chandra Prakash Meena, Prabhdeep Kaur, A. K. Mitra, R. R. Janghel, Anuj Kumar, Gaurav Agarwal, Anand Ranjan, Shivanshu Mittal, Harsh Vazirani, Anuj Kumar Goyal, Siddharth Arjaria, and Ram Nivas Giri for the quality results they produced during the course of work with the authors. The authors further wish to thank Kshitij Verma, Anuj Nandanwar, Anurag Agrawal, Sanjay Kumar, and Pritesh Tiwari for ensuring each and every stage of author submission completed on time. The help provided by them in editing of the manuscript was of a large help at a time when it was needed the most.

The first author would also like to thank his son Apurv Shukla for his support during the course of writing the book. The second author thanks her mother Parvati Tiwari for motivation and blessings in writing the book.

This book required an immense amount of work by numerous people around the globe. The draft was heavily worked over to ensure high quality. The authors wish to thank the entire staff of Taylor

& Francis, Auerbach Press, Glyph International, and all the associated units who worked over different phases of the book. The authors thank all the people who have been associated with and have contributed toward the book.

Above all, we all thank the Almighty for showering us with His constant blessings and love. Without Him, this project would not have been possible.

Anupam Shukla

Ritu Tiwari

Rahul Kala

Authors

Dr. Anupam Shukla is an associate professor in the IT Department of the Indian Institute of Information Technology and Management Gwalior. He has 22 years of teaching experience. His research interest includes speech processing, artificial intelligence, soft computing and bioinformatics. He has published over 120 papers in various national and international journals/conferences. He is referee for 10 international journals including Elsevier, IEEE, and ACM computing and in the editorial board of International Journal of AI and Soft Computing. He also received Gold Medal from Jadavpur University during his postgraduate studies.

Dr. Ritu Tiwari is an assistant professor in the IT Department of Indian Institute of Information Technology and Management Gwalior. Her field of research includes biometrics, artificial neural networks, signal processing, robotics and soft computing. She has published around 50 papers in various national and international journals/conferences. She has received Young Scientist Award from Chhattisgarh Council of Science & Technology in the year 2006. She also received Gold Medal in her post graduation from NIT, Raipur.

Rahul Kala is a student at the Indian Institute of Information Technology and Management Gwalior. His areas of research are hybrid soft computing, robotic planning, biometrics, artificial intelligence, and soft computing. He has published over 25 papers in various international and national journals/conferences. He also takes a keen interest toward free/open source software. He secured All India 8th position in Graduates Aptitude Test in Engineeging-2008 with a percentile of 99.84. Rahul is the winner of Lord of the Code Scholarship Contest organized by KReSIT, IIT Bombay and Red Hat. He also secured 7th position in ACM-International Collegiate Programming Contest Kanpur Regional 2007.

Section I

Soft-Computing Concepts

1 Introduction

1.1 SOFT COMPUTING

From childhood, we have been taught strict rules that we at first believed always hold true. We may have assumed that the world moves by these strict rules. If you ask a child to solve an arithmetic problem, that child can follow the arithmetic rules to find a 100 percent correct solution. But there is another paradigm of thought. In the real world, it is not always possible to be precise in finding answers or solutions to problems. In many situations, natural systems tend to do things that they believe are true rather than things that are compulsorily true. The foundations of defining strict procedures, rules, and methods hence fail if the underlying belief fails. But this gives us an exciting view of a world that is driven by belief, impreciseness, and approximations and that we all have observed numerable times in our daily decision making.

To illustrate this fact, let us take a very simple example. Suppose a bag contains 1,000 pieces of fruit, and each piece can be either an apple or an orange. You are asked to instruct a child to separate the fruits. You might tell the child that the orange ones are oranges and the red ones are apples. But you might then realize that every apple is not perfectly red. Some apples may have lost some of their color over time. So you then might try to explain to the child all the variants of red and orange. You might even try explaining the difference between the odors of the two fruits. You then realize that some apples are highly decolorized or have lost their characteristic odor, so you have already compromised the precision of the child's sorting. Now imagine that there are 100 types of fruits, rather than just two. You would have a great deal of trouble explaining the differences between all the fruits in such a way that the child could precisely separate them. If some fruits do not match the criterion of any fruit, the child might get stuck. This is where imprecision comes into play. In reality, however, any child can easily sort all fruits based on his or her past experiences, beliefs, and confidence.

The example reveals how real life is filled with uncertainty. We usually proceed through this world by accepting the most certain computed fact and regarding it as true. But we can never deny the fact that uncertainty always exists. For example, the next time you get ill and consult a doctor, ask your doctor whether he is fully sure of his diagnosis. After he looks at the methodology of treatment, he might reveal that he took many preventive measures to stop diseases that could occur. Regarding the cure, the inferences were the result of perception rather than a full mathematical proof.

A small look into all the systems might give you a fair idea of the notion being presented here. Uncertainty exists not only in the natural systems with which we are all familiar but also in almost all human-made systems that we encounter. In fact, many of the systems that you appreciate as a marvelous creation of humankind actually use principles of soft computing. This includes speaker recognition, handwriting recognition, face recognition, disease identification, and many more. So the next time you swipe your finger at your fingerprint-secure laptop, be aware that the system is using soft computing. Or the next time you select your hello tune using interactive voice response, know that it also uses soft-computing techniques. And in any system that uses soft-computing techniques, the presence of uncertainties and impreciseness is a given.

But let us see what and how much uncertainty there is. The uncertainty exists in the preciseness of the results. This means that if you ask a soft-computing expert system to get sugar from the market, it might end up getting salt. Or if you asked your system to find the value of 5 divided by 2, your system might give a long number whose value is almost that of the result. But our society loves perfection, so we definitely do not want our systems to be imperfect in any sense. So let us look at how much uncertainty we are talking about. When you go out of your home to a nearby shop, do you consider that you might actually be stabbed on the way or that there might be a bomb blast in which you might lose your life? Most of us are mature enough to believe that such events are so rare as to not occur, and thus we go out of our homes with full confidence. Another way to state this is we are sure that there is nothing unusual in the day, so such events would not take place. The same is the case with soft-computing systems, which have applications in almost every domain.

Another interesting fact that we observe in nature is its complexity. Even the simplest systems that you could imagine are actually very complex. This complexity sometimes reaches the extent that it might not be possible for even the fastest computers on Earth to simulate the system. If you tried to simulate such a system, you would realize the heavy computational load that is involved.

Even machine-controlled systems are not that far from such high computational loads. The well-known example of the traveling salesman problem (TSP) is one case in point. In this problem, n stations are given. A salesman has to start from one station, travel to all of them in some order, and then return to the starting station. The objective is to minimize the total traveling distance.

Many of such problems are very practical in day-to-day life. For example, imagine that you have designed a robot that is supposed to clean n number of places in your home, a problem that is similar to the TSP. This robot senses the surroundings and maps the n places. Then it computes the shortest cycle for implementation. If the number of places is large, the robot might actually end up calculating the shortest distance on one day and cleaning the house on another day. But suppose that the next day, the objects have been moved. This means that the robot must recalculate the shortest path. It can hence be observed that this is not the solution. Rather, the solution lies in finding the best possible path in a given span of time. This would again lead to approximations and uncertainties.

From this example, it is clear that all types of systems cannot run on precise decisions that are the result of hardbound rules. Instead, we need to enter the domain of soft computing. Here we find solutions to all the problems that the systems in our examples faced, with little loss of precision.

Before we continue our journey toward the versatile and unending field of soft computing and its applications, let us first formally define what soft computing is.

1.1.1 What Is Soft Computing?

According to Professor Lofti Zadeh, soft computing is "an emerging approach to computing, which parallels the remarkable ability of the human mind to reason and learn in an environment of uncertainty and imprecision" (Zadeh, 1992b) This definition clearly states the role of imprecision in effective decision making in a finite time.

Definition 1
Soft computing is an emerging approach to computing that parallels the remarkable ability of the human mind to reason and learn in an environment of uncertainty and imprecision.

Natural processes have always been the inspiration behind the design of soft-computing techniques. Soft-computing systems try to imitate, to some extent, the functionality given by natural processes—especially the human brain. It is clear that even if we succeed in automating a small fraction of the natural processes, the results would be enormous. Definition 1 clearly explains the use of uncertainty and imprecision in making systems more robust and scalable. The motivating factor of natural processes and learning from cognition plays a key role in making complex systems possible. This is the beauty of soft computing.

Another commonly used definition of soft computing given by Professor Zadeh states, "The guiding principle of soft computing is to exploit the tolerance for imprecision and uncertainty to achieve tractability, robustness, and low solution cost" (Zadeh, 1996).

Definition 2
The guiding principle of soft computing is to exploit the tolerance for imprecision and uncertainty to achieve tractability, robustness, and low solution cost. (Zadeh, 1994)

This definition clearly states that the processes can be very computationally expensive. In reality, it might not be possible for any system to solve the problem in a finite amount of time. Hence we require soft-computing techniques to give the best possible results in the smallest amount of time. Soft-computing techniques are well suited to handling data for these types of problems. The techniques give the best solutions from the entire pool of solutions possible per the time constraints.

These classes of problems are optimization problems, in which we generate solutions according to the problem and then try to optimize them. Optimization problems may be defined as "finding optimal values for parameters of an object or a system which minimizes an objective (cost) function" (Kasabov, 1998). In optimization problems, we try to generate solutions by varying the solution parameters. The aim is to maximize or minimize the value of the objective function that judges the requirement of the problem or the quality of the solution.

Definition 3
Optimization is about finding optimal values for parameters of an object or a system which minimizes an objective (cost) function (Kasabov, 1998)

Soft-computing techniques have a great deal of relevance in optimization problems, giving very good, time-efficient results. These techniques generate a diverse set of solutions that they can then optimize to return the most optimal solution.

Soft computing, hence, means computation using soft conditions—that is, computing in which errors are undesirable but acceptable to a certain degree. Soft computing sacrifices the preciseness of the solution, but in return it is able to solve problems that either were impossible to solve using traditional methods or were computationally too expensive to solve. Soft computing allows us to model problems that would have been impossible to model due to the complex nature of those problems. This complexity exists not only in modeling; rather, it extends to the domains of space and time. But soft computing optimizes both space and time to solve problems that were unsolvable using traditional means. Soft-computing techniques have thus become a valuable tool for people in a variety of domains and disciplines and are becoming extremely popular due to their high performance, ease of modeling, and ease of use. In the subsequent sections of this chapter, we will discuss these techniques and how they are used.

1.1.2 Soft Computing versus Hard Computing

Now let us divert our attention to hard computing. Unlike soft computing, hard computing deals with precise computation. The rules of hard computing are strict and binding. You are given a procedure to solve a problem. The inputs, outputs, and procedure are all clearly defined. Hence you apply the procedure to get the desired output. Because the rules are so well defined and the inputs so precise, it is possible to get precise answers without any degree of uncertainty. The same results will be returned, no matter how many times you perform the experiment, where you perform the experiment, how you perform the experiment, and who performs the experiment. The results only change if the procedure, underlying assumptions, or rules change. The beauty of such a system is its preciseness.

An example of hard computing is the arithmetic sum. Whenever the inputs are 1 and 2, the output is 3. We can be assured that the answer is correct without worrying about anything else. Most of the laws of physics, mathematics, and science have been modeled using the principles of hard computing. Whenever there is a modeling problem, we solve by taking approximations and assumptions that are believed to hold true. Hence, hard computing models everything as a set of predefined rules that always hold. If these laws were ever to fail, the result would be unimaginable.

Hard computing is thus better than soft computing in that it gives precise answers. But when using hard computing, we need to be aware of the following points, which introduce limitations in the working of hard computing, especially when real life data of enormous size is used.

- **Model complexity and assumptions:** Some problems are very difficult to model. Normally the approach for modeling a problem is to use a set of mathematical equations and variables that mathematically interpret the problem. These equations are then solved simultaneously, keeping the constraints constant to get the final solution. But it is not always possible to model a problem in this way due to its complexity. In such instances, we may opt to use approximations that do not affect the answer much and that are able to give the results. Although we would not make assumptions that would affect the solution much, many problems can still not be modeled due to their complex nature.

  As an example, imagine that you are supposed to solve the problem of character recognition. Here, the input is in the form of a grid on M rows and N columns. Your problem is to identify the character based on this input. Suppose you start making rules to identify each and every character. The number of rules needed to identify only 26 English characters is very large. If you had to create each rule, you would probably become highly confused and leave the work. This means the problem is too difficult for hard computing. Likewise, if we move into the domain of face recognition, the problem increases even more. Thus hard computing is not the method of choice for such complex systems.
- **Computationally expensive problems:** Suppose that you have been able to model the given problem. But then you try giving real life data that are too large in size or dimensionality. "Large size of data" refers to the data required by the system for its initialization or setup, while "dimensionality" refers to the number of parameters on which the output depends. With such large data, it will take the system a long time to compute the final result. But it is natural that the system you are trying to make has to respond in a finite amount of time. Hence such systems would not serve your purpose.

  Imagine that you are asked to optimize a given function when given a set of constraints. You use all sorts of mathematics to try to get the answer, but still the system takes a long time. Even if you get a solution, you may find that other better solutions also exist. As a result, the system fails to perform up to your expectations.
- **Space complexity:** Many systems require a huge amount of memory to perform. It may not always be possible to have such a large amount of memory available. Thus, memory constraints may cause a problem, as they may not allow the system to perform, even if all other factors are favorable.

  Imagine that you are trying to recognize a speaker using point-to-point matching over a set database. You need to make this system on one chip. Your chip may not have enough memory to store all the data sets in any form, especially as the chip memory might be expensive. Hence, the memory constraints would cause another problem.

It should be clear that hard computing, though preferable, is not always able to solve problems. It has a limited scope and cannot be used in all problems. The problems with hard-computing techniques increase exponentially whenever we increase the size of the data or the data's dimensionality.

Soft computing, on the other hand, offers a very simple solution to all the problems. With soft computing, we can solve the problems and obtain good results. Another good feature of soft computing is that we only need to make a model. Tools exist that automatically convert this model into a working code using a graphical user interface (GUI) or simple commands. Soft-computing techniques are now available as in-built libraries in MATLAB®, Java, C, and C++, among others. This makes working with soft computing fun and easy.

1.1.3 Soft-Computing Systems

Soft computing started in the 1990s. Over the past two decades, the field has rapidly matured and found applications in various domains. The field of soft computing is highly multidisciplinary. People specializing in soft computing are studying newer and newer systems and trying to find the solutions to various problems.

The major areas of application of soft computing are given below. It should be noted, however, that the list is much longer and more versatile than this.

- **Biometrics:** These systems deal with the identification and verification of people based on their biometric properties, including speech, face, iris, fingerprint, and so forth. These systems are used for security and automation purposes and are being actively deployed in industry (see Chapters 7, 8, and 9).
- **Bioinformatics:** These systems study and analyze molecular biology using soft-computing tools. Bioinformatics systems have been used to study the complex structures of proteins, which was not possible using any other technique due to their very large size (see Chapter 10).
- **Biomedical systems:** These systems are used to identify the presence of diseases. The diagnosis has varied applications to assist doctors and help them make decisions regarding the presence or absence of diseases. These systems are being actively used for various kinds of analysis, diagnosis, and monitoring (see Chapters 11 and 12).
- **Robotics:** Soft-computing techniques are extensively used for robotic knowledge representation, learning, path planning, control, coordination, and decision making, to name a few. Robotics tends to be a vast area for various artificial intelligence and soft-computing techniques at multiple levels (see Chapter 14).
- **Vulnerability analysis:** Soft-computing techniques are extensively used to model systems in order to find vulnerability. This type of analysis has applications in various domains. For example, in banks, these models are used for credit risk analysis; in information security, for risk analysis; and in social living for social risk analysis (see Chapter 13).
- **Character recognition:** Another exciting field of soft computing is character recognition. Soft-computing techniques identify characters in various scripts that may be typed (or optical) or handwrittens in nature. These systems are extensively used in scanner software, text extractors, and so forth (see Chapter 15).
- **Data mining:** In data mining, we try to extract rules, features, and trends from a large database. Soft-computing techniques are extremely time effective for such purposes. This task is also called knowledge discovery in database.
- **Music:** Soft-computing techniques can also be used for the automatic generation of music, genre recognition, artist recognition, and so forth. These systems can be used to assist composers in generating music or in generating the sounds of musical instruments (see Chapter 17).
- **Natural language processing:** Natural language processing (NLP) refers to the understanding, representation, and conversion of the natural languages used by humans. Soft-computing techniques are being used for all aspects of NLP. The success of these techniques has made communication possible between people and machines, regardless of their original languages.
- **Multiobjective optimizations:** Various natural problems have been modeled as multiobjective optimizations, in which some constraints are given. The problem is to optimize a set of objective functions keeping these constraints valid. Soft-computing techniques offer the best solution to these problems.
- **Wireless networks:** Soft computing is used in wireless networks because it offers the most optimal placement of wireless sensors, intelligent channel allocation, and so on.

Soft-computing techniques have thus made it possible to optimize the performance of wireless networks.

- **Financial and time series prediction:** This is one of the most exciting and heavily studied fields, in which people try to predict the financial timeline and other series w.r.t. time based on a set of parameters. Soft-computing techniques are used to predict the future on the basis of past trends.
- **Image processing:** Soft computing has applications in the field of image processing, including in noise removal, segmentation, and extraction. Image processing is the basic step for any system working with images or videos.
- **Toxicology:** In toxicology, soft computing is being used to predict the affect of drugs on animals.
- **Machine control:** Soft-computing techniques can be used to control the various machines used in factories, including controllers for temperature, speed, trajectory, and so on. This eliminates the need of humans and gives better performances.
- **Software engineering:** Soft-computing techniques may be used for the planning and scheduling of various parts of software engineering projects. They are also used for automatic code generation and testing purposes.
- **Information management:** Soft-computing techniques are used to represent information, to learn from information, and so on.
- **Picture compression:** Another exciting field of soft computing is in picture compression. The compression makes the pictures more compact for storage, handling, and communication purposes (see Chapter 16).
- **Noise removal:** The signals used for various techniques need to first be processed to remove noise. This noise removal is provided by soft computing, which first tries to find the possible amount of noise based on past noises and then generates an antinoise to cancel its effect.
- **Social network analysis:** In social network analysis, the social behaviors of people are analyzed on the basis of social networking sites, forums, and so forth. This field deals with the study of patters and behaviors in these networks.

As is obvious, soft computing is used in a long list of domains. All of these are active fields of research in the modern scientific society. In this book, we study some aspects of some of these fields. Keep in mind that these methods may be generalized for other systems as well. In this book, we attempt to cover as many soft-computing techniques as possible in a diverse set of applications to allow readers to try their hand at any of these fields using any of the methods so they can see the results for themselves.

1.2 ARTIFICIAL INTELLIGENCE

We all know that human beings are intelligent animals. This means that we can perform many tasks that require understanding, perception, and decision making. Through our understanding, we are able to complete tasks in the best possible ways. This intelligence has separated us from machines, which use a definite input, output, and procedure. Suppose you ask your child to get a glass of water; it is likely that your child will do this naturally, without much thought. But asking a machine to do the same work may not be possible. We talk more about humans in comparison to other animals here as they are believed to be species with extraordinary intelligence level. Artificial intelligence (AI) deals with making machines intelligent. In researching and applying AI, we try to induce intelligence in machines by artificial means.

Over the past few decades, researchers have been trying to learn from the way humans work. The quest to create machines that imitate human beings has resulted in many models for solving various problems. Even after so many decades, however, we have been able to imitate humans to only a very brief extent. For the most part, how we learn and perform still remains a mystery. It is fantastic to see how even small children can do such complex tasks as recognizing parents, words, and places so easily. If you tried to get machines to learn the same tasks, you would realize the complexity of the problem. Suppose a child is walking down a road. If some obstacle comes up, that child would move to the side to avoid the obstacle. If you wanted a machine to do the same thing, it would be extremely difficult. The human brain is filled with mysteries that need to be unfolded if we want to learn from them and implement them.

Before we move further in our discussion, let us first formally define *intelligence*. The *Oxford English Dictionary* defines *intelligence* as "intellect, understanding. Intellect is a faculty or reasoning, knowledge, and thinking." Another definition of *intelligence* is "the computational part of the ability to achieve goals in the world. Varying kinds and degrees of intelligence occur in people, many animals and some machines" (McCarthy, 2007). This clearly states our notion of intelligence. Humans are called intelligent because they are able to perceive a situation, comprehend it, look for solutions, and make suitable decisions. This decision-making ability separates us from the "nonintelligent" machines. Intelligence in humans is a natural phenomenon that takes place as a result of lifelong experimentation, learning, and practice.

Definition 4
Intelligence means intellect, understanding. Intellect is a faculty or reasoning, knowledge, and thinking. (Oxford English Dictionary)

Definition 5
Intelligence is the computational part of the ability to achieve goals in the world. Varying kinds and degrees of intelligence occur in people, many animals and some machines. (McCarthy, 2007)

Let us now try to understand artificial intelligence.

1.2.1 What Is Artificial Intelligence?

Artificial intelligence (AI) means adding intelligence to machines artificially so that they become intelligent and behave in ways similar to humans. AI can be formally defined as "the simulation of human intelligence on a machine, so as to make the machine efficient to identify and use the right piece of 'knowledge' at a given step of solving a problem" (Konar, 2000). *Knowledge* is defined as awareness or familiarity. Another definition of artificial intelligence describes it as "the science and engineering of making intelligent machines, especially intelligent computer programs. It is related to the similar task of using computers to understand human intelligence, but AI does not have to confine itself to methods that are biologically observable" (McCarthy, 2007). This definition clearly states that AI techniques are attempting to induce in machines the capability of acquiring knowledge, storing and retrieving knowledge, learning from the available knowledge, comprehending knowledge, and, above all, making decisions.

Definition 6
Artificial intelligence is the simulation of human intelligence on a machine, so as to make the machine efficient to identify and use the right piece of "knowledge" at a given step of solving a problem. (Konar, 2000)

Definition 7
[Artificial intelligence] is the science and engineering of making intelligent machines, especially intelligent computer programs. It is related to the similar task of using computers to understand human intelligence, but AI does not have to confine itself to methods that are biologically observable. (McCarthy, 2007)

AI systems have a great deal scope in making effective expert systems and decision support systems (DSS). Because of the added intelligence, these systems can operate autonomously, which enables them to take end-to-end responsibility of a task. This further alleviates the need of humans in such problems. The need for and use of autonomous machines can easily be seen in robotic systems. For example, robots are being built for rescue operations that are dangerous for humans to perform, as well as for survey operations in inaccessible areas, on other planets, and elsewhere. This saves expense and protects humans from the dangers of such situations. These autonomous robots are a big boon in getting the needed information easily. In addition, the robots can perform in all kinds of conditions and even in unlikely environments.

1.2.2 Problem Solving in AI

The concepts of AI are not restricted to a single set of algorithms. Rather, they extend to the intelligent use of a large number of tools and techniques and to many more algorithms that are designed per the problem requirements. Advances in algorithm design have given us many algorithms that are now being used for problem solving. These advances allow AI techniques to incorporate many diverse fields of algorithms that can be used in multiple ways and on multiple levels.

The most commonly used and most basic set of algorithms are the graph search algorithms. A *graph* is defined as a collection of vertices and edges and may be denoted by $G(V, E)$. Here we are given a set of vertices and a set of edges that connect vertices. We are given a vertex from which we start our search. The problem is to reach the goal vertex by traveling along the edges from vertex to vertex. In the example in Figure 1.1, the edges are $E = \{1, 2, 3, 4, 5, 6\}$, and the vertices are $V = \{(1, 2), (1, 3), (1, 4), (2, 7), (4, 5), (4, 6), (3, 9), (6, 9), (5, 6), (7, 5), (5, 8)\}$. In this example, we are supposed to reach the goal vertex starting from the source vertex. One such solution has been highlighted. In notation, we would present this solution as $1 \rightarrow 4 \rightarrow 5 \rightarrow 8$. The algorithm returns true if the goal vertex is found; otherwise, it returns false. In the same example, the algorithm would return true in the given configuration with 1 as the source and 8 as the destination. However, if we change the source to vertex 2 and the goal to vertex 6, the algorithm would return false, as there is no way to reach vertex 6 by starting from vertex 2.

The number of vertices in a real AI problem may be very large. Thus it may not be possible to generate the entire graph as we did for this example. Instead, the graph is generated as it is explored, and the vertices and edges are added dynamically as the algorithm runs. Using this technique, we are able to save both space and time.

These algorithms have many applications in scheduling, planning, game solving, and other such problems. In each, we model the whole problem as a graph, where every vertex corresponds to a state of the game. To understand this better, let us first explain what is meant by a *state*. A state may be defined as the status of the present situation out of all the possible statuses that the problem can

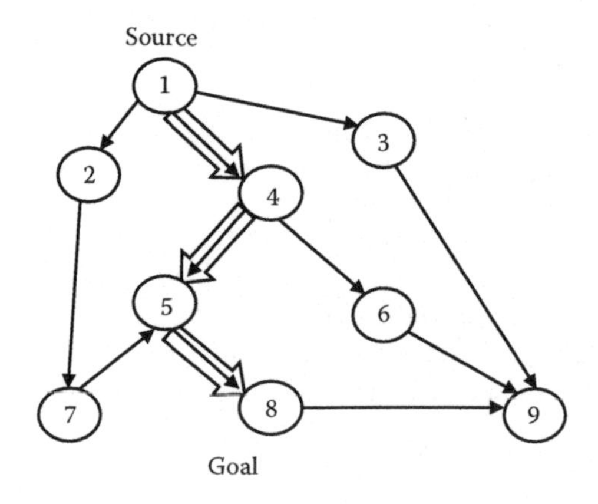

FIGURE 1.1 A general graph.

2	3
1	*B*

FIGURE 1.2a The general game board in a four-puzzle problem.

take. Suppose the problem we are working on can take any of the *n* possible states. Here *n* may be very large. Each of these *n* states is represented by a vertex of the graph. Note that because *n* is very large, the vertices might not be initialized or generated while coding, but they do exist conceptually. In the algorithm, they may be generated whenever needed or whenever explored.

Definition 8

A state may be defined as the status of the present situation out of all the possible statuses that the problem can take.

Now let us come to the edges. An edge exists between two states if it is possible to reach the latter state from the former. Hence if we can make a move of 1 unit and reach a state *j* from a particular state *i*, we would proceed with the making of the edge (*i*, *j*). The edges may be traversed to explore the possible transitions between the states as the moves are made.

To better understand the concepts of graph and search presented above, let us take an example of the famous four-puzzle problem. In this problem, there are four cells. Three of the cells have a number, while the fourth cell is blank (see Figure 1.2a). We can slide any adjacent cell in the place of the blank cell vertically or horizontally. The board is shuffled, and our task is to arrange the numbers in a particular order. In this problem, a state corresponds to a particular configuration of the board. Let us say that we denote a board by <*P*, *Q*, *R*, *S*>, where *P* is the top left cell, *Q* is the top right cell, *R* is the bottom left cell, and *S* is the bottom right cell. Let *B* denote the blank. Possible states are <1, 2, 3, *B*>, <1, 2, *B*, 3>, <1, 3, *B*, 2>, and so on. An edge exists between two states if we are able to reach one state from the other. We know that if there is a *B* is at the bottom right, we can either slide *Q* down or *R* to the right. So if we consider the state <1, 2, 3, *B*>, we can change the state in one move to <1, *B*, 3, 2> by sliding 2 down or to <1, 2, *B*, 3> by sliding 3 to the right (see Figure 1.2b). Note the use of bidirectional arrows in the example. This indicates that the second state can be reached from the first as well as the first from the second. Hence an edge exists from vertex <1, 2, 3, *B*> to <1, *B*, 3, 2> and to <1, 2, *B*, 3> as shown.

The problem is to reach a goal state, such as <3, 2, 1, *B*>, starting from an initial state, such as <1, 2, 3, *B*>. The graph for this is given in Figure 1.2c.

Once the initial source and the final goal are known, the search for the solution can be done using any standard graph search algorithm. We will now discuss the general fundamentals of graph searches. Then we will briefly discuss the various graph searching algorithms.

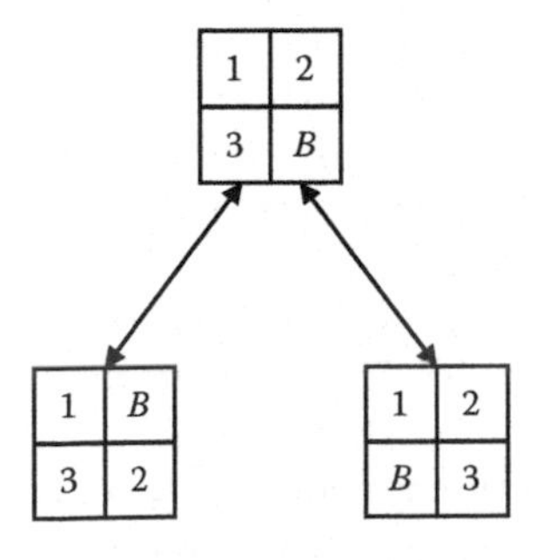

FIGURE 1.2b The game graph for the four-puzzle problem.

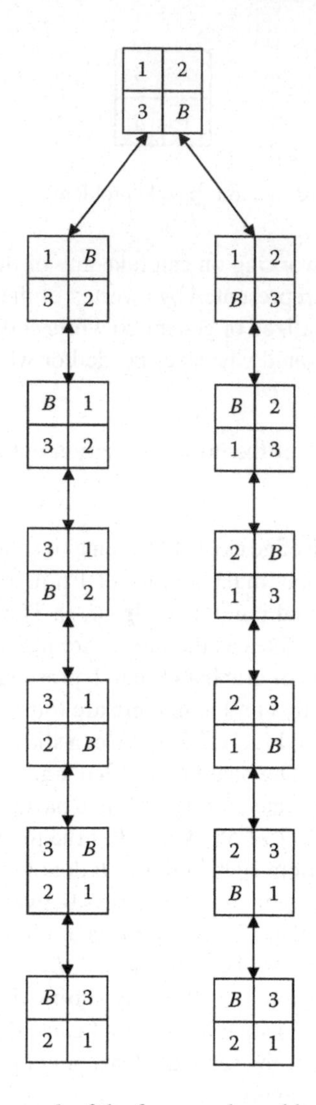

FIGURE 1.2c The complete solution graph of the four-puzzle problem.

A graph search always starts from the source. We maintain two separate lists. One is called the open list, and the second is the closed list. The closed list contains all the vertices (also called *nodes*) that have been visited in the past. The open list stores the vertices that have not yet been visited but that have been discovered. Initially the closed list is empty, and the open list contains only the source node, because the source has been discovered but not yet visited. At all iterations, we select a vertex (say x) from the open list. We remove this vertex x from the open list and expand this vertex. Expansion of a node means to find the vertices next to the vertex. This gives all vertices y such that the (x, y) edge exists. In the Figure 1.2 example, expansion of vertex <1, 2, 3, *B*> would give <1, *B*, 3, 2> and <1, 2, *B*, 3>. If the newly generated vertices are not in an open or closed list, then they are added to the open list. This is because the vertex being added has been discovered but not yet visited. The vertex x is marked as visited and is added to the closed list. In this manner, we proceed until we extract the goal from the open list. If at any point the open list becomes empty, we conclude that it is not possible to reach the goal from the source.

The various graph searching algorithms mainly differ in the way they extract a vertex from the open list. The difference in the strategies used decides the algorithm. Different algorithms work well in different conditions. Choosing an algorithm is much more a question of trying to predict the input. In general, the performance of the different algorithms cannot be compared if the inputs are completely unknown.

The algorithms may be classified into two categories. The first category forms the class of algorithms that do not have any means to evaluate the closeness of a vertex to the goal. The second category can evaluate the possible closeness of a vertex to the goal based on a function known as the *heuristic function*. The word *heuristic may* be defined as "rule of thumb. People develop their own rule of thumb based on their experience acquired over the years." Heuristic functions may be used to "estimate, calculate, uncover, discover, purify, and simplify solutions to problems" (Kliem and Ludin, 1997). Thus, heuristic functions may be defined as functions that "are used in some approaches to search to measure how far a node in a search tree seems to be from a goal" (McCarthy, 2007). The heuristic function is used to measure the fitness or effectiveness of a vertex; the closer a vertex is to the goal, the better would be its heuristic function.

Definition 9

Heuristic means rule of thumb. People develop their own rule of thumb based on their experience acquired over the years. Heuristic may be used to estimate, calculate, uncover, discover, purify, and simplify solutions to problems. (Kliem and Ludin, 1997)

Definition 10

Heuristic functions are used in some approaches to search to measure how far a node in a search tree seems to be from a goal. (McCarthy, 2007)

The following are the most commonly used graph algorithms.

- **Breadth-first search (BFS):** This algorithm first visits the various nodes that are closer to the source and then proceeds to the farther ones. The BFS algorithm works very well if the goal is believed to be very near the source. The BFS gives good results when applied to social network graphs, as it is normally observed that in these graphs, the various nodes lie quite close to one another. In BFS, a node closer to the source is visited first. The open list is implemented by a queue that follows a first-in, first-out (FIFO) structure.
- **Depth-first search (DFS):** Unlike the BFS, the DFS starts by going deep into the graph. It tries to expand nodes and penetrate deeper into the graph. When it is not possible to further expand the nodes at any level, it then moves to expand the nodes of the previous level. The open list is maintained by a stack, which follows a last-in, first-out (LIFO) implementation.
- **Iterative deepening depth-first search (IDDFS):** This type of search follows the principles of the depth-first search, but we limit the maximum depth to a certain limit. This limit keeps increasing as we proceed with the algorithm.
- **Best-first search:** This search selects the best possible nodes from the open list and expands them further. The motivation for this search is that the best nodes would reach the end nodes in the shortest possible time. It continues to select and expand the best nodes until the goal is reached. The nodes are evaluated based on the heuristic function.
- **A* search:** This algorithm is similar to the best-first search. However, instead of optimizing only the probable distance from the goal (heuristic), this search tries to optimize the total path, including the distance traveled to reach the vertex so far from source and the heuristic value of this vertex. The total cost is calculated; this total cost consists of the historic cost and the heuristic cost. The least total cost vertex is selected from the open list.
- **Multi Neuron heuristic search:** This is similar to the A* algorithm. However, instead of selecting the node with the best cost, it selects α number of vertices from the open list. The costs of these vertices vary from best to worst. These vertices are then processed sequentially. This algorithm is believed to work better when the heuristic function may change its

value very sharply between vertices. α is a parameter whose value may be varied to control the algorithm. A value of 1 converts this algorithm to an A* algorithm, while a value of infinity converts this into breadth-first search. The details of this algorithm can be found in Shukla and Kala, 2008a.

One of the most commonly used algorithms for real life problems is the A* algorithm. Other algorithms may be made by making small changes to this algorithm. The A* algorithm is presented in Algorithm 1.1. An example of using this algorithm can be found in Chapter 14.

Algorithm 1.1

A*(graph, source, goal)
Step 1: closed, open ← empty list
Step 2: Calculate cost of open
Step 3: Add source to open
Step 4: While open is not empty
Do
Step 5: Extract the node n from open with the least total cost
Step 6: If n = goal position, then break
Step 7: For each vertex v in expansion of n
Do
Step 8: if v is in neither open nor closed list, then compute costs of v and add to open list with parent n
Step 9: if v is already in open list or closed list, then select the better of the current solution or the solution present already and update the cost and parent of the vertex
Step 10: if v is already in closed list, then select the better of the current solution or the solution present already and update the cost and parent of the vertex and all children from the vertex
Step 11: Add n to closed

1.2.3 Logic in Artificial Intelligence

Just as logic drives many systems in real life, it also has a prominent role in AI. Logic may primarily be seen as a method of knowledge representation. Here we try to represent a system by a set of rules or in a way that can be easily understood and implemented in the machine. Knowledge is of deep interest to system developers because it provides a means for the machine to understand, act, and make decisions and inferences based on a common understanding of the knowledge. It helps erase the gap between human and machine understanding.

Before we go further, let us first give a formal definition of logic: "What a program knows about the world in general, the facts of the specific situation in which it must act, and its goals are all represented by sentences of some mathematical logical language. The program decides what to do by inferring that certain actions are appropriate for achieving its goals" (McCarthy, 2007).

Definition 11
Logic is what a program knows about the world in general, the facts of the specific situation in which it must act, and its goals are all represented by sentences of some mathematical logical language. The program decides what to do by inferring that certain actions are appropriate for achieving its goals. (McCarthy, 2007)

Another term associated with logic is knowledge. *Knowledge* is defined as "a function that maps a domain of clauses onto a range of clauses. The function may take algebraic or relational form depending on the type of applications" (Konar, 2000).

Definition 12

Knowledge is a function that maps a domain of clauses onto a range of clauses. The function may take algebraic or relational form depending on the type of applications. (Konar, 2000)

We now have a fair idea that the motivation is to induce intelligence in machines by introducing some means of knowledge representation. In essence, we are trying to explain logic to a machine that until recently had been doing only trivial tasks. The simplest implementation is called as the production system.

The production system consists of simple if-then clauses. The *if* part, or the antecedent, denotes the condition that must be true for the particular rule to fire. The condition expresses the particular case in which the action would hold true. The *then* part consists of the action or conclusion, also known as the consequent, that results from the rule being fired. The entire knowledge is present in the system in the form of a rule set. These systems use these rules for all operations. Once these rules are ready, we know that all available information has been incorporated into the system in the rules. These systems also have a memory, known as the working memory. This memory stores all the information regarding the system's state. Based on this state, the rules are fired by a rule implementer.

Consider the production rule:

If (*X* marks are more than 80) & (*X* attendance is more than 75%), then (*X* grade is A).

Here the *if* part states all the conditions that if true lead to the action.

1.3 SOFT-COMPUTING TECHNIQUES

In the next few subsections, we state the various concepts that collectively make up the field of soft computing. Soft computing mainly incorporates artificial neural networks, fuzzy logic, and evolutionary computing. New models, called hybrid systems, are currently being built from a combination of these techniques.

1.3.1 Artificial Neural Networks

The artificial neural network (ANN) is an attempt to imitate the functionality of the human brain. The human brain is believed to be composed of millions of small processing units, called neurons, that work in parallel. Neurons are connected to each other via neuron connections. Each individual neuron takes its inputs from a set of neurons. It then processes those inputs and passes the outputs to another set of neurons. The outputs are collected by other neurons for further processing. The human brain is a complex network of neurons in which connections keep breaking and forming. Many models similar to the human brain have been proposed.

ANNs are able to learn past data, just like the human brain. The network learns the past data by repeating the learning process for a number of iterations. ANNs learn the data and then reproduce the correct output whenever the same input is applied. The precision and correctness of the answer depends on the learning and the nature of the data given. Sometimes an ANN may refuse to learn certain data items, while other times it may be able to learn complex data very easily. The precision depends on how well the neural network was able to learn the presented data.

ANNs have another extraordinary capability that allows them to be used for all sorts of applications: Once they have been well taught the historical data, they are able to predict the outputs of unknown inputs with quite high precision. This capability, known as generalization, results from the fact that ANNs can imitate any sort of function, be it simple or complex. This gives ANNs the power to model almost any problem that we see in real life applications.

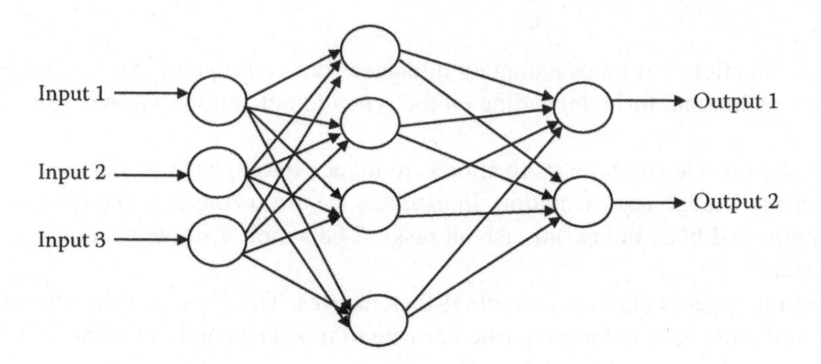

FIGURE 1.3 The architecture of the artificial neural network.

Furthermore ANNs are highly resistant to noise. Thus, if there is any noise in the input data, ANNs are still able to perform appreciably well. This ability makes it possible to make robust applications for ANNs.

Let us now formally define ANNs based on our understanding so far: "The neural network, by its simulating a biological neural network, is a novel computer architecture and a novel algorithmization architecture relative to conventional computers. It allows using very simple computational operations (additions, multiplication and fundamental logic elements) to solve complex, mathematically ill-defined problems, nonlinear problems or stochastic problems" (Graupe, 2007).

Definition 13

The neural network, by its simulating a biological neural network, is a novel computer architecture and a novel algorithmization architecture relative to conventional computers. It allows using very simple computational operations (additions, multiplication and fundamental logic elements) to solve complex, mathematically ill-defined problems, nonlinear problems or stochastic problems. (Graupe, 2007)

The general structure of the ANN is given in Figure 1.3, which shows how the information is processed. Every node takes inputs from the previous nodes and gives its output to subsequent nodes. In this way, there is constant parallel processing by the various nodes. The last node processes the information to generate the final output. From this simple model, many variations can be created (see Chapter 2).

In the next couple of chapters, we will see the underlying reasons why ANNs are able to solve complex problems so easily. We present the complete working of ANNs, as well as the various models that exist. We will also study the application of ANNs to various problems. Finally, we will see how we can use ANNs to solve real life problems.

1.3.2 Fuzzy Systems

Impreciseness is always prevalent in systems. Observations might reveal that you are never able to make discrete rules over a set that always holds precisely true. Rather the rules are always imprecise in nature. We can never work over definite systems that have definite rules. It is from the same observations that fuzzy systems evolved. The impreciseness that is introduced in the system makes them work in a better way. Fuzzy systems are hence being used to model many real life applications.

Before we continue our discussion of fuzzy systems, however, let us first introduce fuzzy sets. In fuzzy sets, every member of the set has a certain degree of belongingness to the set. This degree is called the *membership degree*; the function that decides the membership degree, or belongingness, is known as the *membership function*. This function associates each member of a set with a degree or probability of belongingness. The higher this degree, the more the member is a part of the set.

To illustrate, consider the set good students. A nonfuzzy system would divide the whole class into students who are good students and students who are not good students. If a good student falls below some threshold, that student suddenly becomes a bad student. The fuzzy system, however, would associate some degree by which every student is a good student. Suppose a very bright student has a degree of goodness of 0.9, and a very dull student has a goodness degree of 0.1. If the performance of the good student were to reduce by some amount, there would only be a change in membership degree depending on the extent to which the performance has reduced. The perks and benefits that good students get in the case of fuzzy sets would not suddenly vanish after a small drop in performance, as would be the case with nonfuzzy systems.

Fuzzy systems are implemented by a set of rules called *fuzzy rules*, which may be defined on sets. A nonfuzzy system has very discrete ways of dealing with rules—either a rule fires fully or it does not fire at all, depending on the truth of the expression in the condition specified. On the other hand, in fuzzy systems, the conditions are true to a certain degree and false to a certain degree. The degree of truthfulness and falseness decides the degree to which the rule fires. Hence, every rule is fired to some degree. The outputs of all the rules are aggregated to get the system's final output.

To illustrate, consider a rule that says, "Good students with good attendance get good grades." The grades are a result of the degree of the performance and attendance. These two inputs together decide the grades. Likewise, if there are five such rules, they are aggregated, and the final outcome is computed.

The general procedure of working with fuzzy systems consists of input modeling. Then the membership functions are applied over the crisp inputs to get the fuzzified inputs. Then the rules are applied over these inputs to generate the fuzzy outputs. The various outputs are then aggregated and defuzzified to get the final crisp output. This procedure is given in Figure 1.4.

Fuzzy systems are also known as fuzzy inference systems (FIS). These intelligent systems generate outputs from given inputs using fuzzy logic. These systems are defined as "systems formulating the mapping from a given input to an output using fuzzy logic" (MATLAB Guide).

Definition 14
Fuzzy inference systems are systems formulating the mapping from a given input to an output using fuzzy logic. (MATLAB Guide)

Fuzzy systems continue to be a fascinating playground for experimenting with, working on, and modeling new problems (see Chapter 4).

1.3.3 Evolutionary Algorithm

If you look around the physical world, you might be astonished at how well the various species have adapted over time. Natural evolution has been a key point of motivation for people working in the

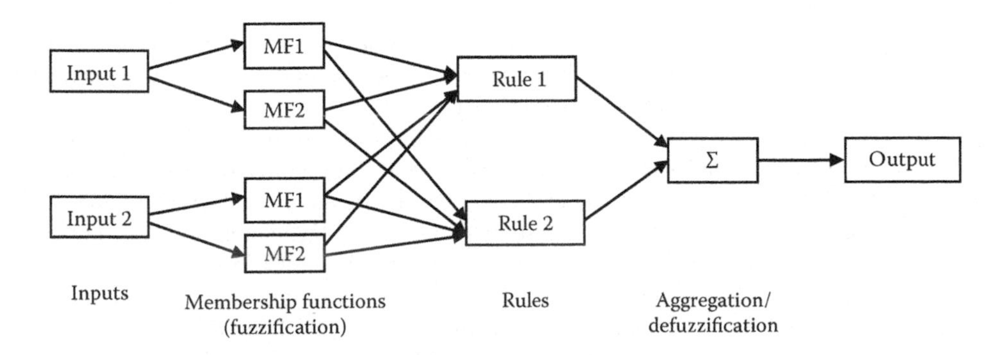

FIGURE 1.4 The architecture of the fuzzy inference system.

field of evolutionary algorithms (EAs), who are trying to imitate the same phenomenon of natural evolution to generate good or optimized solutions. These people are constantly trying to learn from natural evolution and the means and methods by which the best individuals evolve from one generation to the next over time.

Natural evolution results from the fusion of male and female chromosomes to generate one or more offspring. The offspring contain mixed characters of both the male and female counterparts. The offspring may be fitter or weaker than the participating parents. This process of evolution leads to the development of one generation from the previous generation. Over time, the newer generations evolve, and the older ones die out. This survival process goes on and on, following Charles Darwin's theory of survival of the fittest. According to this theory, only the characteristics of the fittest individuals in a population go to the next generation. Others die out. It can easily be seen that both the average quality of fitness and the fittest individual in the population are found as the generations go on evolving.

Evolutionary algorithms work in a similar way. The first step is to generate a random set of solutions to the given problem comprising a population. These random individuals are then made to participate in the evolution process. From these solutions, a few individuals with high fitness are chosen through a method called selection. These individuals are then made to generate offspring. This process of generating offspring by two parents is known as *crossover*. The system then randomly selects some of the newly generated solutions and tries to add new characteristics to them through a process known as *mutation*. Various other operations may also be done by specialized operators. Once the new generation is ready, the same process is applied to that new generation. The fitness of any one solution may be measured by a function known as the *fitness function*.

As we move through newer and newer generations, we find that the quality of solution continues to improve. This improvement is very rapid over the first few generations, but with later generations, it slows. The general structure of simple genetic algorithm, a type of evolutionary algorithm is given in Figure 1.5, and an in-depth analysis and study of this algorithm is in chapter 5.

A formal definition of evolutionary algorithms is "algorithms that maintain a population of structures that evolve according to the rules of selection and other operators, such as recombination and mutation. Each individual in the population is evaluated, receiving a measure of its fitness in the environment" (Spears, 2000).

Definition 15

Evolutionary algorithms are algorithms that maintain a population of structures that evolve according to the rules of selection and other operators, such as recombination and mutation. Each individual in the population is evaluated, receiving a measure of its fitness in the environment. (Spears, 2000)

1.3.4 Hybrid Systems

Each system discussed above has some advantages and some disadvantages. In hybrid systems, we try to mix the existing systems, with the aim of combining the advantages of various systems into a single, more complex system that gives better performances.

Many hybrid systems have been proposed, and many models have been made to better cater to the data. One of the best examples is the fusion of ANNs and FIS to make fuzzy-neuro systems. In this hybrid model, a fuzzy system is built over ANN architecture and is then trained using the back propagation algorithm (BPA). These hybrid systems are highly beneficial in real life applications, such as in controllers. These controllers primarily use the fuzzy logic approach, but optimization of the various parameters is done using ANN training. This combination makes the systems perform with a very high degree of precision. An evolution from this hybrid model is the adaptive neuro-fuzzy inference system (ANFIS), which is similar to the neuro-fuzzy system but is adaptive in nature.

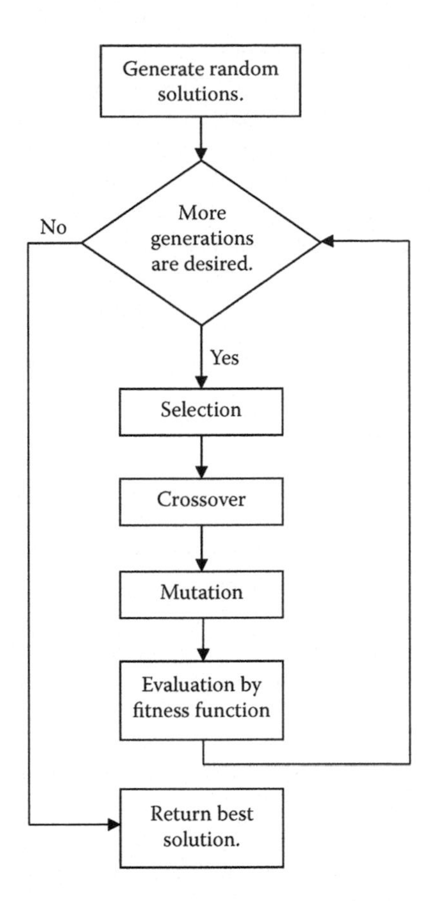

FIGURE 1.5 Simple genetic algoirthm.

Another system that is being used in hybrid models are the evolutionary algorithms. These systems are being added to ANNs to give better performance and obtain the global minima. EAs give the hybrid systems a good performance optimization. They are also being used to make an evolutionary model of ANNs. This system starts from a simple ANN; as time continues, the system evolves the ANN, based on performance, by constantly adding neurons. These systems are also found to give very good performance. In a similar system, EAs evolve the fuzzy system to ensure that the generated system gives a high performance.

The fusion of AI techniques with other algorithms is another common technique. Various models exist, with one system fused to another according to the problem requirements. These systems often cater to the needs of the problem in a better way. They ensure that the performance is good, even as traditional methods fail to deliver good results.

For more information on hybrid models, see Chapter 6.

1.4 EXPERT SYSTEMS

Whenever we need help in everyday life, we consult an expert. An expert is a person who has profound knowledge of the subject; this knowledge is a result of the expert's capabilities and experience. The expert looks at the problems, makes some inferences, and presents the results. As an end customer, we follow the results and assume that the results are true. The expert also continues to update knowledge and to learn from past cases. Experts provide end-to-end solutions to meet our needs. They can make correct decisions, even in the absence or impreciseness of some information.

Expert system (ES) design is motivated by this concept of experts. In this type of design, we try to make systems that can act autonomously. Once these systems are presented with the conditions as inputs, they work over the solutions. The solution given by these systems is then implemented. These systems are a complete design of whatever it might take to draw conclusion once the problem is given. To reach the conclusions may require preprocessing, processing algorithms, or any other procedures. These systems may require a historical database or a knowledge base to learn from.

Expert systems are being built as an end tool to meet the customer's various needs. These tools behave just like an expert who takes care of all the subject's needs. These tools have made it possible for people to enjoy the fruits of soft computing without knowing anything much about the underlying system.

Before we proceed with our discussion of ES, let us formally explain what an expert system is.

1.4.1 What Are Expert Systems?

An expert system may be defined as "knowledge-based systems that provide expertise, similar to that of experts in a restricted application area" (Kasabov, 1998). This definition clearly depicts the quest to make systems that can cater to the user's complete needs, even in the absence of data or in the impreciseness of the data. Another important dimension of an expert system is that of knowledge. Just as an expert works only on the basis of knowledge, the expert systems also require knowledge that needs to constantly update itself. All ES decisions are based on this knowledge. Learning and proper representation of knowledge is an important aspect of these expert systems.

Definition 16

Expert systems are knowledge-based systems that provide expertise, similar to that of experts in a restricted application area. (Kasabov, 1998)

The general architecture of an expert system is given in Figure 1.6, which highlights all the points discussed above. The entire system can be divided into two parts: The first consists of the interface with the user where the required inputs are taken and the required outputs given. The other part is the expert, which processes the inputs to generate outputs and tries to learn new things.

- **User interface:** This is where the user interacts with the system. The user can select options, give inputs, and receive outputs. The user interface is designed to keep all the functionality available in the system in some or the other way.
- **Knowledge base:** This is the pool of knowledge available in the system. Knowledge may be represented in many ways. Standard AI and soft-computing techniques are used for this purpose. This pool of knowledge can be easily queried to solve common problems per the requirements of the user.

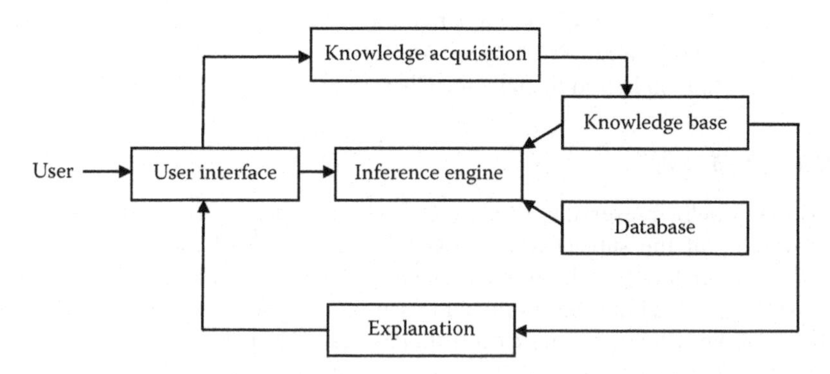

FIGURE 1.6 The architecture of an expert system.

- **Database:** This module is a collection of data. All past or historical data are stored here. The database itself has the potential to generate a huge amount of knowledge once any learning technique, or any other means, is applied over it. Analysis or learning of these data generates the knowledge used by the knowledge base.
- **Inference engine:** This module is responsible for interpreting the knowledge or its inference. It assembles and presents the output to the user in the form desired. This engine controls the system's overall working.
- **Knowledge acquisition:** This module traps data or inputs and adds them to the pool so the system can learn new things from the data that have been presented. Knowledge acquisition is responsible for making the system robust against the newer inputs. This increases the system's performance.
- **Explanation:** This module is responsible for tracing the execution of the entire process. It records how the various inferences were made and why the specific rules were fired. In expert systems, it is responsible for the reasoning behind the generation of the output from the inputs.

Once an ES is designed, we can obtain a robust system that can perform and learn autonomously. The knowledge base and database provide excellent means for drawing relevant conclusions and even for learning. Consider the example of character recognition. The expert system may provide an interface on which the user can draw some character; in return, it identifies that character. The recognized character may be output from the system, or it may be used for any specialized work the software desires.

1.4.2 Expert System Design

In this section, we briefly discuss the design issues of the expert systems. Because these systems must solve the user's needs, the foremost thing to understand is these needs. The basic objective and motive of the ES must be very clear. We further need to think about all the kinds of work that the ES is supposed to perform. The ES must be able to replace the existing system. We need to be clear about the kind of processing that the system needs and about the outputs the system must return for all types of identified user requirements.

The first major task is identification of inputs. All the inputs that can be used should be well known. We need to ensure that all inputs are relevant to the system. We also need to be clear about the way in which these inputs are to be used by the system. There must be a suitable interface for the inputs to be taken from the user. Similarly the outputs must be clearly known. We must work out the number and kind of outputs that the system is expected to produce. The outputs generated must be of some interest to the user or of relevance to the system. There also must be a good interface to present the output to the user.

We need to again look forward for the flow of data. We need to think about which system gets what data, what work the system will perform, what the outputs are, and whether the outputs would be used or processed. Doing all of this will ensure that we are able to meet all the functionalities of the system. All the data needed for a process must be made available so that the system can perform and the functionality can be met.

We further need to work out the correct means for achieving each constituent task. This includes the correct method for storing and processing knowledge, the types of AI and soft-computing tools that need to be used, and so forth. In totality, we must be able to justify the output as the best possible solution considering the problem constraints and the advantages and disadvantages of the system. We present the various techniques and the conditions for them to be used in subsequent chapters.

Once the method has been chosen, we again need to ensure that the ES design is correct. The ES must be able to perform in the given constraints and in the best possible way. It also must be cross verified. Once the implementation is done, it may be checked for performance and errors.

1.4.3 High-End Planning and Autonomous Systems

Now that we know about the building and working of expert systems, let us briefly discuss the highly intelligent autonomous systems, which require a great deal of planning and effective decision making in an autonomous mode. These autonomous robots are designed to do such tasks as chasing thieves, performing rescue operations, surveying unknown lands, and so forth. These robots require AI and soft-computing tools used at various levels to perform the given task without any help from humans.

Robots behave just like humans. Humans have the capability to do multiple things at once. For example, a person might be walking to his room and listening to music, while also answering a phone call to discuss a meeting. The ability of the human brain to do all of this at the same time without problems is a beautiful phenomenon. Unfortunately, most of the systems we develop do not actually incorporate such activities.

When it comes to the field of robotics, various activities need to be performed. These activities vary from higher-end activities, in which strategies must be made, to lower-end activities that deal with simply taking data from sensors. The higher-level activities cater to the needs of the lower-end activities. The following are a few major activities:

- **Data acquisition:** This step deals with the acquisition of data from various sensors, including infrared sensors, global positioning sensors, wireless sensors, or video and audio signals. Each sensor is used for a different purpose, but all of them give valuable information about the surroundings.
- **Processing:** Data, in the form of audio and/or video, must be processed and recognized. Recognition systems are used to extract useful information from the data. The common steps involved are preprocessing or noise removal, segmentation, feature extraction, and recognition. Even raw data from other sensors must be processed to extract useful data and to represent those data in a more useful form.
- **Map building:** This section deals with building maps from the data that have been gathered and processed. The map represents all the surroundings of the robotic world. It tells about the locations of obstacles and traversable paths. It may be represented in various forms, depending on the problem and the constraints. The most commonly used maps are geometric maps, Voronoi maps, topological maps, and hybrid maps.
- **Control:** The ways and means for controlling and moving robots are provided by the robotic controllers, which deal with the task of making the robot move along a desired path. The control uses various techniques to guide the robot and ensure that the robot's physical movement is always according to the desired plan. The controls ultimately drive the motors of the robot wheels to make them move.
- **Path planning:** For the robot to perform a task, we require it to move from some point to another. Path planning deals with the task of finding a path from a given source to a

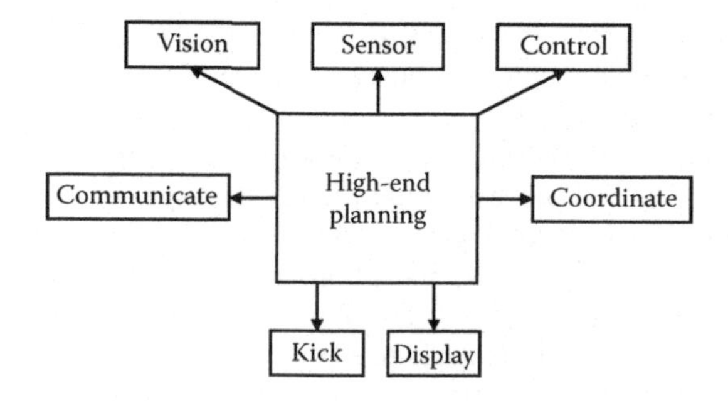

FIGURE 1.7 Software architecture for a robot.

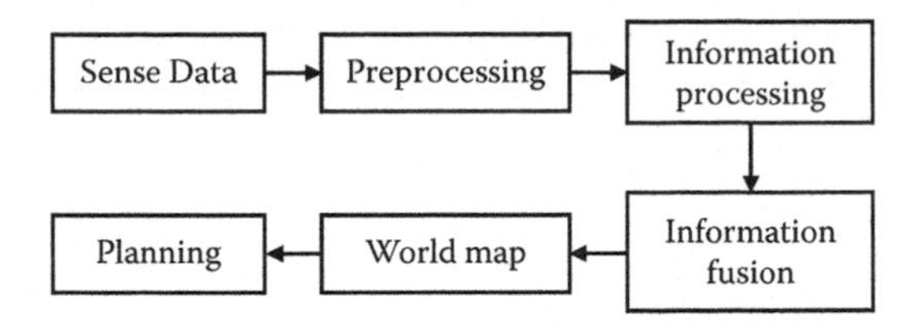

FIGURE 1.8 Architecture of the information processing and planning system in robots.

destination. Various AI and soft-computing techniques are used to find the robot's path. Some of these techniques facilitate guiding the robot in a real-time dynamic environment. If a path is possible, it is found out, and the robot is made to move over the same path.

- **Intelligent planning:** Intelligent planning deals with carrying out the job for which the robot was made. The planning consists of making real-time strategies for the robot. The robot is then guided to act on these strategies. The planning involves various AI and soft-computing techniques to make proper plans and execute them. The feedback mechanism makes the necessary adjustments or even changes the strategy. An example is the Robo Cup, in which robots are made to play a football match autonomously.
- **Multirobot coordination:** Some tasks may require more than one robot operating simultaneously. In such cases, there must be proper planning and coordination among the robots. All robots in a system need to plan and act upon a common strategy. Teamwork and cooperation are the keys to success.

  A picture of a system that requires multirobot coordination is given in Figure 1.7, which shows robots competing autonomously in a football match. The system has been made as a black box without exposing its internal functionality and will not be discussed here. Figure 1.8 shows various tasks performed by the robot for planning.

  The specific problems of path planning and navigation control are discussed in chapter 14.

1.5 TYPES OF PROBLEMS

So far we have been using the word *problems* in a very general context. Looking at various types of problems that exist in the world, we can easily say that there are three main categories of problems for which soft computing is used—classification, functional approximation, and optimization. Every real life application involves one or more of these three problems. Throughout this book, we make frequent references to these problems. Various specialized techniques and soft-computing models exist for each type of problem. First let us explain each category.

1.5.1 Classification

In classification problems, we are supposed to map the inputs to a class of outputs. The output set consists of discrete values or classes. The system is supposed to see the inputs and select one of the classes to which the input belongs. Every input can map only to a single class. Hence, once an input is presented, if the system maps it to the correct class, the answer is regarded as correct. If the system maps it to the incorrect class, the answer is incorrect.

In the case of classification problems, the system tries to learn the differences between the various classes based on the inputs presented. The differentiation among the classes is very simple if there are many differences between the two classes. However, if the two classes are very similar, it becomes very difficult to differentiate.

Before we proceed, let us give a formal definition of *classification*. According to the *Oxford English Dictionary,* classification is "the process of arranging in classes or categories or assigning to a class or category."

Definition 17
Classification is the process of arranging in classes or categories or assigning to a class or category. (Oxford English Dictionary)

The system's performance may be measured as the percentage of inputs correctly classified. It can be observed that the system's performance depends on the number of classes. Classifying an input into two classes is probably much easier than classifying the inputs into 2,000 classes. The output of the system in every case needs to be the specific class to which the system believes that input belongs.

To better explain the concept of classes and classification, let us take the example of speech recognition. Suppose we are making an automatic phone dialer that dials the number we speak. We know that the person speaking will say any one of the ten digits from 0 to 9. Hence, the output for every input can be any one of the ten digits. The system is supposed to identify the digit. Here, we can say that there are ten classes, and the classification problem is to map the input to any of the ten classes.

1.5.2 Functional Approximation

In functional approximation problems, the output is continuous and is an unknown function of the inputs. For every input, there is a specific output in the entire output space. The system is supposed to predict the unknown function that maps the inputs to the outputs. The system tries to predict the same unknown function in the learning phase. Whenever the input is provided, the system calculates the outputs. In these systems, it is very difficult for any input to get the precisely correct output. But the output given by the system is usually very close to the actual output. The closeness depends on the input given and the kind of learning the system was able to perform. The system's performance may be measured by the deviation of the actual output to the desired output. The purpose of such a system's design and learning is to make it possible to imitate the function that maps the inputs to the outputs. Here performance may be measured by the mean percent deviation or root mean square error.

Consider this example: We are trying to make a system that controls a boiler's temperature. The system takes as its input the present temperature and the quantity of fluid in the boiler. The system then outputs the fuel to be added per second to make the boiler work. The more fuel is added, the higher the heat is applied and the higher the temperature. This system is a functional approximation system that tries to map the inputs and the outputs by an otherwise unknown function.

1.5.3 Optimizations

We have already discussed optimization and optimization problems in the introduction. We only elaborate a few points here.

In optimization, we are given a function, known as the objective function. The aim is to minimize or maximize the value of the objective function by adjusting its various parameters. Each combination of parameters marks a solution that may be good or bad, depending on the value of the objective function. Soft-computing techniques generate the best possible parameter set that gives the best values of the objective function considering the time constraints.

1.6 MODELING THE PROBLEM

We talked about design issues in the discussion of expert system designs, where we introduced the design of an effective system. Here we concentrate our attention on the general modeling of the problem and the application of soft-computing techniques to that problem. We also discuss how to

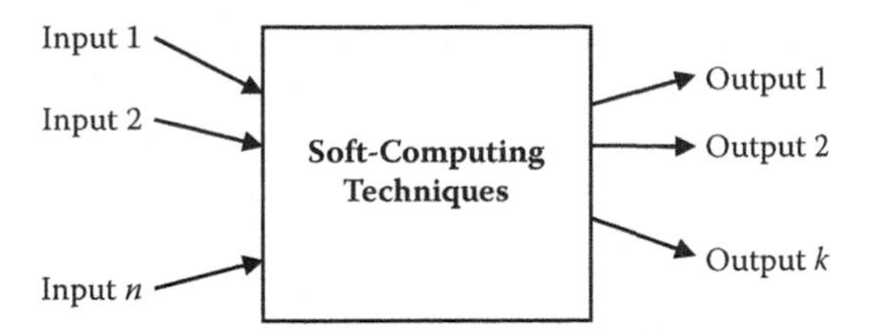

FIGURE 1.9 The black box treatment of soft-computing techniques.

go about problem solving using the soft-computing techniques described here. To begin with, let us first describe the problem.

Soft-computing problems usually involve some situation, and we need to understand that situation to accordingly generate an output. The foremost job is to think about how to represent the entire situation. We try to model the problem in a way that can be given as input to the machine in order to generate output. It is important for the machine to be presented with knowledge, data, and inputs or outputs in a form with which it can work.

Let us look at the system described in Figure 1.9. The inputs need to be preprocessed per the requirements of the problem before being used by the system. Preprocessing makes the inputs more relevant to the system. The soft-computing techniques obtain the outputs from the given input.

To understand better, let us take the example of speaker recognition. We know that the input will be in the form of voice. The output is the identity of the user. But directly taking voice as an input has numerous problems. The first problem the system faces is representation. The voice or speech must be converted to a digital format so the system can understand it; computers do not work with analog inputs. In addition, the range of the inputs must be such that the system can work with it. The second problem involves the amount of data being entered. The voice may be sampled at a high rate, and the amount of data associated with a single word may be too high. The system might not be able to work with such a high amount of data.

One problem with outputs is that it may not be easy for the system to say the speaker's name. There must be a representation such that the system can communicate the identified speaker. Hence the system's output needs to be understandable to the user. In this example, the output may be the speaker number, rather than a name.

We also need to be sure of the soft-computing method being used. We need to ensure that the method works for the data being considered given the problem constraints. The pros and cons of the various soft-computing methods should be analyzed before deciding which model to use.

1.6.1 Input

For any soft-computing technique, it is important to identify the inputs well. Too many inputs may make the system very slow. Hence, we only select those inputs that can appreciably affect the output. If our system has some feature that is not making a fine contribution to the output, that contribution may be neglected from the list of inputs. The input set contains the candidates that affect the output by a reasonable amount. Often while making the system, we may be required to guess the system inputs, because the manner in which different aspects of the problem or different inputs affect the output may not be known at all. The decision of selecting inputs is made using common sense. Let us say, for example, that we are making a system that recognizes some circular object in an image. The color may not be an important input to us, because all objects would be circular whatever color they may be.

As we explained earlier, the system may become very slow and difficult to train if the number of inputs becomes too large. Many times we may be required to adopt feature extraction techniques that extract relevant features from the situation that forms the system's input. In the speaker recognition system, feature extraction consists of power spectral density (PSD), number of zero crossings, resonant frequencies, and so forth. It is impossible for the entire data of continuous speech to be given to the system. On doing so, the system would have faced two problems. First, the training time would have been very large. Second, there would be a problem with performance. We would need too much training data for the system to perform well. So instead we consider the entire system. Calculate PSD, number of zero crossings, resonant frequencies, and so on. Each of these has a limited number of inputs. In addition, because the collection of these features is unique for every speaker of a word, these features make an effective input for the speaker recognition system. Hence it is better to extract these features rather than give the system the complete speech. Note that we did not drop or delete any input or feature here; rather we summarized the entire data onto a few values.

Not that we have given a fair idea of feature extraction, let us provide a formal definition: In feature extraction, "features are essential attributes of an object that distinguishes it from the rest of the objects. Extracting features of an object being a determining factor in object recognition is of prime consideration" (Konar, 2000). The uses of this technique can easily be imagined in the context of practical applications of soft-computing techniques.

Definition 18

Features are essential attributes of an object that distinguishes it from the rest of the objects. Extracting features of an object being a determining factor in object recognition is of prime consideration. (Konar, 2000)

Another term of importance is *dimensionality reduction*. To understand this concept, let us first talk about the terms *dimensionality* and *input space*.

Dimensionality refers to the number of inputs or attributes used by a system in solving a problem. Suppose we have plotted the outputs against the inputs in a graph that has as many axes as the number of inputs and outputs combined. We would end up with a graph that shows the inputs and outputs in a graphical manner, allowing us to study the trends or manners in which the various outputs change with changes in inputs. Using graphs to study these changes is common—corporations make extensive use of graphs to show their profit trends over time. The problem here is that the graph has a very high number of axes (also called dimensionality), and we can only see and understand graphs up to a maximum of three dimensions. This means it is impossible for us to draw these graphs. However, it is still possible to visualize the same information using the best of our visualizing capabilities. To do this, we use a graph known as input space, which represents a high-dimensional space to show the inputs and outputs. From this concept of input space we get the word *dimensionality,* which points toward the number of inputs. We will be referring to this input space throughout the rest of this book.

For the system to work properly, many inputs need to be preprocessed to reduce the final number of inputs. This preprocessing is done by standard dimensionality reduction techniques. Dimensionality reduction maps the large number of inputs to a smaller set of inputs that can be worked with easily. Dimensionality reduction is formally defined as "the process of mapping high dimensional patterns to a lower dimensional subspace and is typically used for visualization or as a preprocessing step in classification applications" (Lotlikar and Kothari, 2000).

Definition 19

Dimensionality reduction is the process of mapping high dimensional patterns to a lower dimensional subspace and is typically used for visualization or as a preprocessing step in classification applications. (Lotlikar and Kothari, 2000)

Note that the specific technique of feature extraction or dimensionality reduction depends on the problem of study. We present some of these techniques in the subsequent units.

Now let us concentrate our attention on the quality of inputs. We know that in almost any soft-computing technique, changing the inputs by a very small amount does not change the outputs by an exceptional amount. In most cases, the change is small, which means the inputs that are very close to each other in value have almost the same output. It may be noted, however, that whenever we use the word *small*, we actually mean a negligible difference; this difference is relative and depends on the problem. In other words, *negligible* may have different magnitudes for different problems. This is a very practical point that can be easily verified. Suppose that in the problem of speech recognition, we experiment n number of times. For each time, the readings may be different. The difference from reading to reading may even exist in consecutive experiments because of the changes in voice, even if the speaker does not change. Hence, in this case, small changes usually will not change the output appreciably.

Another important aspect is that if the same input exists any number of times, the output should always be the same. For the system inputs that we choose, these properties need to hold true in order for the system to perform well. In the example of speaker recognition, if for the set of inputs we have considered two people happen to have all features the same, then our system would fail.

It may also be observed that all soft-computing techniques require that all inputs lie within a finite range of values. It is also better if the inputs given for training are evenly distributed throughout the entire range. This means that all input values must be equally likely to occur in the input. This enables the soft-computing techniques to perform better.

1.6.2 Outputs

Output modeling is as important as input modeling. Outputs must be well known in advance. Because the system can only give valid outputs, not abstract ones, we need to decide the number of outputs and the type of every output according to the problem.

As was the case with inputs, the number of outputs must be limited to keep the complexity of the system low. A very complex system requires a huge amount of time for processing. The amount of data required for training might also be very high. Thus, it is better to limit the number of outputs.

Consider the example of a speaker recognition system. Here the output is the name of the speaker that the system believes is the speaker. The easiest way to deal with this problem is to assign a number to every speaker out of the possible set of speakers. Thus the first speaker would be 1, the second would be 2, and so on. The system then is supposed to output the speaker number (1, 2, etc.). Other methods for modeling the output of such types of classification problems are discussed later.

As was the case for inputs, the outputs must lie between finite ranges of values. It is also desirable for the outputs in the training data to cover the entire output space with equal distribution. This may help the system to perform better.

1.6.3 System

We now know the inputs and the outputs. We next need to build a system that maps the inputs to the outputs. The system may be built using any of the AI or soft-computing approaches that are covered in this text. A combination of two or more of these techniques may also be used to better cater to the needs of the problem. Once the inputs and outputs are known, designing the system using the tools and techniques presented in this text is not a very difficult task.

To get the best efficiency, however, we need to thoroughly understand the technique we are using. We need to know how it works, as well as its merits and demerits. We must use the technique that we are convinced is the best means for solving the problem. In addition, if the existing methods do not perform well, we may need to use a combination of them, along with some intelligent techniques. All of this requires a good understanding of the problem and the systems.

This book describes each method and how it works. Chapter 19 discusses the various techniques that help make an effective system design. The perfect system design may require extensive time, trying newer and newer architectures, experimenting with different scenarios, and working over different parameter values. Only after extensive experimentation may we be able to come up with a good design and validate it against standard solutions.

1.7 MACHINE LEARNING

So far we have talked a great deal about the term *learning* in relation to how a machine tries to learn. Soft computing helps the machine as it tries to memorize the things with which it is presented. This section defines the *machine learning* and all the related terminologies.

It is said that history repeats itself. Hence, it is natural requirement for any intelligent system to learn from the past. If we take a glimpse into the past, we may see many things from which we can learn. The same is true for machines. Machines learn from the past in two ways. Either they are exposed to past happenings to adaptively learn from whatever they experience. In this scenario, there is continuous learning, similar to what happens in human beings. The second form is when a great deal of data is collected and presented collectively to the machine. The machine then takes this pile of data and tries to learn the same.

Whichever method is followed, the end result is a big collection of data. The larger the data, the better the scope of learning by the machine. The data carry hidden facts, rules, and inferences that even the most intelligent humans would not be able to find out easily. The machine, when presented with this pool of data, is supposed to uncover the facts, rules, and inferences so that next time, if the same data occur, it may identify and calculate the correct answer. This capability of the machine is called *machine learning*.

Before we continue our discussion of machine learning, let us first give it a formal definition.

1.7.1 WHAT IS MACHINE LEARNING?

Machine learning is the ability of a machine to learn from the given data and to improve its performance by learning. The motivation is for the machine to find the hidden facts in the data. By doing so, the machine will better be able to get the correct results if the same data are again given as input. The machine summarizes the entire input data into a smaller set that can be consulted to find outputs to inputs in a manageable amount of time.

Thus, machine learning may be formally defined as "computer methods for accumulating, changing, and updating knowledge in an AI computer system" (Kasabov, 1998).

Definition 20
Machine learning refers to computer methods for accumulating, changing, and updating knowledge in an AI computer system. (Kasabov, 1998)

1.7.2 HISTORICAL DATABASE

The main focus in machine learning is to provide a good database from which the machine can learn. The underlying assumption is that the outputs can always be reproduced by some function of the inputs. Hence it is important for the system to be provided with enough data so the machine can try to predict the correct function. To a large extent, the system's performance depends on the volume of data. The data need to be sufficiently large; otherwise the machine will make false predictions that might look correct to the user. In addition, the system must be well validated against all sorts of inputs to ensure that it can cater to the needs of any future input. The diversity of the inputs is also important, and data from all sections of the input space need to be presented.

Let us take the example of a risk assessment system. Suppose the risk of fire at a place depends on the quantity of inflammables at the location and the fire extinguishers available. First we would need a great deal of historical data regarding the standard threat value for different combinations of the two factors. Suppose we have abundant data about cases in which fire extinguishers were few in number and very few data about cases in which fire extinguishers were large in number. Our system might not give a correct prediction for the cases with many fire extinguishers.

The data requirement is met by making a good database. This database is then given to the machine for learning purposes.

1.7.3 Data Acquisition

Now that we know the importance of data—and specifically good data—for the system, let us look at how these databases are built. The standard practice for good machine learning is to make a database for all types of inputs. We use means specific to the problem to collect data for the purpose of machine learning. For example, a speech recognition system requires numerous people to speak the words to be identified multiple times. All the audio clips may then be recorded and stored in the system, and the inputs and outputs may be calculated and stored in the database. In this case, we would need to conduct a survey to collect live data from people.

Many standard problems have standard databases that are available on the Internet. A few examples of these database repositories are the University of California—Irvine (UCI) Machine Learning Repository, the Time Series Data Library (TSDL), Delve data sets, and the UCI Knowledge Discovery in Databases (KDD) Archive. These databases have been built by researchers around the world and donated to be shared under public domain. These databases act as benchmark databases to analyze the performance of any soft-computing technique.

1.7.4 Pattern Matching

We have already discussed the presence of historical data that can be consulted whenever any unknown input is given. We also discussed that the system needs to give the correct answer to the inputs based on the database and the rules it generates. In this section, we briefly discuss another related term—*pattern matching*.

Pattern matching is a collection of methods that can tell us whether and to what degree two given patterns or inputs match. Pattern-matching techniques can hence be used for any recognition system in which we are trying to match any given input to the known patterns available in a database. The pattern that corresponds to the highest match is regarded as the final pattern. These techniques may also be used for verification systems in which the degree of matching decides the authenticity of the person.

For example, in a character recognition system, we have a number of ways in which all the letters can be written. Whenever any input is given, we need to match the input with the available database. This would give us the correct letter that the input represents.

Pattern recognition is a concept similar to pattern matching, which is mostly used for recognition systems. *Pattern recognition* may be defined as "the scientific discipline whose goal is the classification of objects (or patterns) into a number of categories or classes. Depending upon the problem, these objects may be images or signal waveforms or any type of measurements that need to be classified" (Theodoridis and Koutroumbas, 2006).

Definition 21

Pattern recognition is the scientific discipline whose goal is the classification of objects (or patterns) into a number of categories or classes. Depending upon the problem these objects may be images or signal waveforms or any type of measurements that need to be classified. (Theodoridis and Koutroumbas, 2006)

Pattern matching commonly uses AI and soft-computing techniques for the purpose of matching the two sets of patterns. These techniques ensure a timely response by the system, unlike other techniques that are often computationally expensive and hence slower. In most of the real life systems we discuss, soft-computing techniques are applied to extract the rules from the available data; then those rules are used for recognition purposes. This is usually better than matching the input with each and every piece of data available in the database.

Traditional pattern-matching techniques involve statistical methods in which statistical tools are used to determine the degree to which two patterns match. These statistical approaches tend to be very computationally expensive and hence are not commonly used if there are numerous

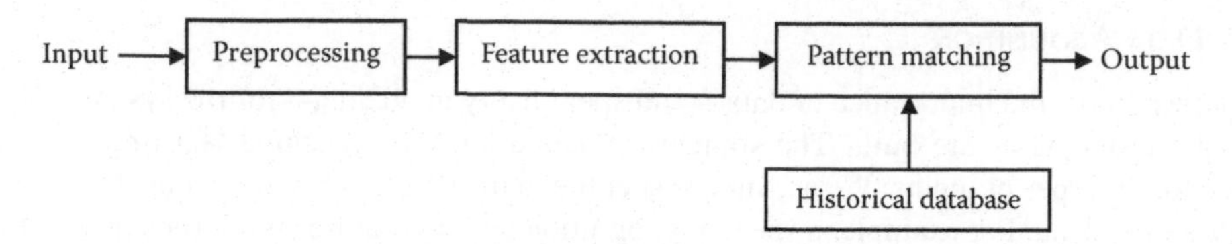

FIGURE 1.10 The various steps involved in a recognition system.

templates available for every kind of pattern in the database. This book does not cover the statistical approaches.

The general structure of the pattern-matching system is given in Figure 1.10, which shows all the steps usually followed by a recognition system. First, the system preprocesses the inputs to remove any noise. The features are then extracted. These extracted features are then matched against a historical database by pattern-matching techniques. The matching results denote the output of the system.

1.8 HANDLING IMPRECISENESS

So far we have talked about the amount of data and about the ways and means of collecting this data. Any sort of experimentation has limitations, and no matter how hard you try, data will always have some amount of impreciseness. This impreciseness can pose a serious limitation to the system. Suppose that your system was supposed to learn data. If there is too much noise, the system may learn the wrong rule, which in turn can make the system behave abnormally.

Any system will fail to perform unless the input is perfect. However, the effect of impreciseness in soft-computing techniques is actually rather low. Soft-computing techniques are appreciably resistant to noise and uncertainties prevalent in the data. They are able to perform well even when placed in highly noisy conditions. The level to which the system can handle noise and uncertainty decides the robustness of the system.

1.8.1 Uncertainties in Data

Uncertainties occur in data when we are not sure that the data presented are correct. It is possible that the data might be wrong. Because of this uncertainty, we cannot decide whether the data should be used for machine learning. If the data are wrong, the training and learning will suffer.

One example of uncertainty is when two sensors get the same data but the data do not match each other. Ideally when two sensors are used for the same reading, the data should match. If the data do not match, then there is uncertainty as to whether we should accept the data. If the data do need to be accepted, then we must determine which is correct.

A related problem is the absence of data. In this case, the data for some of the inputs are completely absent and are not available in the system. Here again, we need to consider how to use this data. A related example is when a sensor is employed to capture some data and the sensor fails to make a reading.

1.8.2 Noise

Noise in the data means that the data present in the database are not the same data present in the system. Rather the data in the database have been deformed or changed to a different value. The amount of deformation depends on the amount of noise. Noise may be present in data for various reasons. It may be due to poor instruments used to measure data or due to mixing of unwanted signals.

Consider our example of speech recognition. If the person speaking is standing where a train can be heard in the background, the voice our system receives would be that of the speaker as well as of the train. Hence the system will not get the correct data and may behave abnormally.

Two very common forms of noise are prevalent in soft-computing systems. The first is impulse noise, in which an attribute, or a set of attributes, is highly noisy for some very small amount of data. The rest of the data has permissible noise. In our speech recognition example, the data being recorded to make the database is from people speaking. If the spoken words of one speaker got mixed with a sudden noise of dropping glass, then this recording would reveal an impulse noise. All other recordings would not have much noise.

The other form of noise occurs when some general noise is randomly distributed through the entire data set. In this case, each and every data set is noisy by some small amount, and the noise is random in nature. Consider the same example of speech recognition. If a fan in the recording room were making noise, this would induce a random noise in all the recorded data sets.

1.9 CLUSTERING

Many times data become abundant in size. Any data-analysis technique has limitations in the amount of data it can handle. Adding more data would result in the system becoming computationally slow. Hence we are often required to limit data within computational limits. Many specialized models of soft computing also face these problems of large data processing. We hence need a technique for restricting the amount of data that the system can serve. This is achieved by clustering the data.

Clustering is an effective way to limit these data. Clustering groups data into various pools. Each pool can then be represented by a representative element of that group; this element has average values of the various parameters. The entire pool may be replaced by this representative, which conveys the same information as the entire pool. In this way, we are able to limit the data without much loss of performance.

Clustering may be formally defined as "the process to partition a data set into clusters or classes, where similar data are assigned to the same cluster whereas dissimilar data should belong to different clusters" (Melin and Castillo, 2005). The distance measure, or the difference in parameters between the two clusters, measures the closeness of the clusters to each other.

Definition 22

Clustering is the process to partition a data set into clusters or classes, where similar data are assigned to the same cluster whereas dissimilar data should belong to different clusters. (Melin and Castillo, 2005)

Clustering is used very commonly in soft computing for the purpose of analysis. Clustering reduces the amount of data so that we can easily visualize, plan, and model the system. In addition, clustering reduces the system's execution time.

Now let us briefly discuss some of the commonly used clustering techniques.

1.9.1 K-Means Clustering

This algorithm takes as its input a number k, along with the data to be clustered, and in return generates k clusters, which is the number of clusters we are supposed to supply in advance. The algorithm divides the data into k clusters. The algorithm is presented in Algorithm 1.2.

Algorithm 1.2

k-Means Clustering (*k*)
C ← random *k* centers
While change in cluster centers is not negligible
Do
    For all elements *e* in data set
    Do
        Find the center c_i, which is closest to *e*
        Add *e* to cluster *i*
    For all centers c_i in *C*
    Do
        C_i ← arithmetic mean of all elements in cluster *i*

To begin, *k* random points are chosen. These points are regarded as the centers of the *k* clusters. The elements are then distributed to these clusters. Any element belongs to the cluster that is closest to it. After all the elements have been distributed in this manner, the cluster centers are modified. Each cluster has a set of elements given to it during distribution. The new cluster center lies at the mean of all the cluster members. All the *k* cluster centers are modified in this manner. We again start with the distribution of elements to the clusters. We repeatedly do this until the changes in the cluster centers become very small or the centers stop changing their positions.

1.9.2 Fuzzy C-Means Clustering

Another commonly used clustering algorithm is the fuzzy C-means clustering. This algorithm is also iterative, which means that the steps are repeated until the stopping criterion is met. The clustering improves with every step. In this algorithm, rather than associating every element with a cluster, we take a fuzzy approach in which every element is associated with every cluster by a certain degree of membership. This is why the algorithm is prefixed *fuzzy*. The degree of membership is large if the element is found near the cluster center. However, if the element is far from the cluster center, then the degree of membership is low. The algorithm is presented in Algorithm 1.3.

For any element, the sum of the degree of belongingness to all clusters is always 1, as is represented in Equation 1.1:

$$\sum_{i=1}^{k} u(e,i) = 1 \tag{1.1}$$

where $u(e, i)$ denotes the membership of the element *e* to the cluster *i*.

The center of any of the clusters is the mean of the points weighted by the degree of membership to this center. The center c_i for any cluster *i* can be given by Equation 1.2:

$$c_i = \frac{\Sigma_e e * u(e,i)^m}{\Sigma_e u(e,i)^m} \tag{1.2}$$

The degree of belongingness of any element *e* to the cluster *i* is the inverse of the distance between the element and the cluster center c_i. This is given by Equation 1.3:

$$u(e,i) = \frac{1}{\text{distance}(e,c_i)} \tag{1.3}$$

The coefficients are normalized and fuzzified with a real parameter $m > 1$ in order to make their sum 1. This degree of membership of any element *e* to cluster *i* can hence be given by Equation 1.4:

$$u(e,i) = \frac{1}{\Sigma_j \left(\frac{\text{distance}(c_k,e)}{\text{distance}(c_j,e)}\right) 2/(m-1)} \tag{1.4}$$

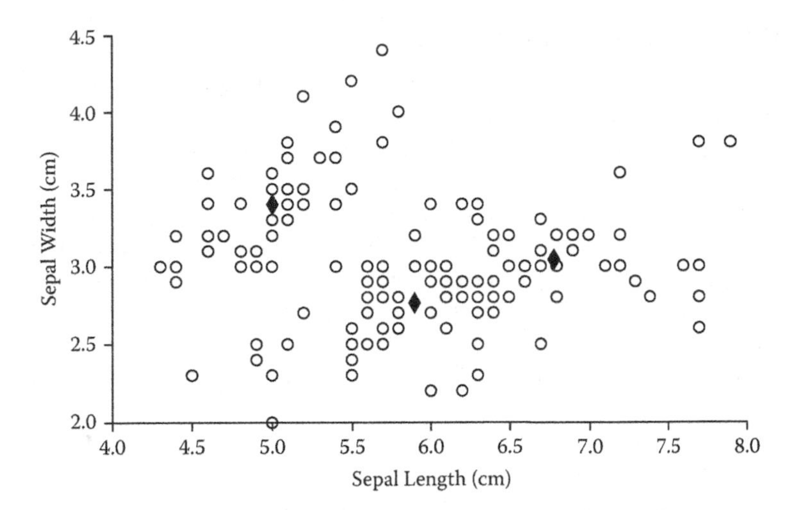

FIGURE 1.11a Fuzzy C-means clustering. *Note*: The black dots indicate centers.

1.9.3 Subtractive Clustering

In Algorithms 1.1 and 1.2, we had to give the number of clusters in advance. In subtractive clustering, however, the number of clusters does not need to be specified in advance. It thus may be used to find clusters when we do not wish to specify the number of clusters; instead, we expect the algorithm to inspect the data and decide the number of clusters on its own. This is a fast algorithm that is used to estimate the number of clusters and the cluster centers in a data set. Due to lack of space in this text and the complexity of the algorithm, we do not study this algorithm in detail.

As an example, we will use the IRIS classification data from the UCI Machine Learning Repository to carry out clustering by these three algorithms. The results are given in Figures 1.11a and 1.11b for fuzzy C-means (FCM) and subtractive clustering. The IRIS data have four features that were measured: sepal length in centimeters, sepal width in centimeters, petal length in centimeters, and petal width in centimeters. The graphs have been plotted between the first two parameters. The solid black circles denote the cluster centers. Note that here, the FCM clustering was intended to give three clusters, and hence it gave 3 clusters. The subtractive clustering gave four clusters on this database.

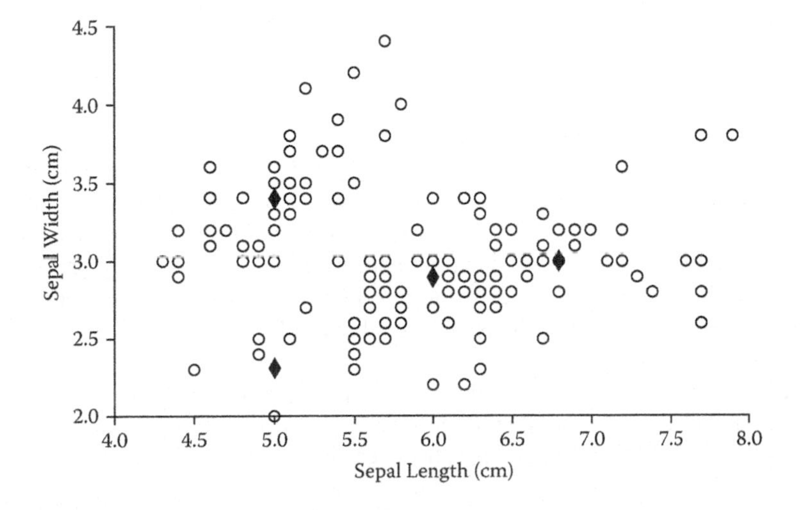

FIGURE 1.11b Subtractive clustering. *Note*: The black dots indicate centers.

1.10 HAZARDS OF SOFT COMPUTING

Every system has a liability associated with it. This is most obvious when the system fails. Every system has certain assumptions and constraints that change with time. It cannot be guaranteed that any system will continue performing without failures, as errors and bugs are prone to every system. Thus, it is possible for a system to generate wrong answers or even crash.

The case of soft computing is even more special. Even after long research, we have not been able to achieve 100 percent accuracy. This means that soft-computing systems are not to be relied upon blindly. Every soft-computing system has some small amount of error that must be considered. Normally such an error is acceptable to the user or application, and hence it does not make much of a difference. But in certain conditions, such an error can prove bad.

Say we make a diagnostic system that detects diseases in humans. Further suppose that these systems replace doctors entirely and take the job of diagnosis entirely. Now suppose a patient is wrongly diagnosed as a result of the system's limitations. The cure the patient takes might be completely irrelevant or it may prove to be disastrous for the patient. The condition is even more tragic if the presence of disease is not shown and the disease actually exists. In any case, the need for doctors to make the final conclusions would prevail.

1.11 ROAD MAP FOR THE FUTURE

Soft-computing techniques have seen research in every domain, ranging from practical applications to theoretical foundations. As a result, these techniques give much better performance and are able to solve problems very efficiently. There is, however, still a great deal of scope for things to be done in the future to enable people to better enjoy the fruits of soft computing and to tap the full potential of soft computing.

Soft computing is still far behind in providing the functionality of the human brain. The brain is still the biggest motivator and challenge for engineers working on soft computing. Both the computational power and performance of the human brain is beyond the reach of our present systems.

The existing systems also have far to go in terms of robustness. Performance is still a major issue for various systems. In addition, so many domains still remain untouched by soft computing, and soft-computing techniques are still in the growing stage in many other fields. Hence, we find a very bright future and much potential for further research into the fields of soft computing. We can surely expect many wonders in the near future as a result of this field.

1.11.1 Accuracy

The best accuracies of well-established systems may revolve around 99 percent to 99.9 percent. These accuracies are achieved when many assumptions have been imposed on a system. The accuracies of some of the more complex problems, however, are much lower. Attempts are being made to build better architectures and to extract better features in order to improve this accuracy. Simultaneous work in all the fields, be it biometrics, bioinformatics, or some other field, are being carried out in the hopes that these systems will one day be able to cross all barriers to achieve 100 percent accuracy.

Systems are also prone to noise. Many different preprocessing techniques have been devised for specific applications to compress a good amount of noise. Along with better preprocessing, better recognition systems might result in good performances in real life applications.

1.11.2 Input Limitations

It is well known that the performance of soft-computing techniques largely depends on the size of the input data set, with higher data sets being a big boon to the system. Presently all systems have a limited data set through which they are trained and through which performance is tested. The limited data-set size gives good results in experimentation but has huge limitations in real life problems. The types of data are also limited in experimentation. In real life problems, the data size needed for a system to perform might be too high.

A speaker recognition system might work well for ten speakers. But for the practical real life problem, if we increase the number of speakers to 10,000, the system might refuse to work at all. This imposes a serious issue on deployment of these systems into the real world.

1.11.3 Computational Constraints

Every system needs to perform under a given time. Even though computation power has increased rapidly over the past few decades, an increase in expectations and performance of soft-computing techniques means we need greater computation power. An exponential rise in the size of data sets and in the number of inputs has also had a deep effect on the computational requirements.

If computation power continues to grow, we may soon have systems that can perform much better in a shorter time. The increase in computation would mean a greater autonomy in the selection of features, data sets, or complexity.

1.11.4 Analogy with the Human Brain

The human brain is one of the most magnificent developments and is responsible for the intelligence of human beings. The history of humankind narrates the wonderful creations that the human brain is capable of producing. We now give a brief glimpse into the kind of system the brain is.

It is estimated that the human brain contains 50 to 100 billion neurons. Of these, about 10 billion are cortical pyramidal cells, which pass signals to each other via approximately 100 trillion synaptic connections. Even the most complex machines that we make, including our super computers, are too small in computation when compared with the brain's massive structure.

The large computation in parallel by the biological neurons is what makes it possible for the brain to do everything it does. It will take a very long time for engineers to build any machine that is comparable to the human brain. But who knows, one day artificial life and artificial humans might turn out to be a reality.

1.11.5 What to Expect in the Future

We have already discussed how performance, robustness, and computation play a major role in changing the face of soft computing in the future. We have even discussed that in the future, soft computing will move from the laboratories and into the practical areas of industry and consumer life. But there is much more that is possible in the very near future.

For years, one of the major problems with soft computing has been its multidisciplinary nature. To make effective systems, either the soft-computing experts will have to master other domains or people from other domains will have to master soft computing. This problem has prevented soft computing from entering many domains.

As a result, there is a considerable amount of work that is yet to be done by soft computing in many domains. Examples include the study of human behavior, the effects of chemicals, or the study of nature at large. Some very positive developments may be seen in the near future when soft computing enters into these newer domains.

In the fields where soft computing has already established itself, the application of newer models may also give some great results. Many of the recently proposed models are yet to find a place in solving various problems. In many of the problems, these models may be adjusted for specific applications.

Making newer and newer models by mixing existing models is another very active field of research. We may soon find many innovative ways to handle the problems that soft-computing techniques are currently facing.

1.12 CONCLUSIONS

Soft-computing tools and techniques have already proved to be a big boon for the industry, where they have been able to solve problems that earlier had no solutions. Soft-computing techniques have found a place in almost every domain and have affected every sphere of human life. The next decade might see a complete transformation in the way society lives and functions, as soft computing moves out of the laboratories and into industry. This book is intended to be a step in the same direction.

CHAPTER SUMMARY

This chapter introduced the field of soft computing. In it, we presented various terms and concepts that are commonly used in the literature to discuss any soft-computing system. The chapter also covered the methodologies, technologies, and concepts of any real life system to fulfill the prerequisite of understanding and making real life systems that we discuss in Section II.

In this chapter, we discussed the importance of uncertainties in systems and proposed the application of uncertainties in making the most complex systems easier to engineer. We provided a formal definition for *soft computing* and described the differences between soft and hard computing. We explained why soft-computing techniques make it possible to develop systems that are impossible to develop with hard computing. We also introduced the multidisciplinary nature of soft computing and gave a brief outline of the various fields in which soft computing is being applied.

We then diverted our attention to the related field of artificial intelligence. We saw how we can model any problem in artificial intelligence and the basic solving techniques it uses. We learned about the graph search algorithm and how it is used to solve common problems. We further presented the concepts of logic and production rules and how they are used to solve real life problems.

We also introduced the various techniques that contribute to soft computing, such as artificial neural networks, evolutionary computing, fuzzy systems, and hybrid systems. We use these techniques for all real life problems. We even talked about expert systems, which are designed for end-to-end customer needs and which behave just like an expert who has complete knowledge of his or her field. We also saw how these systems are designed.

We discussed the various kinds of problems that are solved by soft-computing techniques, including functional approximation, classification, and optimization. We again saw the importance of optimal performance and how to achieve it using soft-computing techniques. We also introduced machine learning and pattern recognition.

We then highlighted the presence of impreciseness and noise and showed how these can reduce a system's performance. We highlighted the importance of systems that can handle impreciseness better, as they are more practical in real life problems.

We also saw the role of clustering in controlling the inputs and bringing them within the computation limits. We discussed the various clustering techniques and presented the hazards of soft computing. Finally, we talked about the future of soft computing in terms of better performance, better systems, and newer domains.

SOLVED EXAMPLES

1. **Solve the water jug problem by using breadth-first search (BFS): You are given a 4-gallon jug and a 3-gallon jug, a pump with unlimited water that you can use to fill the jug, and the ground on which water may be poured. Neither jug has any measurement markings on it. How can you get exactly 2 gallons of water in the 4-gallon jug?**

   Answer: Let us denote a state by <*i*, *j*>, where *i* is the amount of water in the 4-gallon jug and *j* is the amount of water in the 3-gallon jug. At any time, the following moves are possible:

   - Dump out water from the 4-gallon jug.
   - Dump out water from the 3-gallon jug.
   - Pour water from the 4-gallon jug into the 3-gallon jug.
   - Pour water from the 3-gallon jug into the 4-gallon jug.
   - Fill the 4-gallon jug.
   - Fill the 3-gallon jug.

   The initial condition is <4, 0>. The final condition is <2, *>. Starting from <4, 0>, we begin to expand the graph. Any expansion resulting in an already expanded or already seen node is not expanded.

   The solution is given in Figure 1.12.

2. **Assume that the odor of a fruit consists of seven parameters. Each parameter value lies between 0 and 1. Sensors are available that can measure these seven parameters. Each fruit has some characteristic smell that can be used for identification. Calculate the feasibility of the recognition system to be made by (a) rule-based approach and (b) soft-computing approach. Make necessary assumptions wherever necessary.**

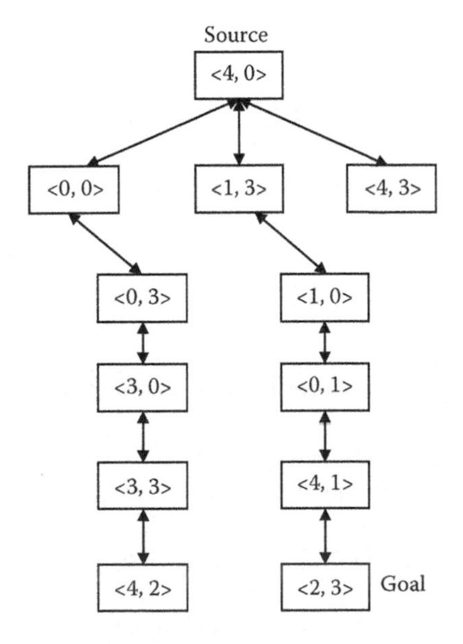

FIGURE 1.12 The solution for the water jug problem.

Answer:

Rule-Based Approach:

Suppose that each of the seven parameters is significant only up to the hundredth decimal place. This means an error of ±0.01 is tolerable.

Thus each of the seven parameters can take values between 0.00 and 1.00 in intervals of 0.01 (100 values).

The total number of combinations of values hence becomes 100^7, or 10^{14}.

In the worst case, we need 10^{14} rules to map each input to some output.

Assuming each rule takes 1 byte of storage capacity, we would be required to store 10^{14} bytes of memory, which is approximately 1000 terabytes.

Further imagine the time required to write this many rules. It might take experimentation of more than a lifetime to make a complete system.

Soft-Computing Approach:

Due to the soft nature of this approach, the system can be made using about 240 training data cases. This would require less than a minute to train and much less storage capacity. The generalization is assumed to be in the system.

3. **Identify the features that may be extracted in a soft-computing system for face recognition.**

Answer: Because it is not possible to give the entire face as input due to the system's limited input capacity, we must use feature extraction.

The following are the possible features that can be used:

- Length of the eye
- Width of the eye
- Diameter of the center of the eye
- Length of mouth
- Width of mouth
- Diameter of the center of the mouth
- Distance between the center points of the eyes
- Distance between the center points of the eye and mouth

Note that these features will be same for a person every time experimentation is done. Image-processing techniques may be used to extract the features. Also note that these conditions are quite robust against changing lighting conditions.

4. **Write approximate rules to differentiate between a straight line and a curve.**

Answer: The following algorithm may be used:

Inputs: A set of consecutive points of the form (x_1, y_1), (x_2, y_2), (x_3, y_3) . . . , (x_n, y_n) such that they form a line or curve.

Output: "Line" if it is a line and "Curve" if it is a curve.

Let the figure represent a line, and let the equation of the line be given by Equation 1.5.

$$(y - y_1) = (x - x_1) * \frac{(y_n - y_1)}{(x_n - x_1)} \tag{1.5}$$

Algorithm

$e \leftarrow 0$
For all points (x, y) in the figure
Do
  Compute y for x by using Equation 1.5
  $e \leftarrow e + | y - y' |$
if e is less than threshold, return "line"
else return "curve"

5. **Comment on the following being used as features for the detection of words spoken by a person (speech recognition system):**

- Amplitude of speech at the end of 0.1 second
- Amplitude of speech at the end of 0.2 second
- Amplitude of speech at the end of 0.3 second
- So on until the nth second

where n is the maximum duration to speak a word.

Answer: The recognition system will face the following problems:

- The speech of different people will be at different speeds. Hence the amplitude after any second may not refer to the same part of the speech.
- For the same person, the duration might change drastically during speech.
- The amplitude highly depends on the speaker. All speakers have different amplitude levels.
- The amplitude varies depending on the speaking condition, mood of the speaker, and so on.

It may be concluded that this would not make a good recognition system.

EXERCISES

General Questions

1. What is soft computing?
2. What role does soft computing play in everyday life?
3. What role can soft computing play in robotics?
4. Explain under what conditions hard computing fails.
5. What is artificial intelligence?
6. How can you use artificial intelligence to solve a maze?
7. What are production rules, and what is their application in artificial intelligence?
8. Explain the various soft-computing techniques.
9. When are hybrid systems preferred over normal systems?
10. Explain the term *expert system*.
11. Take any problem from real life and model it as an expert system.
12. Identify the possible inputs that a system could take while making a system that calculates the risk in stock investment.
13. Give at least five real life examples of (a) classification problems, (b) functional approximation problems, and (c) optimization problems.
14. Explain the concept of machine learning.

15. Based on your understanding, list the parameters on which machine learning depends.
16. What is pattern matching?
17. Derive a mathematical expression to find the degree of matching of two signals.
18. What is the role of soft computing in pattern matching?
19. Define the term *noise*.
20. What role does noise play in the performance of a system?
21. Explain the role of clustering in soft computing.
22. Cluster the data (1, 2, 3), (2, 1, 4), (4, 3, 2), (2, 2, 2), (1, 5, 2), (2, 2, 1), (1, 2, 1), (3, 3, 2), (1, 4, 2), (1, 4, 1) into two clusters using (a) k-means clustering and (b) fuzzy C-means clustering. Compare the results.
23. Express the hazards of soft computing in the field of speech recognition.
24. Identify the applications of a speech recognition system.
25. Comment about the future of soft computing.
26. Identify the applications of subtractive clustering in real life problems.
27. Comment on the applications of soft computing in game playing.
28. What features may be used for the detection of cancer in human beings?

Programming and Practical Questions

1. Use A* algorithm to solve the nine puzzle problem that has 3 × 3 cells with 1 blank cells, using the concepts of four-puzzle problem presented in the text. Assume the source and the goal is given.
2. Assume that the performance of any graph search algorithm is proportional to the number of vertices it visits. Compare the performances of the graph searching techniques presented in solving a maze by taking various types of routes. Then comment on the advantages and disadvantages of the various search techniques. Make suitable assumptions whenever necessary.
3. Randomly generate 10,000 data elements of five dimensions and cluster the data by using (a) k-means clustering and (b) fuzzy C-means clustering.
4. Analyze the time and memory required for the various clustering techniques on increasing the input size.
5. Write rules to differentiate among a square, a triangle, and a circle. Make a program in any language that implements these rules and then check your rules. At what inputs does your program fail? What assumptions did you make?
6. Make a basic system that stores the pictures of characters entered and the name of that character. Whenever a new input is given, the system checks its database. If an exact picture exists, it returns the corresponding character. Can you think of ways to make this system identify small deformed inputs?

2 Artificial Neural Networks I

2.1 ARTIFICIAL NEURAL NETWORKS

As we have noted, a glimpse into the natural world reveals that even a small child is able to do numerous tasks at once. The most complex problems that engineers keep struggling with are so simple for a child to do. Let us look again at the example of a child walking. Probably the first time that child sees an obstacle, he/she may not know what to do. But afterward, whenever he/she meets obstacles, he/she simply takes another route. It is natural for people to both appreciate these observations and learn from them. An intensive study by people coming from multidisciplinary fields marked the evolution of what we call the artificial neural network (ANN).

One of the most important factors that has resulted in a high industrial use of ANNs is their ability to solve very complex problems with great ease. Researchers and developers can use their knowledge and experience to design a simple system that is able to perform wonders. Although the design and implementation of ANNs is simple, a great deal of effort may be required to collect the necessary historical data used for learning. It may also be necessary for the researchers and developers to constantly try newer and newer architectures until they ultimately find an architecture that solves the problem with the best results.

ANNs have a great ability to learn large volumes of data, which enables them to give accurate results whenever the data reappear in the problem. This ability to learn is mainly due to the inherent property of ANNs being able to imitate almost any function. Hence they are great tools for machine learning.

Another important advantage of the neural networks is their ability to give the correct answers to unknown inputs. In most cases, they are able to find the correct answer to unknown inputs with sufficient accuracies whenever that are presented to the system.

A great deal of work over the past few decades has seen very promising architectures, models, and methods for solving the problems that can be solved by the use of ANNs. These networks have become very handy tools for industrial implementations. In fact, ANNs are being used in almost every industry to solve real life problems. Many of the systems we discuss in the next units use ANNs to find the outputs to applied inputs.

ANNs represent a highly connected network of neurons, the basic processing units, that operate in a highly parallel manner. Each neuron does some amount of information processing. It derives inputs from some other neuron and in return gives its output to other neurons for further processing. This layer-by-layer processing of the information results in great computational capability. The general architecture of the ANN is given in Figures 2.1a and 2.1b. As a result of this parallel processing, ANNs are able to achieve great results when applied to real life problems. These networks vary in their complexity, depending on the problem and its requirements. This complexity increases with the number of neurons. As the number of neurons increases, the ANNs become capable of imitating more and more complex functions.

Let us take an in-depth look at this exciting playground for system engineers, beginning with a brief history of ANNs.

2.1.1 Historical Note

The magnificent computing capabilities of the human brain have always been a source of inspiration for people in multiple disciplines. For centuries, people have tried to devise methods to study and

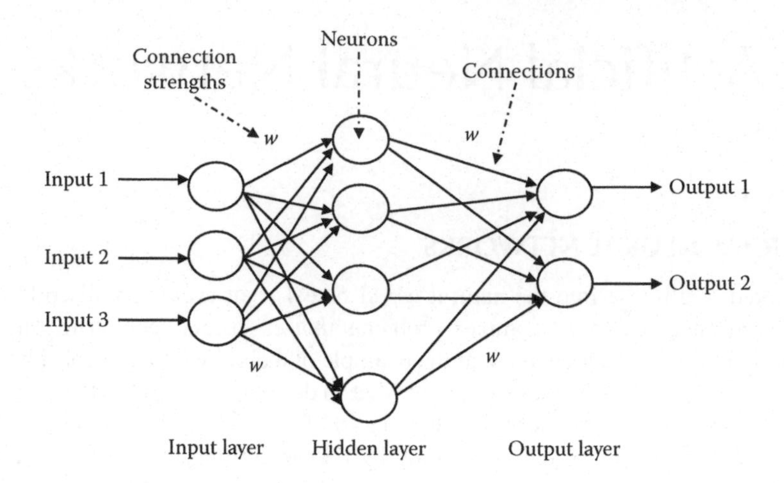

FIGURE 2.1a The architecture of the artificial neural network.

imitate the brain. But it was only relatively recently that research succeeded in getting preliminary results that explained the working of the human brain. In 1943, neurophysiologist Warren McCulloch and mathematical prodigy Walter Pitts studied how neurons worked and were able to design a simple electronic circuit of a simple artificial neural network. They are credited with the McCulloch-Pitts theory of formal neural networks (Haykin, 1994). In 1949, Donald Hebb further extended the work in this field, when he described how neural pathways are strengthened each time they are used (Haykin, 1994). Hebb published a book titled *The Organization of Behaviour*. In the 1950s, when traditional computing began, the field of ANNs began to fall by the wayside. However, in 1954, Marvin Minsky presented his thesis "Theory of Neural-Analog Reinforcement Systems and Its Application to the Brain-Model Problem" and also wrote a paper titled "Steps Toward Artificial Intelligence."

Later John von Neumann invented the von Neumann machine. And in 1958, Frank Rosenblatt, a neurobiologist, proposed the perceptron, which is believed to be the first physical ANN. However, in 1969, Minsky and Seymour Papert described the limited capability of perceptrons in their book *Perceptrons* (Masters, 1993). Between 1959 and 1960, Bernard Wildrow and Marcian Hoff of Stanford University developed the Adaptive Linear Elements (ADALINE) and the Multiple Adaptive Linear Elements (MADELINE) models. In 1986, David E. Rumelhart, Geoffrey E. Hilton, and Ronald J. Williams proposed the back propagation algorithm.

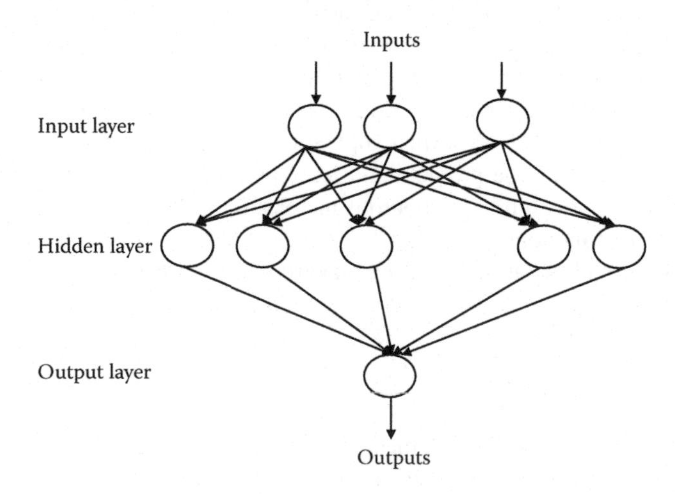

FIGURE 2.1b The schematic diagram of the artificial neural network.

Since the 1990s, the field has grown exponentially. The increase in computing capabilities helped break the shackles of computational infeasibility and has been a major reason for success in the field. People have since begun studying the applications of ANNs in various domains and fields. As a result, ANNs have become very popular industrial tools for all the major real life problems. Further increases in computing capabilities, along with high-quality research and newer models, indicate a very strong future for this field.

2.2 THE BIOLOGICAL NEURON

As already discussed, ANNs are an inspiration from the biological neuron. In this section, we take you through a very complex structure of unimaginable computation. We introduce you to a model that has millions of neurons, all talking to each other at unimaginable speeds. This model is the human brain.

The entire human brain consists of small interconnected processing units called *neurons*. These neurons are connected to each other by nerve fibers. These interconnections make a very complex structure that make it possible for a neuron to pass information to other neurons and take information from other neurons. These fibers enable a huge amount of information to be passed between the neurons. The interneuron information communication makes it possible for multilevel hierarchical information processing, which gives the brain all its problem-solving power.

Each neuron is provided with many inputs, each of which comes from other neurons. The way in which each neuron processes information is still not clear to scientists, who infer that each neuron takes a weighted average of the various inputs presented to it. These weights are the strengths of the interconnection between the two neurons. The weighted average is then made to pass over a nonlinear inhibiting function that limits the neuron's output. The nonlinearity in biological neurons is provided by the presence of potassium ions within each neuron cell and sodium ions outside the cell membrane. The difference in concentrations of these two elements causes an electrical potential difference, which in turn causes a current to flow from outside the cell to inside the cell. This is how the neuron takes its inputs. The difference in potentials between the neuron and its surrounding cell causes the variation of weights. Processing is done by invasion in diffusion current occurs due to the synaptic inhibiting behavior of the neuron. Thus, the processed signal can propagate down to other neighboring neurons.

A neuron has four main structural components: the dendrites, the cell body, the axon, and the synapse (see Figure 2.2). Dendrites are responsible for receiving the signals from other neurons. These

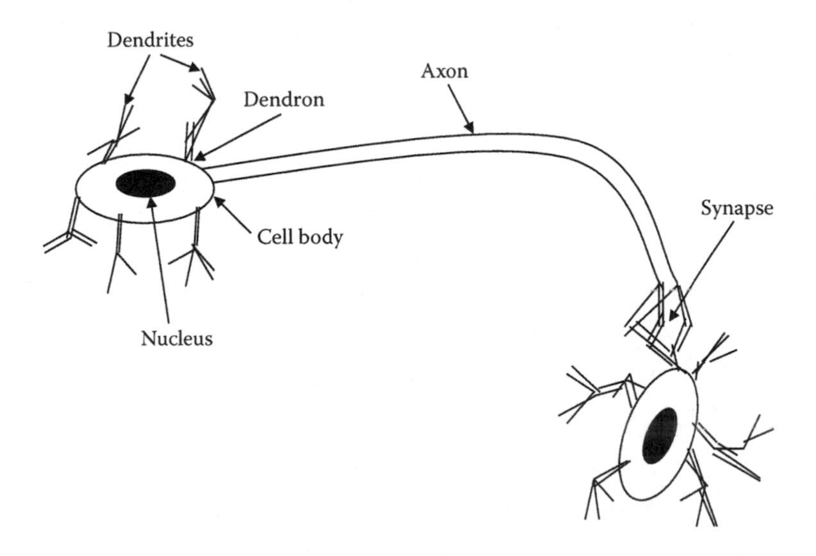

FIGURE 2.2 The biological neuron.

signals are passed through a small, thick fiber called a *dendron*. The received signals collected at different dendrites are processed within the cell body, and the resulting signal is transferred through a long fiber called the *axon*. At the other end of the axon exists an inhibiting unit called a *synapse*, which controls the flow of neuronal current from the originating neuron to the receiving dendrites of neighboring neurons. Axons and dendrites in the central nervous system are typically only about one micrometer thick, whereas some in the peripheral nervous system are much thicker.

2.3 THE ARTIFICIAL NEURON

We have seen that ANNs are a highly interconnected network of artificial neurons. Let us now get acquainted with the artificial neurons. The artifical neuron is the most basic computational unit of information processing in ANNs. Multiple such neurons are employed by ANNs for information processing, with each neuron taking information from a set of neurons, processing it, and giving the result to another neuron for further processing. These neurons are multiple in numbers and contribute to the neural computation that collectively becomes immense, which gives ANNs the power to solve complex, real life problems. These neurons are a direct result of the study of the human brain and attempts to imitate the biological neuron. "The neural network, by its simulating a biological neural network, is a novel computer architecture and a novel algorithmization architecture relative to conventional computers. It allows using very simple computational operations (additions, multiplication, and fundamental logic elements) to solve complex, mathematically ill-defined problems, nonlinear problems, or stochastic problems" (Graupe, 2007).

Definition 23

The neural network, by its simulating a biological neural network, is a novel computer architecture and a novel algorithmization architecture relative to conventional computers. It allows using very simple computational operations (additions, multiplication, and fundamental logic elements) to solve complex, mathematically ill-defined problems, nonlinear problems, or stochastic problems. (Graupe, 2007)

2.3.1 STRUCTURE

The artificial neuron consists of a number of inputs. The information is given as inputs via input connections, each of which has some weight associated with it. An additional input, known as *bias*, is given to the artificial neuron. We study the role of this input in the subsequent sections of this chapter. The neuron also consists of a single output. The output is formed from processing of the various inputs by the neuron. The structure of the neuron is given in Figure 2.3. As stated earlier,

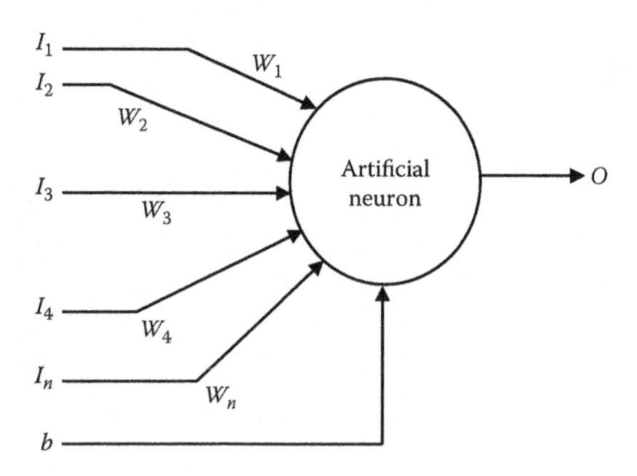

FIGURE 2.3 A single artificial neuron.

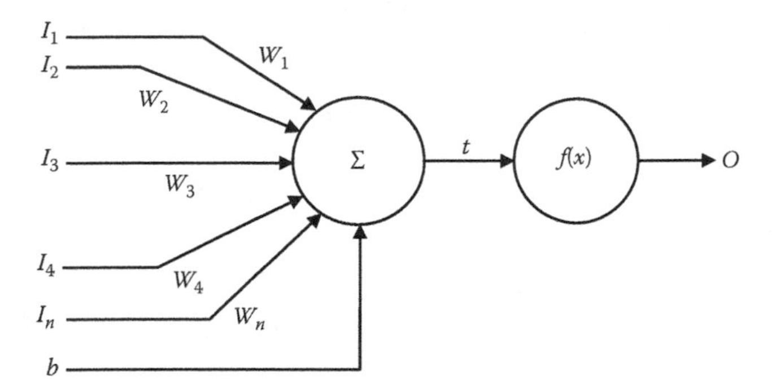

FIGURE 2.4 The processing in a single artificial neuron.

the various inputs come from different neurons or an external source, and the output goes to other neurons or is finally given as system output. This transfer of information takes place via connections that are organized according to the model being used.

In Figure 2.3, the inputs are marked $I_1, I_2, I_3, \dots, I_n$; the weights associated with each connection are given by $W_1, W_2, W_3, \dots, W_n$; b denotes the bias; and the output is denoted by O. Because there is one weight for every input, the number of inputs is equal to the number of weights in a neuron.

2.3.2 The Processing of the Neuron

The artificial neuron does a very basic amount of functionality. The functionality that is performed can be broken down into two steps. The first is the weighted summation, and the second is the activation function. The two steps are applied one after the other, as shown in Figure 2.4.

The first block is the summation block, the role of which is to add together all the given inputs weighted by their connection weights. The bias is added directly and behaves as if the weight of the bias connection were ±1. This block hence acts as an aggregator to reduce all the inputs into a single element. The function of the summation block is given by Equation 2.1:

$$t = \sum_{i=1}^{n} W_i * I_i + b \tag{2.1}$$

The summation forms the input to the next block that is the block of the activation function, where the input is made to pass through a function called the *activation function*. The activation function performs several important tasks, one of the more important of which is to introduce nonlinearity to the network. The nonlinearity enables an ANN to imitate the behavior of complex shapes. Without the activation function, the neural network would only be able to predict simple, linearly separable problems or to imitate linear functions. We discuss the linearly separable problems in the subsequent sections.

Another important feature of the activation function is its ability to limit the neuron's output. This is an important task, because this output is the input for the next neuron. Hence the activation function ensures that the next neuron gets input in the ranges with which it can work.

Many activation functions are available in the literature. Each is suited to a different type of problem. These common activation functions are discussed in the subsequent sections of this chapter.

The mathematical expression of the activation block is given by Equation 2.2. The complete mathematical function of the artificial neuron can be combined from Equations 2.1 and 2.2 to give Equation 2.3:

$$O = f(t) \tag{2.2}$$

where f is any activation function

$$O = f\left(\sum_{i=1}^{n} W_i * I_i + b\right) \tag{2.3}$$

2.3.3 The Perceptron

The perceptron is the most basic model of the ANN. It consists of a single neuron. The perceptron may be seen as a binary classifier that maps every input to an output that is either 0 or 1. The perceptron is given by the function represented by Equation 2.4:

$$f(x) = \begin{matrix} 1 & \text{if } w.x + b > 0 \\ 0 & \text{otherwise} \end{matrix} \tag{2.4}$$

where w is the vector of real-world weights, x is the input, "." is the dot product of the two vectors, and b is the bias.

The perceptron has learning capabilities in that it can learn from the inputs to adapt itself. This adaptation is in the form of adjustment of weights. As a result, the perceptron is able to learn historical data and give a correct answer when those data are presented again.

2.4 MULTILAYER PERCEPTRON

Because the perceptron is a single neuron, it cannot solve all sorts of problems. The limitations of the perceptron can be removed by applying many neurons over a complex network architecture. The application of many perceptrons in various layers gives rise to multilayer perceptrons (MLPs). The computational capabilities of these networks enable them to solve almost any complex problem.

Figure 2.1 presented a very basic ANN architecture, with only a few neurons connected to each other. Real life ANNs, on the other hand, consist of much larger neurons spread over different layers and involving numerous connections. These complex networks cannot be drawn on a piece of paper. However, we are able to present the general architecture of the ANN. Figure 2.5 shows the various components of the ANN.

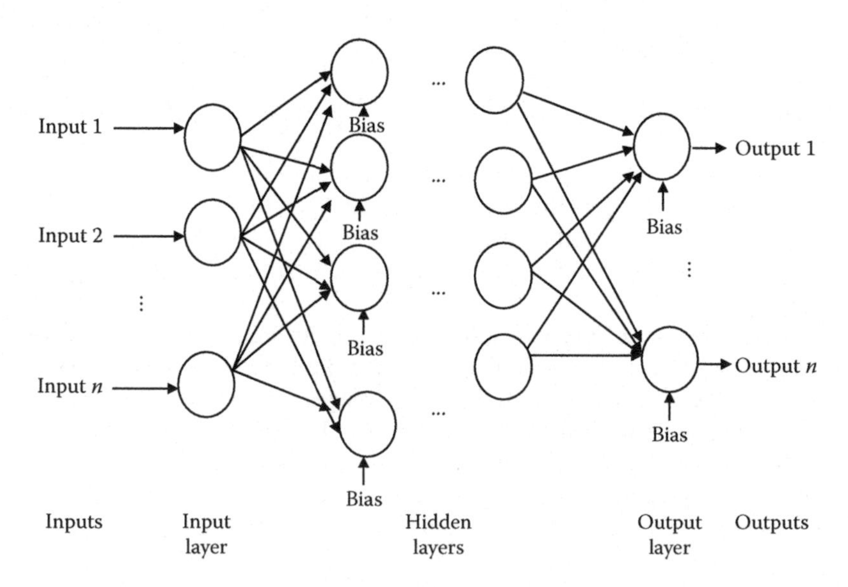

FIGURE 2.5 The general architecture of the multilayer perceptron.

MLPs are formally defined as "networks used to overcome the linear separability limitation of the perceptrons. An MLP consists of an input layer, at least one intermediate or 'hidden' layer, and one output layer, the neurons from each layer being fully connected (in some particular applications, partially connected) to the neurons from the next layer" (Kasabov, 1998). This definition states the basic motive of any intelligent system to have an output corresponding to any input.

Definition 24
Multilayer perceptrons are networks used to overcome the linear separability limitation of the perceptrons. An MLP consists of an input layer, at least one intermediate or "hidden" layer, and one output layer, the neurons from each layer being fully connected (in some particular applications, partially connected) to the neurons from the next layer. (Kasabov, 1998)

In this section, we take you through an in-depth analysis of ANNs, including a discussion of the various concepts and building blocks of these networks.

2.4.1 Layers

The entire architecture of the ANN is in the form of layers, as can easily be seen in Figure 2.5. The number of layers decides the computational complexity of the ANN. A network with fewer layers is always computationally less expensive as compared with a network with a larger number of layers. The complexity of the function imitated by the ANNs also increases as the number of layers increases.

In general, every ANN has at least three layers: the input layer, the hidden layer, and the output layer. Although there is only one input layer and one output layer, there can be multiple hidden layers.

The first layer is the input layer. The number of neurons in the input layer is equal to the number of inputs of the system. Each neuron corresponds to one input.

The second layer is the hidden layer. This layer does the intermediate processing. The number of neurons in each hidden layer may vary. Usually the number of neurons in this layer is larger than the number of neurons in both the input layer and the output layer. The number of hidden layers is usually kept to one in order to control the complexity. We discuss in detail the reasons for this control in Chapter 19.

The last layer is the output layer. In this layer, the number of neurons is equal to the number of outputs that the system is supposed to provide. Each neuron corresponds to a particular output in the system.

Every representation of node in the figure is itself an artificial neuron. Each of these nodes does the work of aggregated addition and activation. The outputs of every neuron are given to the following neuron via the connections, as shown in Figure 2.5.

2.4.2 Weights

All the connections shown in Figure 2.5 correspond to some weight. Weights corresponding to each input are multiplied by the inputs during the aggregation, or weighted addition. The only way to control the system's output is to change the weights of the various connections. The training or learning in an ANN corresponds to the change in their weights. The different combinations of weights for any ANN determines the effectiveness or performance of the ANN.

When we discuss ANNs, we often use the terms *memory* and *learning*. Looking at their performance, it is natural to think of these networks as having some memory and as being able to learn just like the human brain. In fact, memory in the context of ANNs is in the form of its weights. These weights, along with the network architecture, determine the intelligence of the ANN. Hence, whenever we talk of the ANN's ability, we are actually referring to the weights. The ANN of some

TABLE 2.1
Common Activation Functions

No.	Function Name	Function Equation
1.	Identity	$f(x) = x$
2.	Logistic	$f(x) = \frac{1}{1-e^{-x}}$
3.	Sigmoid	$f(x) = \frac{1}{1-e^{x}}$
4.	Tanh	$f(x) = \tanh(x/2)$
5.	Signum	$f(x) = \begin{matrix} +1 & x > 0 \\ -1 & x < 0 \\ \text{undefined} & x = 0 \end{matrix}$
6.	Hyperbolic	$f(x) = \frac{e^{x} - e^{-x}}{e^{x} + e^{-x}}$
7.	Exponential	$f(x) = e^{-x}$
8.	Softmax	$f(x) = \frac{e^{x}}{\sum_i e^{x_i}}$
9.	Unit sum	$f(x) = \frac{x}{\sum_i x_i}$
10.	Square root	$f(x) = \sqrt{x}$
11.	Sine	$f(x) = \sin(x)$
12.	Ramp	$f(x) = \begin{matrix} -1 & x \leq 0 \\ x & -1 < x < 1 \\ +1 & x \geq 1 \end{matrix}$
13.	Step	$f(x) = \begin{matrix} 0 & x < 0 \\ 1 & x \geq 0 \end{matrix}$

problem can be transformed into the ANN of another problem if we copy the weights from the other ANN.

Bias is another modifiable parameter that contributes to the system's performance. For the reasons stated above, bias may usually be taken as a weight when discussing learning. The previous paragraph actually holds true for both weights and bias.

2.4.3 Activation Functions

We talked about activation functions as a fundamental operation of the artificial neuron. We also talked about the application and concept of this function. The various commonly used activation functions are summarized in Table 2.1. The graphs of some of these functions are given in Figures 2.6a through 2.6f.

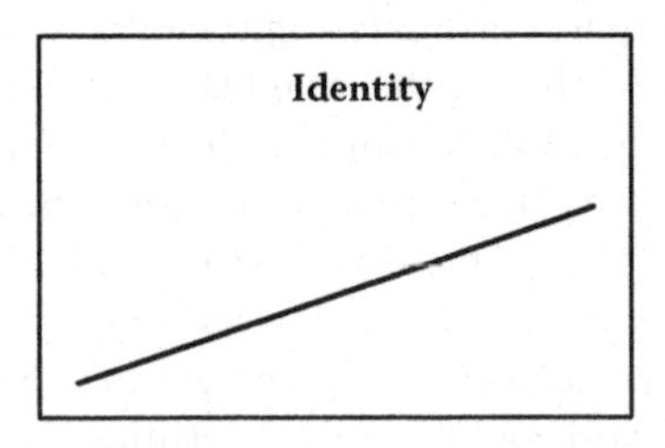

FIGURE 2.6a The identity activation function.

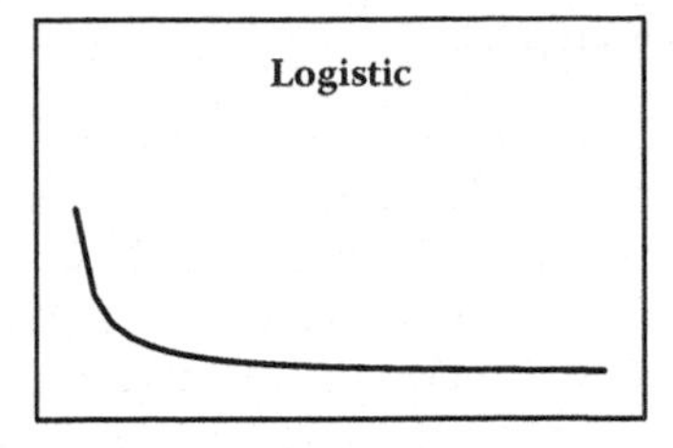

FIGURE 2.6b The logistic activation function.

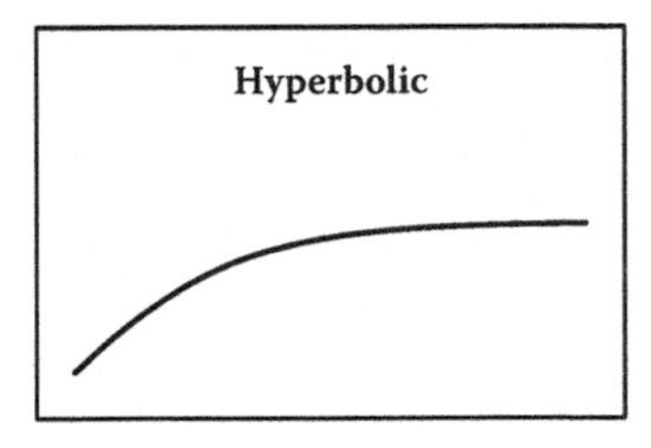

FIGURE 2.6c The hyperbolic activation function.

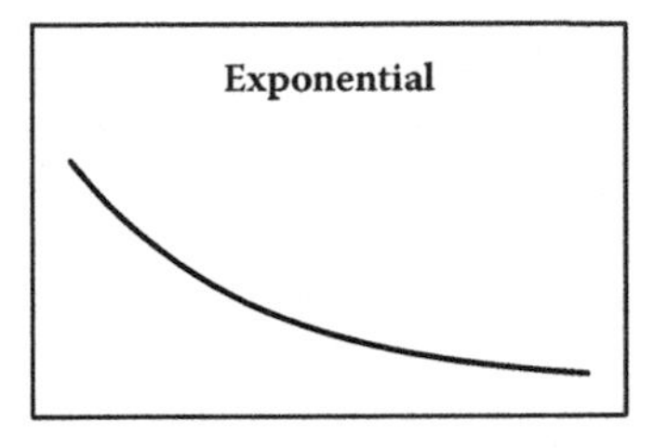

FIGURE 2.6d The exponential activation function.

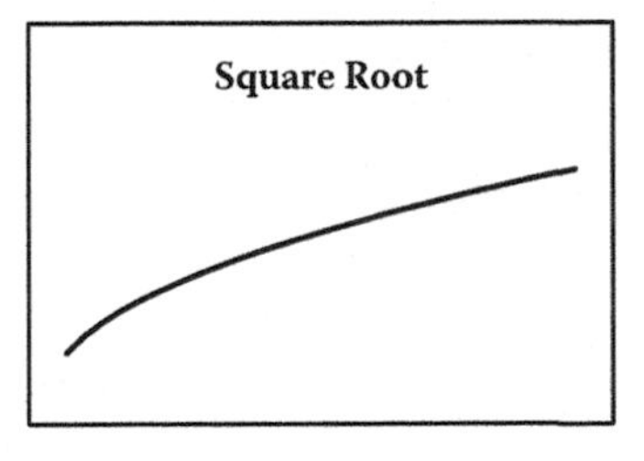

FIGURE 2.6e The square root activation function.

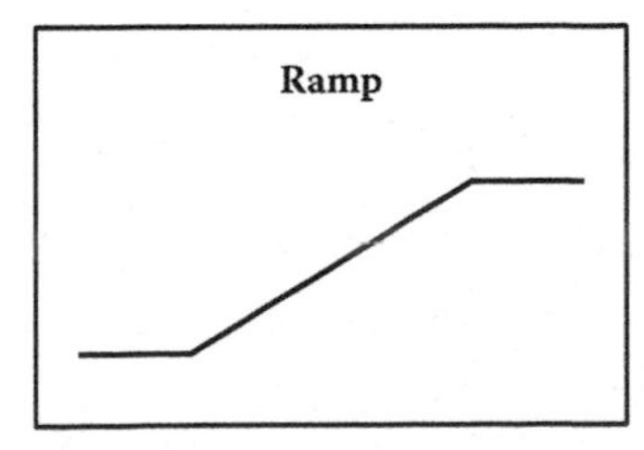

FIGURE 2.6f The ramp activation function.

2.4.4 Feed-Forward Neural Network

In a feed-forward neural network, information flows from the inputs to the outputs, without any cycle in their structure. These simple ANNs are easy to work with and visualize. And yet they still have great capabilities for problem solving. The outputs in these networks are a function of the provided input. The previous outputs have no role in determining the present output.

The non–feed-forward neural network may have cycles in its structure. These networks include recurrent neural networks in which time delay units are applied over the system in cycles, which causes the present outputs to be affected by previous outputs as well as new inputs.

Example 2.1
Calculate the output for a neuron. The inputs are (0.10, 0.90, 0.05), and the corresponding weights are (2, 5, 2). Bias is given to be 1. The activation function is logistic. Also draw the neuron architecture.

Answer:
Using Equation 2.3, we have

$$O = f\left(\sum_{i=1}^{n} W_i * I_i + b\right)$$

where $I_1 = 0.1$, $W_1 = 2$, $I_2 = 0.9$, $W_2 = 5$, $I_3 = 0.05$, $W_3 = 2$, and $b = 1$.

$$f(x) = \frac{1}{1 - e^{-x}}$$

Hence the neuron's output is

$$\begin{aligned} & f(W_1 * I_1 + W_2 * I_2 + W_3 * I_3) \\ & = f(2 * 0.1 + 5 * 0.9 + 2 * 0.05) \\ & = f(0.2 + 4.5 + 0.1) \\ & = f(4.8) \\ & = 1.008 \end{aligned}$$

The neuron architecture is shown in Figure 2.7.

Example 2.2
Consider the network architecture shown in Figure 2.8. All unconnected weights are taken to be 0. Calculate the output for input (2, 4, 1). Assume all bias to be 0. All activation functions are assumed to be identity, for simplicity.

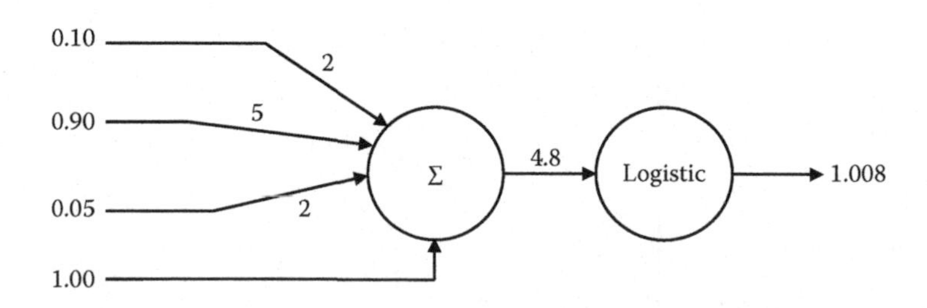

FIGURE 2.7 Figure for Question 2.1.

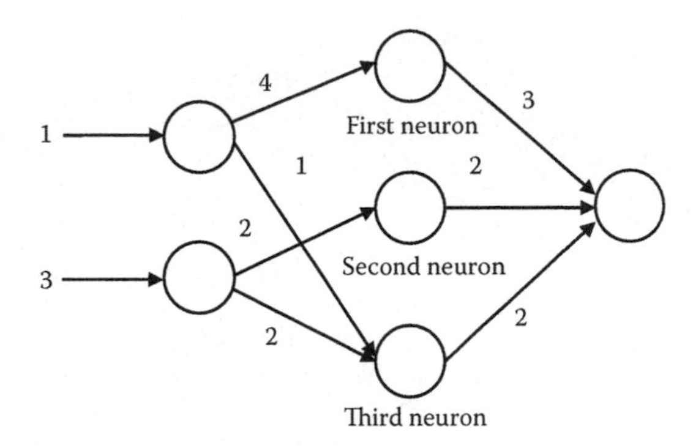

FIGURE 2.8 Figure for Question 2.2.

Answer:
For the input layers, the inputs are 1 and 1. The same become the outputs of this layer.

For the first neuron in the hidden layer, there is only one input—1. The aggregated sum of this neuron is hence 4 * 1, or 4. Because the activation function is identity, the same becomes the output of this neuron.

Similarly the outputs of the second and third neurons are 3 * 2 = 6 and 1 * 1 + 3 * 2 = 7.

The inputs to the output layers are the outputs of the hidden layers. The inputs are 4, 6, and 7. The weights are 3, 2, and 2. Hence the output is 4 * 3 + 6 * 2 + 7 * 2 = 38.

2.5 MODELING THE PROBLEM

Chapter 1 discussed quite a bit about the problem and problem modeling in soft computing. The same principles apply to ANNs. With ANNs, it is preferable for inputs with similar parameters to have almost similar outputs. No two test cases should have the same inputs but different outputs. The entire input space should be covered by the training data to ensure that the system can handle all kinds of inputs that a real life system may encounter.

Sometimes transformations may be applied, depending on the problem. These transformations change the shape of the input space and transform it to a shape that is much easier for the ANN to work with. As a result, the training data, and the trends represented by the training data, get well placed in the input space and start obeying the laws of good soft-computing inputs. These transformations change the way the ANN behaves. The problem resulting from the application of transformations may be better suited for training purposes by the ANN. We do not discuss these techniques much in this book.

2.5.1 Functional Prediction

Functional prediction is one type of problem handled by ANNs. This type of problem usually has a continuous output for every input. In this case, the ANN is made to predict an unknown function of the form $y = f(i_1, i_2, i_3, \ldots, i_n)$, where $i_1, i_2, i_3, \ldots, i_n$ arc thc various inputs.

If we plot a graph between the output and the inputs, we would have a complex surface with a graph spanning across n dimensions. For simplicity, let us assume that the system to be predicted is $f(x, y) = \sin(X^2 + Y^2)$. The plot is given in Figure 2.9.

The training data consist of some points given on this graph. The ANN acts in a way similar to human perception. If we were given some points in this graph, we would try to complete or extrapolate these points using smooth curves. We have two ways to complete the curve in order to find the value of the function at unknown points. In the first approach, we try to see the training data in the vicinity of this point or input. Based on these data, we try to extrapolate the function and find the output at some points. Such a method is realizable as well; consider Figure 2.9 again. If we had sufficient points

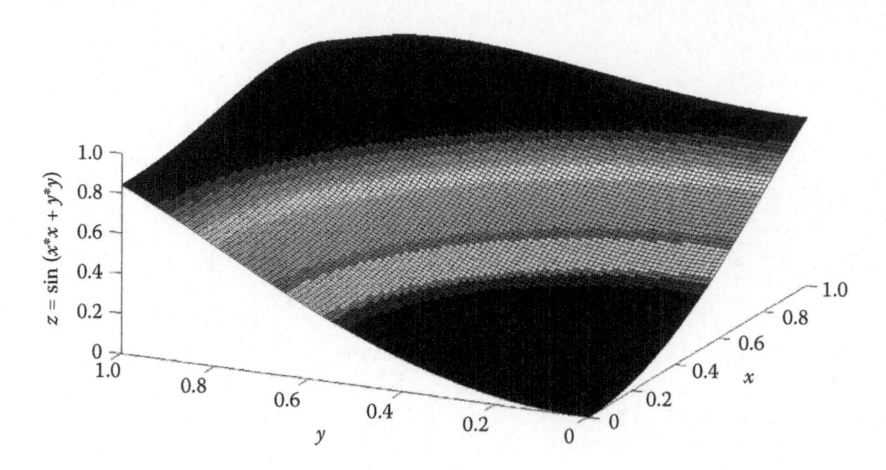

FIGURE 2.9 The function $f(x, y) = \sin(X^2 + Y^2)$.

available in the training data set near the uppermost part of the curve, we might not even consider looking at the lower half of the curve. Rather we would try to predict the output based on only these points. This technique forms a localized way of dealing with problems in which the output only depends on nearby training data cases. However, these systems face problems in the presence of noise.

The other way to complete the curve is to look at all the training data sets and, based on these sets, form a global curve that gives the least error or deviation with the training data set points while also maintaining its smoothness. Once this curve for the entire input space has been plotted, we can find the output at any point. In this globalized approach, the output of a point depends on all the inputs in the training data sets. This technique is much more resistant to noise but is very difficult to learn.

The same happens to be true with ANNs that try to predict or construct the surface to complete the whole curve. The data requirements for training system may hence be easily understood by the readers.

2.5.2 Classification

Classification is the second type of problem handled by ANNs. In these problems, the ANN must classify the inputs and map them to one of the many available classes. The output in these problems is discrete and gives the class to which the input belongs.

An important factor in these problems is the number of classes. More the classes, the greater the complexity of the algorithm. To give a better understanding of this concept, we plot these problems on a graph. Suppose that the data we want to classify depend upon two parameters. We plot this on a two-dimensional graph, with the classes X and O marked in the graph by x and o (see Figure 2.10).

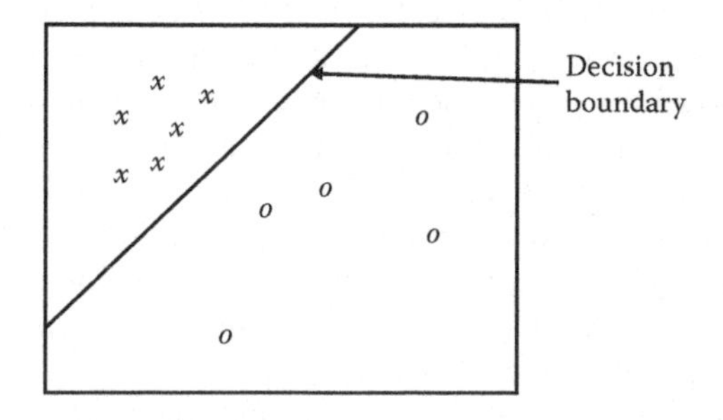

FIGURE 2.10 The classification problem.

In this example, the ANN is supposed to divide the entire space into two distinct regions. The first region belongs to the class *X*, and the second region to the class *O*.

The line in the figure is called the *decision boundary*. This line draws the boundary between the different classes. The decision boundary may be in the form of a line, oval, or any other simple or complex shape. If the decision boundary can easily be drawn in the multidimensional input space of the problem, then the problem can easily be solved by the ANN. Our hope is that the ANN is able to imitate the line, and thus separate the classes, with the same ease as that we have drawing the decision boundary.

ANNs work by predicting the decision boundaries in the *n*-dimensional input space. Say the decision boundary can be most appropriately represented by the equation $y = f(i_1, i_2, i_3, \ldots, i_n)$, where $i_1, i_2, i_3, \ldots, i_n$ are the various inputs. It is clear from our understanding of coordinate geometry that the value inside the curve would be less than 0 and that outside the curve would be greater than 0. This principle is used to find whether the points lie inside or outside the curve. This is done for all classes and is packaged in one ANN for simpler design and implementation.

2.5.3 Normalization

As discussed in Chapter 1, it is preferred that the inputs and outputs in any system lie between a fixed and manageable upper and lower limit. This avoids the problem of ending up with a computation above the programming tool's limitations at any of the intermediate levels. This further ensures that all the inputs to any of the neurons in any of the stages are within the acceptable limits of the neuron.

We therefore fix the upper bounds and the lower bounds for all the inputs to fall between some minima and maxima. This process is done by the technique of normalization, which is a means for mapping an input with a known minimum and maximum range to a range that is desired.

Suppose that any of the input attributes in the problem have a lower limit of x_{min} and an upper limit of x_{max}. We want to map the input within the limits of y_{min} and y_{max}. The general formula for computing the corresponding output *y* for any input *x* is given by Equation 2.5:

$$y = y_{min} + \frac{y_{max} - y_{min}}{x_{max} - x_{min}} * (x - x_{min}) \tag{2.5}$$

Say we want our inputs and outputs to lie between –1 and 1. Now $y_{min} = -1$ and $y_{max} = 1$. Equation 2.5 thus modifies to Equation 2.6:

$$y = -1 + \frac{2 * (x - x_{min})}{x_{max} - x_{min}} \tag{2.6}$$

Note that in the MATLAB® implementation of ANN, the input ranges must be specified at the time of network creation. For details, please refer Appendix A.

2.5.4 The Problem of Nonlinear Separability

Refer back to Figure 2.10, which presents a simple model in which an ANN must classify a set of data into two classes. We see that the separation could easily be made by a simple line. Consider a perceptron with two inputs *x* and *y* and a bias *b*. The output of this model would be as given by Equation 2.7:

$$o = f(w_1 * x + w_2 * y + b) \tag{2.7}$$

TABLE 2.2
XOR Problem

A	*B*	*Y* = *A* XOR *B*
0	0	0
0	1	1
1	0	1
1	1	1

Say, for simplicity, that the activation function is identity, as shown in Equation 2.8:

$$o = w_1 * x + w_2 * y + b \tag{2.8}$$

Because this equation can represent any line, we can train the perceptron to solve the problem of Figure 2.10 by adjusting the weights w_1 and w_2 and the bias b.

Imagine we are given an exclusive-or (XOR) problem, in which we must train a neural network to learn the XOR output given in Table 2.2. This problem can be plotted as shown in Figure 2.11. Because no single line is able to classify the sequence, the problem cannot be solved by a perceptron. This problem with perceptrons ultimately led to the use of multiple neurons in a layered manner. The XOR problem can easily be solved by MLPs, because they have the capability to imitate any complex function.

The concept of nonlinear activation functions further helps in generating complex nonlinear functions that are imitated by the ANN. This principle also finds an application here.

2.5.5 Bias

As already mentioned, bias is the additional input given to ANNs. During testing, they are treated just like any other input and are modified according to the same rules as the inputs. Bias is believed to give the system more flexibility. Biases can be used to add means of adjustment to ANNs, and hence become instrumental tools in adjusting the ANN at the time of training.

Let us take the example of a simple, one-neuron system. Say it has two inputs x and y. The output of this system would be given by Equations 2.7 and 2.8.

Say the problem is classificatory in nature. From our understanding of the decision boundary, we know that the equation represents the decision boundary. We can easily conclude that without bias, we cannot solve the problem given in Figure 2.12. We can also see from Figure 2.12 that changing the bias gives greater freedom to the decision boundary to adjust itself according to the problem's requirements. The bias moves the decision boundary forward (or backward) to keep the slope

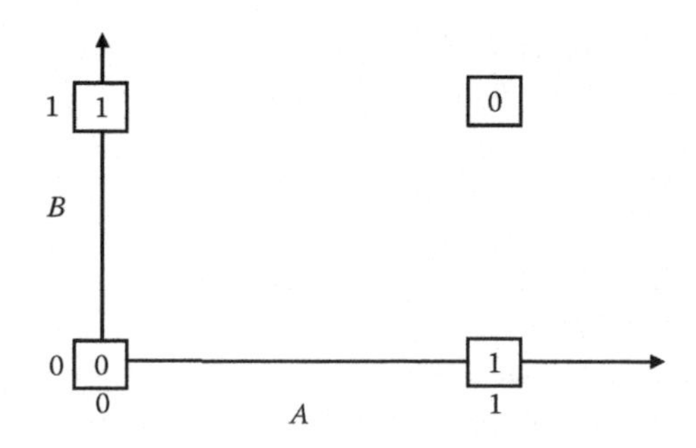

FIGURE 2.11 The XOR problem.

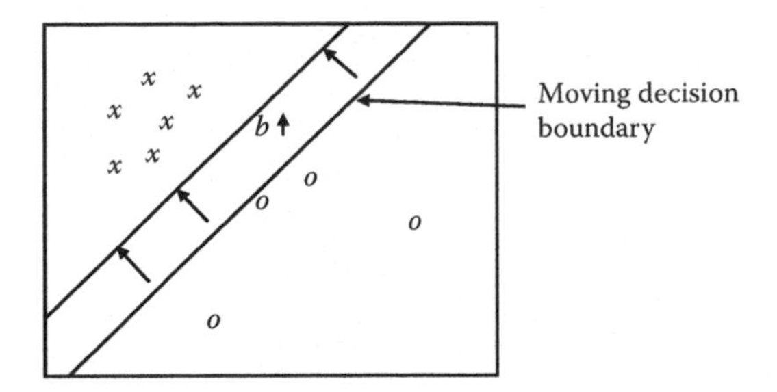

FIGURE 2.12 The role of bias in moving the decision boundary.

constant. (The effect of changing weights w_1 and w_2 may be visualized by the readers themselves; they would cause the slope of the line to change.)

2.6 TYPES OF DATA INVOLVED

We know that ANNs need a good amount of data to train themselves. More the data, the better the system's performance. The data in this approach are generally divided into three sets, each of which is discussed below. The distribution of data among these sets is the work of the designer and largely depends on the quality and quantity of the data available in the historical database.

- **Training data:** These data are used by the training algorithm to set the ANN's parameters, weights, and biases. Training data make up the largest set of data, comprising almost 80 percent of the data.
- **Testing data:** This data set is used when the final ANN is ready. Testing data, which are completely unknown, are used to test the system's actual performance. If the performance is poor, we may have to reconfigure the ANN, or we might find that it would be best to use another technique. The testing data set consists of about 10 percent of the data.
- **Validation data:** These data are used for validation purposes during the training period. At any point during training, we measure the error on both the training data and the validation data. If the error in training data is decreasing but on validation data is increasing, we stop the training to prevent the problem of overgeneralization (a problem discussed in Chapter 19). The size of this data set is about 10 percent of the entire data set.

2.7 TRAINING

We have discussed the fact that ANNs are able to learn the data presented in a process known as *training*. Here we describe how to train the ANN so it can adjust itself in such a way that it gives a correct answer whenever the problem is again given to it.

The training algorithm tries to adjust the various weights (and biases) and set them to a value such that the ANN performs better at the applied input. Thus, the entire training process is a means of finding the right combination of weights and biases for which the ANN performs at its best. Note that each ANN might not be able to train itself well for all architectures. This performance and training depends on the number of hidden layers, as well as on the neurons in these hidden layers. Changing the architecture cannot be performed during training. The training algorithm does not play with the network architecture. Hence ANN designers try various combinations and numbers of hidden layers and hidden layer neurons to train the network repeatedly in order to get the configuration for the best performance. Because this entire activity is manual, it may take a long time and a great deal of experience before a designer is able to select a final design. A good ANN architecture

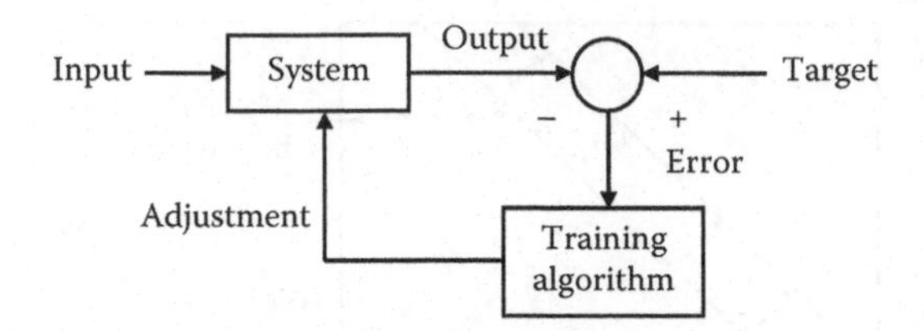

FIGURE 2.13 The basic block diagram of supervised learning.

gives the best performance in the least number of layers and the least number of neurons. (For more information, refer to Chapter 19.) This performance is measured using the testing data set.

Before we discuss the training algorithm further, let us discuss the important terms used in the training process.

2.7.1 Types of Learning

When discussing ANNs, *learning* refers to the learning of the data presented. This learning may be broadly classified into three categories.

- **Supervised learning:** In this kind of learning, both the inputs and the outputs are well determined and supplied to the training algorithm. Hence whenever an input is applied, we can calculate the error. We try to adjust the weights in such a way that this error is reduced. The error may be reduced as a result of repeating the application of the input. Supervised learning takes its analogy from the way a person might learn by being assisted by someone who keeps correcting the mistakes. As a result, the mistakes are reduced to an appreciable extent. Supervised learning is used for most real life applications, as we shall see in the upcoming chapters. The basic block diagram of this learning is given in Figure 2.13.

  In this chapter, we will study only supervised learning. The training method, and the algorithm that we present below, is a supervised learning technique. The other learning techniques are discussed in Chapter 3.
- **Unsupervised learning:** In this type of learning, the target outputs are unknown. The inputs are applied, and the system is adjusted based on these inputs only. Either the supporting weights of the problem are added or the dissipative nodes are decreased. In either case, the system changes according to the inputs. The result is that the system gets ready to give the correct answer whenever the same input is applied. The unsupervised learning algorithm only needs to calculate the direction of improvement of weight given by the output. The basic block diagram for unsupervised learning is given in Figure 2.14.
- **Reinforcement learning:** As its name implies, this type of learning is based on the reinforcement process. In this system, the input is applied. Based on the output, the system either gives some reward to the network or punishes the network. If the answer was in the

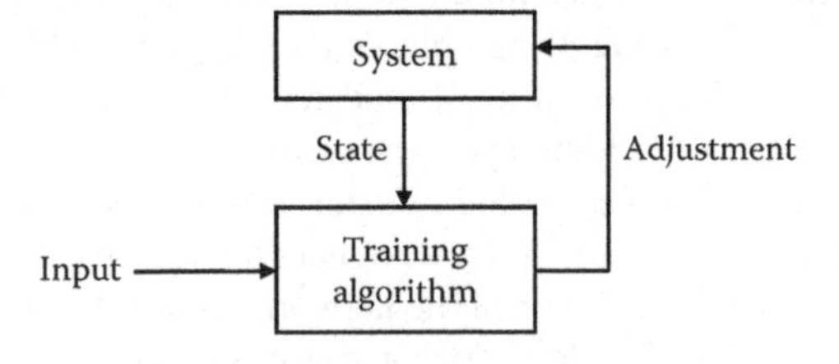

FIGURE 2.14 The basic block diagram of unsupervised learning.

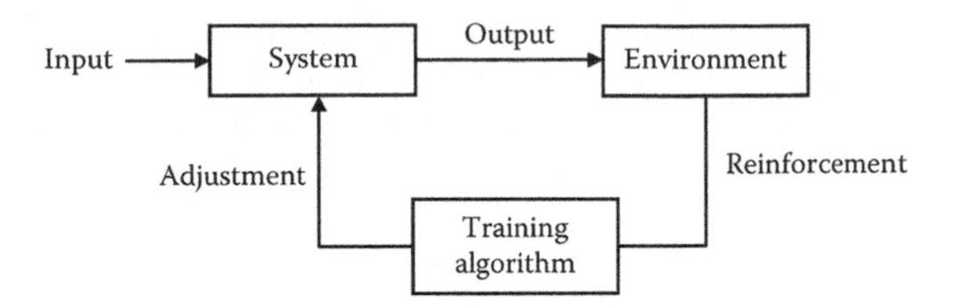

FIGURE 2.15 The basic block diagram of reinforcement learning.

correct direction, the result is a reward, and vice versa. In this learning technique, the system tries to maximize the rewards and minimize the punishment. The basic block diagram is given in Figure 2.15.

2.7.2 The Stages of Supervised Learning

We now know all the basic parts of the ANN architecture. We also have a fair idea about the working of the ANN whenever an input is given. To train the ANN, we need to apply the input, get the output, and accordingly adjust the weights and biases. ANNs are used for training as well as for testing. Training deals with giving inputs to the network, noting its errors, and making related modifications. This involves a forward step in which the input is applied and the output is retrieved and a backward step in which the error is calculated and the relevant changes are made. In testing, however, there is only a forward path.

The application of inputs and the calculation of outputs both require the flow of information from one neuron to another until the final output neuron calculates the final answer. This flow of information is forward, from the input to the output, and it happens layer by layer. This step in the training process is called the *feed-forward step*.

Once the output to the applied input is known, the error can be calculated. The task now is to adjust the weights according to the error. This error is propagated backward from the output layer to the input layer. In this manner, all the weights and biases are adjusted, starting from the output neurons.

The general diagram of the two stages is given by Figure 2.16.

2.7.3 Error Function

Error, a measure of the ANN's performance, helps us in analyzing the ANN. A good ANN has minimum error. As such, a good training algorithm moves the ANN from high error configuration to a lower error configuration. The objective is to minimize the total error associated with the problem.

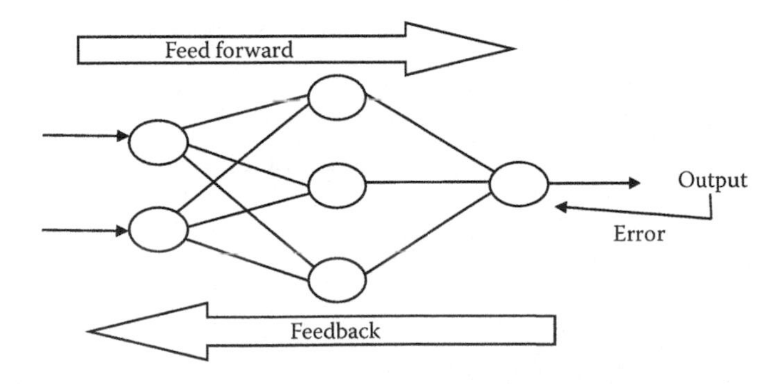

FIGURE 2.16 The feed forward and the feedback stages in the artificial neural network.

An error in ANN is measured by using error measurement functions. These error functions represent the net error from all the inputs that are in the training or validating data set. Various types of error functions may be used, depending on the problem and choice of the designer. A few of the major, commonly used error functions are given below. Each error function performs as conveyed by its meaning.

- Mean absolute error
- Mean squared error
- Mean squared error with regression
- Sum squared error

An ANN's error is a direct result of the adjustments in weights and biases carried out in the ANN. The only way to effect the error during real-time training is to change the weights and biases. Every combination of weights and biases has some error associated with it. This information can be plotted graphically in a multidimensional graph, where the number of dimensions is the number of weights and biases, plus 1 to represent error. These graphs, which are known as error surface graphs or configuration graphs, represent the error for every combination of weights and biases. We discussed in Chapter 1 that the input spaces and graph are highly dimensional in nature and hence are impossible to draw and very difficult to visualize. The same is true for error graphs, but to an even greater extent. The number of weights and biases are themselves so large that we may find it difficult just to count them. Now imagine a graph that has this high number of axes or dimensionality. However, the training algorithm is supposed to find the minimal error by examining this graph. Thus, it would be advisable to understand the graph before proceeding with the text. We also refer to this graph numerous times throughout the text.

One such error graph is given in Figure 2.17, which shows two weights—$weight_1$ and $weight_2$. Error is plotted for all combinations of $weight_1$ and $weight_2$ on the vertical axis.

2.7.4 Epoch

ANN training is done in a batch-processing mode, in which a batch consists of all the training data set cases. This complete pack is presented to the system for training. After the system finishes processing all these inputs, the same batch is applied again for another iteration of updating the weights and biases. We continue in this manner, presenting the system with the same pack of inputs and outputs. A single iteration of applying all the inputs, or the input package, is known as the *epoch*.

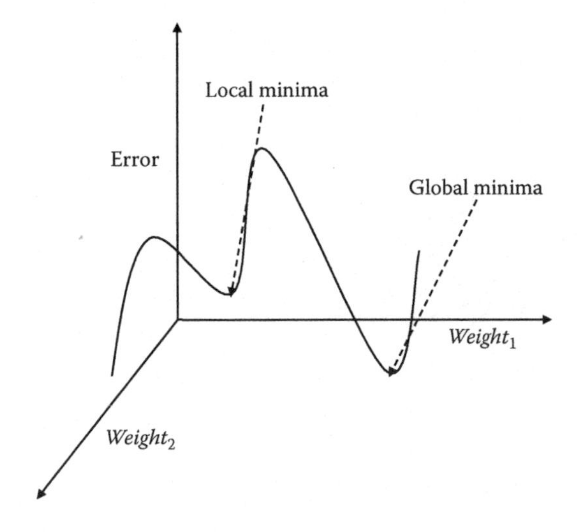

FIGURE 2.17 The error graph.

At every step of this process, the ANN learns to some extent. As a result of applying the same inputs multiple times, the system is able to train itself with permissible error. After multiple epochs, the system is said to be fully trained. The number of epochs required for the system to train itself varies according to the problem and the training data.

2.7.5 Learning Rate

The learning rate is the rate by which the system learns the data. A larger learning rate means that a system would be able to learn the data very quickly. On the other hand, whenever the learning rate is low, the system would learn very slowly. The learning rate is denoted by η, a fraction that measures between 0 and 1.

A high learning rate makes the system learn fast, but it gives larger total errors. A low value of learning rate, on the contrary, takes a lot of time to train, but the total error of the system is low. Suppose you are watching a movie with a very high frame rate. You would not be able to detect very small defects that the individual pictures in the movie may have. However, if you were watching a film at a slow frame rate, you would be able to see smaller defects. The same is true for ANNs. The higher learning rates force an ANN to try to jump to a region where error is expected to be less. However, at the higher rate, it may not be able to jump to the correct place where a minimum lies. If the rate are lower, however, the ANN may take a while to reach the minima, but it would certainly reach it eventually.

Consider Figure 2.17. At higher learning rates, the ANN would jump from a point closer to the minima in the direction of the minima. This would take it to a point ahead of the minima (which has error more than the minima). This would again result in the ANN trying to jump to the minima again (this time in the opposite direction) in the direction of the minima. This would again take it ahead of the minima. The ANN would thus keep oscillating between points in an attempt to reach the minima. At lower learning rates, however, the ANN would advance slowly and gradually toward the minima and would eventually reach it.

Now consider Figure 2.18, in which the errors are represented by plotting the error contours. Contours are the regions having the same value as the error function. A contour line of a function of two variables is a curve along which the function has a constant value. In this figure, each contour line shows all points that have the same error value. Training with a low learning rate gives training as depicted in Figure 2.18a; a large value of learning rate is depicted in Figure 2.18b.

Definition 25
A contour line of a function of two variables is a curve along which the function has a constant value.

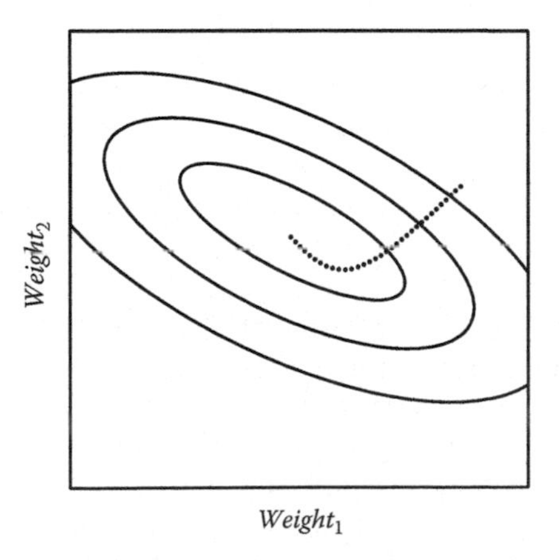

FIGURE 2.18a Error contours and learning at a small learning rate.

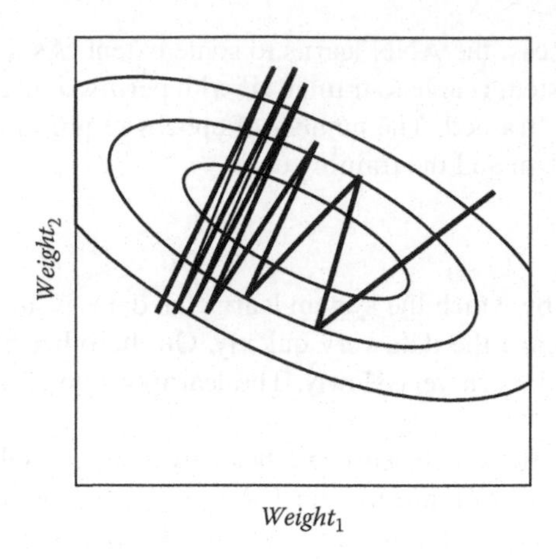

FIGURE 2.18b Error contours and learning at a large learning rate.

2.7.6 Variable Learning Rate

In the previous section, we saw that the learning rate must be chosen correctly. A too high or too low value will not perform well. Each value of learning rate has own problems and limitations.

One solution for choosing the correct learning rate is to keep the learning rate as a variable. A variable learning rate allows a system to be more flexible. The learning rate at the start is kept high. As the training continues, however, the learning rate is gradually reduced. The concept of a variable learning rate is analogous to the frame rate example we mentioned earlier. Suppose there is an action scene in your movie. You might consider watching the scene at a lower frame rate so as to get the complete details. Or you may fast forward through most of the scenes because they do not appeal you. In either case, you are trying to vary the frame rate per your regions of interest in the movie. More details regarding this concept are discussed in Chapter 19.

2.7.7 Momentum

The training algorithm is an iterative algorithm, which means that at each step, the algorithm tries to move in such a way that the total error is reduced. After several iterations, the algorithm might find the system close to a point where the error is quite small.

Looking at Figure 2.17, we clearly observe that there are two valleys in the plotted error surface. The first valley is shallow, and the second one is deep. The deeper valley corresponds to a lower error, and hence to a better configuration of the ANN, than the shallower valley. The deeper valley is known as the *global minima*, which is the least error in the entire surface. The shallower valley is called the *local minima.*

During training, the ANN keeps moving to a place where the error is reduced. The decision of the direction of movement is made by analyzing the surface and its slope. It is natural for the ANN to get trapped at some local minima, which may reduce its abillity to reach a global minima. This problem is given by Figure 2.17. Because the training algorithm is trying to move in such a way that error is reduced, it keeps following the local minima. But it can easily be seen that if it continues to move in the same direction, it would eventually attain the global minima.

This problem is solved by introducing a term called *momentum.* The momentum keeps pushing the training algorithm to continue moving in the previous direction, making it possible for the training algorithm to escape out of the local minima. The meaning of *momentum* in this case is analogous to the meaning of *momentum* in the physical world. For example, a ball moving has momentum that keeps it moving in the same direction.

2.7.8 Stopping Condition

The training algorithm may train itself infinitely, trying to minimize the error more and more. In any real life problem, however, we have to establish when the algorithm should stop training the ANN. The algorithm stops according to the stopping condition.

Normally one or more of the following criteria are used as stopping conditions:

- **Time:** The algorithm may be stopped when the time taken to execute exceeds the threshold.
- **Epoch:** The algorithm has a specified maximum number of epochs. Upon exceeding this number, the algorithm may be stopped.
- **Goal:** The algorithm may be stopped if the error measured by the system reduces to more than a specific value. It may not be useful to continue training after this point.
- **Validating data:** If the error on validation data starts increasing, even if there is a decrease in the testing data, it would be better to stop further training.
- **Gradient:** Gradient refers to the improvement of the performance or the lowering of the error in between epochs. If the algorithm fails to improve the performance or lower the error appreciably between epochs, the training may be stopped. Here it is assumed that further training would not result in much of a difference in performance.

In general, it is not preferred to train a network more than a specified amount. If we continue training, the network might assume more complex functions than the simple ones required of it. This effectively reduces performance in unknown data. This problem is known as the problem of overgeneralization. A good way to avoid overgeneralization is to use validating data separate from the training data. These concepts are presented in more detail in Chapter 19.

2.7.9 Back Propagation Algorithm

The previous section discussed the general concepts of training. Here we discuss one of the algorithms used to train the ANN. This algorithm, called the back propagation algorithm (BPA), is a supervised learning algorithm that calculates the error and then back propagates the error to previous layers. These previous layers then adjust their weights and biases accordingly, and the configuration is thus changed.

The changing of the weights and biases follows the approach of steepest descents, which is explained in the next section. For now we will gather all the concepts and present them as a formal algorithm that is is given by Figure 2.19.

The complete working and an in-depth analysis of the ANN is not used in real life problems and is hence outside the scope of the book.

2.7.10 Steepest Descent Approach

Steepest descent is a commonly used approach for controlling the output by a controllable variable. The objective is to reach a local minima of the function, given any position at which the system is presently located. This approach runs on the function position at any instant and modifies the controlled variable such that the function moves closer to the local minima in the next run.

Steepest descent, also known as gradient descent, may be defined as "the basic learning rule of the adaptive network in which the gradient vector is derived by successive invocations of the chain rule" (Melin and Castillo, 2005).

Definition 26

Steepest descent is the basic learning rule of the adaptive network in which the gradient vector is derived by successive invocations of the chain rule. (Melin and Castillo, 2005)

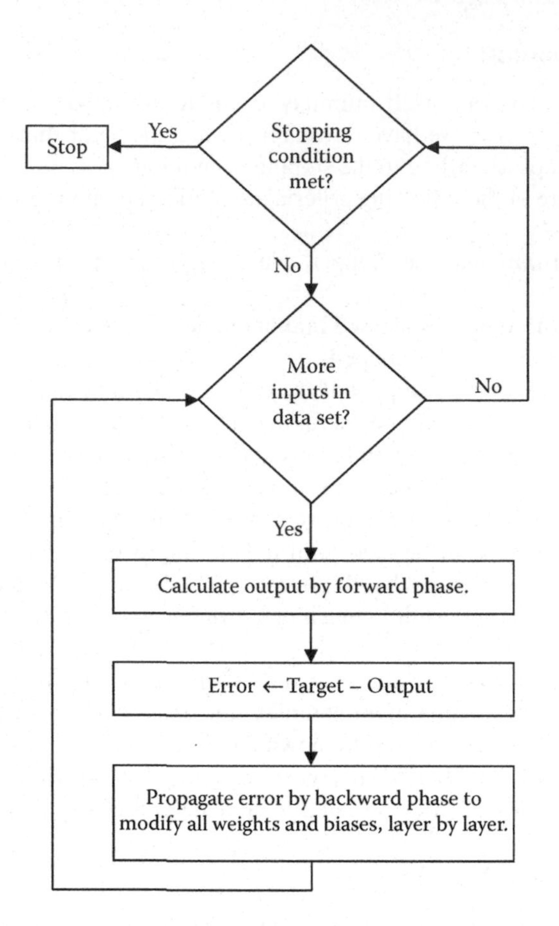

FIGURE 2.19 The flow diagram of the back propagation algorithm.

In the context of ANNs, the error may easily be visualized as a function for weights and biases for the same inputs. Here we work in the context of error graphs (configuration graphs) given in Figure 2.17. The goal is to minimize the error, or reach a point in the graph that corresponds to the deepest point in a valley. This forms the function that the steepest descent is supposed to minimize. The weights and biases are the variable parameters. The slope is then calculated, and the system is moved accordingly.

It may hence be seen that whenever the ANN is trained by the BPA, the error at the first few epochs changes drastically. Then, as it proceeds, the change becomes more and more gradual. As we reach the last few epochs, the error function behaves as if it were almost constant. Further training hardly has any effect on the error. A typical curve of the error plotted against the epoch is shown in Figure 2.20. Note that the error plotted on the y-axis is in a logarithmic scale, for better visibility.

2.7.11 Mathematical Analysis of the BPA Expression

The first step in the BPA is the forward phase, where the output from the system is calculated. This step has already been discussed in earlier sections. Because the final desired target output and the output from the system are known, the error can easily be computed.

For the output neurons, which decide the system's final output, the error δ may be calculated by the expression given in Equation 2.9:

$$\delta = F' * (O_{desired} - O_{actual}) = O_{actual}(1 - O_{actual})(O_{desired} - O_{actual}) \tag{2.9}$$

where $O_{desired}$ is the desired target output and O_{actual} is the actual output obtained by the system.

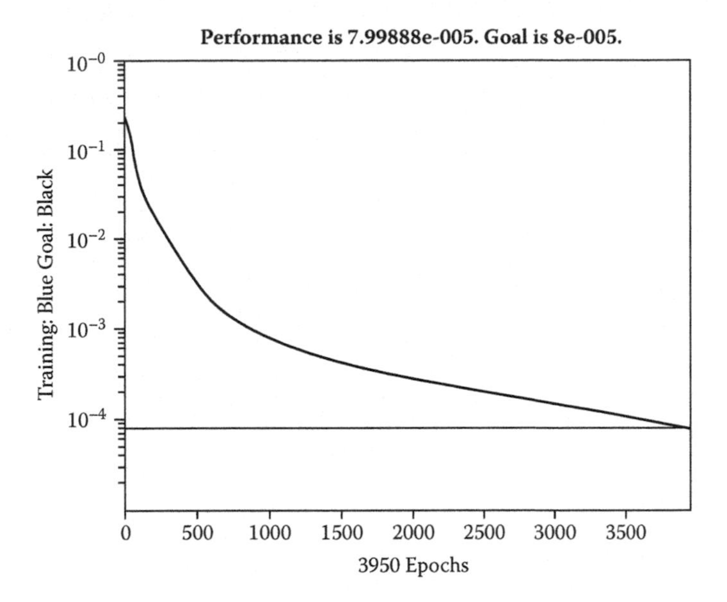

FIGURE 2.20 Error versus epochs while training with the back propagation algorithm.

Now we need to modify the weights. The weight from the neuron p at layer $k-1$ to neuron q at layer k is denoted by $W_{p,q,k}$ (see Figure 2.21a). The change in the corresponding weight is denoted by $\Delta W_{p,q,k}$. The expression of $\Delta W_{p,q,k}$ is given in Equation 2.10, and the final weight is given by Equation 2.11.

$$\Delta W_{p,q,k} = \eta \delta_{q,k} O_{p,j} \tag{2.10}$$

$$W_{p,q,k}(n+1) = W_{p,q,k}(n) + \Delta W_{p,q,k} \tag{2.11}$$

where $W_{p,q,k}(n)$ is the weight from neuron q at layer k to neuron p at layer $k-1$ at the nth iteration, $\delta_{p,k}$ is the error generated at neuron q in layer k, $O_{p,j}$ is the output of neuron p positioned at layer j, and η is the learning rate.

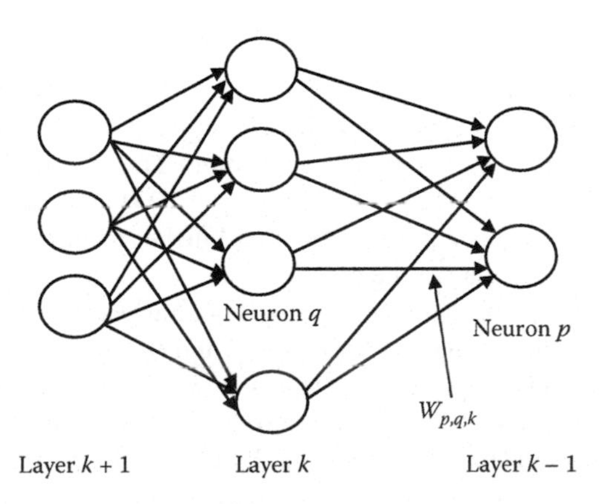

FIGURE 2.21a The notation of weights in ANN for the back propagation algorithm.

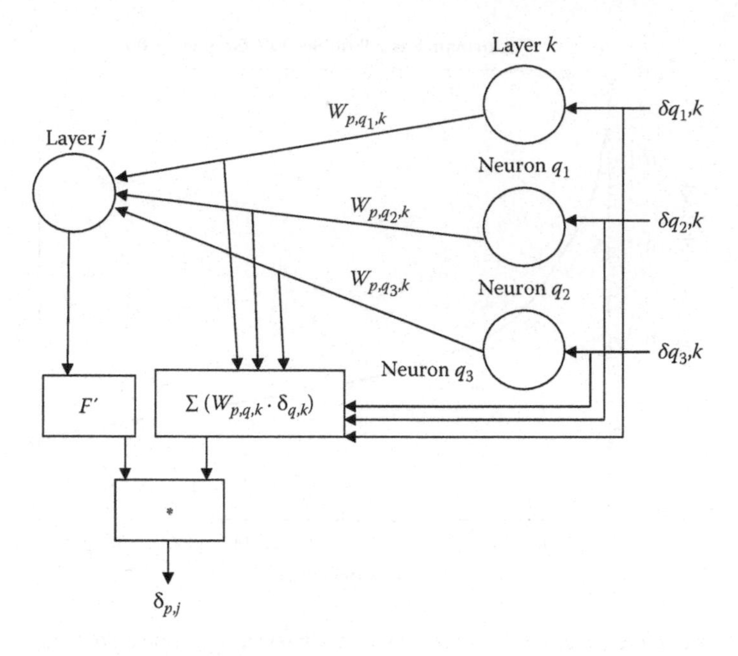

FIGURE 2.21b The computation of δ_p at layer j.

The error of neuron p at layer j is given by Equation 2.12:

$$\delta_{p,j} = O_{p,j}(1 - O_{p,j})(\Sigma\, \delta_{q,k} W_{p,q,k}) \tag{2.12}$$

where $q\ \varepsilon\ \{q_1, q_2, q_3\}$, as given in Figure 2.21b.

2.8 ISSUES IN ANN

In this section, we discuss some of the important aspects that must be considered in an ANN. These aspects play a key role in any real life problem and must be studied for a good ANN design. Other issues will be covered in Chapter 19.

2.8.1 Convergence

Usually, whenever we train an ANN, the error keeps decreasing as the epochs increase. The decrease in error is more drastic at the start and less so at the later epochs. The change in error decreases, and the error function in the later stages behaves as a constant. This is referred to as *convergence*. We say that the ANN error has converged to a point. Figure 2.20 shows one such system in which the error function has converged to an error of about 10^{-4}. Further training would not have resulted in any appreciable decrease in the value of the error.

It may however be noted that the error function will not always converge to some point. It is also possible that the ANN may start behaving abnormally and the error may increase indefinitely. In this case, the ANNs would not give a decent performance. Such actions may be caused by various reasons, including large input and output ranges, large learning rates, large momentum, complex network architectures, or too much noise in the data. It may also be possible that the problem cannot be solved by an ANN because the function to be predicted might be too complex or the wrong features, or missing features, were selected for the problem.

2.8.2 Generalization

Generalization is the ability of the ANN to give correct answers to unknown inputs. Generalization is a measure of how well a system has been trained. Generalization is measured by the performance of the system for the testing data set. The ANN's generalization capability should hence be as high as possible.

Many times, however, the system fails to train or the performance is too low. This often results from the system being unable to generalize the problem or to form general interpretations from the given parameters. The problem may be due to not enough neurons or a bad input. Whenever the ANN is able to generalize, however, the performance increases and the error decreases.

2.8.3 Overgeneralization

Another problem associated with ANN is overgeneralization. In this situation, the ANN performs very well for the training data but fails to perform well for the validation data. This problem may be caused by excessive neurons, excessive training duration, or other similar reasons.

Overgeneralization occurs when the ANN, by its training algorithm, starts imitating very complex functions in place of the simple ones. The complex functions fit the training data points well, which causes the training data error to keep reducing. But the case for validation data is different: Even around the points of the training data, the ANN starts behaving abnormally, depending on the complexity of the function it has assumed. Consider Figure 2.22. This figure shows that the training data sets are the points of intersection of the two curves. The figure shows the actual curve, a curve with an apt number of neurons, and a curve with a very large number of neurons. The complexity of the function is the highest for the curve with a large number of neurons. Hence this function does not give good performance.

The solution to the problem of overgeneralization is to take a separate set of data, called validation data, and stop training as soon as the error starts increasing in the validation data set, as shown in Figure 2.23.

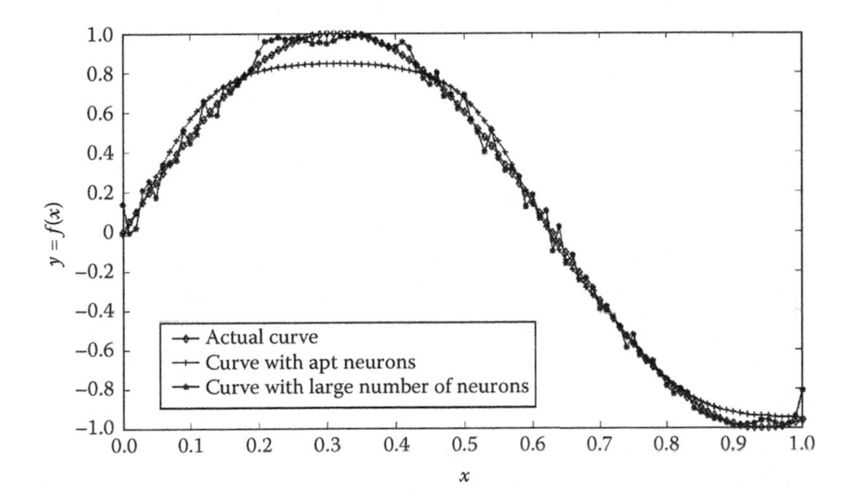

FIGURE 2.22 The overgeneralization problem.

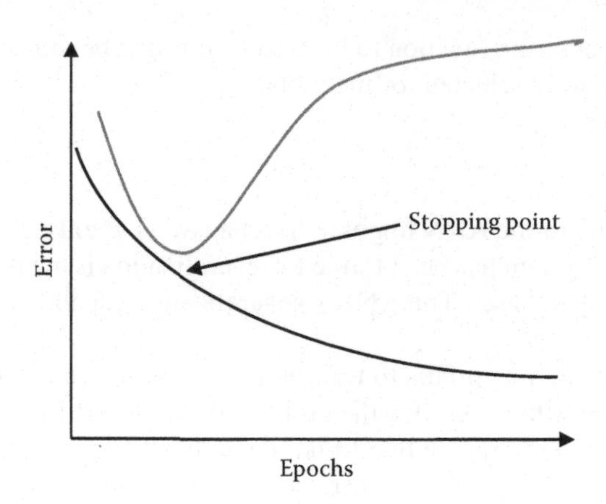

FIGURE 2.23 Early stopping with the help of a validation data set.

2.8.4 Complexity

The number of layers and the number of neurons in the ANN must always be limited. An increase in the number of layers increases the complexity of the ANN by a great extent. With our knowledge of ANNs and the BPA, we can well imagine the computation increase in every addition of a layer. The same is true for the addition of neurons. These additions also have the problems of overgeneralization, as discussed above.

Thus, it is best to have just one hidden layer. If the system does not train, however, we may have two hidden layers. The number of neurons in any case must be the minimum that can make the ANN reach the goal.

2.9 EXAMPLE OF TIME SERIES FORECASTING

We now look at an example of time series forecasting to make clear all the concepts of ANN and their practical usage. The problem in this example is a general class of problems that try to predict the future based on past trends. It takes as input the past values of the variable or series being monitored. Then the ANNs are applied to predict the future value.

This problem has huge practical applications and can be applied over various domains. It may be used to predict the number of customers at any store, the agricultural yield, or even the stock market behavior. If future outcomes are known in advance, people can make the necessary arrangements to deal with the situation. For the stock market prediction, as an example, people may consider where to invest and actually make huge profits out of the system.

ANNs are good means for solving these problems because they can imitate the huge, complex functions that occur in many real life series. In addition, ANNs are much more resistant to noise in the data and hence are better suited for real life data, which are prone to such noise.

2.9.1 Problem Description

The time series consists of a series of observations in the form of $(x_1, x_2, x_3, x_4, \ldots)$. The problem is to predict the future values of the series based on past values. The problem assumes that the future values are a function of past values. It assumes that it is possible to get the future values based on the past values as the series repeats itself.

The performance of the series is measured by the root mean square error, given by Equation 2.13. Here error e_i is measured as the difference between the value given by the system and the actual value at time i.

$$MSE = \sqrt{\frac{\sum_{i=t+1}^{t+L} e_i^2}{L}} \tag{2.13}$$

A very common tool for series analysis is autocorrelation, which is given by Equation 2.14.

$$r_k = \frac{\sum_{t=1}^{s-k} (x_t - \bar{x})(x_{t+k} - \bar{x})}{\sum_{t=1}^{s} (x_t - \bar{x})} \tag{2.14}$$

where r_k is the correlation coefficient, x_t is the value of the sequence at time t, $\bar{x}$ is the average value of x in the particular time frame being considered, and S is the size of the series.

Autocorrelations can be useful for decomposition of the main features of the time series, such as the trend and seasonal effects. A trend is the constant increase or decrease in the value of the observed series along with time. In the case of sales prediction, this trend may be due to inflation, increasing demands, and so forth. The seasonal effects occur in series and repeat in the time series. In the case of sales prediction, these seasonal effects include seasonal products such as air conditioners.

The specific problem that we take here is from the quarterly S&P 500 Index for 1900 to 1996, which comes under the category of financial series (Makridakis, Wheelwright, and Hyndman, 1998). The data for this come from the Time Series Data Library.

2.9.2 Inputs

The time series prediction problem uses the concept of a time window, which is a collection of numbers in the form $<I_1, I_2, I_3, \ldots, I_k>$, where each I_i represents the ith input to the system. Hence there are k inputs in all. The input I_i denotes the value at a time lag of I_i units. Thus if we are trying to calculate the output at time t, the inputs would be the value of the sequence at time $t - I_1$, $t - I_2$, $t - I_3, \ldots, t - I_k$.

One of the major problems is to decide the time window. If we increase the size of the window, the performance will increase. But this would require more training data, as well as more computational time. On the other hand, if we make the window smaller, the performance will decrease.

Another important aspect is estimating the time lags in the time window—in other words, estimating the input I_i to be used as inputs to the system. We need to judiciously decide the lags that affect the output of a system. When we use the inputs as $<I_1, I_2, I_3, \ldots, I_k>$, we actually mean that the output at the tth time will depend on output at times $t - I_1$, $t - I_2$, $t - I_3, \ldots, t - I_k$. The actual value can only be obtained with prior knowledge of the seasonal and trend characteristics or by trial-and-error approach.

In our example, we took the time window of $<I_1, I_{12}, I_{13}>$. The inputs were given to the system after a log operation. This preprocessing exploited the differences between the inputs and resulted in better recognition as compared with the original inputs.

2.9.3 Outputs

The system's output was the value of the series at the tth time. We repeated the experiment for all ts to predict/regenerate the complete sequence (starting from $t = 15$, because before that all previous inputs were unknown).

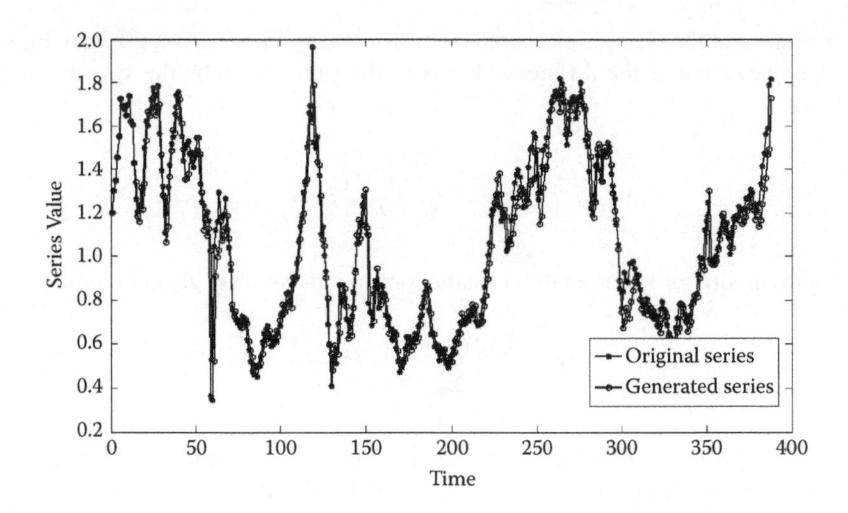

FIGURE 2.24 The time series forecasting experiment.

2.9.4 Network

The network was formed by trial and error. After numerous tries, we got the optimal results by the following network configuration in which we tried changing only the number of neurons in the hidden layer.

Number of hidden layers: 1
Number of neurons in hidden layer: 20
Activation functions of all neurons: Tan Sigmoid (tansig)
Training method: Train Gradient Descend (traingdm)
Learning rate: 0.1
Momentum: 0.01
Max epochs: 2,500
Goal: 0.01

Note that the goal was not met, even upon reaching the maximum epochs during training.

2.9.5 Results

The results of the algorithm are given in Figure 2.24. The data on the left consist of the training cases, while the data on the right consist of the test data. The results are further summarized in Table 2.3.

TABLE 2.3
Results of Time Forecasting Experiment

No.	Case	No. of Data	Correlation	Root Mean Square Error
1.	Training data	185	0.956963	0.113043
2.	Testing data	188	0.959198	0.108290
3.	Total	373	0.958757	0.110673

In this example, we used ANN to solve a real life problem and got good results. More practical guidelines can be found in Chapter 19. Other real life cases using ANN are available in Section II.

2.10 CONCLUSIONS

The ANN can be seen as an instrumental tool for solving virtually any real life problem. ANNs have revolutionized industry because of their huge impact and ease of use. Recent years have seen strong research and very promising models that further strengthen the impact of these networks. Simultaneous work in various domains is being carried out to exploit every advantage of ANNs and to improve their performance even more. ANNs are definitely the technology of the future. At the same time, however, it should be noted that the limitations of ANNs must be determined and solved accordingly. The rich literature in the ANN knowledge bank makes it a very exciting field to study and experiment with.

Although these networks are easy to use, it does take experience and knowledge to build an effective ANN design. Many young researchers may not bother much with the role of various parameters and network complexity. In real life systems, however, these play a great role in making a robust system that gives a very high performance. Designing a good ANN is more of an art than a technical concept. It requires comprehension and patience to try various designs until being with performance.

Equally important is the selection of inputs and outputs. An attempt to train an ANN with the wrong features may not succeed. Even if it succeeds, however, it would never run on a real life problem with real life data. Thus we need to think broadly about the feasibility of the selected input and output parameters before even trying to train the system.

ANNs, if used cautiously, are an excellent means for solving emerging industrial problems.

CHAPTER SUMMARY

This chapter presented the artificial neural networks as effective tools for solving real life problems. Various issues, advantages, and disadvantages related to the use of ANNs were discussed. These networks derive their motivation from basic biological neurons, which are the processing units that collectively make up the brain. This chapter also compared the power and architecture of the human brain with that of the ANN. The chapter introduced and discussed the working of the most basic single artificial neuron, as well as the model and workings of the perceptron.

The more complex ANNs were then presented, including the multilayer perceptron (MLP). We discussed how these ANNs are able to solve complex real life problems. The chapter covered the various terms used in the field of ANNs, including the meaning and role of neurons, layers, activation function, weights, bias, and so forth.

We also discussed how to model the problem and proceed with an ANN solution toward the same. This process involves normalization of the inputs and outputs as a means of basic preprocessing. This was followed by a discussion of the training of ANNs. Learning is categorized into three types: the supervised learning, unsupervised learning, and the reinforcement learning. We also presented the basic terminologies of training, such as learning rate, momentum, variable learning rate, and so forth. A supervised learning algorithm called the back propagation algorithm was also discussed and analyzed.

The last part of the chapter discussed the major issues that must be observed in any ANN design. These issues include convergence, generalization, overgeneralization, and network complexity. We saw the need to make suitable measures to handle problems associated with these issues.

In the end, we took a real life problem of time series forecasting and solved it by using ANN.

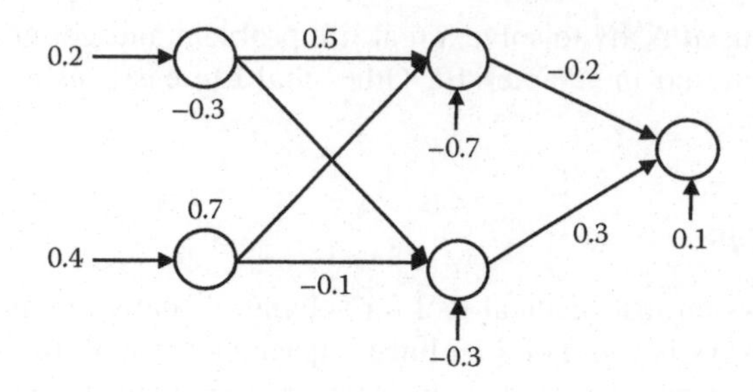

FIGURE 2.25a Figure for Example 1.

SOLVED EXAMPLES

1. **For the network given in Figure 2.25a, calculate the output. All activation functions are hyperbolic given by Equation 2.15. Biases, if present, are indicated.**

$$f(x) = \frac{e^x - e^{-x}}{e^x + e^{-x}} \tag{2.15}$$

Answer: The answer is self-explanatory in Figure 2.25b, in which the numbers indicate the value flow.

The final answer is –0.1709.

2. **It is said that every intelligent ANN has a number of neurons less than the total number of inputs in the training data. Explain.**

Answer: In every real life ANN, the numbers of neurons is much lower than the number of inputs in training data.

Consider an ANN with the following configuration:

Number of hidden layers: 1

Number of neurons in hidden layer = Number of input test cases (say n)

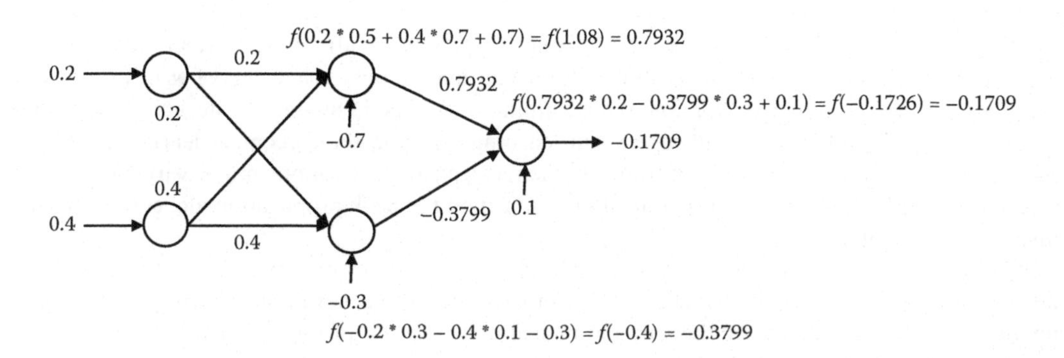

FIGURE 2.25b Figure for Answer 1.

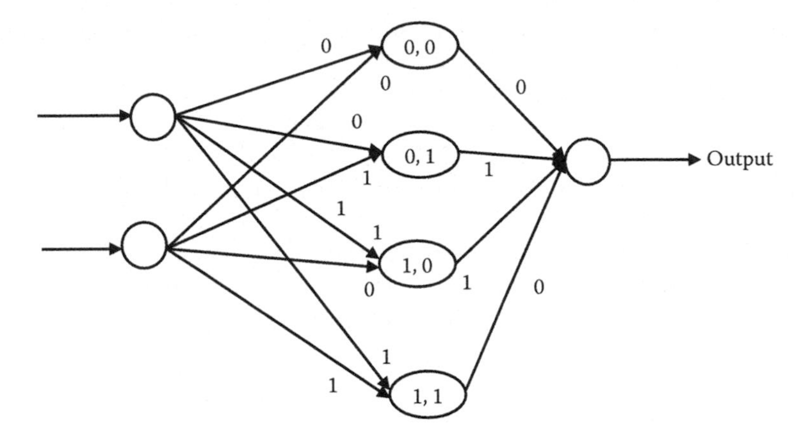

FIGURE 2.26 Figure for Example 2.

Let weight w_{ij} from the input i to the neuron j of the hidden layer store the input parameter of the jth input test case.
Let weight w_j from the neuron j of the hidden layer to the output store the output of the jth input test case.
Let us suppose that in place of multiplying the input by the weight, we instead take the difference for all weights from input to hidden layer.
Let us suppose the activation function to be as given by Equation 2.16

$$f(x) = \begin{matrix} 1 & \text{if} \quad x = 0 \\ 0 & \text{otherwise} \end{matrix} \tag{2.16}$$

We can see that the ANN does not require training and is able to give correct answers to all the test cases.

Now, take the example of the XOR problem given in Table 2.2. The network is shown in Figure 2.26. The input case is written inside the neuron for simplicity of understanding.

Let the inputs be denoted by (x, y), where x is the input to the first input neuron and y is the input to the second input neuron. Suppose the input is (1, 0). Neuron 1 will give an output of $f(|1 - 0| + |0 - 0|) = f(1) = 0$. Neuron 2 will give an output of $f(|1 - 0| + |0 - 1|) = f(2) = 0$. Neuron 3 will give an output of $f(|1 - 1| + |0 - 0|) = f(0) = 1$. Neuron 4 will give an output of $f(|1 - 1| + |0 - 1|) = f(1) = 0$.

So the final answer is $0 * 0 + 1 * 0 + 1 * 1 + 0 = 1$, which is correct.

It may be seen for any of the cases that only one neuron is activated, and the others are passive. The output corresponding to the activated neuron is given as output. It may even be seen that the generalizing power of these networks is 0; they cannot give outputs to any of the unknown inputs. However, if needed, suitable modifications may be made to incorporate this behavior as well.

Hence it can be stated that if an ANN is generalizing or trying to frame rules, the number of neurons would be much less as compared with the training data set size, depending on the ANN's generalizing capability.

3. **How do you determine the number of hidden layers and the number of hidden layer neurons for any problem in ANN design?**

Answer: There is no direct methodology to determine the number of hidden layers and the number of neurons. The number can only be determined by regular experimentation. The following points, however, may be useful in deciding the number.

1. The number of layers is always to be kept as one. If the network does not train at all, the number of layers may be increased to two.
2. The hidden layer should have the least number of neurons required to train the system.
3. Start the experimentation with a small number of neurons.
4. If the system does not train, increase the number of neurons.
5. Do not force training by allowing training to run for a long period of time.
6. Do not increase the neurons abnormally.
7. Keep the neurons always less than the number of input test cases.
8. Neurons are, in general, higher than the number of inputs and outputs.
9. Rules of thumb, as suggested by many literatures, do not always work.

4. How are ANNs able to handle noise in data?

Answer: ANNs are known to be able to handle noise in data very well. The presence of noise to a certain extent does not have a big effect on performance.

To understand, let us take a simple ANN that is used for functional approximation. Consider input data set (I, O), where I denotes the inputs, and O the outputs. Let us say that k of these inputs had some large noise in them. Let us say also that there was a general small amount of noise in all the data.

During training, the ANN is actually trying to form a function from the given values. When any of the k inputs are met, they fail to fit well in the graph. Hence the ANN may avoid them in the natural process (assuming k is much less than *total input size*). In addition, the general noise is suppressed in the ANN's attempt to accommodate all the functions.

Readers may consider this analogous to the system in which a regression formula is applied for curve fitting. Eventually, all the inputs are reframed to match the curve that comes by regression analysis.

EXERCISES

General Questions

1. What is the role of the activation function in ANN?
2. What role does bias play in ANN?
3. What is the effect of the number of neurons and the number of hidden layers in ANN?
4. Compare and contrast ANNs with the human brain.
5. What is the effect of learning rate in training an ANN?
6. Explain the different learning techniques in ANN.
7. What would happen if the number of neurons in the hidden layer were too low?
8. What is the role of training data input size on the learning time required? Does this size depend on the type of data?
9. What would be the effect of keeping momentum 0 in an ANN?
10. Comment on the factors that affect the time required for training the ANN.
11. What is overgeneralization?
12. Comment on conditions when ANN should not be used for problem solving.
13. How is performance measured in an ANN?
14. State the back propagation algorithm.
15. What is the use of the variable learning rate in ANN? How is it better than a constant learning rate?
16. How do you save the ANN from being trapped in local minima?
17. Differentiate between local and global minima.

18. What is the steepest descent approach?
19. How important, according to you, is the role of initial weights and biases in training the ANN?
20. What is meant by an epoch?
21. How can the problem of overgeneralization be solved?
22. What would happen if the amount of data to train an ANN were too low?
23. What role do activation functions play in the training of the ANN?
24. Can we incorporate methods of adding and deleting neurons and hidden layers during training so that the system not only trains itself but also decides its own architecture?
25. Suppose the system discussed in Question 24 were possible. Identify the advantages and disadvantages.
26. Imagine a program that starts with one neuron in a hidden layer. It tries to train the system. If the system does not train, it adds another neuron. This process continues until the network is trained. Identify the advantages and disadvantages of this strategy. Will this strategy work?
27. Comment on the statement, "Weights are the memory of the ANN."
28. Is it necessary that the error of ANN always reduces as the epochs increase? Why or why not?

Practical Questions

(Use any ANN simulator to answer these questions.)

1. Randomly generate 500 training samples for any assumed function that takes two parameters. Train the ANN on these data. Now give unknown inputs and measure the performance of the system.
2. In the same data that were used for Practical Question 1, add 1% noise to each piece of data and repeat the experiment. Did you observe any change in performance? Repeat the experiment for 5%, 10%, 15%, 20%, 25%, ..., 50% noise to see how the performance changes.
3. In the same data that were used for Practical Question 1, change the number of hidden layers and neurons to observe the effects of overgeneralization. Change the learning rate to observe reduced performance by ANN oscillation.
4. Make an ANN that takes a Fibonacci number and generates the next Fibonacci number in the series.
5. Try to generate random data of five dimensions and an associated random output. Try training an ANN with this data. Now keep increasing the number of inputs in the input data and note the number of neurons required in the hidden layer to meet the goal. Compare this change in the number of neurons with the growth when the input data size is increased for Practical Question 1.
6. Solve the n-input XOR problem by ANN, where n is known. Comment on the increase in the number of neurons required as n increases.
7. Download any time series from the UCI Machine Learning Repository and solve it using ANNs.

3 Artificial Neural Networks II

3.1 TYPES OF ARTIFICIAL NEURAL NETWORKS

Chapter 2 provided an in-depth analysis of the ANN model and its training in the supervised mode. The chapter introduced the back propagation algorithm (BPA) for machine learning and defined various parameters commonly used in ANN terminology. The ideas presented in that chapter denote just one of the many types of ANNs that are actively used in various problems.

The abundant research in the field of neurocomputing, as well as the active inputs from associated fields such as the cognitive sciences, has led to various ANN models and learning techniques. Whereas Chapter 2 offered an in-depth analysis of problem solving with ANN and BPA, this chapter broadly explores several other models and methodologies to introduce you to the infinite possibilities and real life applicability of ANN.

The major models studied in this chapter are radial basis function networks, learning vector quantization, recurrent neural networks, Hopfield neural networks, and Kohonen's self-organizing maps. Each of these models, in its own way, maps inputs to outputs with the help of inferences from historical databases by machine learning.

We studied the supervised learning in the previous chapter. In this chapter we present the other two types of learning, i.e., the unsupervised learning and reinforcement learning. These learning are the basis of every neural network model that we study.

3.1.1 Unsupervised Learning

In unsupervised learning (UL), the system is given only the inputs to learn; the output is unknown to the system. Therefore, the system has to learn or train itself without knowing the final outcome, which is why a UL system does not learn about its performance and the errors. This type of learning is very different from supervised learning, in which the whole learning was based on the output—in other words, the output drove the system although true. It may be hard to imagine that the system is being trained without having the final answers. Thus, unsupervised learning can be defined as the technique of machine learning when the inputs are known and the outputs are unknown.

Definition 27
Unsupervised learning is the technique of machine learning when the inputs are known and the outputs are unknown.

UL has a deep analogy to the natural systems. Much of the learning in human beings is in the absence of a strict trainer who keeps repeating the questions so that the system is able to train itself, as would happen in supervised learning. In reality, children learn a great deal on their own without any formal training. Let us take the example of a child walking. Nobody tells him how to differentiate between the paths and the obstacles. Instead, the child keeps walking or playing. When he encounters an obstacle, he might initially collide, but later he is able to identify his surroundings and act accordingly.

The same is true for UL. Even upon initially seeing the input, the system can draw many conclusions. The system in this technique of learning tries to identify the inputs by analyzing a relation or pattern. In machine learning, the inputs always repeat themselves. Hence, identifying the input is of great significance in learning. The system can identify similar kinds of inputs, even during the training phase. This capability gives the power to a system using the UL technique to give the correct output once an input is applied. In short, the system analyzes the input and identifies the possible patterns. It then generates the output based on the common attributes those kinds of inputs possess.

To study the role of inputs in UL techniques, let us look at another example. Suppose we need to make a real life system that identifies the digit spoken by an individual. Based on the system requirements, we identify the relevant parameters and make the needed database. If the choice of features is good, the inputs with digit 0 would be located at some distant place of the input space, the digit 1 at some other place, and so on. When the system is given any historic input, it visualizes the inputs. It then tries to demarcate between the inputs to decide the possible space at which that input lies. Every input close to this input is identified accordingly and tried to learn similarly. Because the digits are assumed to be reasonably apart, the UL system is able to link them by the inputs.

Two common examples of unsupervised learning are self-organizing maps (SOMs) and recurrent neural networks, which are studied later in this chapter.

3.1.2 Reinforcement Learning

In reinforcement learning (RL), the exact output is unknown (as with supervised learning), but we come to know the goodness of a solution. The system is made to do a set of actions a number of times and receives a reward for every action that it performs. This reward depends on the goodness of the solution. If the solution is bad, then a punishment or a negative reward may be given. Thus, every action is associated with some reward or punishment, depending on the final outcome. The reward or punishment is dynamic and changes with respect to the environment. In other words, if there is a change in environment, an action that resulted in a reward can later result in a punishment. Through these rewards and punishments, the system learns about its performance. Other than these rewards or punishments, the system is not provided any other information for learning. Thus, the system tries to find a solution in which the rewards are maximized and the punishments are avoided.

Reinforcement learning is when the learner does not explicitly know the input-output instances, but it receives some form of feedback from its environment. The learner adapts its parameters based on states of its actions (Konar, pg 427)

Definition 27

Reinforcement learning is when the learner does not explicitly know the input-output instances, but it receives some form of feedback from its environment. The learner adapts its parameters based on states of its actions (Konar, pg 427)

The RL technique has many applications in fields related to control. In these fields, the learning is applied to control a machine. Reinforcement learning hence is used in robot path planning and control. In these problems, the environment is known. Every time there is some collision or unwanted action, the system is punished. If, on the other hand, the system is able to guide the robot toward the goal, the system is rewarded, with the reward depending on the path quality.

The system uses a problem-specific function, called the reinforcement function, to give the reinforcements, or the rewards and punishments. Another important concept is the environment that defines the physical conditions. As mentioned earlier, the reinforcement always depends on the environment. The final outcome is the action or actions that optimize the reinforcement. This is also called the *policy for reinforcement maximization.*

RL techniques are still in the developing phase and have not been applied to real life problems to a big extent. Therefore, we do not discuss this learning technique any further in this book. Rather, in the next few sections, we discuss some of the major ANN models that have found a variety of uses in real life problems.

3.2 RADIAL BASIS FUNCTION NETWORK

The radial basis function network (RBFN) is another type of ANN that performs unsupervised or supervised learning. These networks, which are known for their ability to model complex scenarios in a very simple architecture, are used for classification, series prediction, control applications, and so forth. They can perform both supervised and unsupervised learning, depending on the way the network has been modeled and used. The RBFN uses radial basis functions (RBFs) as activation functions, which allows it to easily solve the problem or perform a mapping of the inputs to outputs. This section addresses the various features of the RBFNs one by one.

3.2.1 CONCEPT

RBFNs make use of data known as *data centers* or *data guides*, which serve as neurons of the only hidden layer that these networks have. These data centers may be taken from the input data only or from elsewhere in the multidimensional input space. These data centers act as guiding factors for the RBFNs.

To fully understand RBFNs, it is best to visualize the input space where the outputs are plotted against the inputs in a high-dimensional graph. In this way, it is possible to visualize the plotting of training data that are known to the system before the application of unknown inputs. Because these data may be too large and unworkable for computational reasons, the system needs to start by summarizing the data. The resultant summarization is the data centers. Each data center represents an area of similar input and hence similar output. Now we have an input space consisting of a finite number of points already plotted. The next task is to use the data centers to extrapolate and plot outputs in the entire input space. These data centers are consulted whenever the decision regarding the output is to be taken for any of the inputs presented. Each data center contributes to the system's final output by trying to induce the characteristics of the type of output it represents to the final output. Hence, RBFNs may be visualized as systems in which many centers are quarreling to influence the final output and make it closer in magnitude to the output represented by them. The influence of each center depends on the distance of the unknown point from the center—the larger the distance, the smaller the influence. In other words, every center is more powerful in its locality. Thus, for better performance, it is best to have these data centers distributed in the entire input space, especially for critical inputs.

Another way to look at RBFNs is through the ANN layered model approach. In this case, RBFNs exploit the factor of nonlinearity. ANNs extrapolate the input space mainly by using the known inputs. Hence, these systems act as regressors. The better the system covers the entire space using smooth curves, the better the expected performance. Conventional ANNs use highly nonlinear functions as activation functions, and these functions keep changing with time. As a result, conventional ANNs often have a problem with generalization. RBFNs, on the other hand, use a nonlinear function at one layer that is kept constant. In addition, experiments show that the choice of this function does not have a remarkable effect on the system's performance. The other activation functions for RBFNs are simple linear mapping. As a result, RBFNs achieve greater performance than other ANNs over specific problems.

Conventional ANNs are prone to get stuck at the local minima. To avoid this, we apply various methods, such as genetic optimizations and momentum. However, doing so makes conventional ANNs highly computationally expensive, and the performance can never be guaranteed. These problems do not exist in RBFNs, which are better in regression as compared with conventional ANNs.

RBFNs mainly work with huge amounts of data. These networks can better handle the large input sizes because they are computationally less expensive and capable of complex modeling.

RBFNs also involve more data centers or hidden-layer neurons than do conventional ANNs. The effect of adding neurons does not add the same complexity as the addition of hidden-layer neurons in conventional ANNs, because the constant structure of the activation function limits the complexity. This may even be visualized in the input space: The addition of centers means a greater flexibility to adapt the outputs, especially at some regions of the space that lie close to the center. However, this does not strictly mean that we would be able to imitate any functional behavior or more complex functions, as is possible with conventional ANNs.

3.2.2 Network Architecture

The basic architecture of RBFNs is similar to conventional ANNs, which have one hidden layer. The first layer is the input layer, where the inputs are present. The second layer is the only hidden layer in the network. The last layer is the output layer, which gives the final output.

The hidden layer consists of a number of data centers, as discussed earlier. In RBFNs, these data centers do the work of neurons in ANNs. Each neuron uses a nonlinear activation function, which adds nonlinearity to the system, thus enabling the system to model complex shapes. The hidden neurons are connected to the output by weights. The output neurons simply take the weighted average of the inputs given to it. This becomes the final answer. The general structure for RBFNs is shown in Figure 3.1.

3.2.3 Mathematical Analysis

RBFNs work just like any ANN. The input is applied, and then the various hidden neurons are given the inputs. These neurons perform the summation; then the output is passed through the activation function. The output nodes do the aggregated addition, which generates the system's final output.

Upon the application of any input, each neuron takes the norm (usually the Euclidian norm) of the input with the neuron itself. The norm for any input is its distance from the data center in the input space. Each data center possesses a location in the input space. Thus, the system's first task is to find the norm, which is denoted by the function $\| x \|$, by calculating the distance between the applied input and this center. The norm is given in Equation 3.1:

$$h_i = \|x - c_i\| \tag{3.1}$$

where c_i is the ith data center or the ith neuron, x is the input, and h_i is the aggregated sum.

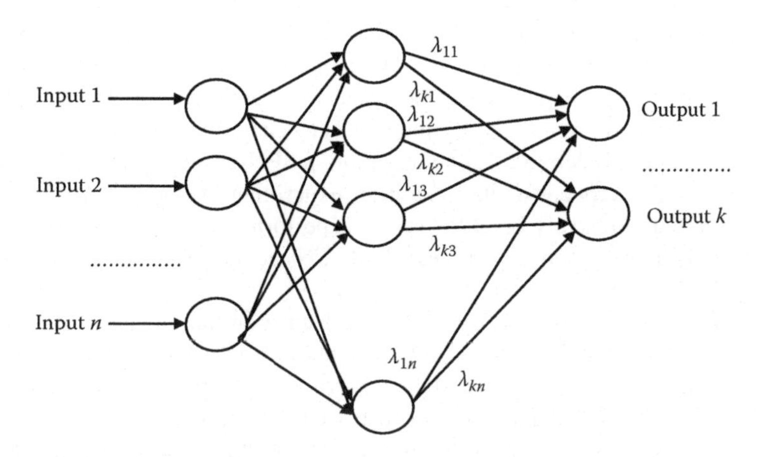

FIGURE 3.1 The architecture of the radial basis function network.

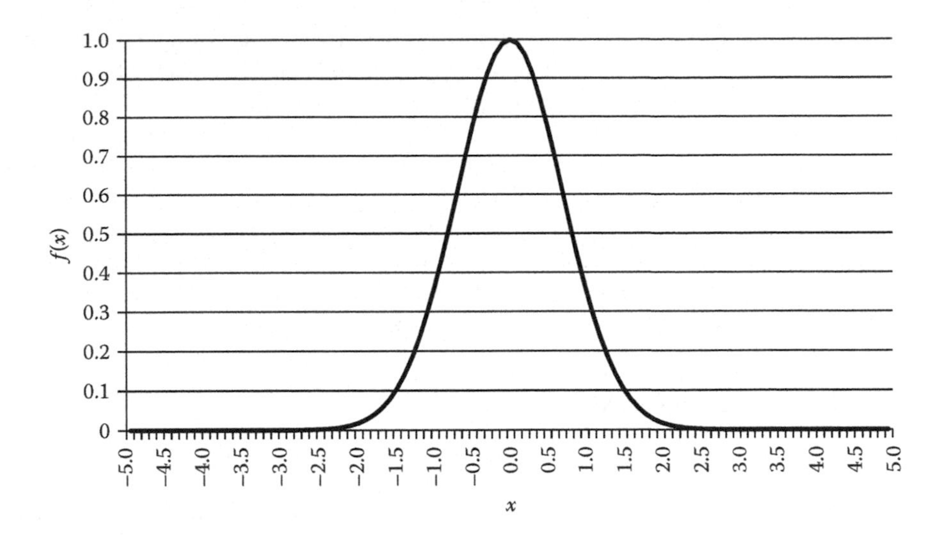

FIGURE 3.2 The graph of Gaussian function.

The nonlinear activation function, or the basin function, used in these networks is usually Gaussian. The Gaussian function is given by Equation 3.2:

$$\rho(x) = e^{-\beta x^2} \tag{3.2}$$

The graph of Equation 3.2 is shown in Figure 3.2. Here the peak is always found at the mean position corresponding to an input of 0.

The calculated norm of distance by Equation 3.1 is intended to decide the influence of the data center in determining the final output. This occurs in the second step, where we apply the activation function. As we go away from the data center, the influence must decrease. This is achieved by using the Gaussian activation function given by Equation 3.2.

The summation h_i is passed through the activation function. Hence, using Equations 3.1 and 3.2, we get Equation 3.3:

$$\rho(\| x - c_i \|) = e^{-\beta\|x-c_i\|^2} \tag{3.3}$$

Now the task is simple: We know the output of every data center; this is the kind of output it advocates. We also know the influence of this data center based on its distance from the unknown input. We now simply take a weighted average. The output that each data center advocates is available in the form of the weight between the hidden layer and the output layer. The output layer simply takes a weighted summation over these inputs, as is represented in Equation 3.4:

$$v_j = \sum_{i=1}^{n} \lambda_i e^{-\beta\|x-c_i\|^2} \tag{3.4}$$

This equation denotes the final output of the jth output neuron.

3.2.4 Training

The training in RBFN is done to adjust the various parameters introduced in the mathematical model. The adjustment of these parameters gives the precise network that is able to solve the real-life problem. In ANN terminology, these parameters comprise the weights between the input layer and the hidden layer and between the hidden layer and the output layer. In data center and input space terminology, these parameters are the data centers and the output advocated by the data centers. Sometimes the activation function constants may also be modified.

The RBFN may use a variety of training methods. Which adjust the system parameters to reduce error. Many statistical approaches have been used that view training as an optimization of a function using statistical tools. The back propagation algorithm with steepest descent approach may also be used. Some of the other commonly used techniques for training include assigning each weight basis function, numerical approaches, interpolation, least squares approach, orthogonal least squares using Gram-Schmidt algorithm, gradient descent or back propagation, and expectation maximization algorithm. Note that we do not discuss these techniques in this book.

3.3 LEARNING VECTOR QUANTIZATION

The learning vector quantization (LVQ) is a means of supervised learning. The LVQ algorithms have a variety of uses in classification problems. In these problems, the various inputs need to be separated according to decision boundaries. Thus, this algorithm works on the principle of Hebbian learning.

3.3.1 Concept

The general concept of LVQs comes from Hebbian learning, which is an understanding of the human brain and the learning associated with it. Hebbian learning is derived from Hebb's postulate, which states, "When an axon of cell A is near enough to excite a cell B and repeatedly or persistently takes part in firing it, some growth processes or metabolic changes take place in one or both cells such that A's efficiency, as one of the cells firing B, is increased."

LVQ introduced the novel concept of *neighborhood*. For any classification problem, if we know the output at some input, it is natural that the output at the neighboring points would be the same. Consider, for example, the speaker recognition system. Suppose our database says that a set of parameters with the duration of speech of 16 seconds corresponds to Speaker B. Now suppose we get an unknown input with the same parameters but the duration changed to 16.1 seconds. It is natural to believe that it was the same Speaker B who took a little longer this time. Hence, for any classification problem, we may make many inferences if we know the output class at the nearby inputs.

Consider now the input space, which marks the various classes by labels in a high-dimensional graph of inputs. Again we know beforehand the classes corresponding to the training data. We can visualize this information as already being marked in the graph. However, these data would be too large, covering large areas of the graphs. We cannot handle such large volumes of data due to computational and memory constraints. Now suppose that these data are, by some intelligent means, reduced to a set of new points called *centers* or *guides*. In this input space, whole classes are well determined, which means the input graph is now much lighter and easier to handle. Each center denotes a class that is a valid output for the input on which it lies, as well as for the nearby inputs. Now whenever any unknown data are applied, the system tries to find the closest center where the output is well known and associates itself with that center. If the problem is nicely modeled and the inputs are good, we would be able to find some center very close to it which would decide its answer or the class to which the input maps.

Taking these networks as strict ANNs and following the strict ANN methodology, we can easily visualize the work of each of center as neurons. The activation is much loosely defined which models process such that one can find the minimum distance in the input space to each neuron.

With this basic concept, we come to the very important concept of learning. Learning in a Hebbian approach consists of constantly evaluating the effects of the neighboring neurons. All these effects decide the activity of every neuron. The combined activity of all the neurons gives the final network. Thus, the learning in this case is supervised: The actual output is known, and the adjustments in the activities of the various neurons are calculated by the error in output.

This approach is also referred to as the winner-takes-all approach, because we first study the activities of all the neurons, and then we decide the winning neuron. The activity of this winning neuron is adjusted according to the answer. This approach may be visualized as a very apt learning strategy for any classification problem in which the winning neuron is the point in the input space to which the applied input associates and output the neuron's associated class. The continuous adjustment of the neuron activity and the effects of those neurons upon the neighboring neuronal activity in multiple epochs is done performed by the system. This adjusts the various weights, or parameters, of the network in order to better adapt the network to the problem, which allows these networks to perform very well for classification problems.

3.3.2 Architecture

The whole architecture of LVQs consists of three layers: the input layer, the classification layer, and the output layer. The inputs are presented in the input layer.

The classification layer consists of a number of neurons per class. Thus, there need to be an equal number of or more neurons in this layer than the number of classes. The total number of neurons depends upon the number of neurons used per class, which in turn depends on the ease by which the various classes may be separated. Each neuron has a class associated with it. In addition, each neuron in these networks denotes some point in the high-dimensional input space. These neurons represent regions where the class possibly lies. The neurons may be visualized as the subclasses of every class. Each class has one or more subclasses, and similarly these networks have one or more neurons per class.

The relation between the class and the subclass is an interesting subject to study. Let us look again at the input space where these neurons are placed, or need to be placed, as centers or guides. Ideally the various classes must reside at completely distinct regions in the input space—in other words, the interclass separation must be high. The various training data sets that correspond to a particular class must reside as close as possible—in other words, the intraclass separation must be low. But this does not happen. The various classes may lie close together and may even intermingle with one another in the input space. Thus, we need extra neurons or centers placed in such a way that the classes can be separated. In other words, we need to break the region in the input space that is occupied by a single class into smaller regions that do not intermingle with other foreign classes. These two regions can now act as two separate classes and are hence referred to as subclasses.

The third layer consists of only one neuron per class. This layer takes as input the output of the various neurons and tries to sum up the outputs per class, so that every class has only one output. The second layer gives output only to the subclass that lies closest to the given input vector. It does this by using a hard limiter activation function that activates only the neuron that lies closest to the given input.

The architecture of these networks is similar to the one shown in Figure 3.1.

3.3.3 Mathematical Modeling

Suppose that an input i has been applied. The LVQ begins by calculating the distance of this input i from all the other neurons that are present in the second layer. This distance, also called the *norm*, denotes the distance of separation between the neuron, or the guide, and the unknown input. This norm is given in Equation 3.5:

$$a_i^l = \| W_i^1 - I \| . \tag{3.5}$$

where a_i^l is the distance between the input and the ith neuron. The weight W_i^1 is the position of the neuron.

The next task is to find the neuron, or guide, that lies closest to this input in the input space. The activation function simply selects the minimum distance of all a_i^l. This is the output of this layer and is given by Equation 3.6:

$$a^l = \min\{a_i^l\} \tag{3.6}$$

Once the neuron that lies closest to the input is known, we need to find the class that corresponds to this neuron. This class is the system's final output. This lookup is done by the second layer, which is the layer that simply keeps track of the classes. All the subclasses per class are mapped to a single class, and the network uses a matrix of weights to track the various classes. The weight matrix W^2 of this layer is a mapping between the subclasses and the classes. The columns in the matrix represent the subclasses, and the rows represent the classes. Thus a unit element in this weight matrix may be represented by Equation 3.7:

$$\begin{aligned} w_{ij}^2 &= 1, \quad \text{if subclass } j \text{ belongs to class } i \\ &= 0, \quad \text{otherwise} \end{aligned} \tag{3.7}$$

Because every neuron can be a part of only one class, each column has exactly one 1 corresponding to the class to which it belongs.

The final answer is the class that has the minimum distance, and this is the output of the entire network.

3.3.4 Training

Training in an LVQ has two phases. In the first phase, we calculate the correct neuron that is activated whenever an input is applied. Because only one neuron is activated by the concept of LVQ, we get a single answer. The second phase is the modification of the parameters, or weights, of the LVQ. The parameters are in the form of the positions of the various neurons, or guides. The weights are modified to make the network move in a direction such that the error is reduced.

Upon application of an input i, let us assume that neuron k was activated. We know that the input belonged to class l. This naturally means that neuron k also belonged to class l. The W^2 matrix is modified, if needed, to reflect this change and now shows 1 for the element corresponding to class l for subclass k.

Next we modify the W^1 matrix. Modification of this matrix depends upon whether the output's predicted class was the same as or different from the desired class. Hence, the modification rule relies on the ability of the network to classify the input correctly or incorrectly.

Based on these concepts, two algorithms have been proposed for learning in the LVQ.

3.3.4.1 LVQ 1 Algorithm

In this algorithm, only the winning neuron is modified; no change made to any other neuron. The modified neuron moves toward the input or away from the input, depending on the condition.

Suppose the network identified the output class correctly. In this case, the strategy is to move the winning neuron more toward input i. This is achieved by adding a small parameter in the direction of the input, as represented by Equation 3.8:

$$W_k^1(t+1) = W_k^1 + \alpha(t) * (i - W_k^1(t)) \tag{3.8}$$

If, however, the network does not identify the class correctly, we try to move the weight of the winning neuron a little away from input i, as shown in Equation 3.9:

$$W_k^1(t+1) = W_k^1 - \alpha(t) * (i - W_k^1(t)) \tag{3.9}$$

For the other neurons that are not the winning neurons, the weight remains unchanged, as shown in Equation 3.10:

$$W_k^1(t+1) = W_k^1(t) \tag{3.10}$$

where t denotes time, and the parameter $\alpha(t)$ is the learning rate; $\alpha(t)$ always lies between 0 and 1. The higher the value of this parameter, the larger the changes that are made and the more the neuron would move. This parameter continues to reduce with time. At larger times, the system becomes more stable and converges to some definite state.

3.3.4.2 LVQ 2.1 Algorithm

This algorithm modifies two neurons in place of one. Out of these two neurons, one belongs to the correct class and the other to the wrong class. In other words, one of these neurons belongs to output class l and the other does not.

In addition, input i for this algorithm must lie within the window formed between the midplanes of the two selected closest neurons. Let w denote the width of this window. Let j and k denote the neurons to be modified . Let the distance of input i to neurons j and k be d_j and d_k, respectively.

We say that input i falls in a window of width w if Equation 3.11 holds:

$$\min\left(\frac{d_j}{d_k}, \frac{d_k}{d_j}\right) < s \tag{3.11}$$

where s is given by Equation 3.12:

$$s = \frac{1-w}{1+w} \tag{3.12}$$

The window concept is shown in Figure 3.3. In this scenario, it is necessary that input i lies within the window as shown in Figure 3.3. The modification of the weights is given by Equations 3.13 and 3.14: The neuron that corresponds to the correct output class is modified by Equation 3.13, while the neuron that corresponds to the incorrect output class is modified by Equation 3.14.

$$W_k^1(t+1) = W_k^1 + \alpha(t) * \left(i - W_k^1(t)\right) \tag{3.13}$$

$$W_k^1(t+1) = W_k^1 - \alpha(t) * \left(i - W_k^1(t)\right) \tag{3.14}$$

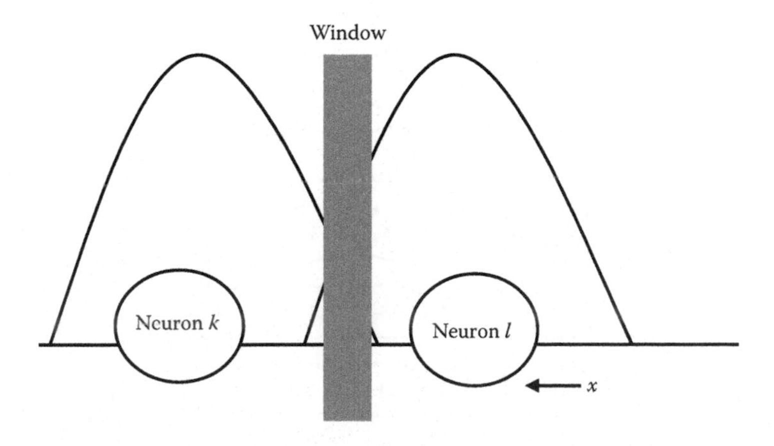

FIGURE 3.3 The window in the learning vector quantization 2 learning algorithm.

3.4 SELF-ORGANIZING MAPS

The self-organizing map (SOM) is quite similar to the LVQ. SOMs are used for learning in an unsupervised mode. In this mode, the inputs are known, but the outputs are unknown. SOMs are known for forming features in the whole input space, which helps in learning. SOMs also offer a good means of clustering the input data into distinct clusters that lie over the SOM's feature map.

3.4.1 CONCEPT

SOMs, which are used for unsupervised learning, make use of the concept of feature maps. The entire high-dimensional input data can be mapped on a two-dimensional map called the *feature map*. In this map, the closer regions are closer in the high-dimensional input space and vice versa.

Each node in this feature map represents some point in the input space. In this feature map, the neighboring points of a point *p* are the points that lie close to it. It is natural that these points would lie close to point *p* in the input space as well. The network, just like in LVQ, tries to find the neuron that is closest to the input that is applied. This neuron is activated, and the others are not activated. Thus there is only one neuron that wins, using a winner-takes-all approach.

The modification in weights follows a Hebbian rule, in that the adjustment of the activity has an effect on the neighboring neurons. This effect loses its impact on moving farther from the wining neuron. Thus if we adjust a neuron, we would also be adjusting its neighbors. The magnitude continues to decrease as we move farther from the input and toward the distinct neighbors.

The feature map is the most interesting and novel concept introduced by SOM networks. The training and use of SOM involves mapping up of the highly dimensional and vast input space into these two-dimensional finite feature maps. Mapping refers to giving a 2-dimensional representation with finite and discrete size to the input space. This limits the data into computational limits, which is a prime goal of learning. We cannot distribute the input space uniformly because the training data might not be uniform. Doing so would lead to some very inactive regions of the feature map. The distribution needs to be done judiciously to maximize the effectiveness of every cell of the feature map while still covering the entire input space.

3.4.2 ARCHITECTURE

The SOM has a simple two-layer architecture. The first layer is the input layer, which is where an input is applied. The second layer is the classifying layer. This layer consists of the feature map, also known as the self-organizing feature map (SOFM), which comprises a set of neurons that are usually arranged in a two-dimensional matrix. Each neuron represents a point in the input space. The task of the classifying layer is to find the neuron that lies closest to the input presented. The SOFM is shown in Figure 3.4.

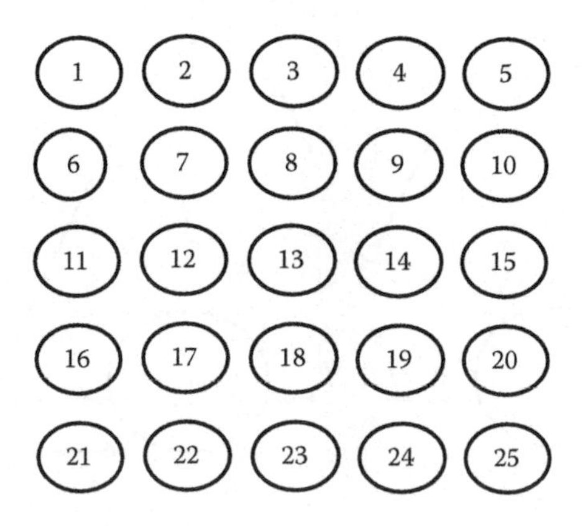

FIGURE 3.4 The self-organizing feature map.

3.4.3 Mathematical Analysis

The concept of SOM's mathematical framework is the same as that of the LVQ. The first step in the SOM is to compute the neuron that lies closest to the given input. This is done by calculating the norm given in Equation 3.15 and selecting the minimum distance, as given in Equation 3.16:

$$a_i = \|W_i - I\| \tag{3.15}$$

where W_i represents the weight, or the position, of the ith neuron in the input space And a_i is the output of the ith neuron.

$$a = \min\{a_i\} \tag{3.16}$$

The class corresponding to the output class of the winning neuron a is regarded as the final network output.

3.4.4 Training

The SOM's training procedure consists of calculating the class that corresponds to the network and modifying the network's parameters, or weights. Calculation of the correct class with a given input i, which was discussed in the previous section, happens by the selection of the closest neuron or the winning neuron. Modification of the winning neuron's weight at any time t is given by Equations 3.17 and 3.18:

$$W_k(t+1) = W_k + h(t) * \left(i - W_k(t)\right) \tag{3.17}$$

or

$$W_k(t+1) = (1 - h(t))W_k + h(t) * i \tag{3.18}$$

where W_k represents the weight, or position, of the kth neuron in the input space, t is time, and i is input. The modification uses the parameter $h(t)$. The parameter also accounts for the closeness of the neuron to the winning neuron, which is useful for the training of neighboring neurons.

We change the weights not only of the winning neuron but also of the neighboring neurons according to Equations 3.16 and 3.17. This change depends on the distance of the neuron from the winning neuron. This distance is measured by the parameter $\alpha(t)$, which is much less for neighboring neurons that lie close to the winning neuron. This parameter continues to decrease as we move away from the winning neuron. Neighboring neurons are defined as those neurons that lie within a certain predefined radius r of the winning neuron, as given in Equation 3.19. The modification of weights is done for all neighboring neurons $N(r)$.

$$N(r) = \{k: d_{jk} \le r \text{ for all } k\} \tag{3.19}$$

where $N(r)$ denotes the neighboring neurons, j denotes the winning neuron, and d_{jk} denotes the distance of neuron k from neuron j.

The parameter $h(t)$, given by Equation 3.20, accounts for the reducing impact of training as we move farther from the winning neuron.

$$h(t) = \alpha(t) * e^{-\frac{\|W_j - W_k\|^2}{2\sigma^2(t)}} \tag{3.20}$$

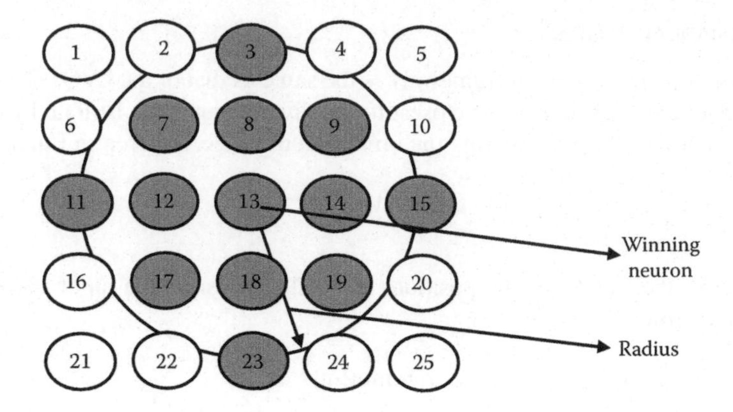

FIGURE 3.5 The neighborhood of the winning neuron in self-organizing maps.

where $W(j)$ corresponds to the weight of the winning neuron, W_k corresponds to the weight of the neighboring neuron, and $\alpha(t)$ is the learning rate.

The learning rate continues to decrease with time. The neurons lying within the circle are in the neighborhood. The concept of neighborhood is depicted in Figure 3.5.

3.5 RECURRENT NEURAL NETWORK

Recurrent neural networks are a special type of ANN in which the output of one or more neurons is fed back into the network as inputs, forming the input for the next iteration. These networks offer a good means of machine learning in which past outputs may potentially affect the next outputs. For this reason, they are used in various problems. Recurrent neural networks have good capabilities for predicting output based on past outputs and inputs.

3.5.1 Concept

Because recurrent neural networks have additional inputs that come from the output of the neurons, they are also known as time delay networks. This means that the output of any neuron at time t is the input given to the system at time $t + 1$. Or, in general, the output is delayed for m units of time. This output is then given as input to the network at time $t + m$. Whenever we talk of *time* in these networks, we refer to discredited time, which means 1 unit of time is the time taken by the system to process one input case.

The additional inputs given to the system help better adapt the system for certain problems, such as series prediction, functional approximation, and recognition systems. For these types of problems, each output can have an effect on the next output, which is why recurrent networks give better performance.

3.5.2 Architecture

The architecture of these networks is similar to that of the ANNs. In conventional ANNs, however, neurons connect one layer to the next layer, with no evidence of an internal cycle being formed. The existence of cycles is possible in recurrent neural networks. The existence of cycles implies that the information keeps cycling between the neurons. The external inputs and outputs ensure that this information is added with new inputs, as well as affecting the other outputs. The processing of information in this cycle takes place in a way similar to what occurs in conventional ANNs. However, the cycle iterates the information only once for each unit of time. The presence of the cycle in the

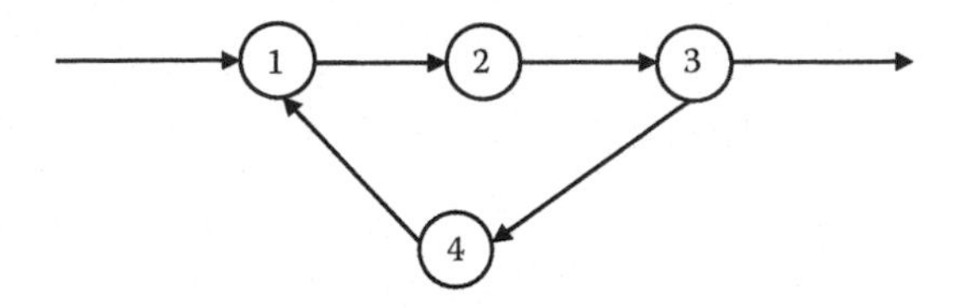

FIGURE 3.6 The cycle in the recurrent artificial neural network.

network is shown in Figure 3.6 where each circle denotes a neuron. This general architecture is shown in Figure 3.7.

3.5.3 Training

The training of these networks is similar to the training in conventional ANN training. The standard back propagation algorithm may be used for the training of these networks. Training involves calculation of outputs, determination of error, and back propagation of error to preceding layers to adjust for weights and biases.

3.6 HOPFIELD NEURAL NETWORK

Hopfield neural networks take their inspiration from the feedback system in biological neurons, in which continuous feedback from the various nodes helps guide the output of the systems. Hopfield neural networks make use of this same phenomenon by using the recurrent neural network structure discussed in the previous section. This enables the Hopfield nets to adapt themselves and give the correct results to the unknown inputs.

There are two types of Hopfield neural networks: the binary Hopfield neural network and the continuous Hopfield neural network. The binary Hopfield neural network has binary outputs that are either 0 or 1. The continuous Hopfield network, on the other hand, has outputs that can be any value from 0 to 1. We only discuss the binary Hopfield neural networks in the rest of the text.

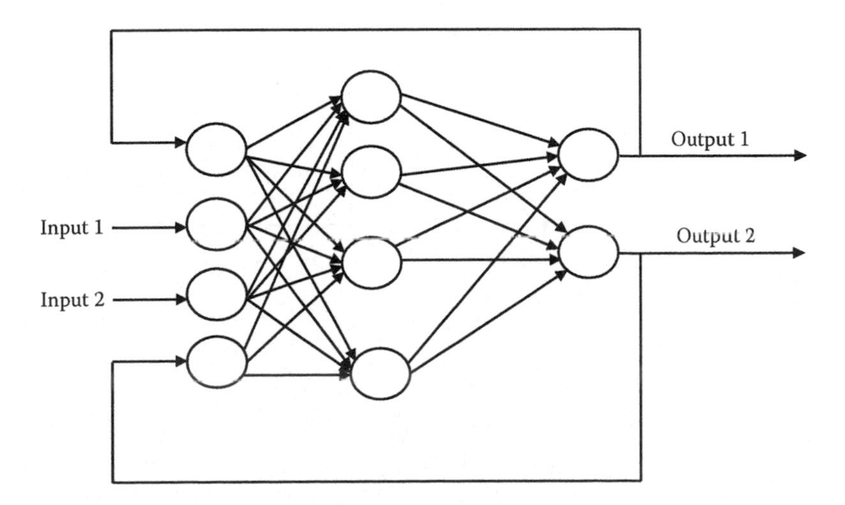

FIGURE 3.7 The architecture of the recurrent artificial neural network.

3.6.1 Concept

The basic concept in these networks is that of feedback. Every neuron is provided with feedback from the other neurons. These feedbacks are incorporated by every neuron while calculating its output. Thus, in Hopfield nets, the final output of any one neuron is the summation of the input plus the feedbacks from the various neurons. The input is usually referred to as the external input, while the feedbacks are referred to as the internal inputs.

These feedbacks and inputs are then passed over the activation function to get the network's final output. From this point in the system on, the Hopfield networks behave as any other conventional ANN, with the continuous feedback driving the networks toward the correct output.

These feedbacks make the Hopfield networks perform much better than conventional ANNs. A single hidden layer Hopfield network can imitate the working of multilayer ANNs because of the various states possible from the internal input feedbacks. In general, an n neuron system in a binary Hopfield neural network has 2^n states possible. Every state here corresponds to an output combination of the hidden layer neuron, or simply the internal input.

3.6.2 Architecture

In general, the architecture of these networks is the same as that of any ANN. The difference, however, lies at the level of the neuron architecture. The neurons in these networks are fitted with extra internal inputs in the form of feedbacks from the various other neurons. The past outputs of these neurons are used as inputs in the subsequent times. These feedbacks act as extra internal outputs that drive the system at every unit of time.

The general architecture of every neuron is shown in Figure 3.8. Each neuron i has a feedback input weight w_{ij} that comes from any other neuron j. Any neuron does not give feedback to itself (in other words, $w_{ii} = 0$). The final summation is performed and passed to the activation function.

The general architecture of the layer is shown in Figure 3.9, which clearly shows that every output is fed back into all the other neurons except the originating neuron. Thus, the feedbacks are processed as inputs by the neurons. The output of a neuron at time t is given as input to the other neurons at time $t + 1$. This feedback continues to drive the system as t increases.

3.6.3 Mathematical Modeling

This section discusses the mathematical formulation of the Hopfield ANN. The input to any neuron is given by Equation 3.21:

$$Z_i(t+1) = \sum_{i \neq j} W_{ij} * y_j(t) + I_i \tag{3.21}$$

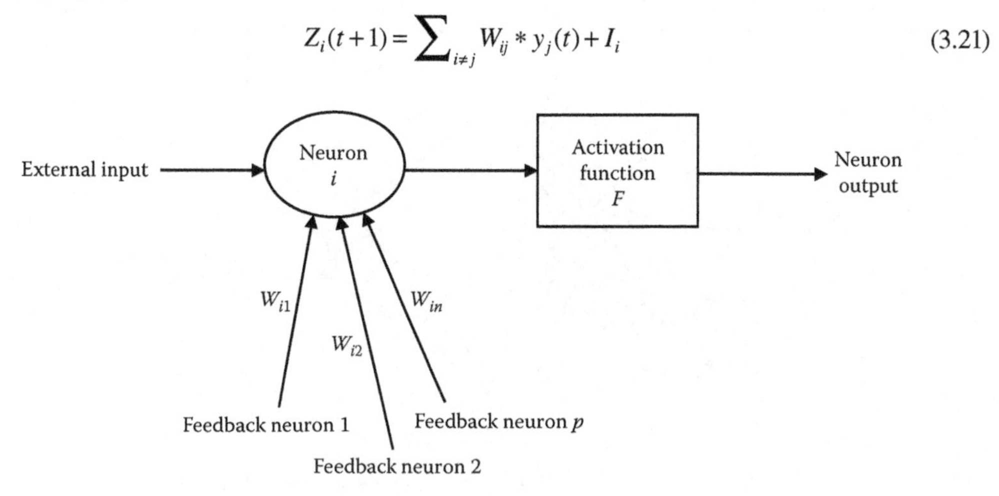

FIGURE 3.8 The structure of a neuron in a Hopfield neural network.

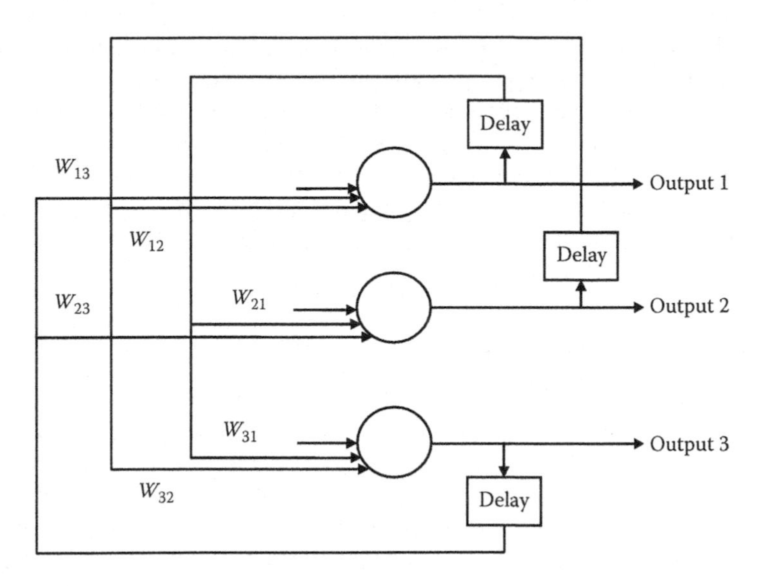

FIGURE 3.9 A hidden layer of the Hopfield neural network.

where y_j denotes the output of the neuron j at any time t, W_{ij} is the associated weight, and I_i denotes the external inputs given to the system.

Output Z_i passes through the activation function $F(x)$. The activation function for a binary Hopfield network is given by Equation 3.22:

$$\begin{aligned} F(x) &= 1, \quad \text{if } x \geq Th \\ &= 0, \quad \text{if } x < 0\ Th \end{aligned} \tag{3.22}$$

where Th denotes the threshold.

The result of the activation function is given in Equation 3.23:

$$\begin{aligned} Y_i(t+1) &= 1, \quad \text{if } Z_i > Th_i \\ &= Y_i(t), \quad \text{if } Z_i = Th_i \\ &= 0, \quad \text{if } Z_i < Th_i \end{aligned} \tag{3.23}$$

This is the output of the neuron.

The interrelation between the various neurons follows the same equations as that of the conventional ANN.

3.6.4 Training

The Hopfield neural networks do not need separate training methods. The entire procedure of adjusting weights can be done by having a look at the inputs themselves. The equation of the weight matrix W in the Hopfield neural network is given by Equation 3.24:

$$W = \sum_{i=1}^{L} (2 * x_i - \overline{1}) * (2 * x_i - \overline{1})^T \tag{3.24}$$

Once the weights have been set according to Equation 3.22, we can use the network to directly test on unknown inputs. The whole input vector is given one by one to the system. The system takes the first input and starts processing according to the mathematical model. The processing naturally goes on with time, as the outputs are refed into the system as feedbacks. This keeps the system busy for a few iterations. After some iteration, the change in output is reduced, or the output converges to some point. This output is regarded as the final output. Then the next element in the input vector is applied. This process continues until all input vectors have been applied to the network. The number of outputs is hence equal to the number of inputs, and every input corresponds to some output.

3.7 ADAPTIVE RESONANCE THEORY

Adaptive resonance theory (ART) was proposed to cope with the problem of forgetting. ANNs are prone to forget previously learned rules when they are not revised for a long time, even as the ANN continues to learn newer rules. This forgetting poses problems for conventional ANNs, because the old rules keep fading with time. ART provides a good means of memorizing new rules without losing information from the older rules. Thus the ART networks are able to preserve the network rules.

Two very important words that require special attention in these networks are *adaptive* and *resonance*. Resonance refers to the state of these networks that allows them to learn from the new inputs without losing the older ones. The matching of new inputs to previously learned states constitutes the fundamental concept of these networks. Adaptive refers to the inherent nature of these systems to change or adapt themselves according to the presented inputs. The entire network adapts to changes in the inputs at the time of learning. This feature enables it to learn new rules without forgetting the old ones. This is similar in nature to the way human beings learn and retain information.

3.7.1 Concept

The learning of new patterns without forgetting the older ones is called *plasticity*. The need for plasticity arises when we need to learn many patterns over time. ANNs usually tend to forget older patterns as they are exposed to newer ones, which leads to problems in recognizing older patterns. If we observe the biological system, however, this problem does not seem to exist. We are able to remember friends, memorize contact numbers, and so forth, even though we may have not revised that information for a very long time. Observing the biological system was the motivation for making systems that can preserve the rules or information even as that system continues learning newer rules.

ART networks are unsupervised neural networks that employ neurons to map the inputs to outputs. Each neuron represents some point in the input space. However, these neurons are saved from being driven out in due course of time by taking preventive measures. This ensures that the information regarding the old rules is not lost.

3.7.2 Architecture

The architecture of the ART consists of two major layers. The inputs are given to the first layer, and the classified output may be collected from the second layer. Processing takes place between these two layers.

The first layer, where inputs are applied, is also called the *comparison layer* (for reasons discussed later). The weights between the two layers are responsible for the network's entire memory. These weights are represented in a weight matrix that maps every input to every neuron. The rows of this matrix represent the prototype; each row represents some point in the multidimensional input space, and the number of rows is equal to the number of neurons, or prototypes. As the inputs go from the first layer to the second, they are multiplied by the corresponding weights. In other words, the product of the input and the weight matrix is taken. This input is then given to the second layer.

The second layer is also called the *recognition layer* or the *competitive layer*. This layer does the work of recognition or classification. It selects only one class—the class that has the largest match according to the inputs given to the layer. The output of this class is made 1, and all the others are made 0. Hence, we are able to classify the output.

Per the basic concept of the ART, some constraints must be applied before any modification or learning can occur. These constraints are applied at the architectural level. The architecture uses on an orienting subsystem that resets the training of any specific input if that input does not match well. This feature helps in plasticity control, as discussed earlier. Thus, the orienting system orients the system either by letting it change or by inhibiting the change.

3.7.3 Training

The training of the ART consists of passing the inputs one by one to the network and modifying the weight or the prototype vector. The basic motivation is the same as for any classifying network in which we try to move a network so that it performs well the next time the same input is given. It is believed that this trains and tunes the network for enhanced performance.

In ART, the inputs are first applied at the input layer. The weights are multiplied by the inputs and passed from the first layer to the second. The output goes to the second layer, which is the classification layer. Here the outputs are decided according to their closeness. Only one class is decided, and it is regarded as the final class. According to the produced output, the corresponding prototype vector is moved toward the input in an attempt to reduce error in the next run. The vector is modified at this stage.

Once the class is known, the algorithm follows the backward phase, and the matching is done by the first layer. The prototype vector is matched with the applied input. If the change is too small, then the training continues. If, however, there is a large gap between the prototype and the applied input, the orienting system sends a reset signal to reset the system to its initial stage, nullifying the last change. The inhibitor is set high for this winning neuron to inhibit the neuron from being activated in the next run. The corresponding prototype is temporarily deleted. The training is then repeated for this input but with the previous winning neuron deactivated to make sure the prototype does not get destroyed in the process of training when a highly unmatching input is applied. For such inputs, a new prototype is usually selected, and the same is tuned. This system is continued as all inputs are processed for the training of the algorithm. The selection goes on and on, unless we are able to find a vector that matches nicely with the input.

The general architecture of the algorithm is shown in Figure 3.10. Figure 3.11 shows the training algorithm.

Due to lack of space, we do not study the ART in detail as a part of this text.

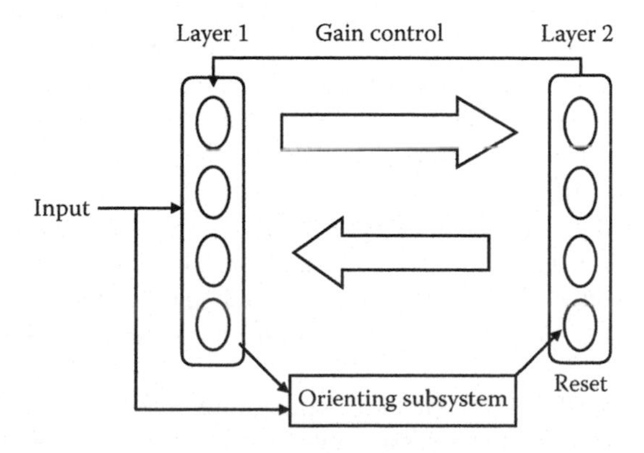

FIGURE 3.10 Basic adaptive resonance theory (ART) network architecture.

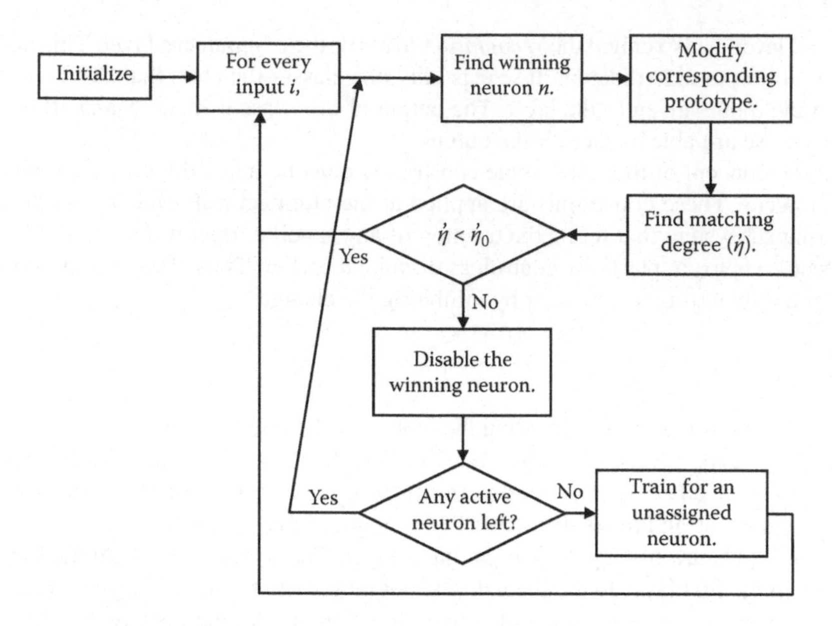

FIGURE 3.11 The learning algorithm in ART networks.

3.8 CHARACTER RECOGNITION BY COMMONLY USED ANNs

In this section, we apply the studied algorithms to the problem of character recognition to give you a practical insight into the use of the concepts presented in this chapter in a real life problem. The solutions are applied to identify the characters in the scanned document or image. To do this, We first have a look at the problem, and then we try to solve the problem using the various methods presented.

3.8.1 Problem Description

Character recognition identifies the character. The character may be obtained from scanned documents, embedded images, or real-time scanning such as the input from a connected pen. Microsoft's Tablet PC is an exciting innovation in this field. Other examples of such innovations include scanning of text documents, data extractors, and robot-vision-based guidance.

Character recognition is of two types: offline character recognition and online character recognition. For example, in the Tablet PC (an online character recognition tool) the system scans the pen's motion and notes the pen's statistics at various instances of time while the character is being written. Hence, the system has a great deal of information about the manner in which the character was made. In offline character recognition, however, the entire character is given as an input image, and the system is then supposed to identify the character from that single image.

The problem with character recognition, as with any recognition system, involves preprocessing, segmentation, feature extraction, and recognition. In preprocessing, the system tries to remove the noise that might have spoiled the image. In segmentation, the entire image is broken down: text is broken down into paragraphs, lines, words, and characters. The most basic entity that remains is the character itself. This character is then given as input to the feature extraction, in which the features of the character are extracted. Finally, in the recognition phase, the final recognition of the character is complete.

In our discussion, we will only be dealing with the recognition phase. We will try to solve the problem of recognition using the various tools and techniques presented in this chapter. It is assumed that the preprocessing and segmentation have already been done.

3.8.2 Inputs

Let us take as the input the entire image. Feature extraction has not been applied for this problem, thus it is natural that the solution should face related problems of learning due to high-dimensional data. Testing has been done on a separate testing data, which were generated by occultation of reversal of some bits in the training data.

M x N inputs are given to the system, where M and N are the dimensions of the image. Each pixel is represented by one input. Every input is either 0 or 1, depending on the presence of ink at that point. The size is taken as 6 x 6 for optimal training. A higher size would require more input cases for training and thus slower processing. A lower size may not be able to train the system, as differentiation between images would be difficult.

3.8.3 Outputs

The situation presented is a typical classification problem. The issue is to correctly identify the class to which any input belongs. The system will then give a single output to every system. Which is the class to which the input belongs, or the identified character.

3.8.4 Solution by Radial Basis Function Network

The first algorithm we used to recognize the input character was RBFN. The inputs were given to the RBFN, the outputs were numbered 1 to 10 for every character, and the network was created.

We found that the network showed 100 percent accuracy in both testing and training data. (The testing data were created by changing a couple of bits in the training data.) To test the robustness of the network created, we occluded the letter more and more. The algorithm gave correct results to even the input shown in Figure 3.12 for the input "A."

3.8.5 Solution by Learning Vector Quantization

The next algorithm used was the LVQ. We found that the network again showed 100 percent accuracy in both testing and training data. The total number of neurons in the hidden layer was fixed to 5. A learning rate of 0.01 was used. The learning function used was *learnlvl*. The goal was 10^{-3}, and the epochs were 1000. The training curve is shown in Figure 3.13.

Before we applied LVQ, it was necessary to convert the outputs to indices in order to make every output a class. In our example, there were 10 classes. The output was thus a 10×10 matrix [A] 10×10 (please use subscripts for 10×10), where every element a_{ij} was the input test case i belonged to class j. The details of this can be found in Appendix A.

The position of the weights is given in Figure 3.14a before training. Initially all weights were at the middle of the whole high-dimensional cube (only three dimensions are shown). Upon training, this data changed to that shown in Figure 3.14b. The inputs may be visualized in this cube.

The result on the test of occultation was again found to be the same. The algorithm gave correct results to the problem given in Figure 3.12.

0	0	0	1	0	0
0	0	1	0	1	0
0	1	0	0	0	1
0	1	1	1	1	1
0	1	0	0	0	1
1	0	0	0	0	1

FIGURE 3.12 Input corresponding to letter A.

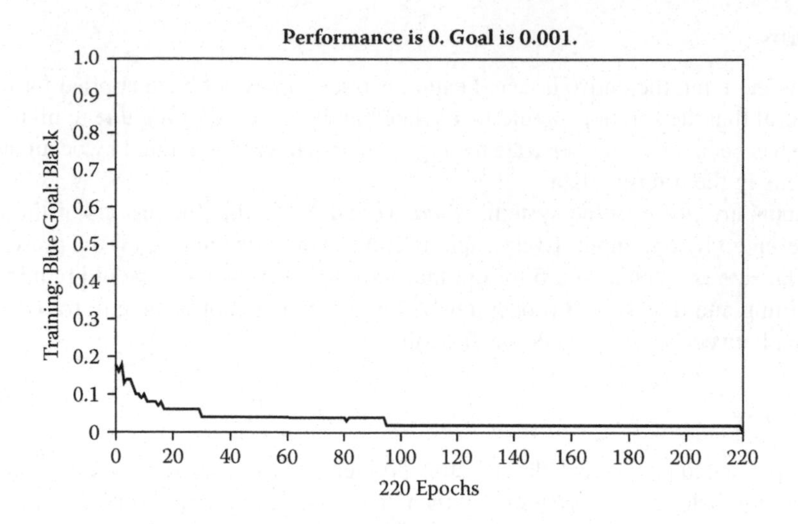

FIGURE 3.13 The training curve for learning vector quantization (LVQ).

3.8.6 Solution by Self-Organizing Map

The next algorithm applied was SOM. We again found that the network showed 100 percent accuracy in both testing and training data. In this case, the codebook used for the SOM was of dimension 4 x 4. The network was trained for 1000 epochs.

The plots of the weights at the beginning and end of the training are shown in Figures 3.15a and 3.15b. Initially all the weights were at the center of the input space, but they moved apart as the training continued.

The result on the test of occultation was again found to be the same. This algorithm gave better results to the occultation and could withstand an even higher change of bits.

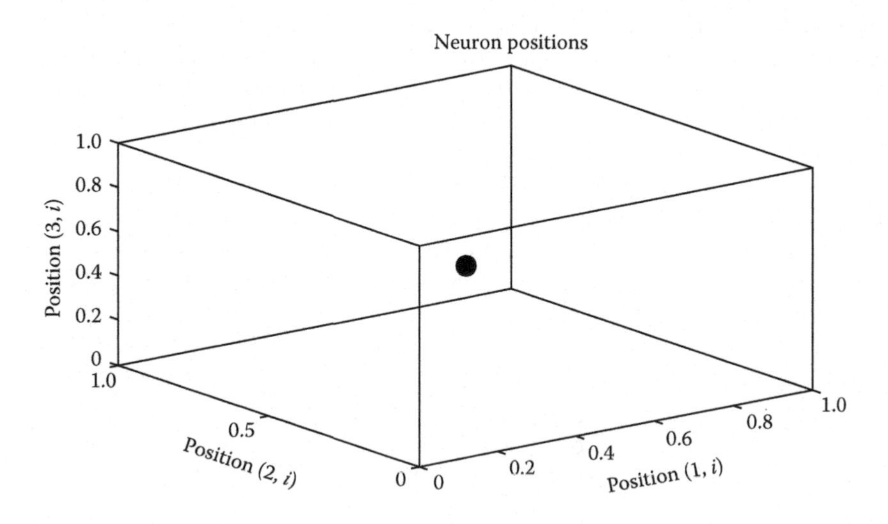

FIGURE 3.14a The initial placement of weights in LVQ before training (only three dimensions are shown).

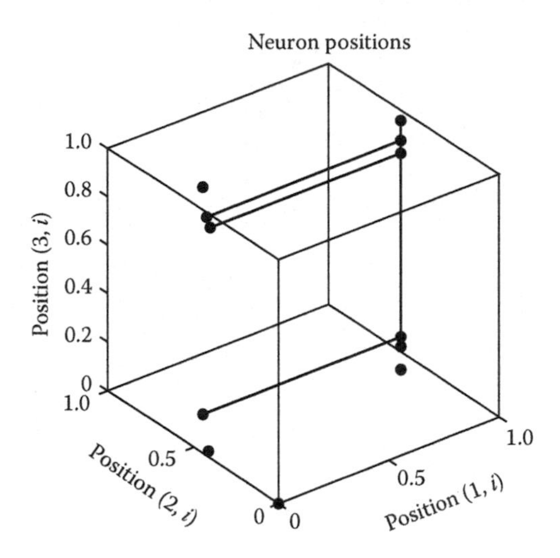

FIGURE 3.14b The placement of weights in LVQ after training (only three dimensions are shown).

3.8.7 Solution by Recurrent Neural Network

The next algorithm applied was recurrent neural network. We used the Elman model for experimental purposes. We again found that the network showed 100 percent accuracy in both testing and training data. The network consisted of 14 neurons in the hidden layer, with *tan sigmoid* (*tansig*) and *log sigmoid* (*logsig*) activation functions of the hidden and output layers. The epochs were set to 1500, the learning rate to 0.01, and the momentum to 0.1. The goal was set to 10^{-5}. The training graph is shown in Figure 3.16.

In this instance, we needed to convert the inputs and outputs to sequences before giving them to the network. Similarly, outputs collected from inputs had to be converted back to

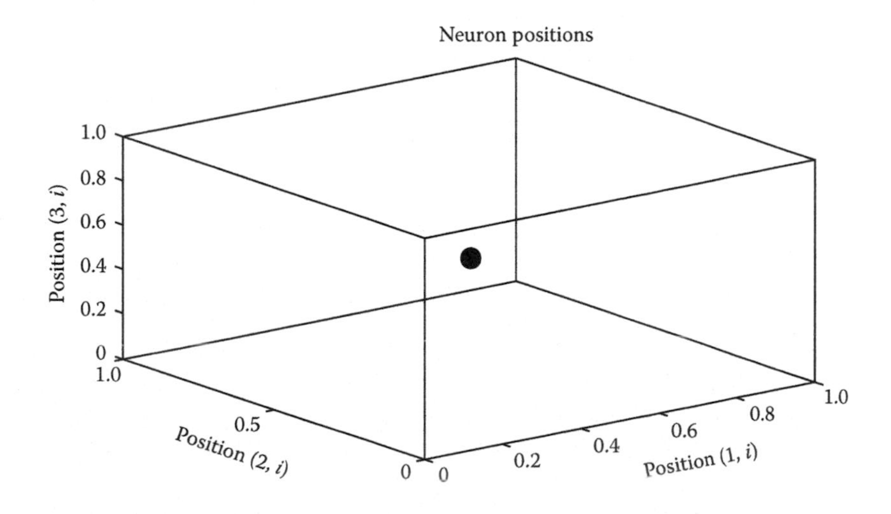

FIGURE 3.15a The position of the weights in the input space before training (only three dimensions are shown).

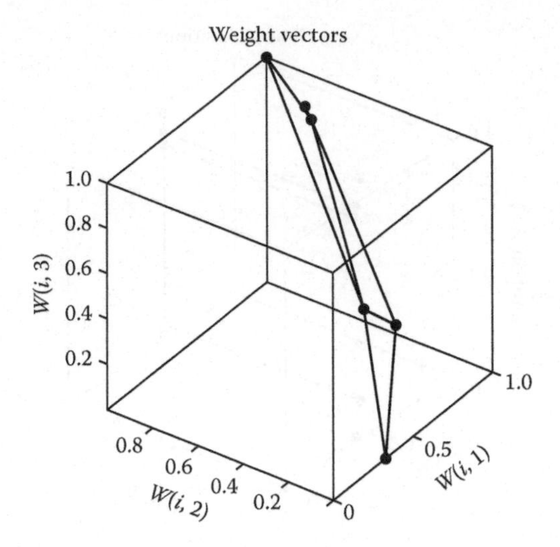

FIGURE 3.15b The position of the weights in the input space after training (only three dimensions are shown).

numbers. This is explained in Appendix A. The result on the test of occultation was again found to be correct.

3.8.8 Solution by Hopfield Neural Network

To the same problem and inputs, we applied Hopfield neural networks. MATLAB® was used as a platform. We again found that the network showed 100 percent accuracy in both testing and training data. We observed that the Hopfield neural network stabilized at the input points given to it for training. It also stabilized for the presented test data and reached the nearest training data. The result on the test of occultation was again found to be correct, even with these ANNs.

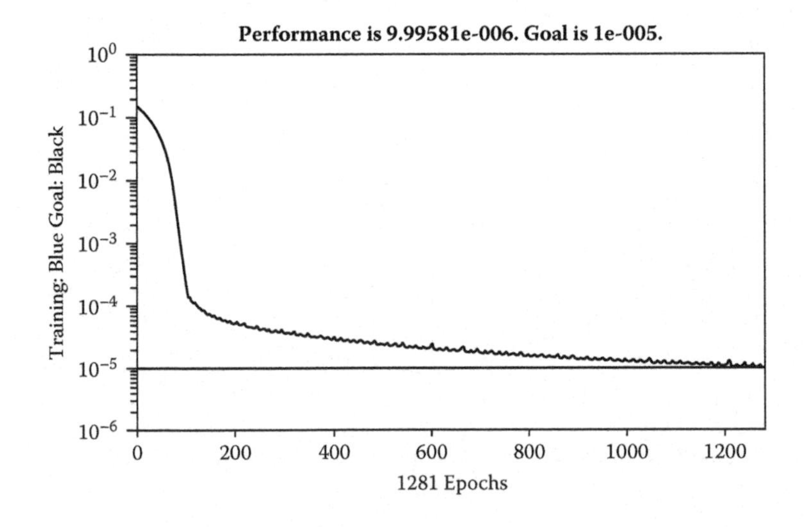

FIGURE 3.16 The training curve for recurrent Elman networks.

CHAPTER SUMMARY

This chapter offers a basic guide for the application of various kinds of networks in practical real-life applications. We studied the various designs and architectures possible for artificial neural networks. Each network had its own way of dealing with the problem presented. Many were ideally suited for functional prediction problems, in which the role of ANN is to predict a function that maps the inputs to the outputs. Others were more suited for classification problems, in which the ANN is supposed to classify the data into classes. Various types of learning were presented, including unsupervised learning as well as reinforcement learning. The manner, or principle, by which these types of learning memorize inputs was also discussed. We also saw how these two types of learning differ from supervised learning.

The first specialized ANN model introduced was the radial basis function network. The basic concept behind this network, and its associated mathematical model and learning, was introduced. We discussed how these networks are able to give high performance. We similarly introduced the other ANN model—learning vector quantization—and discussed its principles, architectures, and mathematical modeling. Other models discussed include recurrent neural networks and Hopfield neural networks. We also discussed self-organizing maps and adaptive resonance theory.

Toward the end of the chapter, the real life example of character recognition was studied. We discussed how we solved the character recognition problem using each of the discussed methods. Each method gave a very high performance to the applied inputs.

SOLVED EXAMPLES

1. Comment on localization and generalization in the context of ANNs in terms of design and training time.

Answer: There are two important, related terminologies whenever we talk about the context of machine learning. The first is generalization, in which the network tries to see the entire network and make some generalized rules that are believed to be true for all cases. The output of a point in input space is governed by all the points given for training. Because the network must analyze all the points to frame generalized rules, the network training is slower and requires multiple epochs. But the storage requirements (or neuronal requirements) is much lower in these networks—a few nodes can store rules for the entire network. All the cases or regions where these rules apply can be replaced by these few neurons. Hence, the network does not have to store a great deal of information. The same happens if we try to make an ANN predict any common function, such as $f(x) = \text{Sin}(x)$ in the region of 0 to π.

The other terminology is localization, which may sometimes be considered to be the counterpart of generalization. In localized networks, the output at a point is only dependent on the surrounding, or neighboring, points given at the time of training. The simplest or most localized example would be to calculate the neighboring points of any given point and take their weighted sum. In such a network, the storage requirements are fairly high, because we need to store almost the entire training data. The training requirements, however, are fairly low, and the network should either not require training or train fairly early.

It is not completely in the hands of a designer to decide the level of generalization. Depending on the type of training data, ANN may completely reject training when it tries to train using the concepts of generality. It would demand a greater number of neurons, which would lead to high memory requirements and loss of generality. Functional prediction ANNs require higher generality than classifiers.

The ANN with BPA is one of the most generalizing algorithms that specially suits the requirements of functional prediction problems. RBFN, which is less generalized as compared with ANN with BPA, loses generality as the number of neurons or data centers is increased. The same is true for SOM and LVQ, which are all already quite localized. Generality of ART, recurrent neural networks, and Hopfield networks also follow similar rules.

2. Solve the problem of character recognition as given in Section 3.8 by ANN with BPA.

Answer: Because the problem of character recognition is classificatory in nature, it would be justified to make some adaptations in the BPA before using it to solve the problem. Here we keep the inputs the same as discussed in Section 3.8. The output in this case is the class to which the input belongs.

The modification we make to the traditional BPA is the addition of a new layer. This layer ensures that only a single class is returned by the system as an output. This layer is given n number of inputs, where n is the number of classes into which the data can be classified. The output is the class to which the applied input belongs. The n number of inputs to this layer are the outputs of a traditional ANN with BPA. Here each output lies between –1 and 1. The higher the output, the greater the probability of the input belonging to the same class. In other words, output i denoting class i should be 1 for all inputs that belong to class i and should be –1 for all other inputs. The last added layer selects only the output with the maximum value in the preceding layer and outputs its class number. This ensures that only one class is selected as the final answer.

The network is shown in Figure 3.17, in which the first block is the traditional ANN with BPA. The second layer outputs the class corresponding to the maximum output given from the preceding layer.

Suppose the input was A. The output for the training of this letter is <1, –1 , –1, –1, –1, –1, –1, –1, –1, –1>. This denotes a complete probability of occurrence of A and no probability for the occurrence of other characters. The output corresponding to B is similarly <-1, 1, –1, –1, –1, –1, –1, –1, –1, –1>.

Using this model, we train and test the system. The network had a hidden layer with 40 neurons and an output layer with 10 neurons, each corresponding to the 10 letters. The epochs were fixed to 1000, and t goal was set to 10^{-2}. The learning rate was 0.01, and momentum was 0.1. The results again showed an accuracy of 100 percent in both training and testing data. When tested for larger noise, the network showed poor performance with many errors. The training curve is shown in Figure 3.18. The network was trained in 800 epochs.

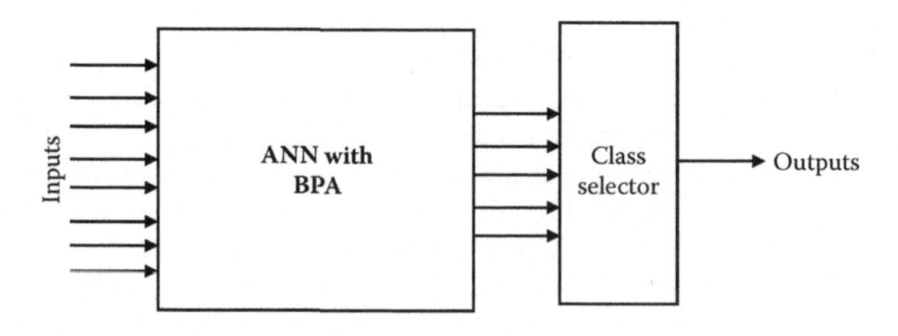

FIGURE 3.17 The artificial neural network (ANN) model for Question 2.

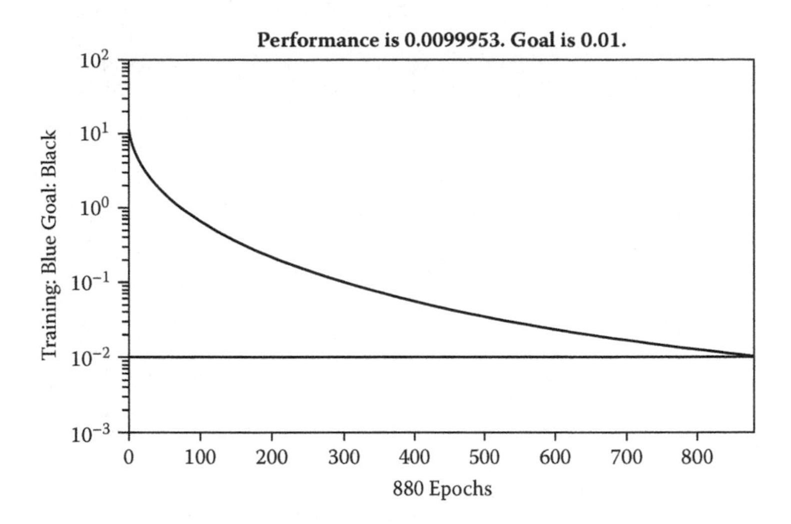

FIGURE 3.18 The training curve for the character recognition problem with ANN and back propagation algorithm (BPA).

3. Write a short note on adaptive learning.

Answer: The algorithms discussed so far were trained using the concepts of batch training, in which the training inputs were applied one after the other. The whole process was repeated in multiple epochs. This learning solved problems with good performance in the testing data set, but in reality this may not be the case with a good system. In actual systems, all the inputs may not be available in one go. There may be a need to train using online data, which may be available in a number of ways. In addition, in such systems, it may not be necessary to remember the historical data, as these systems deal with learning new data and forgetting old data. This kind of learning, known as adaptive learning, has applications in game playing, robotics, and other similar applications.

The BPA and various other ANNs have been designed to incorporate adaptive learning by including a change in the learning methods and their mathematical foundations.

4. What is the role of the number of codebook vectors or neurons in LVQ? Explain with an example.

Answer: As discussed in Question 1, the increase in the number of codebook vectors results in a loss of generality. This loss would take less training time and would make the entire system localized. In classification problems, the loss of generality has no major impact in training the system, as would be the case with functional prediction problems.

We take the example of classification in the Integrated Risk Information System (IRIS) database from the University of California, Irvine (UCI), Machine Learning Repository. In this example, there were a total of 150 data sets. We took 90 for training and 60 for testing. The number of neurons was varied from 1 to 110. The percentage of accuracy for the various neurons for the training data, testing data, and total are given in Figure 3.19.

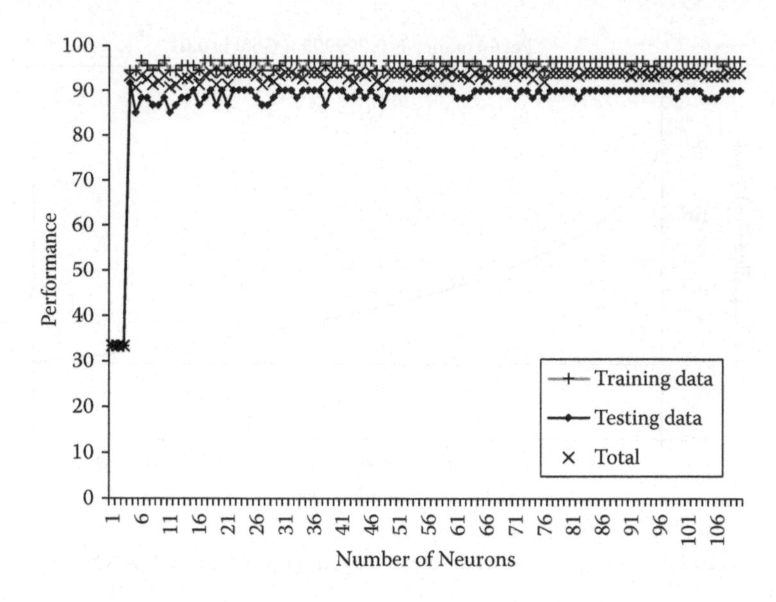

FIGURE 3.19 Performance versus the number of neurons.

EXERCISES

General Questions

1. What is the difference between supervised and unsupervised learning?
2. What is reinforcement learning, and how is it different from supervised and unsupervised learning?
3. Under what conditions is unsupervised learning better than supervised learning?
4. Why is there a need for so many ANN architectures and algorithms?
5. Explain the concept of time delay with reference to recurrent neural networks.
6. Explain how SOMs can be used for clustering.
7. Explain the concept behind LVQ.
8. Which is the best ANN design for the purpose of classification? Prediction?
9. Why is ART coined with the word *resonance*?
10. Explain the basic difference between BPA and learning in SOM.
11. Explain the difference between SOM and LVQ. Which is better in what context?
12. Explain the Hopfield neural networks as classifiers.
13. What would be the effect of increasing the number of codebook vectors in LVQ?
14. Explain the working of ART.
15. Why is training not required in certain ANNs?
16. Explain the concept of neighborhood in the case of SOM. What role does it play in performance?
17. Explain the term *competitive neural networks.*
18. Explain the term *winner takes all.* How does it differ from its counterpart? Which is better?
19. Explain the concept of window and window width in the case of LVQ.
20. Classify each ANN design into supervised and unsupervised learning.
21. Explain the concept of convergence in ART.

PRACTICAL QUESTIONS

1. Generate random data and use SOM for the purpose of classification. Compare the results with k-means clustering and fuzzy C-means clustering.
2. Use the different ANN designs for the classification of IRIS data. The data set may be taken from the UCI Machine Learning Repository.
3. Use RBFN for the functional prediction problem of time series analysis presented in Chapter 2. Compare the results with that of BPA.
4. What is the role of the number of neurons in SOM?
5. Using the data in Practical Question 2, try to introduce noise to study the effect of addition of noise in various designs of ANN.

4 Fuzzy Inference Systems

4.1 FUZZY SYSTEMS

The concepts of soft computing that we introduced in Chapter 1 form the guiding principles of fuzzy logic. With fuzzy logic, the system is made to give the most probable output to any kind of input based on the predefined rules. Fuzzy systems are a modified form of rule-based approach, in which rules are applied to find the output of any input. In place of strict rules applied over the classes or sets, however, the rules are much softer in the manner in which they are applied. This not only gives a scope for the function to perform well in the presence of uncertainties and noise, but it also makes it possible to obtain more realistic systems that imitate natural behavior. Fuzzy systems are used in numerous real life industrial applications, including biomedical engineering and robotic control. The success of fuzzy systems in such varied domains clearly speaks to the effectiveness of these systems, which form an integral part of soft computing. Fuzzy logic is the natural choice when modeling systems that have predefined rules governing their behavior.

Other interesting use of fuzzy systems is in classification and pattern recognition problems, where they are able to easily determine the output class that the input corresponds to. The fuzzy nature of these systems serves as an instrumental tool in finding the output class by a set of rules that may be framed by looking at the training data.

Fuzzy systems get their name from the uncertainty or probability they associate with the various stages of functioning as they calculate the outputs from the applied inputs. The rules governing the behavior of the fuzzy system are based on the classes of inputs. These rules simply denote that certain types of inputs have a certain types of output. With fuzzy systems, we study the inputs and outputs in groups or classes. The novel concept behind fuzzy systems is that the the input belongs to a class by a certain degree or a certain probability. This concept is further used to work over the rules to come up with a certain answer to the problem. The probability-based association of the fuzzy systems allows them to imitate various systems that could not have been built using traditional approaches. These systems hence act as a boon in the implementation of the rule-based approach by soft-computing techniques.

Fuzzy systems are entirely rule driven. Mapping of the inputs and outputs is accomplished by the rules, which are specified during the design. The different rules affect the output in their own way. In other words, the rules try to find the output according to their own understanding of the system for which they are made. The final result is the output calculated by the combined effect of all the rules and is given as output of the system. It is very likely that this output was not the result of any of the rules; rather it is the result of all the rules being put together.

Just like any system, the fuzzy system maps the inputs to the outputs. This mapping is derived from various rules that are fuzzy in their implementation. The rules are written in the form of normal English rules, which can easily be framed after a study of the system. As we have explained, the output is the combined effect of all these rules put together. Note that the different rules do not behave in similar manner to one another. Some rules may result in a high output, while others may result in a low output. The aggregation of the output predicted by all these rules computes the final answer to the unknown input that was given to the system.

In the subsequent sections of this chapter, we discuss the various issues and concepts of the fuzzy system, including its design and its usage in real life applications.

4.2 HISTORICAL NOTE

The history of fuzzy logic goes back to the days of Aristotle and the binary logic representing true and false, which began the development of logic in the history of humankind. Multilogic also evolved about the same time, but not to a very good extent. It was not until 1964 that Lofti Zadeh introduced the concept of fuzzy logic, when he introduced a formal method of dealing with and problem solving with fuzzy sets. The field attracted the attention of numerous researchers worldwide and initiated a great deal of work in this field. Fuzzy logic then joined the application domain, where it has been used in numerous systems and consumer applications, including washing machines, camcorders, and microwave ovens, to name just a few.

4.3 FUZZY LOGIC

In this section, we introduce the concept of logic and, hence, fuzzy logic. We discussed logic in Chapter 1, where we saw how logic is used for problem solving. In this section, we give an in-depth analysis of the same and then move to a discussion of fuzzy logic.

4.3.1 Logic

Every mapping of the inputs to the outputs is done using a set of guidelines, or functions, that are the inherent properties of the system being considered. This mapping forms the basis of logic. We must figure out the knowledge that is available in the system and then determine how to store it in a usable manner. We try to represent the system by a set of rules or in a way that can be easily understood and implemented by the machine. Knowledge is of deep interest to system developers, as it provides a means for the machine to understand, act, and make decisions and inferences based on the common understanding of the general people. This knowledge removes the gap between human and machine understanding.

Recall from Chapter 1 the definition of *logic*: "What a program knows about the world in general, the facts of the specific situation in which it must act, and its goals are all represented by sentences of some mathematical logical language. The program decides what to do by inferring that certain actions are appropriate for achieving its goals" (McCarthy, 2007). We also defined *knowledge* as "a function that maps a domain of clauses onto a range of clauses. The function may take algebraic or relational form depending on the type of applications" (Konar, 2000).

Logic is used to make machines intelligent and to empower them with the ability to make decisions. Logic makes it possible for machines to take input and act in the desired manner. Machines are able to do this because they follow a set rules that denote the knowledge assembled or represented in the system.

The rules are simple if-then clauses. The *if* part, also called the *antecedent*, denotes the condition that must be true for the particular rule to fire. The condition expresses the particular case in which the action would hold true. The *then* part, known as the *consequent*, consists of the action or the conclusion that results from the rule being fired. The entire set of knowledge is mapped onto this rule set. The systems then use these rules for all operations. Once these rules are ready, we know that all available information has been incorporated into the system in the rules. These systems also have a memory associated with them called the *working memory*, which stores all the information regarding the state of the system. Based on the state, the rules are fired by a rule implementer.

Consider the rule:

If (X marks are more than 80) & (X attendance is more than 75%), then (X grade is A).

Here the *if* part states all the conditions that if true lead to the action.

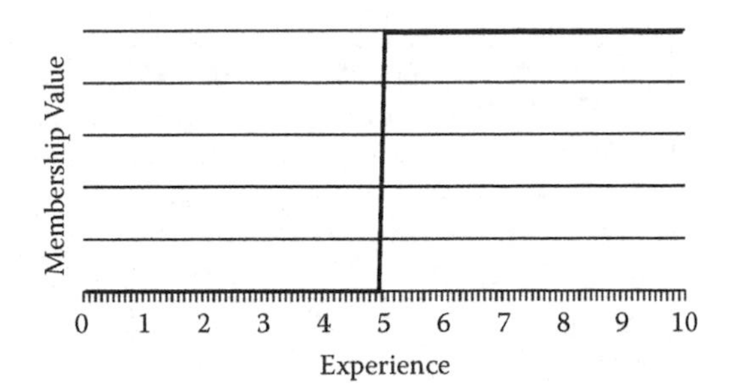

FIGURE 4.1a The variable *experience* for the accident risk problem.

4.3.2 Problems with Nonfuzzy Logic

Now from the concept of logic, we move to the concept of fuzzy logic. It is clear from the example above that any rules specify certain conditions or antecedents for the corresponding actions or consequents to be activated. This means that either the action will take place or it will not. If the condition were the set of conditions joined by logical operators, then the same concept holds true. Once again, the various conditions are evaluated using the state of the system and are worked using the logical operators. This decides whether the final condition will be true or false. If the final condition is true, the corresponding action is activated. Thus, in the above example, both the statements must be true for the system to perform the action.

In the real world, however, this might not give a very realistic picture of the entire system. Consider this example:

> If (driver experience is high) & (road is bad), then (accident risk is moderate).
>
> If (driver experience is low) & (road is bad), then (accident risk is high).
>
> If (road is good), then (accident risk is low).

In this problem, we have a system that is trying to find the risk of accident by taking the inputs of driver experience and road condition. But driver experience, road, and accident have been defined in quite abstract terms. Suppose that driver experience of more than or equal to 5 years is *high* and less than that is *low*. Further suppose that road condition is measured by a road index that lies between 0 and 1. A *bad* road means an index of less than or equal to 0.4, while a *good* road means an index larger than 0.4. Further let us suppose that accident is measured as a probability. A *high* accident means a probability of 0.7. *Moderate* means a probability of 0.4. *Low* means a probability of 0.2. This is summarized in Figures 4.1a and 4.1b. Figure 4.1a depicts the two levels of experience as *low* and *high*, while Figure 4.1b depicts the two levels of road as *bad* and *good*.

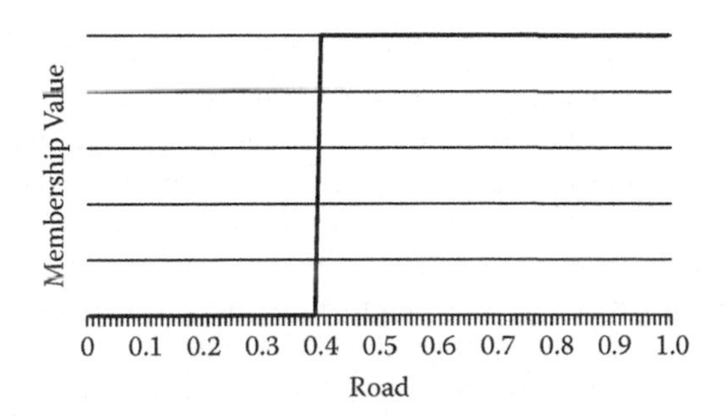

FIGURE 4.1b The variable *road* for the accident risk problem.

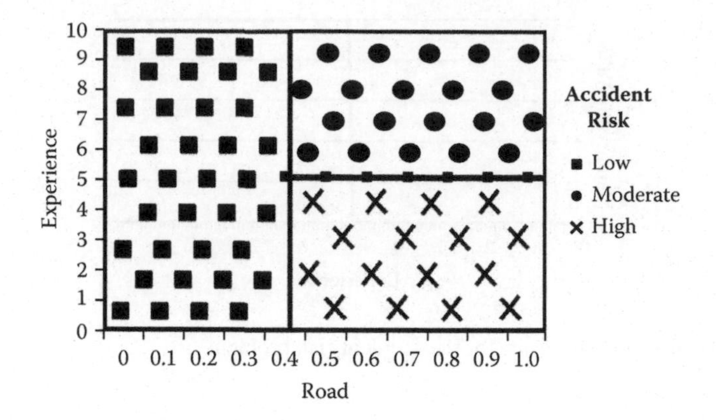

FIGURE 4.2 The output of the nonfuzzy system for the accident risk problem.

Working with these rules, it may easily be seen that for a person with 5 years of experience driving on a road with road index 0.5, the accident probability is *moderate* (or 0.4). This sounds reasonable. But suppose that under the same conditions, we have a driver with an experience of 1 day less than 5 years. In such a case, the accident probability suddenly becomes *high* (or 0.7). Thus with a decrease of just 1 day of experience, we see a sudden increase in the probability of an accident. This is the unrealistic nature that fuzzy systems are good at modeling. The output for the various inputs is given in Figure 4.2, where the three levels of accident are represented by three different symbols.

The unrealistic nature of the above system can be better solved by using fuzzy logic, as we shall see later in this chapter.

Let us consider another model of solving the same problem using a nonfuzzy system. This time we use simple mathematical functions to map the output to the inputs. This system of problem solving is commonly known as a *human logic system*. The system discussed here is its very basic version and is given by Equations 4.1 through 4.3.

$$accident_{total} = accident_{experience} + accident_{road} \tag{4.1}$$

where $accident_{total}$ is the total probability of an accident (output), $accident_{experience}$ is the accident probability due to experience, and $accident_{road}$ is the accident probability due to road conditions.

$$\begin{aligned} accident_{experience} &= 0, \quad \text{if } experience \geq 10 \\ &= (1 - experience/10)/2, \quad \text{otherwise} \end{aligned} \tag{4.2}$$

$$accident_{road} = (1 - road)/2 \tag{4.3}$$

Analyzing this system, we can easily see that if the driver does not know how to drive (*experience* = 0), the probability of an accident for a very good road (*road* = 1) is 0.5. This means there is only a 50 percent chance that there will be an accident. In reality, this would be more than 90 percent, because whenever you give a car to a new driver, that person is always accompanied by an experienced driver because accidents are very likely. The same is also true in the case that the road is very bad. The surface of this function is given by Figure 4.3.

These two problem-solving methods are used together in nonfuzzy systems. Even these systems find interesting applications and can be adapted well for modeling problems. However we do not study these nonfuzz systems in this text. Simply by looking at these two examples, we can easily see that nonfuzzy systems have problems. In the rest of this chapter, we will see how fuzzy systems can solve these problems with better system design.

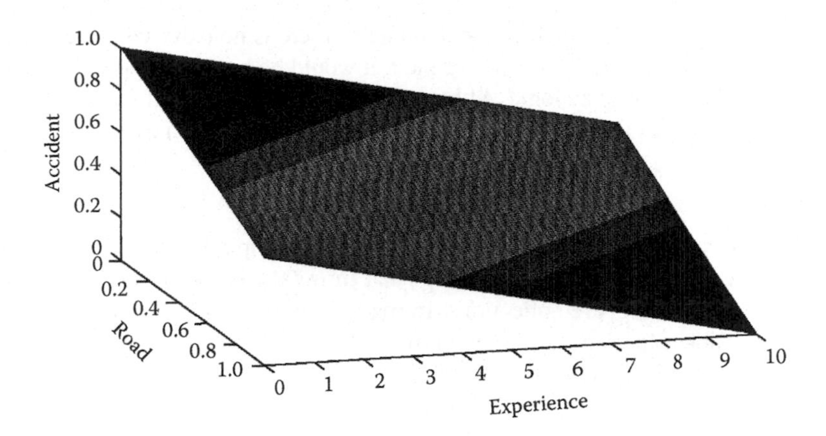

FIGURE 4.3 The surface plot of *accident risk* in the second example.

4.3.3 Fuzzy Logic

In this section, we introduce the concept of fuzzy logic and learn how it is different from normal logic. This discussion forms the basis of our coverage of fuzzy inference systems, which are entire systems that use fuzziness to map inputs to outputs.

In our earlier example, we saw that every input belongs to one and only one class—for example, *experience* could be either *high* or *low*. This restriction is the basic reason for the problems encountered in nonfuzzy systems. In fuzzy logic, however, this restriction is changed or generalized.

In fuzzy logic, every input belongs to every class. The degree of association of the input to the various classes varies. In other words, the input belongs to the different classes by different degrees of associations. This association may be very strong to some class but weak for other classes, or the association may be moderate for all classes. Hence in fuzzy logic, we would never say that the input *i* is high, low, moderate, and so on. Rather we would say that the input *i* is high to some extent, moderate to some other extent, and so forth. The higher the degree of association of the input to some class, the more characteristics of that particular class it implements.

In our example of the road, we find that under fuzzy logic, *experience* can be *high* and *low* at the same time. Thus the driver's experience may be high to the extent of 80 percent and low to the extent of 10 percent. This means the driver's behavior closely follows the behavior of experienced drivers, but the 10 percent association indicates that this behavior to some extent follows the behavior of inexperienced drivers. Hence when we apply the rules using the specialized operators that we study next, the output is the aggregation of both effects. This gives good results when applied over real life cases.

4.3.4 When Not to Use Fuzzy

A fuzzy approach is not the best approach for all types of problems; therefore we need to study the problem to be solved before applying fuzzy approach. In this section, we discuss some types of problems for which fuzzy logic should not be applied.

Suppose we have identified the system. We know the inputs and outputs, but we do not have a clear idea of the rules that map the inputs to outputs. In this instance, fuzzy logic may not be the best approach. Fuzzy logic follows simple English rules that must be known for a system to have an effective design. In the absence of these rules, the performance might be poor, or we may have to apply many efforts to study the patterns of inputs and outputs in search of the rules.

Consider the natural systems of physics, in which the bounding equations are well known and perform well. In this situation, fuzzy systems may not perform as well as the already established mathematical equations. Say the situation is that of a car moving at speed v and acceleration a.

Further suppose that the condition is ideal, which means there is no external force, friction, and so forth on the car. If we are to find the speed at time t, it would be better to apply the standard mathematical equation rather than fuzzy logic. Although in the same problem, if we introduce additional constraints that mathematics finds it very difficult to cater, the problem may become fuzzy.

4.3.5 Fuzzy Sets

We have already discussed the concept of degree of membership and various classes in terms of fuzzy logic. We now formalize the same concept that forms the basis of fuzzy sets. According to the theory of mathematics, sets are collections. In road example, we may regard experience to be a collection of all possible experiences (in years) that the driver has. This may be any value greater than or equal to 0 and may be represented by Equation 4.4.

$$experience = \{z: z \geq 0\} \tag{4.4}$$

In a fuzzy approach, we represent each element of a set with a certain probability. This is shown as a/b, where a denotes the element of the set and b denotes the degree of membership of a in the set. Consider the set of high experience, which would be given by Equation 4.5:

$$high_{experience} = \{z/\mu(z) : z \geq 0\} \tag{4.5}$$

where we assume that the degree of membership of z in the set is given by the function $\mu(z)$. Thus it is natural that the degree of membership will increase as z increases, because as experience increases, the driver will more closely follow the characteristics of an experienced driver.

4.4 MEMBERSHIP FUNCTIONS

In the previous section, we talked about the degree of membership, which denotes the belongingness of any value to any input. Every element is denoted with a certain degree of association that is given by a function known as the membership function (MF).

The MF takes as a single input the element whose membership needs to be found and returns the membership degree of that input. The function may be denoted by $\mu(z)$, where z is the element.

Any input may have one or more membership functions associated with it. In our road example, we have the input *experience* associated with two MFs—*high* and *low*. There are no set guidelines as to how many MFs make an ideal system; the choice usually lies with the designer's implementation. Having an idea of the rules or how the system works may play an important role in deciding the number of MFs.

The member functions for the two classes of experience—*low* and *high*—are given in Figures 4.4a and 4.4b, respectively.

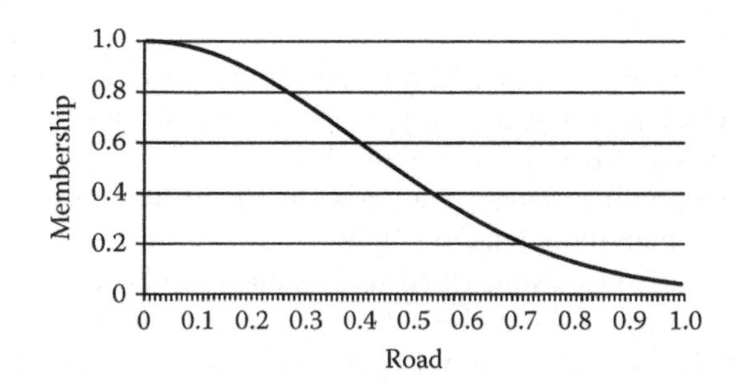

FIGURE 4.4a The membership function *low* for road in the accident risk problem.

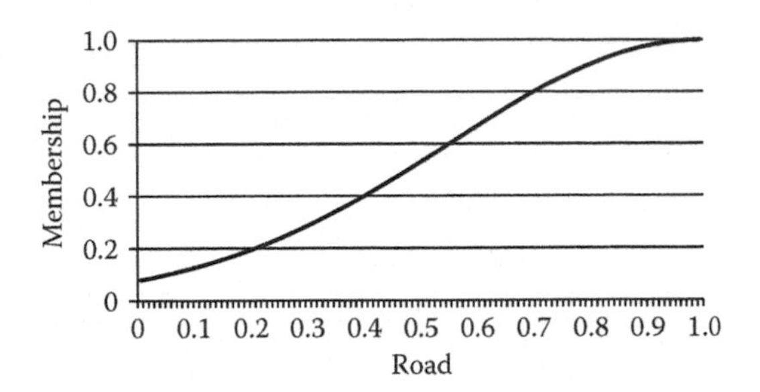

FIGURE 4.4b The membership function *high* for road in the accident risk problem.

It can clearly be seen from Figure 4.4a that as long as experience is low, membership is high. This means that the lower experience values have a high similarity with the class *low*, with the highest being when *experience* is 0. This situation is desirable, because if we make any rule for the class of low-experienced drivers, these people would be more likely to exhibit the properties of the rule. As we keep increasing experience, the membership value keeps decreasing and ultimately reaches 0 at an experience of 10 years. The converse would be true for Figure 4.4b.

MFs are defined by the system designer according to the problem. Normally designers prefer to use standard membership functions, which have been used in numerous problems. We now discuss a few of these membership functions.

4.4.1 Gaussian Membership Functions

The Gaussian MF depicts the Gaussian curve, given in Figure 4.5a. This widely used membership function denotes either a sharp Gaussian decrease or a sharp Gaussian increase in the membership value. The Gaussian MF is given by Equation 4.6:

$$f(x,c,\sigma)=e^{\frac{-(x-c)^2}{2\sigma^2}} \tag{4.6}$$

where c and σ are parameters that may be adjusted to control the behavior of the function, x is the given input, and c is the input for which the membership value is maximum (or 1) or the center of the curve.

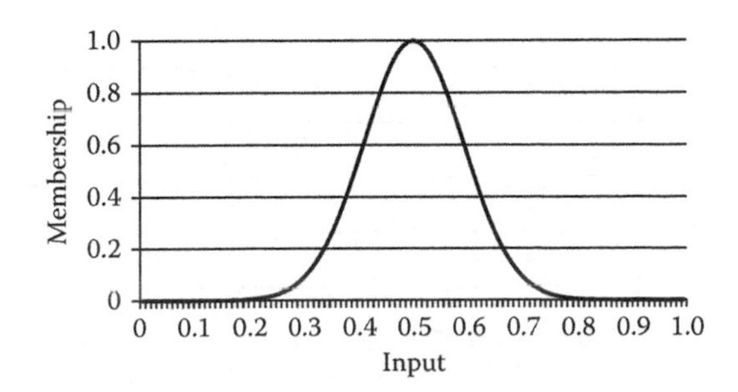

FIGURE 4.5a Standard membership functions: Gaussian.

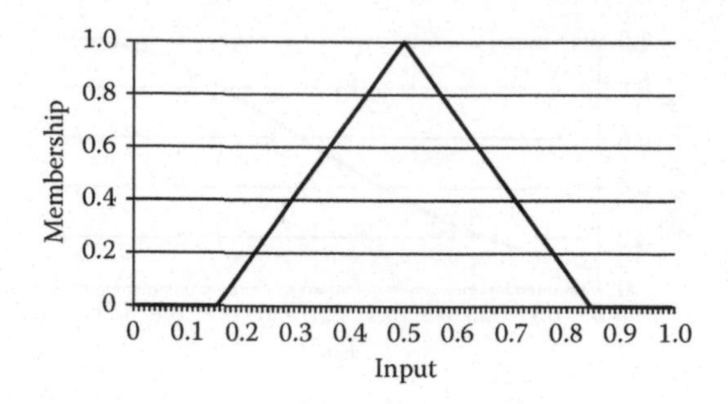

FIGURE 4.5b Standard membership functions: Triangular.

4.4.2 Triangular Membership Function

This function, which denotes a straight-line decrease or increase in the membership value, is used in situations where there is a simple linear degradation or upgradation of the membership value. The curve of this function is shown in Figure 4.5b and in Equation 4.7.

$$f(x,a,b,c) = \begin{matrix} 0 & \text{if } x \le a \\ \dfrac{x-a}{b-a} & \text{if } a \le x \le b \\ \dfrac{c-x}{c-b} & \text{if } b \le x \le c \\ 0 & \text{if } c \le x \end{matrix} \tag{4.7}$$

where a, b, and c are parameters such that $a \le b \le c$. The membership value is 0 until it reaches point a. From that point, the membership value starts increasing and touches a maximum of 1 when it is at point b. It then starts decreasing until it reaches 0 at point c. From c onward the membership value is 0.

4.4.3 Sigmoidal Membership Function

The sigmoidal MF, which depicts the sigmoidal function, is given by Figure 4.5c and by Equation 4.8.

$$f(x,a,c) = \frac{1}{1+e^{-a(x-c)}} \tag{4.8}$$

where a and c are parameters.

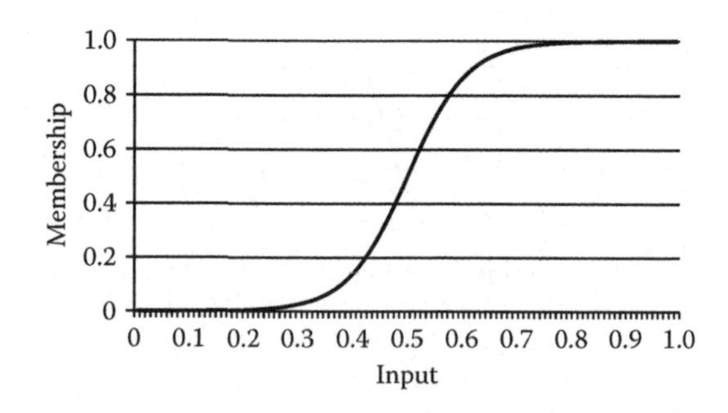

FIGURE 4.5c Standard membership functions: Sigmoidal.

4.4.4 Other Membership Functions

Other standard membership functions are summarized in Table 4.1.

The motivation behind the use of fuzzy sets is that we must be able to implement a traditional rule-based approach. Using MF, we can determine the degree of association of any element to any of the classes of inputs or outputs. This empowers us to replace all conditions of the form "if a is i,

TABLE 4.1
Commonly Used Membership Functions

S. No.	Name	Equation	Graph
1.	Generalized bell-shaped	$f(x,a,b,c)=\frac{1}{1+\left\|\frac{x-c}{a}\right\|^{2b}}$	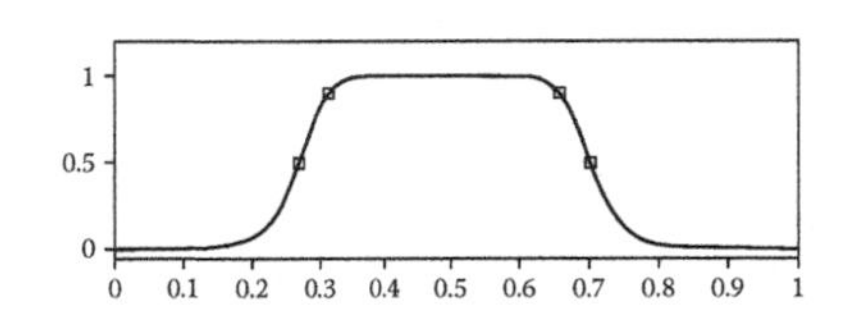
2.	Gaussian combination	$f(x,\sigma,c)=e^{\frac{-(x-c)^2}{2\sigma^2}}$	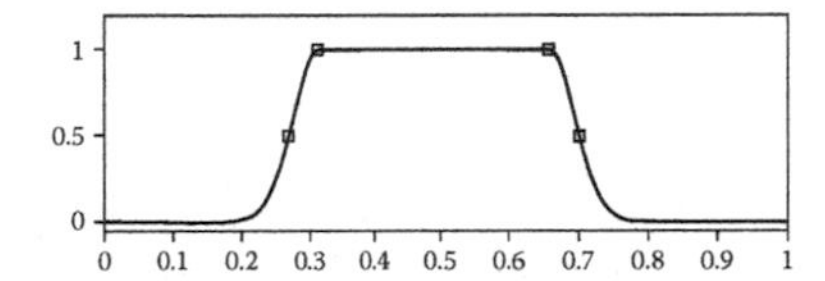
3.	Difference sigmoidal	$f(x,a_1,c_1,a_2,c_2) = \frac{1}{1+e^{-a_1(x-c_1)}}-\frac{1}{1+e^{-a_2(x-c_2)}}$	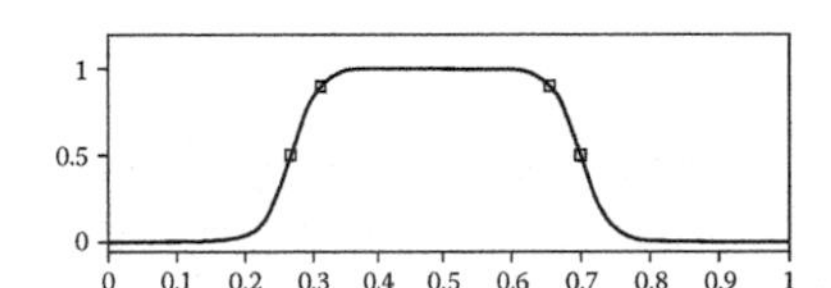
4.	Product sigmoidal	$f(x,a_1,c_1,a_2,c_2) = \frac{1}{1+e^{-a_1(x-c_1)}}*\frac{1}{1+e^{-a_2(x-c_2)}}$	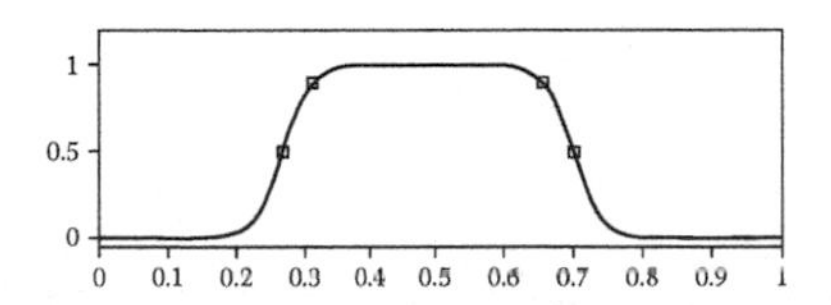
5.	S-shaped		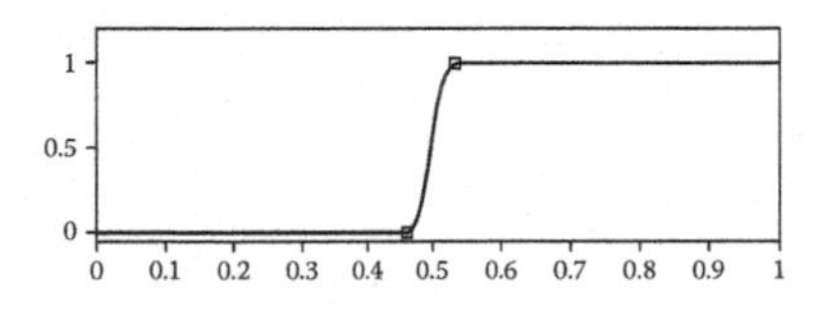

(continued)

TABLE 4.1 (CONTINUED)
Commonly Used Membership Functions

S. No.	Name	Equation	Graph
6.	Z-shaped		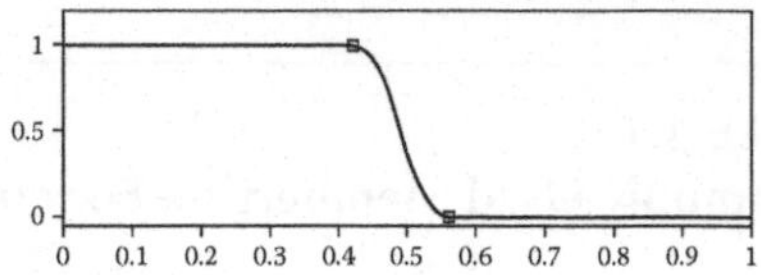
7.	Pi-shaped	$f(x, a, b, c, d) = \text{S-shaped}(x, a, b) * \text{Z-shaped}(x, c, d)$	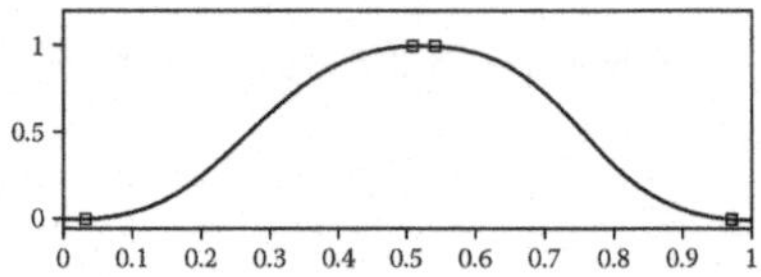

then b is j" with their membership values. We concentrate only on the antecedents here. Suppose the condition reads, "If driver is highly experienced." We can easily replace this condition with the membership value given by the membership function of $high_{experience}$. In the later sections of this chapter, we will replace other operations of the rule-based approach until we have the entire fuzzy-based inference engine ready.

4.5 FUZZY LOGICAL OPERATORS

In this section, we study the fuzzy way of dealing with logical operators. Any condition in a rule-based approach may carry a number of logical operators. These operators must be evaluated to get the value of the entire condition. Consider the condition given by Equation 4.9. Here the various conditions are joined using the logical operators *AND* and *OR*. Any operator may also be applied with a unary operator *NOT.*

$$c = (x_1 \text{ AND } x_2) \text{ AND NOT } (x_3 \text{ OR } x_4) \tag{4.9}$$

where x_1, x_2, and x_3 represent the various conditions.

The various operators according to the rules of Boolean algebra follow the precedence order *NOT, AND, OR*, with *NOT* having the highest precedence.

Any condition may ultimately be represented using the generalized form given in Equation 4.10:

$$c = [\text{NOT}]\ x_1 \text{ op } [\text{NOT}]\ x_2 \text{ op } [\text{NOT}]\ x_3 \text{ op } x_4 \ldots\ldots \text{ op } [\text{NOT}]\ x_n \tag{4.10}$$

where $x_1, x_2, x_3, \ldots, x_n$ represent the various conditions, each condition is of the form $y_i = f_j$, f_j is the membership function, y_i is the variable, [NOT] means that its presence is optional, and op stands for AND/OR.

We saw in the previous section that in fuzzy systems, each condition $x_1, x_2, x_3, \ldots, x_n$ denotes some value. This value is the degree of association of the variable to the particular class and is given by the governing membership function.

In this section, we model how the logical operator handles fuzzy arithmetic. Each operator takes two membership values (or just one in the case of NOT) and returns as a result the membership value according to the operator's logic. Various conditions joined by logical operators may be handled in a manner similar to how we handled the logical operators in Boolean algebra. Similar to boolean operators, fuzzy operators have the rule of precedence, associative law, commutative law, etc.

4.5.1 AND Operator

AND is a binary operator that takes two inputs and returns a single output. It may be represented by Equation 4.11.

$$c = x \text{ AND } y \tag{4.11}$$

In Boolean algebra, the functioning of AND is given by Table 4.2. The output is true (or 1) only if both of its inputs are 1; otherwise it is 0. In a logical sense, this means that the operator returns a high only when the first *and* the second inputs are high.

The fuzzy AND does not have as its inputs 0 or 1. Instead it has a continuous range of values from 0 to 1. In fuzzy systems, we usually take the AND operator as the *min* or *product*, both of which have their conventional meaning and are represented by Equations 4.12 and 4.13:

$$c = \min\{x, y\} \quad \text{(for a min system)} \tag{4.12}$$

$$c = x * y \quad \text{(for a product system)} \tag{4.13}$$

Observe that in both the cases the inputs and outputs are bounded between 0 and 1 and that the system follows the outputs of the Boolean algebra system when given Boolean inputs. We take two sample graphs for the variables x and y. The resultant graph generated by the fuzzy AND operator using both the *min* method and the *product* method are given in Figure 4.6a, while Figure 4.6b shows their binary equivalents.

4.5.1.1 Realization of Min and Product

It may be interesting to observe the behavior of the AND operators in the inputs given in Figure 4.6. Looking at Figures 4.6a and 4.6b, we may easily observe a very strong correlation between the fuzzy and the nonfuzzy counterparts.

Consider the system with which we are supposed to find the vulnerability of intrusion at some location. We know that for this intrusion, an intruder must break two security doors, one after the other. After that, the intruder may exploit the system. A simple fuzzy rule might say, "If ($door_1$

TABLE 4.2
Truth Table for AND

x	y	$c = x$ AND y
0	0	0
0	1	0
1	0	0
1	1	1

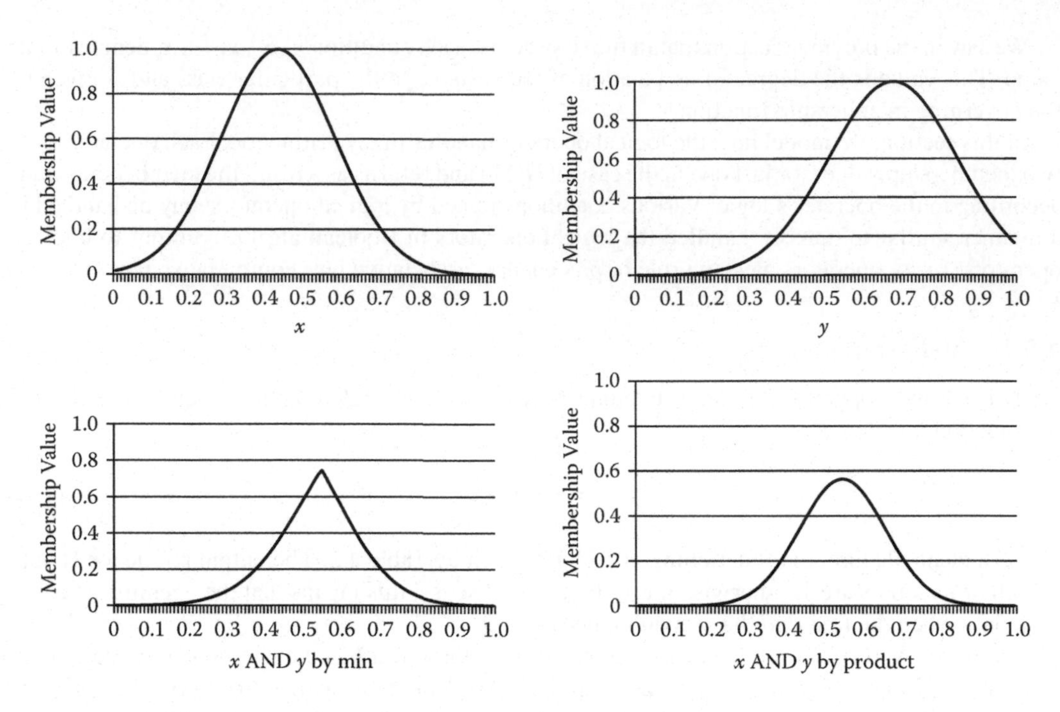

FIGURE 4.6a The AND logical operator in fuzzy arithmetic.

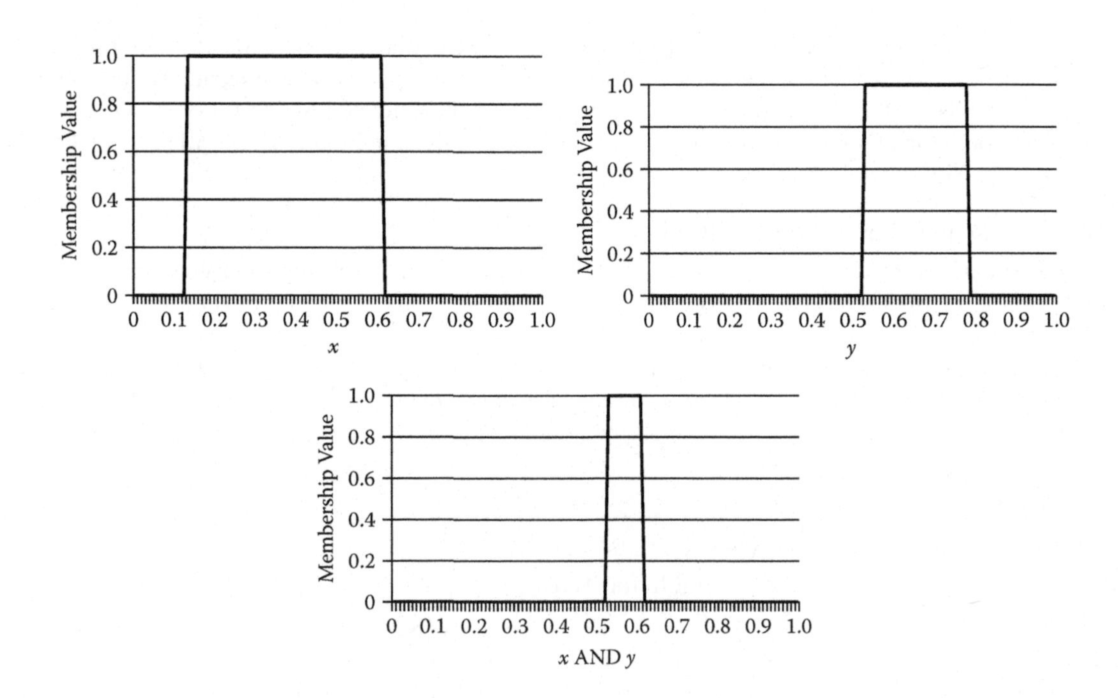

FIGURE 4.6b The AND logical operator in Boolean (nonfuzzy) arithmetic.

security is poor) AND ($door_2$ security is poor), then (*intrusion* is high)." The AND may be resolved by the use of a *product* operator. A mathematical counterpart may also suggest that the same problem may be solved by the use of probability, where the probability (P) of intrusion is given by Equation 4.14. This sounds very similar to the use of *product* as the AND operator.

$$P(intrusion) = P(door_1 \text{ is passed}) * P(door_2 \text{ is passed}) \tag{4.14}$$

Consider another example. Suppose you are traveling on a road and need to measure the comfort of traveling. The comfort depends on the road condition and the vehicle condition. The general rule may be framed as, "If (*vehicle condition* is bad) AND (*road condition* is bad), then (*comfort* is poor)." In this system, it may easily be observed that if the road has too many curves and traffic, no matter how good the vehicle is, the drive would not be comfortable. In addition, if the vehicle is in very bad shape, the drive would not be comfortable. In such a case, it may be seen that the comfort behaves as the minimum of the two factors. We assume that the comfort is measured by asking the person traveling, who gets dissatisfied when either of the conditions is bad and thus reports the drive uncomfortable. We further assume that if the person is traveling on a dirt road, it would not make any difference whether he travels by a very expensive car or a normal car, since he would not enjoy the drive in any case.

4.5.2 OR Operator

OR is another binary operator that takes two inputs and returns a single output. It may be represented by Equation 4.15. In Boolean algebra, the functioning of OR is given by Table 4.3. The output is true (or 1) if any of its inputs are true (or 1); otherwise it is 0.

$$c = x \text{ OR } y \tag{4.15}$$

In fuzzy systems, we take the OR operator as the *max* or the *probabilistic or*. Both have their conventional meaning and are represented by Equations 4.16 and 4.17.

$$c = \max\{x, y\} \qquad \text{(for a max system)} \tag{4.16}$$

$$c = x + y - x * y \qquad \text{(for a probabilistic OR system)} \tag{4.17}$$

Again observe that in both cases, the inputs and outputs are bounded between 0 and 1 and that the system follows the outputs of Boolean algebra when given Boolean inputs. The graphs for the OR operator are given in Figure 4.7a, while Figure 4.7b shows their binary equivalents.

4.5.2.1 Realization of Max

Consider the same vulnerability analysis system in which we measure the intrusion risk. Consider that the same two doors are not sequential this time, but parallel. In this case, the intruder may break

TABLE 4.3
Truth Table for OR

x	y	$c = x$ OR y
0	0	0
0	1	0
1	0	0
1	1	1

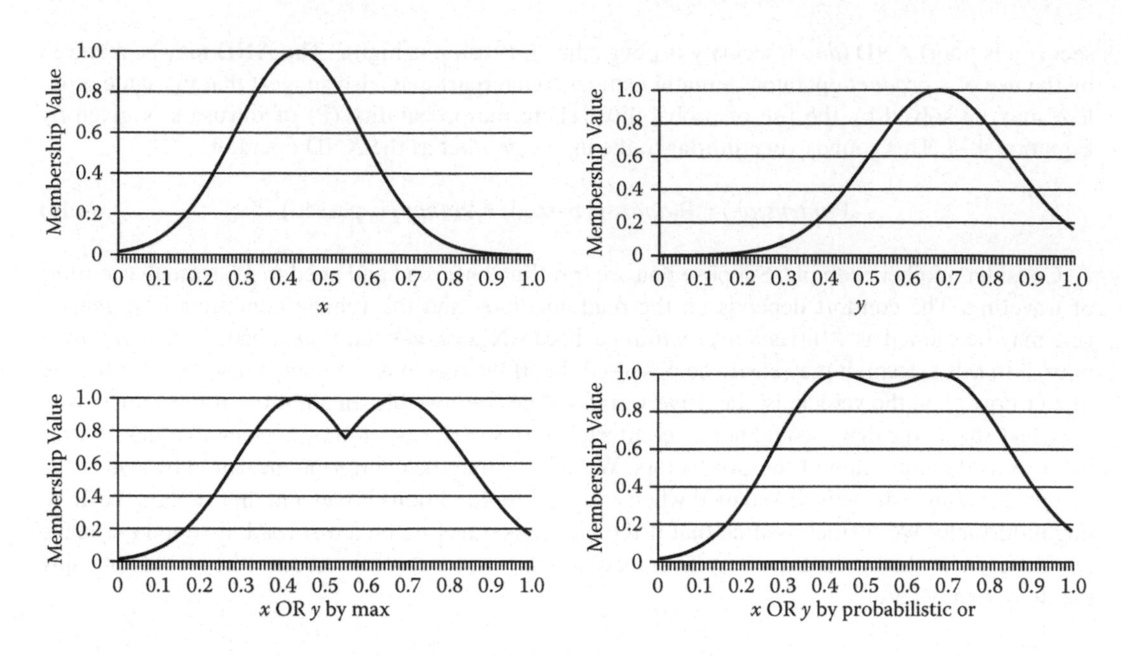

FIGURE 4.7a The OR logical operator in fuzzy arithmetic.

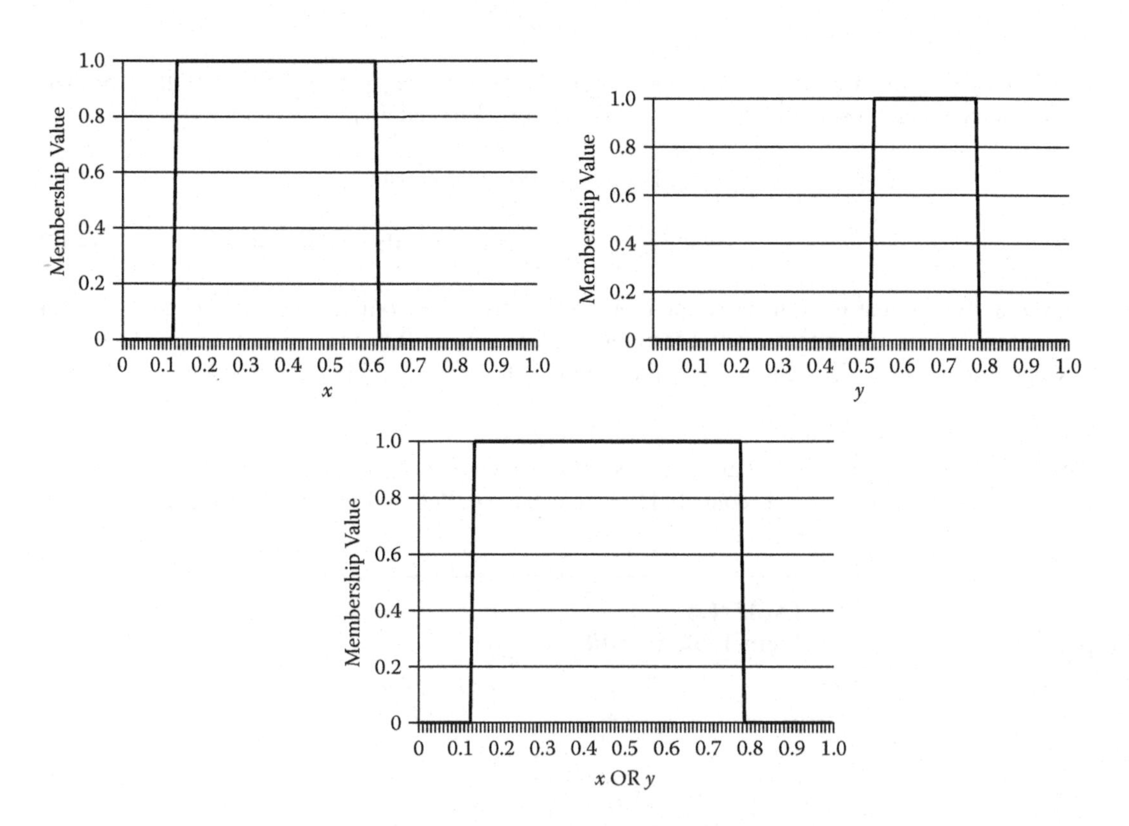

FIGURE 4.7b The OR logical operator in Boolean (nonfuzzy) arithmetic.

TABLE 4.4
Truth Table for NOT

x	c = NOT x
0	1

any one of the doors to exploit the vulnerability. In such a situation, we may write the fuzzy rule as, "If ($door_1$ security is poor) OR ($door_2$ security is poor), then (*intrusion* is high)." Considering the system from the intruder's point of view, he would first select the door in which the intrusion is most likely; this is the door with the least security or the highest chance of intrusion. He would then break the security for intrusion. This situation behaves in a similar way to the *max* operator.

4.5.3 NOT Operator

NOT is a unary operator that takes one input and returns a single output. It may be represented by Equation 4.18. In Boolean algebra, the functioning of *NOT* is given by Table 4.4. The output is the reverse of the input.

$$c = \text{NOT } x \tag{4.18}$$

In fuzzy systems, the NOT operator does exactly the same thing—reversal—as represented by Equation 4.19. The graphs for the NOT are given in Figure 4.8a. Figure 4.8b shows their binary equivalents.

$$c = 1 - x \tag{4.19}$$

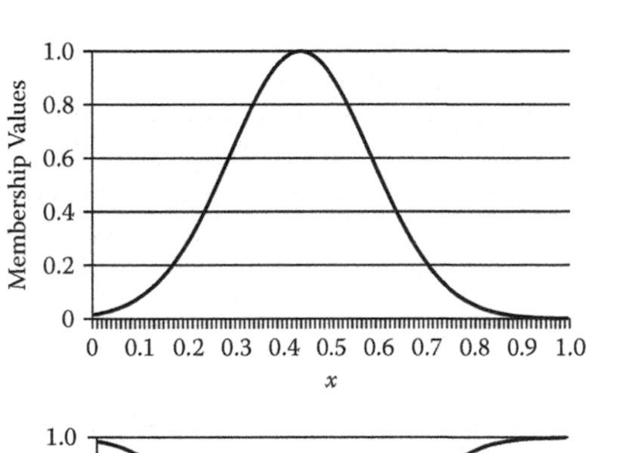

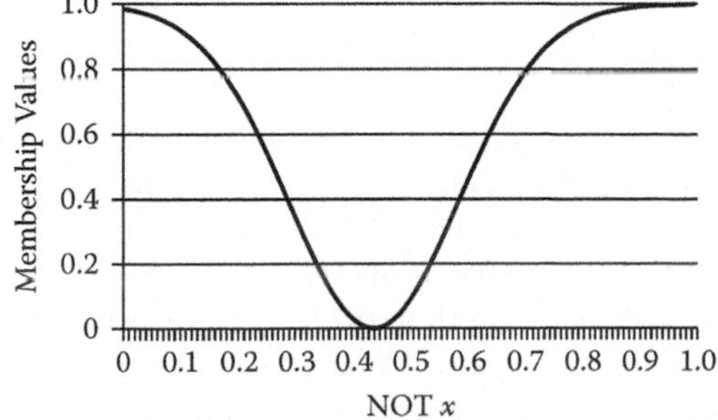

FIGURE 4.8a The NOT logical operator in fuzzy arithmetic.

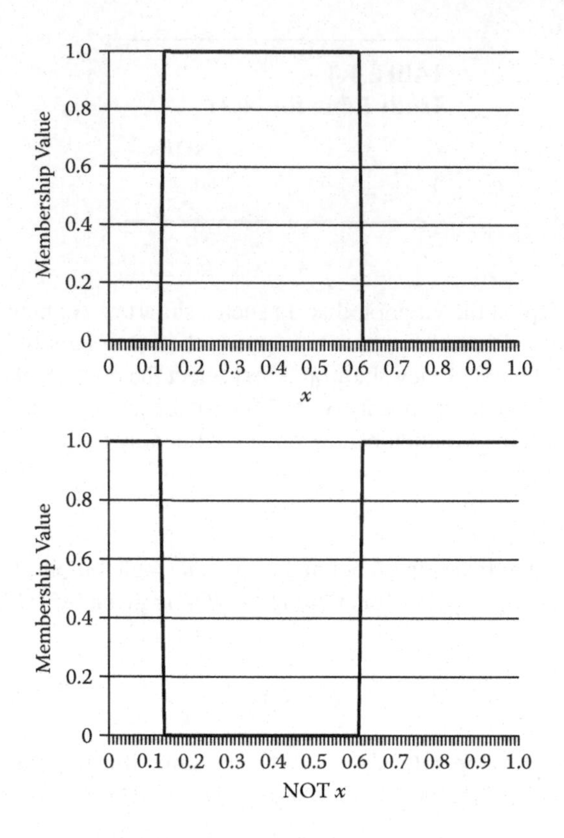

FIGURE 4.8b The NOT logical operator in Boolean (nonfuzzy) arithmetic.

4.5.4 Implication

As we proceed through the text, we will advance toward the conversion of any general rule-based approach to a fuzzy approach. So far, we have reduced any rule to the form "if x, then $y_1 = c_1$ and y_2 and c_1 and y_2," or "$x \rightarrow y$". We now resolve the THEN operator ($\rightarrow$), which is known as the *implication operator* and is given by Equation 4.20. We may even consider a much more generalized manner in which a rule may be written considering all the inputs and outputs. The general way of representing such a rule is given by Equation 4.21.

$$x \rightarrow y \tag{4.20}$$

$$\begin{aligned} &\textbf{if } [\text{NOT}]\ x_1 = f_1 \text{ op } [\text{NOT}]\ x_2 = f_2 \text{ op } [\text{NOT}]\ x_3 = f_3 \text{ op } x_4 = f_4 \ldots \text{ op } [\text{NOT}]\ x_n = f_n \\ &\quad \textbf{then } y_1 = f_1 \text{ AND } y_2 = f_2 \text{ AND } y_3 = f_3 \text{ AND } y_n = f_n \end{aligned} \tag{4.21}$$

where $x_1, x_2, x_3, \ldots, x_n$ are the input variables, f_j is the membership functions, [NOT] means that its presence is optional, op stands for and/or, and $y_1, y_2, y_3, \ldots, y_n$ are the output variables.

The nonfuzzy systems have a series of rules in the "if . . ., then . . ., else." format. Whenever the condition is true, the corresponding statements are executed, and the resulting output may be operated. This procedure, however, does not work in fuzzy systems, where no condition is true or false. On the contrary, the truth is always to some degree in the interval 0 to 1. Hence we need formal methods to carry out implication and, as we shall see later, to combine the results of various rules.

The AND represented in the left part of the expression in Equation 4.21 is different from the one given on the right side. The AND on the left side represents the combination of the various factors

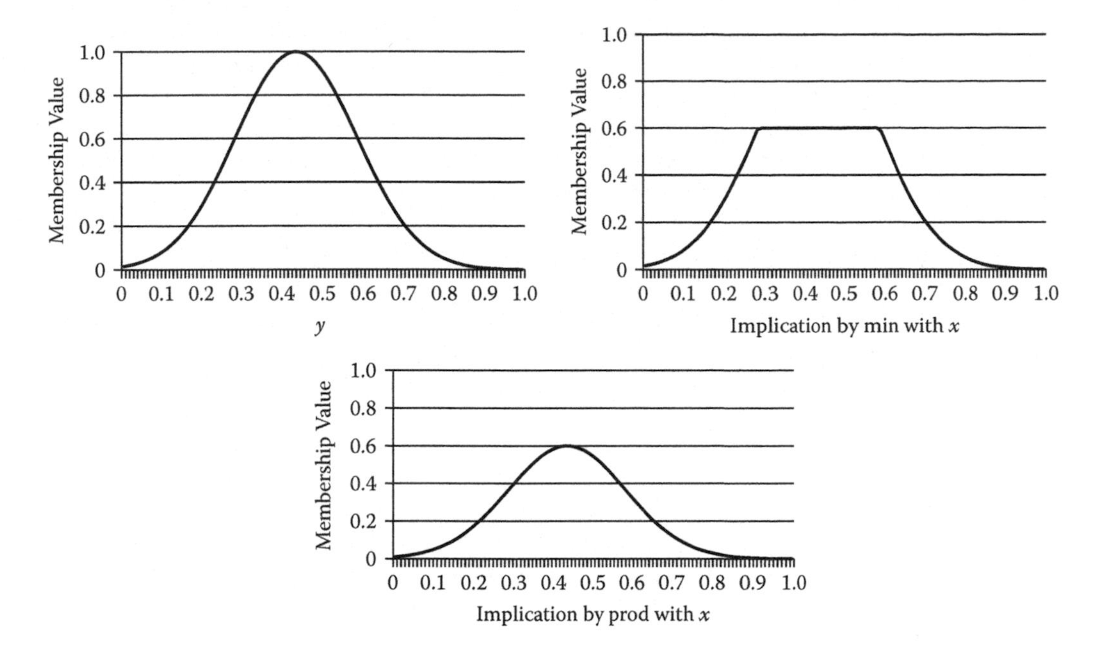

FIGURE 4.9 The *min* and *product* implication operators ($x = 0.6$) in fuzzy arithmetic.

to represent the rule. They are combined using the logical equivalent fuzzy operators. On the other hand, the AND on the right is not combined; it represents the same conditions affecting the various output variables. They are all independent of each other and are treated separately while working.

Similarly the = sign occurs both on the left and the right of THEN. These two = are also different. The former is the check for equality, which checks the closeness of the input to any membership function. The latter is an assignment, where we try to assign the output to a membership. We will now see how this assignment is done.

Implication is a binary operation that takes two inputs and returns a single output. Implication in the case of fuzzy systems is normally performed by the *minimum* or the *product* function. This is the same operation that we used in the *AND* operator and is given by Equations 4.22 and 4.23. The graphs are the same as shown in Figure 4.9.

$$x \rightarrow y = \min(x, y) \quad (4.22)$$

$$x \rightarrow y = x * y \quad (4.23)$$

where x represents the final calculated value by the application of the various logical operators and y represents the selected membership function of the output variable. We get a single membership value for a given input. The same value is used for the purpose of calculation. The output is the entire membership function graph. The operation is applied by a single membership value on the entire membership function.

Unlike the AND operator, the implication function is not the combination of conditions according to the laws of the logical operators. Rather it tries to do an assignment. Due to its way of functioning, this operation is sometimes known as *chopping* if using the *min* operator. It chops off the regions in the membership graph of the output variable. Similarly the operator may be called *squashing* when working with the *prod* operator, as this squashes the entire graph into a lower length graph.

We know that if the condition is $x \rightarrow y$, it means that we are trying to associate the output of the variable x by that represented by the MF y. If we assume that the condition given to the left of THEN was completely true—that is, it has a membership of 1—then the output would exactly follow the membership graph of the selected MF of the output variable. Suppose the condition is, "If road is bad, then accident is high." We can say that if there is only this condition in the entire system and the road is given to be *bad* (bad with high membership), the output would be a *high* accident (high with a large membership degree). This needs to be true, because the output has to follow the rules. In case of *bad* road, it should give a *high* accident.

If we reduce the membership degree, however, the effect of this rule also reduces, which means we have an idea of the output, but we are not that sure of the output. For this reason, we minimize the output's membership degree. The lower membership degree signifies that the confidence, or belongingness, of the output to that class is low. In the same example, if there is lower membership of the rules, the accident would remain *high*, but its membership value would reduce. This means we are not that sure of the accident being *high*. This is exactly what the implication operator does.

4.6 MORE OPERATIONS

We have converted a significant amount of the rule-based approach to model it on the lines of fuzzy logic. In this section, we proceed with our discussion of other operations, including aggregation and defuzzification. After discussing these operators, we put everything together to produce the final model that will be the fuzzy implementation of the rule-based approach, also known as the fuzzy inference engine.

4.6.1 Aggregation

In the previous section, we saw how an if-then clause can be used to find the output for any class. We saw that the output class was identified and its member function was operated according to the value of the condition. This gave us a membership function that was the output of the class for that rule.

In any fuzzy system, numerous rules exist. This means by using knowledge of inputs and systems so far, we are able to obtain a set of functions for each and every rule for every output class. To complete the system, we need a means for deriving the final output from these individual functions. This will enable us to use the various rules to generate the final output. The final output is affected by each rule and by the decided membership functions.

The work of aggregating all the rules together to form a single output is done by the aggregation operator, which may be visualized as a summation of the various rules to get the final output. This summation represents an MF that is the combination, or the aggregation, of the constituent MFs.

We mainly use three kinds of functions for the aggregation: *sum, maximum,* and *probabilistic or.* In whichever of these methods we follow, the final outcome must always be between 0 and 1. This is with regard to the property of the membership function. The graphs for all three of the aggregation for three different rules x, y, and z are given in Figure 4.10.

4.6.1.1 Realization of Sum and Max

We use sum, max, and probabilistic or for the purpose of aggregation. In this section, we present the novelty behind the use of these functions and how they are catering to the needs of the functions for the purpose of effective fuzzy system design.

The first function used is *sum.* The motivation behind this operation is simply the way in which one would normally handle multiple rules. The effect of the different rules is simply added. Any value greater than 1 is taken to be 1 itself. Suppose you have to decide between fast driving or slow driving. You leave the decision to different people who will decide it for you. Normally you would add the number of people suggesting fast driving and the number of people suggesting slow driving. This is the voting mechanism, which works on the principles of addition.

Similarly assume that you asked the different people the same question. This time it was a heterogeneous group of people with different levels of understanding. All try to answer the question

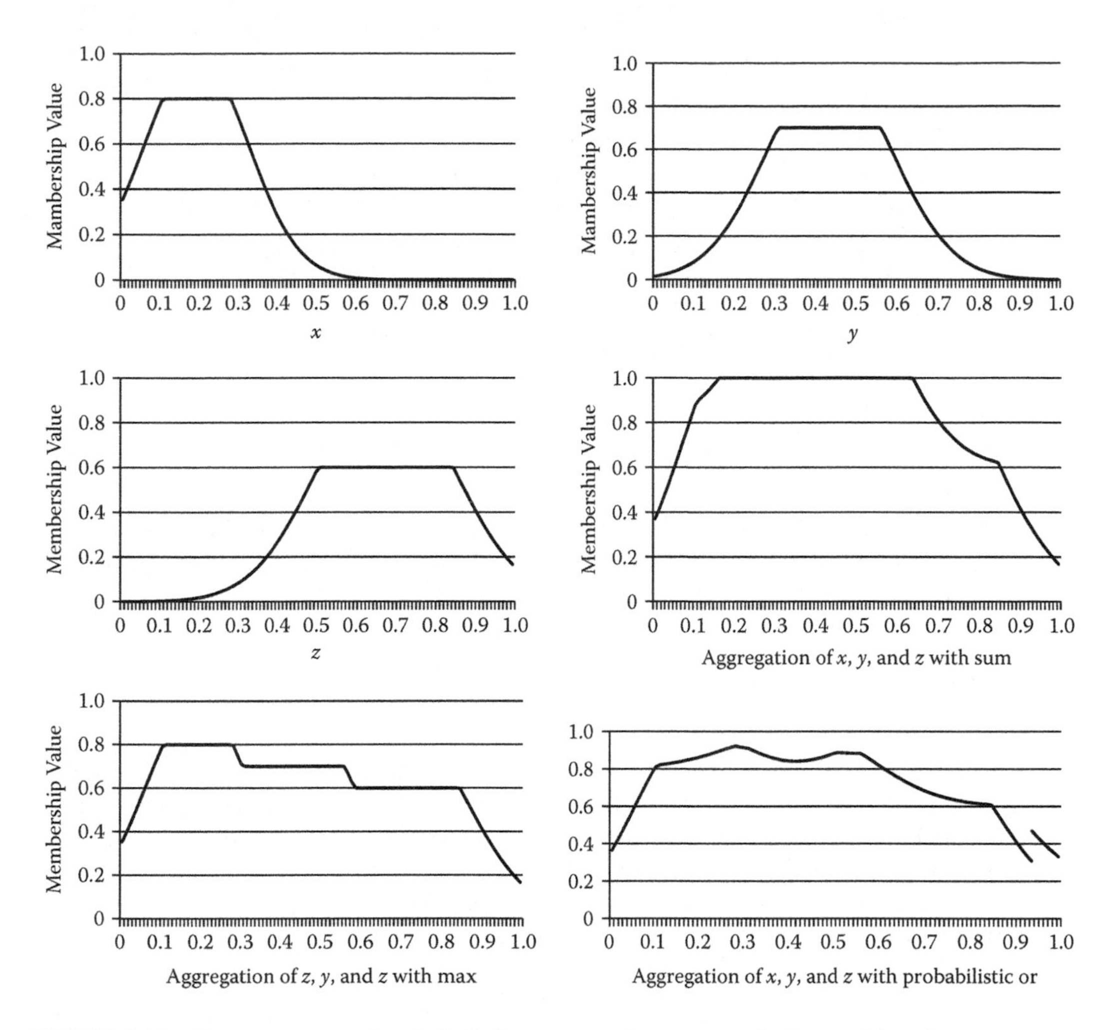

FIGURE 4.10 The *sum*, *max*, and *probabalistic or* aggregation operators in fuzzy arithmetic.

according to their own understanding. In this case, you might trust the person who is supposed to be the best or who has the greatest confidence in her responce. You might follow her decision blindly. This is the motivation behind the use of the max operator.

4.6.2 Defuzzification

So far we have the aggregated output as a result of applying the various rules. Now we need to return the crisp output, or the numeral output, that the system is expected to give. This is done by the defuzzification operator. This process converts the calculated membership to a single numeric output for each output variable. In concept, this is the opposite of the fuzzification function, in which we converted the numeric inputs into membership degrees by the use of membership functions. The defuzzification process is the last step that gives the final output of the system.

Defuzzification is applied to the obtained membership degrees to generate the crisp output. The process of defuzzification involves analysis of the entire membership function to find the most optimal value according to the logical or problem requirements.

Various methods are used to defuzzify the outputs. The most prominent methods are centroid, bisector, largest of maximum (LOM), mean of maximum (MOM), and smallest of maximum (SOM). Here all functions have their usual meaning.

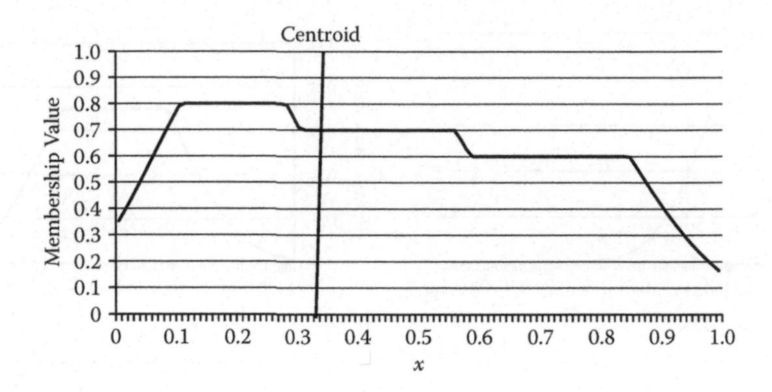

FIGURE 4.11a Defuzzification by centroid.

Centroid: The centroid finds the centroid of the total area represented by the membership curve. The centroid is a concept similar to that of the center of mass for a body. The centroid may be calculated by Equation 4.24:

$$o = \frac{\int x_i * m_i * dx}{\int m_i * dx} \tag{4.24}$$

where o is the final defuzzified output, x_i is the range of values of the output variable, and m_i is the corresponding membership function. Figure 4.11a shows a sample membership curve and the corresponding defuzzified output calculated by centroid method.

Bisector: The bisector finds the bisector of the total area represented by the membership curve. The area bisector divides the whole membership function area into two equal halves, as given by Equation 4.25:

$$\int_{x_1}^{o} m * dx = \int_{o}^{x_2} m * dx \tag{4.25}$$

where o is the final defuzzified output, x_1 and x_2 are the ranges of output, and m is the corresponding membership function. Figure 4.11b shows a sample membership curve and the corresponding defuzzified output calculated by bisector method.

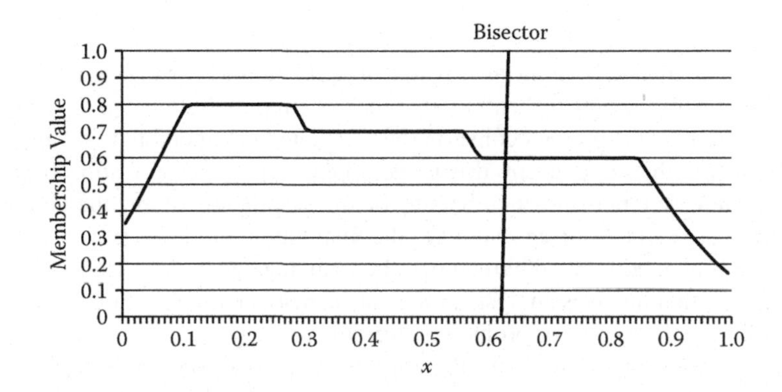

FIGURE 4.11b Defuzzification by bisector.

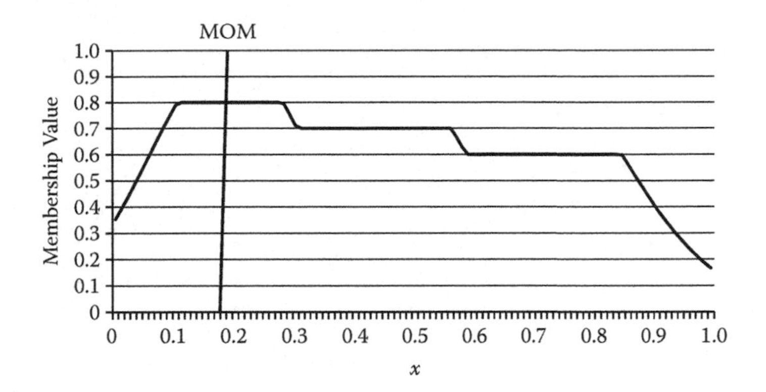

FIGURE 4.11c Defuzzification by mean of minimum.

MOM: The mean of maximum is the average maximizing at which the membership function is the maximum. This method works similar to the *max* operation in implication in that it tries to obtain the solution with the maximum membership degree. The general equation is given by Equation 4.26:

$$o = \frac{\int x * dx}{\int dx} \tag{4.26}$$

where o is the final defuzzified output and x covers all values of the output range where membership is maximum. Figure 4.11c shows a sample membership curve and the corresponding defuzzified output calculated by MOM.

SOM: The smallest of maximum is the smallest value of the output variable at which the membership function is the maximum. This method explores all the values of the output variable where the maximum membership degree is found and then gives the smallest of those values. The general equation is given in Equation 4.27:

$$o = \min(x) \tag{4.27}$$

where o is the final defuzzified output and x covers all values of the output range where membership is maximum. Figure 4.11d shows a sample membership curve and the corresponding defuzzified output calculated by SOM.

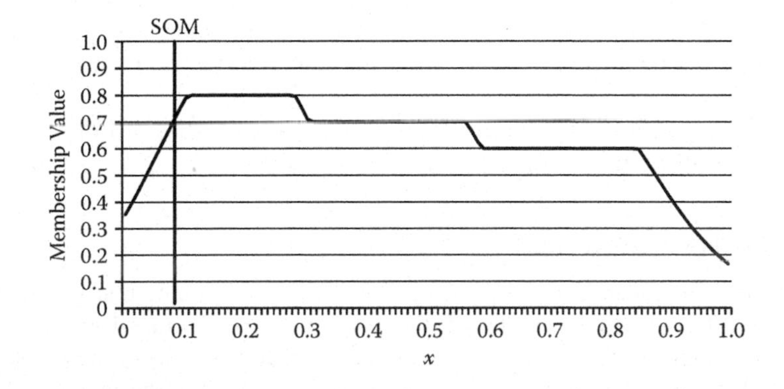

FIGURE 4.11d Defuzzification by smallest of maximum.

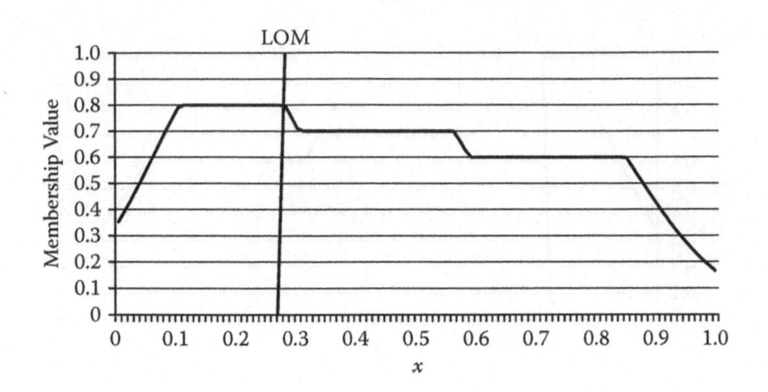

FIGURE 4.11e Defuzzification by largest of maximum.

LOM: This is same as SOM, except that we select the maximum value. The equation is given in Equation 4.28 and the graph is in Figure 4.11e.

$$o = \max(x) \tag{4.28}$$

where o is the final defuzzified output and x covers all values of the output range where membership is maximum.

4.7 FUZZY INFERENCE SYSTEMS

In the previous section, we learned about the means and methods with which we can convert any general rule-based system into a fuzzy logic–based system. The motivation behind this task was to make use of rules that might be generally known in a system to model the system that is driven by these rules. Because the rules were known beforehand, it is natural for the system to follow the desired outputs. This gives rise to a complete system that can be used to model the complexities in real life problems.

In this section, we present a step-by-step approach to how the system finds the correct output to any problem. We cover all the concepts that were presented earlier to engineer a complete system. We also learn how the various parts of the system perform, one after the other, to give the correct output from the inputs.

The fuzzy inference system (FIS) is an intelligent system that is built to give the correct outputs to the known and unknown inputs. The outputs are mapped to the inputs by a set of rules that are cautiously framed after a study of the system's input and output behavior. FIS has a great ability to change the common-language description consisting of rules into a complete system. FISs are hence good at modeling real life problems once we know the common characteristics or the general rules of the system.

We start by discussing the general methodology and characteristics of FIS. We then provide a step-by-step guide to working with these systems.

4.7.1 Fuzzy Inference System Design

This section covers the general design principles of the fuzzy systems that so far we have been discussing in general. A good fuzzy system design needs to correctly map the inputs to the outputs, which is done by designing the correct rules and the correct adjustments of those rules.

The major task, as with any soft-computing system, is identifying the inputs and outputs. The inputs are decided based on the system, rather than on the approach. Hence, fuzzy logic does not play that important a role as far as selection of inputs and outputs is concerned. The important aspect regarding selection of the inputs and outputs is that they must lie within a finite range and their variation must be known. Knowing the variation is helpful at the time of setting up the MFs of the inputs and outputs. In addition, we must know the manner in which the various inputs relate to the outputs. We cannot take any randomly distributed data and try the fuzzy approach. Instead we must know the general guiding rules that relate the inputs to the outputs. This is how fuzzy systems differ from artificial neural networks. In artificial neural networks, we simply give the inputs and the outputs to the network so it can make the rules on its own.

Along with inputs and outputs, the design consists of selecting the correct rules that relate the inputs to the outputs. These rules should be known in general, which means we must have an idea of how the inputs map to the outputs by the application of the various rules. In real life systems, we generally need to know the manner in which the output behaves upon the increase or decrease of any particular input, as this helps when framing the rules. Thus we can say that when a particular input combination is low, the output is low, and vice versa. The rules may be specifically studied by looking at the system's behavior. In most real life applications, we try to correlate the change in the values of input variables to that of the output variables in the presence of multiple inputs and outputs.

The rules are the driving factors of fuzzy systems. A system with defective rules will not be able to perform very well, especially in the presence of a high amount of data. Even a system with correctly designed rules may be further optimized by the adjustment of the different parameters.

The fuzzy systems also depend on the selection and the correct parameterization of the MFs. Fuzzy systems may be fine-tuned by adjusting the parameters of the various MFs of inputs and outputs. Because application of different types of MFs may often have a deep impact on the fuzzy system's performance, we must clearly identify the input as well as the type of data that the system would encounter. Based on this, the MF may be selected. Doing so is more of an art and experience rather than a deep knowledge of fuzzy systems.

4.7.2 The Fuzzy Process

In this section, we study the step-by-step process that maps an input to the output. We have already studied the various steps involved. Here we present a complete picture of the system. Let us return to our road example. Consider the following three rules:

R0: If (driver experience is high) & (road is bad), then (accident risk is moderate).

R1: If (driver experience is low) & (road is bad), then (accident risk is high).

R2: If (road is good), then (accident risk is low).

This system takes in two inputs—driver experience and road. Let each input be rated on a scale of 0 to 1. The system gives one output—the risk of the accident. Let the risk of the accident be measured on a scale of 0 to 1. The higher the value of the input, the higher the risk of accident. Suppose we apply any arbitrary inputs x and y. We explain the process for getting the final output from this system. It is assumed that the system has already been designed.

Fuzzification: The process starts with the fuzzification of the inputs. In this step, we calculate the value of the degree of membership for each input to each of the needed classes by using the associated membership functions. This step is done for each input and for each MF per the requirement of the rules. In our road example, while solving for R0, we would first have to find the degree of membership of x to *high* and of y to *bad*. Suppose the membership functions of a *high experience* and *bad road* are μ_A and μ_B, respectively. From the fuzzification, we get two membership degrees. The first denotes the membership of x to the class of *high experience*, or $\mu_A(x)$, and the second is the membership degree of y to the class of bad *road*, or $\mu_B(y)$. We can clearly see that each input is associated with some class, and the corresponding membership degree is thus found out.

Logical operations: We must combine the various logical operations in each and every rule. This is done using the fuzzy logical operators that we studied in the earlier sections of this chapter. Each logical operator takes one or two operands, and each operand is a number. The answer is another single number that is the result of the logical operation over the operands. In the road example, the *min* method may be applied over $\mu_A(x)$ and $\mu_B(y)$ to give a single number that may replace the left part of the *then* expression. The number is actually the membership degree of the inputs to the particular rule that is being considered.

Implication: In the implication step, we assign the output variable some value or membership degree. This membership degree is in the form of a graph or a set of values for every output, as we saw in the previous section. The implication is applied to each output variable, resulting in a membership graph per output variable. In our example, implication on condition R0 would result in a graph being made for the output variable *accident*.

Aggregation: In the aggregation step, we combine the different rules to study their combined effects by using the operators we discussed in the previous section. Aggregation combines all rules into one. This results in the combination of the different membership graphs to generate a common membership graph. In our example, the three rules—R0, R1, and R2—are combined to produce a common graph that will then be further processed.

Defuzzification: This step completes the fuzzy system by giving back the output as desired by the system—that is, a crisp, or numeric, output. Defuzzification is carried out using any of the operators we discussed in the previous sections. It converts the graph we obtained in the aggregation step into a numeral that is given as the output. This step is done for each output variable in the system.

4.7.3 Illustrative Example

Consider once again the road example. Suppose the membership functions of the two input variables *experience* and *road* and one output variable *accident* are as given in Figures 4.12a, 4.12b, and 4.12c, respectively. The input variables *road* and *experience* have two membership functions. The only output variable *accident* has three membership functions. This matches the rules we have considered.

Suppose we apply an input of 2.5 to *experience* and 0.4 to *road*. Now we want to study the system's output. First we must to fuzzify the inputs. If we are solving for rule R0, we must try to find the membership degree of the 2.5 input to *high experience*. This comes out to be 0.1605, as shown in Figure 4.13a. Similarly for an input of 0.4 for *bad road*, the degree of membership of MF *bad*

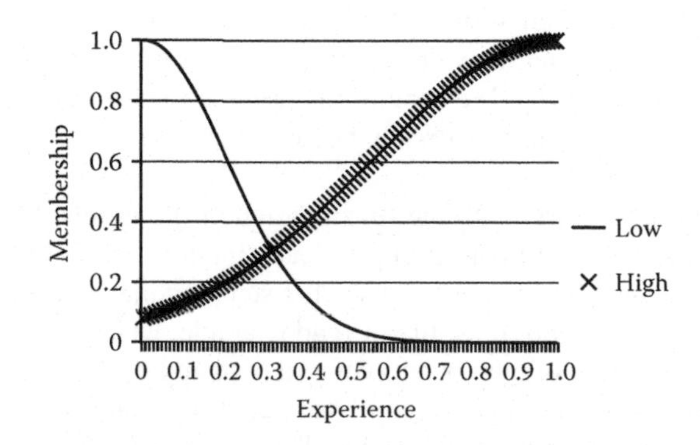

FIGURE 4.12a The membership functions for *experience* for the accident risk problem.

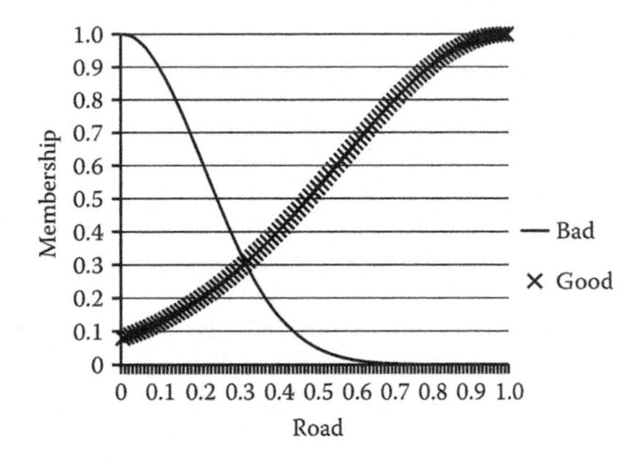

FIGURE 4.12b The membership functions for *road* for the accident risk problem.

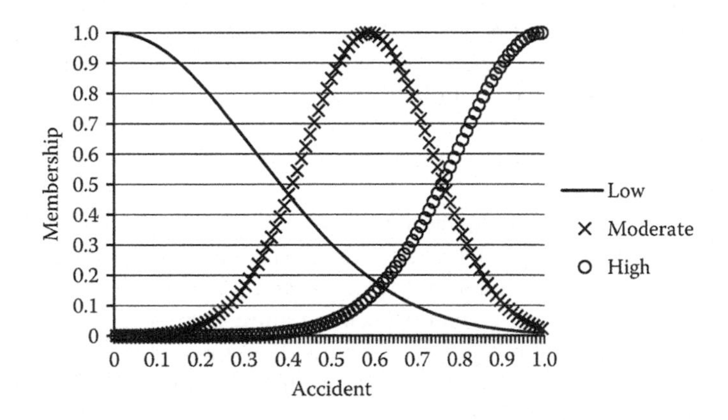

FIGURE 4.12c The membership functions for *accident* for the accident risk problem.

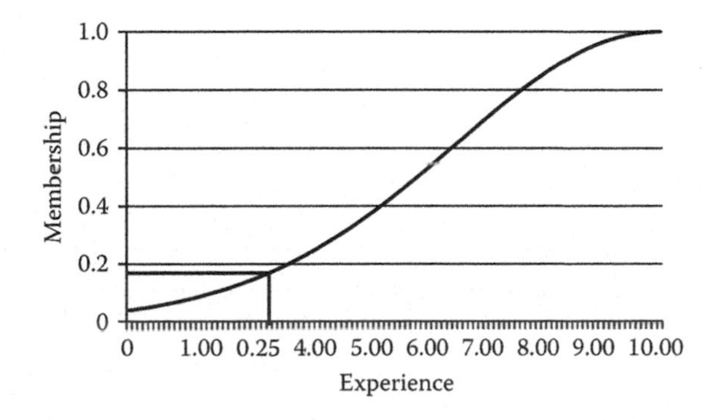

FIGURE 4.13a Fuzzification for rule 0 for *experience* in the accident risk problem.

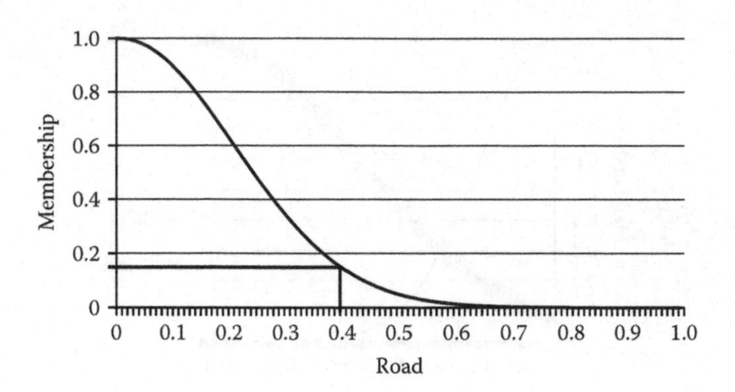

FIGURE 4.13b Fuzzification for rule 0 for *road* in the accident risk problem.

comes out to be 0.1443. This is given in Figure 4.13b. The values for the rest of the rules may be calculated in a similar manner.

We now apply the AND operator between the two values of 0.1605 and 0.1443. Suppose that the *min* operator is used for AND. This would give the result given in Equation 4.29. The values for the rest of the rules may be calculated in a similar manner.

$$y = \min(0.1605, 0.1443) = 0.1443 \tag{4.29}$$

The next step is implication, which we need to carry out along the point $y = 0.1443$. For rule R0, the result is given in Figure 4.14. The other rules may be handled in a similar manner.

Then we use aggregation for the three rules R0, R1, and R2. Suppose we are using *max* as the method of aggregation; the resultant graph is shown in Figure 4.15.

At the end, defuzzification of the resultant graph is done to yield the net result for the output *accident*. In this example, we used centroid as a defuzzification operator. The final answer comes out to be 0.385, as shown in Figure 4.16.

In this way, we can get the output from any set of inputs. The complete FIS system is given in Figure 4.17. It denotes the manner in which the output is mapped to the inputs.

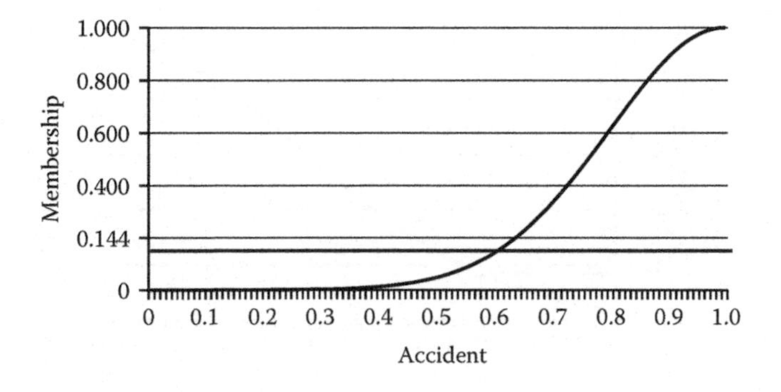

FIGURE 4.14 The implication operator.

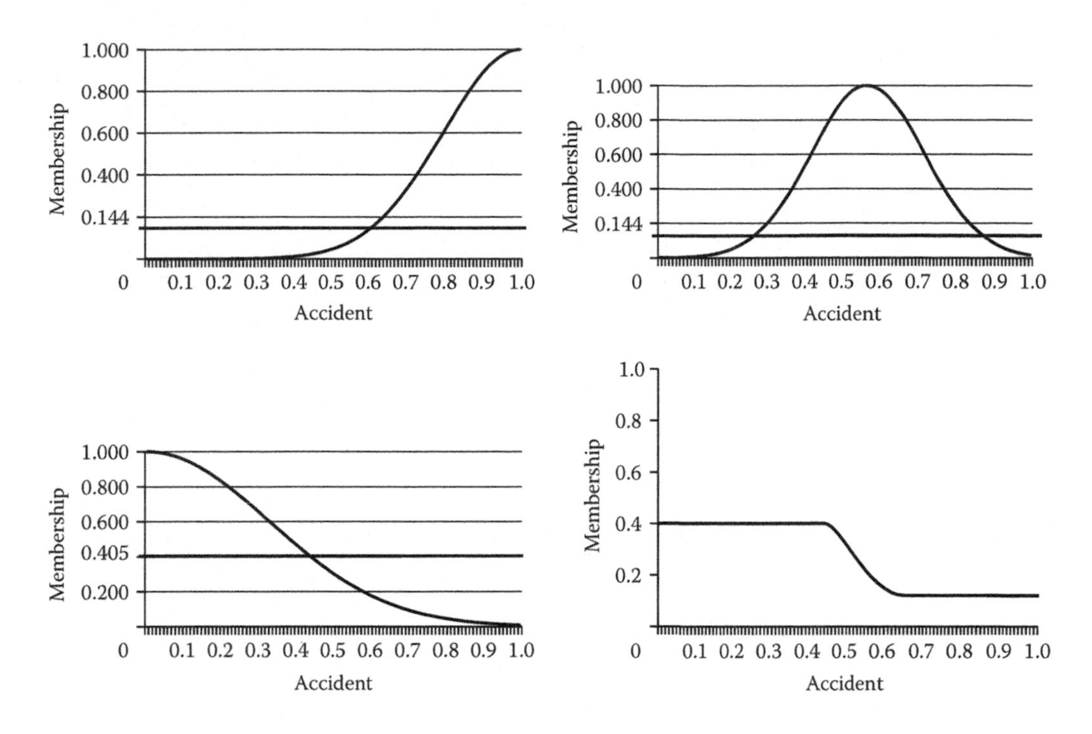

FIGURE 4.15 The aggregation of the three rules in the accident risk problem.

4.7.4 Surface Diagrams

To get a better understanding of the system and its behavior, we often use surface diagrams. A surface diagram is a multidimensional representation of the entire system. It tries to show the entire input space. However, because the input space is highly dimensional, it cannot be represented on the screen because we cannot see more than three dimensions. One of these dimensions is fixed to show the output. The output is hence plotted against any two inputs taken on the other two axes. The surface diagram shows the effect of changing these inputs on the output. The other inputs must be kept constant. Figure 4.18 plots the surface diagram of the road problem.

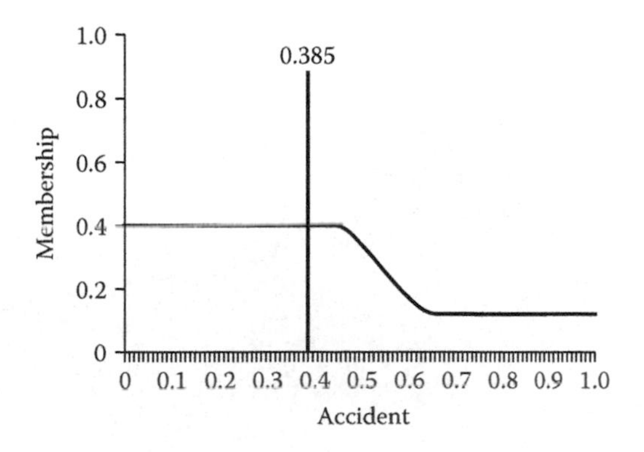

FIGURE 4.16 The defuzzification of the output in the accident risk problem.

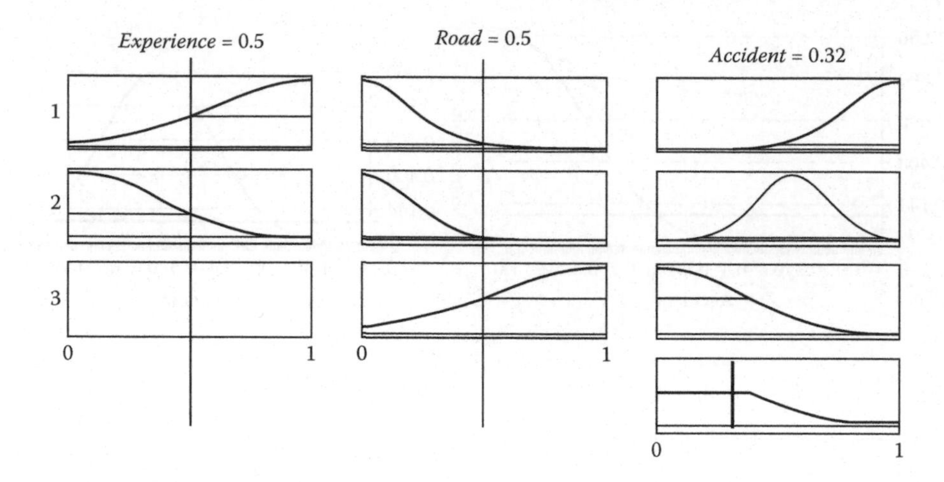

FIGURE 4.17 The complete fuzzy inference system for the accident risk problem.

4.8 TYPE-2 FUZZY SYSTEMS

Our entire discussion so far has focued on the initial model of fuzzy logic. Professor Zadeh proposed another model known as type-2 (T2) fuzzy logic. The initial model that we discussed earlier was henceforth called type-1 (T1) fuzzy logic. The T2 fuzzy logic system (T2 FLS) uses the T2 fuzzy sets (T2 FS) for operations. T2 FS is a more generalized form of sets that can model the fuzziness to an even greater extent. In this section, we briefly discuss the T2 fuzzy sets and T2 fuzzy systems.

4.8.1 T2 Fuzzy Sets

From our discussion so far, we know that the T1 FS denotes the fuzziness or impreciseness present with any input. We use MFs to measure the belongingness of any input to any of the membership classes. We even plotted these MFs on a graph to see the membership values for different inputs.

The T2 FSs are a higher level of abstraction that denote the uncertainties associated with MFs. Hence they may be referred to as the fuzzy fuzzy models, because they denote the fuzziness in the T1 fuzzy model. The graph of any trivial membership function is given in Figure 4.19a. Now suppose the left corner of the triangle depicted in the graph is not well known. Say it can lie anywhere in the small region, as shown in Figure 4.19b. Now suppose that the same uncertainty exists in

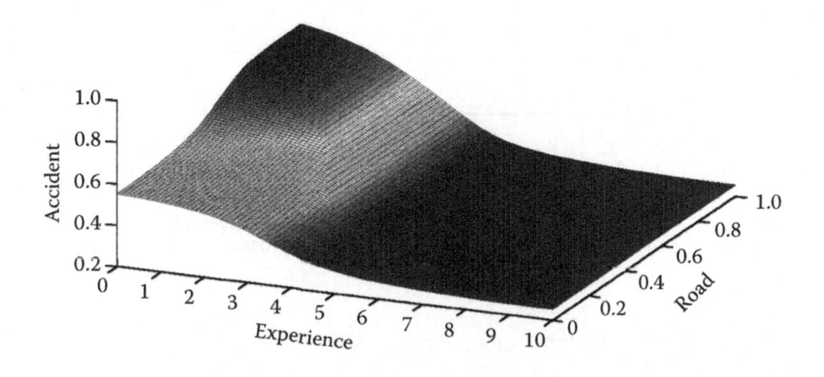

FIGURE 4.18 The surface diagram for the accident risk problem.

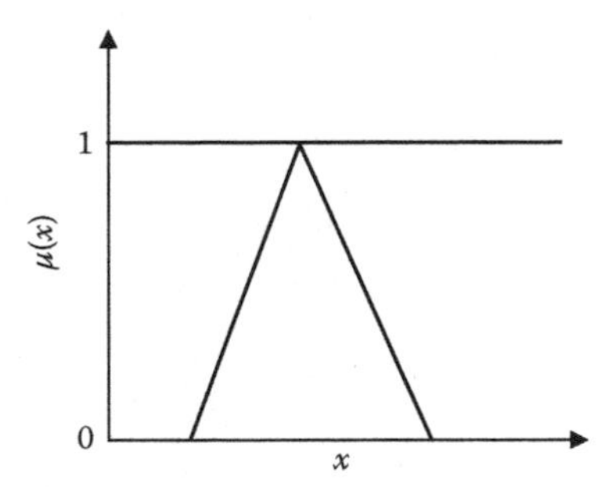

FIGURE 4.19a The type-1 membership function.

the entire curve, as is represented in Figure 4.19b. Here the points required by the T1 FLS can lie anywhere in the associated region. We cannot be sure of where these points will lie.

Suppose that you are to find the membership of any input x to any class. In T1 FLS, the natural approach would be to use the associated membership function and calculate the value—say, $\mu'(x)$. The same may be seen from the membership function graph, where the membership values for the different inputs were plotted. However, the same is not true for a T2 FLS, in which you would first take the input x and consult the graph only to find the range of values within which the membership degree can lie (see Figure 4.19b). The membership degree $\mu(x)$ associated with the input is fuzzy. This fuzziness of the fuzzy system forms the concept of T2 FIS.

Because the membership function $\mu'(x)$ in case of T2 FS is fuzzy, the fuzziness must have some value. This fuzziness denotes the degree of certainty in the T1 FS equivalent MF. Hence we may represent the MF in the case of T2 FLS as $\mu(x,u)$, where u is the point over which the membership value must be calculated. It may easily be visualized that this function represents a three-dimensional graph showing the final membership value x and u. The membership value may even be denoted by the grayness of the curve—that is, higher membership value curves may be darker than their counterparts.

Figure 4.20 shows a T2 FS membership function that we represented by $\mu(x,u)$. Let there be a total of N membership functions. This means that u can take N values corresponding to each of these N membership functions for the corresponding x. These values may be denoted as $u_1 = MF_1(x)$, $u_2 = MF_2(x)$, $u_3 = MF_3(x)$, . . ., $u_N = MF_N(x)$.

The graph given in Figure 4.19b presents a uniformly colored curve. It may hence be interpreted that the probability or membership value is equal for all points in the curve. Let this value be 1. Hence if a point lies inside the region, its membership value would be 1; otherwise 0. A three-dimensional equivalent of this graph would be a graph having a discrete vertical axis where points

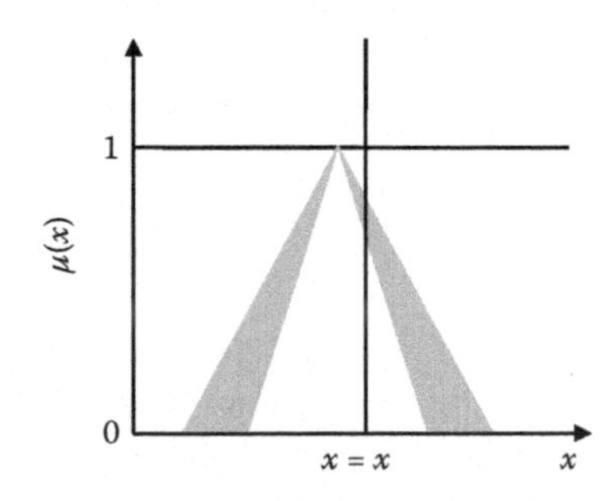

FIGURE 4.19b The type-2 (T2) membership function.

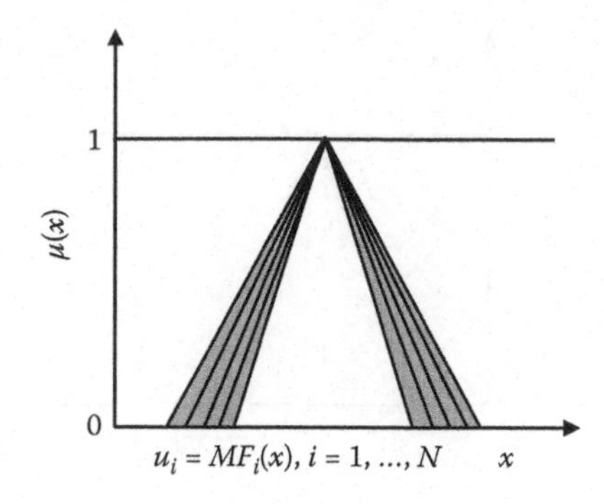

FIGURE 4.20 The membership functions in T2 fuzzy sets.

may only lie at a value of 0 or 1. These form a very special type of systems known as the interval type-2 fuzzy sets (IT2 FS). For these sets, all $\mu(x,u) = 1$.

The IT2 FS is commonly used. In fact, most T2 FLS applications use IT2 FS. The latter's underlying assumption results in lesser computation and complexity, making it possible to use these systems for many applications. Their counterparts, unfortunately, require a great deal of time due to the large computation involved.

For better understanding, we usually represent the membership curves in two-dimensional maps only. This depiction is similar to the one shown in Figure 4.19b, where the nonzero membership areas are colored. The curve so obtained is called the footprint of uncertainty (FOU), which depicts the areas and their associated membership degrees in a two-dimensional graph.

4.8.2 Representations of T2 FS

The T2 FS is commonly represented using vertical slice representation or wavy slice representation. The vertical style is a simpler representation and is more commonly used for computational purposes. The wavy slide representation, however, is more commonly used for theoretical purposes. We study each on its own.

If we slice the three-dimensional membership plot at any value of x, we would get a two-dimensional figure with axes of $\mu(x,u)$ and u, where x is a constant across which the plot was cut. This plot is called the vertical slice at any particular x.

The vertical slice representation uses these properties to represent the fuzzy system. Here we monitor two membership functions—the upper membership function (UMF) and the lower membership function (LMF). The UMF contains the maximum u for any fixed x, while the LMF contains the least (see Figure 4.21).

The wavy slice representation uses embedded MF in its representation. This representation is also known as the Mendel-John Representation Theorem. The embedded fuzzy system is a general curve that is within the least and the maximum values. It may easily be seen that the union of all such curves gives the FOU. The embedded membership function is depicted in Figure 4.22.

4.8.3 Solving a T2 FS

The basic approach for solving a T2 FS is quite similar to the T1 FS we discussed earlier. The system takes in crisp inputs. The process of fuzzification is carried out. The fuzzified inputs fire the rules. The results of various rules are aggregated to form the fuzzified output. In a T1 system, the fuzzified output had to be defuzzified. However, in a T2 system, we need to apply the additional operation of type reduction before the output can be defuzzified.

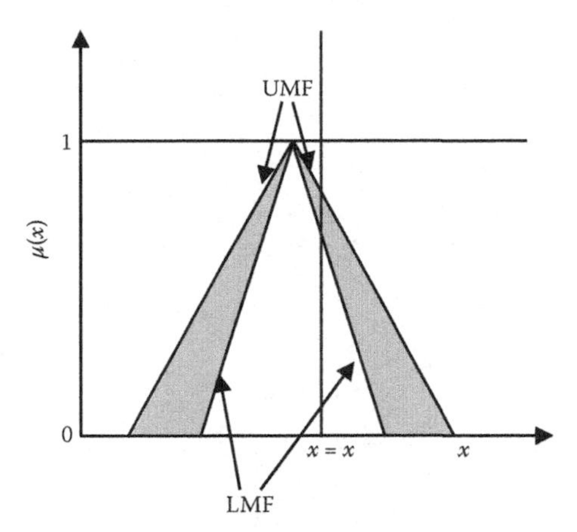

FIGURE 4.21 The lower membership function (LMF) and upper membership function (UMF).

One of the major differences between the T1 and T2 systems is how they deal with the rules. In T2 fuzzy sets, every input and every output MF is in turn an MF. Hence the rules need to be worked out accordingly. The added uncertaininty increases the complexity to a very large extent. Every combination of MF for every rule to be fired must be handled separately, and the outputs are later combined.

For simplicity, consider the simplest rule, having one antecedent and one consequent. Let the rule be, "If x is P_1, then y is Q_1," where P_1 and Q_1 are in turn made up of MFs. Let us say that P_1 is made up of membership functions P_1^1, P_1^2, P_1^3, . . ., P_1^{np}, and Q_1 is made up of membership functions Q_1^1, Q_1^2, Q_1^3, . . ., Q_1^{nq}. Now every combination of P_1^i and Q_1^i must be worked out separately. This way we would be able to deal with the rules in a similar way that we dealt with them in T1 systems. The results of all the combinations are then combined. There will be nP x nQ combinations possible, where nP and nQ represent the number of MFs of the MF P_1 and the MF Q_1, respectively. The firing of rules is diagrammatically shown in Figure 4.23.

The implication in a T1 FS consists of taking the minimum of the antecedents and slicing the output MF at the same level. The T2 FS is, to a reasonable extent, the same, except that this operation must be done for both the UMF and the LMF (see Figure 4.24).

The type reduction (TR) step converts a T2 FS into a T1 FS. This step may be performed by the center of sets (COS) mechanism, which makes use of the Karnik-Mendel (KM) algorithm. After the TR, the defuzzification operation may be carried out, or we may alternatively use COS defuzzification.

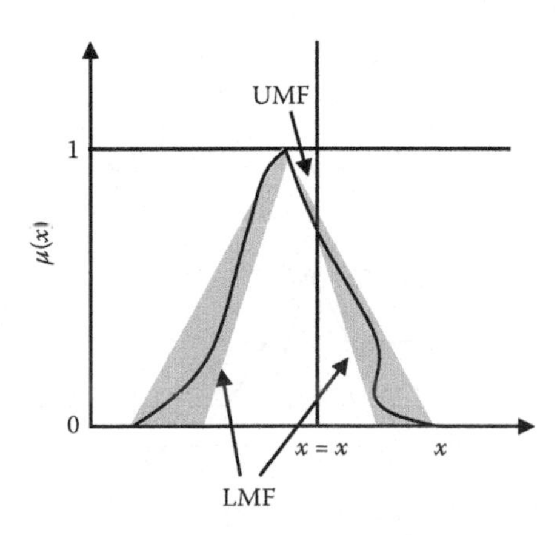

FIGURE 4.22 The embedded fuzzy sets.

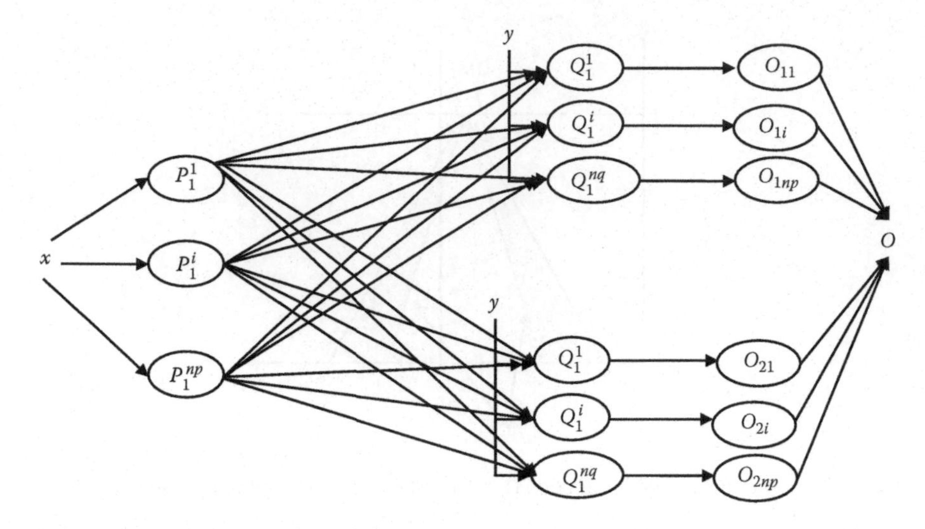

FIGURE 4.23 The rule-firing methodology.

4.9 OTHER SETS

So far we have studied fuzzy sets. But two other sets, based on similar ideas, are commonly used in real life applications—vague sets and rough sets. Readers may want to recollect the set fundamentals, as well as the basics of fuzzy sets, before proceeding with the text.

4.9.1 Rough Sets

We have only considered the sets in which some fuzziness of an element belongs to the set. The rough sets have evolved as tools to better analyze experimental data. These data suffer from noise as well as another major problem—sometimes some values in these data may be totally absent. Thus the rough set is a generalized concept in which the existence of any element in a set is vague. This means that the existence of that element cannot be determined. The rough set theory defines a

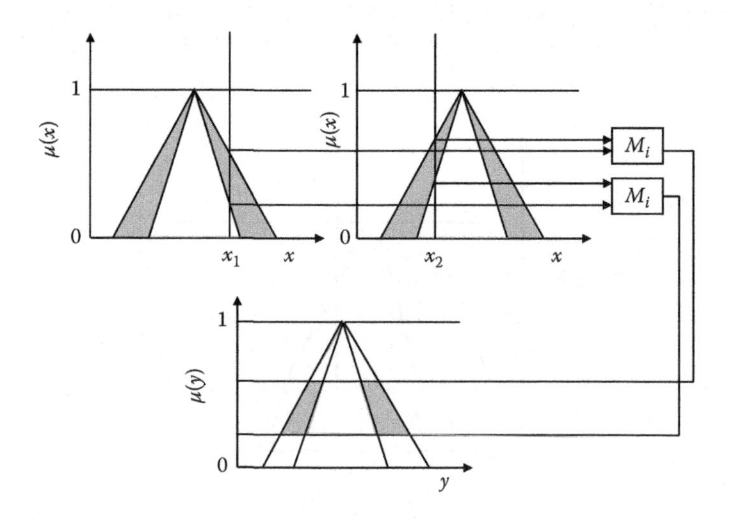

FIGURE 4.24 Implication in T2 fuzzy systems.

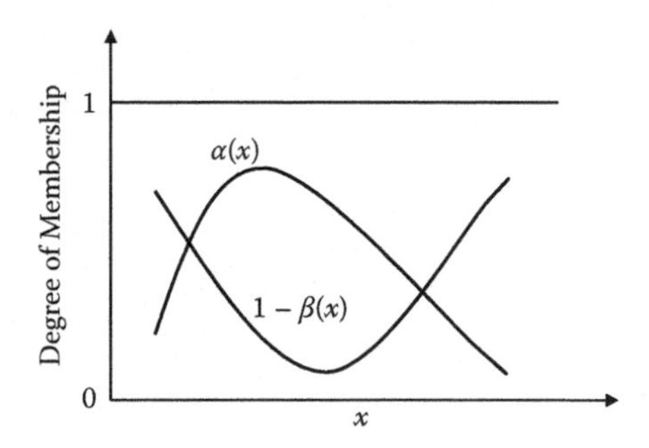

FIGURE 4.25 The vague set membership function.

boundary that consists of elements whose existence in a set is vague or not precisely known. If the boundary is of zero width or does not contain any element, the rough set is called a crisp set and becomes a traditional mathematical set with no impreciseness.

4.9.2 Vague Sets

Vague sets (VS) are sets in which each element has both a degree of trueness and a degree of falseness associated with it. This means there is some degree to which the existence of an element is true and some degree to which it is false. The sum of the two degrees is not necessarily 1, as was the case with fuzzy logic. As a matter of fact, the sum is always less than or equal to 1. Both degrees lie between 0 and 1. The membership of any element x in a VS may hence be represented as $<\alpha(x), 1 - \beta(x)>$, where $\alpha(x)$ denotes the degree of trueness, and $\beta(x)$ denotes the degree of falseness. It is evident that $\alpha(x) + \beta(x) \leq 1$. The membership function in the case of VS may be plotted as shown in Figure 4.25.

4.9.3 Intuitionistic Fuzzy Sets

Intuitionistic fuzzy sets (IFS) are a similar concept to that of the VS. In IFS, two degrees of membership are associated with any element of the set. The first degree measures the membership of the element, and the second measures the nonmembership. It is denoted by $<\mu(x), V(x)>$, where $\mu(x)$ measures the degree of membership and $V(x)$ the degree of nonmembership. The plot for this membership function is given in Figure 4.26.

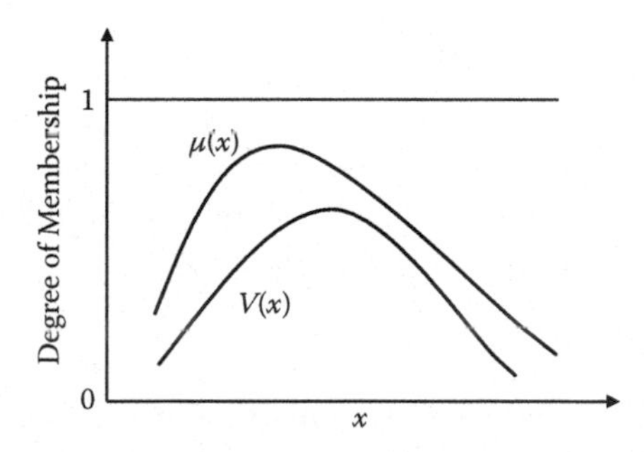

FIGURE 4.26 The intuitionistic fuzzy sets membership function.

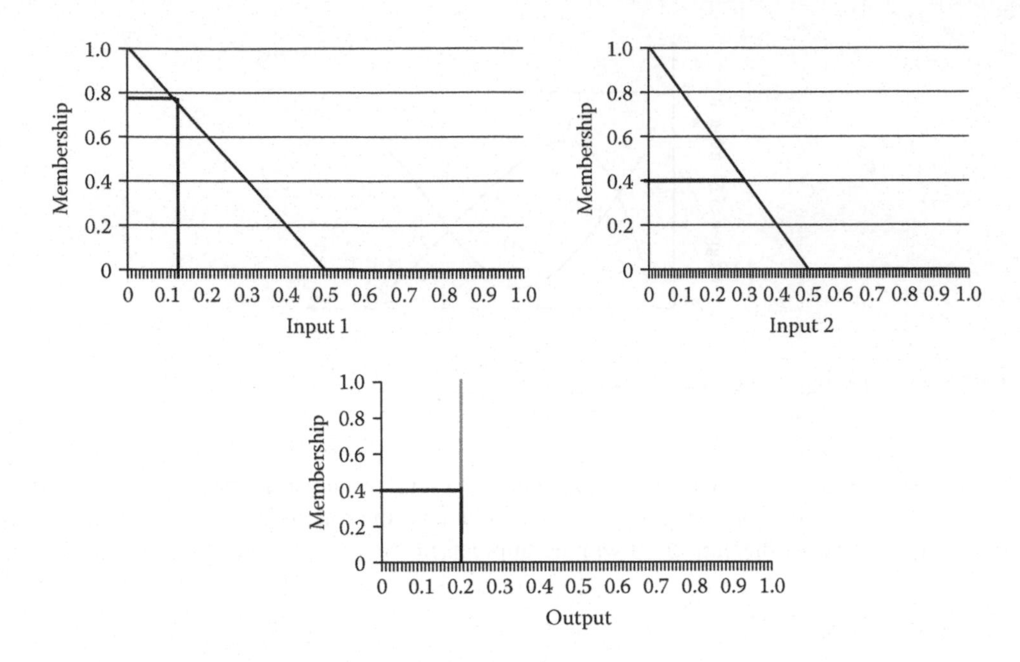

FIGURE 4.27 Implication in the Sugeno fuzzy inference system.

4.10 SUGENO FUZZY SYSTEMS

The T1 FIS that we have discussed is known as Mamdani FIS and is widely used in real life applications. There is another FIS known as Sugeno FIS or Takagi-Sugeno-Kang FIS. This model is same as that of the Mamdani FIS; however the output membership functions in Sugeno can only be linear or constant. The rule in such a system is of the form, "If x is P, then y is Q," where Q is a constant or crisp number. This rule represents the zero-order Sugeno FIS. In these systems, the implication method is simply multiplication or minimum, and the aggregation operator includes outputs of the various rules. Figure 4.27 shows the implication operation in such systems. The aggregation is given in Figure 4.28. Defuzzification in these systems is simply the weighted mean, as shown by dotted line in Figure 4.28.

First-order Sugeno FIS is a more generalized FIS. In this system, rules may be of the form "If x is P, then $y = a * x + b$," where a and b are constants. This is a similar concept to the zero-order system, except that the output can move about in a linear fashion. The higher-order Sugeno systems are computationally very expensive and hence not used in real life applications.

4.11 EXAMPLE: FUZZY CONTROLLER

Among the numerous applications of fuzzy logic, we have controllers. Fuzzy logic has found immense applications in such systems, where we try to control the output of a machine to attain some predefined output catering to the machine's constraints. To fully understand fuzzy logic, we take the example of a fuzzy controller used in robotic control.

4.11.1 Problem Description

Robotic fuzzy controller is used to move robots from a source to a destination, or goal. This problem has relevance in the field of robotics, which applies intelligent systems to make a map and decide the path. Then it becomes the duty of the robotic controller to move the robot following the desired path.

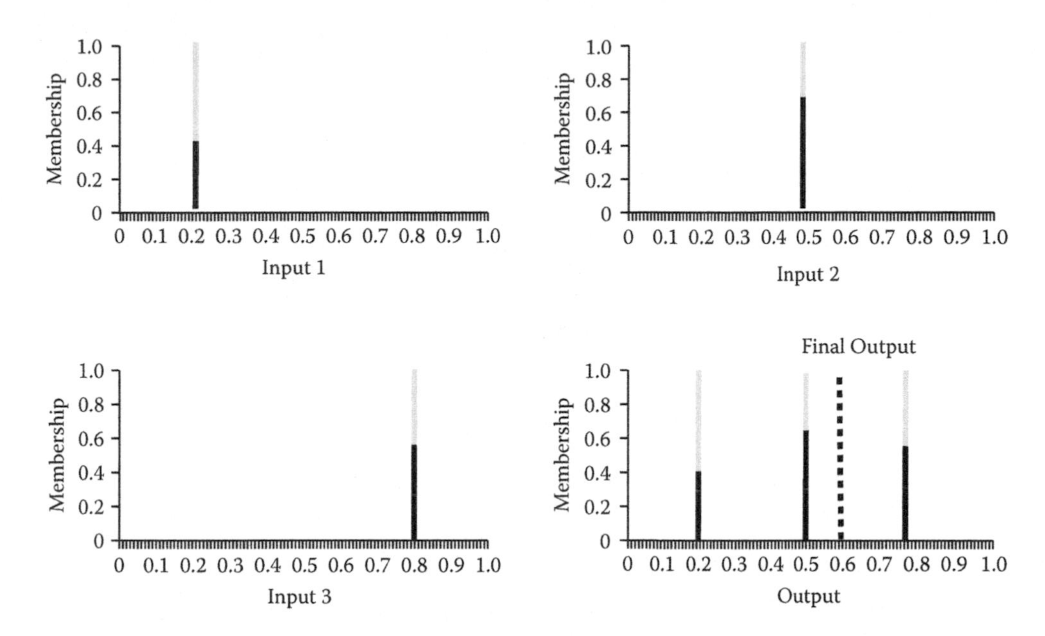

FIGURE 4.28 The aggregation and defuzzification in Sugeno fuzzy inference systems.

The robot basically consists of wheels and motors that drive the wheels. All robotic movements are governed by wheels. For example, a carlike robot rotates by the rotation of its wheels. Considering these facts, the robot cannot make every possible move. It is able to turn only by a certain amount of angle. Furthermore a very sharp turn would make the journey very unsmooth and would further require a reduction of speed. This is undesirable.

For the sake of simplicity, we assume that the robot's speed is constant and cannot change. We further assume that no obstacles exist anywhere in the map. Under these constraints and assumptions, we must move the robot.

The robot in this example is more of a carlike robot. It can only move forward. Of course, in such a robot, backward motion is possible, but that is seldom used in real life situations or in experimental purposes. In addition, we can turn the robot in both a clockwise and a counterclockwise direction by any desired amount. As in a car, the turning of robot is done by turning the wheels at the required angle.

4.11.2 Inputs and Outputs

At any time, the robot's motion depends on the angle and the goal. These form the inputs to the system. The angle α is the angle by which the robot must turn in order to face the goal. This angle is measured by taking the difference between the robot's current angle φ and the angle of the goal measured by the robot's current position θ. The result is always between –180 degrees and 180 degrees, as shown in Figure 4.29.

The other input is the goal, or the distance between the robot's current position and the position of the goal. The distance is normalized by multiplication of a constant so that it lies between 0 and 1.

The system has a single output—the angle by which the robot may be turned at the next move. The robot then physically moves and turns by this angle in its next move. This angle may be positive or negative, depending on whether the desired move is in a clockwise or counterclockwise direction. As we proceed, this angle usually gets smaller and smaller, because over time the robot orients itself in the direction of the goal. After this, the robot just needs to move toward the goal or march in a straight line toward the goal.

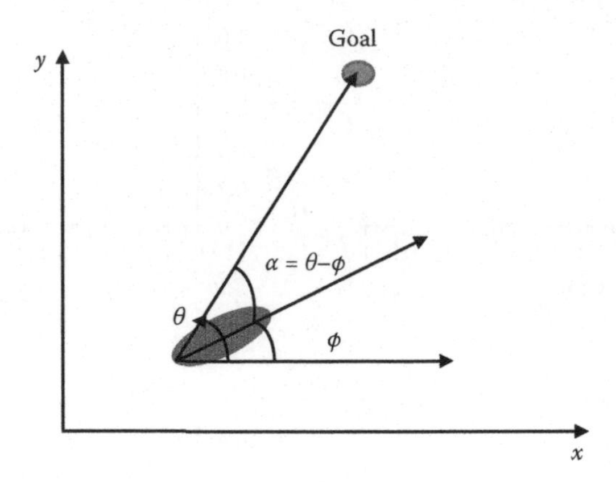

FIGURE 4.29 The measurement of angle α in the robotic control problem.

4.11.3 Membership Functions

After deciding the inputs and outputs, we form the membership functions for each input and output, as given by Figures 4.30a, 4.30b, and 4.30c for *angle*, *goal*, and *output turn*, respectively. All the membership functions used are Gaussian, except for the extremes of the angle, which are trap membership function in nature.

It may be seen from these figures that any MF starts from the midpoint, or extreme, of the neighboring MFs. This helps us frame the rules. At the time at which some input corresponds to the maxima of some MF, it also happens to lie at the minima of other MFs. It may hence be seen that at these inputs, only one MF is active with a membership value of 1, while all others are inactive with a membership value of 0. The output corresponding to this input may be the precise output at the consequent of the rule. When the rule is evaluated, there happens to be a direct mapping of the input that has a membership value of 1 to the output that may again have a membership value of 1. Hence using this mechanism, we can create a type of lookup between known inputs and outputs, and we can perfectly match the desired output using rules. As we deviate slowly from this point, the other MFs start getting active and start influencing the output.

Consider the input angle. At the time when $\alpha = 0$, only one membership function (called *no*) is active. This MF has a value of 1 at $\alpha = 0$. The corresponding output can be mapped to the membership

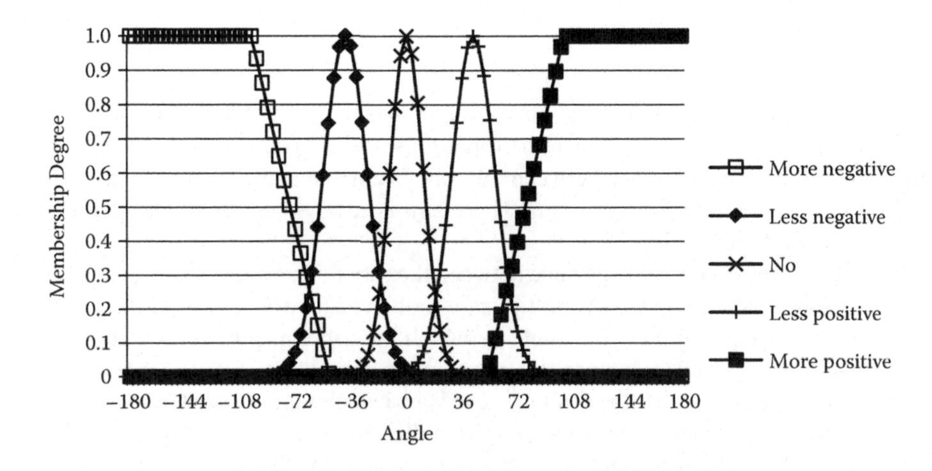

FIGURE 4.30a Membership functions for *angle* in the robotic control problem.

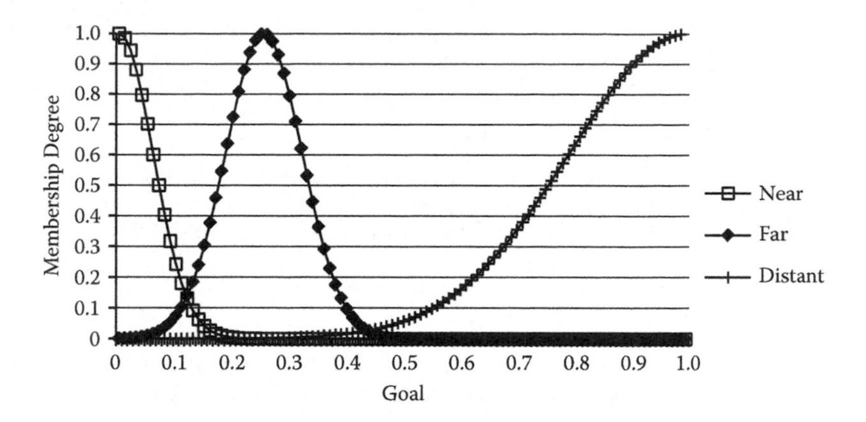

FIGURE 4.30b Membership functions for *distance* in the robotic control problem.

function *no* of the output, which also has a value of 0 as the output with only one active MF. Hence on application of $\alpha = 0$, this rule gives a perfect output of 0, which is desirable.

The *angle* is composed of five MFs that cover the range from –180 degrees to 180 degrees. These are named (from least to right, according to Figure 4.30a) *more negative* (*moren*), *less negative* (*lessn*), *no difference* (*no*), *less positive* (*lessp*), and *more positive* (*morep*). Except for the two extreme MFs, these are more or less equally distributed. The inspiration for this comes from the following fact: Suppose two conditions—$\alpha = 150$ and $\alpha = 100$. Even though the difference in value of α is very large, the turn for both these cases would be around the maximum possible value that is comfortable. This is because in any other case, the robot would take too long to orient itself. Hence it can be seen that there is almost no difference in the output, even for a large change in inputs.

Again, as discussed above, the MF at $\alpha = 0$ stands for *no turn*. This is the region where we require making no turns or very small turns, as the robot is almost facing the goal. The region between these is covered by one MF on each side (*lessp* and *lessn*), which has been cautiously placed around 45 degrees because it was easy for us to visualize the preferable turn around this angle.

Similarly the input *distance* has three MFs: *near*, *far*, and *distant*. The *distant* MF follows the same philosophy as the extreme MFs of the input *angle*. If the distance is large, we would prefer not to make turns so as to avoid sharp turns. It does not make much difference if the goal is so distant that the robot cannot see or so near that the robot might just see; the output is not much affected. The MF *near* covers the region where the goal is so near that we need to make very sharp turns to reach it.

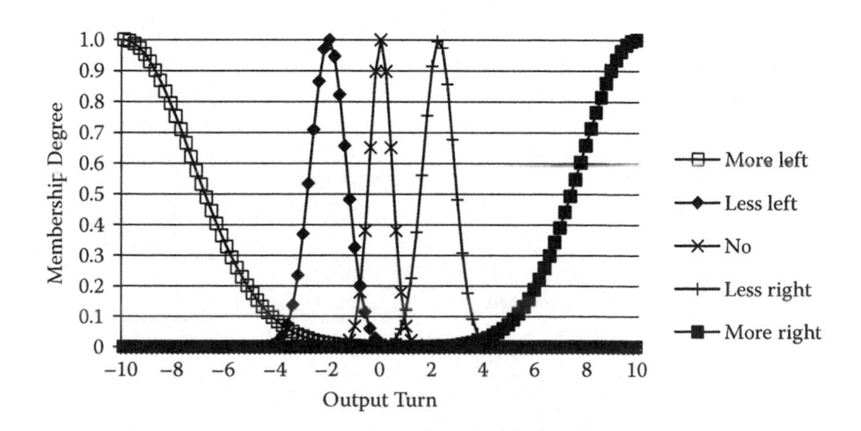

FIGURE 4.30c Membership functions for *output turn* in the robotic control problem.

Rule1: If (α is morep) then (output is morer)
Rule2: If (α is lessp) then (output is lessr)
Rule3: If (α is no) then (output is no)
Rule4: If (α is lessn) then (output is lessl)
Rule5: If (α is moren) then (output is morel)
Rule6: If (α is not morep) and (goal is distant) then (output is no)
Rule7: If (α is not moren) and (goal is distant) then (output is no)
Rule8: If (α is lessp) and (goal is near) then (output is morer)
Rule9: If (α is lessn) and (goal is near) then (output is morel)

FIGURE 4.31 The fuzzy rules for the robotic control problem.

If we do not do so, we might miss the goal. The length of this MF is less that may be attributed to the philosophy of motion. The MF *far* covers cases where the goal is within the comfortable regions.

The output MF is divided into five regions: *more left* (*morel*), *less left* (*lessl*), *no turn* (*no*), *less right* (*lessr*), and *more right* (*morer*). The extreme MFs are larger in coverage, as this avoids too large of turns being taken, which would be undesirable. The location of each MF matches the desirable angle of turn of the MFs of the input angle.

4.11.4 Rules

Rules form the basis of the FIS, because the behavior of the system largely depends on the rules. We use the same understanding and philosophy of inputs that we used to form the MFs. However, although we discuss MFs, rules, and results sequentially in this book, the design of any fuzzy system does not perfectly go in that order. In fact, it happens in iterations of these steps, in which we first form MFs, then frame rules, then simulate the system, and finally see results. Afterward the system's errors and shortcomings are noted, and accordingly the MFs and rules are modified.

We framed a total of nine rules for the system. These rules, which are given in Figure 4.31, may be understood from our understanding of the inputs and the MFs.

4.11.5 Results and Simulation

The model we made was validated and tested by a simulation engine. The general approach followed was that we first calculated the input needed by the FIS according to the present conditions. This was then entered into the FIS to get the angle of the next move. This angle was then implemented in the next move. This procedure was repeated until the goal was reached. This simulation is shown in Figure 4.32.

Based on this simulation methodology, the path traced by the robot for various runs with different initial positions and angles is given in Figures 4.33a through 4.33d. In each figure, the robot is moving upward. Hence the lower point in the path is the initial position and the upper point is the final position. The angle of the robot is the direction in which it initially moves.

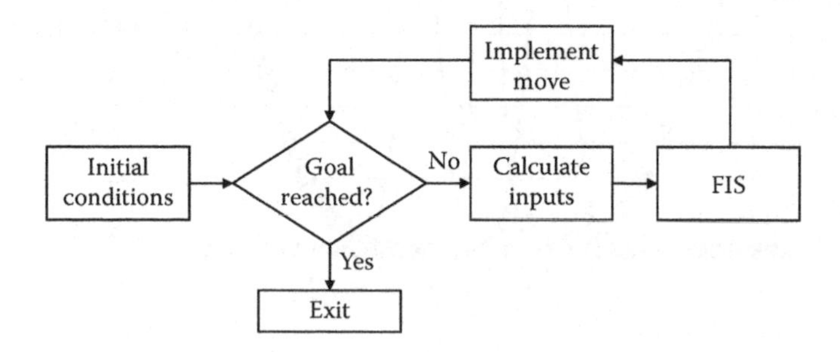

FIGURE 4.32 The simulation process in the robotic control problem.

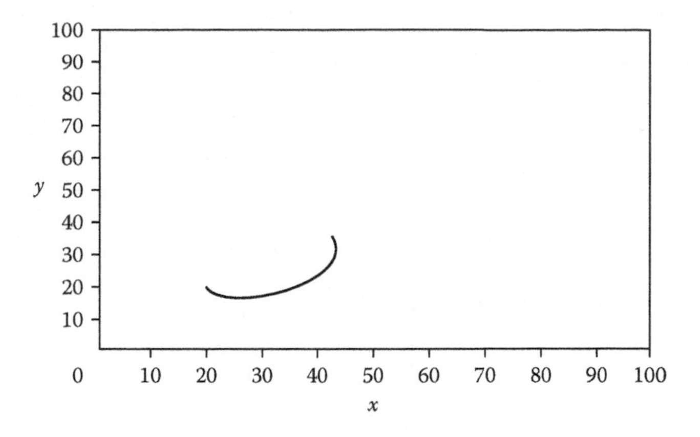

FIGURE 4.33a Path traced by robot at first run.

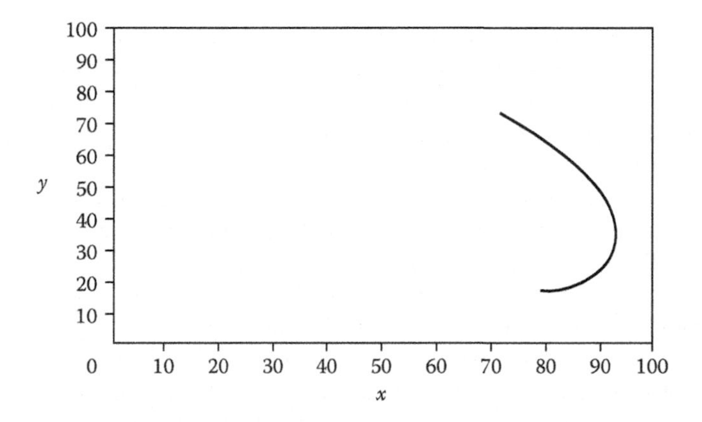

FIGURE 4.33b Path traced by robot at second run.

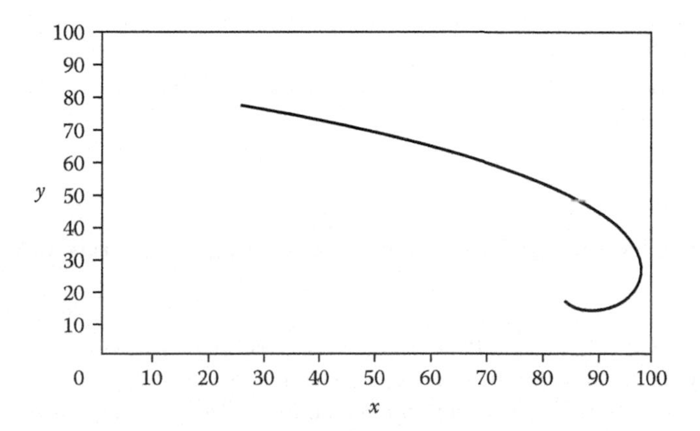

FIGURE 4.33c Path traced by robot at third run.

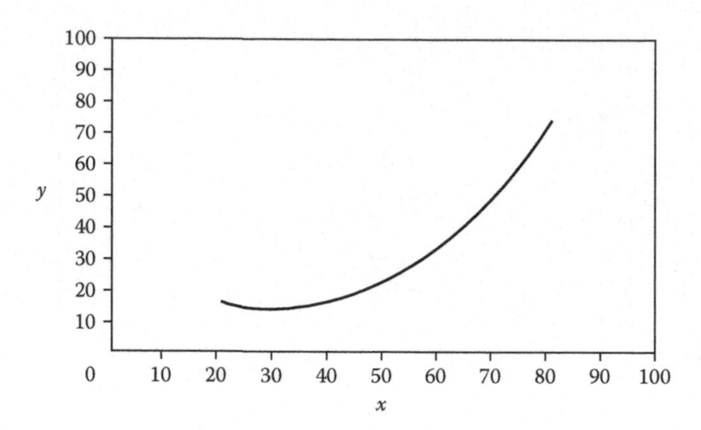

FIGURE 4.33d Path traced by robot at fourth run.

CHAPTER SUMMARY

This chapter presented an in-depth analysis of fuzzy systems. We started our discussion with fuzzy logic, where we discussed the basics of fuzzy logic and its difference from normal logic. Fuzzy and nonfuzzy systems were then compared and contrasted. The loopholes of nonfuzzy systems formed our motivation for using fuzzy systems. Fuzzy sets were another major topic that formed the basic foundation of the chapter.

To develop and understand a fuzzy system, the essential parameters of a rule-based approach and its fuzzy counterparts were presented. This enabled modeling of the fuzzy systems along the lines of a basic rule–based approach that is well understood and commonly used. Fuzzy membership functions enabled us to determine the degree of membership or belongingness of an element to a fuzzy set. Numerous commonly used fuzzy sets were illustrated.

The next topic of discussion was the fuzzy logical operators. Here we studied the fuzzy counterparts of various logical operations, including AND, OR, NOT, and implication. The other fuzzy operators included aggregation and defuzzification. Then fuzzy inference systems were presented. We studied the manner in which FIS maps the inputs to the outputs, as well as other design issues of fuzzy systems.

Another topic of discussion was the type-2 fuzzy system. These systems have been shown to model impreciseness or fuzziness better than the type-1 fuzzy systems. We also moved from the fuzzy sets to the other sets—namely, rough sets, vague sets, and intuitionistic fuzzy sets.

At the end of the chapter, a real life example of a fuzzy controller was built using the fuzzy inference system. This system could move a robot from the known initial position to a final position by making a smooth transition in its path.

SOLVED EXAMPLES

1. Discuss the general methodology of problem solving using fuzzy inference systems (FIS).

Answer: Problem solving in FIS is an iterative process in which we make a model and keep modifying the model until satisfactory results are achieved.

First we must study the problem and decide the inputs and outputs. We discussed the ways and means to do this in Chapter 1. Once the inputs and the outputs are known, the next step is to decide the membership functions. Initially it would be preferable to go with a limited number of MFs, rather than crowding the model with too many. Another important

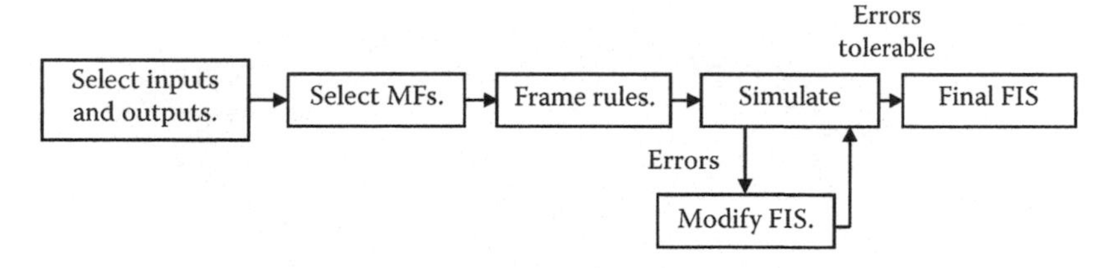

FIGURE 4.34 Problem solving in fuzzy inference systems.

aspect is the placement of the MFs. These are normally placed by developing an insight into the kind of input the system would give or the kind of output that is expected from the system. In the input side, suppose that the lower inputs are very likely and appear most of the time, while the higher inputs are very rare. In such a case, more MFs may be placed at the lower regions. Also suppose that the output is expected to drastically change for small changes in input at the lower inputs. At the higher inputs, however, even a large change in input produces a small change in output. In such a case as well, it would be advisable to place more MFs at the lower region.

The MFs are usually placed at points around which the corresponding output is fairly known. This helps map the input to the output according to the rules. In addition, an MF usually reaches its maxima of 1 when the previous MF reaches its minima of 0. At this point, there is only one active MF with a membership value of 1. This further allows us to easily map the inputs to outputs by rules. Next the rules are framed using common sense.

Once this model is ready, it is tested by the simulation engine, by known inputs, or by common sense. The discrepancies and errors are noted. Now we must modify the model accordingly. Many times the wrong outputs may result from the fact that we did not consider many cases and hence did not frame rules for these cases. In many other cases, the errors may be due to the wrong placement of the MFs. These issues may be fine-tuned according to the requirements. If the errors are not due to either of these reasons, we may consider adding up the MFs and replacing MFs in regions where the output was wrong.

The modified FIS is again tested; this process continues until we get the desired output for all inputs. This process is given in Figure 4.34.

Suppose we already know some inputs and outputs. Reiterating in search of the most optimal solution is a time-consuming step that requires a great deal of patience and energy. These problems require some algorithm to perform these tasks and find the most optimal structure. We will see in Chapter 6 that the neuro-fuzzy systems are an effective way of doing such things. We will also see the application of genetic optimizations as another solution.

2. **Compare an artifical neural network (ANN) with FIS for problem solving.**

Answer: Both ANN and FIS are intelligent systems. They are used to give correct outputs to the inputs presented. The ANN learns from the historical database itself, whereas the FIS must be tuned manually so that it imitates the historical database, if available.

Knowledge exists in every intelligent system. Through this knowledge, the system is able to map inputs to outputs or, in other words, give the correct output for the inputs. This knowledge needs some kind of knowledge representation and usage. In the case of ANNs, this knowledge exists in the form of weights between neurons. In the case of fuzzy systems, knowledge is in form of rules.

ANN training usually happens with great ease in that the rules used to map the inputs to the outputs are simple enough. This training is when the bulk of the data is in agreement by

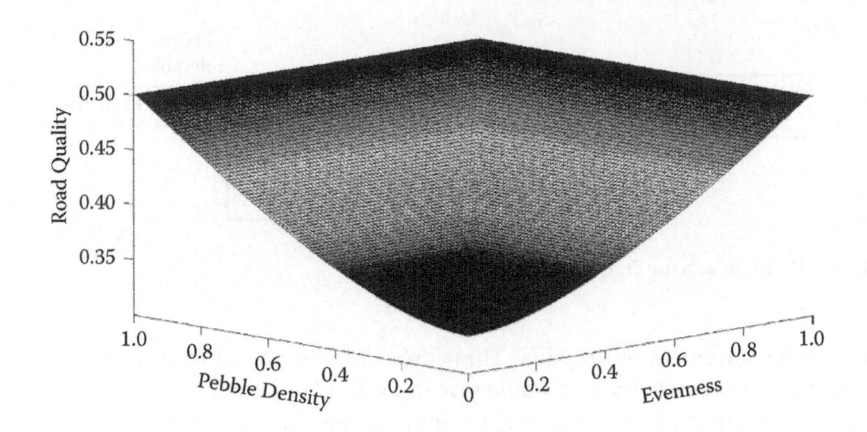

FIGURE 4.35a The surface for Example 3 in the case of FIS with one membership function.

rules, without any great anomalies. If the rules are not simple, however, the data training requires a great deal of effort and may be fulfilled with the addition of neurons. Every kind of data cannot be trained by ANN. We normally try to train the ANN with more neurons or for more time to get a decent performance. But using more neurons or more time may not do well in the testing data due to loss of generalization, as discussed in Chapter 2. In short, the ANN training is simple for most simple and clearly defined problems.

The same is true for the FIS. If an FIS does not perform well, we may make necessary modifications. If we are unable to get high performance, we may add more rules or MFs, though this would not be required in most of the simple problems. Simpler problems would give good performances even with a low number of rules and MFs.

Hence the neurons and layers in ANN are similar to the rules and MFs in FIS.

The FIS however needs a fair idea of the initial rule and an understanding of the system. This helps frame the correct rules, which are very necessary for system. This is not the case with ANN, where rules are automatically extracted.

3. **Suppose that the quality of road Q is measured on a scale of 0 to 1. Two factors affect Q: the average evenness of the road and the average pebble density. Both inputs can be measured by standard practices and are normalized to lie between 0 and 1. Both have an equal effect on Q. Design a fuzzy system for finding Q. Is there any other way to solve the problem?**

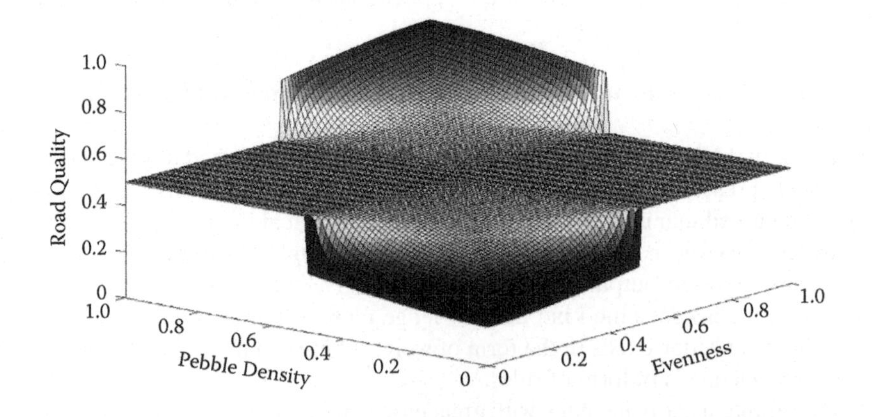

FIGURE 4.35b The surface for Example 3 in the case of FIS with three membership functions.

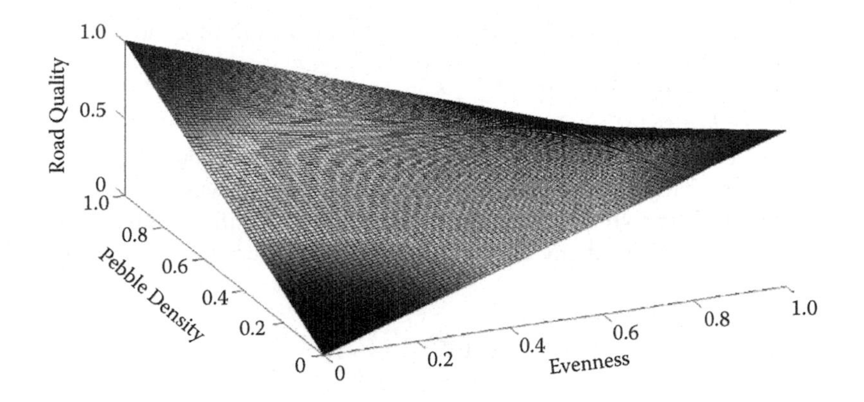

FIGURE 4.35c The surface for Example 3 in the case of nonfuzzy systems.

Answer: Let the average evenness of the road be given by R and the average pebble density be given by P. It is given that both inputs equally affect the output and are equally distributed over their input ranges. The fuzzy model that is generated has one MF per input and output. Let the MF of R be r, P be p, and Q be q. All the MFs are triangular in nature, with maxima at the input corresponding to the input/output of 0 and minima corresponding to the input/output of 1. The system has one rule: "If (R is r) and (P is p), then (Q is q)." The surface model of the resultant system is shown in Figure 4.35a. When we change the number of MFs to three and number of rules to three, the resultant surface is given in Figure 4.35b.

This problem can also be solved by a simple mathematical equation, as given by Equation 4.30:

$$Q = R * (1 - P) + P * (1 - R) \tag{4.30}$$

The plot of this surface is given in Figure 4.35c. Although it is natural that a fuzzy approach would ultimately lead to this equation, using FIS can add computational overheads to the system.

From Figures 4.35a through 4.35c, we can make two major inferences

- Adding rules and MFs makes the plot more complex. The same behavior was exhibited by ANNs.
- Many simple problems can be solved by very few rules and MFs.

EXERCISES

General Questions

1. Compare and contrast fuzzy sets with (a) vague sets and (b) rough sets.
2. What is the difference between binary logic and fuzzy logic?
3. What are membership functions?
4. Compare a production rule-based problem-solving approach with fuzzy logic.
5. What are rules in an FIS?
6. Compare and contrast Sugeno and Mandami type-1 fuzzy logic systems.
7. Compare and contrast Type I and Type II fuzzy logic systems.
8. Suppose $a = 0.05$ and $b = 0.3$. Calculate (with respect to fuzzy logic) (a) a AND b, (b) a OR b, and (c) (NOT a) AND (NOT b). Make suitable assumptions wherever necessary.

9. The following identities hold in Boolean logic. Do they hold in fuzzy logic?
   (a) a OR $b = b$ OR a
   (b) a OR (b OR c) = (a OR b) OR c
   (c) NOT (a OR b) = (NOT a) OR (NOT b)
10. Consider the FIS given in Section 4.7.3. Find the output for the input <0.23, 0.87>.
11. Does the order of firing rules affect the output in FIS? Why or why not?
12. What is the role of defuzzification in FIS? Name a few defuzzification methods.
13. Why is there a need for so many defuzzification methods, when any one of them can be used for the same purpose?
14. Explain type-2 fuzzy inference systems.
15. What do we mean by aggregation in an FIS?
16. What are surface diagrams? Why are they needed?
17. Explain fuzzy controllers.
18. What is the difference between fuzzification and defuzzification?
19. What is a crisp number?
20. What is the role of attaching weights to rules in FIS?

Practical Questions

1. Imagine a car moving toward a wall at some speed. You are supposed to stop the car before it crashes against the wall. Simulate this problem using a fuzzy controller.
2. Suppose the health of a person depends on his age and medical history. Make an FIS of this system. Try using various AND, OR, implication, and defuzzification methods.
3. In the solution generated in Practical Question 1, study the effect of adding or deleting membership functions of each input and output. What is the least number of MFs required for a desirable behavior of the system?
4. Make an FIS that ranks (assigns scores to) cricket-playing nations. What parameters do you consider? Make suitable assumptions wherever necessary.

5 Evolutionary Algorithms

5.1 EVOLUTIONARY ALGORITHMS

Evolution is a novel system that has led to the creation of great systems in the natural and artificial world. The systems we see today show tremendous performance, in part because they have adapted to and evolved from the changing environment over time. Evolution is a predominant phenomenon in most natural systems. In artificial systems, evolution is attributed to the study of the success of this phenomenon in its natural counterparts. We see that these systems keep getting better over time as a result of continuous adjustment, adaptation, and learning.

Looking at any species in the natural world can give us a clear idea of evolution. The most complex species is the result of continuous evolution over time. Evolution started with a simple species. Over time, these species evolved into newer and newer species, each more complex and better suited to the changing environments than the last. In general, each new species performs much better when compared with the earlier ones. This makes it possible for the continuous generation of new species from older ones that keep performing even better. The results are the most beautifully adapted species that we find today. The best example is of human beings, who have outperformed everyone in terms of intelligence.

Inspired by the natural systems, the artificial systems are also trying to learn from evolutionary principles. The systems now model a problem in such a way that the solution evolves over time. This makes possible the generation of various kinds of solutions, which keep improving with time. The artificial systems being built are better adaptive to the changing environment.

Optimization has always been one of the major problems in the field of computation. In these computations, we must optimize the solutions to any problem. If the number of possible solutions were constant, it would make sense to find all the possible solutions and then compare their performances. In reality, however, the number of solutions is so large that any machine would not be able to generate them all in a finite duration of time. The large number of possible solutions has always made researchers think about and come up with newer means of finding the most optimized solutions to problems in a finite duration of time.

Optimization problems always have multiple solutions with varying degree of goodness or accuracies. The best solution is the one that has the most optimal (maximum or minimum) value of the objective function. The objective function is a means of measuring the goodness or accuracy of a solution to these optimization problems. In other words, the aim of optimization is to minimize or maximize the value of the objective function. This is achieved by varying the parameters or variables upon which the objective function depends. Every set of values of variables corresponds to some level of goodness as measured by the fitness function. The final output of the algorithm is the value set of these variables, which gives the most optimal value of the objective function.

Optimization problems are also studied as search problems. The goal is to search for the best possible combination of values of the parameters that gives the most optimal objective value of the objective function. The total number of possible combinations of values for the various parameters is infinite, which is what makes the problem interesting. We must search for a point in the entire world where the extent of the world can be infinite. To better understand the problem, let us understand a space known as *search space*. Imagine a very high-dimensional space that has as many dimensions as the number of variables in the optimization problem plus one. In this space, each dimension corresponds to a variable, and the last dimension corresponds to the value of the objective function for the values of these variables. We take this axis as the output axis, pointing vertically upward. Now we plot the surface of the objective function in this space. This would be a surface in the high-dimensional space, which is the search space. The plotted surface would be filled with valleys and hills as the objective function value increases or decreases. The problem of optimization may now be viewed as the problem of finding the deepest valley in this search space; hence it is a search problem.

Consider the case in which we need to minimize the total revenue of any firm. We are given the various parameters and constraints of the firm. The largest possible revenue that satisfies all the constraints is the best, or the most optimized, solution to the problem.

Evolutionary computation has many applications in the optimization problems, which keep generating solutions with each newer generation obtaining solutions better than those obtained by the previous generation. Some of these applications include robotics, functional optimizations, and character and other recognition systems, to name a few.

5.2 HISTORICAL NOTE

Work in the field of evolutionary computation started in 1950s and the 1960s, when researchers tried to model systems inspired by evolution. In the 1960s, I. Rechenberg introduced *evolution strategies*, a method he used to optimize real-value parameters for devices such as airfoils. Later *evolutionary programming* was developed by L. K. Fogel, A. J. Owens, and M. F. Walsh. Here candidate solutions were represented as finite-state machines, which were evolved by randomly mutating their state-transition diagrams and selecting the fittest. Since then, these machines are being used for the problems of optimizations.

Other work in the field include that done by G. E. P. Box, G. J. Friedman, W. W. Bledsoe, H. J. Bremermann, J. Reed, R. Toombs, and N. A. Baricelli. In the 1950s and 1960s, computer simulations and controlled experiments were used by Baricelli, A. S. Fraser, F. G. Martin, and C. C. Cockerham. In the 1960s, John Holland led to the development of genetic algorithms. Holland studied adaptation to find a means of incorporating this natural adaptation into machines. In 1975, he published *Adaptation in Natural and Artificial Systems*. I. Rechenberg, L. K. Fogel, A. J. Owens, and M. F. Walsh also proposed various models for evolutionary programming. Genetic algorithm today has been greatly modified since it was initially proposed by J. H. Holland.

5.3 BIOLOGICAL INSPIRATION

As discussed earlier, evolutionary computation, or evolutionary algorithms (EAs), is an inspiration derived from its biological counterparts. In this section, we discuss the evolution process in the biological world so we can better understand evolution in the artificial systems by genetic algorithms (GAs), which we discuss in the next section. Note that the strong correlation between the artificial and the natural evolution systems has resulted in many terminologies of the artificial systems being borrowed from the natural evolution system.

The GA is inspired from the survival and reproductive systems of the natural species. A species consists of populations. These populations comprise the set of individuals of the species at that point in time. The capability of the species may be taken as the average capability, or fitness, of all the individuals of that species. An individual, in this case, represents a single entity, or creature, in this population. The different individuals in the population sexually interact to generate newer individuals. Thus, the sexual interaction among the individuals of a population results in a newer set of individuals of a newer population. In natural systems, the generation of newer populations is in the form of generations. A generation sexually interacts to generate the next generation. It is generally observed that the average fitness of the newer generation is better than that of the preceding generations, and thus the newer generations can better adapt to the changing environment, as theorized by Charles Darwin's survival of the fittest. His theory states that only the traits of the fittest individuals pass from one generation to the next, while the traits of the others die out. In other words, the better characteristics are retained from one generation to the next, whereas the poorer characteristics are killed in the process.

Any species is internally composed of cells. These cells are, in turn, made up of chromosomes. The chromosomes, which are strings of DNA, are the individual's identity. The chromosomes and DNA decide the behavior, characteristics, and other properties of that individual. Internally every chromosome is divided into genes. Each gene encodes a particular region. Genes may be combined in numerous ways to generate various kinds of chromosomes or features in the individual. The different kinds of encoding for a gene for a trait are called *alleles*. Each gene is located at a particular locus, or position, on the chromosome. Many organisms have multiple chromosomes in each cell. The complete collection of genetic material (all chromosomes taken together) is called the organism's *genome*.

Two more important terms are *genotype* and *phenotype*. *Genotype* refers to the particular set of genes contained in a genome. The genotype decides all the characteristics of the individual. If two people have the same genotype, they may be considered identical. The *phenotype* consists of the set of observed characteristics of the individual. The genotype decides these characteristics, whereas the phenotype is a collection of observed features in any order or manner or representation.

Most sexual organisms have a paired chromosome structure called a *diploid*. During sexual reproduction, there is a recombination of chromosomes from the two parents. This recombination is also referred to as *crossover*. In this step, exchange of genes takes place between the two parents to form a gamete, or a single chromosome. These gametes pair up to create a full set of diploid chromosomes. There is often a change of nucleotides from the parents to the child. This change is referred as *mutation*. Normally the change is the result of copying errors.

Another concept in the natural systems is that of *fitness*, or the ability of the individual to survive in the entire population. A high-fitness individual means a high chance of survival. Fitness is also a measure of the individual's reproductive power, or fertility.

5.4 GENETIC ALGORITHMS

In this section, we give a brief overview of the genetic algorithms. Much of our discussion is restricted to the traditional GA, commonly known as the simple genetic algorithm (SGA). The GA is an effective way to solve real life problems based on evolutionary ideas. It is the most commonly used technique of evolutionary algorithms. In this section, we study the basic algorithm and its working. The details of the various steps are then discussed in the rest of the chapter.

5.4.1 Concept

We talked much about the concept of GA when we discussed its biological counterparts. When we compare the biological systems with systems of the SGA, we find that many of the concepts are the same.

The GA starts with a set of solutions or populations. Each individual in the population set represents a solution to the problem. The nature of the solution, however, varies. Some solutions may be very good, while others are very bad. The goodness of the solution varies from individual to individual within the population. To measure the goodness of the solution, we use a function called the *fitness function*, which must be made according to the program logic.

Initially the population set contains all randomly generated solutions that are initially generated. This is called the *first generation* of the population. Then the various genetic operators are applied over the population. The genetic operators help in the generation of a new population set from the old population set. The new population set is formed by the genetic reproduction of the parents to form the children. Various other genetic operators are also applied that help generate the new population set. These specialized operators help in better optimization of the solution.

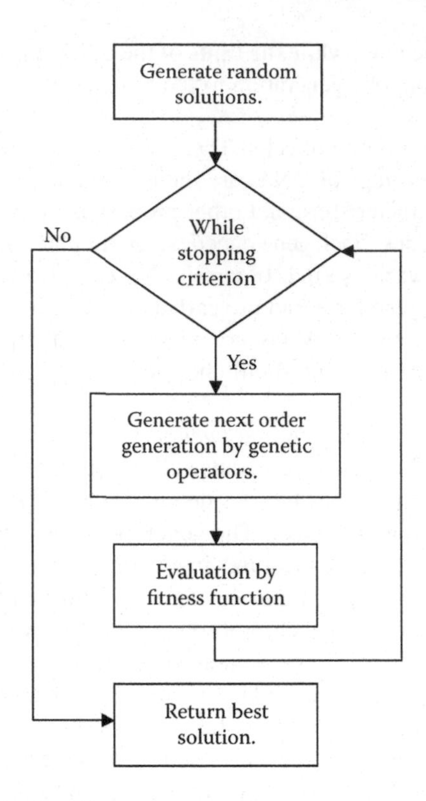

FIGURE 5.1 The simple genetic algorithm.

In this manner, we generate higher-order generations from the lower order. The fitness of any solution or individual may be measured by the fitness function. A higher value of the fitness function means better solutions. The fitness of the population is measured by the average fitness of all the individuals that make up the population. It is observed that the fitness keeps increasing as we move forward with the algorithm. The higher-order generations are more fit when compared with the lower-order generations. This is true for all best fitness individuals in the population, average fitness in the population as well as the worst fitness in the population. The GA results in improvisation as the generation proceeds. This improvisation is very rapid in the first few generations. After that, however, the algorithm more or less converges to some point to give the most optimal solution to the problem.

The general algorithm of GA is given in Figure 5.1. We discuss each step of the algorithm, as well as the other details of the algorithm, in the next subsections. The basic algorithm follows an iterative approach and continues to generate newer generations. For the most part, the solution keeps improving with time until it finally converges or the improvement in the solution is unnoticeable. There must be a stopping criterion at which the loop terminates. This step may be based on various criterions that we will study later.

The iterative nature of these algorithms is a very important aspect, especially with respect to soft computing. We know that the search space is infinite and cannot be explored to generate solutions; the time is always infinite. The beauty of this algorithm is that it finds the best point in this infinite search space as early as possible. The amount of search space traversed in the algorithm's run time may be a very minute proportion of the total search space. The iterative approach is thus a very useful tool that gives valuable solutions per the time constraints.

5.4.2 Solution

The GA is applied to the optimization problems, in which we are required to optimize an objective function by varying some variables or parameters. The real life problem first must be converted to this optimization problem before GA is applied over it. Whenever we talk of *problem* in this section, we are referring to this optimization problem. Solution is an individual in the population set that represents a set of parameters or variables of the problem to be optimized. The solution may be good or bad, depending on the fitness measured by the fitness function.

Consider the problem of functional optimization, in which we are given a set of input variables. We have an objective function that we must optimize, which is the goal of the problem. In addition, we have a set of constraints that every solution must follow. An example of this problem is given by Equation 5.1. In this problem, a solution may be represented by two values, one corresponding to the value of a and the other corresponding to the value of b. The general form of any solution may be taken as (a, b).

$$\text{Maximize: } a * b \tag{5.1}$$

Constraints:

$$0 \leq a, b \leq 5$$

$$a + b > 5$$

The solution may be feasible or infeasible. Infeasible solutions usually disobey some constraint of the problem and are hence not valid solutions. Infeasible solutions cannot be returned as the final solutions by the GA and are usually not accepted in the solutions at all. Thus whenever these solutions try to enter into the population set, they are rejected. We rather repair them to make them feasible and then allow them to enter the population set. Another possibility for dealing with infeasible solutions is to assign them a very low fitness value. According to the theory of survival of the fittest, it may be assumed that they would then die out with the passage of time. In the above problem, the solution (2, 10) is infeasible because it does not obey the constraint $0 \leq a, b \leq 5$.

So far we have represented the solution in an easy-to-understand format. This type of solution, known as the *phenotype solution*, may be used for computation of fitness or other comparisons or operations. The GA, however, may require a different representation of the solution, in this case the *genotype solution*, to perform and function.

The genotype solution is usually a sequence of bits or numbers. Each bit may be either 0 or 1. The whole phenotype solution is represented in such a way that we may represent it in the form of continuous bits. The entire bit sequence is the solution in the genotype form. The conversion of the solution into this format of bit string comes from the classic theory of GA. The various genetic operators were designed assuming the individuals were a series of bits. We shall later see that representation of the solution in the form of a continuous series of numbers (not necessarily binary) also works and often gives better performance.

In our example, we represented a solution in the form (a, b), where each a and b can take any value between 0 and 5 (6 values in total). To represent a, we require a minimum of three bits ($2^3 = 8 \leq 6$). Similarly to represent b, we require at least three bits. The entire solution may hence be represented as a sequence of six bits, with first three bits denoting a and the next three bits denoting b (see Figure 5.2). The solution (2, 3) is represented as $a = (010)_2 = 2$ and $b = (011)_2 = 3$.

Another way of representing the solutions is directly by their numbers. The other genetic operators can be modified to handle the solutions with a numeral representation, which saves the

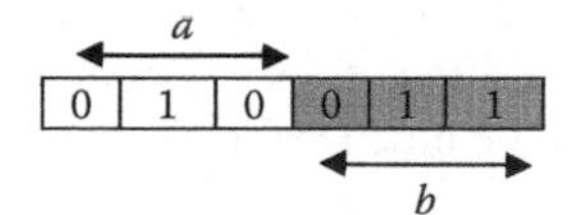

FIGURE 5.2 The genotype representation of a solution.

overhead of continuously converting binary to decimal and vice versa. We will see in the later sections some specific genetic operators for this type of implementation that are widely used for their ease of handling.

In general, for any problem that we need to optimize with the help of GA, we first must represent it in the form of a solution. Once this has been done, we can apply the various genetic operators to optimize and find the most optimal solution. The use of the other GA operators is quite a trivial task. Hence one of the major design and implementation issues is a good solution design that deals with the manner in which we convert the solution to a sequence of bits or numbers.

The solutions may take different representations. If we try to train an ANN by GA, the solution represents the different weights and biases if the structure is kept constant. If the network is allowed to change, we must work to again with some means of network representation. In many cases, the solution may be represented in the form of a graph. Similarly in many game-playing problems, the solution is in the form of two-dimensional tables. Once the phenotype has been represented, it should be easy to convert it into a genotype. We will look into the diverse means of representing the solution in the next units, where we discuss real life applications.

5.4.3 Initial Population

One of the first steps in the algorithm is to generate an initial population, or the starting population that ultimately leads to the generation of newer and newer generations. Mostly the initial population is generated randomly. We know the total length of the solution's genotype representation. The standard practice is to generate random series of 0's and 1's to fill this solution. This generates highly random solutions. Due to the random nature, all kinds of solutions are generated, some with high fitness value and some with poor fitness value.

Imagine you are asked to hunt in some area for treasure that has been hidden by prehistoric people. You may take one of two options to find the treasure. The first is to delegate a large number of people. This step is important because you need a large number of people to go to the diverse areas of the region where the treasure may lie. A large number of people will be able to cover a very large area at once, thus maximizing the search operation. But we also know that hunting for such treasures requires a lot of cooperation and coordination. Hints, clues, and findings need to be shared among all searchers. This information helps guide each person hunting for treasure to think of the likely places where that treasure lies and then to hunt at that location. This is where the very large number of people creates a problem. A great deal of time will be wasted in communicating among the hunters. They may not be able to guide each other effectively. For this reason, you may often wish to go with the second option. This is to use smaller, well-coordinated groups.

Recall our notion of the search space, which we introduced earlier in this chapter. The problem is to search for a point in this highly dimensional graph of infinite possibilities. The various individuals are, in simple terms, the agents (the hunters) who try to find the global minima (the treasure). The hints are in the form of fitness values of the individuals. Say the fitness value where an individual is delegated is very poor. This naturally means it is not a good area in which to search. In our analogy, it is as if some hunter had said, "This area is bad; please do not come here." Coordination is in the form of interactions among the individuals and the global planning that we would see later. Based on this analogy, we must try to analyze the role and importance of individuals and population size.

The number of individuals in the population, or the population size, must also be specified. A very high value would result in a lot of computation in generating one population to the next, but it would give a lot of scope to the individuals to generate all kinds of diverse solutions. This may even result in early convergence or the generation of very good solutions early. If, however, this number is kept very low, the generations would be much less time consuming to calculate. However, it would not be able to support high diversity.

The initial solution does not play a very vital role in the GA's solution generation, as long as the initial population contains individuals of all kinds. Each bit, in some way, represents or contributes to characteristics of the solution. These characteristics may be good or bad. Good characteristics would make the fitness value of the solution high, and vice versa. The entire working of the GA tries to generate different combinations of these characteristics. If some characteristics that may be very good for the solution are completely missing from the population, optimization would suffer. Hence having all kinds of solutions in the population is the prime requirement for any GA to work effectively.

This idea can be understood by looking at our analogy of the treasure hunt. The various hunters must collectively cover the entire search space as much as possible to ensure that no part of the area goes unnoticed. Every attempt, or possibility, of hunt must be exploited. A very serious problem occurs if a large area does not have a hunter in the vicinity. It might be possible that this was the place the treasure, or the global minima, was present. Thus individual characteristics are analogous to the area or region in the search space where the search operation is happening.

5.4.4 Genetic Operators

The conversion of one generation to the next is done by means of genetic operators. In this section, we briefly look into the various genetic operators and the means by which conversion takes place. In general, the conversion to a higher generation improves the population's fitness, although there is no assurance of the same. The conversion is guided by the natural system.

1. **Selection:** The first operator applied is selection. Here we select the pairs of individuals that will reproduce to generate newer populations. Selection forms pairs of individuals that will later cross-breed to generate individuals. Selection is an important step, because it decides the parents that are responsible for the individuals being generated. Thus a number of strategies for the selection process have been developed. All try to cross-breed good solutions in the hope of generating good individuals in the resulting populations.

   Selection is done on the basis of the individual's reproductive capability, or fitness value. An individual having a higher fitness has a high reproductive capability and thus is likely to be selected numerous times. The weaker individuals, on the other hand, may not be selected at all for reproduction. This is also prevalent in natural systems, in which a powerful animal mates more often than does a weaker one. This selection is desirable because the more that animal mates, the more offspring it would produce, resulting in a larger proportion of this individual's offspring in the entire population set. Because those offspring come from a powerful parent, they will be strong. Hence the next generation represents a very powerful community as a whole. This same process is imitated by the selection operator.

   Selection is shown in Tables 5.1a and 5.1b, in which a random selection has been applied.

TABLE 5.1A
Population Pool

Phenotype	Genotype
(1, 1)	001001
(2, 1)	010001
(3, 2)	011010
(2, 2)	010010
(2, 4)	010100
(4, 2)	100010

TABLE 5.1B
Selection Operation

Phenotype	Genotype	Phenotype	Genotype	Crossover Point	Ind. 1 Phenotype	Ind. 1 Phenotype	Ind. 2 Genotype	Ind. 2 Genotype
(1, 1)	001001	(2, 1)	010001	2	000001	(0, 1)	011001	(3, 1)
(2, 4)	010100	(3, 2)	011010	5	010100	(2, 4)	011010	(3, 2)
(2, 2)	010010	(4, 2)	100010	4	010010	(2, 2)	100010	(4, 2)

2. **Crossover:** Selection results in the formation of pairs of parents of the entire population set. The next operation they need to perform is crossover, or the genetic operator that results in the mixing of two parents to generate individuals of the new population. Crossover is done by taking some bits from the first parent and other bits from the other parent. One of the commonly used crossovers is the single-point crossover, in which we randomly select a point in the chromosome. The bits to the left of this point are taken from the first parent and the bits to the right are taken from the other parent. This generates new individuals in the new population.

   Crossover is motivated by the intrinsic hope to form offspring that takes the best characteristics from the two parents. It is hoped that the new individuals generated from the crossover will take the better characteristics from the parents and leave out the bad ones. In the natural systems, this crossover happens when children are brighter than their parents. Note that out of the entire pool of individuals, only one may find the global minima. This would serve our purpose; it would also attract the other individuals to come toward minima. Hence even if the probability of such a phenomenon to occur is very small, it would be worth taking a try. As an example, of the billions of people in the world, we have only a few Nobel Laureates. These people are strong enough to inspire and drive the rest of us. It may, however, be noted that the various characteristics are always independent entities here. It is not as easy as saying that the child takes good eyesight from the mother and intellect from father. Rather they are interdependent, and only the combination of their values decides the final fitness of the newly generated individual.

   Using our example of functional optimization, the crossover of two random parents is shown in Figure 5.3. The crossover in this case is applied after the second position. The exchange of bits can be seen in the figure. The parents in this case are (2, 3) and (1, 0). The crossover across the second position yields the solutions (3, 0) and (0, 3).

   The result of applying the crossover to the selected pool given in Table 5.1b is given in Table 5.2. Here each pair contributes two individuals to the new population. Each crossover operation has a crossover being applied at some distinct crossover point.
3. **Mutation:** Crossover results in the mixing of characteristics between two individuals. Many characteristics may be completely absent in the solution pool, even if many of those

TABLE 5.2
Crossover Operation

Phenotype	Genotype	Phenotype	Genotype
(1, 1)	001001	(2, 1)	010001
(2, 4)	010100	(3, 2)	011010
(2, 2)	010010	(4, 2)	100010

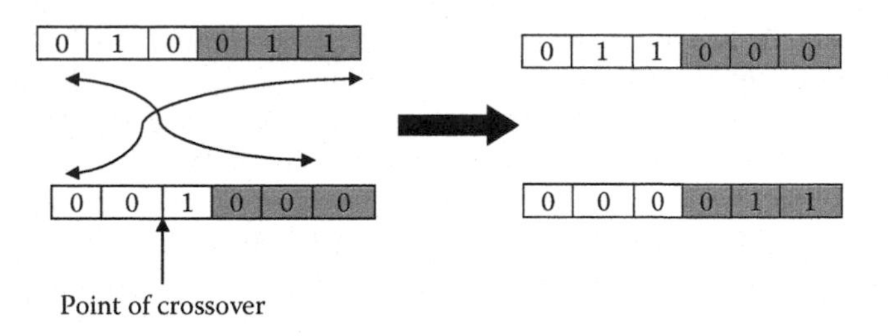

FIGURE 5.3 The crossover genetic operator.

missing characteristics may have been desirable for good solutions. This absence may be due to the fact that these characteristics were never generated, or that these characteristics somehow got deleted. The term *characteristic* here refers to the bits in the genotype representation. These characteristics need to be introduced into the population. Mutation is the process by which we try to introduce random characteristics into the population. We do this done by flipping bits. A random bit in the individual is selected and is simply swapped during mutation. Thus if it represented 1, it now represents 0, and vice versa. This mutation is the characteristic that we have artificially introduced into the system. In general, if this characteristic is good, it will stay in the solution and survive. If the characteristic is bad, however, it will die out over time.

This whole process is like an experiment. In this case, we saw the population pool comprising the individuals. We made an inference that some type of characteristic may be absent in the present population pool, so we artificially removed an individual, genetically infused the characteristics into it, and then put it back in the population pool. Now we make experimental observations: This artificially synthesized individual interacts with the other individuals and tries to pass the inserted characteristic. If the characteristic is good, then all the individuals benefit, and the new characteristic becomes a demand in the population pool, and all the new individuals in later generations further spread the new characteristic. In all, the experiment is termed successful. However if the characteristic were bad, it would decrease the fitness of the artificially synthesized individual. As a result, the reproductive capability of that individual would suffer, and it would not be able to transfer the characteristic much. If it were able to pass the characteristic, the fitness of the new individuals would also reduce. Eventually all these individuals would die out over time, leaving the other, fitter individuals to dominate the population pool.

This mutation is given in Figure 5.4, where it may be seen that (2, 3) transforms to (3, 3).

4. **Other genetic operators:** There are various other genetic operators that are used for various specialized purposes. Each aids in the generation of a good population set. Each genetic operator, in its own way, contributes to optimization by adding the possibility of the generation of good solutions. We will study these other genetic operators in the next section.

Now we may clearly understand the communication and coordination that we mentioned in our analogy of the GA to the treasure hunt operation. A hunter at a site that is highly likely to be the site

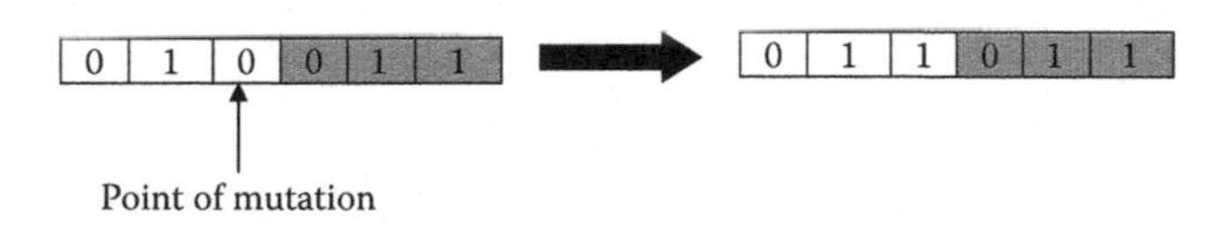

FIGURE 5.4 The mutation genetic operator.

where treasure is found must communicate this fact to the others. Hunters at locations that are likely not the site of treasure might naturally be very interested in joining this first hunter and helping him search his site. This is analogous to crossover. At the same time, some hunters might consider roaming about and searching for new areas. These also need to communicate their findings to others in case they happen to explore something interesting. This is analogous to mutation. The selection is nature driven. Because every hunter is updated about the location and hints of every other hunter, they are all always attracted to go to the best possible place discovered so far. This is because of the high likeliness of the treasure at that location. Sooner or later, they might all end up searching for the treasure at the same place, because they would have collectively unveiled enough information to make the probability of treasure very high at some place. This would be a strong attraction to any hunter.

5.4.5 Fitness Function

The quality of the solution or of any individual is measured by the fitness function. This function decides how good the solution is. In general, a good solution will have a high value of fitness function, whereas a bad solution will have a low value. Sometimes the converse relation may also be applied. Using the fitness function, we may be able to rank the solutions from good to bad, which is very helpful in the selection operator. The fitness function gives a quantifiable way for us to measure the solutions and the behavior of the algorithm. The fitness function is hence very helpful for knowing how the algorithm is behaving over time and over generations.

Based on the fitness of the solutions as given by the fitness function, we are generally interested in three kinds of fitness. The first is the best fitness. The best solutions have the best value of the fitness function out of the entire population set. The best fitness gives us an idea of the algorithm's performance or of the most optimal solutions so far. At the time that the stopping criterion is met, the best solution is used as the final solution of the GA. The second type of fitness is the average fitness, which gives us an idea of the average solution in the population set. The third type is the worst fitness, which is the worst solution in the entire population set. In general, all three fitnesses increase (or optimize) with time. Toward the end, they usually attain some optimal value and converge.

As an example, the objective function $a * b$ may be taken as the fitness function. The larger the value of the fitness function, the better the performance. This would make the solution better. The fitness values for some individuals are given in Table 5.3. In this example, we have assigned a value of –1 to all the infeasible solutions.

5.4.6 Stopping Condition

The GA proceeds in an iterative manner and runs generation by generation. This procedure goes on and on, and the solutions generally keep improving. However, we must be able to decide when we

TABLE 5.3
Fitness Value of Some Individuals

Phenotype	Genotype	Fitness
(1, 1)	001001	–1
(2, 1)	010001	–1
(3, 2)	011010	6
(2, 2)	010010	–1
(2, 4)	010100	8
(4, 2)	100010	8

would be willing to terminate the program. This is done by the stopping criterion, which decides when the program will be terminated. The best solution at that point would be regarded as the final solution.

There are numerous ways to set the stopping criterion:

- If the solutions being generated are sufficiently optimized and further optimizations may not be helpful, we may wish to stop the algorithm. This is when the goal that was set before training has been met.
- If the time taken by the algorithm or the number of generations exceeds a certain number, we may wish to stop.
- If the improvement by subsequent generations or time is too low and the solution seems to have converged, we may choose to stop the algorithm. This is also known as *stall time* or *stall generation*.

5.5 FITNESS SCALING

We discussed the selection operator in the previous section. Selection tries to select the parents that can generate good solutions in the next generation population. Selection relies heavily on the fitness values of those individuals. The solutions with higher fitness need to be selected more often than the solutions with lower fitness values. This necessitates a fitness function to help in selection. However, fitness-based selection has some problems. It is likely that in such systems, the higher fitness individuals be selected in very large numbers. This would mean the new generation would be dominated by a few individuals, and the convergence would be very fast. In practice, however, we may only get local minima and not global minima from this system. These problems with fitness give rise to the need for additional methods that can be used for effective selection.

Imagine that in a population, the most powerful animal mates so much that the world becomes filled with the offspring generated by it. The high fitness of these offspring would make other individuals die, by the rule of survival of the fittest. The offspring generated would be highly similar, because they come from similar parents. This would kill the individuals that had good characteristics but were still in the improving phase. In the analogy of hunting treasure, this would mean that a hunter got so excited that he called all others to hunt at a specific area. Now everyone is hunting at a place that is good but may not be the best; the best place might go unexplored.

Scaling is one such mechanism for effective selection. In scaling, the individuals in the population are rescaled according to their fitness values. The scaling of fitness values gives rise to the expected value of the individuals. This new value will then help the selection operator to select the parents. The expected value denotes the fitness of the solution in the new scale.

In this section, we take a brief look at some of the commonly used fitness scaling methods. All of these methods take the fitness value of the solutions and scale them according to the needs.

5.5.1 RANK SCALING

In rank scaling, the individuals in a population are ranked according to their fitness value. Each individual is given a rank from 1 to N. The expected value of any individual is on the basis of rank. The expected value of any solution in this method is given by Equation 5.2:

$$ExpValue(i,t) = Min + (Max - Min)\frac{rank(i,t) - 1}{N - 1} \tag{5.2}$$

where $ExpValue(i, t)$ denotes the expected value of individual i at time t; Min is the minimum most expected value, or the value for the solution with first rank; Max is the maximum most expected value, or the value for the solution with rank N; and N is the number of individuals in the population.

In equation (5.2), we are basically normalizing the ranks between a Min and a Max value. This may be selected based on the data and the problem under consideration. Rank scaling rescales the individuals according to their rank.

5.5.2 Proportional Scaling

Proportional scaling scales the individuals directly on the values of the fitness function. The expected value is directly proportional to the individual's fitness. This may be represented by Equation 5.3.

$$ExpVal(i, t) = \alpha\, fit(i, t) \tag{5.3}$$

where α is any constant and $fit(i, t)$ is the fitness value.

Proportional scaling has problems when the fitness of the individuals is not in a very good range. It also has problems when the fitness value of a few individuals is comparatively very large, which causes the next generation to be dominated by a few individuals and leads to fast convergence, as discussed earlier.

5.5.3 Top Scaling

Top scaling selects a top few individuals and scales them to the same value of $1/N$. All other individuals are scaled to a value of 0. In other words, only the top specified number of individuals is considered capable of participation in selection. The rest are rejected during scaling. All the selected individuals have equal probability. The number of individuals selected is specified as some proportion of the total population α. This scaling method is given by Equation 5.4:

$$ExpVal(i, t) = \begin{cases} 1/N, & \text{if } i \text{ is within } 100 * \alpha\% \text{ of individuals in the population} \\ 0, & \text{otherwise} \end{cases} \tag{5.4}$$

5.6 SELECTION

We discussed selection and its role in GA in the previous section. The basic objective of selection is to pair up individuals in the population for crossover. Numerous selection techniques are available for this purpose. A good selection technique maximizes the chances of selection of good individuals. In this section, we focus our attention on some of the commonly used selection techniques.

5.6.1 Roulette Wheel Selection

In roulette wheel selection, the probability of selecting any individual is directly proportional to that individual's expectation value. Hence the more fit solutions have a greater chance of being selected again and again. This makes the fitter individuals participate more and more in the crossover. The weaker individuals get eliminated from the process. The simplest way of implementing this is by the roulette wheel. Each individual in the population represents a section of the roulette wheel. The higher the individual's fitness, the larger its area on the roulette wheel, thus increasing the probability of the selection of this individual. To perform selection, the roulette wheel is spun N times, where N is the number of individuals in the population. After each spin, the reading of where the wheel stops is noted. The wheel is naturally more likely to stop at individuals whose fitness value is large.

For implementation purposes, the roulette wheel is implemented by first summing up all the expectations of all individuals. Let this sum be T. We then select any random number r between 0 and T. We move through each and every individual and sum their fitness. We keep moving until the summated value is less than r. The final selected value is the one that makes the sum exceed r, as given below:

Roulette Wheel (*P*)

$T \leftarrow$ Sum of fitness of all solutions
$r \leftarrow$ Any random number between 0 and T
$s \leftarrow 0$
for every individual p in population P
do
  $s \leftarrow s + fit(p)$
  if $s \geq r$
  then return p

5.6.2 Stochastic Universal Sampling

The roulette wheel implementation faces one problem: It is likely that the high-fitness individuals will keep getting selected, and the lower ones will not get selected at all. In such a case, the population would result in only a few solutions being repeated over again and again, which would kill diversity. To prevent this situation, we have another reimplementation of the roulette wheel. In this type, there are N selectors in place of one. This makes it possible to simultaneously select N individuals. The wheel is rotated only once, and the N populations are selected. This implementation is called stochastic universal sampling (SUS).

```
ptr = Rand(); /* Returns random number uniformly distributed in [0,1] */
for (sum = i = 0; i < N; i++)
    for (sum += ExpVal(i, t); sum > ptr; ptr++)
        Select(i);
```

(Mitchell, 1999)

where *ExpVal*(i, t) denotes the expected value or the fitness function for the individual i at time t.

5.6.3 Rank Selection

Rank selection is similar to that of the roulette wheel, except that the individuals are scaled according to their ranks. The higher fitness solutions have a higher chance of being selected. However, this method removes the dominance of a few high-fitness solutions, as might be the case with the roulette wheel. In roulette wheel selection, it is possible for the entire selected population to consist of only a few individuals that get selected again and again. Using rank selection, however, even if some individual has an exceptionally large fitness value, it would not be selected again and again, because the selection is on the basis of rank that are uniformly distributed rather than fitness value.

5.6.4 Tournament Selection

Tournament selection further solves the problem of a solution being excessively selected and dominating the population. This method of selection is inspired by the tournament, or the competitive method. In this selection, we first select any two random individuals from the population. We also select a random number r between 0 and 1. If r is found to be greater than a predefined number k (approximately 0.75), then the lesser fit individual is selected; otherwise the individual with the higher fitness is selected. This is given below:

Tournament Selection (k, P)

$i_1, i_2 \leftarrow$ any two individuals from population P
$r \leftarrow$ any random number between 0 and 1
if $r > k$
    if $fit(i_1) > fit(i_2)$, return i_2; else return i_1
else
    if $fit(i_1) > fit(i_2)$, return i_1; else return i_2

We may even generalize this algorithm by taking more individuals in the competition. In such a case as well, the fittest individual is selected with some high probability. The number of individuals to be selected for the tournament per selection is called the *tournament size*, and this number must be specified.

In this case, we are selecting candidates and making them compete with each other. The competition is performed on the basis of fitness of the individuals. Normally the higher fitness individual wins. However, there is also some probability for the reverse to happen, which makes it possible for the higher probability solutions to be present in larger number. The lower fitness solutions are eliminated in the next generation in the same way. Because the population is not dominated by some particular solution, the rapid convergence does not happen, thus making it possible to reach a global minima.

5.6.5 Other Selection Methods

Based on the same principles, various other selection methods have also been proposed. These methods perform the selection process in their own way. Some of the other commonly known methods are steady state selection, remainder selection, random selection, and uniform selection.

5.7 MUTATION

Mutation is the addition of new characteristics into the population. With mutation, we randomly try to inject newer characteristics into the population. If these characteristics are good, they will survive and continue into the next population. If they are undesirable, however, they will be eliminated in subsequent iterations. The mutation results in adding newer areas to the GA's search space.

Mutation is carried out on a very few individuals, because these are random characteristics that can drive the population anywhere. If the number is kept very high, the whole GA would behave almost random in nature, because all individuals would be deformed by the mutation operation. Too much randomness in the operation and the direction of the crossover and other operators would be lost.

The amount of mutation to be carried out is specified by the mutation rate r_m, a number between 0 and 1. This number is usually kept low to keep a limited mutation in the algorithm. We iterate through each entity in a solution in the population. Each entity has a probability of r_m to be selected for mutation. The mutation operator is applied over the selected entity sets. In this section, we discuss the different mutation strategies that are applied by GA.

5.7.1 Uniform Mutation

Uniform mutation is the most common mutation operation. In this operation, the value at the selected location is simply flipped. Thus if a 0 is present in the location, it changes to 1, and vice versa.

However, if the solution is represented in the form of numbers, the way to handle the mutation operation is somewhat different. In this operation, the number at the selected location is replaced by a random number. The random number selected must lie within the allowed range.

5.7.2 Gaussian Mutation

Gaussian mutation happens in the case of the numeral representation of the solution, where the solution is represented by simple numbers. In this type of mutation, we add a randomly generated number to the existing value at the selected location. The random number is taken from the Gaussian distribution with a center at 0. This normally generates very small values of either sign, which makes small changes to the solutions. Sometimes big changes are also produced, however, that cause a big change in the solution.

5.7.3 Variable Mutation Rate

The mutation rate may be changed during the running of the algorithm. This is known as a variable mutation rate. This mutation is sometimes used to control the behavior of the algorithm. A high mutation rate, as discussed earlier, is the sign of randomness, which might be a good phenomenon at the start of the algorithm, when we are trying to explore newer and newer possibilities. At the end of the algorithm, however, the mutation rate may be kept small to allow the algorithm to converge. In this manner, we may vary the mutation rate according to some principle.

Other details about the use of mutation and the other practical issues associated with it are given in Chapter 19.

5.8 CROSSOVER

In crossover, we mix two parents to generate a new individual by taking some genes from the first parent and some from the other. The crossover operation, as discussed earlier, tries to generate fitter solutions by combining the characteristics of the parents. Crossover is applied in various ways, as we discuss in this section.

Crossover also has an associated probability called the *probability of crossover*, or p_c. This probability decides the total percentage of populations that will go to the next generation as a result of the crossover operation. This probability is usually kept high because crossover has much less randomness in it. The solutions in crossover are more or less predictable and along the lines of the parents. Hence this probability is kept high to allow the algorithm to converge at the correct time. The rest of the individuals in the new population come from mutation or other genetic operators as discussed elsewhere in this chapter.

5.8.1 Single-Point Crossover

The single-point crossover is the simplest type of crossover. We select any random number between 0 and n, where n is the length of the chromosome. This point is called the crossover point. The genes to the left of the crossover point are taken from the first parent, while those to the right are taken from the second parent to form the new individual.

If the solution is present in the form of phenotype, the same analogy may be used. This crossover was presented in Figure 5.3.

5.8.2 Two-Point Crossover

Many times the one-point crossover does not give very good results. In such instances, the two-point crossover is used instead. In this crossover, we select two crossover points in place of the one that was used in one-point crossover. The new solution is made by taking genes from between the two crossover points from one parent and the rest of the genes from the second parent to generate the new individual. This type of crossover results in passing on more diverse characteristics from the two parents. The two-point crossover is shown in Figure 5.5.

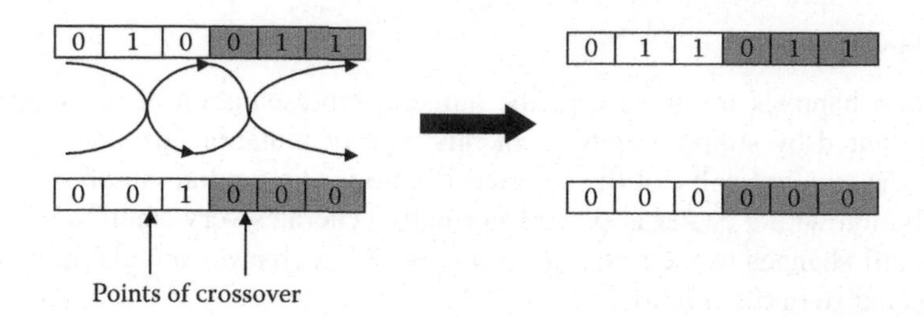

FIGURE 5.5 The two-point crossover genetic operator.

5.8.3 Scattered Crossover

One of the major problems with the two aforementioned crossover techniques is that they are position oriented, which means positioning has a great deal of relevance in deciding the crossover's efficiency. This limits the type of individuals that can be generated. A solution to this is the scattered crossover, which is position independent. This means that the position and the ordering of the bits does not carry importance.

In this technique, we first create a random vector of binary values. The length of this vector is the length of the individuals. Each element of this vector is either 0 or 1, which denotes the parent. If the vector contains 0 at any position, the gene corresponding to this position is taken from the first parent, and vice versa. Thus position has no relevance in this; and hence this solution is position independent. This characteristic is important because of the highly interconnected nature of the genes and the attributes in deciding the fitness.

5.8.4 Intermediate Crossover

The intermediate crossover is usually applied over the numeral representation. In this case, the value of any gene of the new individual is an intermediate value that is taken from the two parents. This value is calculated by Equation 5.5:

$$c = p_1 + r * \alpha * (p_2 - p_1) \tag{5.5}$$

where c is the gene of the child that we need to find out, p_1 and p_2 are the corresponding genes of the two parents, r is any random number between 0 and 1, and α is a fixed ratio. The obtained gene will lie in the interval of the parents if the ratio α is between 0 and 1.

5.8.5 Heuristic Crossover

The heuristic crossover is another operator used primarily between parents with numeral representation. This method is similar to the intermediate method. However, in this method, the gene of the new individual lies on the line drawn between the genes of the two parents. It is preferred to keep the generated gene a small distance from the fitter parent and farther from the worse fit parent.

5.9 OTHER GENETIC OPERATORS

We have already studied some of the major genetic operators used in GA. In this section, we state some additional operators that are commonly used for various specialized purposes. These operators add more flexibility to the working of the GA.

5.9.1 Eliticism

Eliticism selects some predefined proportion of individuals with highest fitness from one generation and passes it straight to the next generation. This operator saves these high-fitness individuals from being destroyed by crossover or mutation and makes them available in the next generation. This operator also ensures that the best populations will be carried forward, which in turn ensures that the globally best individual will be selected as the final solution. Without this operator, the generated individuals may be of fitness lower than the fitness of the best fit individual of the preceding population. This would result in lowering the best fitness value across the two generations.

5.9.2 Insert and Delete

The insert operator allows the insertion of new individuals in the population pool. This makes the entire algorithm flexible to accepting new individuals. Based on the solution, at any point, we can generate new individuals with high fitness. These individuals may be added to the pool by using the insert operator.

Similarly, the delete operator allows deletion of some individuals from the population set. We may wish to delete certain individuals that have a low fitness value. This operator may be used in such circumstances.

5.9.3 Hard and Soft Mutation

Many times the mutation operator may be split in two. One part, or the soft mutation, brings about small changes in the individual. Because the changes are small, big changes might not be expected in the behavior of these solutions. This operatore may be used to control the system convergence or for general addition of new characteristics.

The other operator, or the hard mutation, is used to carry out big changes in the individual. This mutation may completely transform the individual. It brings major changes in the solution's behavior. This mutation may be used very rarely to add very new characteristics to the system.

5.9.4 Repair

Many times, the GA may generate infeasible solutions. The repair operator may be used to repair the infeasible solutions before they are stored in the population pool. This operator converts an infeasible solution into a solution that could be feasible. For this, we need to study the problem constraints and examine the individual. We make as slight modifications as possible in order to make the solution feasible. The fitness value of this newly generated solution cannot be guaranteed.

5.10 ALGORITHM WORKING

We have already seen how the various genetic operators are used in the working of the algorithm and in generating new, high-fitness individuals. In this section, we look at how the algorithm works and converges in order to find a good solution.

Any individual in a GA is represented in the form of a string of bits or numbers. This string can be seen as the building blocks of the individual, where each block contributes to some observable characteristic of the individual. If we assume that we followed a genotype representation of the individual, where each cell contains either 0 or 1 in the solution, the population pool can be assumed to be a pool consisting of numerous occurrences of 0's and 1's. As the algorithm starts, there are random 0's and 1's in the population pool, which contains almost all types of solutions. At this point, no order can be observed in the solutions.

However, as we go further with the algorithm, we can observe characteristics in the solutions that lie in the solution pool. Imagine there is a position i where the occurrence of 1 is very favorable and where the occurrence of 0 is unfavorable. As the algorithm proceeds, the number of individuals with a 1 at position *i* keep increasing, while those with a 0 at that position keep decreasing. After a few iterations, all the individuals in the population pool will have 1 at position *i*. In this way, the algorithm converges. This is true for all positions in the individual representation; the more prominent positions converge early. In these positions, all unwanted characteristics or entries die out soon. The nonprominent positions take time to converge.

We can look at the population at any time in form of schemas. A schema is the representation of the solutions in the population. The schema contains a series of 0's, 1's, and *'s. The length of the string is the same as the length of the chromosome in the representation of the individual. The presence of bit 1 at some location in the schema means that all the individuals following this schema have 1 at this location. Similarly 0 means that all subsequent individuals have an occurrence of 0 at that location. The presence of * means that the individuals can have either 0 or 1. Hence a schema of 1**0**0* can stand for 10000000, 10100100, 11000101, and so forth. But it cannot stand for 00000000 or 10110000.

At the start of the algorithm, the schema is very general, consisting of only *'s. It may be represented as *******. As the algorithm proceeds, the schema will have definite values for some of the prominent positions. This means that the presence of the opposite bit was making the fitness low, and hence those were eliminated. Now the schema will be of the form 1***0**. This denotes that all the individuals in the population have a 1 at the first position and a 0 at the fourth position. As we continue, the schema gets more deterministic. After some more iterations, it may be represented by the form 10**0**.

Another way to look at this is through a hypercube. Think of a chromosome of length *n* as a collection of bits. This chromosome can be represented in an *n*-dimensional graph, with the points lying on the extremes of the axes. The search space for a three-dimensional graph is given in Figure 5.6a, where each corner of the cube denotes a valid solution of varying fitness. This denotes the whole search space of the GA. Any point searched by the GA would lie at the corner of this *n*-dimensional cube. We start with a very big search space of dimension *n*. As we proceed, we keep getting certain values at specific positions. This shortens the GA's search space by eliminating some dimensions, making it possible to get the needed optimal solutions.

The same search space may be plotted in the numeral representation form, where the number of axes is the number of numerals or variables in the solution. For a problem involving two variables, this space is shown in Figure 5.6b, where the various individuals may be visualized as points in the space. Every point denotes a feasible or infeasible solution. Furthermore every point has some value of fitness associated with it. Earlier we visualized this idea in the form of hills and valleys in the space, with the objective being to find the deepest valley.

Initially the individuals are randomly distributed in this search space. As a result of the uniform random initial solution generation, they cover almost the entire search space. Some individuals are

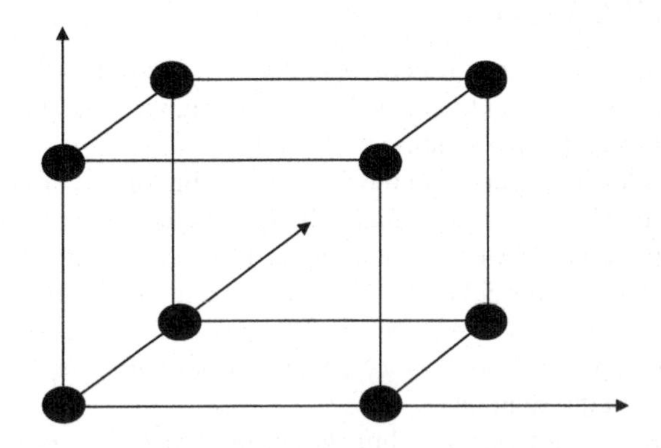

FIGURE 5.6a The genotype representation of search space as a hypercube.

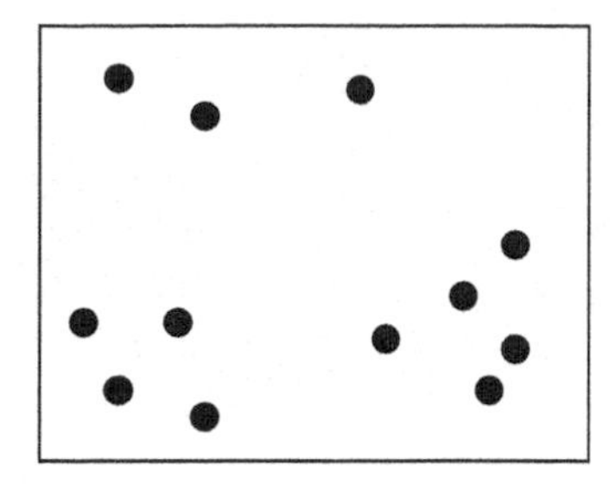

FIGURE 5.6b The phenotype representation of search space.

located at areas where the fitness is good, while others are located at areas where the fitness is bad. As the algorithm runs from one generation to the other, the solutions at the good places start attracting the individuals at the bad places. The attraction results in the movement, or jump, of the bad individual toward the good individual. The new individuals that are formed lie closer to the good individual.

During the initial iteration of the algorithm, the better proportions of individuals attract all other individuals. As a result, two things happen. The first is that the entire population starts congregating to a few prominent areas in the search space. This is of great help, as now many individuals search for global minima at an area that has high fitness. The second is that the entire search space keeps reducing, because the individuals completely leave the areas where they were originally located to join other individuals for a search at likely locations. As these individuals move, the search space gets smaller and smaller. At the later stages of the algorithm, the individuals with high fitness values converge when they move or jump in the direction of the fitter solution. In this way, all the individuals start converging at some point in the search space. As the generations continue, all individuals converge, with the search space reduced to a point.

But there is another major concept here, and that is of addition of mutation. Mutation throws the individuals in random directions by some very small to small amounts and, in turn, expects the individuals to find good regions with high fitness values. The magnitude, or distance, by which the individuals are thrown is usually low and depends on the mutation rate used. These solutions try to explore new areas and to search for the presence of areas with good fitness values. If they succeed in doing so, all the individuals start following in this new area. In this case, the search space keeps reducing as the iterations proceed. This can be seen by plotting or observing the areas where the solutions are found. Some regions would soon become absent from the solution pool, and no individual would be found in these areas.

5.10.1 Convergence

It is evident from the previous section that the algorithm converges as we proceed with the generations. Convergence occurs when the fitness function value behaves almost constant, even upon further generations. The convergence depends to a good extent on the problem and the kind of effect the change in parameter brings. For most simple problems, the convergence is relatively early. More complex problems, however, may require more generations to converge.

Along with convergence, it is important that the algorithm reaches global minima instead of stopping at some local minima. The global minima can only be explored when the algorithm generates diverse solutions. Trying to attain fast convergence may sometimes lead to local minima in place of global minima—this is undesirable.

We use Rastrigin's function to test the convergence of GA. The curve, because of its characteristic nature, is difficult for attaining global minima. Many techniques get trapped at some of the local minima and find it difficult to attain global minima. Rastrigin's function is given by Equation 5.6, and graph is given in Figure 5.7a. Figure 5.7b shows the local and global minima. The mean, best,

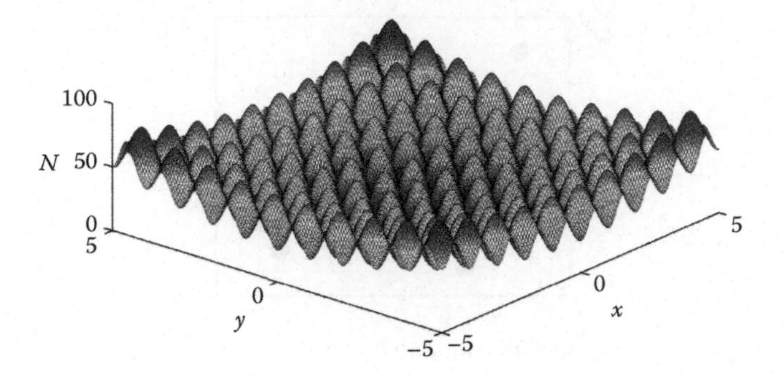

FIGURE 5.7a The surface of the Rastringen's function.

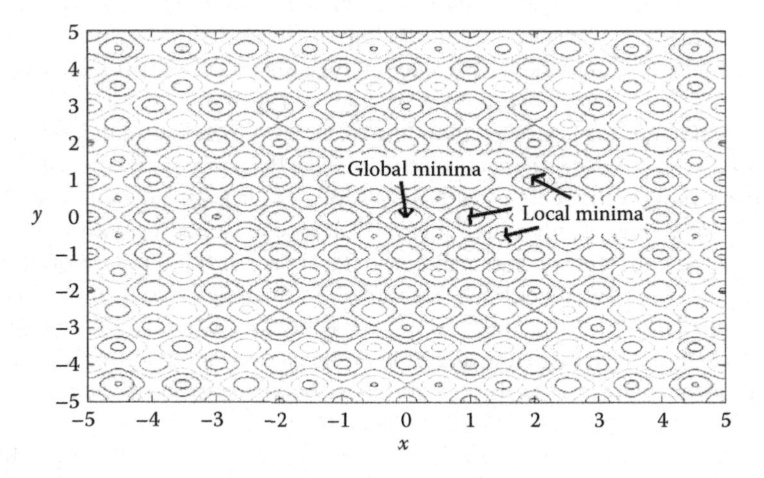

FIGURE 5.7b The contours of the Rastringen's function, showing global and local minima.

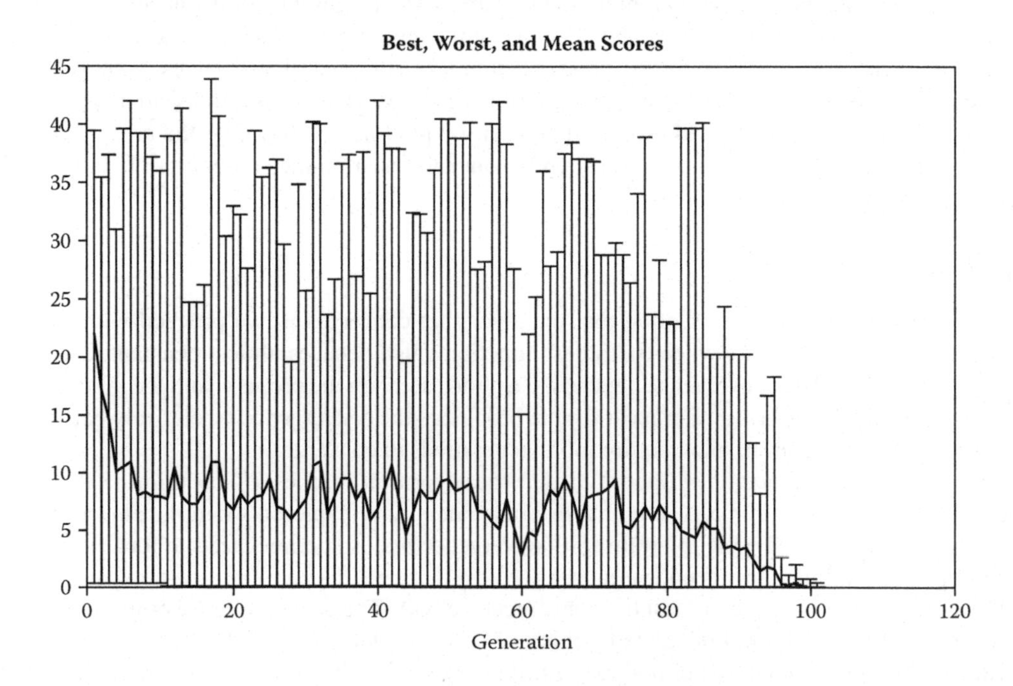

FIGURE 5.8a The convergence of best, worst, and mean fitness individuals in genetic algorithms.

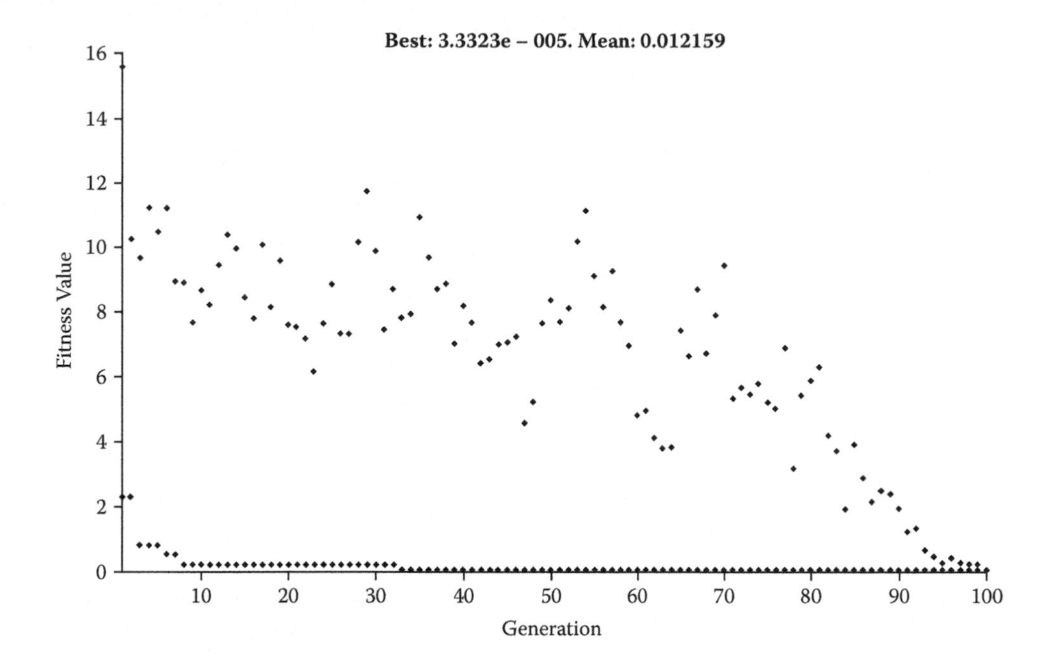

FIGURE 5.8b The convergence of mean fitness in genetic algorithms.

and worst values in different generations of the GA are shown in Figure 5.8a. Figure 5.8b shows the mean and best fitness values per generation. The convergence may be seen in the two figures.

$$\text{Ras}(x) = 20 + x_1^2 + x_2^2 - 10\,(\cos 2\pi x_1 + \cos 2\pi x_2) \tag{5.6}$$

5.11 DIVERSITY

Consider the numeral or bit string representation of any individual. It consists of a set of features that characterize the individual. This information may be plotted on the search space, as discussed earlier. All such individuals in the population may be plotted in this graph.

Diversity is the average distance between the individuals in a population. This distance may be measured as the geometric, Manhattan, or any other distance between the features of the corresponding individuals in this n-dimensional map. Diversity is said to be high if the average distance between the individuals is high. A high distance means that the individuals are far apart in this n-dimensional map and cover almost the entire map. On the contrary, if the individuals in the population are close together, the diversity is low, which means the individuals are concentrated at some region of the map. With low distances, there might not be any individual at some locations. Figure 5.9a shows a space with high diversity, while Figure 5.9b shows a space with low diversity.

Diversity is of vital importance to the GA. High diversity means a large search space for the algorithm to find the global minima. Consider a GA over any problem. Initially all the solutions are fairly far apart, and hence the diversity is high. As the algorithm runs, the individuals start moving in search of the minima. Each iteration repositions these individuals within the population map. As the algorithm runs, more and more solutions would start concentrating near the minima. As time goes on, more and more individuals would start concentrating around the global minima. This process keeps iterating until time expires or the stopping criterion is met. The concentration of individuals toward the minima is accompanied by a drop in density.

Although it helps in convergence, it is usually not good that individuals completely vanish from some regions of the graph. This is because the position where the majority of the individuals are

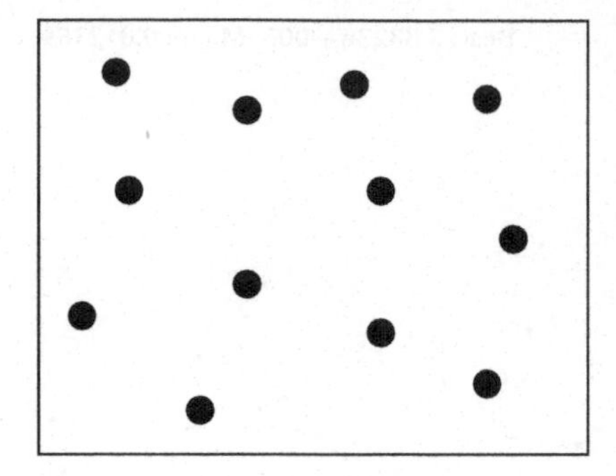

FIGURE 5.9a High population diversity in search space.

located may be the local minima. The region from which all individuals have vanished may actually contain the global minima. But if no individuals are left in that region, it would be impossible to reach the global minima, and it would be very possible for the local minima to be returned by the algorithm.

This situation may be seen in another way as well. The absence of individuals from some regions of the graph means the deletion of certain features. These features are deleted when they result in undesirable fitness, which occurs when all the individuals carrying that feature have an undesirable fitness value. But this does not necessarily mean that the feature was bad. It could be that the good fitness values were somewhere that could not be explored. If the feature were not deleted, the better fitness value might have been explored, which could have resulted in reaching the global minima.

An analogy of this can again be derived from the natural systems. Suppose different communities in a region share different characteristics. The performance of a particular community in some job would be better than in others. This would motivate us to find individuals from that community for the task. However, because of local or global transactions, it can never be guaranteed that the best individual would come from the particular community. Hence the other communities, even though they have lesser probability, must also be considered at all times.

5.12 GRAMMATICAL EVOLUTION

So far we have only discussed GA, in which the length of the chromosome is kept constant. But this is a problem in numerous applications, where the length may change. Some examples are the automatic generation of code or the evolution of a variable ANN or fuzzy system, where we cannot

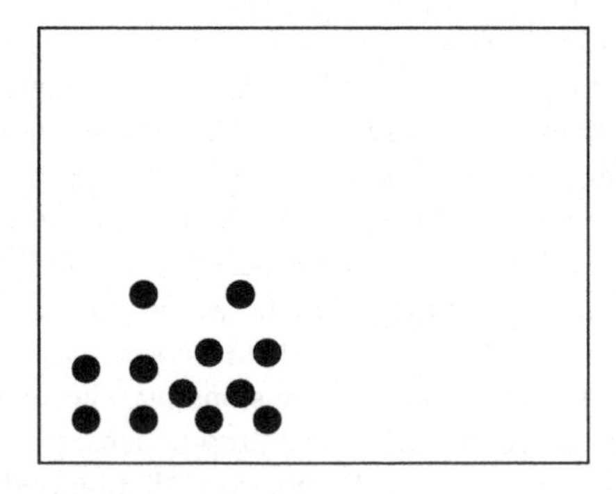

FIGURE 5.9b Low population diversity in search space.

fix the length of the chromosome. Hence we need some mechanism to use the power of GA over these problems. This is done by a related concept known as grammatical evolution (GE). GE is an effective means for applying GA in a situation where the chromosome length varies.

In GE, the chromosome is always represented in terms of a long sequence of integers. This representation can be mapped to the needed phenotype representation by the use of simple rules. This representation further enables us to carry out all GA operations on this long sequence of integers, because the length is now fixed. These integers then need to be converted back into the phenotype representation for evaluation of the fitness function. The conversion process completes the requirement of all the GA operations.

We next explain the process of converting the sequence of integers back to the symbolic form. Before that, however, we introduce the concept of grammar, which will help us understand this conversion.

GE uses the notion of Backus-Naur form (BNF) to represent the language. BNF carries all the information of the language syntax in the form of various rules. The grammar is a specification of the terminating characters, the nonterminating characters, the starting character, and so on. Rules govern the generation of the grammar. The grammar may thus be represented by $\{N, T, P, S\}$, where N is a set of nonterminals, T is a set of terminals, P is a set of production rules, and S is the start symbol.

The following is an example of a grammar:

N = { <expr>, <biop>, <uop>, <bool> }
T = { and, or, xor, nand, not, true, false, (,) }
S = { <expr> }

P can be represented as:

(A) <expr> ::= (<expr> <biop> <expr>)
| <uop> <expr>
| <bool>

(B) <biop> ::= and
| or
| xor
| nand

(C) <uop> ::= not

(D) <bool> ::= true
| false

Every rule has a number of choices associated with it. Looking at the grammar, we can easily see that rule A can have three choices: <expr> ::= (<expr> <biop> <expr>), <uop> <expr>, and <bool>. Similarly rule B can have four choices. The number of choices for rules C and D are one and two, respectively. All of these are summarized in Table 5.4.

To convert any sequence of integers into the symbolic form, we use the following procedure. At the end, we get back the symbolic form, or the phenotype representation, which can then be used for the needed purposes.

We first take the first integer c contained in the sequence. At the same time, we take the starting symbol S from the grammar. We then need to select the rule that is to be fired. In order to do so, we first look at the list of choices available for the rule associated with S. Say that S was <expr>. We have three choices available: <expr> ::= (<expr> <biop> <expr>), <uop> <expr>, and <bool>.

TABLE 5.4
Number of Choices Available From Each Production Rule

Rule	Number of Choices
A	3
B	4
C	1
D	2

We calculate the number of these rules in the list; let this number be *r*. We select the *i*th choice from this list, where *i* is given by Equation 5.7. *S* is now completely replaced by the selected choice.

$$i = c \bmod r \tag{5.7}$$

At any general point of time, the expression in hand would be a collection of terminating and nonterminating characters, with some operators. We select the first nonterminating symbol. We again follow Equation 5.7 to replace the nonterminating symbol with the selected choice.

Consider the following rule from the given grammar—that is, given the nonterminating <biop>, which describes the set of binary operators that can be used. There are four production rules to select from. As can be seen, the choices are effectively labeled with integers counting from 0.

(B) <biop> ::= and (0)
| or (1)
| xor (2)
| nand (3)

If we assume the integer being read at some point of time from the sequence list is 6, then 6 mod 4 = 2.

We would select rule (2), or XOR. In other words, <biop> is replaced with XOR. Each time a production rule has to be selected to transform a nonterminal, another codon is read. In this way, the system traverses the genome.

The chromosome consists of a sequence of integers or codons that are read one after the other. Likewise the solution is modified by the application of the rules. It is possible that many codons may not be read and that the solution may consist of only terminating characters. In such a case, further operation is stopped, and this forms the final solution. It may even be possible that all the codons may be read and still nonterminating symbols will exist in the solution. In this case, the solution may be neglected, given the lowest fitness value, or completed by default terminating symbols. Many times, the chromosome may be reread from the start to facilitate the completion of the solution.

5.13 OTHER OPTIMIZATION TECHNIQUES

So far we mainly have concentrated on GA, which is the most traditional and universally used technique, especially for optimization purposes. However, other techniques exist, more or less along the same lines as GA. Like GA, these other techniques have primarily been inspired by the natural processes. In this section, we discuss the two most commonly used techniques besides GA— particle swarm optimization (PSO) and ant colony optimization (ACO).

5.13.1 Particle Swarm Optimization

PSO is a technique inspired by the natural process of flocking birds, which is also known as swarm intelligence. It derives intelligence by studying the social or group behavior of the animals. Imagine a group of birds must search for some food. They all search in random directions, starting from random positions. Each bird knows how close it is to the food, because the birds are coordinated with each other. Hence for any bird, it would be natural for it to move toward the bird that is closest to the food of all the birds. It would also be guided by its own experiences so far in the search process.

The PSO is a similar algorithm. Recollect our discussion of the search space with the graphical plotting of the populations on an n-dimensional graph. We discussed that as the time passes, the solutions keep concentrating nearer some point that is supposed to be the global minima. In PSO, we use the same ideology, along with the inspiration from the birds.

As the individuals moved in GA to find the global minima, so too do the individuals move in PSO in search for minima. In GA, however, the movement was on the basis of mutation and crossover operators. In PSO, we follow a more mathematical, systematic approach to find the solution. The movement of individuals in PSO is guided by a set of equations that are made from our inspiration of the flocking birds.

Each individual or particle in PSO has a particular speed at which it is moving. Furthermore each particle knows the position in which it presently lies. The fitness function is a measure of how close it is to the goal. Each particle knows the best possible position it encountered during its path from the start. Hence the history is saved. This history acts as a local guide for the particle to estimate the best position and to then move to that position.

In addition, all the particles know the particle that is closest to the goal (or that has the minimum fitness value). All the particles would hence try to move toward this particle in order to lower their distance from the goal, or minima. This acts as a global planner to guide all the birds to move toward the position that may be taken as the global minima.

The movement of the particles updates both their speeds and the positions. Particle movement is governed by the local and global planner. In PSO, this is modeled by a set of equations. Let the speed of any particle i at any iteration t be v_i^t. Let the particle be at position x_i^t at this iteration. At the next iteration $t + 1$, the position and the speed of the particles get updated as given by Equations 5.8 and 5.9:

$$v_i^{t+1} = v_i^t + c_1 * r * (x_{bi}^t - x_i^t) + c_2 * r * (x_g^t - x_i^t) \tag{5.8}$$

$$x_i^{t+1} = x_i^t + v_i^t \tag{5.9}$$

where c_1 and c_2 are constants that are usually both fixed as 1, r is any random number between 0 and 1, x_{bi}^t is the best solution in the path of particle i until iteration t, and x_g^t is the globally best solution until iteration t.

In this manner, the positions and speeds keep updating with each iteration. The particles keep moving toward the globally best solution and continue to concentrate and move there. In such a way, the algorithm finds the solution. At any time, some particle may land at a point that was better than the globally best point known. In such a case, all other particles start traveling toward this new point.

The PSO algorithm is given in Figure 5.10. As in GA, we first generate an initial set of solutions. These solutions are distributed in the entire search space. Each particle is assigned a random velocity. Now we use Equations 5.8 and 5.9 to update each particle. Another task performed at all iterations is to calculate the local best and globally best positions for all particles. In this manner, the particles keep moving. The algorithm stops when some prespecified stopping condition is met. The maximum velocity of any particle is fixed. If the velocity during any operation results in a velocity more than v_m, it is limited to v_m.

5.13.2 Ant Colony Optimizations

The other optimization studied here is the ant colony optimization (ACO). Like the PSO and GA, this algorithm is also inspired by the natural process—in this case, the process of ants working in their colonies. This process is simulated in the ACO algorithm.

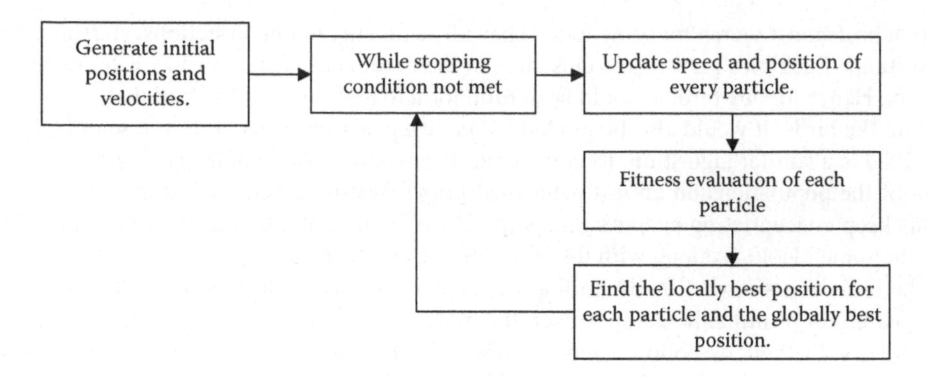

FIGURE 5.10 The particle swarm optimization algorithm.

The ants in this algorithm are expected to move about the search space. In their motion, the ants, or solutions, build the partial solution to the problem to be solved. At any time, numerous moves are possible; the decision of which move is made follows a probabilistic approach. As the ant moves, it keeps building this partial solution. This is the basic fundamental of the algorithm.

The move of any ant is guided by two factors—trails and attractiveness. The trails are the entire experience of the ant in the course of its travel. This information is stored in a cumulative form in what is referred to as the *pheromone*. This pheromone information is used to direct the search of future ants and thus needs to be modified or updated at the ant's each and every move. Attractiveness, also known as the heuristic value, is associated with the node and represents the quality of the node.

The ants move in the graph according to the probability decided by the trails and the attractiveness. As the ant moves, it keeps building the partial solution. Its move is always such that the ant does not land on a vertex that it has visited before. For this purpose, all visited vertices are stored in memory.

Once all the ants have completed their tour, the pheromone on the edges is updated. First we decrease the pheromone by a small percentage to avoid excess pheromone being deposited at some edge over time. Then the edges receive pheromones depending on the quality of the solutions.

Another concept used in ACO is that of the demon actions. This concept is optional in ACO. Demon actions are used for all centralized actions that cannot be performed by a single ant. These are used according to the needs of the specific problem being considered. One of the commonly used demon actions is to run a local optimization search. Another task that may be performed here is an update of the global information.

The general ACO algorithm is given in Figure 5.11.

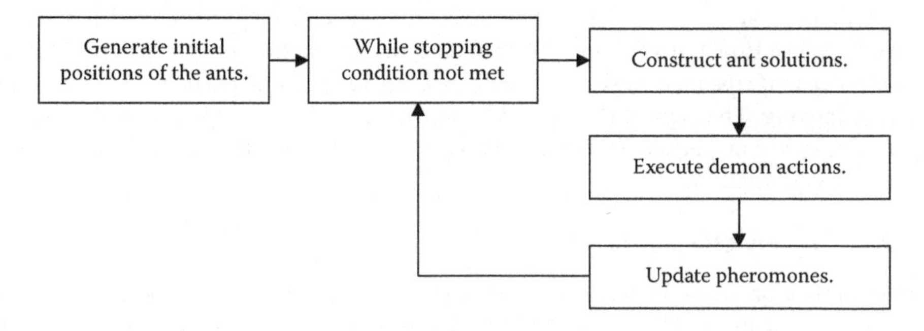

FIGURE 5.11 The ant colony optimization algorithm.

5.14 METAHEURISTIC SEARCH

In this section, we discuss some more metaheuristic searching techniques that are also commonly used for numerous purposes and optimizations. These include the k-nearest neighbor approach, simulated annealing search, taboo search, hill climbing, and so on.

The k-nearest neighbor is another search algorithm that starts from some location in the graph and tries to reach the goal. The readers may consult our earlier discussion on A* algorithm for the details of search algorithms. At any node n, there is a set of nodes that we can traverse from the current node. These nodes are called the neighbourhood of n, and they have an edge connecting to n. The edges have different weights, which denote the cost of traveling from the current node to the next node. In a k-nearest neighbor approach, we select the best k moves out of all the moves possible. These k moves are then executed. In this way, the graph keeps getting explored, and we get the final answer.

Another similar approach is the simulated annealing (SA), which derives its analogy from the way in which metals are cooled and frozen into a minimum energy crystalline structure. This analogy is used to prevent the algorithm from getting trapped into some local minima. This algorithm makes some small changes to the current node n in order to arrive at some other node. If the new node is better, then the move is executed. If, however, the selected node is not better, the move is executed with some probability p. In this algorithm, each node n has an energy value given by the fitness function $f(n)$. Furthermore there is another parameter introduced called temperature (T).

The probability of the move being executed, in case it leads to a solution with an undesirable fitness, is given by Equation 5.10, which is inspired by the annealing process

$$p_{\Delta E} = \frac{1}{1+e^{\frac{\Delta E}{T}}} \tag{5.10}$$

where ΔE is the difference in fitness of the current and the next node.

T starts high and slowly decreases as the algorithm runs, because the behavior of the algorithm must be random at first so that it can search more locations. As the time goes on, the algorithm needs to converge to the minima; hence the temperature is reduced.

5.15 TRAVELING SALESMAN PROBLEM

To completely understand GA, we apply it to a widely studied problem called the traveling salesman problem (TSP). In the literature, this problem has been solved using a variety of methods. TSP is used as a benchmark problem for various methods to be studied, compared, and analyzed. We first study the problem and then develop a solution for it.

5.15.1 Problem Description

A salesman has been given the task of selling products. He must sell to a total of n houses. The salesman is supposed to go to each and every house. The starting house is given. The salesman must then return to the starting house at the end. In addition, every house is reachable from every house, but the distances between the houses are all different. The salesman must travel in such a way that the total length of the path traversed is minimal.

This problem is known to be an Non-Deterministic Polynomial Time (NP)-hard problem, which cannot be solved in polynomial time. In other words, the solution to this problem exponentially increases with the increase in input size. For even moderately large inputs, these problems may take years to solve, which, of course, is unacceptable.

Suppose there are seven houses, as shown in Figure 5.12. Numerous paths are possible. The most optimal path is shown in black in Figure 5.12. The dotted path forms another solution.

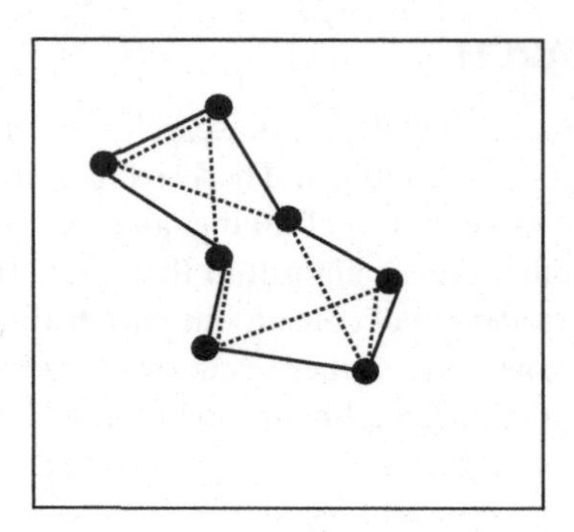

FIGURE 5.12 The various paths in the traveling salesman problem (TSP).

The traditional methods do not work for these problems. Instead we have to use soft-computing techniques. These techniques do not necessarily give the most optimal output; however, the output generated is highly optimal in a finite or desirable amount of time. In this section, we develop a solution to this problem using GA.

We are now ready to give a formal description of the problem and to model how the GA can be used over it. Let there be n number of houses that the salesman must visit. Let each house i have coordinates (x_i, y_i). Further suppose that the distance between two houses is the geometric distance between them. We need to find the sequence of houses that the salesman must traverse so that the entire path length is the shortest.

Let the starting house be s. The salesman must travel to all n houses. Let the order in which he travels be $<s, a_1, a_2, a_3, a_4, \ldots, a_{n-1}\, s>$. For this problem, we represent any solution by $< a_1, a_2, a_3, a_4, \ldots, a_{n-1>}$. The actual path traced by the salesman would be $s \rightarrow a_1 \rightarrow a_2 \rightarrow a_3 \rightarrow a_4 \ldots a_{n-1} \rightarrow s$.

5.15.2 Crossover

We do not use a conventional crossover operator in this example, because it may yield infeasible solutions. We hence modify the crossover operation to suit this problem. Let there be 5 houses that the salesman must visit. Let us assume that the crossover operation is to be applied to two individuals <1, 4, 3, 2, 5> and <2, 5, 3, 1, 4>. Suppose a single-point crossover is to be applied across position 3. In such a scenario, the solutions generated by a conventional crossover operator would be <1, 4, 3, 1, 4> and <2, 5, 3, 2, 5>. Clearly both are infeasible because the first doesn't visit houses 2 and 5, and the second does not consider houses 1 and 4.

In this problem, between any two individuals, a normal crossover is first applied. Then we iterate the newly formed individual to ensure that it does not contain duplicate entries. If a duplicate entry of any integer is found, its value is replaced by the first unused integer from the phenotype representation. In this way, the solution becomes stable.

5.15.3 Mutation

As discussed above, mutation also results in an infeasible path. For the same reasons, we modify the mutation operation in the same manner for this problem.

5.15.4 Fitness Function

As discussed above, for any problem, the total path traced by the salesman is $s \rightarrow a_1 \rightarrow a_2 \rightarrow a_3 \rightarrow a_4 \ldots a_{n-1} \rightarrow s$. The fitness function for this problem is the sum of the geometric distance between all sets of points. Note that the first and the last points are constants that are given to us initially.

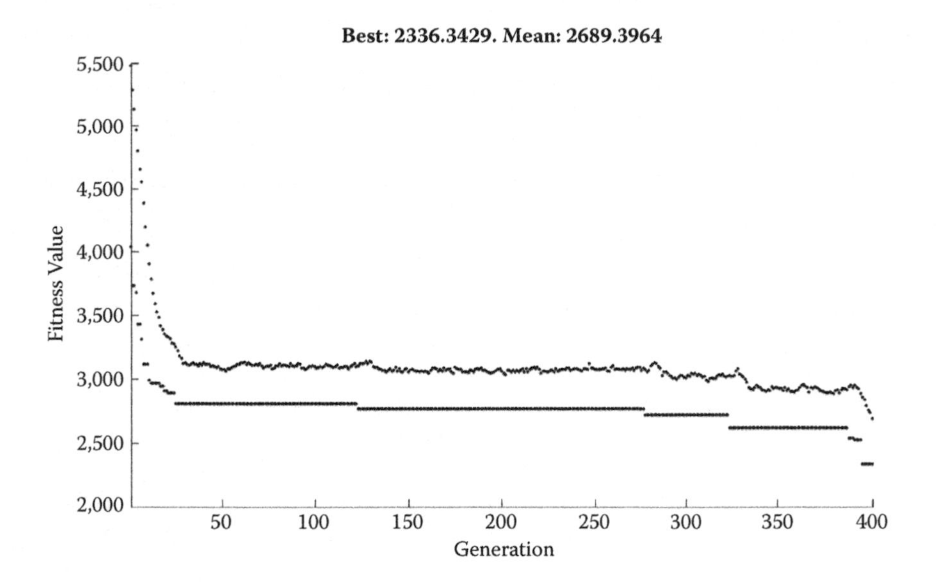

FIGURE 5.13 The mean and minimum fitness during optimization of the genetic algorithm for TSP.

5.15.5 Results

We simulated this problem. We took a map of the size 500 × 500 and generated 20 random initial points to act as cities. An additional point was generated to act as the starting point, which has a fixed value of (250, 250).

The double-vector method of population representation was used. The total number of individuals in the population was 1,000. A uniform creation function, rank-based scaling function, and stochastic uniform selection were used. The elite count was 2. Modified single-point crossover was used. The program was executed for 400 generations. The final path length received was 2,336 units.

The training curve showing the average and mean values is given in Figure 5.13. The path traveled by the salesman is given in Figure 5.14. The path is highly optimal. We also received the final output in a finite amount of time, proving that the algorithm easily solved the problem.

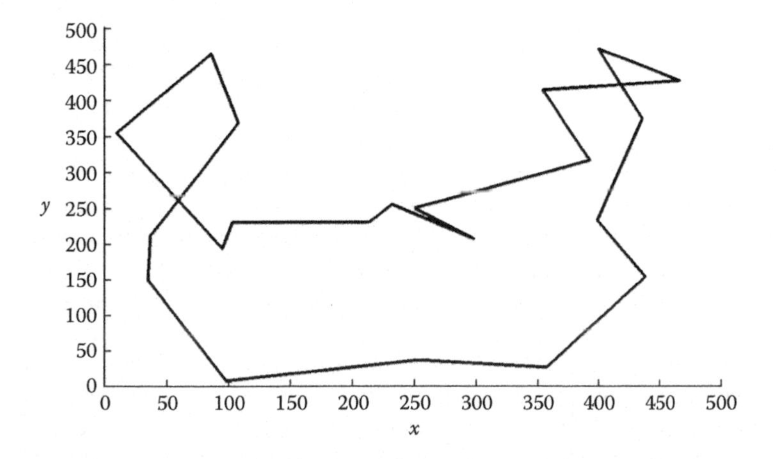

FIGURE 5.14 The result of TSP by genetic algorithm.

CHAPTER SUMMARY

This chapter presented evolutionary algorithms. First the biological inspiration behind evolutionary computing was discussed. This was followed by a discussion of the genetic algorithms. GA is an effective tool for problem solving and optimizations. Problem solving with GA involves problem encoding, which incorporates the genotype and the phenotype representations. There are various other genetic operators that drive the algorithm and aid in the generation of effective solutions. The GA follows an iterative approach in which the solutions keep improving with time.

Selection was one of the major operators we discussed. GA uses numerous selection techniques, including roulette wheel selection and tournament selection. The other major operators used are crossover and mutation. We discussed the crossover and mutation rate, as well as the variable crossover and mutation rate. The fitness of any individual in the algorithm is measured with the help of a fitness function. We discussed the role, need, and working of the fitness function as well. The GA theoretical foundations, convergence, and working were also explained in the chapter.

The other major area presented was the use of other methods for optimization. These methods include the particle swarm optimization (PSO) and the ant colony optimization (ACO). Both are excellent means for solving real life applications. The last part of the chapter consisted of metaheuristic search techniques, such as simulated annealing and *k*-nearest neighbor. Toward the end, we used GA to solve the very famous travelling salesman problem (TSP). The algorithm was able to give very good results to the problem in finite amount of time.

SOLVED EXAMPLES

1. **What is convergence in genetic algorithms? Discuss the role of mutation and crossover in convergence.**

   Answer: If we plot the fitness of population (usually mean or best) against the generations, we would see that fitness changes rapidly at the start of the algorithm. Toward the higher generations, the fitness behaves more or less constant. This is the time when individuals in the GA are very near the global minima and are moving toward the global minima very slowly. At these higher generations, we say that the GA's fitness has converged to the point of a global minima, and further optimizations may not be beneficial. Although we use the word *global* here, it can never be assumed that the GA has converged to a global minima and not local minima.

   Mutation and crossover are two contrasting operators when discussed in reference to GA. Crossover inspects all the individuals in the search space and, based on these, tries to find a region of optimal fitness. This region is usually within the domain covered by these individuals. Crossover drags individuals closer to the suspected points of minima. As a result, the search space keeps getting shorter and shorter, and in this way, the solution converges very sharply.

   Mutation does exactly the opposite. It tries to explore newer and newer areas in the hope of finding optimal points. This results in an increase in the search space and the placement of individuals at newer regions that may be outside the search domain where the individuals have so far been placed. This obviously delays convergence.

   The contrasting nature of the two operators plays a role when we talk about global and local optima. The fast convergence of the crossover might make it converge at some local optima, which is undesirable. In this situation, we would need a means to make the algorithm search for the global minima. This means is done by mutation. If mutation is kept at 0, the convergence is very fast in GA. However, this often leads to a local minima. At the same time, a very high value of mutation and low crossover might result in the algorithm acting in a random manner. In this case, the training graph would reveal very uncertain trends in the population's mean fitness value. Thus the convergence would not be met unless the crossover were high enough.

2. Suggest a strategy for fixing mutation and crossover rate in GA.

Answer: Mutation in GA is responsible for exploring newer areas in the search space, while crossover is responsible for converging the individuals to a minima (see Example 1). Hence we need a balance between the two tasks. Accordingly, by studying the convergence of the fitness versus generation graph, the mutation and crossover may be fixed. If the graph reveals a very fast convergence, the crossover rate may be reduced, and the mutation may be increased. Likewise, if the algorithm reveals a very slow convergence, the crossover may be increased, and the mutation may be decreased.

According to the same principles, we have the concepts of variable mutation and crossover rates. These are an attempt to modify the constants in real time by studying the fitness trends. The variable scheme decreases the crossover rate and increases the mutation rate, and vice versa, in case the algorithm seems to have converged at some point. This ensures that the system makes the best use of the time for execution and tries its best to find a balance between the exploration for the global minima and the convergence in order to go deep into the most optimal value in minima. The convergence of the algorithm denotes that the individuals are in similar areas and hence are trying to find the exact place of the minima. In such a case, we might consider exploring new areas to look for the possibility of global minima somewhere else in the search space. At the same time, if there is too much randomness, we might consider concentrating on some minima. This is achieved by adjusting the crossover and mutation. Many other schemes are also possible for fixing these parameters.

3. Comment of localized crossover in evolutionary algorithms.

Answer: *Localization* is a term that can easily be seen in the natural systems. Any person interacts mostly with the people from his or her locality. This interaction influences people. As a result, the locality keeps driving itself and developing over time. Every individual of the locality helps others develop and specialize according to the local conditions. In the entire population (world), there are many such localities that develop over time. These numerous localities coexist, and their collection makes the world population.

The optimal development of a locality cannot take place by this method alone. The locality needs to be shown and given exposure to the outside as well. In the natural systems, this happens as a result of various global movements that result in global interactions between people from different localities. This helps further improve the localities.

Our inspiration for the GA was the human evolution system. Here the same phenomenon happens, in which individuals interact with neighboring individuals. Then there are global interactions as well.

Coming into the GA world, in a localized crossover, individuals are allowed to crossover only with neighboring individuals. Here the neighborhood is defined by the closeness of the individuals. This closeness is measured by the geometric, or Euclidian, distance. There is also a global crossover operation, in which the crossover takes place at the global level. A proper balance is maintained between the two, which is very helpful for maintaining the population's diversity.

4. What is the need for soft and hard mutation, and what is their relevance in GA?

Answer: Many times mutation is classified under two separate heads, soft and hard. Soft mutation brings in soft changes to the species, while hard mutation makes very prominent changes in the species. As a result, soft mutation is carried out with a relatively high probability as compared with hard mutation, which happens with a very low probability.

Hard mutation places the individuals at some unknown locations in the search space. This is because of the prominence of changes it brings to their location. The species lands at some completely new place and then becomes part of the population there. If the place results in optimal fitness, new individuals will join along with the generations. If, however, this results in a bad fitness, even this individual may die out over time. Because the location may be completely new, there is no general idea of the performance of this newly explored area.

Soft mutation, on the other hand, makes only few changes in the position of the individuals in the search space. These changes are random in nature. Soft mutation may be seen as a mechanism by which individuals are picked up and placed at any area nearby. The fitness is usually expected to be around the average fitness in the area. In addition, the area is not completely new for the algorithm, as it has already been explored to some extent.

Mutation is responsible for the search of new areas in the search space. At the same time, it must not affect the convergence too much (see Examples 1 and 2). This explains the relatively high probability of soft mutation and the very low probability of hard mutation. According to the roles they play in crossover and search for optima, these probabilities are justifiable.

5. What are infeasible solutions? State and compare the means for handling infeasible solutions in GA.

Answer: In feasible solutions, the solutions do not obey the problem constraints. Say the problem says the value of some x needs to be less than 10. The GA generates a solution that has x as 15 as an infeasible solution.

There are two major ways to handle infeasible solutions. The first is to assign the infeasible solutions a very low fitness value. Because GA believes in survival of the fittest, these infeasible solutions would be deleted from the population in some generations.

The second way is to employ a repair operation in GA. The repair would result in correcting the inappropriate values and assigning feasible values per the logic. This converts an infeasible solution into a feasible solution and hence solves the problem.

The choice between the two methods is largely dependent on the problem. If the problem has a very high occurrence of infeasible solutions, the repair operation may be preferred. Also if the crossover of two good feasible solutions can give an infeasible solution, the repair method may be better, because we cannot expect a very large number of solutions being deleted as would be the case with the fitness method. If again and again infeasible solutions keep creeping in large numbers, they would affect the total populations and hence cannot be entertained.

If, however, the number of infeasible solutions is few and has limited occurrence, the fitness function may be used. This method would result in deletion of the infeasible solutions as the generations increase.

6. Explain the concept of GA with species and sexual selection.

Answer: This is another method of implementation of GA in which the entire population is divided into males and females. Females are the individuals scattered throughout the population with a higher level of fitness. Males are the rest of the individuals. The number of females is quite less when compared with the males. The crossover always takes place between the males and the females. The crossover is a multiparent crossover operation. The power of a male in crossover depends on his distance from the female. The females also prefer to crossover with the males who lie in the neighborhood. The new population is generated in the pool. The males, females, and the fitness values are updated for the next generation. Thus this method preserves the population's diversity.

QUESTIONS

GENERAL QUESTIONS

1. What is optimization? How is it carried out in GA?
2. What is the relevance of specifying initial population in GA?
3. What role do elite children play in GA?
4. Explain the concept and importance of variable crossover rate.
5. Compare and contrast various metaheuristic search techniques.
6. Compare and contrast (a) genetic algorithms and particle swarm optimization and (b) genetic algorithms and ant colony optimizations.
7. What is diversity in GA? What role does it play in convergence of GA?
8. Explain the concept of generation in GA.
9. What are the similarities and differences in the workings of GA and of ANN?
10. What is the effect of increasing or decreasing the number of individuals in population on various performance parameters in GA?
11. Compare roulette wheel selection with random selection.
12. What are the various ways of doing crossover operations in GA?
13. What is the use of having insert and delete operators in GA? Suggest possible problems where these operators might find application.
14. It is said that the lower fitness individuals get killed in GA. How is it done?
15. Explain a few methods for selection operation in GA.
16. Why are the methods of selection, crossover, and so forth called operators?
17. What is swarm intelligence? Is it different from particle swarm optimizations?
18. What is the difference between the phenotype and the genotype representation of individuals in GA?

PRACTICAL QUESTIONS

1. Take different synthetic functions and optimize them using GA. What is the effect of changing crossover and mutation rate? Repeat the experiment for Rastringen's function.
2. Recollect the data that you took for clustering in Chapter 1. Cluster the same data by GA. Do you get the same results as obtained by other methods?
3. Consider the traveling salesman problem. Observe the effects of changing the number of generations and elite count. What would happen if the elite count were kept to 0?
4. Recollect the fuzzy controller we designed in Chapter 3. Assume a few standard test cases with known inputs and outputs. Optimize the generated fuzzy controller by using GA.

6 Hybrid Systems

6.1 INTRODUCTION

So far we have discussed numerous algorithms and solved a number of related real life applications. However, many real life applications are so complex that they cannot be solved by the application of a single algorithm. This situation has necessitated the need to develop algorithms that mix two or more of the studied algorithms. The choice of algorithms depends on the needs and characteristics of the problem. These mixed systems, called hybrid systems, further help in solving problems to a reasonably good extent and in achieving higher performances.

Hybrid systems use two or more specialized algorithms in an interconnected manner. These algorithms complement each other and solve the problems faced by the other algorithms. Hybrid systems as a whole present a robust system that can handle real life inputs even better that single algorithms. In the future, the size and dimensionality of data is expected to increase tremendously. In addition, most of the systems will have moved from the laboratories into industry. Thus hybrid systems will be a great boon and will provide promising solution for the growing needs.

Recently, due to an increase in the need and application of hybrid systems, these systems are being experimented with, studied, and deployed even more extensively. As a result, many new hybrid models have been proposed, and many modifications have been suggested for the existing hybrid models. This increase in the trend of hybrid systems is both in the theoretical domain, where researchers are improving the algorithms, as well as in the application domain, where these algorithms are being adapted to handle real life problems.

Hybrid algorithms make extensive use of artificial neural networks (ANNs), genetic algorithm (GA), and fuzzy logic (FL), along with heuristics and other artificial intelligence (AI) practices to solve real life problems. The individual algorithms have already been discussed in Chapters 1 to 5. This chapter is devoted to the numerous ways to integrate these algorithms to make even better systems. The immense possibilities in which this integration can be carried out are a very fascinating feature of the hybrid systems.

Before we continue our discussion of hybrid systems and how the individual systems are fused together, we will first look again at the individual systems. In this case, we will discuss the strengths and weaknesses of the individual systems so we can have them in mind when discussing how to fuse them into hybrid systems.

Then we will put the different constituents together to make effective systems. In this chapter, we mainly study six systems:

- Adaptive neuro-fuzzy inference systems (ANFIS), which use neural networks to learn fuzzy systems
- Fuzzy neural networks, which are ANNs that take fuzzy inputs
- Evolutionary ANNs, which use genetic algorithm (GA) to evolve ANNs
- Evolutionary fuzzy systems, which use GA to evolve fuzzy inference systems (FIS)
- Rule extraction in ANNs, which extract fuzzy rules from a trained ANN
- Modular ANNs, which introduce the concept of modularity in ANNs

6.2 KEY TAKEAWAYS FROM INDIVIDUAL SYSTEMS

The ANN, GA, and FL are known for their ability to model and solve problems. In this section, we discuss the key advantages and disadvantages of each system.

6.2.1 Artificial Neural Networks

ANNs are known for their ability to generalize and learn from past data. They can approximate almost any function without having any predefined idea of the problem or its solutions. ANNs are an inspiration from the human brain, which has unimaginable powers. The highly parallel nature of ANNs further encourages their use in real life applications and systems. Furthermore ANNs are resistant to noise to a certain extent.

However, ANNs are limited in that they need a high amount of training data. If there is not enough training data, the ANN's performance is very bad. These data also must be distributed evenly throughout the search space for effective performance. Another problem is that the parameters (number of neurons, hidden layer) must be manually fixed. A change in these parameters can bring a paramount change in the system's performance.

6.2.2 Fuzzy Systems

Fuzzy systems, as is clear from the name, are known for impreciseness. They can solve problems very effectively by adding impreciseness to the systems. Fuzzy systems are an excellent means of problem solving when we have an approximate idea of the rules that map the inputs to the outputs. The size of the training data needed to validate the system does not need to be large, because the rules govern the outputs. The computational requirement in these systems is fairly low.

However, fuzzy systems cannot be effectively used when the rules governing the mapping of inputs to outputs is not well known. Another major problem is that the parameters must be manually fixed and manipulated to achieve higher accuracies. A change of parameters may make a paramount difference in the system's performance.

6.2.3 Genetic Algorithms

GAs are very good optimizers. They can be used for almost any kind of optimization, which is of a great value in real life problems. The optimization powers of GAs make them very handy tools for problem solving. GAs can be used to optimize various parameters and to solve many problems in real time; these solutions may otherwise not be possible in finite time. GAs are also used for various search-related operations. However, GAs are often computationally expensive when used for numerous problems.

6.2.4 Logic and AI-Based Systems

Logic in the form of production rules and AI methods is also employed in hybrid systems. Logic and AI-based systems provide precise logic and rules that drive the output. They provide exact outputs in the required computational time. However, they fail when the dimensionality of data is large or when the time required is larger than what we can afford.

6.3 ADAPTIVE NEURO-FUZZY INFERENCE SYSTEMS

The adaptive neuro-fuzzy inference system (ANFIS) belongs to the class of systems commonly known as neuro-fuzzy systems. Neuro-fuzzy systems combine the powers of ANN with those of fuzzy systems. As discussed earlier, the basic motive of such a combination is to take the best features of both to create a more robust system.

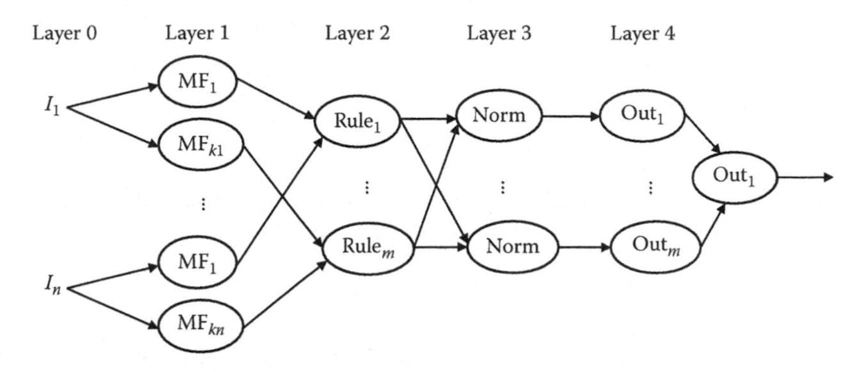

FIGURE 6.1 The general architecture of ANFIS.

In this section, we take the learning capability of ANNs and the logical approach of fuzzy logic to get a system that is commonly known as the ANFIS or adaptive neuro-fuzzy inference system. Here *adaptive* means that the system adapts to the problem's requirements. Chapter 4 explained that fuzzy systems alone usually have problems because the parameters must be adjusted by the designer in a trial-and-error manner. This procedure leads to long cycles of system modification and evaluation. This process goes on and on until decent levels of accuracy are met, which means the designer must have a lot of patience when attempting to design an effective system. In many instances, designers completely fail to adjust the system's parameters. Obviously this is a big hindrance in the use of fuzzy systems. This is especially true in case we have some training data available.

ANNs, on the other hand, have great powers of learning. They can learn large volumes of data by adjusting their parameters. This means ANNs have very high performance. However, unlike fuzzy systems, ANNs do not work on the concept of rules, a concept that is often desirable in terms of design. ANFIS combines these complementary benefits to make a complete system.

6.3.1 General Architecture

In this section, we study the general architecture of ANFIS, which is primarily a fuzzy system. It does all the tasks of a fuzzy system, in the same order. The innovation of the ANFIS comes from the fact that we can model the FIS in the form of an ANN, as is depicted in Figure 6.1.

Figure 6.1 represents both ANN and FIS. The ANN representation is used for learning purposes, while the FIS representation is used for evaluation and testing purposes. In this section, we discuss the role and uses of the different layers of FIS.

6.3.1.1 Layer 0

In this layer, the inputs are presented to the system. Apart from applying the inputs, this layer does not perform any other function.

6.3.1.2 Layer 1

In this first layer of the ANN, fuzzification is carried out, which is why this layer is sometimes called the *fuzzification layer*. The fuzzification in FIS is carried out by applying the inputs to the membership functions (MFs). In this layer, the ANN's activation function is the MF of the FIS. More formally, let the activation function of the neuron that connects the *i*th input to the *j*th MF of the input be $\mu_{ij}(x)$. According to figure, this would receive an input of I_i. Let the output produced by this neuron be O_{ij}, where O^1_{ij} is given by Equation 6.1:

$$O^1_{ij} = \mu_{ij}\,(I_i) \tag{6.1}$$

The commonly used MF $\mu(x)$ is given by Equation 6.2:

$$\mu(x) = \frac{1}{1 - \left|\frac{x-c}{a}\right|^{2b}} \tag{6.2}$$

where a, b, and c are parameters. We discuss the role and uses of these parameters in the next section. Note that Equation 6.2 represents the MF we used in Chapter 4.

6.3.1.3 Layer 2

This layer represents the AND operation that is found in the antecedents in any fuzzy rule. Each neuron of this layer takes its various inputs from the preceding layer. It returns just one output, which is usually the product of the inputs. Recall that one implementation of the AND operator that we used in FIS was the product of the inputs. The activation function of this neuron is given in Equation 6.3. Here this neuron takes as its input the fuzzified inputs from the first layer (O_{ij}^1). The output is the final "ANDed" result (W_k^2) of the neuron corresponding to the kth rule.

$$W_k^2 = \Pi O_{ij}^1 \tag{6.3}$$

6.3.1.4 Layer 3

The third layer is called the *normalization layer.* The output (W_k^2) that we received from Layer 2 is actually referred to as the rule's firing power. This output denotes, or symbolizes, the relative importance of the rule. This is similar to what we discussed and implemented in the implication operator in Chapter 4. Every rule k has some weight (firing power) W_k^2 attached to it. This layer normalizes the inputs according to applied rules and weights given by Equation 6.4:

$$W_k^3 = \frac{W_2^2}{\Sigma_i W_i^2} \tag{6.4}$$

6.3.1.5 Layer 4

This layer calculates the individual output for every rule. For simplicity, we have taken only one output variable. In the case of multiple outputs, the additional output variables would be added in parallel to the system. In other words, this layer performs the role of the implication operator that we studied in Chapter 4. The output of this layer, however, is much generalized and is given by Equation 6.5:

$$O_1^4 = W_k^3 * \left(\sum_i a_i * I_i + b\right) \tag{6.5}$$

where a_i and b are parameters and I_i is the input from the previous layer. We will study the role of these parameters in the later sections of this chapter.

6.3.1.6 Layer 5

This layer simply takes the weighted sum of the outputs of Layer 4 to give the final answer of the system.

6.3.2 Problem Solving in ANFIS

In the previous section, we discussed how to represent an FIS in the form of an ANN. Doing so enables us to use the existing back propagation algorithm (BPA) to train the system in supervised learning mode. In this section, we develop a general problem-solving methodology for ANFIS.

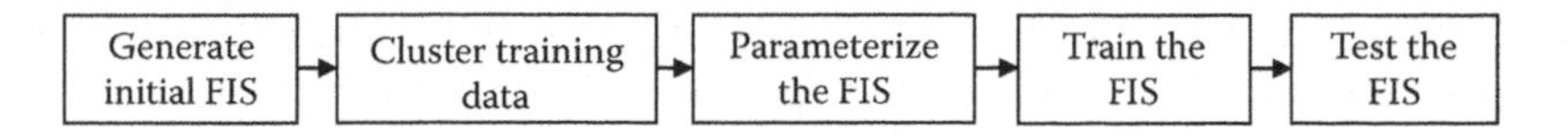

FIGURE 6.2 The problem-solving methodology in ANFIS.

The motivation of ANFIS is to train the FIS using neural network architecture. We start with an initial FIS for the purpose of training. We then represent, or implement, that FIS in the form given by Figure 6.1. This is then parameterized and trained using an algorithm similar to BPA. Training data may be clustered for faster training. The general problem-solving methodology in ANFIS is given in Figure 6.2. We discuss the various steps of the algorithm in the next subsections. We avoid a very mathematical and detailed discussion of this topic, due to a shortage of space.

6.3.2.1 Initial FIS

As stated above, we start by generating an initial FIS, which may be further trained by the system. If the initial FIS is not generated, we may use a randomly generated FIS with predefined inputs and a predefined number of MFs per input. All MF combinations make rules for the randomly generated FIS.

6.3.2.2 Clustering Training Data

In some cases, the data may be too large for the network to train. Training with a large amount of data may result in the training time being too large. For this reason, the training data are often clustered to reduce it to a manageable size. These data may then be trained by the training algorithm. We may use any clustering method for this, including fuzzy C-means clustering or k-means clustering.

6.3.2.3 Parameterization of the FIS

The next step is to add parameters to the generated FIS model. These parameters have a special relevance to the FIS in that they help the ANN implement the ANFIS to fine-tune the system. Hence the system's performance should improve with an increase in the number of parameters, which, in turn, increases the degrees of freedom associated with the FIS. We use generalized versions of all MFs in the system. We further induce weights wherever possible. All this helps make the FIS more generalized to the same problem.

6.3.2.4 Training

A training algorithm is used to adjust the parameters that we introduced in the previous step. This training algorithm finds the most optimal set of values for which the outputs generated by the system follow the training outputs as closely as possible with the least errors.

6.3.2.5 Testing

Finally, the modified FIS is ready to handle the known and unknown inputs. The FIS may be used directly for the implementation.

6.3.3 Training

The training adjusts the parameters of the FIS. This adjustment greatly improves the system's performance. The training algorithm's output is the fixed parameters that make the FIS attain such high performance. Two very commonly used training algorithms are the back propagation algorithm and hybrid learning.

6.3.3.1 Back Propagation Algorithm

The BPA in ANN was discussed in Chapter 2. A similar mechanism is followed to train the ANFIS. Here the inputs are applied, and then the outputs are calculated. The error is measured as a difference of the actual and the observed output. This error is back propagated and is used to modify the various parameters. This modification follows a steepest descent approach.

6.3.3.2 Hybrid Training

Hybrid training is a special algorithm that tries to train the ANFIS in two stages: the forward stage and the backward stage. In our earlier discussion of the layer evaluation in ANFIS, we introduced numerous parameters to the system. These parameters can be classified under two separate heads. The first are the *premise* parameters, which are found on the left side of the implication. These parameters are nonlinear in nature. The second classification is the *consequent* parameters, found on the right side. These are linear in nature.

In the forward stage of the hybrid training algorithm, we use the least-squares method to find the error. We modify the consequent parameters in order to minimize error. During this stage, the premise parameters are kept as constants. In the backward stage, the reverse is true: The errors are propagated backward, and the premise parameters are updated. A steepest descent approach is used for the modification.

6.3.4 Types of ANFIS

ANFIS may be classified into various types, depending on the rules or logic used. We briefly discuss the different classifications in this section. Based on the logic in which the rules are formed or connected, the ANFIS is divided into two major types:

- **OR type:** This system uses OR logic for the various rules that are framed. The OR logic connects the MFs in an OR manner.
- **AND type:** This system connects the rules with the help of the AND operator, as we discussed earlier.

Furthermore the ANFIS may be divided on the basis of the governing logic:

- **Mamdani type:** This system is based on the Mamdani logic.
- **Logical type:** Here the FIS uses a logical working of the rules.

6.3.5 Convergence in ANFIS

The training graph of the ANFIS, in which we plot the errors against training iterations, behaves in a similar manner to the training graph of the ANN. While training in ANFIS, we see that as the iterations grow, the error gets smaller, and the outputs of the inputs given to the ANFIS closely follow actual values. This is largely because of the gradient descent approach that is used in both these algorithms. Hence the training curve of the ANFIS actually converges. This convergence, both in terms of number of iterations and time, depends on many parameters, including the complexity of the FIS, the learning rate, and so on. Again recall from Chapter 2 that we usually use three data sets. The first is for training, the second for validation, and the third for testing. In this way, we were able to induce ANN over FIS and to optimize the FIS parameters. Doing so reduced the necessity of using trial and error to make an FIS. Furthermore the resulting FIS gives high performance while also following the properties of the FIS.

6.3.6 Application in a Real Life Problem

So far we have studied the principles and workings of the ANFIS. In this section, we apply ANFIS to solve a real life problem—that of glass identification (Box and Jenkins, 1970). This problem asks us to find the type of glass. The parameters that affect the decision are measured.

The inputs to the system have nine attributes: the refractive index (RI), sodium (Na), magnesium (Mg), aluminium (Al), silicon (Si), potassium (K), calcium (Ca), barium (Ba), and iron (Fe). All metal inputs are unit measurement, weight percent in corresponding oxide. The output denotes the type of glass: building windows float processed, building windows non–float processed, vehicle windows float processed, vehicle windows non–float processed (none in this database), containers, tableware, and headlamps.

The database contains 214 data entries. Out of this, we used 173 (~80%) for training, 21 (~10%) for testing, and 20 (~10%) for validation.

All inputs were normalized to lie between 0 and 1 before application of ANFIS. Each set had inputs from almost all classes. The ANFIS was made to generate an FIS containing three MFs per input and output. Grid partitioning was used to evolve the initial FIS. A hybrid training mechanism was used for 30 iterations.

The system, after training, gave a net accuracy of 76.6300 percent by correctly identifying 164 out of 214 data items. The performance on only the training data set was 83.2370 percent, with the correct identification of 144 out of 173 classes.

One thing that we would be interested in is the closeness of the desired output to the actual output. This is shown in Figure 6.3 for the training, testing, and validation data sets. The various inputs are plotted consecutively on the x-axis. For each input, two points are plotted: the actual output and the desired output. Ideally these two points should coincide, but this happens only when the training is perfect with zero errors. In reality, these points do not coincide, and their separation visually depicts the system error per input.

Figure 6.4 shows the surface diagram obtained by varying the first two inputs (all others being constant). This figure graphically depicts the input space. For the output to be plotted against a maximum of two dimensions, we are required to keep the other outputs constant.

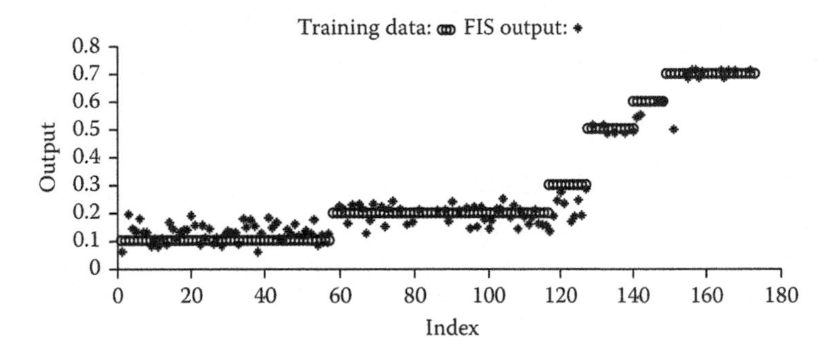

FIGURE 6.3a The closeness of the training outputs in ANFIS.

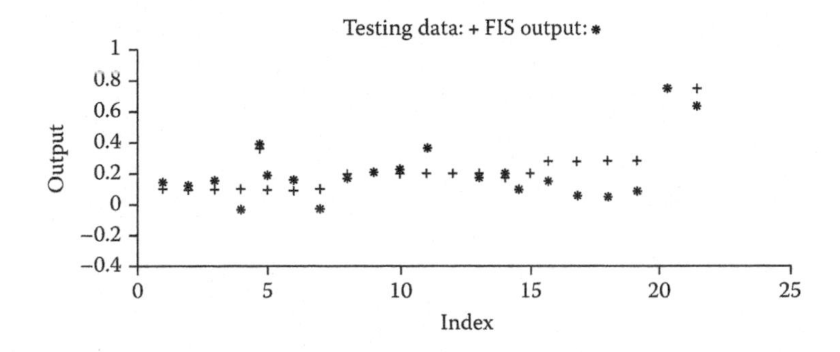

FIGURE 6.3b The closeness of the validation outputs in ANFIS.

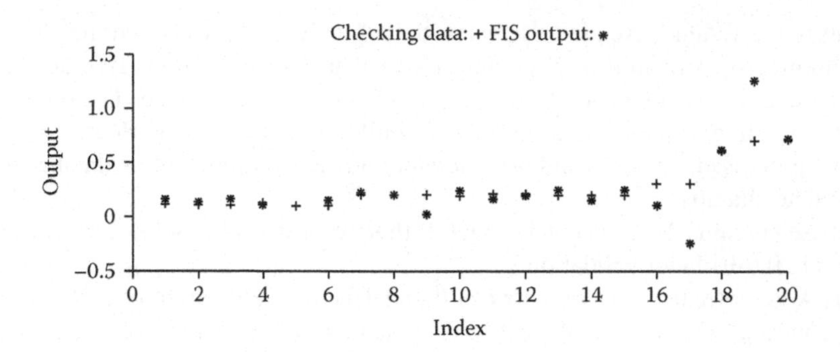

FIGURE 6.3c The closeness of the testing outputs in ANFIS.

6.4 EVOLUTIONARY NEURAL NETWORKS

We studied ANN training earlier in this text. The ANN using BPA as a training algorithm still has shortcomings. It is quite likely that BPA results in some local minima in place of global minima. We also must specify the initial parameters before the learning can start. These shortcomings pose restrictions on the use of ANNs. The GA, on the other hand, is known for its ability to optimize. In this section, we fuse this capability of the GA with the ANN's ability to train.

This hybrid solution overcomes many of the problems that arise with ANN training. The solution generated by the GA is more likely to be global minima, in contrast to the training provided by the BPA. Furthermore the GA can be used to generate the complete ANN. In such a case, we need not specify any initial parameter for the ANN, as the GA would be able to completely evolve the ANN.

We study the problem of the ANN's evolution on two levels. First we try to evolve an ANN that has a fixed structure. Then we generalize this solution to evolve the structure of the ANN, as well as the weights and other parameters.

6.4.1 Evolving a Fixed-Structure ANN

In this case, the ANN's structure is fixed which needs to be specified by the user beforehand. The GA fixes the values and the weights and biases of the various nodes. In other words, the GA optimizes the network parameters for better performance.

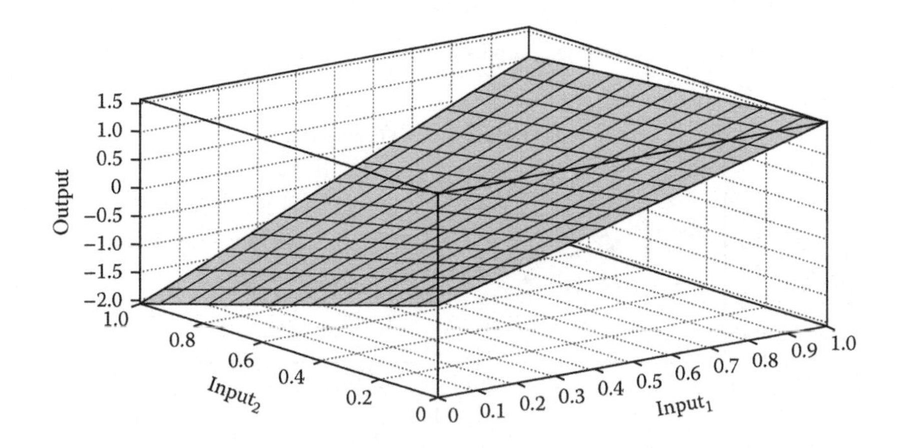

FIGURE 6.4 The surface diagram for the problem.

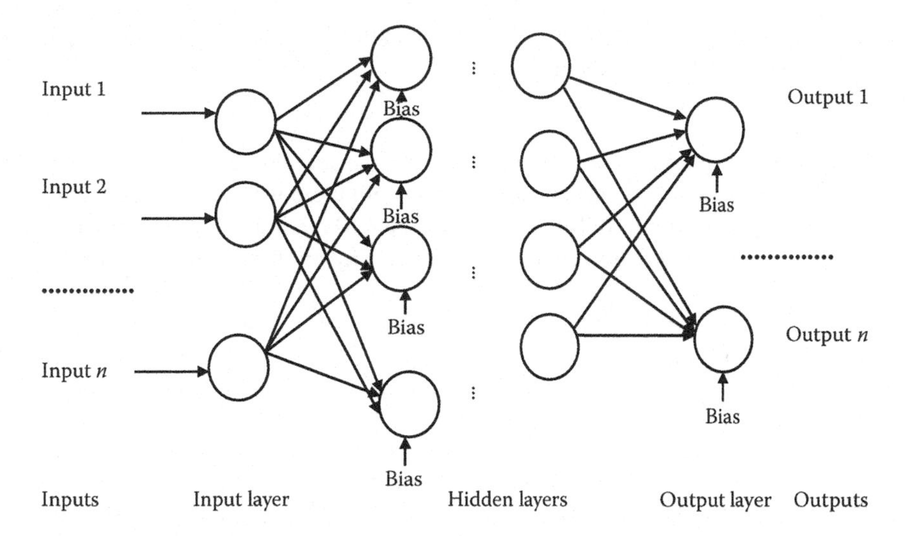

FIGURE 6.5 The general architecture of artificial neural networks (ANNs).

An ANN is a collection of various neurons arranged in layers. The last layer is always the output layer. There may also be many hidden layers between the input layer and the output layer. Each layer has one or more neurons. In a fully connected approach, every neuron of layer i is connected to every neuron of the forward layer $i + 1$ by some weight. This weight is adjusted during training. In addition, every neuron has some bias associated with it. The general structure of the ANN is given in Figure 6.5.

6.4.1.1 Problem Encoding

For simplicity, we only consider the case of one hidden layer. Let there be I inputs, H neurons in the hidden layer, and O outputs. This network structure may be represented by $[I * H * O]$. This system has $I * H$ weights between the input layer and the hidden layer and $H * O$ weights between the hidden layer and the output layer, making the total number of weights $W\ I * H + H * O$. Furthermore the number of biases is equal to the number of neurons. Each hidden layer and each output layer neuron has an associated bias. Hence the total number of biases is $H + O$. All of these weights and outputs are the system parameters that need to be optimized. This means that for a single-layer ANN, $I * H + H * O + H + O$ parameters must be optimized.

Problem encoding consists of arranging these parameters in a linear array, or the phenotype problem representation. The population may be represented using any double-vector or bit-string representation.

6.4.1.2 Genetic Operators

All standard genetic operators (see Chapter 5) are used, including selection, crossover, eliticism, mutation, and so on. The genetic operators ensure creation of good individuals from one population to the next.

6.4.1.3 Fitness Function

The fitness of any individual in the population is measured by the fitness function. The fitness function here consists of the ANN and its training data set. In the fitness function, we first create and initialize an ANN according to the various parameters that the GA generated in the individual. These parameters were extracted from the individual and were then used to set the weights and biases of the ANN.

Then the training data set is passed through the ANN. The ANN's performance against this data set is measured to determine the net fitness value of the GA that needs to be maximized (or the negative performance to be minimized). Hence every time the GA demands the measurement of fitness value of some individual, the ANN is created, and the value is measured by the performance. This interfaces the GA with the ANN during training.

6.4.1.4 Testing

Once the GA reaches its optimal state and terminates per the stopping criterion, we get the final values of the weights and biases. We then create the ANN with these weights and bias values; which is then regarded as the most optimal ANN to result from the ANN training. We can then use the evolves ANN for testing. Note that validation data are not necessarily required in this type of training.

6.4.1.5 Experimental Verification

In this step, we experimentally verify the evolutionary ANN by using it to solve a real life application. In this case, we solve the Box and Jenkins glass furnace problem (Box and Jenkins, 1970) using evolutionary ANN.

The Box and Jenkins gas furnace data consist of 290 measurements of the gas furnace system. The input measurement $u(k)$ is the gas flow rate into the furnace, and the output measurement $y(k)$ is the concentration of the outlet gas. The sampling interval is 9 seconds (s). We wish to determine a fuzzy model of the gas furnace system. In our simulations, we assume that $y(t) = f(y(t-1), y(t-2), y(t-3), y(t-4), u(t-1), u(t-2))$.

The gas furnace problem is a functional prediction problem. Of the 290 instances in the data set, we set 232 (~80%) for training and 58 (~20%) for testing. Each instance has 10 attributes. The output contains seven classes.

We use a single hidden layer ANN with a structure of [10 * 13 * 1]. Initially we form this structure using trial and error in ANNs. We then apply GA for parameter optimization. The weight matrix consists of 10 * 13 weights between the input layer and the hidden layer, 13 * 1 weights between the hidden layer and the output layer, and 13 hidden layer biases and 1 output layer bias, for a total of 157 variables for the GA.

In GA, we use the double-vector method of population representation. The total number of individuals in the population is 10. We choose a uniform creation function, rank-based scaling function, and stochastic uniform selection methods. The elite count is 2. We then use single-point crossover. The program is executed for 100 generations. The crossover rate is 0.7.

The training curve showing the average and mean values is given in Figure 6.6. The root mean square error (RMSE) for the trained ANN is 0.080614 on training data and 0.0919 on testing data.

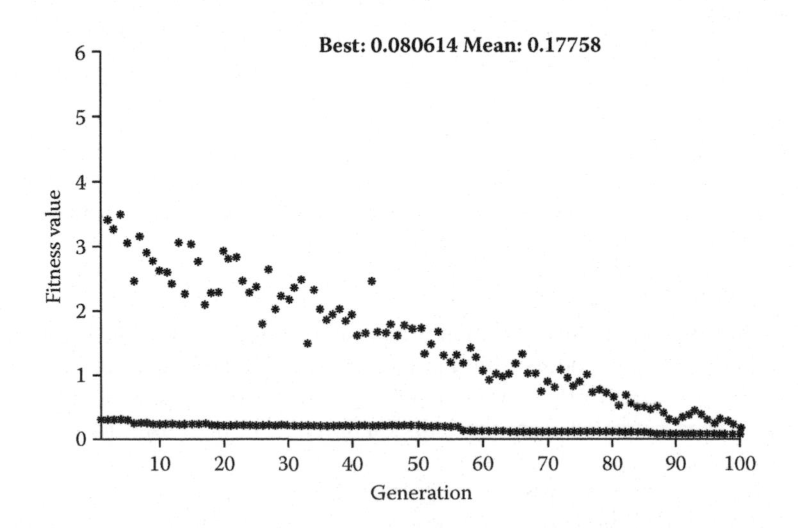

FIGURE 6.6 The training of ANNs by genetic algorithms.

This clearly shows that the GA trained the ANN well and the resultant ANN has achieved high performance.

6.4.2 Evolving-Variable Structure ANN

In the previous section, we saw an interesting application of GA in ANN in which we trained an ANN with the help of GA. However, that was only one part of the solution; the designer would still have to waste a great deal of time and energy in deciding the correct architecture for the system. Furthermore the generated solution might not reach the global minima because of the wrong architecture being selected. This poses a serious problem and limitation with ANNs.

In this section, we solve this limitation by using GA to completely evolve ANN. The use of GA as proposed in this section not only generates optimal values for different weights and biases but also evolves the correct architecture for the ANN. This not only relieves the designer from the task of experimenting again and again to find the correct architecture, but also ensures that the most optimal architecture will evolve. This further means that the solution will be highly generalized.

The basic principle or ideology behind the problem remains the same. Here we will train an ANN with the help of a GA. Because we need not specify any parameter about the ANN, we may say that we will evolve an ANN with the help of GA. We first look at one of the most crucial aspects of the problem—problem representation or encoding. We then move to the other aspects of GA, which are more or less the same as those discussed in the previous section.

Problem encoding in this case may be done in two ways: direct encoding or grammatical encoding. Both are discussed below. We use the same notations as in the previous section for the rest of our discussion.

6.4.2.1 Direct Encoding

In direct encoding, the ANN's entire structure is encoded directly into the chromosome. The chromosome now represents two things. The first is the connection scheme that exists in the ANN, while the second is the actual weights of the ANN. Both may be appended one after the other to make the complete chromosome. Suppose the system has a total of N neurons. This is the size of the ANN. Our task is to use GA to find the correct connection (or architecture) among these N nodes and the actual weights wherever the connections exist.

The connections between the neurons are denoted by the connection matrix, which is of size $N \times N$. The matrix contains a 1 at cell a_{ij} if there is a connection from neuron i to neuron j; otherwise, it contains a 0. In this manner, we can denote any ANN having any form of structure. Consider the matrix given in Figure 6.7a, which corresponds to the network architecture shown in Figure 6.7b. It may further be verified that Figure 6.7b is the only network possible by the use of the matrix given in Figure 6.7a. The encoding is hence unique.

	1	2	3	4	5	6	7
1	0	0	1	0	1	0	0
2	0	0	0	1	1	0	0
3	0	0	0	0	0	0	1
4	0	0	0	0	0	1	0
5	0	0	0	0	0	1	1
6	0	0	0	0	0	0	0
7	0	0	0	0	0	0	0

FIGURE 6.7a The representation of connections in an ANN.

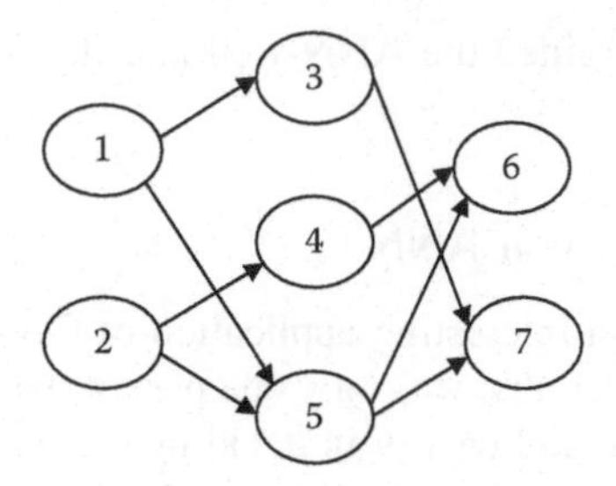

FIGURE 6.7b The ANN corresponding to Figure 6.7a.

The actual weights between the neurons may be stored in another similar matrix in which each cell contains the weights between the corresponding neurons. These weights may be ignored if no connection exists between the neurons as depicted by the connection matrix.

In this way, we implement a direct encoding. To make a chromosome from this encoding, the two matrices may be represented linearly and appended one after the other. We may follow a bit-string representation or a double-vector representation for implementation.

6.4.2.2 Grammatical Encoding

Grammatical encoding is a different style of coding that does not directly store the connections and weights. Instead, it uses grammar to store the information necessary to generate the architecture whenever needed. The grammatical rules are framed and used whenever the conversion from the genotype to the actual model must be done. This grammar-based encoding is very useful, because it reduces the size of the string. This type of encoding has added features to the convergence and speed of the GA and is hence very useful.

The GA frames its own specialized grammar to help in the notation and evolution of the ANN. For further details of this notation, refer to our discussion of grammatical evolution in Chapter 5.

6.4.2.3 Fitness Function

The fitness function in this method is the same as we discussed in the fixed-architecture model of the ANN. In this case, however, the ANN's architecture is variable, which means we must construct the ANN entirely by using the information supplied by the individual solution to the GA. We first form the architecture and then assign the weights to the different connections and biases. The rest of the procedure is the same as discussed above.

The experimental verification of this method may be done in a manner similar to that described in the previous section.

6.4.3 Evolving Learning Rule

Another exciting fusion of the ANNs with GA is done in the domain of learning rules. Here we use GA to evolve learning rules for the ANN. This is another area where hybrid systems have resulted in enhanced capabilities of the system. We combine the learning capability of the ANN and the available learning base with the optimizing capabilities of the GA. The result is that the GA can enhance the ANN's learning by telling it the exact adjustments that must be made during back propagation of the errors.

In a conventional ANN with BPA, the learning follows these steps: We first apply the inputs to the network to get the corresponding outputs. The error is calculated by the difference between the current output and the desired output. Per the back propagation fundamentals, this error is back propagated through the previous layers to adjust their weights. This weight adjustment is typically done using the steepest gradient approach.

Systems using the evolving learning rule follow the same mechanism. The only difference is in the modification of the weights. From our understanding of the BPA, we know that the amount of modification made to any weight in a single-layer ANN is a function of the activation function of the following variables:

a_i, the activation of input i
o_j, the activation of output unit j
t_j, the training signal on output unit j
w_{ij}, the current weight on the link from i to j

The change in weight may hence be represented as given in Equation 6.6:

$$\Delta w_{ij} = f(a_i, o_j, t_j, w_{ij}) \tag{6.6}$$

Here we assume that the learning rule should be a linear function of these variables and all their pairwise products—that is, the general form of the learning rule is as given in Equation 6.7:

$$\Delta w_{ij} = k_0 (k_1 w_{ij} + k_2 a_i + k_3 o_j + k_4 t_j + k_5 w_{ij} a_i + k_6 w_{ij} o_j + k_7 w_{ij} t_j + k_8 a_i o_j + k_9 a_i t_j + k_{10} o_j t_j) \tag{6.7}$$

where the k_m $(1 < m < 10)$ are constant coefficients and k_0 is a scale parameter that affects how much the weights can change on any cycle.

In these systems, we use GA to evolve the values of the constants k_m's. The encoding of these constants may be done in a double-vector or bit-string format.

6.5 EVOLVING FUZZY LOGIC

In this section, we use a combination of FIS and GA for the purpose of problem solving. FIS are known for their ability to create systems based on lingual rules and to handle uncertainty. They introduce nonlinearity to make effective MFs or systems that are much closer to real life systems. The GA is a very good tool for optimization purposes in FIS.

One of the major disadvantages of FIS is that the rules must be clearly defined because they govern the mapping of the inputs to the outputs. Thus the rules must be complete in all respects, which require a great deal of patience and experience on the part of the system designers. One small mistake in creating the rules can have a huge impact on the system's performance. The same is true for the MFs. We must cautiously decide the number of MFs, as well as the values of their parameters, as this also has a deep impact on the performance of the FIS. Many times, even if the system performs, the performance is not always optimal. In other cases, the designer must constantly make adjustments to search for the right set of rules and parameters. This poses a serious limitation to the use of FIS. Despite poor performance, however, there are good chances of the system being trapped at some local minima.

In this system, we use the GA's optimization powers in the FIS, similar to how we used GA for the training of ANNs. We follow the same ideology in this section. As in the previous section, we divide our discussion under two heads. The first covers the evolution of the FIS with fixed rules. The second is the entirely generalized form, in which we evolve the entire FIS along with the structure.

6.5.1 Evolving a Fixed-Structure FIS

We first evolve a fixed FIS. Because much of the discussion is similar to that of the evolutionary ANN, we mainly address problem encoding here.

Because the structure of the FIS is fixed, the number of MFs for all inputs, outputs, and rules has already been decided. We must now optimize the parameters of all the MFs. Suppose the FIS contains n inputs and m outputs, where each n input and each m output will have some MFs. Let us assume that any input i has n_i MFs. Similarly let us assume that any output o_i has m_i MFs.

The MFs can be of any shape and size and can be at any location. They may have any number of parameters. For simplicity, we assume that all MFs are Gaussian in nature, which means they have two parameters through which we decide the spread and location of the Gaussian function. Through this, we can easily calculate the total number of parameters that the GA is supposed to optimize.

All these parameters are represented in a single array with values one after the other. We may represent these parameters using either a double-vector or a bit-string mechanism.

The individual contains the FIS. Whenever the fitness of any individual is to be measured, we first initialize an FIS. Then we use this FIS to get the system's performance for some training inputs. This performance becomes the GA's fitness value. In this manner, the GA connects to the FIS.

One problem that this system faces is that of feasibility. Many solutions generated by the FIS may be infeasible due to a number of reasons. One possibility could be when a *high* MF of any input/output aligns itself before the *low* MF. Another possibility could occur when some MFs try to cover a large part of the input region. These solutions are naturally undesirable. While the system designer was working, he or she would have avoided such illegal states through common sense; the GA, however, would not know to avoid generation of these illegal states. Hence the performance might be low.

We can make use of this problem to create even better system design. For example, we could give all parameters a relatively small space within which they can optimize themselves, instead of wandering through the entire search domain. This would make the GA's search space much more concrete and would hence result in enhanced system performance. Limiting the search space, however, should be done cautiously, because a very narrow search space might miss the location of the global minima. On the other hand, a very large search space might suffer the same problem of the generation of a large number of infeasible solutions.

After the entire training has been completed by reaching the specified stopping criterion, the next step is to use the trained parameters to generate the final FIS. For this, we take the optimized values of the parameters that the GA returns. We then generate the final FIS using these parameters as the values of the MF parameters.

6.5.1.1 Experimental Verification

We study this method by applying it to a small problem. Consider the problem of a robotic controller that we studied in Chapter 4. We saw that a great deal of emphasis was placed on the decision of the MFs and their parameters. Here we optimize these parameters using evolutionary FIS.

The FIS used has the same structure as the one we used in the problem in Chapter 4. Recall from that chapter that the problem was to make a robot reach the goal starting from a given source. The system had two inputs—the angle and the goal. The angle α is the angle by which the robot must turn in order to face the goal. This angle is measured by taking the difference between the robot's current angle φ and the angle of the goal measured by the robot's current position θ. The result is always between –180 degrees and 180 degrees. The other input is the goal, or the distance between the robot's current position and the position of the goal. The distance is normalized by multiplication of a constant, so that it lies between 0 and 1. The system has a single output: the angle by which the robot may turn at the next move.

The input angle has five MFs. The first and the last of them were trapazoidal MF (trapmf) in nature, whereas all others are Gaussian MF. The other input (distance) has three MFs associated with it, all of which are Gaussian in nature. The output has five MFs, all of which are Gaussian in nature.

The five MFs of input angle take a total of ten parameters, with each MF taking two parameters. Although the two extreme MFs of the input angle take four parameters according to the equation, the other two parameters are constant; hence we do not modify them. Similarly the three MFs of the input distance take six parameters and the five MFs of the output take ten parameters. This makes the total number of variable parameters 26. Thus the total number of variables in GA is 26.

Rule 1: If (α is morep) then (output is morer)
Rule 2: If (α is lessp) then (output is lessr)
Rule 3: If (α is no) then (output is no)
Rule 4: If (α is lessn) then (output is lessl)
Rule 5: If (α is moren) then (output is morel)
Rule 6: If (α is not morep) and (goal is distant) then (output is no)
Rule 7: If (α is not moren) and (goal is distant) then (output is no)
Rule 8: If (α is lessp) and (goal is near) then (output is morer)
Rule 9: If (α is lessn) and (goal is near) then (output is morel)

FIGURE 6.8 The fuzzy rules for the robotic control problem.

The system has nine rules, which may be understood from our understanding of the inputs and the MFs. The rules are given in Figure 6.8. The GA's fitness function consists of three benchmark problems, given in Figure 6.9a, b, and c. The source, goal, and initial angle at which the robot is facing are indicated in the figure.

We use two factors to measure the performance. The performance P for any single test case i is hence given by Equation 6.8:

$$P_i = L_i * T_i \tag{6.8}$$

where L_i denotes the length of the path for test case i and T_i denotes the maximum turn taken by the robot. Both factors are normalized to lie between 0 and 1.

The total fitness F for any individual is the sum of performances in the three benchmark maps, as given in Equation 6.9:

$$F = P_1 + P_2 + P_3 \tag{6.9}$$

The GA optimizes the various parameters using the double-vector method of population representation. The total number of individuals in the population is 20. A uniform creation function, rank-based scaling function, and stochastic uniform selection method is used. The elite count is 2. Single-point crossover is used. The program repeats for 100 generations. The crossover rate is fixed at 0.8. The mean and best fitness of the population at various generations are shown in Figure 6.10.

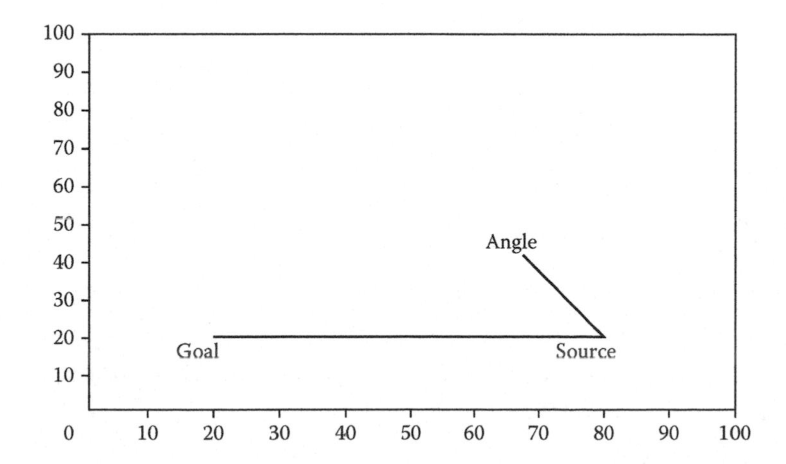

FIGURE 6.9a The first benchmark map used to measure fitness in genetic algorithms for the robotic control problem.

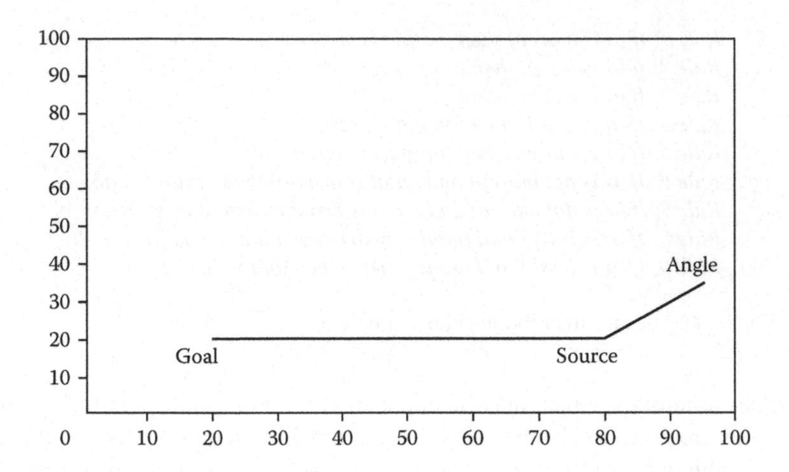

FIGURE 6.9b The second benchmark map used to measure fitness in genetic algorithms for the robotic control problem.

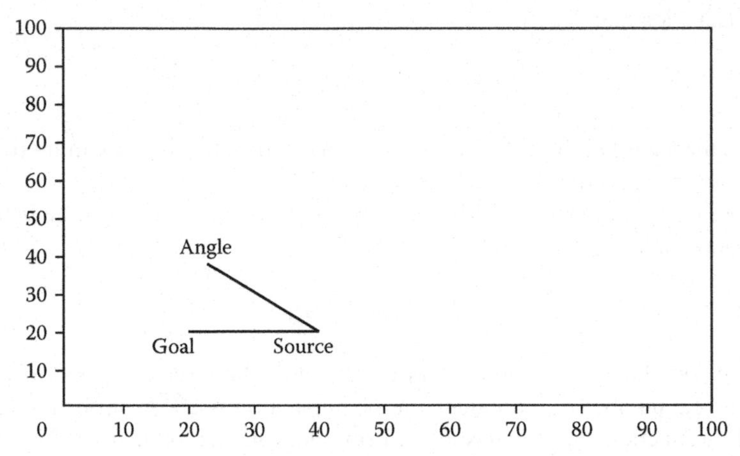

FIGURE 6.9c The third benchmark map used to measure fitness in genetic algorithms for the robotic control problem.

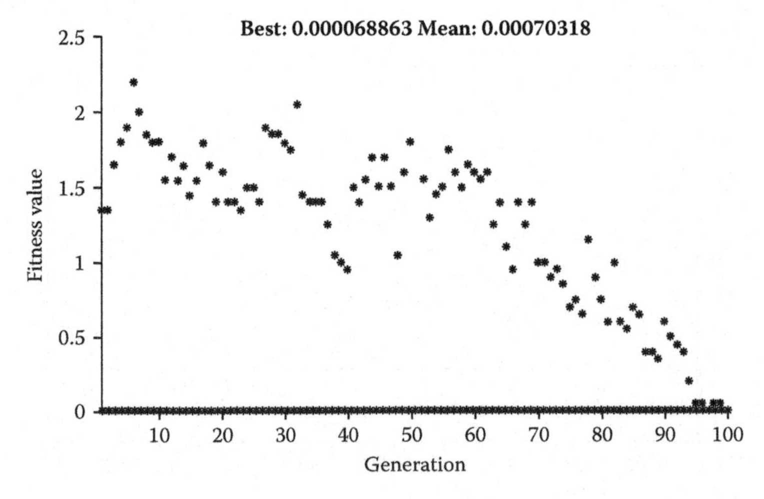

FIGURE 6.10 The best and mean values at various generations of the genetic algorithm in training for the robotic control problem.

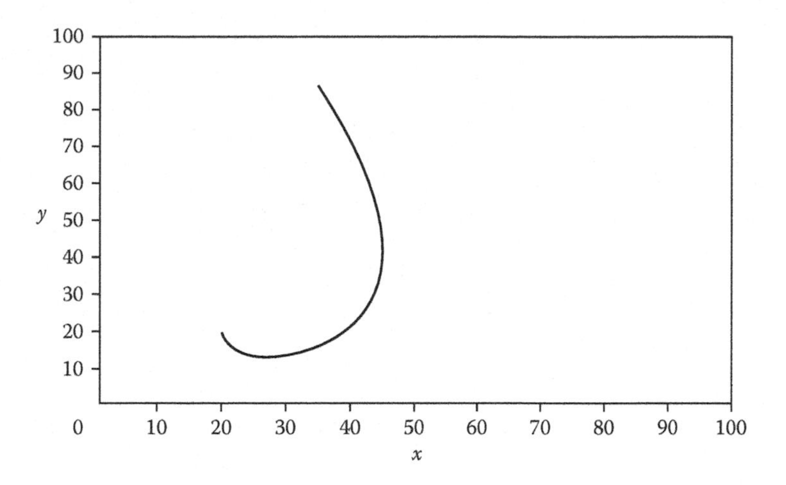

FIGURE 6.11a The path traced by a robot at various combinations of source, destination, and initial angle: First run.

Based on this simulation methodology, the path traced by the robot for various runs with different initial positions and angles is given in Figures 6.11a through 6.11d. In all the figures, the robot is moving upward, so the lower point in the path is the initial position and the upper point is the final position. The robot's angle is the direction in which it initially moves.

Thus using GA we can optimize the performance of the FIS that we built in Chapter 4. More constraints or preferences may be added during fitness evaluation for results optimized for specific requirements. This would make the FIS perform very well per the specific requirements. GA helps attain global minima for efficient solution generation.

6.5.2 Evolving a Variable-Structured FIS

In the previous section, we optimized a fuzzy system to a good extent. But the solution proposed may not be very optimal, because the designer must still do the work to select membership functions and frame rules. The fuzzy system's design still remains a very tedious activity. In addition, because

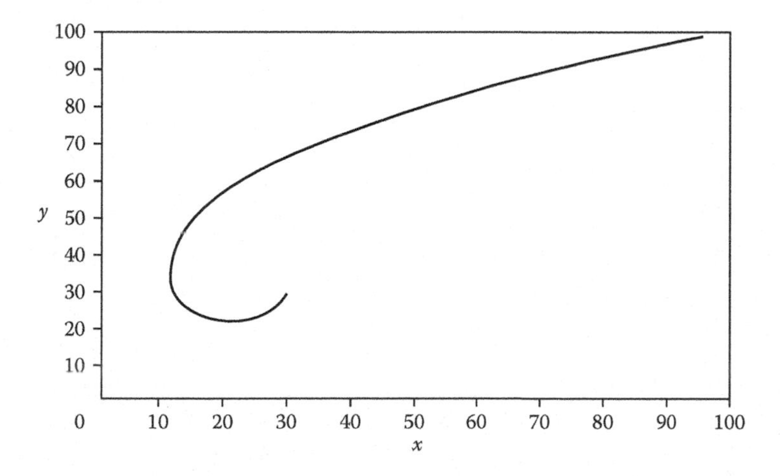

FIGURE 6.11b The path traced by a robot at various combinations of source, destination, and initial angle: Second run.

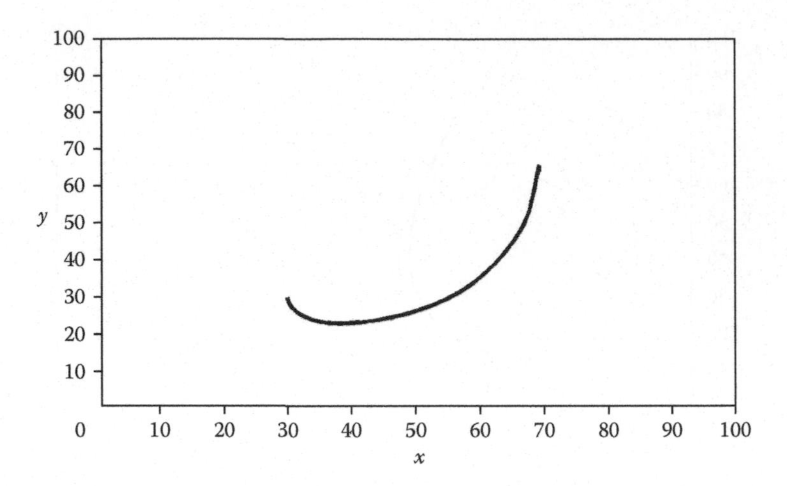

FIGURE 6.11c The path traced by a robot at various combinations of source, destination, and initial angle: Third run.

the rules and MFs were framed by trial and error, it is very likely that the solution generated is not globally optimal.

In this section, we further remove these limitations by making a very flexible system. This system will be able to evolve the complete FIS without having any earlier idea of the kind of problem. Thus the FIS generated is likely to be the most optimal structure. The process of evolution for the FIS is similar to the way we evolved the ANNs with a flexible architecture.

For this system, we mainly follow a direct encoding scheme in which the entire architecture is directly represented as a chromosome or an individual of the GA. This architecture is then applied by the genetic operators and the process of evolution goes on and on. The chromosome can be internally divided into three parts: the MF, the MF parameter values, and the rules. These three parts append each other to make the entire chromosome. This entire chromosome is a collection of numbers that may be encoded as double vector or bit string for implementation purposes.

The first part of the chromosome is the MF. Here we designate the number of MFs for both the inputs and the outputs. The number of MFs per input/output is fixed to a maximum value of α,

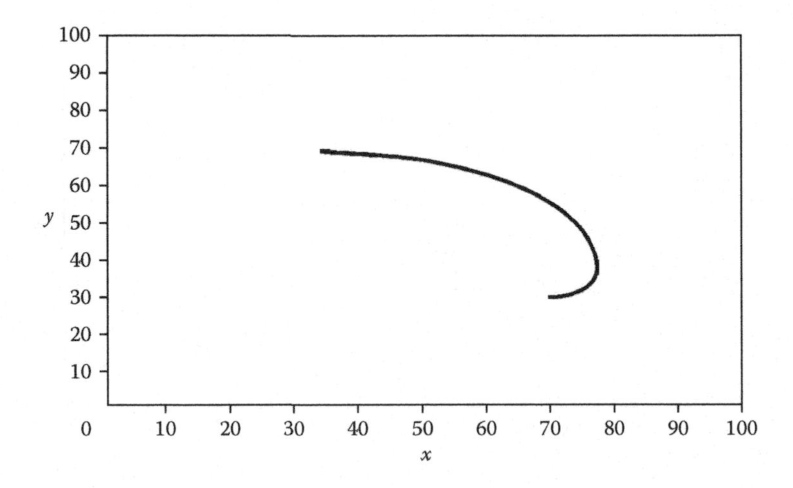

FIGURE 6.11d The path traced by a robot at various combinations of source, destination, and initial angle: Fourth run.

which means the number of MFs per input/output can be anywhere from 1 to α. This is represented as straight integers, with one integer corresponding to each input or output. Thus if there are n inputs and outputs, there would be a room for all n integers in the chromosome, and all these would have legal values between 1 and α.

The other part of the chromosome is the MF parameters. Here we assume that we can completely denote an MF by its MF and its extreme locations in the input graph. One way to deal with these locations would be to vary them according to the type of MF. Thus there would be three values per MF in the order of type, left end, and right end. Here we make room for all α MFs per input/output. If an MF does not exist, there may be any arbitrary values stored in the corresponding locations. Thus if there are n inputs and outputs, there would be $3 * n * \alpha$ numbers reserved in the chromosome to represent the MF values, all of which may not be important. Another major point of concern here is that the MFs must be sorted internally for the algorithm to work. This comes from the natural intuition that the MF *low* can never be higher than the MF *high*.

The last part of the chromosome consists of the rules. Any rule in an FIS may be represented by Equation 6.10:

$$\begin{aligned}&\text{If } (input_1 \text{ is [NOT] } MF_a) \text{ and } (input_1 \text{ is [NOT] } MF_b) \ldots (input_k \text{ is [NOT] } MF_z),\\&\text{then } (output_1 \text{ is } MF_p) \text{ and } (output_2 \text{ is } MF_q) \ldots\end{aligned} \tag{6.10}$$

where NOT is an optional construct.

Now we discuss the mechanism for representing the same rule in the form of a sequence of integrals. Every $input_i$ can be associated with any one of the corresponding MFs that belong to Input $_i$. Suppose this $input_i$ has a corresponding entry of MF_t in the rule. In other words, the rule has a term ($input_i$ is MF_t). We replace this word by the integer t. Similarly if the term is ($input_i$ is not MF_t), we replace it with the integer $-t$. Suppose that $input_i$ does not participate in the rule at all. In such a case, we write 0 to the contribution corresponding to $Input_i$. The same is true for the outputs, except that the NOT construct is not present at the outputs. Hence negative integers are not allowed in the rules.

We can completely replace the rule with the integer string $<t_1\ t_2\ t_3\ t_4 \ldots o_1\ o_2>$, where any t_i is the MF associated with $Input_i$ and any o_i is the MF associated with output i. Any integer can be positive, negative, or zero. Because there are n inputs and outputs, this string will have exactly n integers. Furthermore each integer can be only within the range of $-\alpha$ to α.

Let us suppose that the system can have a maximum of β rules. Each rule would be a collection of n integers. This means that the rules can occupy the maximum space of $n * \beta$ in the chromosome. It is not necessary for all rules to be occupied. In fact, each individual may have many unoccupied rules. The empty rules may be stored with any arbitrary value. The total number of rules that are actually present is denoted at the start.

We have discussed the maximum space needed for any chromosome to be implemented and the fact that the extra segment per the MF or rules may be given any arbitrary value. The needed integers would lie somewhere inside this chromosome at their respective positions. The appended representation of all the constituents makes up the complete chromosome.

The other operators of GA, such as mutation, selection, and crossover, may be applied, as long as the constraints, such as the maximum number of MFs per input/output and the ordering of MFs, are kept in mind. If an individual generated as an initial solution or during mutation or crossover does not adhere to these rules, we apply a simple repair operation to make it a valid chromosome of the GA.

The same problem of robotic control that we discussed and solved in the previous section may be solved using this mechanism as well. Upon solving, we would see that the FIS generated is optimal per the user's demands and requirements. With this, we end our discussion on evolutionary systems.

6.6 FUZZY ARTIFICIAL NEURAL NETWORKS WITH FUZZY INPUTS

We have already modeled the FIS by ANN architecture and saw that the FIS could be made to learn. We did this using the neuro-fuzzy systems. In this section we will combine the powers of ANN with that of fuzzy logic (FL). We do so to make the ANN benefit from the FL by making it capable of handling uncertain or imprecise inputs. This impreciseness is a natural occurrence that we take from real life examples. This has led to the modeling of impreciseness or fuzziness in almost every problem that we find. The models accounting for this impreciseness can handle data much better than the other systems.

6.6.1 Basic Concepts

The ANN that we build in this section will be able take fuzzy inputs, which means we can give any type of input—fuzzy or nonfuzzy. The inputs given to the system are handled in a manner similar to any ANN, but with due respect given to fuzziness in every stage. For this procedure, we define some new operators and methods for working with fuzzy inputs to give the same functionality that the individual tasks give to integers in a conventional ANN. The outputs generated by the ANN are also fuzzy in nature, which means we can never say with confidence that we have the final output; the output always has some degree of uncertainty associated with it.

The uncertainty in real life applications may be in any number of forms. Imagine we are doing an experiment with five sensors taking different readings. Unfortunately one of the sensors is broken and does not give a reading. This situation adds uncertainty in the system. We have two options. We may choose to neglect the reading and discard the recorded data, but this would result in loss of information. Or we could choose to work with the data, keeping in mind the effect of uncertainty at every step. In this situation, we would not only save data from being lost, but we would also get an idea of the system's output and performance. Note that even though the other sensors gave readings, their reading can never be guaranteed, because there is always some uncertainty.

In this section, our main focus is to use all our understanding of conventional ANNs to study the effect adding fuzziness to these systems. We will see how the various mathematical operators need to be modified to handle this fuzziness. Then we present the actual picture of the system.

6.6.2 Fuzzy Arithmetic Operations

ANN calculation of an output for any input or learning using BPA uses some arithmetic operators both to predict the output and to learn. Hence the major task before us is to form fuzzy equivalents of these arithmetic operations. Once this is done, we can simply replace the nonfuzzy operators from the ANN with the modified fuzzy operators.

Recall our discussion about fuzzy sets, in which we learned that every member of a fuzzy set has two important pieces of information. One is its attribute or its value, and the other is the degree of certainty or membership of this member. Say the output is 5 with a membership value of 0.5. This means we have 50 percent confidence that the output will be 5. We use this same concept throughout the rest of our discussion here. This information may also be represented by the notation p/x, where p is the membership degree and x is the value. Thus, in our example, the representation would be 0.5/5.

Addition: Addition of two fuzzy sets simply adds the corresponding values. The answer is the arithmetic addition of the corresponding values. The membership, or certainty, associated with this answer is the minima of the two membership functions. Of all the different combinations possible that make the same sum, the maxima of this minima is taken as the final value. In this way, we get both the value and the membership function.

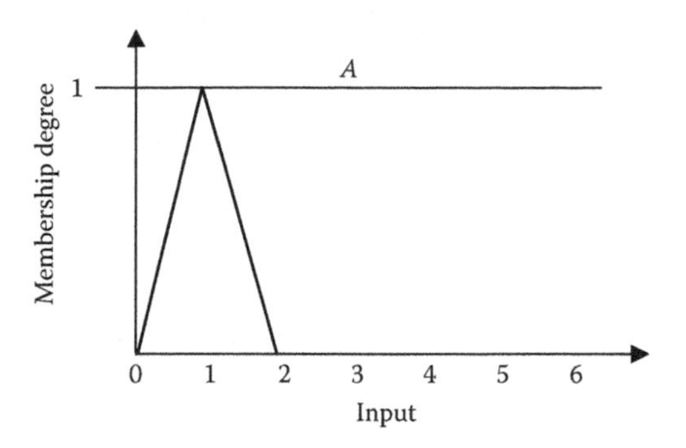

FIGURE 6.12a The mathematics of fuzzy sets: Input *A*.

The addition is given by Equation 6.11. This operation may also be seen in Figure 6.12c using the inputs given by Figures 6.12a and 6.12b.

$$\mu_{A+B}(z) = \max\{\mu_A(x) + \mu_B(y)/z = x + y\} \tag{6.11}$$

Product: This operation is similar to the addition commonly used in ANN. In product, we multiply the two values to get the final output value. The membership associated with the output is calculated in a similar mechanism to the one used for addition. Note that if the fuzziness is removed, or the probability of occurrence is kept at 1, the fuzzy arithmetic converts to conventional arithmetic. The product operation is given in Equation 6.12. This is shown in Figure 6.12d over the inputs given by Figures 6.12a and 6.12b.

$$\mu_{AB}(z) = \max\{\mu_A(x) \cap \mu_B(y)/z = xy\} \tag{6.12}$$

Function: Here we study the effect of adding any nonlinear function or any function in general. Function is a fundamental operation used in ANN as the activation function. The value after application of the activation function operation is the same as the function being applied to the original value. This keeps the system consistent with the nonfuzzy system or conventional ANN.

The application of a function over fuzzy input is given by Equation 6.13. The operation is shown in Figure 6.12e for the activation function 1/(1 + exp(–x)) over the inputs given by Figures 6.12a and 6.12b.

$$\mu_{f(Net)}(z) = \max\{\mu_{Net}(x)/z = f(x)\} \tag{6.13}$$

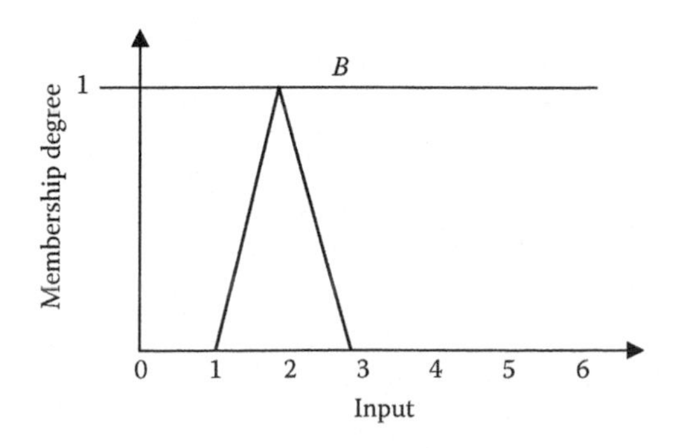

FIGURE 6.12b The mathematics of fuzzy sets: Input *B*.

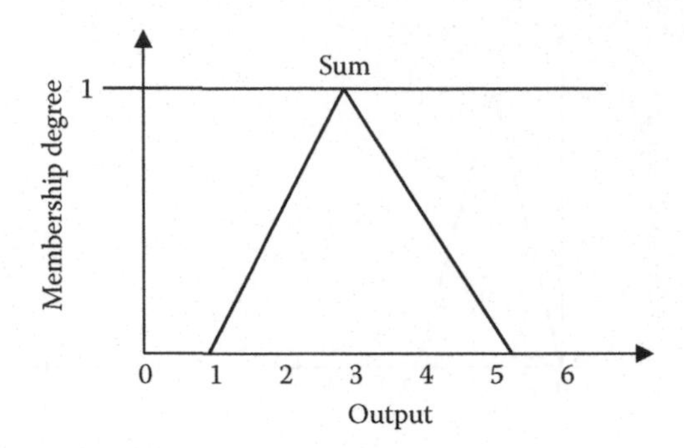

FIGURE 6.12c The mathematics of fuzzy sets: Sum operation.

6.6.3 ALPHA Cut

In this section, we discuss another concept associated with fuzzy algebra. This concept is known as ALPHA (α) cut. The operations of fuzzy algebra are usually done in a section of the input that has high membership value. This is also known as level sets. The h level set of a fuzzy number consists of all elements whose membership value is greater than h. This forms a closed interval in the input. Suppose that X is the fuzzy set; the h level set would be given by Equation 6.14:

$$[X]_h = \left[[X]_h^L, [X]_h^U\right] \tag{6.14}$$

The complete mathematics of the fuzzy logic is outside the scope of this book, but it can be done using the concepts presented. The BPA uses the error function, derivatives, and so forth, in addition to the presented operators. Without going deeper into the other operators, we present the modified BPA here.

6.6.4 Modified BPA

To learn the available database, the fuzzy neural networks use the learning algorithm, which is a modified form of the BPA. Here we apply the fuzzy inputs, look at the fuzzy output, and calculate the fuzzy error that is passed backward (or back propagated) to the previous layers. As with BPA, training is done in a batch-processing manner and is terminated based on some stopping criterion. All of this is carried out for a number of α cuts, as shown in algorithm.

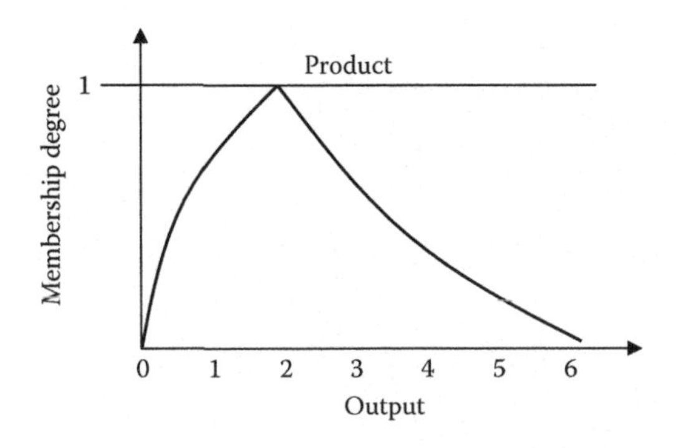

FIGURE 6.12d The mathematics of fuzzy sets: Product operation.

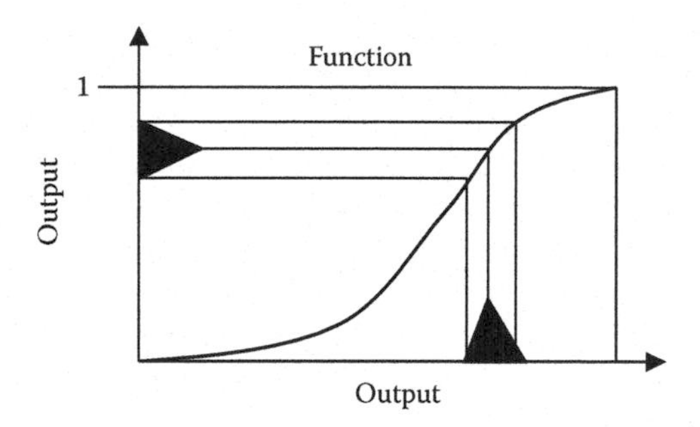

FIGURE 6.12e The mathematics of fuzzy sets: Function operation.

Learning Algorithm in Fuzzy ANN with Fuzzy Inputs

Step 1: Initialize the network
Step 2: While stopping condition not met
Step 3: For all $h = h_1, h_2, \ldots, h_n$
Step 4: For all input, output pair i in data set
Step 5: Calculate fuzzy output o
Step 6: Calculate fuzzy error e
Step 7: Back propagate and adjust fuzzy weights according to error e

6.7 RULE EXTRACTION FROM ANN

We learned about problem solving using ANN in Chapters 2 and 3. We saw that ANNs are great tools for learning historical data and predicting correct outputs from unknown inputs. This capability of ANNs enables them to be used in a variety of applications. We even discussed some of these applications. ANNs are a perfect choice of system design, because they do the entire work of rule formulation, learning, and so forth on their own, relieving designer from all these jobs.

6.7.1 Need of Rule Extraction

The extraordinary capability of ANNs often leads to the conclusion that they are black boxes. This means they simply take inputs and give outputs, without revealing much about the process, rules, or means of doing so to the user. Thus the designer has no knowledge or information whatsoever about how the output was achieved. Skilled and experience designers may be able to interpret something by looking at the training curve, network architecture, or training data set. But still, there is no way to understand how the inputs map to the outputs.

While studying expert systems in Chapter 1, we discussed the importance of rules. In Chapter 2, we stressed the fact that the weights in an ANN play the role of rules. These weights decide the mapping of the inputs to the outputs because the learning involves adjustment of the weights. In one sense, the weights are just a collection of real numbers. But in fact, they denote a long list of real numbers that would not make much sense to a person.

Hence it is clear that we need some means for interpreting the weights of the ANN. The motivation is not to accept the ANN as black boxes but to make some sense of them so that designers or users can better understand the system. This will further help us understand the relation between the inputs and the outputs.

Fuzzy systems are known for their ability to work in accordance with lingual rules that may be framed according to common sense. These systems map the inputs to the outputs according to these rules, which by nature are very clear and easily readable by humans. This fact not only helps us understand the relation between inputs and outputs, but also gives us a means for predicting outputs to unknown inputs. If an input does not give a correct output in an ANN, we cannot do anything much about it. In FIS, however, we may be able to reanalyze the rules and find the faults. These faults may due to any reason, including unframed rules that result from neglecting some cases. The easily understandable nature of the FIS helps a great deal in understanding system and design.

In this section, we fuse the discussed capability of fuzzy rules into the ANN. This is called the extraction of rules from the ANN. Although any form of rules may be extracted from the ANN, we limit our discussion here to the fuzzy rules we discussed in Chapter 4. These rules are always extracted from a trained ANN, because this is the point in the system when the rules have been framed or learned by the ANN. Here we take an ANN and convert it into a set of fuzzy rules that behave in a similar manner to that of the ANN in both known and unknown inputs.

6.7.2 System Inputs, Outputs, and Performance

In order to completely represent the system, we need a collection of fuzzy rules. These rules have the same inputs as those of the ANN. Furthermore the rules have the same outputs as those of the ANN. The MFs are variable according to each input and output and may be framed in a variety of ways in order to get the best set of extracted features. These fuzzy rules may be of the form given below:

If ($Input_1$ is [NOT] MF_{a1}) and ($Input_2$ is [NOT] MF_{b1}) . . ., then ($Output_1$ is MF_{p1}) and ($Output_2$ is MF_{q1}) . . .
If ($Input_1$ is [NOT] MF_{a2}) and ($Input_2$ is [NOT] MF_{b2}) . . ., then ($Output_1$ is MF_{p2}) and ($Output_2$ is MF_{q2}) . . .
If ($Input_1$ is [NOT] MF_{a3}) and ($Input_2$ is [NOT] MF_{b3}) . . ., then ($Output_1$ is MF_{p3}) and ($Output_2$ is MF_{q3}) . . .
If ($Input_1$ is [NOT] MF_{an}) and ($Input_2$ is [NOT] MF_{bn}) . . ., then ($Output_1$ is MF_{pn}) and ($Output_2$ is MF_{qn}) . . .

This system has a total of n rules. As we learned in our discussion of ANN, a good ANN is one that gives the best performance in the least number of neurons. Furthermore as we learned in our discussion of FIS, a good FIS gives the best performance in the least number of rules and MFs per input and output. The same is true here. Good extracted rules imitate the ANN in the best possible way in the least number of rules and MFs. Hence the extracting algorithm prefers to keep the number of rules and MFs limited when extracting rules. This is called the property of *comprehensibility.*

The other factors that contribute to the quality of the rules are *accuracy* and *fidelity*. The *accuracy* measures the correctness of the outputs given by the extracted fuzzy rules to unknown inputs. *Fidelity* indicates how well the extracted fuzzy rules imitate the behavior of the ANN.

6.7.3 Extraction Algorithms

Various algorithms have been proposed for the extraction of rules from a trained ANN. Over the past few decades, research in this field has resulted in algorithms that try to optimize the performance of the extraction to give better extracted rules in various ways. Here we study the broad category of extracted rules. The rule extraction algorithms are usually classified in two broad domains: decomposition algorithms and pedagogical algorithms.

Decomposition algorithms try to decompose the whole ANN into rules. They express the conditions under which the output or the hidden units in the ANN are activated by a set of logical rules. These rules collectively make the system, where the individual rules are aggregated to get the resultant outputs. This type includes the algorithms KBANN/M of N and RULEX.

Pedagogical algorithms follow a pedagogical approach to learning or evolving the fuzzy rules. The entire ANN is seen as a black box in which the various rules are tested for performance. These rules are framed on the algorithmic logic. This type includes the learning algorithm of Craven and Shavlik (1994).

Numerous techniques are applied in both these approaches, including the use of genetic algorithms, heuristics, and other intelligent systems.

6.8 MODULAR NEURAL NETWORK

Modular neural networks (MNNs) are modular implementations of ANNs. The basic motivation is to divide the entire complex task or problem into much more manageable problems. This is done in a modular manner. Hence MNNs are very useful in problem solving because they make many problems realizable that were not possible in the earlier architectures.

6.8.1 Need for MNNs

The rapid increase in the use of ANNs and the very high accuracies received from them offer encouragement for the use of ANNs to solve various problems. Real life problems, however, may be very complex. As discussed in Chapter 1, real life applications become more difficult to handle as the size or dimensionality of the training data increases. When this is the case, the different soft computing approaches face difficulties. These approaches also have difficulty performing well when the mapping of the inputs to the outputs is in a complex manner and not easily realizable even for ANNs. The rapid increase in the demands of ANNs indicate a great need for solving these problems. We also discussed in Chapter 1 that computational and memory constraints pose a serious limitation to the working of ANNs or of any soft-computing method in general. MNNs, however, can solve these problems by introducing the novel concept of modularity.

If we try to solve a complex problem with ANNs, we face various problems, including a limited time required for training. In addition, ANNs often refuse to train due to high complexity. In any case, the performance of the ANN for complex problems is not very high. This performance further reduces with addition of noise or unknown data. Another problem is that ANNs tend to forget past data when they are exposed to new data, especially when the data are not presented for long. These problems are solved by the use of MNNs, which use a modular design for solving the problem in parallel.

6.8.2 Biological Inspiration

The idea behind modularity was inspired from the natural systems. Modularity exists in the human brain. Every part of the brain has some specific task. One part is responsible for hearing, another for thinking, and so on. In the brain, there is a natural integration of information that enables it to act in a highly sophisticated manner. The brain's modularity in at various forms and in various levels.

6.8.3 Modularity in ANN

Modularity is the breaking up of a complex task into a set of simpler tasks. Each task may be done by some independent or dependent component or system. The basic objective of modularity is for numerous systems to collectively perform some task better than any single system. This means that the problem being considered needs to be modular in nature as well. In other words, it should be possible to divide the entire problem into sets of problems in such a manner that they can be handled by different systems. These divided problems are known as *modules*.

The modules of the ANN may be divided into two categories. In the first, known as *soft modularity*, there is some dependency among the various divided problems or modules. These dependencies are communicated during the run and need to be met for the entire algorithm to continue. In the second modularity, known as *hard modularity*, the various modules work entirely independent of one another. They all perform autonomously to solve their part of the problem or module.

If we look at the conventional ANN, we can easily see that the entire architecture can be broken down into modules. Furthermore the various tasks in the problem can be broken down. This gives a great deal of scope for the division of ANNs into modules. The manner in which this division is performed differs from problem to problem.

6.8.4 Working of the MNNs

The MNN first divides the entire problem into smaller problems, or modules. The parts of the problem are then solved separately by each module, with each module finding its own respective results. After the various modules have been solved, they are communicated to the central component, called the *integrator.* The integrator divides the problem into the subproblems and hands this information to the various modules. It also takes the results of the various modules and integrates them, or adds them up, to get the final solution.

The working of the MNN is similar in concept to the divide-and-conquer approach. This approach believes in dividing the task into a set of related smaller tasks. This division is known as the divide stage. These related smaller tasks are solved using a recursive approach. The solution to each is then combined to get the final solution. This integration is known as the conquer stage.

Hence MNNs are very actively used in the field of research and development. The work in MNNs lies at two levels. The first level is the division of the problem into subproblems. The second level is the integration of the various subproblems and the formulation of the final result from the individual results.

The division of the problem into subproblems, or modules, is one of the first tasks that is performed. This step is done during the design of the MNN. The division may be carried out in a variety of ways, such as adaptive resonance theory (ART), hierarchical networks, or ART with back propagation (ART-BP) algorithms.

The integration deals with the combination of various results. Earlier models used addition or the mean to combine the results of various modules. Other nonlinear approaches were also used, including Baysean networks, Dampster-Shafer belief theory, and fuzzy modeling. We will study some of these techniques in the next section.

6.8.4.1 ART-BP Network

This ANN contains the unsupervised ART for dividing the problem into constituent modules or ANNs. The training of this ART is done in an unsupervised manner, and it mainly clusters the training data and divides it into various modules. Each ANN is trained using a supervised BPA. Each ANN is an ART network. The training here goes on in parallel among the various ANNs, which improves the training speed. The basic block diagram of an ART-BP network is given in Figure 6.13.

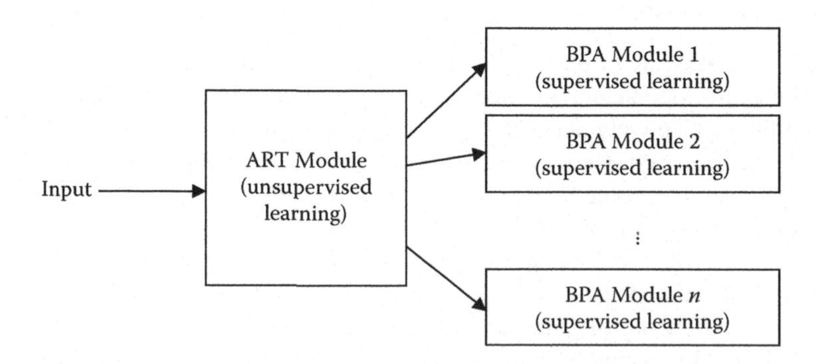

FIGURE 6.13 The block diagram of the adaptive resonance theory by back propagation network.

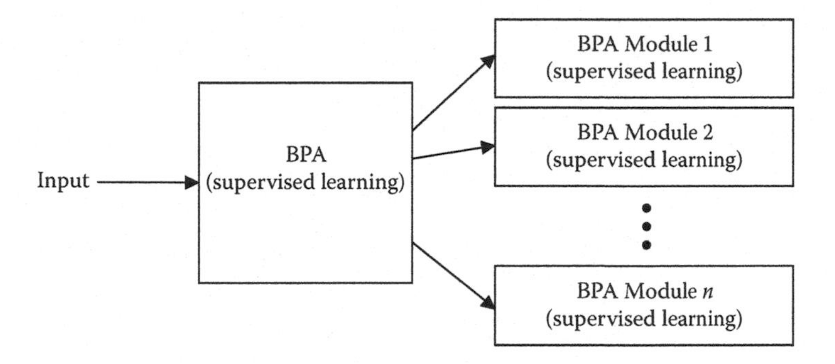

FIGURE 6.14 The block diagram of the hierarchical network.

6.8.4.2 Hierarchical Network

This network contains a BPA at the two hierarchies. In all other terms, it is similar to that of the ART-BP algorithm. However, here the first BPA does the minor work of classifying whatever the problem demands. The information at this level is very vague and not accurate. The final solution is generated by the next ANN, which uses BPA to get the final answer. The basic block diagram of a hierarchical network is given in Figure 6.14.

6.8.4.3 Multiple-Experts Network

With this network, the basic motivation is to solve the same problem using a variety of experts or ANNs. Each expert gives the output according to its analysis. Then the final output is calculated using all these outputs. The differences in the various ANNs may be introduced by varying the number of neurons, architecture, or initial weights. The decision of combination is made according to the weighted mean. An additional parameter of *gait* is inserted into each ANN's output. This parameter helps control the contributions of the various ANNs. Gait weighs each expert and decides its influence in the system's final decision. Some experts are assumed to be good and are thus given higher weights, whereas others may be given smaller weights. Fuzzy reasoning or any other method may also be used for integration. The basic block diagram of these networks is given in Figure 6.15.

6.8.4.4 Ensemble Networks

Ensemble networks consist of multiple networks that are identical in nature. Each network solves the problem according to its own parameters. The combination of the results to generate a final

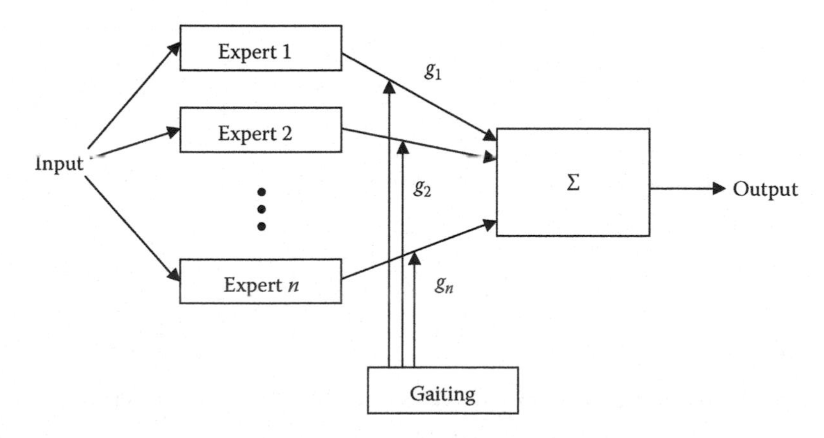

FIGURE 6.15 The block diagram of the multiple-experts network.

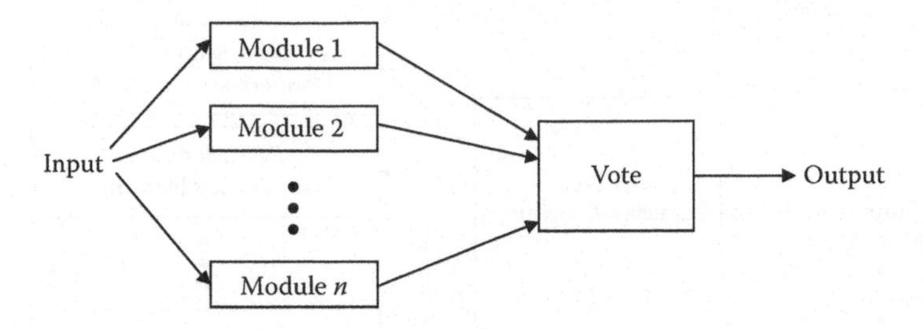

FIGURE 6.16 The block diagram of the ensemble networks.

solution is done by the principle of majority vote. In this case, each network nominates one class that it believes the input belongs to. The result is based on the class that gets the maximum votes. Sometimes weighted average may also be used. The block diagram of ensembles is given in Figure 6.16.

6.8.4.5 Hierarchical Competitive Modular Neural Network

Hierarchical competitive modular neural networks (HCMNNs) contain a hierarchy of ART networks. The lower hierarchy (also called the stem ART) has a low level of detail and classification. The higher level (also called the leaf ART) has a relatively higher detail. This higher ART maps the input to the ANN using supervised BPA. In this manner, the entire problem is tackled in a hierarchical manner. The basic block diagram is given in Figure 6.17.

6.8.4.6 Merge-and-Glue Network

In this network, we first divide the task based on heuristics according to the problem's logic. This distributes every problem to the networks. The different networks may be trained in parallel. After all the networks have been trained, the different networks combine to form a single, large network using the weights and architectures of the individual networks. Learning can then be resumed in this larger network, starting from the network's error configuration. We also add some more neurons, which are supposed to discover more features as they are exposed to more training data. These neurons are known as *glue*. The basic working of this type of network is given in Figure 6.18.

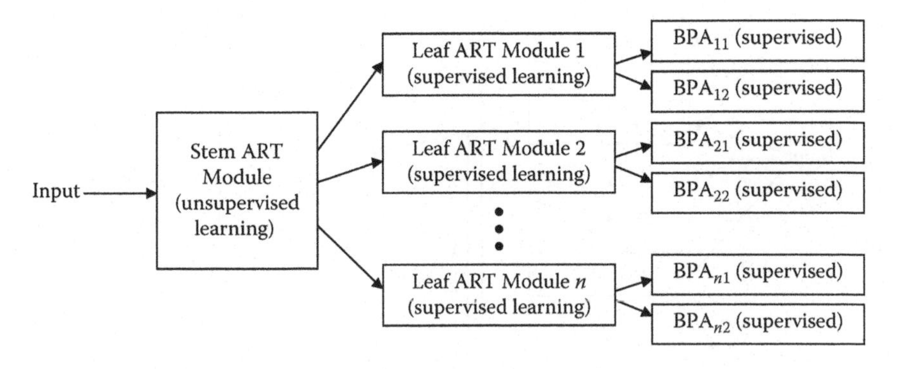

FIGURE 6.17 The block diagram of the hierarchical competitive modular neural network.

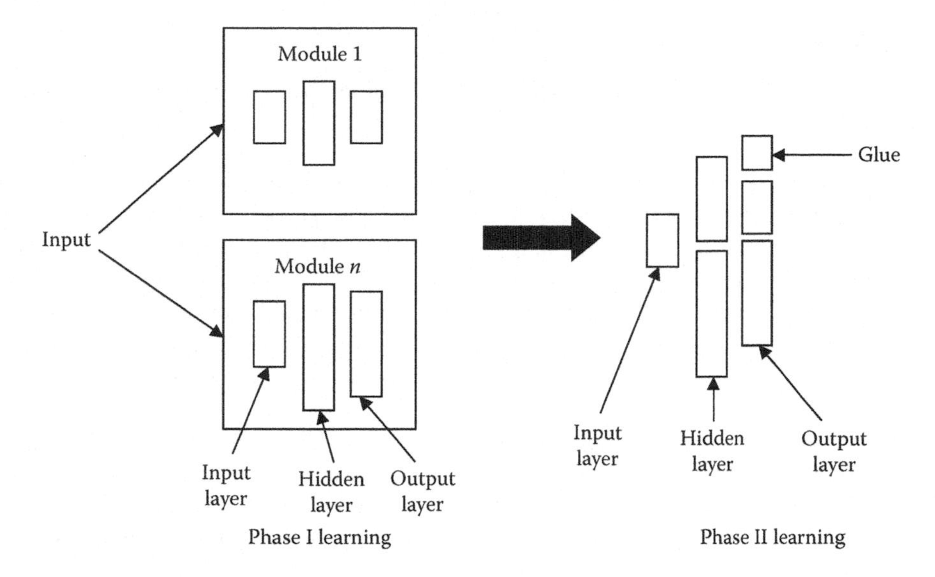

FIGURE 6.18 The block diagram of the merge-and-glue network.

CHAPTER SUMMARY

In this chapter, we presented the intensely useful systems that are built by fusing the properties of previously studied systems, including artificial neural networks, genetic algorithms, fuzzy inference systems, artificial intelligence, and logic. We explained the basic properties of these systems that can be exploited to generate robust systems. These systems are known as hybrid systems. The need for and usefulness of these hybrid systems in real life applications carries a special importance of its own. In this chapter, we presented numerous hybrid systems, their concepts, and workings.

The first system discussed results from the fusion of FIS with ANN, which generated the ANFIS. The FIS was represented in a neural network architecture, which enabled the application of ANN training to the generated FIS. We further discussed how the ANFIS could be used to optimize the FIS parameters, thus yielding a good system.

The genetic algorithm (GA) was then used to evolve ANNs, in a process known as evolutionary ANN. The system was discussed under two heads: fixed architecture and variable architecture. We saw how these systems could be used to attain global optima to reach the highest level of performance.

A similar system is one that uses GA with FIS. This time, the GA was used to evolve FIS. Again we saw that the application of GA led to the evolution of an FIS that showed great performance. The application of GA further ensured that the system would reach global optima.

The next topic of discussion was fuzzy artificial neural networks. These ANNs combined with the power of fuzzy logic to introduce the novel concept of impreciseness. These systems had fuzzy weights, fuzzy inputs, and fuzzy outputs. We studied and analyzed fuzzy algebra; using this, we converted the back propagation algorithm (BPA) to work over fuzzy numbers.

We then studied how we could extract rules from the ANNs. The importance of this task, and the ways and means to do so, was presented. The final topic discussed was modular artificial neural networks (MNNs). This section included a discussion of the need for these networks and their applicability. Numerous algorithms that implement MNNs and that are widely used in real life applications were presented.

Throughout the text, we looked at various real life problems and solved these problems by the discussed methods.

SOLVED EXAMPLES

1. In MNNs, is it preferable to have a large number of modules or a low number of modules? Explain.

Answer: For this example, we consider the MNN in which the problem is divided and subdivided in each hierarchy. In this manner, the solution is generated.

The increase in hierarchy, or the number of modules, is a great boon to execution time. As the modules increase, the running time gets smaller and smaller. This is true for both the testing phase and the training phase. The time factor plays an even more vital role in the training phase, where time drastically reduced with an increase in the number of modules. This often makes it possible to get the networks train which would not have been possible in any other manner.

The increase in the number of modules, however, also has a negative effect. The generalizing capability of the ANN continues to decrease as we increase the number of modules. The ANN will continue to be localized to a set of inputs in the input space, which is an undesirable effect for any ANN. Furthermore an increase in the number of modules means an increase in the system's memory requirements.

Suppose we keep increasing the number of modules to the degree that the number of modules is equal to the number of inputs. Now every module has effectively one input to train and test. The unknown inputs find the closest modules or the closest of the known inputs. This highly localized ANN is undesirable.

Thus when deciding the correct number of modules, it is necessary to strike the right balance among training and testing time, memory requirements, and the generalizing capability.

2. How do you handle the problem of missing data in experiments using fuzzy ANNs?

Answer: Fuzzy ANNs can take both fuzzy and nonfuzzy inputs, which allow them to handle the experiments that have missing values. This can happen in both training and testing phases.

Consider an experiment with n attributes. An entry in this experiment may be represented as $<x_1, x_2, x_3, \ldots, x_n>$. These form the inputs to the ANN. The ANN may be used for functional approximation between the inputs and the outputs. In general, the inputs would be well determined, and we would be able to work with them using fuzzy or nonfuzzy ANNs.

Now suppose that the observed value for an input i is missing from the database. This input would be in the form $<x_1, ?, x_3, \ldots, x_n>$, where ? denotes the missing value. Suppose that this attribute had legal values in the range of a to b. We replace the missing value entirely by this range in order to give the fuzzy input to the ANN. The input given to the ANN would be $<x_1, [a\ b], x_3, \ldots, x_n>$.

This method would work both in the training and the testing phase. Furthermore the uncertainties in the output can be handled in a similar manner.

3. Explain the importance of controlling the number of neurons in evolutionary ANN.

Answer: We know that any good ANN solves the problem given with the highest performance and the least number of neurons and layers. The evolutionary ANN does not respect this fact. Hence it may often generate large networks that may not perform well when testing data. This would result in reduced performance.

Therefore it is necessary to devise a means for restricting the number of neurons from being too large in size. At the same time, however, we need to respect the fact that for many problems, the optimal solution is only when the number of neurons are kept quite high, perhaps due to the complexity of the problem. The system needs to generate large networks in these solutions.

Various modifications to the earlier discussed model are necessary to cater to these needs. We must start with the networks that have low neurons as the initial solution. This would control the network from growing too much. In addition, very large networks being generated may be reframed or repaired at the start. The restriction size may increase as the time increases. Another way to handle this could be to introduce an extra penalty for large networks. Further mutation and crossover need to be modified to keep the size of the network from growing rapidly.

4. It is better to have as less as possible number of rules being extracted out of the ANN. Similarly the number of MFs and antecedents need to be limited. Explain.

Answer: As was the case with ANNs, in which we saw that the fewer the neurons, the greater the generalizing capability. The same happens to be true with FIS. Here the number of rules needs to be as small as possible. The more rules, the more difficult it will be for us to conceptualize and understand the system. Similarly, the generalizing power decreases as the rules or MFs increase.

Consider the system with n training data cases. Each is of the form $<I_1, I_2, I_3, I_k, O_1, O_2, O_3, O_l>$, where there are k inputs and l outputs. For every input we frame a rule "If ($Input_1$ is MI_1) and ($Input_2$ is MI_2) and ($Input_3$ is MI_3) . . ., then ($Output_l$ is MO_1) and ($Output_l$ is MO_1) and ($Output_l$ is MO_1) . . ." where each MI_1, MI_2, MI_3, . . ., MO_1, MO_2, MO_3 is an MF with centers at I_1, I_2, I_3, . . ., O_1, O_2, O_3, . . ., respectively and of width such that every MF extends from the center of the maxima of the previous MF and goes to the maxima of the next MF. This would make a system of n rules. Clearly the performance would be 100 percent for the training data.

This system, however, may not perform well on the testing data. Furthermore it is highly localized in nature.

QUESTIONS

General Questions

1. What are hybrid systems? What is the advantage of using hybrid systems in place of non-hybrid systems?
2. What is the role of production rules and artificial intelligence (AU) in hybrid systems? Give an example.
3. In what sense is ANFIS better than (a) ANN? (b) FIS?
4. What are evolutionary ANN and FIS? Compare and contrast.
5. What is the use of having fuzziness in ANNs?
6. What do we mean by modularity in neural networks? Under what circumstances is it beneficial?
7. What is the use of extracting rules from ANNs?
8. Compare and contrast the system of rule extraction from ANN to that from ANFIS.
9. Name a few MNNs. Briefly explain the working of each.
10. Why is it better to have a low number of rules extracted when extracting rules from ANNs?

11. Suggest a method for implementation of evolutionary ANNs. Is there a possibility of generating infeasible solutions? If yes, how would you deal with those solutions?
12. Compare the generalizing capability of the MNNs with that of the ANNs.
13. How is fuzzy arithmetic different from conventional arithmetic? Discuss the major operators of fuzzy arithmetic.
14. Explain the grammatical evolution of ANN.
15. Compare the search spaces of the two means of evolution of variable-architecture ANNs.
16. Explain the working of the various layers of ANFIS.
17. Why are modifiable parameters kept high in ANFIS? Does it affect the training data size requirements?
18. Based on the discussed systems, explain how the ANN, GA, and FIS complement each other.
19. Explain the characteristics of good rules extracted from a trained ANN.
20. What is the advantage of fuzzy ANNs over conventional ANNs?

Practical Questions

1. Download any database from UCI Machine Learning Repository and solve the problem using the various hybrid methods discussed in this chapter.
2. Consider the iris database we used for clustering in Chapter 1. Now cluster the same data by dividing them into the least number of clusters such that the average distance of every data element d_i from its corresponding center c_i is always less than some threshold α: $| d_i - c_i | < \alpha$. Use evolutionary algorithms.
3. Write grammatical rules for the evolution of (a) ANN and (b) FIS by grammatical evolution. Test these rules on any real life application.

Section II

Soft Computing in Biosystems

7 Physiological Biometrics

7.1 INTRODUCTION

On an average day, you may encounter a number of people and identify each by face. Even when you meet new people only briefly, you can remember them, at least for a short time. A person calls you from behind, and you may know who it is without turning around. If you spend enough time with certain people, it is highly likely that you will start noticing peculiar facts about them, such as "he walks very fast" or "she stammers." The fact is clear: We are able to identify people by their face, habits, speech, and numerous other means. This identification is very natural to us. Every day we learn and identify using all these features in a completely online and adaptive system. We do not need to sit with a file every morning, trying to recollect all the faces. It just happens on its own. Learning a new face does not mean we forget older ones. The actual system in application is very simple: We look at someone and immediately remember him, to whatever extent it may be. Subsequent meetings might result in even better remembrance. This occurs even though someone's features change with time. A person will not look the same 20 years from now as he looks right now. A person's voice will change in numerous ways over time. In practice, we do not really realize all this. We are able to take the big and small changes that take place over time and adjust our recognition systems accordingly. All this happens throughout our lifetime.

No machine has been able to imitate the behavior of recognition to this same extent. Even the best-developed automated systems have much lower scalability and accuracies as compared with natural systems or human beings. The extreme potential for recognition in humans and other species is certainly a big mystery. It is also an inspiration for all of us.

This inspiration, coupled with the need for automation, has led to efforts in making automated systems that recognize people by biometric features, in manner similar to how humans recognize features. These systems, or applications, are the point of discussion in this chapter and in the next. In these chapters, we study making systems that see a face and automatically recognize it. Or, similarly, these systems may hear a voice and identify or verify the speaker.

The application of and need for such systems are immense. They are being extensively used for both recognition and identification. These systems not only make programs faster by eliminating the need of human intervention, they also offer more rigorous security.

In this chapter, we first take a brief look at the characteristics and basics of the system. We then move on to the job of building these systems. We develop a face-recognition system, followed by a similar system for the iris. This chapter is dedicated to building the physiological biometric systems. In the next chapter, we continue our discussion, extending it to behavioral biometric systems. Then we further extend our discussion on multimodal biometric systems that use more than one biometric modality for the purpose of recognition. Each modality is explained in the next section.

7.1.1 WHAT IS A BIOMETRIC SYSTEM?

Biometrics is the personal or physical characteristics of a person. These characteristics belong to a person as a physical or behavioral entity and include the face, eye, fingerprint, and speech. Thus biometric systems use biometric identities for the purpose of identification or verification.

The biometric features of every individual are distinct; no two people in the world have the same biometric feature. This means that each individual in the world has a fingerprint that is possessed by only him and no one else in the world. Likewise speech is a feature that no two people in the world have in common. These ideas have led to continuous efforts to identify others by these biometric identities. At the same time, these biometric identities have become the legal proof of convicting someone.

In most cases, these biometric identities can be easily matched by humans. For example, it is simple to see two photos and decide whether each shows the same person. Similarly it is easy to listen to two voices to judge whether they match. This task, however, is not so easy in the case of signatures, fingerprints, or even iris, which require specialization in order to identify them.

Today automated systems are able to identify people on any of these biometric identities. Furthermore these biometric identities have been fused together for an even more enhanced system.

7.1.2 Need for Biometric Systems

Security is a prime area of concern. The number of cases in which security is violated and a crime is executed is on the rise. The infringement of security has resulted in a reconsideration of our security systems. Security systems should be such that all the possibilities of infringement are ruled out. The saving of a single attack can save the lives of thousands of people and families.

Looking at past cases of security information, we can see that the traditional methods of ID cards, user names, and passwords are very easy to break. These systems do not guarantee security against many possible infringements. This fact motivates using features that are built into humans as a means of secure identification. These features identify the person and cannot easily be copied by any other person. This level of security is achieved by the use of biometric features. A thief may be able to snatch a personal identification number (PIC), security code, or passwords. He may also be able to forge ID cards and other legal documents. But it would be nearly impossible to copy someone's face or speech. The biometric systems use features that are normally not visible or observable. Thus fooling a person using biometric identification would be much simpler than fooling a machine using the same biometrics for authentication purposes. For the same reasons, we find immense use of fingerprint, face, and other such biometric identification systems at the security checks for all major secured places.

The other use of biometrics comes from the motivation of automation. Automatic biometric systems can carry out tasks that would have required a human presence in the past; and then, even with a human presence, it would have taken a very long time to complete. Imagine the valuable time that biometric systems would save if classrooms were fitted with cameras and face detectors. The amount of time spent in attendance would be reduced to zero and would add security against proxies. The use of these systems for security check saves time. Imagine if your mobile could tell you the person standing in front of you. This might help you in locating people easily.

7.2 TYPES OF BIOMETRIC SYSTEMS

Biometrics is usually classified under two heads, or modalities: physiological biometrics and behavioral biometrics. The combination of two biometric modalities leads to another very interesting type of biometric system: multimodal biometric systems.

7.2.1 Physiological Biometric Systems

Physiological biometric systems use static or active biometric characteristics, or the physiological biometrics, for the purpose of identification and verification. These characteristics remain constant and do not change with time. In other words, they can be captured in a unit time by the capturing device. Examples of these biometrics include face, iris, fingerprint, ear, retina scan, DNA, and palm vein. Because they are independent of time, these systems are sometimes considered simpler to capture and work with. The duration for which the characteristic is captured does not play any role in these systems. This chapter is entirely devoted to these physiological biometric systems.

7.2.2 Behavioral Biometric Systems

Behavioral biometric systems use dynamic characteristics that change with respect to time. These systems have to record the characteristic over a specific time frame in order to do the recognition. The entire recording needs to be sampled at certain time frames, depending on the computational time and characteristic. The sampling must always be with a frequency greater than twice the maximum frequency according to the Nyquist criterion. Examples of these biometrics are speech, keystroke, gesture, and gait. Time is an additional dimension that is added to the system. As a result, these systems may sometimes be more difficult to handle. In general, however, we do not often work in the time domain, and hence the affect of time as an added dimension is not very prominent. We look into the behavioral biometric systems in Chapter 8.

7.2.3 Fused Biometric Systems

The individual biometrics may not always give good or secure systems. These systems face various problems. They are much more difficult to train, and sometimes they do not train at all if the data are too large. In addition, these systems may not give a good performance. We often require an even larger accuracy, which these systems may fail to give. To solve these problems, we usually mix two or more biometric features to make fused biometric systems. These systems combine the best features of the individual biometric systems; the resultant system is much more scalable and is able to solve the problem of recognition or verification. The fused biometric systems reach an accuracy that is larger than the individual biometric systems. We discuss these systems in greater detail in Chapter 9.

7.3 RECOGNITION SYSTEMS

In Chapter 1, we discussed recognition systems. The systems that we present later in this chapter are recognition systems that do the task of recognizing a person. Let us quickly brush up on the concepts studied earlier before we make these systems using the artificial intelligence (AI) and soft-computing techniques.

Recognition systems use a device that captures raw data. This device may be a camera for face recognition, a microphone for speech, or a fingerprint reader for fingerprints. The device returns the captured data. These data need to be preprocessed to remove noise and for other problem-specific purposes. The data are then segmented to extract the good part, or the needed part. These segmented data still cannot be used for recognition, because they carry too much information; soft-computing methods would not be able to work with it all. Therefore we extract useful features from the data that are presented. The extracted data are then used for all kinds of recognition.

Recognition systems use a historical database for training. This database contains various instances for every person. It supplies useful information to the system, allowing the system to learn about different people and their characteristics. Internally the system adjusts its parameters so that it gives good performance during the learning process. The training database has extracted features that are used for the purpose of training.

We have been using two important systems in this text: verification and identification. Verification systems are the ones in which a person claims some identity. The role of these systems is to verify or check whether the identity is correct. The system then returns a true if the claim is correct and a false if the claim is incorrect. Hence this system acts more or less like a biometric password, which restricts the entry of possible intruders and allows the entry of all genuine people.

In identification systems, the person's identity is to be found out. The list of choices is already available in the database. The system is asked to find out who the person is. It may be possible that the person's identity does not match with anyone in the database. The system may return the person that it identifies, which may not be correct, or it may return its inability to identify the person.

The adding of people to the database is known as the process of enrollment. In any of the systems, a person needs to be enrolled before the system is asked to verify or identify. The enrollment process involves the characteristics of the person being given to the database and the database being asked to learn the same.

7.4 FACE RECOGNITION

The first biometric system that we build here is the face-recognition system. This system is given the face of a person and is asked to identify the person. The input may be captured by a camera. The captured input goes through the same cycles that we discussed earlier so that it can be used for recognition purposes. The face, because of its easy visualization and natural similarity, is one of the most widely studied systems. It has numerous exciting characteristics that attract the interests of people. In the application domain, face-recognition systems are widely used because of their easy installment and use as recognizers. The need for face-recognition systems is growing rapidly with the increase in need for added security.

Face recognition is a classification problem. The input is to be mapped to one of the many possible classes. The number of classes in the output depends on the number of people registered in the database. Accordingly the complexity of the problem increases as the number of people or the number of classes increase. Each class represents an individual who is a potential person whose identity is being recognized.

One of the major problems associated with face recognition is that of dimensionality. We might think that once the entire picture is given to any classifier for recognition, the solution is simple. We might relate this to the character-recognition systems we studied, in which the entire character was fed into the classifier (i.e., ANN). In reality, however, it is not always so easy because of the high dimensionality associated with the problem. Let us say that the image we are trying to give as input has already been shrunk to the size of 100×100 pixels. Further suppose that we have converted the image from color to black and white. Each pixel can be denoted by a number between 0 and 1 to denote the amount of whiteness. Now suppose we give the same as an input to the ANN. This would mean we would need a total of 10,000 inputs in the ANN. For simplicity, even if we assume one output class and the number of inputs to be as small as the number of outputs, the system would have more than 10^8 weights. Imagine the amount of computation time for even one single forward pass of the system. Likewise, imagine the time required for the training. The number of training data sets would be of the order of weights. This places a large demand for the training data set as well. It may hence be seen that such a system is impossible to make.

With respect to the above discussion, we realize the need for feature extraction or other dimensionality reduction techniques. These techniques form the crux of this chapter. We will primarily be discussing two techniques that are commonly used in real life applications: principal component analysis (PCA) and regularized linear discriminant analysis (R-LDA). We apply both of these techniques in the task of extracting features from the image that is given as an input to the system. This results in a finite or manageable number of features that we may then use as an input to the classifier.

The class of solution that is obtained using these techniques is known as the appearance-based approach. In this approach we work over the actual pixels that make up the image. The features that are extracted are decided on the basis of these pixels. This approach extracts features that are pixels and that can be seen; thus the name. These methods are easy to work with and visualize. However, as the lighting or the human pose changes, there is a complete change in the picture as well as in the features extracted using these approaches. This poses a serious limitation to these methods.

In these two approaches, we will convert the two-dimensional image matrices into single-dimension input vectors by concatenation of the pixels, one after the other, row by row. Each image will be passed through the dimensional reduction and will then be used as standard input for the classifier.

After studying these two methods, we will look at another widely used mechanism for feature extraction—the morphological method. We will use the morphological method to extract the features from a face that is given as a straight input to the classifier. The face when seen as a picture may change a number of times due to various parameters. If you take two photographs of the same person, there will be various differences between the two images. If this were not the case, it would be simple to do point-to-point mapping of the image to the images stored in the historical database. Illumination is one of the major factors that cause changes in images. Illumination causes the whole image, or some of its parts, to grow brighter. As a result, the effective value of the image's pixels changes.

Similarly facial expressions cause change in the image acquired. A person might be smiling in one image, neutral in another, or sad in another. The expressions cause changes in the relative position of the various features, which disturbs the entire image. A similar factor is the position of the person when the image is taken. Position also plays a key role in the changes that the face acquires. The orientation of the camera often disturbs the entire image. In other cases, the person may be placed in slightly different ways or may be facing a different direction than that faced in the different photographs. Other major factors that may cause changes are accessories (glasses, beards, etc.), age, and backgrounds.

To make a comprehensive system that handles all these aspects, we select and extract the features that are more or less constant under any given circumstance. For this we use morphological methods, in which the various features present in the face (such as eyes and lips) are identified and the distance between them are extracted and used for recognition. These distances are constant against most of the factors discussed above.

The classifier that we implement in this section is the artificial neural network (ANN). We first used the conventional ANN with the back propagation algorithm (BPA). Later in this chapter we implement radial basis function networks (RBFNs). These two form very characteristic classifiers. The ANN with BPA has very good generalizing powers that try to form relations or rules from the training data and then use those rules during testing to map the inputs to the outputs. These ANNs try to imitate the output patterns by imitating the functions that classify the data. RBFNs, on the other hand, do the same task but by a kind of codebook approach, in which the known data points lie scattered in the entire input space and their effect at the various points is used to map the inputs to the outputs. RBFNs also try imitating the problem, but here the solution is based on the codebook vectors or data guides and their effects at various places calculated during training. They can hence easily solve any localized problem where the output at a place primarily depends upon the outputs of the nearby inputs.

We first study the three dimensionality-reduction techniques: PCA, R-LDA, and morphological methods. Then we study the two classifiers: ANNs with BPA and RBFNs. Then we apply a combination of dimensionality reduction with classifiers and present the results in various cases. For experimentation, we discuss the work of Shukla et al. (2009f, 2009g).

7.4.1 Dimensionality Reduction with PCA

Principal component analysis (PCA) is the first technique we use for the purpose of dimensionality reduction. The PCA selects the best components from the given components. The entire high-dimensional vector may be supposed to be made up of components, where each component has a different level of relevance to the entire vector. Suppose these components are sorted in order of decreasing levels of relevance. The top few components have very high relevance and are thus very important. The bottom few, on the contrary, are not very important and may be neglected. The main job of the PCA is to break up the entire high-dimensionality vector into these components. Then the task becomes very simple. We may pick up the required number of components starting from the top. The rest of the components are believed to be not as important and may be neglected. In other words, it is assumed that the information carried out with the top components is enough to represent the entire vector, and the bottom components may not add much information.

The entire PCA is applicable due to the redundancy in information in the various dimensions or attributes. We have a very high-dimensional vector in hand. Analysis of these dimensions reveals that a great deal of redundancy exists in the information provided by the various attributes. In other words, many attributes together would convey more or less the same thing. In such a case, we would be justified to replace these attributes with a single attribute. This would not cause much loss in information, even though it would result in a heavy reduction in dimension. Note that the similarity between two vectors is noted by measuring their correlation. A high correlation means that the attributes measure almost the same thing. Correlation between any two vectors always lies between 0 and 1.

This redundancy is what the PCA considers when reducing the vector size and extracting the components. The PCA tries to remove all attributes that are redundant. This will naturally result in a great reduction in dimensionality. The process of component extraction keeps extracting distinct components that reveal a great deal of information by themselves that had not been revealed so far. Initially we extract the component that has a high degree of correlation to the other attributes. We then continue in this manner, extracting the component with a high correlation to the other attributes and a low correlation to the already extracted components. This ensures that we extract interesting attributes that convey very good information. As we keep extracting the components, we observe that the interestingness is reduced. The newly extracted components have a high degree of correlation to some attribute that has already been extracted. Hence it does not convey anything much.

In this way, the components are extracted from the given vector. In order to fully represent the original vector, we must extract as many components as the dimensionality of the vector. This will ensure no loss of information. We can effectively represent the entire vector in a comparatively smaller dimension by a little loss of information. This is an acceptable loss and is usually preferred, because it makes it possible to solve a problem that was otherwise unsolvable.

Again the amount of information loss by the selection of any number of components depends entirely on the amount of redundancy present in the problem. If the level of redundancy is very high, there would be very little loss, even upon selecting a very few number of components. On the other hand, the loss in information would be very large if the redundancy were not there. In other words, if the data set has a very large redundancy, then only a few top components may be selected. If, however, there is virtually no redundancy, we may effectively select all the components. The correlation may be taken as a measure of the similarity or the redundancy in the attributes.

This property of selecting the best components by exploiting the redundancy in data allows PCA-based systems to attain a very high level of performance, even by the use of limited attributes. This method finds a great deal of application not only in face recognition, but also in numerous other domains and applications of soft computing.

We now look at the method again from the point of view of application. We present the manner in which PCA is applied, along with its mathematical foundation. But first let us revise some formulas and concepts that will be used in the rest of the section.

7.4.1.1 Standard Deviation

Standard deviation (SD) is a measure of the deviation that the data have in general with respect to the mean data. SD can be calculated by using the formula given in Equation 7.1:

$$s = \sqrt{\frac{\sum_{i=1}^{n}(X_i - \bar{X})^2}{n-1}} \tag{7.1}$$

where X_i is any data element, $\overline{X}$ is the mean, and n is the number of elements. The square of the SD is known as *variance*.

7.4.1.2 Covariance

Covariance is the deviation of two attributes or features with respect to each other. This measures the effect of one attribute on the other, as given by Equation 7.2:

$$s = \sqrt{\frac{\sum_{i=1}^{n}(Y_i - \overline{Y})(Y_i - \overline{Y})}{n-1}} \tag{7.2}$$

7.4.1.3 Covariance Matrix

The vector that is to be used has a number of attributes. Every pair of attributes will have a covariance between then. This is the matrix where every cell C_{ij} is denoted by the correlation between the attribute i and the attribute j. Note that $C_{ij} = C_{ji}$.

7.4.1.4 Eigen Vectors

These vectors are scaled by the application of linear transformations of the vector; however, they are not moved. The eigen vectors of any two vectors are always perpendicular to each other. These vectors always follow Equation 7.3. The eigen value is the value associated with the eigen vector, and it denotes the amount of scaling that is performed.

$$Ax = \lambda x \tag{7.3}$$

where A is the vector, λ is the eigen value, and x is the eigen vector.

So far we have concatenated the two-dimensional image to convert it into a single-dimension vector. Suppose we have N such images in the training database. We take a vector $X = (x_1, x_{2,}\ldots, x_i,\ldots, x_N)$, where each x_i denotes an image of dimensions $p \times q$. Let n be the dimensions of each image ($= p \times q$).

To apply the PCA, we first subtract the mean from each data element. This returns a set of elements whose mean is 0. We then calculate the covariance matrix, which measures the covariance between any two pairs of elements. The next step is to determine the eigen vectors and the eigen values of this covariance matrix. The eigen vectors try to approximate as many points as possible. Next we select the maximum eigen values by inspecting the eigen vector and the eigen values. The maximum eigen value element is the principal component. We then select the top few eigen values. These are the components that have a very high significance. The feature vector is then made, which contains all the selected or extracted components. In this manner, we get the feature vector that is the most important part of the PCA algorithm. In the end, we simply use this vector to multiply the vector containing historical data to get the reduced or shortened system. The mathematical notation of this method is given below.

The aim of PCA is to project the original image vector into a feature vector such that the dimensionality of the image vector is reduced. This mapping, or projection, gives us the shortened image vector with limited dimensionality. Suppose that W is the feature vector, X is the original image vector, and Y is the vector with reduced dimensionality; the relation between the variables is given by Equation 7.4. It is assumed that there is a linear mapping from the original image vector to the reduced dimensionality vector. The whole task is thus to calculate the feature vector W.

$$Y = W^T X \tag{7.4}$$

In our case, the vector Y is of size $m \times N$, where m is the required number of dimensions that are to be given to the classifier and N is the total number of images in the training data set.

The transformation, or the feature vector, is a vector of size $n \times m$, where n denotes the total dimensionality of the image vector. The vector X is the matrix of vectors in the training data set that we already have.

The feature vector matrix W contains the largest possible eigen vectors that are extracted out by analysis. The various eigen vectors are inserted one after the other in the feature matrix, column by column. In this manner, the collection of the m largest eigen vectors are the columns of the feature matrix W. Recall that the eigen vectors are calculated using Equation 7.3.

As discussed above, to calculate the eigen vectors and the eigen values, we first subtract the mean form all the attributes in the image vector. This transforms the original image vector into a vector that has 0 mean. This vector is taken for the computation of the eigen vectors for the rest of the steps. This matrix is given by Equation 7.5. Equation 7.6 denotes the mean μ_i for any attribute or row of matrix X.

$$Z = X - \mu \tag{7.5}$$

$$\mu_i = \frac{1}{N}\left(\sum_{j=1}^{N} x_{ij}\right) \tag{7.6}$$

We then need to calculate the covariance matrix of the vectors. In the preceding section, we learned the means for calculating the covariance matrix for multiple attributes. In this case, we have in the image vector multiple attributes and multiple values per attribute. Using these, we need to calculate the covariance matrix that will help us identify the low-variance attributes. We then try to linearly approximate the attributes using attributes that can replace the bulk of the attributes from the high-dimensional image vector. This replacement results in dimensionality reduction. The covariance matrix may be defined by Equation 7.7. Readers may expand this matrix in terms of individual attributes and their corresponding values in order to study the formula as a standard variance formula, which we studied earlier in this chapter.

$$\begin{aligned} C_X &= (1/n)\, Z\, Z^T \\ &= C_X = (1/n)(X - \mu)(X - \mu)^T \end{aligned} \tag{7.7}$$

where C is the covariance matrix. The size of matrix T is $n \times N$, which means that the covariance matrix is a square matrix of size $n \times n$.

Because this is a square matrix, we may easily calculate the eigen vectors from it. The eigen vectors with the maximum eigen values are selected and inserted into the feature matrix W.

Ideally the covariance matrix calculated should be a diagonal matrix. This means a covariance of 1 of an attribute with itself. On the other hand, the covariance must be 0 between any other pair of attributes. Because this may not always be the case, however, we try to diagonalize the matrix to get the best results from the feature decomposition. This method of extracting components is achieved by decomposition of the eigen vector.

Here we assume an orthogonal matrix P that gives a matrix Y as a linear combination of the P and the feature vector Z. Here P is the matrix of the principal components. This can be given by Equation 7.8:

$$Y = PZ \tag{7.8}$$

The covariance matrix C_Y is a diagonal matrix given by Equation 7.9:

$$C_Y = (1/n)\, Y\, Y^T \tag{7.9}$$

On substituting Y from Equation 7.8, we can calculate the matrix C_Y to be given by Equation 7.10:

$$C_Y = P\,C_X\,P^T \tag{7.10}$$

This gives us the exact principal components to be used in the feature vector W.

In most applications, the PCA is used as a black box, in which the algorithm extracts the components and returns the same. The extracted components have a great deal of relevance to us when we use the extracted principal components as an input to the classifier. This makes very scalable and high-performing systems with only a little loss of generality. These systems make it possible to solve real life problems in a natural way that otherwise would not have been possible.

7.4.2 Dimensionality Reduction by R-LDA

The R-LDA is the regularized form of Fisher's discriminant. It is a type of linear discriminant analysis (LDA).

The problem that comes with using the PCA is that it does not consider the output to which an input belongs. Instead, it selects the principal components only on the basis of the attribute or the input values from the training data set. This may result in the deletion of some very important features due to the lack of consideration of the output. Consider an image in which two inputs are very close to each other. These two inputs, however, belong to two distinct classes A and B. If we apply PCA, it is very likely that it will find these two inputs redundant. In reality, however, both inputs convey very different information and both are of use to the system. Hence it would not be good to neglect the output. These problems occur especially in classification problems. Here, too close inputs belonging to different classes convey a great amount of information that must be known and retained and that cannot be generalized.

Fischer's LDA is a method that solves this problem by accounting for the presence of different classes in the training data set. This method adapts the PCA for classification by taking into account the presence of multiple classes. Each class is dealt with separately, and the covariance is measured for both the same class and the mutual covariance between the different classes. The basic aim of this algorithm is to maximize the value of the objective function given by Equation 7.11:

$$J = \frac{|W^T S_B W|}{|W^T S_W W|} \tag{7.11}$$

where S_B stands for the scatter matrix in the same class and S_W is the scatter matrix in between classes.

This formula is derived from our earlier formula for the PCA. But we have separated the different classes to handle each separately. The scatter matrices are always proportional to the covariance matrices and do not induce any change in output. The two scatter matrices are given by Equations 7.12 and 7.13:

$$S_B = \sum_c (\mu_C - \bar{X})(\mu_C - \bar{X})^T \tag{7.12}$$

$$S_B = \sum_c \sum_{ic} (x_i - \mu_C)(x_i - \mu_C)^T \tag{7.13}$$

The rest of the procedure is the same as in the PCA and may be used in the same way.

We now study the regularization in Fisher's LDA. This LDA faces a serious problem known as small sample size (SSS). If the dimensionality of the data is very high, which is usually the case, the number of training samples in the training data set need to be very high for effective training or tuning of the system parameters. If the training data set size is not large, the system may not train well. Usually the training data sets are much more limited as compared with the problem's dimensionality. This puts a very serious limitation on the training of the system.

The LDA is usually able to solve this problem to some extent by adding a regularization parameter. This parameter controls the covariant matrix and hence helps solve the SSS problem. Regularization reduces the high variance related to the eigen value estimates of the within-class scatter matrix at the expense of potentially increased bias. The tradeoff between the variance and the bias, depending on the severity of the SSS problem, is controlled by the strength of regularization. The equation for regularization is given by Equation 7.14:

$$J = \frac{|W^T S_B W|}{|\eta W^T S_B W + W^T S_W W|} \tag{7.14}$$

where all symbols have meaning as discussed in Equation 7.11. η is the regularization parameter, which lies between 0 and 1 and which controls the level of regularization. If we keep η to be 0, the equation reduces to the conventional equation in which there is no regularization. If, however, we keep the regularization parameter as 1, Equation 7.14 converts to Equation 7.15, in which there is a string effect of regularization.

$$J = \frac{|W^T S_B W|}{|\eta W^T (S_B + S_W) W|} \tag{7.15}$$

7.4.3 Morphological Methods

In this section, we study how to apply the morphological methods, or the morphological analysis (MA), to the problem of face recognition, which is a highly multidimensional, complex problem to solve. We extracted the features in the previous section using PCA and R-LDA. Here we extract the features from the image of the face, but with a different strategy. But first we give a general definition of the term *morphology* as it is used here.

The term *morphology* comes from the classical Greek word *morphe*, which means "shape or form." In this method, we study how the different shapes are formed, combined, and exist in any given simple or complex pattern. We break the problem into numerous parts or objects and then study the shapes, forms, and properties of each. The art lies both in breaking the pattern and in the study of the broken objects. Morphology does not restrict itself to face, images, and so forth. Rather it spans much further, making MA a universal tool for every kind of problem.

The objects in question can be physical (e.g., an organism, an anatomy, a geography, or an ecology) or mental (e.g., word forms, concepts, or systems of ideas). The study or analysis of these objects is not a very easy task because of the numerous problems associated with them. One of the major problems is that the objects are nonmeasurable in nature. Numerical facts and figures can easily be fed into any system, but even simple objects need to be quantified or measured using some metrics for use in any system. This causes a serious problem in the systems, where we have to develop means for extracting the usable information in limited dimensions.

The other problem is that these objects are very difficult to break down, identify, or resolve. Doing so requires additional processing and complex algorithms. Uncertainties exist in every step, making this work even more difficult. If you were asked to separate the human eye from an image, you may not be certain about the exact pixel where the eye ends and the rest of the face begins. Different people may interpret this point differently. Furthermore these systems cannot be easily modeled in a meaningful way.

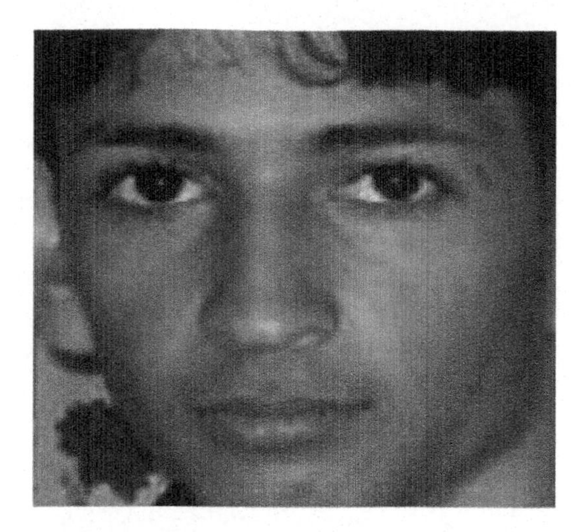

FIGURE 7.1a Processing of the face image: Gray image.

The MA comes into play for these kinds of problems. In this section, we analyze the different issues that are more or less similar for the various problems. Based on this, we then try to solve the problems by keeping the loss of information or the uncertainty as small as possible.

We now apply the fundamentals and principles of MA to the problem of face recognition. It must now be clear that our approach to extracting the features will include separating out the various parts of the face and studying them accordingly. In our case, we will try to separate and identify the two eyes, the nose, and the lips, as these features are relatively easy to identify. Various image-processing tools are used to carry out the various operations or steps in identifying the features of the image.

The input to the system is a color image. This image is first converted to the equivalent grayscale image. This results in a single-color vector per pixel and is much easier to handle than the color images. The gray image is shown in Figure 7.1a.

The contrast in the image to be segmented may be different from the contrast in the background image. We calculate the image's gradient by studying the changes in contrast within the image. We next apply the Edge and Sobel operators to calculate the threshold value. This gives us the binary gradient image. The resulting image is shown in Figure 7.1b. The processed binary gradient mask image

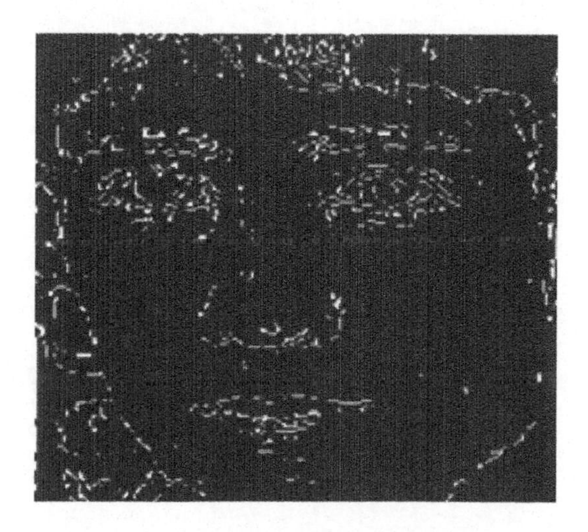

FIGURE 7.1b Processing of the face image: Binary gradient mask.

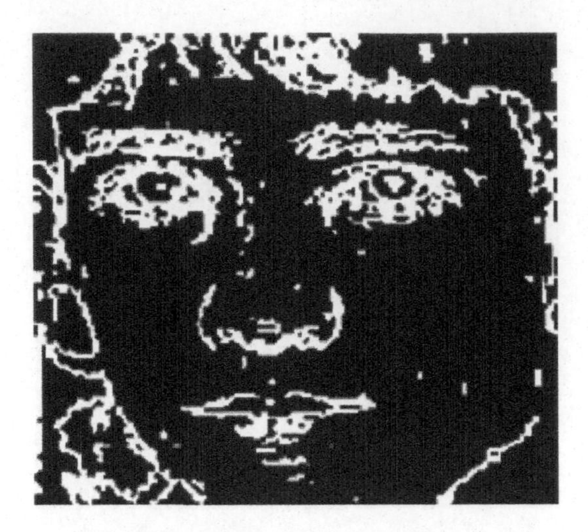

FIGURE 7.1c Processing of the face image: Dilated gradient mask.

still shows lines of high contrast. The linear gaps are removed by applying linear structuring elements, such as dilating the binary gradient image. The dilated gradient mask is shown in Figure 7.1c. This image is very similar to the human perception of the face. The next task is to fill the holes by applying region filling. The resulting binary image is shown in Figure 7.1d. Extraction of eight-connected sets of pixel components is then done to suppress light structures connected to the image border. This is shown in Figure 7.1e. Filtering, thinning, and pruning are then implemented, and the resulting image is the segmented image, as shown in Figure 7.1f. The segmented image is then superimposed with the initial gray image. The resulting outlined initial image is shown in Figure 7.1g. Finally Figure 7.1h shows the final obtained image.

These steps enhance the various features that we were supposed to extract. The features include eyebrows, eyes, nose, and mouth. These features are distinctly visible from the surroundings, and the unwanted background and features have been removed from the image. This makes the task of feature extraction very simple.

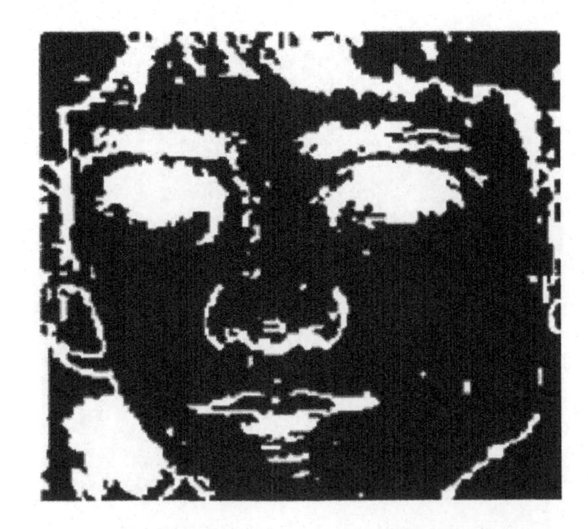

FIGURE 7.1d Processing of the face image: Binary image with filled holes.

FIGURE 7.1e Processing of the face image: Cleared border image.

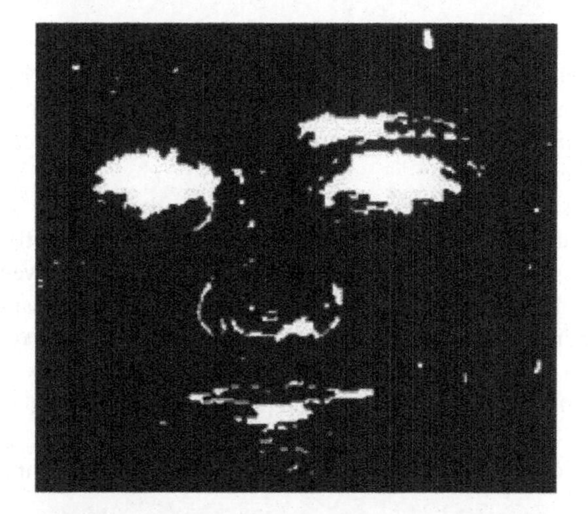

FIGURE 7.1f Processing of the face image: Segmented Image.

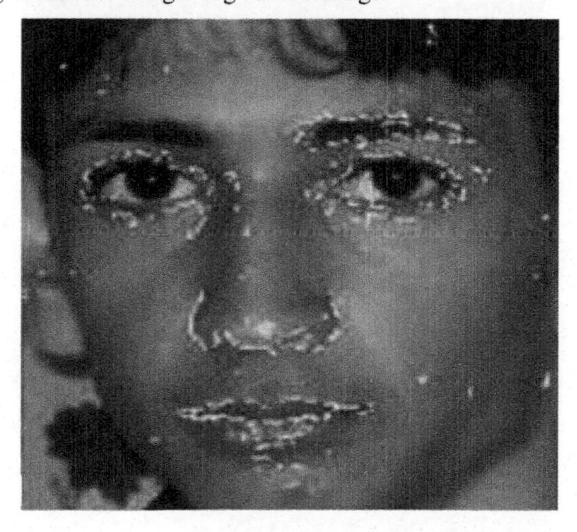

FIGURE 7.1g Processing of the face image: Outlined Original Image.

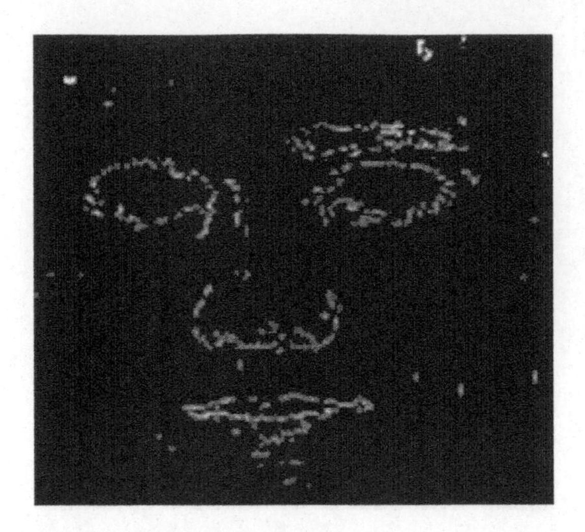

FIGURE 7.1h Processing of the Face Image: Final Image.

The next task is to extract the features that may be used as an input for the classifier for identification. For this, we determine the normal center of gravity (NCG) in the various identified facial features. NCGs are the intensity-weighted centroids of each extracted feature. The NCG is shown with white dots in Figure 7.2.

The other feature that we extract to give to the classifier is the distance between the various features. This includes the distance between the eyes, the distance between the eyes and the lips, the distance between the eyes and the nose, and so on. These distances are given in Figure 7.3.

All the inputs, or the features, are extracted from the input to the classifier. The classifier does the rest of the job. The inputs are finite and good. These inputs are different for each person and remain constant for each person, no matter how many times the picture is taken and processed. In this way, we can expect the classifier to give a good performance. Note that the quality of the inputs determines the classifier's efficiency. The number of inputs, their variations, and noise levels all play a great role in the behavior of the classifier and in the efficiency of the entire system.

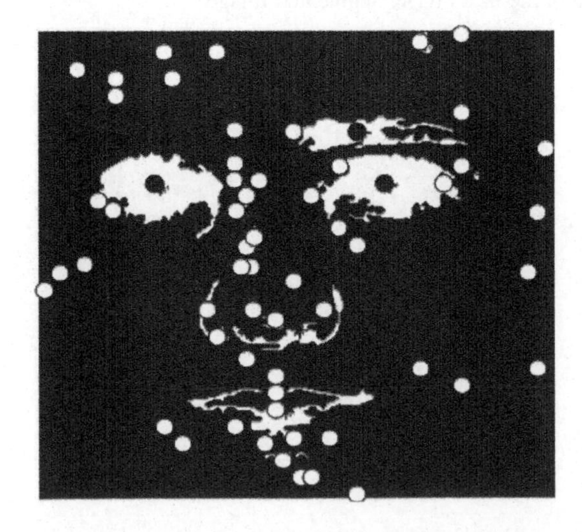

FIGURE 7.2 Determination of normal center of gravity (shown with white dots).

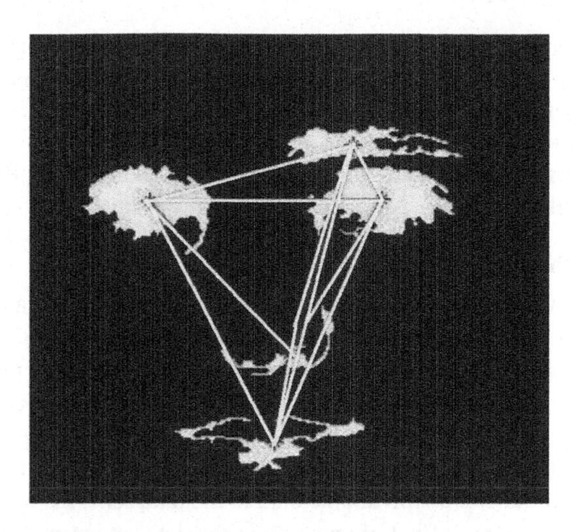

FIGURE 7.3 Calculation of Euclidean distances between various facial parts.

7.4.4 Classification with ANNs

The next job is classification, in which we are supposed to find the correct class to which a given input belongs. In other words, we need to find the correct person whose face has been given in the database. The classifier that we use here is the artificial neural network (ANN). We have used two types of ANNs separately: ANN with BPA and radial basis function networks (RBFNs). Here we discuss the role of these ANNs as classifiers.

Both types of ANNs use a classification model. The inputs that are given to these ANNs are obtained in the same way that they were obtained for PCA, R-LDA, and MA. The number of inputs is different for each technique. The inputs for the first two methods are decided by trial and error, in which we try to take an apt number of inputs that give high performance but that do not take too long to train.

The number of inputs must be in accordance with the number of training data sets available. It would not be good to take in too many inputs if the training data set were too small and we were getting sufficient efficiency with a lower number of inputs. As we increase the number of inputs, the number of neurons in the input layer increases. This means a similar increase in the number of neurons in the hidden layer. If we look from the point of view of connections, for every increase in a neuron, the number of weights of interconnections increases by a large number. This, in turn, increases the number of parameters that need to be tuned by a large amount. These large numbers of parameters require a large training data set in order to train optimally. But this data set is not available. Hence this would result in a suboptimal performance that may not give good efficiency.

The number of inputs in the morphological methods is finite and hence can be used. These inputs include the Euclidian distances discussed in the previous section. These form good inputs to the system because they remain more or less constant regardless of how the image is captured and processed. This makes the interclass distance in the input space as small as possible, which is a needed characteristic in an ANN. This further ensures that the classification work is easy.

The system's outputs are based on a classification model in which there are as many outputs as there are classes. Each output denotes a class. The output numerically stands for the probability, or chance, for the given input to belong to the class being considered. Suppose a system has a record of n different users. There would thus be n different outputs for every input I of the form $<o_1, o_2, o_3, \ldots, o_n>$, where any general output o_i denotes the probability of the input I to belong to the class i. The probability can be a maximum of 1 and a minimum of –1. A probability of 1 denotes that an input definitely belongs to the same class, while a probability of –1 means the input definitely does not belong to the same class.

Hence whenever we apply any input, the system gives us n number of outputs. We now select the maximum number of outputs; the class corresponding to that number is returned as the answer. Ideally one of the n outputs must be 1 and the rest must be –1. But in reality, this would not be possible due to obvious reasons. Hence we select for the maximal output.

The inputs and outputs are common to both the ANN with BPA and the RBFNs. We now discuss each one with regard to the problem.

7.4.4.1 The ANN with BPA

The ANN with BPA is the first classifier we use over the problem. This classifier performs the task of classification by generalizing the class and the class boundaries. It tries to predict the regions in the input space where each concerned class is present or is influential. As a result of the training, it is successful in making some function that separates the region in the input space where class is present from the rest of the region. ANN with BPA is known for its generalizing capability. Generalization is the concept used in training as well. Here the ANN tries to look at the global input space; the decision at any point is the result of the globally available data in the data set. This makes these systems more resistant to noise in general, as the effect of any data in the data set is limited in nature.

To understand the role of ANN with BPA in classification, consider the input space. The inputs and outputs are plotted in the form of a graph. The graph's dimensionality is the number of inputs plus 1 for the output. The output represents a surface in the input space. The output surface shows a binary surface that takes a high value at regions that lie in the class and a low value at regions that lie outside the class.

These ANNs use BPA as the training algorithm, which attempts to set the values of the unknown parameters in such a way that the total error for the training data set is reduced. The BPA fixes the values of the system parameters or the weights in the same respect. It tries to keep the boundary that separates the class regions from the nonclass regions as optimal as possible. If any data set belongs to the class, the BPA tries to modify the parameters such that this training data set lies in the class region as comfortably as possible. This means the surrounding areas are also likely to lie in the region. The converse is true: If the data set does not belong to the class, the BPA tries to completely remove the data set from the region accommodated by the class. This means the parameters are adjusted in such a way that this point is nowhere near the region accommodated by the class. This may again mean that the surrounding points are not in the region occupied by the class.

We now discuss the role of the number of neurons in the hidden layer. As discussed earlier, as we increase the number of neurons in the hidden layer, the number of parameters in the system increases as well. This makes the system even more flexible. As we discussed, if an input belongs to some class, then the surrounding inputs may also belong to the same class. This is because the ANN would take some region in the transition between the input space that belongs to a class and the input space that does not belong to the class. This transition is not sudden; rather it requires some space in the input space graph.

If the number of neurons is far too many, the ANN will be very sensitive. This would enable it to make very rapid or sharp transitions in very little space, due to the flexibility introduced by the large number of system parameters. This gives the ANN the ability to imitate any function, as complex or unrealistic as possible.

The converse is true: If the number of neurons or parameters is less, the ANN is restricted to attaining complex architecture or imitating complex functions. It can only make a gradual transition from the class-occupying region to the non-class-occupying region. This inhibits a very sensitive structure. Note that a very sensitive structure is highly unlikely, because the ANN can now imitate anything one imagines; hence the performance in the unknown inputs is very bad.

7.4.4.2 RBFN

The radial basis function network is the second classifier that we implement. Unlike the ANN with BPA, which we saw is highly generalized, a different role is played by these ANNs. RBFNs are highly localized in nature. Thus we may expect a better performance, as the classification problems are known to be highly localized in nature.

RBFNs select some training data sets, which act as the guides for the rest of the training and testing data. The output to these guides or data centers can be assumed to be perfect. These guides are spread all across the input space. Each guide has parameters that are adjusted in the learning phase for the most optimal performance. Each guide has an effect on the entire input space. The effect decreases as we move farther from the guide. The rate at which this decrease takes place is determined by the guide's parameters.

To know the output for an input, the influence of all the guides is taken into account, and the final output is evaluated accordingly. The result is more dominated by the guides that were very close to the unknown inputs and very less dominated by the ones that were far away. In the training phase, if the input belongs to the same class, the guide tries to increase its output, and vice versa.

Hence the localized nature of the algorithm may be easily seen. The output is mainly dominated by the guides that are very close to the input. This makes the system an excellent choice for classification problems.

7.4.5 Results

To test the system, all the discussed feature extraction methods, as well as the classifiers, were implemented and tested over the grimace face database. This database contains 360 color face images of 18 individuals, with 20 images per person. The database has images that vary in emotions, lighting, and position.

7.4.5.1 PCA, R-LDA, and MA with ANN and BPA

For the PCA and R-LDA, 180 images (50%) were selected for the training data set and 180 images (50%) were selected for testing. These corresponded to 5 images per person out of the 10 images available. For MA, 120 images, corresponding to 8 for each person, were used for training and the remaining 120 images were used for testing.

We first extracted 20 features using PCA. Similarly 14 features were extracted for the same database using R-LDA. In the case of MA, the number of inputs consisted of 6; these were the Euclidean distances. These features were given to the ANN and RBFN as inputs in both the training and the testing stages. The number of output classes was 18.

The configuration of ANN with BPA for these methods, used for training and testing, is shown in Table 7.1. Training graphs of ANN with BPA applied to the PCA and R-LDA preprocessed training set are shown in Figures 7.4 and 7.5. The results with the use of ANN and BPA are given in Table 7.2.

TABLE 7.1
Artificial Neural Network with Back Propagation Algorithm Configuration

ANN with BPA Configuration	PCA	R-LDA	MA
Input vector nodes	20	14	6
Number of hidden layers	2	2	2
Number of neurons (hidden layer 1, hidden layer 2, and output layer)	30, 35, 18	29, 29, 18	31, 37, 15
Transfer functions (hidden layer 1, hidden layer 2, and output layer)	Tansigmoid, tansigmoid, linear	Tansigmoid, tansigmoid, linear	Tansigmoid, tansigmoid, purelin
Learning rate	0.001	0.001	0.0001

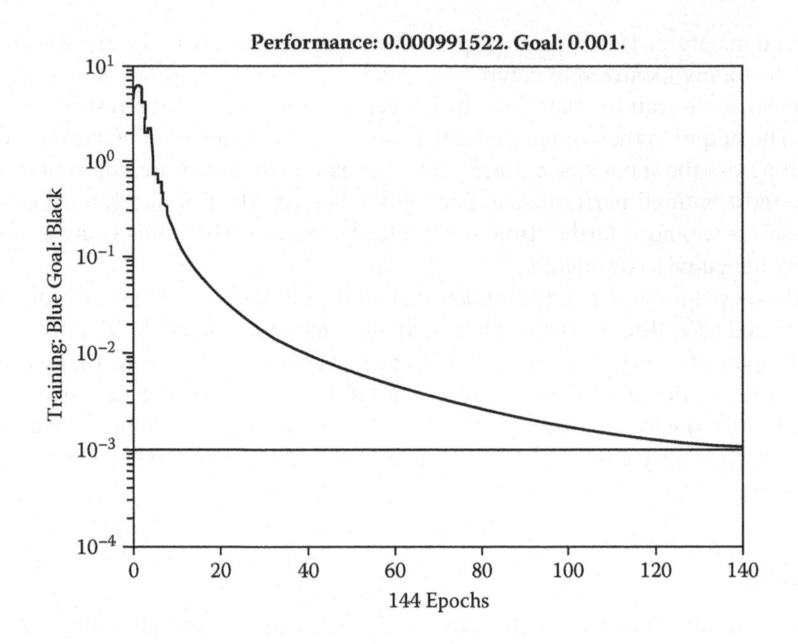

FIGURE 7.4 Learning of ANN with BPA after preprocessing by PCA.

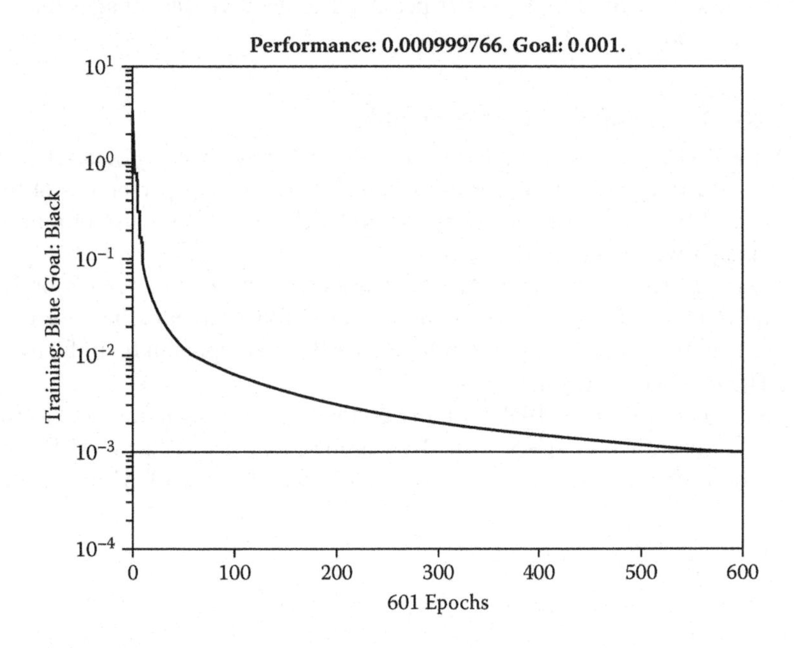

FIGURE 7.5 Learning of ANN with BPA after preprocessing by R-LDA.

TABLE 7.2
Recognition Using the Methods with ANN and BPA

Methods	PCA+BPA	R-LDA+BPA	MA
No. of error images	13	11	9
Recognition rate	92.77% (167/180)	93.88% (169/180)	93.33% (112/120)

TABLE 7.3
RBFN Neural Network Configuration

RBFN Configuration	for PCA	for R-LDA
Number of radial basis layers	1	1
Number of neurons (input layer, radial basis layer, and output layer)	20, 135, 18	14, 169, 18
Spread	0.4	0.8

7.4.5.2 PCA and R-LDA with RBFN

The inputs were same with RBFN as they were for ANN with BPA. The configuration of RBFN with PCA and R-LDA, used for training and testing, is shown in Table 7.3. RBFN has a single hidden layer that comprises 135 neurons for PCA and 169 neurons for R-LDA input vectors. Training graphs of RBFN applied to the PCA and R-LDA preprocessed training sets are shown in Figures 7.6 and 7.7. The results from the use of RBFN with PCA and R-LDA are given in Tables 7.3 and 7.4.

Each ANN took a different amount of time for training input feature vectors. Training of the RBFN was faster as compared with ANN with BPA. This may be accounted for by RBFN's localized nature.

We may further deduce that RBFN's performance was better in both cases. This shows that a localized ANN performed better, which is usually expected in a classification problem because of its localized nature. Furthermore in the problem, R-LDA gave a better performance as compared with PCA.

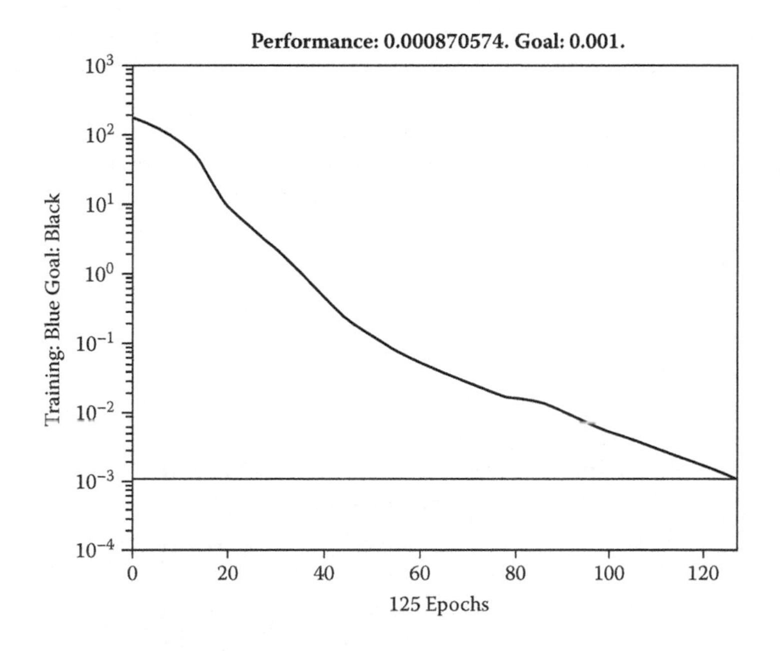

FIGURE 7.6 Learning of RBFN after preprocessing by PCA.

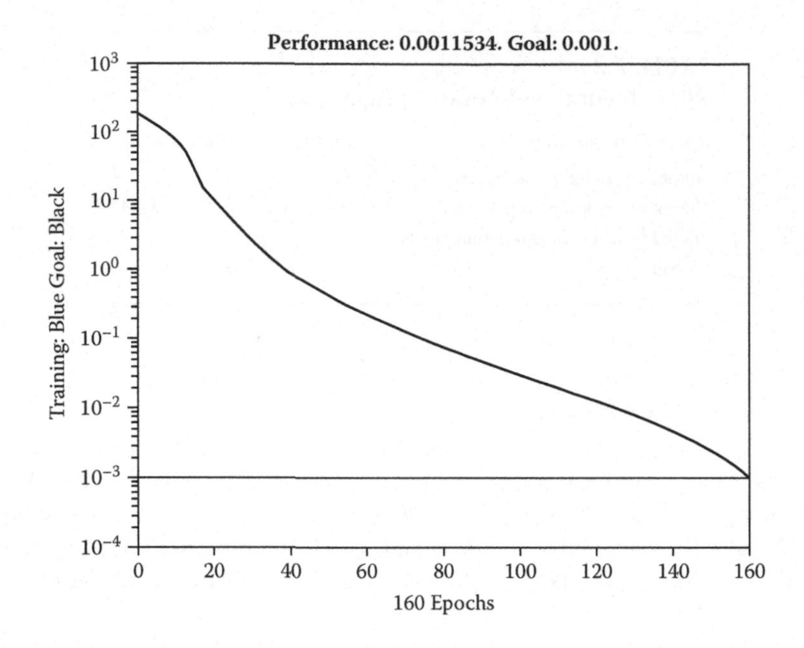

FIGURE 7.7 Learning of RBFN after preprocessing by R-LDA.

7.4.6 Concluding Remarks for Face as a Biometric

In this section, we discussed three important feature extraction techniques—PCA, R-LDA, and MA. These techniques extracted features from the face. The extracted features were then used on two classifiers—ANN with BPA and RBFN. We saw that under all combinations of feature extraction mechanism and classifier, we were able to solve the problem of face recognition. All methods gave handsome performances, and we were able to identify the first biometric under discussion.

As with any classification problem, our example problem was found to be much localized in nature. As a result, the RBFN's performance was better than its counterpart ANN with BPA under all cases. Out of PCA and R-LDA, it was found that the R-LDA performed slightly better than PCA. Furthermore the MA was independent of expression, lighting, or other conditions and gave good performance in the presence of these factors.

TABLE 7.4
Recognition Rate using PCA and R-LDA with Radial Basis Function Network

Methods	PCA+RBFN	R-LDA+RBFN
No. of error images	10	7
Recognition rate	94.44% (170/180)	96.11% (173/180)

7.5 HAND GEOMETRY

In this section, we take another very exciting physiological biometric feature: hand geometry. We will try to identify people by the peculiarity of their hands. Hand geometry is usually difficult to visualize as a means of identification, because humans do not usually use this as a feature for recognizing another person. We identify someone when we see her or hear her. But it is hard to imagine recognizing someone based only on her hand.

Yet the human hand is filled with a great deal of information that leaves room for a large amount of variation between two hands. People may have different lengths of fingers, different palm widths, and so forth. If we observe the hands of any two people, it is highly unlikely that they would have the same values for all features. Normally one might expect a natural deviation in some or the other features, because these features being exactly the same is very unrealistic and has almost zero probability. This is the motivation behind the hand geometry biometric identification systems.

The basic approach in this section is as in the previous section. We will first discuss some techniques for preprocessing and feature extraction. Then we will build a classifier that does the job of mapping the inputs into classes.

In the precious section, we discussed the problem of identification in which we were given an entirely unknown input and were asked to find out who the person was. In this section, we take a slightly different approach. Here we deal with the problem of verification, which means we need to verify the claimed identity of the person. We can make two decisions. One is that the claimed identity is correct, and the other is that the claim is false.

The hand is a very interesting biometric feature. The input in hand geometry may be taken by the use of scanners. The scanners scan the palm and convert it to a digital format or to the form of an image that may be used for further processing. In the past, pegs were used to orient the human hand on the scanner. These pegs, however, caused unintentional effects by disturbing the shape of the hand. Two main weaknesses of pegs are that the pegs deform the shape of the hand silhouette and that users might place their hands incorrectly.

One of the major differences between the palm and the face is in dimensionality. As compared with face, the hand represents a biometric feature of lower dimensionality. The various features that we will be considering in this section are the length and widths of the fingers. The recognition consists of identifying the various points on the human hand. Once these points have been identified, we will then use the Euclidean norm for measuring the lengths, widths, and other features of the human hand. These features will be used as straight inputs to the classifier that will use these features for the purpose of classification.

Geometrically significant landmarks are extracted from the segmented images of the palm. These are used for hand alignment and for the construction of recognition features. The features extracted are resistant to orientations. We get this property by applying the contour extraction algorithm, which extracts and corrects the contours.

For the classification of the extracted data, we use ANNs with BPA. These networks, as discussed in the previous section, can be modeled to work with classification problems. As discussed earlier, they do a fairly good job at learning the training data sets and identifying those regions in the input space that are inside the class and those that are outside the class. This makes the system work for the testing data and for the data that are completely unknown to the system.

The hand-recognition biometric has a very special domain of application. Consider applications such as access control for immigration, border control, dormitory meal plan access, and so forth. In these cases, we primarily require carrying out only verification; not identification. Very distinctive biometrics, such as fingerprints and iris, may not be acceptable. In addition, good frictional skin contact is required for fingerprint imaging systems, and a special illumination setup is needed for iris or retina-based identification systems. Both of these may be too expensive. In these cases, hand

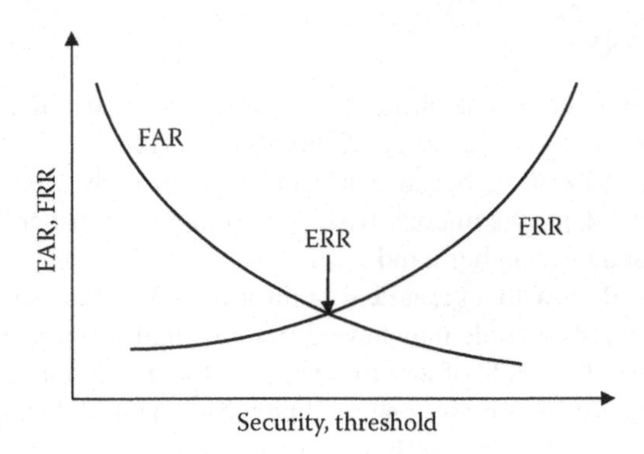

FIGURE 7.8 FAR, FRR versus security, threshold.

geometry finds an application. Furthermore hand geometry is ideally suited for integration with other biometrics—in particular, fingerprints. For instance, an identification and verification system may use fingerprints for (infrequent) identification and may use hand geometry for (frequent) verification.

To measure the effectiveness of any identification system, we must measure two standard error-denoting metrics: false rejection rate (FRR) and false acceptance rate (FAR). FRR is the rate or probability by which a genuine person is rejected access. This means that a genuine user went for a security check, and the system identified him as an intruder. FAR, on the other hand, measures the rate or probability by which a nonlegitimate user is given access. This means that an intruder came for a security check and was given access to the system. Both phenomena are undesirable and hence lead to poor system performance.

The way to control these factors is by changing the system parameters. Two major parameters may be changed. The first is the rejection threshold. This is the minimum confidence that a system must have in any person for the system to grant the access. If the confidence falls below this level, the user is identified as an intruder, and access is not given. An increase in this parameter would obviously result in the FRR being increased and the FAR being decreased. The other parameter is the security parameters, such as like lighting conditions, absence of dust, and other physical parameters that may lead to lesser error and better verification.

Hence it is clear that FAR and FRR are inversely related. If we were to plot a graph between these factors with respect to the security or the threshold, it would be as shown in Figure 7.8. The intersection of FAR and FFR is the point at which accept and reject errors are equal. This is known as the equal error rate (EER) or the crossover error rate (CER). Low EER/CER scores indicate high levels of accuracy.

We now discuss each step involved in the process of identification in hand geometry. Per our methodology of the recognition systems, we will first study image acquisition. Then we will discuss image preprocessing and feature extraction. We then study the classifier, in which we will be using ANN with BPA. Finally we present the experiments and results. These discuss the work of Shukla and Sharma (2008l).

7.5.1 Image Acquisition

The first task that needs to be performed is image acquisition. This step comprises a light source, a flat-bed scanner, and a black flat surface used as a background. The user is asked to place his hand on the scanner, with the hand touching the scanner. The user can place his hand freely, without the need of pegs to fix the hand's position. During scanning, we must ensure that the subject's fingers do not touch one another and that the hand lies flat and stays on the flat surface of the scanner. The scanner scans the image of the hand. Randomly placed hands are scanned with a dark background using 150 dot per inch (dpi) scanning quality. The image captured is a 640 × 480 pixel color photograph in JPEG format. Only the right hand images of users are acquired.

7.5.2 Image Preprocessing

We follow more or less the same procedure for image processing that we used for the face. The given image is a color image. Hence we first convert it into a grayscale image. This allows us to work better with the image.

7.5.2.1 Filtering

A median filter is applied to remove noise in the image.

7.5.2.2 Binarization

We convert the image into a pure black-and-white image such that the hand is white and all other backgrounds are black. To do this, we follow the mechanism of thresholding, in which all pixels above a threshold value are assigned white and the rest are assigned black. This is given by Equation 7.16:

$$\begin{aligned} I'(x, y) &= 1, \text{ if } I(x, y) \geq \text{threshold} \\ &= 0, \text{ otherwise} \end{aligned} \tag{7.16}$$

where I' denotes the black-and-white image and I denotes the original gray image.

The threshold value is automatically computed using Otsu's (1979) method. Then the border of the hand's silhouette is smoothed using morphological opening and closing. Because of the black background, there is a clear distinction in intensity between the hand and the background. Therefore, the image's histogram is bimodal. The result is shown in Figure 7.9.

7.5.2.3 Contour Detection

The next task is to find out the contour between the hand and the background. This effectively separates the hand from the background. We apply the Laplacian of Gaussian operator for this

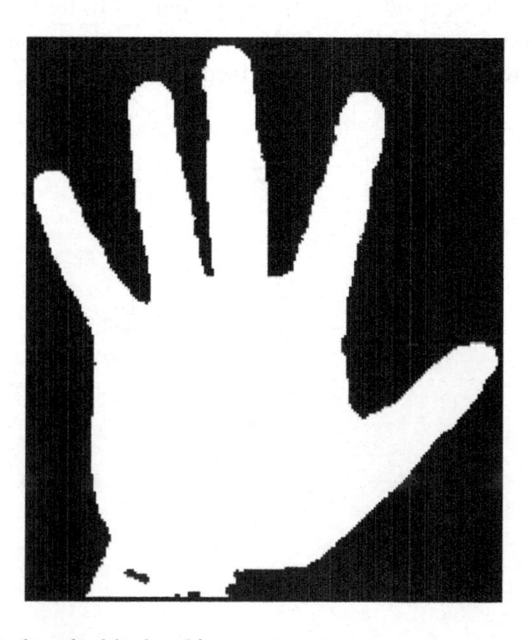

FIGURE 7.9 Binarized black-and-white hand image.

purpose. This operator finds the regions of high change in intensity in order to detect and separate the different contours.

7.5.3 Feature Extraction

The next step is to extract the features from the obtained image vector. We extract the good features from the image. The extracted features will be directly used as inputs for the classifier system. The commonly extracted features for the problem of hand geometry include thumb length, first finger length, middle finger length, ring finger length, wrist length, thumb base width, first finger width, middle finger width, ring finger width, little finger width, hand perimeter, and hand surface. All the parameters mentioned here have their usual meanings.

To extract these features, we must locate the landmark points in the given image vector. These points will enable us to easily extract the various features needed. Based on the features mentioned, the important landmark points for us are the fingertips and the valley points of the various fingers. From this we may simply calculate the various distances that form the input for the classifier. Here the fingertips refer to the topmost position of each finger, while the valley points refer to the bottommost positions, which are found in between two fingertips.

To calculate the landmark points, or the fingertips and valley points, we first find a reference position in the given image. The reference point we take here is the middle point of the wrist. The wrist may be easily seen in the image, as it is at the bottommost location. We traverse from left to right to find the extreme points of the wrist and then take their mean to get the reference point. The reference axis is defined as the line from the reference point to the middle fingertip landmark.

Comparing the distances with those of other neighboring points on the hand contour, the fingertips are the points that have the most distance, and the valley points, the least. By this mechanism, we are easily able to differentiate between the fingertips and the valleys. We then calculate the various lengths per the requirements of the feature to be extracted. The extracted features used here are the lengths of each finger and the widths of each finger at three locations.

7.5.3.1 Finger Baselines

The baselines are the lines that connect the finger valleys. These lines originate from one valley, cross the finger, and go to the next valley, as shown in Figure 7.10 for a single finger. We first create the baseline for the middle finger; this is the line connecting the second and third valley points. The finger baselines of three of the fingers are obtained by connecting the valley points that are on

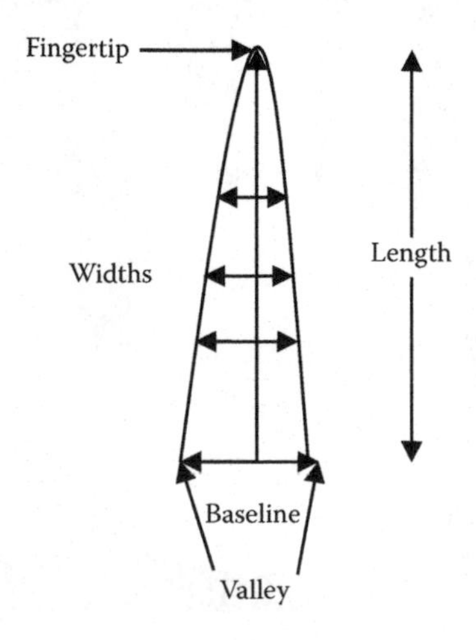

FIGURE 7.10 The features in a finger.

both sides of that particular finger. However, for the others, each has only one adjacent valley point. Thus, the other valley points are assumed to be on the opposite side of the finger in such a way that its distance from the fingertip is the same as that of the known valley point and the fingertip.

7.5.3.2 Finger Lengths

The finger lengths are measured as the distance between the fingertips to the middle points of the finger baselines. These finger lengths are shown in Figure 7.10.

7.5.3.3 Finger Widths

Finger widths are the widths of a finger measured at three locations, as shown in Figure 7.10. The first is measured at the middle of the finger length; the second, at one-third the finger length; and the last, at two-thirds the finger length.

7.5.4 Classification by ANN

In this problem, we use ANN with BPA as a classifier in a similar manner that we used in face recognition. The same fundamentals of generalization hold true here as well. The ANN with BPA is again able to solve the problem and give good results. Recall that here the problem is of verification rather than identification. Thus the ANN has a single output that denotes the confidence of the ANN of the applied input being the correct person. This confidence ranges from –1 to 1. A 1 denotes a complete certainty that the person is authentic, while a confidence of –1 denotes that the person is definitely a fraud.

We also must define a threshold confidence level. This is the minimal confidence that the ANN must have for any output for the ANN to regard the person as authentic. If the output is below this threshold, the ANN decides against the person's authenticity and regards him as a fraud.

Let us consider that the output of the ANN is o. Let t be the threshold. The system output is always 1 or 0, where 1 denotes that the person is genuine and 0 denotes that the person is a fraud. The system output may be given by the Equation 7.17:

$$\begin{aligned} Y &= 1, \text{ if } o \geq t \\ &= 0, \text{ otherwise} \end{aligned} \tag{7.17}$$

Using this equation, we may verify the identity of every individual in the database.

7.5.5 Results

The system was trained and tested against a database that consisted of 61 users. For each user, multiple scans of hands were collected for both the left and the right hands. The left and right hands for every person are very different from each other, and hence these were taken as hands of completely different persons. Four hand images for every person were selected randomly for training, and another two hand images were used as testing data.

The features extracted for some of the people are shown in Table 7.5. There are a total of 13 features, from L1 to F8 and L9 to 113. The explanation of these features is given in Figure 7.11. The training curve for the system is given in Figure 7.12.

The system performance is measured against the standard FRR and FAR ratios as given by Equations 7.18 and 7.19:

$$\text{FRR} = \frac{\text{Number of rejected genuine claims}}{\text{Total number of genuine accesses}} \tag{7.18}$$

$$\text{FAR} = \frac{\text{Number of accepted imposter claims}}{\text{Total number of imposter accesses}} \tag{7.19}$$

TABLE 7.5
Features Extracted from the Samples

	Measurement of Hand Geometry Features (in Pixels)												
Person	**F1**	**F2**	**F3**	**F4**	**F5**	**F6**	**F7**	**F8**	**F9**	**F10**	**F11**	**F12**	**F13**
A	497	605	683	620	502	170	143	165	131	148	136	145	129
B	494	589	637	599	485	154	143	156	136	153	130	130	126
C	466	618	656	603	537	142	128	154	119	144	124	139	115
D	437	571	644	572	486	128	106	136	117	125	109	113	102
E	485	604	692	632	501	143	125	156	132	145	132	143	128
F	485	604	646	604	494	150	121	136	119	144	128	136	127
G	481	549	615	580	455	139	125	136	116	128	108	125	104
H	424	663	706	671	577	154	118	144	116	146	120	135	114
I	504	592	639	586	509	150	116	153	124	142	120	121	117

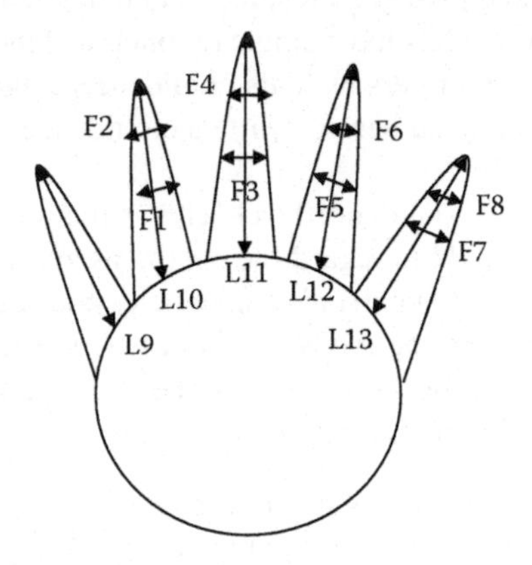

FIGURE 7.11 The features extracted from the hand.

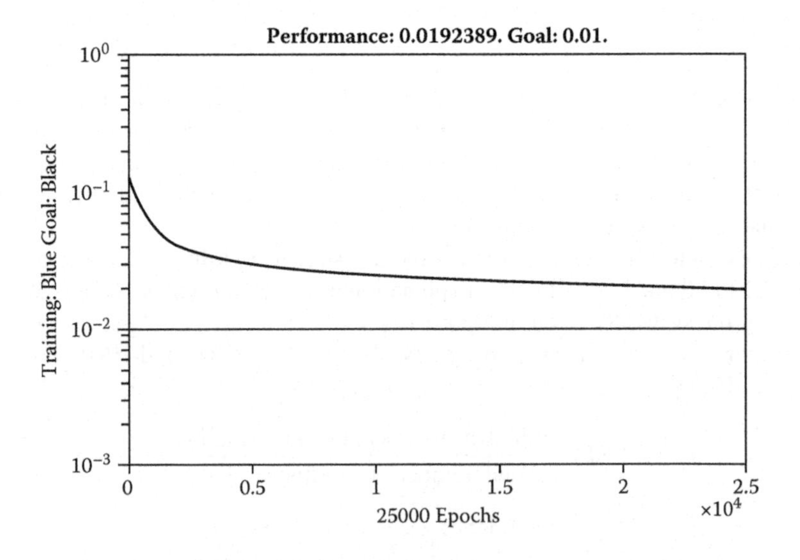

FIGURE 7.12 The training curve during hand recognition.

The built system was tested. The FAR for the system came out to be 4.2828, and the FRR came out to be 4.0000.

7.5.6 Concluding Remarks for Hand as a Biometric

In this section, we discussed a good means of using a biometric identity for the purpose of verification. Hand geometry was used as a novel and unconventional way of verification. In this system, the hand images were scanned and were used for the extraction of various features. These features were used by a verifier built using ANN with BPA. The system error was measured with the help of FAR and FRR. We saw that we could vary this error by varying the system's threshold. The best possible threshold was chosen that returned the least error in terms of both FAR and FRR.

To test the system, the images of 61 users were scanned, with multiple images per individual. Using these images as inputs, the features were extracted, and the resulting system was tested using ANN with BPA. The performance showed FAR and FRR of the system to be 4.2828 and 4.0000, respectively.

The present system offered a set of good features that were extracted for the given system. The other biometric features may be explored in the future. In addition, the fusion of hand geometry with fingerprinting or any other recognition system may be implemented. Such a system may give even greater accuracy and performance.

7.6 IRIS

The third biometric identity we discuss is the iris. The iris has always been a source of attraction for people. This has led to the idea that the iris forms a characteristic feature of humans that can be used for identification. The human iris is an annular region between the pupil (generally the darkest portion of the eye) and the sclera. In the entire human face, the iris remains the most characteristic identity, and it has numerous very exciting and characteristic features. If you closely observe the eyes, or look closely at the picture of an eye, you will likely see the possibilities for its use as an identification system.

In this section, we study the iris, which happens to be one of the most integral parts of the eye. The biomedical literature suggests that irises are as distinct as fingerprints or patterns of retinal blood vessels. Some of the characteristic features of the iris are freckles, coronas, stripes, furrows, and crypts.

The immense identification potential in the human iris is clear from the fact that a study of more than two million different pairs of human irises shows that no iris patterns were identical in even as much as one-third of their form. Even genetically identical eyes—for example from twins—have different iris patterns.

It is easy to imagine why the iris is so good for identification. Imagine seeing someone you know whose face was covered except for the eyes. In most cases, you would be able to identify the person. This might not be the case, however, if the converse were considered. The identification would be very poor, or not possible at all, if we only saw the lips of a person or only the ears. This alone gives us enough reason for realizing that the iris is a very promising biometric identification identity.

The iris is another feature with high dimensionality. This necessitates the study of good features in the iris, as well as the identification and the extraction of those features. Following the principles of any soft-computing system, these features must be robust and must not change appreciably when they are measured multiple times in multiple scans. The image-processing concepts are used for the purpose of identification and extraction of the features.

As with face recognition, the method of capturing the eye is with the help of cameras. In face recognition, we advocated that the cameras should be placed at a convenient location in which the system would be able to identify and extract the face for the purpose of verification or identification. For the iris, however, this may not always be necessary. The cameras used to scan the eye are generally placed so that they can easily capture the person's eye. This makes the task of separating the eye from the rest of the body relatively easy.

Iris recognition faces various problems that make it difficult for the classifier to make verification. Some of the commonly faced problems in iris verification and identification are the presence of eyelashes and eyelids, the iris's nonelastic deformation as the pupil changes size, head tilt, cyclovergence of the eye, and so forth.

In this case, detection of the corner is used as a means of extracting the features from the human iris. We use a mechanism known as the canny edge detection scheme to detect the iris in the eye's digital image. The corners in the iris image are detected using a covariance matrix of change in intensity along rows and columns. All detected corners are considered to be features of the iris image. For recognition, the corners of both the iris images (the scanned image and the data set) are detected, and the total number of corners that are matched between the two images are obtained. The two iris images belong to the same person if the number of matched corners is greater than the threshold value.

In this section, we study the iris in detail and discuss the various features that make good inputs for identification and verification. We then study the ways to extract these features. The manner of extraction largely follows the steps involved in the precious two cases of the face and the eye. We then make a classifier that verifies and identifies per the system requirement. This again will be by the use of ANN with BPA.

7.6.1 Human Iris

In this section, we give a brief description of the human eye. Much of the terms discussed in this section are only for the general awareness of readers. We will not be requiring these terms for the recognition of the iris.

The iris is a thin circular diaphragm that lies between the cornea and the lens. The various characteristics in the human eye are shown in Figure 7.13. The iris is perforated close to its center by a circular aperture known as the pupil. The iris controls the amount of light entering through the pupil. It does this by adjusting the size of the pupil. The iris is about 12 mm in diameter. The pupil size can vary from 10 percent to 80 percent of the iris diameter.

The iris consists of a number of layers. The lowest layer is called the *epithelium*, which contains dense pigmentation cells. The next layer is the *stromal layer*, which contains blood vessels, pigment cells, and the two iris muscles. The density of stromal pigmentation determines the color of the iris. The iris surface may be divided into two zones that may be easily differentiated by their color. The outer zone is the ciliary zone, and the inner zone is the pupillary zone. These two zones are divided by the collarette, which appears as a zigzag pattern.

Formation of the iris begins during the third month of embryonic life. In the first year of life, the unique pattern on the surface of the iris is formed. Pigmentation of the stroma takes place over the first few years of life. Formation of the unique patterns of the iris is random and not related to any genetic factors. Pigmentation of the iris determines its color and is dependent on the genetics. As

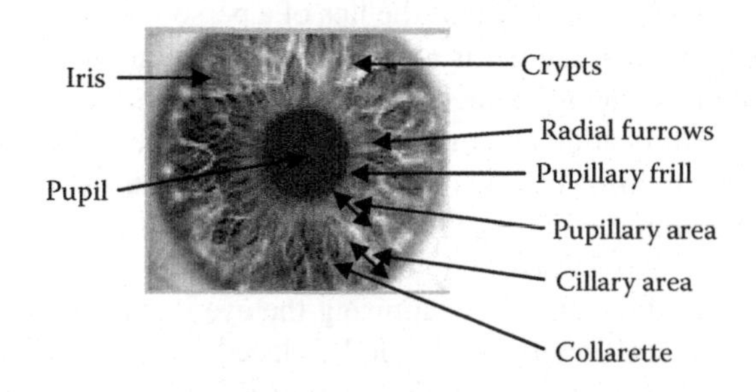

FIGURE 7.13 A front-on view of the human eye.

a result of the epigenetic nature of iris patterns, the two eyes of an individual contain completely independent iris patterns. For this same reason, identical twins possess uncorrelated iris patterns.

As can be observed, the human eye has numerous distinctive features that may easily be used for identification purposes. The recognition or verification of the iris may be performed by scanning either of a person's two irises. Thus the iris is a natural choice for biometric identity. This, however, also requires special capturing of the eye from a close vicinity.

The iris has the great mathematical advantage that its pattern variability from person to person is enormous. In addition, the iris is well protected by the eyelids, which suppress many of the possible variations. The image of the iris is relatively insensitive to angle of illumination, due to its planar nature. Being a distinctive organ, the iris may be easily isolated from the entire image of the face by using standard image-processing techniques. In addition, the iris remains reasonably stable over time, which is not the case for face or hand, which are both bound to change with age.

We now discuss the various steps involved in the process of iris verification. Because these steps are largely the same as discussed in the previous two sections, we will be brief in our discussion to avoid repetition.

7.6.2 Image Acquisition

The iris is extracted from the image taken of the human eye. For this we use a camera that is placed at an approximate distance of 9 cm from the user's eye. This gives the best results, as the iris is easily visible and separable. The distance between the user and the source of light is preferred to be about 12 cm. The image that is taken must have good resolution and sharpness and must be under good lighting conditions. These features play a more important part as compared with face, due to the nature of the iris.

7.6.3 Preprocessing

The image requires preprocessing. We apply various operations to the image in order to extract the necessary features. We apply the various image preprocessing steps as discussed in the case of hand geometry—gray conversion, binarization, and edge detection. The steps for a sample iris are given in Figures 7.14a through 17.4c.

The edge detection is the stage at which we detect the edges of the objects and try to separate them. Here we make use of the canny edge detection algorithm. This algorithm first makes the image smooth by applying a Gaussian filter. This also removes any noise from the image. Then the gradient magnitude and the orientation are calculated, using finite-difference approximations for the partial derivatives. Then we apply nonmaxima suppression to the gradient magnitude. Finally we use a double-thresholding algorithm to detect and link edges. The details of this operator are, however, beyond the scope of the book.

7.6.4 Feature Extraction

We now introduce the concept of corners and use it to extract the corners from the system. The preprocessed image is processed to extract the corner points by using a covariance matrix of change in intensity at each point. We use a window matrix of size 3 × 3 that is circulated in the entire image. This determines whether any point is a corner point. The covariance matrix M is given by Equation 7.18:

$$M = \begin{bmatrix} \sum D_x^2 & D_x D_y \\ D_x D_y & \sum D_y^2 \end{bmatrix} \tag{7.18}$$

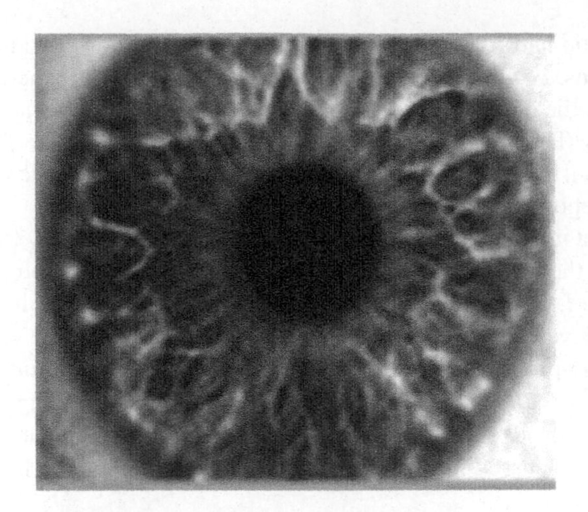

FIGURE 7.14a Gray image of the iris.

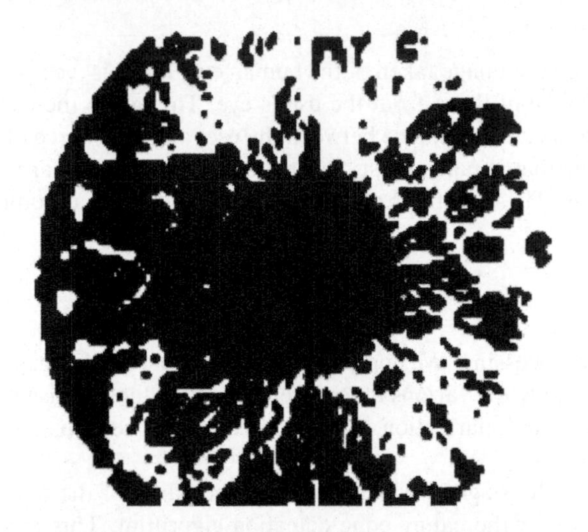

FIGURE 7.14b Binary image of the iris.

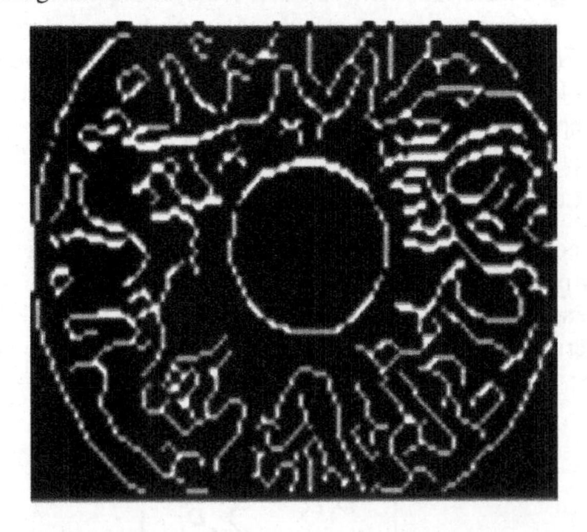

FIGURE 7.14c Detected edges of the iris.

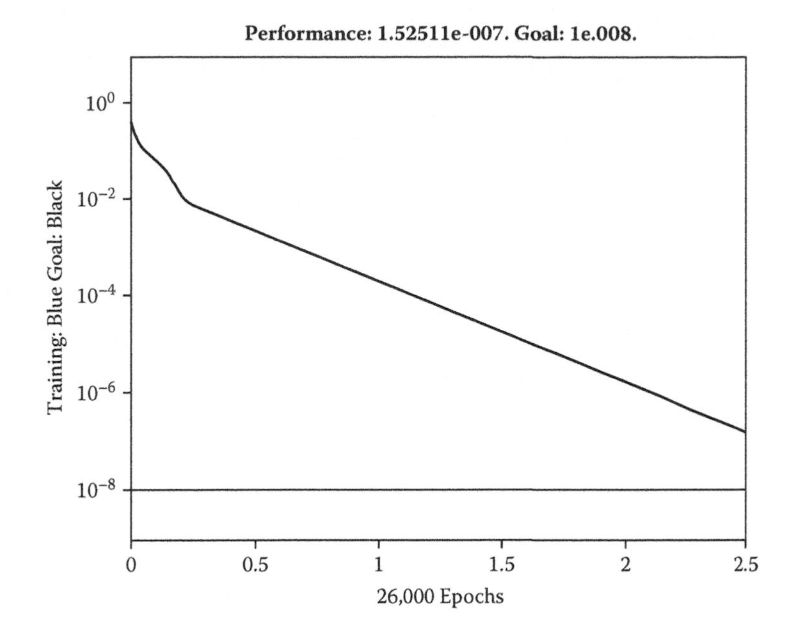

FIGURE 7.15 The training curve for the iris problem.

where D_x is the change in intensity when measured along the columns and D_y is the change in intensity when measured along the rows.

The point is considered to be a corner point if the eigen value of M is found to be greater than some threshold.

7.6.5 Results

The system was trained using a database of 120 individuals whose eyes had been captured with the help of a camera. This database was used by the system for verification purposes. MATLAB® functions were used for the various image-processing tasks to extract the features. ANN with BPA was used for training and testing based on these features. The training curve is shown in Figure 7.15.

The values of the threshold were fixed by trial and error in order to give the least errors to both FRR and FAR. The system after training and testing gave an FRR of 5 percent and an FAR of 4 percent. These are acceptable errors in the system and prove that the system was able to carry out the task with acceptable accuracies.

7.6.6 Concluding Remarks for Iris as a Biometric

In this method, we mainly followed an approach similar to that used for both the hand geometry and the face recognition. This method used ANN with BPA for verification of the individual. The system used the corner detection approach to find the corners in the image of the eye. These extracted corners were the features that the ANN was trained with. The preprocessing of the eye image involved gray conversion, binarization, and edge detection. The system that was developed could easily solve the problem of verification. When we applied these mechanisms over the database of irises, we got an acceptable FRR and FAR, which shows that the iris can be used as a good biometric identity.

The placement of cameras is a potential problem in iris verification. Also, eyelids pose a serious limitation to the system, restricting its efficiency. These issues need to be addressed in the future. The fusion of iris detection may be combined with face recognition for even greater accuracies.

8 Behavioral Biometrics

8.1 INTRODUCTION

Behavioral biometrics deals with active features for the purpose of biometric identification. Activation features change with time and are hence termed *behavioral*. In this chapter, we will analyze and identify the behaviors associated with some features for the purpose of identification. The behavior may be in the form of sound, as is the case with speech; motion, as is the case with signature; and so on.

The basic philosophy of these systems is exactly the same as for the physiological biometric systems. Thus we follow the same principles of acquiring data, preprocessing and segmenting it, and then using the extracted features for recognition or verification with the help of classifiers. The same mechanisms are followed, and similar discussion and treatment of the problems are adopted in these biometric systems.

The addition of time, however, adds a new dimension to the problem and leads to many more techniques, methods, and ways in which we may deal with the systems. This addition of time as a new dimension, however, also leads to various problems and issues that are unique to these systems. We look at all of them in detail in our discussion of the various behavioral biometric systems.

Behavioral biometric systems use the concept of sampling as an additional step in their working. This means we need to decide certain sample frequency. Data are then captured at this particular frequency. Say the sample frequency decided is f Hz. This means we would acquire data after every $1/f$ seconds. Consider the speaker-recognition system, in which speech is recorded through microphones and saved in the system. The recording involves the signal's amplitude at time intervals of $1/f$ seconds. This time decides the system's storage capacity and later the processing time required by the system. A very small sampling time or a large sampling frequency may mean that the system runs very slowly and takes a lot of space to store and process. On the other hand, a very large sampling time or a small sampling frequency may result in the loss of certain features, or even features that get totally missed by the system. Hence sampling time must be fixed accordingly. Usually sampling frequency is kept high enough so that time and storage constraints are not major problems in biometric identification or verification.

Readers might confuse a behavioral biometric system with a system in which samples are recorded over time and a bigger classifier is used to take in features extracted from all the individual samples. This theory may seem to be correct, but it actually has numerous issues. One of the major issues is of dimensionality. We saw in Chapter 7 that dimensionality plays a key role and requires measures for its reduction so that any classifier may work. Adding numerous samples would result in a very large increase in dimensions, which would make the system very complex and hence untrainable.

Another problem associated with the approach is that it can never be ascertained that the features occur at the same time each time the recording is done. The different features usually tend to occur at different times, depending on various known and unknown causes. Consider, for example, a speaker-verification system in which the person is supposed to say words X. She may sometimes say the first part of it very fast and the rest of it reasonably slowly, or vice versa. This puts a limitation on the system. We will see later in this chapter that working in the frequency domain solves this problem and results in better efficiencies. However, even the frequency domain has problems. Instead we work at a domain that is a mixture of time and frequency.

In this chapter, we study two very exciting systems: the speaker recognition system and the signature verification system. Both systems are extensively used in real life applications. The former uses a very characteristic property of speech, which has been highly studied and developed. The other has a very high relevance and practical importance due to traditional use. With these examples, we will see how the problem of behavioral biometrics is solved and designed.

8.2 SPEECH

Right from childhood, we hear so many sounds. We keep associating voices with actions, events, people, and so forth. As we grow up, we start learning the art of associating sounds with numerous things. This learning gives us the very extraordinary power of perceiving things just by hearing. In turn, this means that speech is a very essential part of human life that many times conveys more that what sight can do. Speech holds so much information in itself, which helps us in various ways in our daily life.

For example, if someone coming from behind calls us, we can readily identify him by voice. Similarly if we chat with someone by phone, we are able to make many inferences, even though we have never seen the other person. It is common for people to guess the sex, age, or even the mood of the other person by the way he or she speaks. Likewise, we are able to guess the singer of a song or say the voice of our beloved ones very easily. This all means that sound truly carries something very extraordinary in itself.

In theory, sound is just like any other signal. It has amplitude that changes over time. It is composed of many frequencies. Sound possesses all the properties and factors that any signal holds. Analysis shows that it comes under the class of very complex signals. Even the most basic sound signal, like a heartbeat, changes very rapidly in an unpredictable manner when plotted against time. This makes sound a very difficult signal to visually study or analyze. It also puts forward the notion of the sound signal being of a very large dimensionality.

Our discussion in Chapter 7 focused on the features that can easily be seen and, hence, appreciated. The situation in this chapter is different. Sound is only heard; it cannot be seen. As a result, it may often be difficult to visualize or appreciate the systems. It may even be difficult to determine the features or attributes that the system may have. We discuss the various standard features used for speech analysis later in this section. We use different visualizations of the same data to determine the effectiveness and the validity of the factors that we use for identification or verification.

We have learned that speech signals have numerous uses in natural systems. This fact has motivated the use of speech to make excellent systems that can convey a great deal of information. The rich and diverse features embedded in every speech signal are a good source for extracting unique pieces of information from a system. Here we do the same job of extracting the information embedded in the speech signal.

We again use speech for the purpose of identification. As we have learned, speech is one of the biometrics that is extensively used for identification and verification. For example, it may be used at entrance gates for security checks or for the identification of the person speaking. Speech may also be used as a biometric password for online services.

The best thing about speech is that it is very easy to capture. All that is needed is a microphone, which is relatively cheap and universally available. This is one reason speech has become the choice of millions of researchers and practitioners worldwide. In addition, speech has an added advantage in that it is easy to process through the use of any speech- or signal-processing-related software.

Speech, however, is very prone to noise, often in the form of background noise, which is always present in the environments where we live. In many places like markets, busy streets, and so forth, this background noise rises to such an extent that it is almost impossible to make any type of identification, even with the best systems. However, even if you are somewhere that is completely noise-free, the noise from various factors may still creep in and disturb the whole system. For these reasons, it is necessary to preprocess the recorded speech before it can be used for any kind of identification. This preprocessing step mainly includes noise removal, which in the case of speech may usually be done by a low-pass filter.

The problem of Speaker recognition has been solved using two very common classifiers: artificial neural networks (ANN) with the back propagation algorithm (BPA) and adaptive neuro-fuzzy inference systems (ANFIS). ANN with BPA has a very close resemblance to the human brain. Just as the brain can easily solve and perform the tasks that we wish to solve, ANN with BPA can do the same. In addition, as we have shown throughout this text. ANN with BPA has great learning and generalizing powers, which again makes a very strong case for their use as classifiers.

An easy way to visualize this system is in the form of rules. When a system tries to identify someone, it uses each and every characteristic, feature, and behavior of that person. In this way, the system is related to a rule-based approach, in which the rules play a major role in mapping the inputs to the outputs. For this problem, we use neuro-fuzzy systems. These systems use ANN to train the fuzzy systems and are very useful if some of the system's inputs and outputs as well as some of the rules are known.

In this chapter, we deal with speech in multiple ways. First we study and extract some features from the signal to identify the age and sex of the speaker. This is done by applying both ANN with BPA and the neuro-fuzzy system. We also discuss the very interesting area of multilingual speaker recognition. For this system, we use a modified modular neural network (MNN) approach. We then use wavelet analysis to extract another set of very unique features that forms another easy and efficient system of identification. In this chapter, we address the problem of features and classification separately. The general methodology that we follow is given in Figure 8.1.

Any combination of the feature set with any of the classifiers makes a complete system for the task to be accomplished. We first discuss the various features of speech that make up good identification features. We then study the various identification systems. The results of combining the various features with the various classifiers are presented at the end of the discussion. This section describes the results and findings of Shukla and Tiwari (2008c, 2008i, 2008m, 2007a, 2006), Shukla et al. (2008b, 2009e), and Agarwal, Prateek, et al. (2009).

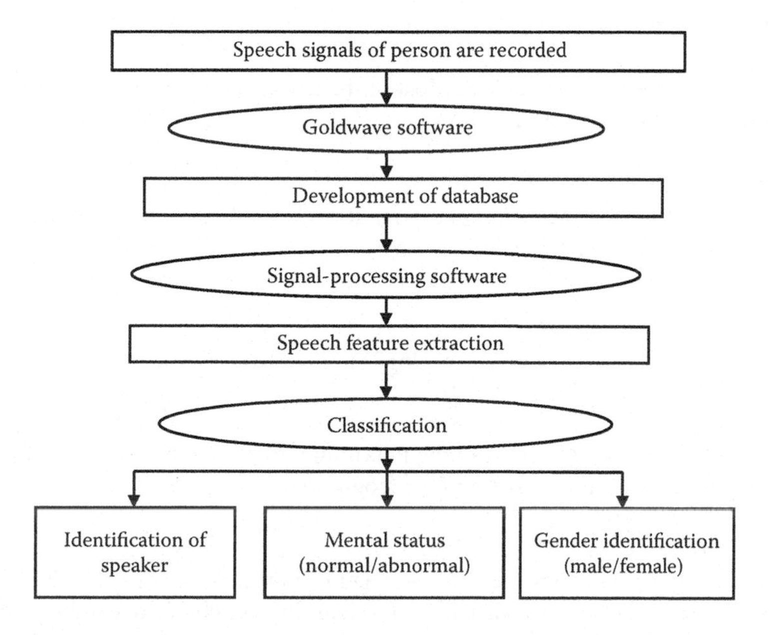

FIGURE 8.1 The general methodology of using speech as a biometric.

8.2.1 Speech Input

Input is captured with the help of microphones and is then given to a computer that stores it in the form of a WAV file. The analog speech signal that comes from the microphone is digitized at a frequency of 16 KHz. The determination of 16KHz as the proper frequency is obtained from the Nyquist criterion, which states that to avoid any loss of information, a signal must be sampled at a frequency that is double the bandwidth. Speech signals are normally up to 3 KHz, but some of the fricative sounds of the speech can go up to 5 KHz. To comfortably accommodate the worst cases, we keep the sampling frequency at 16 KHz. Remember that a very high sampling frequency would increase the file size and the processing time. On the other hand, a very low sampling frequency might result in a loss of information.

Once this recording has been made, it requires preprocessing and segmentation. In these steps, we remove noise from the recorded sample and extract the needed part from it. For example, if sentences were recorded, we would break each sentence into individual words. Similarly, the silence between words and other such features must be cut. We use GoldWave software for these purposes.

8.2.2 Speech Features

In this step, we extract the relevant, or good, features from the speech signal. These features form the input for whichever classifier the system uses. The various speech features must be constant when measured again and again in multiple samples from the user. Furthermore they must vary as much as possible among different people.

The following are the commonly extracted features for speech:

- Maximum amplitude, or cepstral analysis
- Peak and average power spectral density
- Number of zero crossing (NZC)
- Formant frequency
- Time duration.
- Pitch and pitch amplitude

We deal with some of these features below. Most of the features may be extracted using any signal-processing tool and then be used straight away.

8.2.2.1 Cepstral Analysis

In this nonlinear signal-processing technique, we first apply the Fourier transformation to the signal sequence. Then the real and complex logarithmics of the resulting signal are taken, followed by application of the inverse Fourier transformation. This analysis is represented by Equation 8.1:

$$Y = \frac{1}{2\Pi}\int_{-\Pi}^{\Pi} \log[X(e^{j\omega})e^{j\omega}\, d\omega \tag{8.1}$$

In this analysis, we may be interested in the maximum amplitude of the signal that returns after the cepstral analysis to be used as a feature for recognition.

8.2.2.2 Power Spectral Density

Power spectral density (PSD) helps analyze the signal's power or energy as a function of its frequency. PSD shows how the power is distributed among the various frequencies. One such case is shown in Figure 8.2. The PSD may be calculated in numerous ways. One of the ways is by the use of Welch Formula.

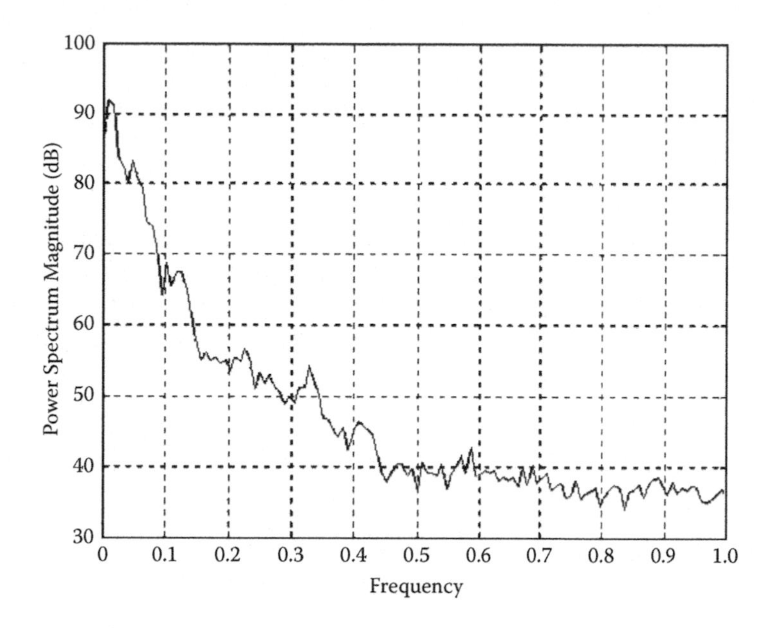

FIGURE 8.2 The power spectral density versus the frequency graph for a speech signal.

Here we extract the average PSD per frequency to make a very good input for the recognition system. The peak PSD may also be measured in the same manner.

8.2.2.3 Spectrogram Analysis

This technique is used to analyze the signal's spectral density with respect to the time. Here we use time-dependent analysis to study and analyze how the power per frequency varies with time. This analysis is done by plotting the spectrogram that represents time on the x-axis and the frequency on the y-axis. The amplitude or energy is plotted on the z-axis. For the sake of representation in a graph, The amplitude is represented by shades of gray. The spectrograph is shown in Figure 8.3.

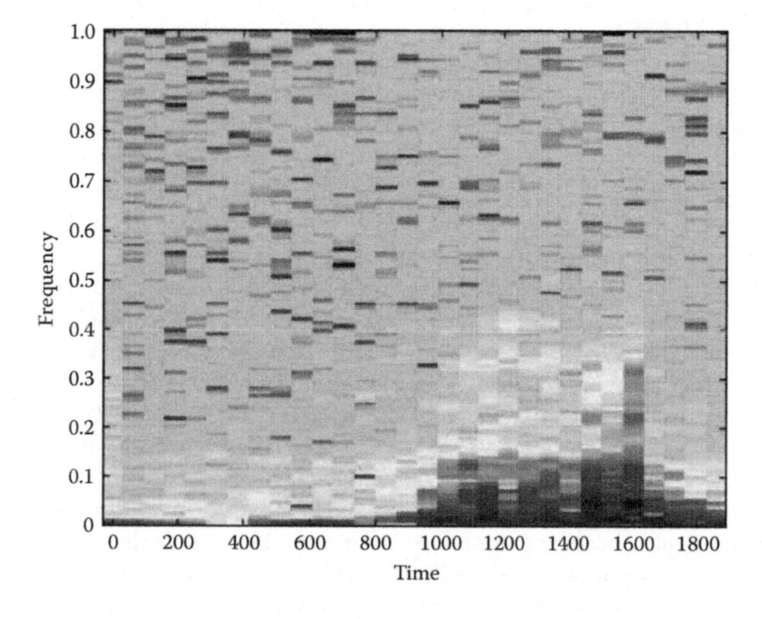

FIGURE 8.3 The spectrogram of a speech signal.

8.2.2.4 Number of Zero Crossings

NZC denotes the number of times the signal crosses the *x*-axis (or the time axis). A crossing is defined as a change in amplitute from positive to negative or from negative to positive. These crossing are calculated and used as features.

8.2.2.5 Formant Frequencies

The voice signal produced by each person always consists of a base frequency. This frequency is known as the *fundamental frequency*. The frequency for each person is determined by numerous factors in the mouth. The resulting frequencies consist of a number of harmonics along with the base frequency. These harmonics are known as the *formants*.

8.2.2.6 Time

In this case, this denotes the total time taken by the signal. For example, if the recording were of a person saying some word, we would measure the total time the person takes to complete that word. Time is used as a direct feature for identification

8.2.2.7 Pitch and Amplitude

The pitch measures the signal's frequency. We use Fourier analysis to measure the most dominant frequency. Amplitude measures the signal's height. We may use the maximum or the mean amplitude as features for this system.

8.2.3 Wavelet Analysis

In the previous section, we presented one set of features that could easily be extracted using the signal-processing application and commands. In this section, we present another set of features that can be used for both identification and verification. We use wavelet analysis, which studies the entire signal and extracts a prespecified number of features called *wavelet coefficients*. These coefficients are then extracted and used as input for the classifier. In this section, we mainly study the wavelet analysis technique. We also present two more similar analysis techniques—Fourier analysis and short-time Fourier analysis—that may be used in place of wavelet analysis for recognition purposes.

8.2.3.1 Fourier Analysis

This technique breaks down a signal into its constituent frequencies. A signal that was originally in the time domain is converted entirely into the frequency domain. The result of Fourier analysis is a set of frequencies that the original signal is made up of. If we were to add all the frequencies returned by the Fourier analysis, we would get our original signal. The technique breaks up signals in order of their dominance in the given signal. In other words, the first few frequencies are predominantly present in the signal, and thus these mainly decide the signal. The last few frequencies, however, are not so important and may be neglected.

Because a Fourier analysis returns an infinite number of frequencies, we need to specify some threshold above which we will accept frequencies. This number is usually selected on the basis of the number of inputs our system can handle. Selecting too many frequencies might result in the system becoming very complex and not able to train itself. Taking too few frequencies might result in the system being unable to classify due to too little intraclass separation. This is why the number of frequencies to be selected must be decided judiciously. The same situation is true for the other analysis techniques (i.e., short-time Fourier and wavelet analysis).

Hence in Fourier analysis, we break the given signal into a set of frequencies and select the major frequencies. The signal is nearly reconstructed from these selected frequencies but without much loss of information. This makes Fourier analysis a good technique for feature extraction.

However, Fourier analysis has a major drawback. While representing any signal in the frequency domain, we lose all the information about time, which is a major factor in a number of cases. Some

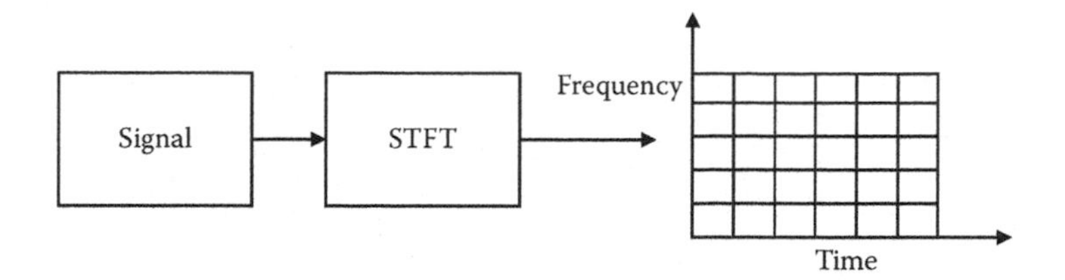

FIGURE 8.4 The short-time Fourier transform in time and frequency domains.

signals consist of short-duration behaviors taken at specific times that may not be noted by Fourier analysis. This often results in the loss of good information that could have helped during classification. These short-duration behaviors include drift, trends, abrupt changes, and beginnings and ends of events.

8.2.3.2 Short-Time Fourier Analysis

Short-time Fourier analysis techniques solve the drawback that we discussed above in Fourier analysis. With this type of analysis, we introduce some information about time in the analysis by specifying a window of a constant timeframe. The Fourier analysis is then applied inside this window. This transformation of the signal into both time and frequency is known as a short-time Fourier transform (STFT). Thus this type of analysis caters to the twin needs of frequency and time, with the frequency information being supplied within the windows of time. The original signal is mapped into a two-dimensional matrix in which the x-axis denotes the time and the y-axis denotes the frequencies. As the window moves across the original signal, the frequency constituents are supplied for every time box. In this way, the matrix is filled from left to right, as shown in Figure 8.4.

The STFT approach fixes the time and frequency in such a way that the best information from both is obtained with the least loss of information. The window's size plays a key role in fixing the role of the time and frequency components. A correct match between the time and frequency is required to ensure coverage of the maximum amount of information with the least number of coefficients. Thus if the window were too small, the information would be dominated by time. Similarly a too wide window would lead to the dominance of frequency, as is the case with Fourier analysis.

In many cases, however, it may be very difficult to fix the window. Some regions may require a smaller window, while others may require a larger window. In such cases, it becomes difficult to optimize the system. Hence a fixed-size window often fails to cater to the needs of time and frequency and is thus a limitation with STFT.

8.2.3.3 Wavelet Analysis

Wavelet analysis improves the limitations of short-time Fourier analysis. In wavelet analysis, the size of the window can be kept variable. In other words, the window is allowed to expand or collapse at different regions or different times of the signal. This gives the system much-needed flexibility for analyzing the signal and deciding the best possible window size to optimize system performance. Wavelet analysis allows us to use long time intervals when we want more precise, low-frequency information and shorter time intervals when we want high-frequency information. This technique is represented by Figure 8.5. As is shown in Figures 8.4 and 8.5, wavelet analysis works on a time-scale region rather than a time-frequency region.

In wavelet analysis, the signal is broken down during wavelet decomposition into approximation and detail. That approximation may then be further broken down at the second level into approximation and detail. In this manner, the approximations continue to be broken down. For n-level decomposition, there are $n + 1$ possible ways to decompose, or encode, the signal. As this breaking down continues, the level of details increases, The details at later stages carry much less information as compared with the initial few details. Therefore we must choose a threshold and stop the

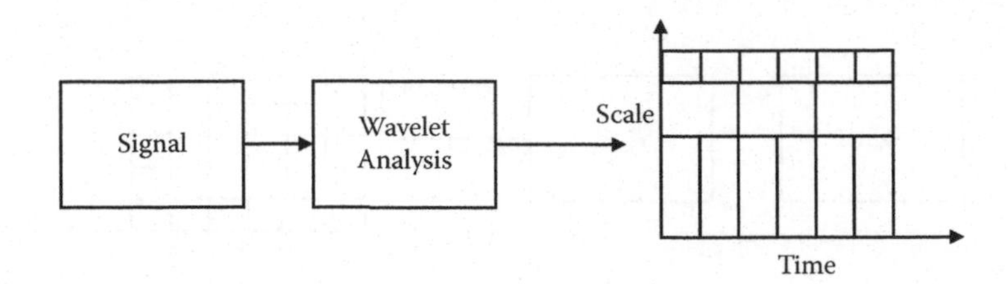

FIGURE 8.5 The wavelet analysis.

process of decomposition after this threshold is met. The decomposition is shown in Figure 8.6, in which $S = A_1 + D_1 = (A_2 + D_2) + D_1 = ((A_3 + D_3) + D_2) + D_1$.

8.2.4 ANN with BPA

After studying two sets of feature-extraction methods, we now come to the other major part of the problem—the task of identification using the extracted features. We have used ANN with BPA, modular neural network (MNN), and neuro-fuzzy systems for this task. In this section, we study the ANN with BPA. The next two sections are devoted to the use of MNN and neuro-fuzzy systems.

The use of ANN with BPA in this problem is similar to its one in Chapter 7, in which we used a classification model of ANN. The same principles are used here. For the tasks that deal with the prediction of age and so on, a single output predicts the needed output. In this way, we can easily use ANN using either the classification or the functional prediction models. In either case, the inputs must be normalized before the system can use them to calculate the system.

In Chapter 7, we spoke a great deal about the use of ANNs as a classifier, with special regard to practical real life problems. In this chapter, we spend provide only a brief discussion of the role of ANN with BPA in functional prediction problems. This discussion will be helpful when we complete the system design and discuss the results.

For any classification problem, it is generally assumed that the data should be as close together as possible in the input space for the intraclass entities and as far apart as possible for the interclass entities. For a functional prediction problem, however, the converse is true: The data must be diverse and carry good information on trends. In other words, the values of the outputs at various locations of the input space must be known. This requires that the data be normally distributed across the input space. In this problem, however, we use both classification and functional prediction with the same data and the same extracted features.

Any real life problem has a high dimensionality that represents millions of possible options for various inputs. This high dimensionality puts a serious demand on the training data set to be rich

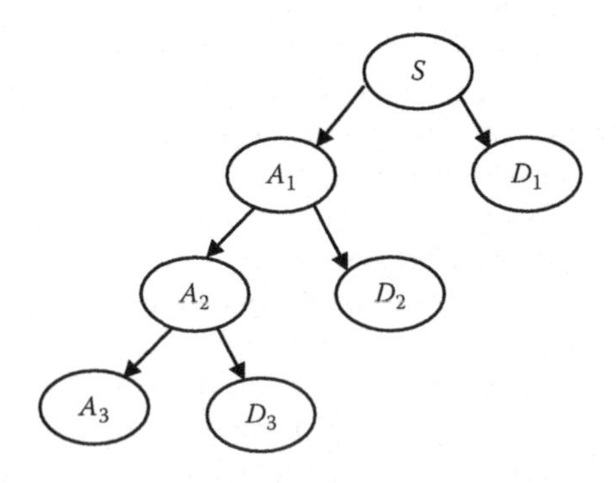

FIGURE 8.6 The decomposition in a wavelet transform.

and large. But a rich and large training data set is very difficult to implement when designing a system. Most of the data available for training might not cover diversity, and getting good data might be too expensive in terms of time and money. As a result, we must work over the system with the help of available inputs number and quality of the inputs.

Out of all the possibilities for any one problem, only a few values may occur in reality. This limits the input space by a reasonable amount. Out of all these values, the ones that actually occur commonly might be even fewer. Sometimes the input lies scattered over a few areas of the entire possible input space. In such a situation, the ANN with BPA does its best to find patterns and generalize the system globally. Thus if the system were to receive data that are very distinct from the data sets used for training, it might not get the correct output. On the other hand, if a system receives input whose nearby locations are with known output, it is highly likely that the output will be computed correctly. This is because the training over that region fine-tuned the network well for the inputs belonging to those regions. For this to occur, we assume that the input attributes are complete and are the ones that affect system performance. In addition, the training had to have been done judiciously, with the best possible architecture in terms of generality. Note that the ANN's generality plays a big role in the multilevel generality of the system. The ANN is further able to learn as much as possible with the limited test cases.

8.2.5 ANFIS

The second classifier and predictor we study here is the ANFIS, which is essentially a fuzzy system that is trained using an ANN learning procedure. This training is possible because the fuzzy inference systems (FIS) can be modeled in the same way as ANN. In this section, we spend time discussing the role and working of ANFIS in regard to practical real life problems. ANFIS is commonly used when we can figure out some rules that drive a system. ANFIS provides an excellent means for tuning the resultant system and enabling it to match the available data set. This capability of ANFIS makes it an easy choice for use in many systems.

However because of its very strong optimization powers, ANFIS is also being used to solve problems whose rules are completely unknown. ANFIS is able to determine the rules starting from nothing. It first assumes an initial all-connected model that forms all possible input combinations in the generated membership functions. These combinations are then matched to the closest available output membership function. Modifications take place in multiple iterations, and then the values are fixed. For this reason, ANFIS is becoming popular for use, even when the system might not be traditionally rule driven.

The ability to use ANFIS without any initial idea of the rules works in numerous cases. We can easily understand why this is so if we look back to the previous section at our discussion of the practical uses of ANNs. The input lies at some locations of the input space whose outputs are clearly known. The ANFIS can then easily map these inputs to outputs and learn the mapping technique as rules. The inputs near the training data must closely follow it. Again, the ANFIS does not take the time to learn, adapt to, or imitate this behavior, as it just needs to know or learn the level of change that must be produced with respect to distance. The inputs far away from this have the same problem in that we have an input to test but there is no nearby input from the training data set. These inputs, as in the ANN, may get almost correct output due to the generalizing capability of ANFIS or the generalizing capability of the rules. More details on this topic are presented in Chapter 19.

However, the problem of classification and functional prediction is in no way simple. Even though the outputs at a point may be influenced by the correct influencing forces, the problems of dimensionality and limited data size still exist. The large dimensionality induces more parameters that require more training samples. But these data are always expensive and limited in size. Hence there are always constraints that make it difficult for the ANFIS to determine the effect of the rules at various places or the importance of those rules.

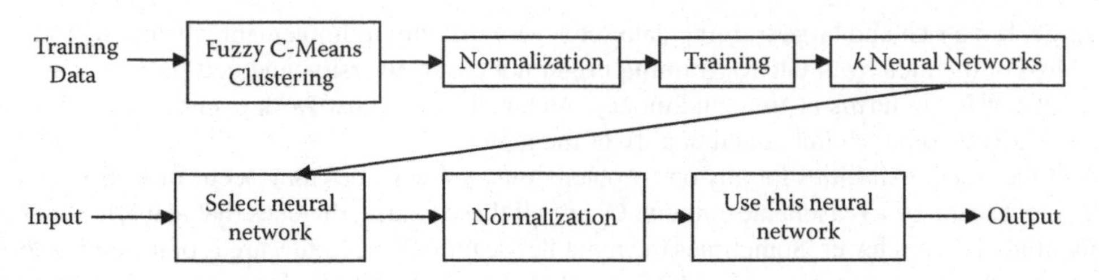

FIGURE 8.7 The hierarchical training algorithm.

8.2.6 Modular Neural Network

The third kind of solution is obtained with the help of modified modular neural networks (MNNs). In many real life applications, enormous research results in massive training data sets. We learned earlier that training large data sets may result in a system refusing to train or taking too much time to train. For these situations, we use MNNs, which apply modularity over ANNs for enhanced performance. We again use a hierarchical model, in which the training data set is first clustered so that each cluster has its own ANN. When an unknown input is given to the system, the system first finds the cluster to which the input belongs. Then the input is processed by the system's individual neural network. The major steps of the algorithm are given in Figure 8.7.

In this algorithm we make use of fuzzy C-means clustering to classify the inputs and allot them the ANNs. This divides the entire input space clearly into regions and allots each region a distinct ANN. Each ANN is trained independently using the members that belong to it, making it possible for the k ANNs to train simultaneously in parallel. This process saves time and results in better learning with fewer errors. Fuzzy C-means clustering creates cluster centers and cluster boundaries, which are used to allocate clusters to unknown inputs. These inputs can be determined by a simple rule stating that every input chooses the ANN or cluster that is closest to it. During testing, whenever an input is given, the best cluster that is closest to the input is selected. We then process the input by this neural network. The logical structure of the algorithm is given in Figure 8.8.

The number of clusters or the number of ANNs, to be used in the MNN must be decided upfront. This value depends on the level to which we want the system to be distributed. The use of too many ANNs would make the system more localized, and the generalizing capability could be lost. By every cluster that we increase, there is some loss of generalizing capability. This would lead to the ANN not knowing what is happening at the other part of the input space. As a result, any kind of optimization the ANN makes would be only on the basis of the area that it is enclosing or that has

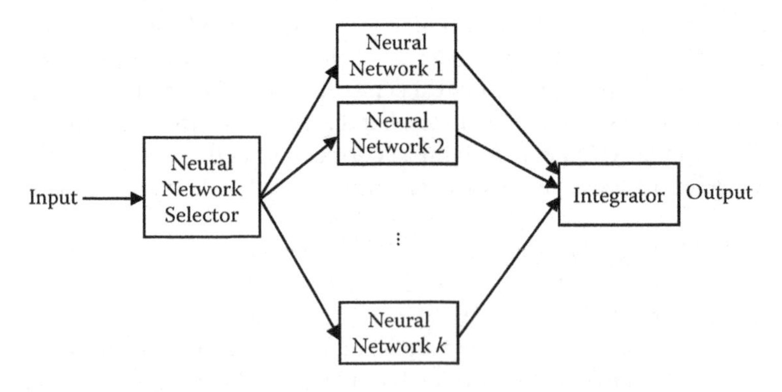

FIGURE 8.8 The hierarchical structure of the training algorithm.

been allotted to it, making all the decisions local. The generalizing is lost when we split the input space into multiple ANNs so that the ANNs are restricted to only a smaller region.

On the other hand, increasing *k* would mean that we would get better training efficiencies, resulting in much fewer errors. Each ANN would have only that many data item sets to learn; it would not be overburdened with undue data items. We also would be able to reach a higher number of epochs. When the error is decreasing appreciably, the increase in the number of epochs is a desirable feature that results in better system performance. The training time would also be reduced, which be a great boon to the systems. Recall that we are considering a system that has far too many inputs in the training data sets; therefore methods for reducing the training time are important in making the system applicable in real life.

A good balance among the generalizing capability, the training time, and the training performance must be made. These factors must be accounted for when specifying the number of clusters, which determines the system's efficiency and performance.

8.2.7 Systems and Results

We now know the various features and classifiers that may be used in any combination. We have also introduced the various problems that need to be solved. To fully understand and get practical experience with real life applications, let us choose the features and classifier to solve one of the problems. Studying how to do this will give us knowledge of the real life applications associated with speech as a biometric, while also enabling us to justify the aforementioned issues with each soft-computing technique.

8.2.7.1 ANN with BPA

We first develop an identification system by using ANN with BPA. We use the first set of discussed features, which include time duration, number of zero crossing, maximum cepstral, average power spectral density, pitch amplitude, pitch frequency, PSD, and four formant frequencies (F1 through F4). The problem to be solved is the identification of a person's identity. A total of 11 features are used. Data for 20 speakers are used for training and testing. The inputs for some of the speakers are shown in Table 8.1.

For training, the goal is fixed at 0.01, momentum at 0.5, and learning rate at 0.01. The training curve for the problem is shown in Figure 8.9.

After we are done identifying the speakers, we look at the problems of sex determination, identification of age group, and mental state. In all of these, the same inputs are used along with ANN with BPA. We first analyze the inputs manually to find out if there is some possibility or trend by which the specified problems can be solved.

First, we look at the sex determination problem. In general, we observe that the formant frequencies are higher in females as compared with the males. The same trend is true for pitch. In peak PSD and average PSD, however, the reverse is true, with males having higher PSDs as compared with females. Maximum amplitude of a male speaker is higher than that of a female speaker.

In the problem of identifying the person's age group, four age groups are considered: 16–30, 30–50, 50–60, and over 60. The general analysis is first done on the attributes, revealing that the pitch frequency is higher for young speakers as compared with older speakers. Average spectral density, peak spectral density, and maximum amplitude all increase with age; however, for people over 60 years, it is less. The same trend is true for number of zero crossings and time duration of utterances. Formant frequency of speakers decreases with age.

The final experiment was identification of the speaker's mental state. This is classified as either normal or abnormal. The speaker's mental status is related to duration of pauses, pitch, formant frequency, amplitude, and time duration. If the duration of pauses is more than the specified limit, per the rules of acoustical linguistics, the speaker's mental status will fall under abnormal conditions. Pitch and formant frequencies of a normal speaker are more than those of an abnormal speaker.

TABLE 8.1
Features for Some Speakers

Speaker	Time Duration	No. of Zero Crossing	Max Cepstral	Average PSD	Pitch Amplitude	Pitch Frequency	Peak PSD	Formant Frequency			
								F1	F2	F3	F4
01	11,928	15	2.0619	0.004484	0.01191	0.91281	0.03125	0.03125	0.10938	0.19531	0.29688
02	11,952	12	1.9008	0.002518	0.01192	0.96094	0.03125	0.039063	0.17188	0.25781	0.30469
03	10,240	2	1.4760	0.000286	0.01191	0.98438	0.039063	0.039063	0.13281	0.19531	0.35156
04	12,213	8	1.3320	0.000156	0.01200	0.96875	0.046875	0.039063	0.14063	0.19531	0.35938
05	10,302	6	1.8464	0.001032	0.01184	0.92345	0.039063	0.046875	0.09375	0.21094	0.30469

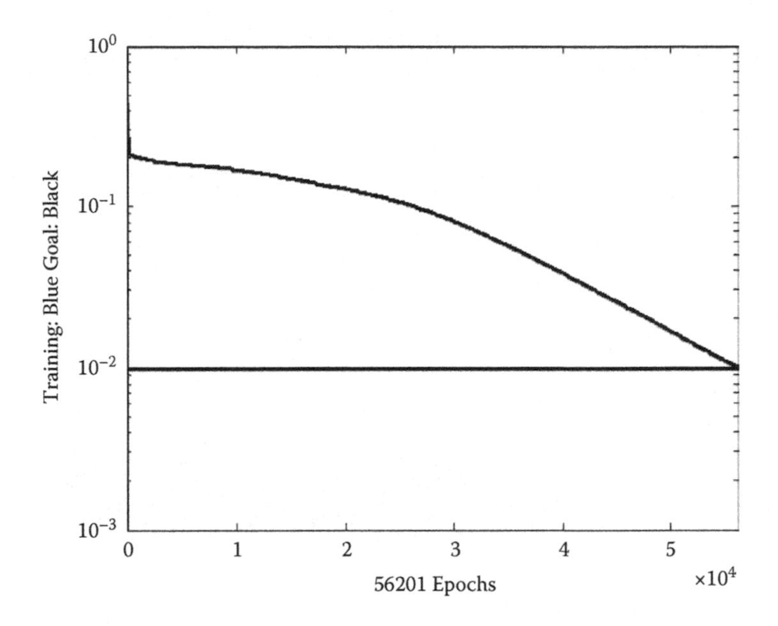

FIGURE 8.9 The training curve for artificial neural network with back propagation algorithm for speech.

Maximum amplitude of an abnormal speaker is greater than that of a normal speaker. Duration of the utterance of the word for a normal speaker is very small as compared with an abnormal speaker.

In all the experiments of this example, we use the variable learning rule. ANN is trained for 50 speakers and tested on 50 speakers. The training graph and the learning rate for different epochs in the problem of sex determination are shown in Figure 8.10a. Similarly Figures 8.10b and 8.10c show the graphs for the problem of age group and mental state determination. In all cases, the ANN was able to achieve high accuracies and thus solved the problem well. Hence we could use ANN with

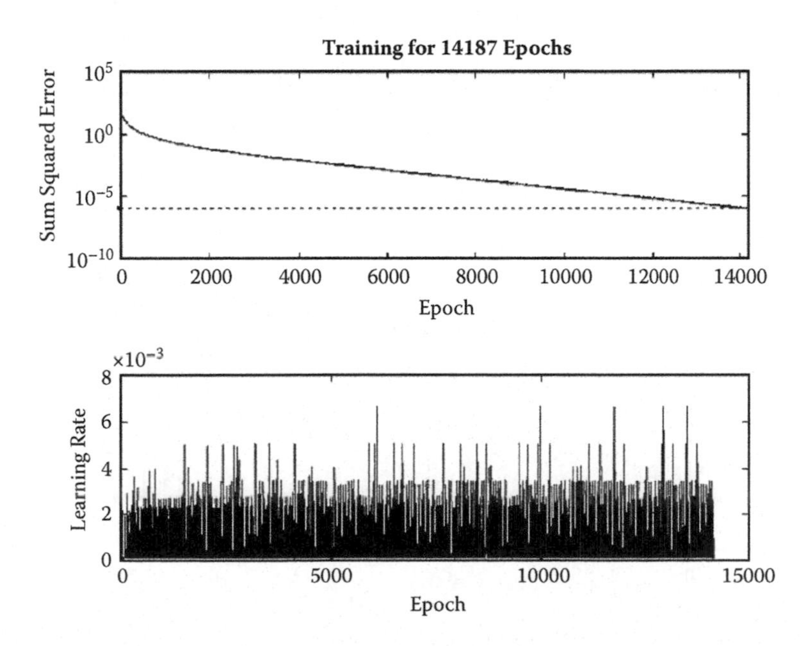

FIGURE 8.10a The training graph for the gender identification problem.

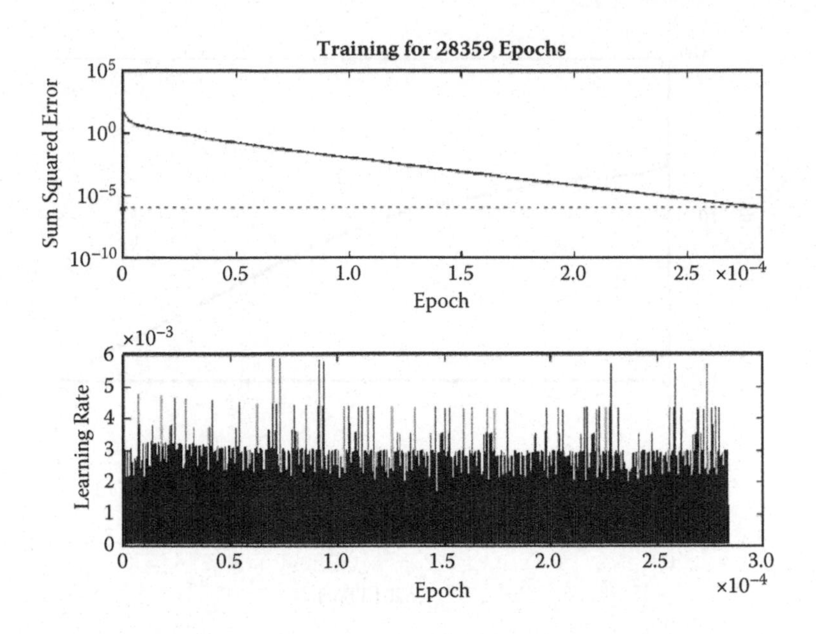

FIGURE 8.10b The training graph for the age group identification problem.

BPA for the problems of sex determination, age group determination, mental state determination, and speaker authentication.

8.2.7.2 Neuro-Fuzzy System

In this section, we solve the same set of problems that we solved in the previous section. But this time we use neuro-fuzzy systems. The same set of features that we used in the earlier section is used in this system. For the development of the classifier, we use the ANFIS editor of MATLAB®.

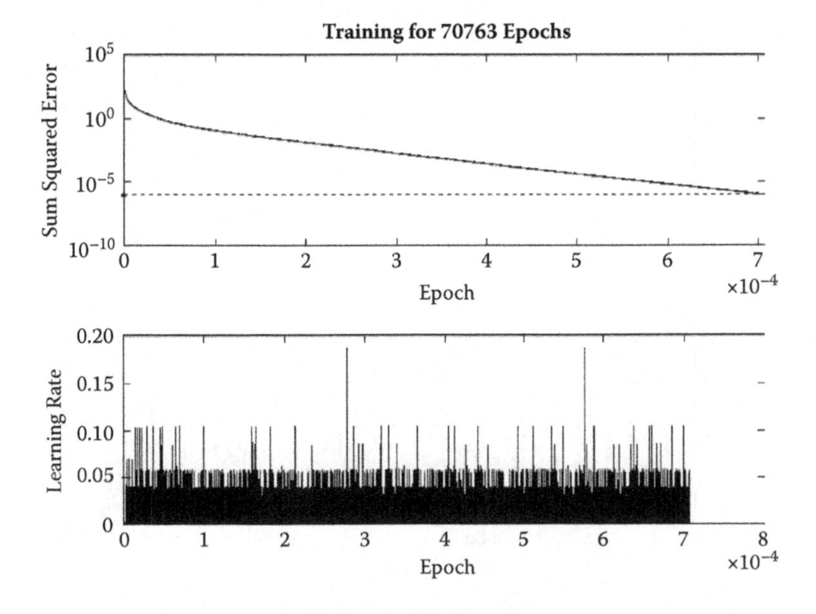

FIGURE 8.10c The training graph for the mental state identification problem.

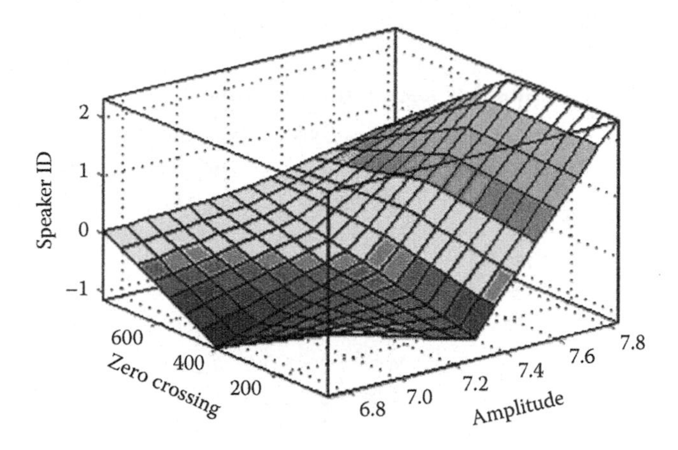

FIGURE 8.11 Speaker authentication surface diagram for zero crossing and amplitude.

In all the cases that we consider here, the model is tested for two cases: speaker independent and speaker dependent. We have two sets of data, open and closed, consisting of inputs from the same speakers and different speakers, respectively. The model is trained using some of the speech features of some of the speakers from the entire database of 50 speakers and using some words from the entire database of 36 words. Then the system is tested for another set of speakers from the open and closed sets, depending on the case. In this way, an attempt has been made to develop a speaker-independent system (the open set) and to compare its performance in terms of a recognition score with the speaker-dependent system (the closed set).

In all the cases, the input membership functions (MFs) used are triangular, trapezoidal, generalized bell, Gaussian, Gaussian2, pi MF(pimf), difference of two sigmoidal MF(dsigmf) and product of two sigmoidal MF (psigmf). A comparison of these MFs has been performed. And the Sugeno fuzzy inference system (FIS) has been used. FIS training has been done using the hybrid optimization method.

The first problem is of speaker identification. For this, we use four features: zero crossings, average PSD, amplitude, and pitch. The surface graph after training is given in Figure 8.11.

The model is trained by speech features for five different speakers. The training data set is shown in Table 8.2. The testing data represented in Table 8.3 is then performed for two cases as follows:

- *Case 1: Speaker-independent system (open set):* The speech data used for testing the model is from the open set of ten speakers chosen at random. The output MF type is constant. The result shows that for Case 1, the recognition score varies from 80 percent to 100 percent and depends on the type of input MF. Table 8.4 shows the details of the input function, error after 20 epochs, average testing error, and efficiency for the training and testing data for the output MF type (constant). Figure 8.12 plots the results of testing a speaker-independent system and shows the difference between the actual output and the desired position. To get a better recognition score, another attempt is then made using a different output MF type (linear), keeping the input membership function the same. The results are given in Table 8.5. The results show that the recognition score does not change, which signifies the input data's robustness.
- *Case 2: Speaker-dependent system (closed set):* The speech data used for testing the ANFIS model is from a closed set of 10 speakers chosen at random. The model is configured for the input and output MF as used in Case 1. Tables 8.4 and 8.5 show the details of the input function, epoch 20 error, average testing error, and efficiency for the training and testing data for the constant and linear output MF types. The recognition score is 100 percent.

TABLE 8.2
Training Data for ANFIS for Speaker Recognition Problem

Word	Speaker	Number of Zero Crossings	Average PSD (in dB)	Amplitude (CEPS.)	Pitch (in Hz)	Logical Value
Word 1	1	167	14.3163990	7.4423	103.630	0
	2	383	26.9000790	7.6895	96.796	0.25
	3	157	51.4011360	7.8329	95.048	0.50
	4	135	30.1214800	7.5097	100.280	0.75
	5	575	0.9795378	6.9015	119.450	1.00
Word 2	1	186	26.0394220	7.2388	102.060	0
	2	42	5.0929102	6.8439	109.930	0.25
	3	250	17.4231570	7.5522	103.950	0.50
	4	652	10.9843770	7.4904	106.890	0.75
	5	351	6.6334056	6.7282	108.410	1.00
Word 3	1	778	27.5996440	7.4132	98.860	0
	2	420	4.7895334	7.1098	109.280	0.25
	3	180	40.848630	7.5232	98.780	0.50
	4	547	36.936820	7.5484	98.304	0.75
	5	475	1.8309492	6.9738	111.480	1.00

Figure 8.13 shows the results for testing the speaker-dependent system. The tables reflect that the epoch 20 error and average testing error are less if the output MF type is constant and the input MF remained unchanged.

The next problem is to determine the mental state, for which we use five features: time duration, zero crossing, average PSD, amplitude, and pitch. The surface graph after training is given in Figure 8.14. The speakers are given logical values of 0 for normal and 1 for abnormal. A value less than 0.5 in testing signifies a normal person, while a value greater than 0.5 signified abnormal. Once again, we have achieved an efficiency of 80 percent to 100 percent, depending on the membership function used. The results for the speaker-independent and speaker-dependent systems are shown in Figures 8.15a and 8.15b.

Now we take the problem of sex determination, for which the factors are zero crossing, amplitude, average PSD, and pitch. The surface graph after training is shown in Figure 8.16. For this problem, 0 represents male and 1 represents female. Thus a value less than 0.5 in testing signifies a male person, and vice versa. Again we have achieved an efficiency of 80 percent to 100 percent,

TABLE 8.3
Speaker Data for the Testing of ANFIS

Word	Speaker	Number of Zero Crossings	Average PSD (in dB)	Amplitude (CEPS)	Pitch (in Hz)	Logical Value
Word 4	1	59	22.2946050	6.9768	102.190	0
	2	139	6.3087162	6.9041	105.860	0.25
	3	55	26.4870500	7.0516	99.954	0.50
	4	81	26.7639990	7.0733	100.960	0.75
	5	421	0.9057146	6.5066	116.000	1.00

TABLE 8.4
Results for Output Membership Function type 'Constant' with various Input Membership Functions

		Training Data		Testing Data	
Input MF Type	Epoch 20 Error	Average Testing Error	Efficiency (%)	Average Testing Error	Efficiency (%)
trimf	9.85E-06	9.66E-06	100	0.54526	20
trapmf	3.87E-06	3.87E-06	100	0.25129	80
gbellmf	6.36E-06	6.53E-06	100	0.36568	40
gaussmf	6.82E-06	7.02E-06	100	0.38494	60
gauss2mf	2.61E-06	2.61E-06	100	0.21143	80
pimf	1.10E-05	1.10E-05	100	0.16622	60
dsigmf	2.35E-06	2.40E-06	100	0.21763	80
psigmf	2.35E-06	2.40E-06	100	0.21763	80

depending on the membership function used. The results for the speaker-independent and speaker-dependent systems are shown in Figures 8.17a and 8.17b.

8.2.7.3 Modular Neural Networks

For the next solution, we use the MNN for classification purposes. The problem being solved is of multilingual speaker recognition. This system is able to recognize speakers when the data set consists of words spoken in a variety of languages. The database consists of one sentence recorded from 32 speakers (19 male and 13 female) in 8 different languages. The sentence is worded in such a way that in each word, every consonant succeeds a vowel and vice versa. The reason behind this is that whenever we pronounce any letter, a vowel sound is always generated. A total of 904 words are recorded in 8 different Indian languages. The features used for this system are Cepstral coefficient, number of zero crossing, average PSD, and length of file.

When the traditional approach of using only one ANN is applied to the problem, the performance is as low as18 percent. Per the algorithmic requirements, the data were first grouped into 20 clusters. These data were then distributed among the 20 ANNs. The architecture and details of the various ANNs are shown in Table 8.6. Each ANN belonged to a different cluster. The training curve for one of the ANNs is shown in Figure 8.18. The efficiencies of single ANNs are given in Figure 8.19.

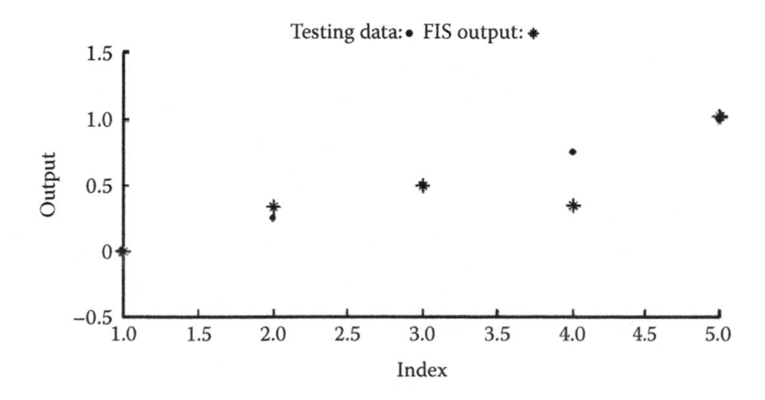

FIGURE 8.12 Results after testing the speaker-independent system using ANFIS.

TABLE 8.5
Results for Output Membership Function type 'Linear' with various Input Membership Functions

		Training Data		Testing Data	
Input MF Type	**Epoch 20 Error**	**Average Testing Error**	**Efficiency (%)**	**Average Testing Error**	**Efficiency (%)**
trimf	3.24E-06	3.24E-06	100	0.39654	20
trapmf	7.63E-06	7.63E-06	100	0.25779	60
gbellmf	1.70E-05	2.38E-05	100	0.27950	40
gaussmf	1.03E-05	1.03E-05	100	0.35046	60
gauss2mf	1.29E-05	1.29E-05	100	0.19440	80
pimf	5.77E-06	2.11E-05	100	0.18111	60
dsigmf	1.40E-05	1.40E-05	100	0.18753	80
psigmf	3.20E-05	3.20E-05	100	0.18493	80

From the 20 clusters of 904 input data, the system's average performance is 95.354 percent, with 42 errors. The resulting system solved the problem that a single ANN was unable to solve. In addition, solving the problem took much less training time as compared with solving it using a single ANN. Hence it is clear that a single MNN is able to perform better and is able to solve the problem of multilingual speaker recognition.

8.2.7.4 Wavelet Coefficients with ANN and BPA

As our final experiment, we use the wavelet coefficients with ANN and BPA for the task of speaker recognition. We recorded data from 20 speakers who were asked to say five words each: "ab," "is," "baar," "aap," and "apne." The speakers repeated these words and were recorded. MATLAB's wavelet analysis kit was used. The number of levels was specified to be six, giving us six different values found from wavelet analysis. The results of some of the inputs of wavelet analysis are given in Table 8.7. The ANN had one hidden layer with 48 neurons in it.

When the algorithm was executed on the collected data, we got a performance of 97.5 percent using 20 speakers and 40 test cases (39 correctly identified). This clearly shows that the algorithm works well and gives correct results on almost all inputs.

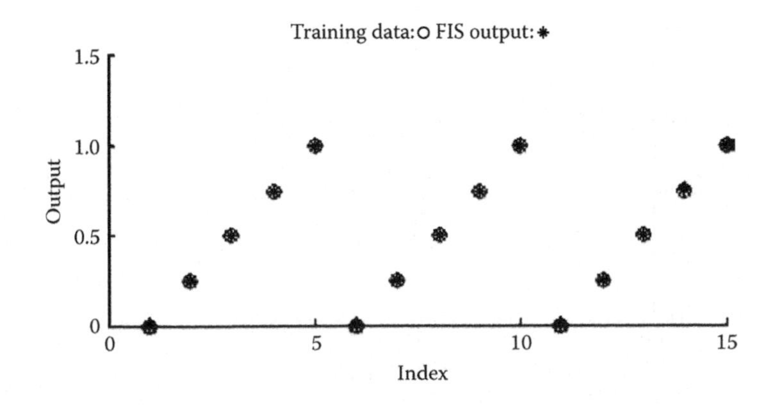

FIGURE 8.13 Results after testing the speaker-dependent system using ANFIS.

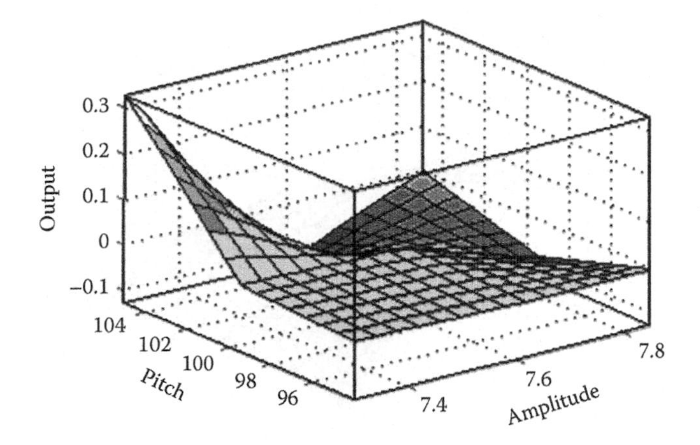

FIGURE 8.14 Mental status determination surface diagram for pitch and amplitude.

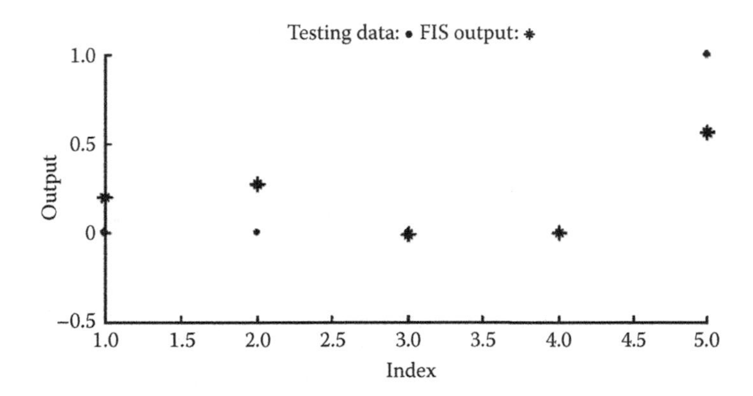

FIGURE 8.15a Results for the mental state determination for the speaker-independent system.

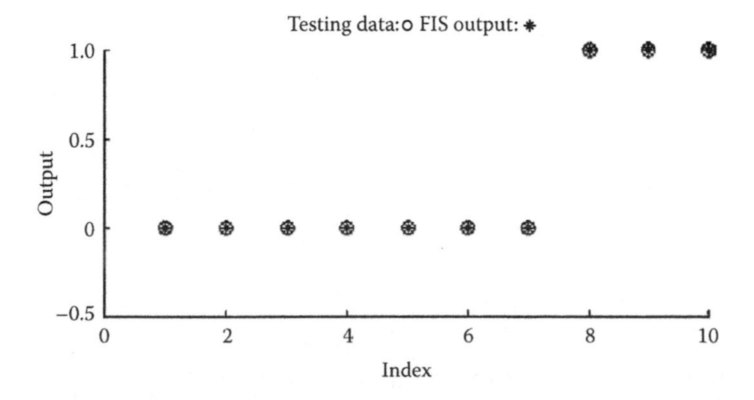

FIGURE 8.15b Results for the mental state determination for the speaker-dependent system.

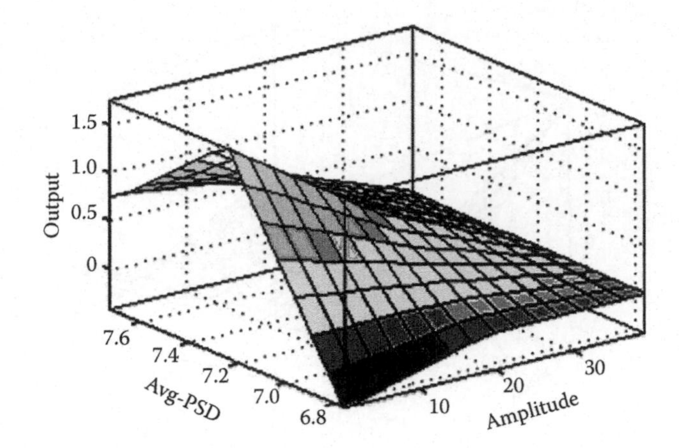

FIGURE 8.16 Gender determination surface diagram for average PSD and amplitude.

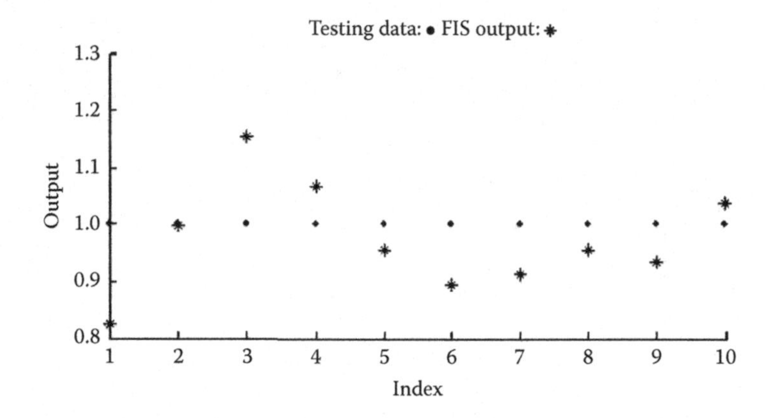

FIGURE 8.17a Results after testing the speaker-independent system.

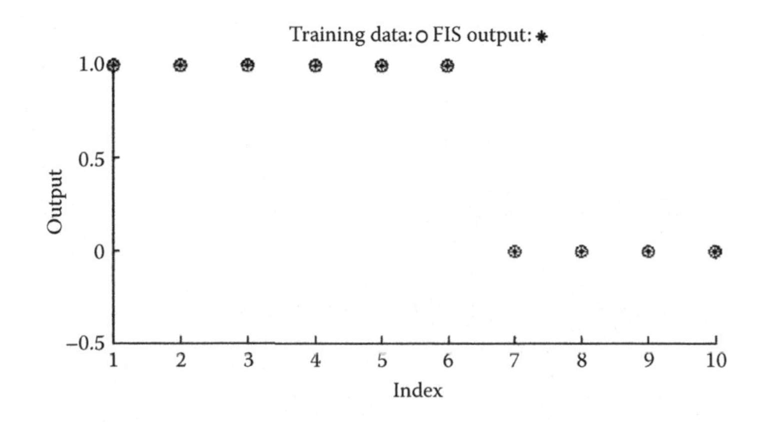

FIGURE 8.17b Results after testing the speaker-dependent system.

TABLE 8.6
Architecture of the various Artificial Neural Networks

Cluster No.	Error Goal (δ)	No. of Neurons/Layer	No. of Targets	Momentum (α)	Max. Epochs
1	0.0120	28	31	0.25	40,000
2	0.0110	52	58	0.26	40,000
3	0.0140	15	19	0.26	25,000
4	0.0100	40	44	0.26	50,000
5	0.0090	56	58	0.28	50,000
6	0.0120	44	55	0.28	25,000
7	0.0090	52	56	0.28	50,000
8	0.0160	20	25	0.24	50,000
9	0.0130	20	28	0.24	50,000
10	0.0080	78	80	0.27	50,000
11	0.0100	42	45	0.25	50,000
12	0.0118	36	45	0.25	22,000
13	0.0118	40	50	0.25	30,000
14	0.0110	27	29	0.28	50,000
15	0.0080	71	73	0.26	50,000
16	0.0100	36	42	0.25	50,000
17	0.0100	33	37	0.25	40,000
18	0.0100	24	27	0.24	50,000
19	0.0100	52	57	0.27	30,000
20	0.0100	38	43	0.26	50,000

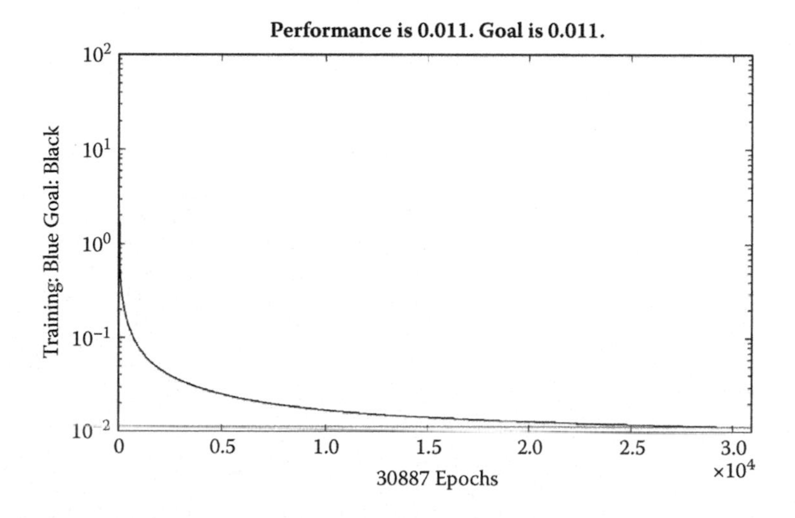

FIGURE 8.18 The training curve of an artificial neural network.

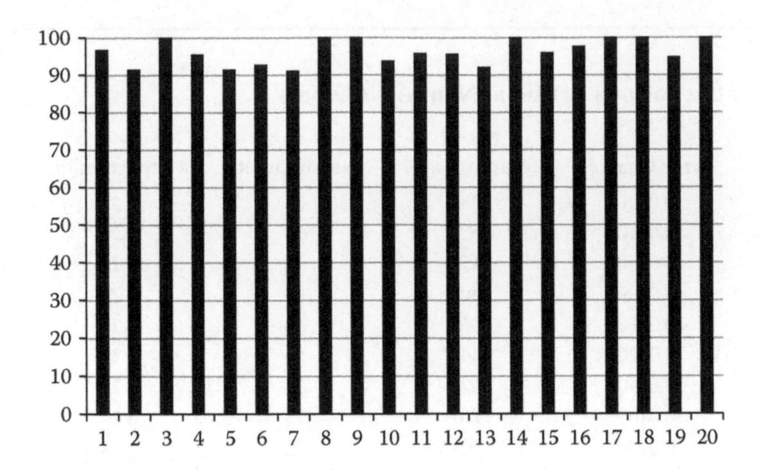

FIGURE 8.19 Cluster versus efficiency graph for various artificial neural networks.

8.2.8 Concluding Remarks for the Speech Biometric

In this section, we adopted a multidimensional approach toward problem solving with the biometric of speech. We first presented various features that could be extracted from speech. At one end, we studied features such as number of zero crossings, PSD, and so forth. On the other end, we learned about the wavelet transform, STFT, and Fourier transform. We then discussed the various classifiers that could be used for the problem, including ANN with BPA, neuro-fuzzy, and MNN. We also studied the various problems that can be solved using speech as a biometric, including speaker identification as well as age group, sex, and mental state.

We used various combinations of problems, features, and classifiers and presented various cases of problem solving. In this manner, we obtained a multidimensional view of the use of speech as a biometric. In all cases, we were easily able to solve the problem by the stated methods and achieved good efficiencies. In every case, we built a system that could effectively be used in real life applications.

TABLE 8.7
Results of the Wavelet Analysis

S. No.	Speaker	Word	D1	D2	D3	D4	D5	A5
1	A	1	1,825	3,642	7,275	14,542	29,075	1,825
2	A	2	1,811	3,614	7,220	14,432	28,855	1,811
3	A	3	1,904	3,799	7,590	15,172	30,335	1,904
4	A	4	1,873	3,737	7,466	14,924	29,840	1,873
5	A	5	1,845	3,682	7,355	14,702	29,395	1,845
6	B	1	1,842	3,675	7,341	14,673	29,338	1,842
7	B	2	1,799	3,590	7,171	14,334	28,659	1,799
8	B	3	1,852	3,696	7,384	14,760	29,512	1,852
9	B	4	1,851	3,694	7,380	14,752	29,495	1,851
10	B	5	1,843	3,678	7,347	14,685	29,362	1,843
11	C	1	1,835	3,662	7,316	14,623	29,238	1,835
12	C	2	1,848	3,687	7,365	14,722	29,435	1,848
13	C	3	1,883	3,757	7,505	15,001	29,994	1,883

8.3 SIGNATURE CLASSIFICATION

Signature is the second biometric feature that we consider in this chapter. The use of signature for the authentication purposes is a very classic task. Signatures have evolved in such a way that they can be used as a written proof of something by some person. Even today, signatures are used at all levels of organizations, departments, and institutions. Thus the signature forms an integral part of the working of any society. For this reason, there is a need for systems that can identify and validate a person's signature. This system would enable us to automatically verify any claim by the use of a developed system. Furthermore the system would be able to stop all fraud and foged documents and in turn help in increasing the security against illegal access.

Signatures are another biometric known to be unique to every individual, which is why so many people use signatures for all major authentication systems. We all start with simple ways of writing or signing our names. As we grow and mature, our handwriting and signatures start gaining characteristics. This evolution results from various reasons, including education, change in behavior, experience, and so on. As adults, our handwriting is unique to each of us. It can easily be observed that the handwriting and signature of any two people rarely match, even in the case of twins with the same upbringing.

In particular, our signatures become very characteristic in nature because of the manner in which they are written. Our signatures denote the natural way in which we would write. There is a scope for everything, including jerks, styles, and other characters. Furthermore when we sign our names, we do not think about legibility or grammatical or syntactic regulations. This all makes signatures very unique, meaning they can easily be used as a biometric identity.

Experts are able to match signatures by noting the characteristic regions, the style of writing, the various curves and their locations, and so on. They are trained in how to differentiate the original signature from fake signatures, even though the signatures may look much the same. In the case of the machines, however, this may not be possible. The problem lies in devising suitable characteristics or features that a machine can use for the purpose of recognition and verification.

Signature recognition and verification are usually studied under two separate categories: offline and online. In the offline approach, the entire signature is available as an input to the system. We have no idea of the manner in which the person signing made the signature. In other words, the person signs and then the signature is fed into the system. The input typically consists of scanned images of the original signature. A common example of this technique is a person who signs a legal document and then sends it to someone else, who has to verify the signature. In this case, the person would scan the document and send it as an input to the system for the purpose of verification.

In the online approach, the person signs directly into the system. This technique typically uses of light pens, which are pointing devices that connect to the system. It requires a transducer to capture the signature as it is written. The online system produces time information, such as acceleration (speed of writing), retouching, pressure, and pen movement. The various movements of the signature may be recorded at various time stamps so that we get added information. A common example of the online approach is a person who makes his signature using the special input device, for the purposes of authentication.

Note that although we talk a lot about signatures and automation, we are in no means referring to the concept of digital signatures. These are an entirely different concept used widely in the context of information security. We are also not dealing with the issues of information security, whether in this chapter or in the entire book. In the systems we describe, we are not trying to secure or authenticate online information flow; rather the motivation is to identify or authenticate the physical person using the biometric identity. These are two completely different domains that are often thought to be the same.

Another thing about good characteristics of the features that we have been discussing throughout the text is the high interclass and low intraclass separation. This characteristic is also true in the case of signatures. The interclass separation is the average distance between the different speakers in input space. The intraclass separation is the average distance between the various instances of same speakers in the input space. Many of the features that you may visualize in the case of signatures face the

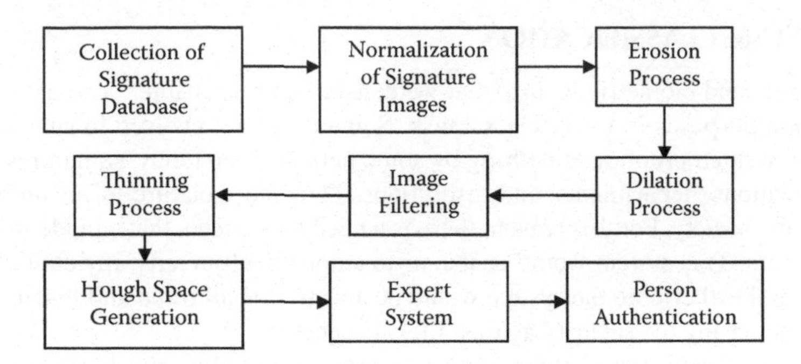

FIGURE 8.20 The general methodology of the signature verification system.

problem of low interclass separation. Namely, the signature of the same person varies to very large extent from signing to signing. This difference may be due to mood, conditions, comfort, and so forth. In fact, a person signing two times, one right after the other, may make changes between signatures. This obviously creates a problem in the system. This may be even more problematic than similar issues with face recognition, where we saw how various features that remain constant over region.

In general, the features of a signature are classified into two main groups. The first group contains the global features, which describe or identify the signature as a whole (i.e., the global characteristics of the signature such as the width and the baseline). These global features are less sensitive to the effect of noise. The other group incorporates the local features, which represent a portion or a limited region of the signature (e.g. critical points and gradients).

We follow the same guidelines for a recognition system that we have been following throughout the text. Figure 8.20 shows the general method typical to this problem, including the individual methods used for each stage. We first discuss the preprocessing stage. Then we move onto feature extraction. Finally, we briefly discuss classification, followed by the experimental results. Because image acquisition here is trivial and has already been discussed numerous times in this text, we do not cover it here.

8.3.1 Preprocessing

After image acquisition, the first step is to normalize the image to fix the range of all the values of the image pixels between 0 and 1. The original image is shown in Figure 8.21a, while the normalized image is shown in Figure 8.21b. Next we apply thresholding, or binarization, to make the entire image black and white. The value of the threshold must be judiciously chosen. The binarized image is shown in Figure 8.21c.

Then we apply the processes of dilation and erosion to free the image of noise and to remove any backgrounds that may be present in the image. Erosion is a process that changes near-white pixels to white pixels, which shrinks the edges of the drawing. This step removes all objects whose size is smaller than the threshold. Hence erosion, also known as shrinking, removes all noise and backgrounds. Dilation is a similar process that changes the near-black pixels to black pixels. This thickens the lines of the image or the signature. All unconnected components are connected and all holes in the drawing are filled. This step is also known as growing, boldening, and expansion. The image after erudition and dilation is shown in Figure 8.21d.

The final step is to detect the edges in the image. Figure 8.21e shows the image after edge detection.

FIGURE 8.21a Signature preprocessing: Original image.

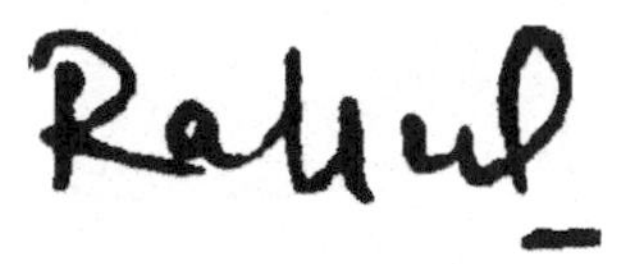

FIGURE 8.21b Signature preprocessing: Normalized image.

8.3.2 Feature Extraction

In this problem, we use an analysis technique similar to Fourier and wavelet analysis for feature extraction. This technique is called the Hough transform. The main objective is to represent the entire signature with as few coefficients and as little loss of information as possible for the sake of feature extraction or to limit the problem's dimensionality. The Hough transform represents the entire image or drawing vector in the form of coefficients. From these coefficients we may select the few best. Note that this transform works only for black-and-white images.

The Hough transform works by trying to detect or extract straight lines from the presented graphic. Although this transform may also be applied in higher orders, where circles or more complex figures are used in place of lines, doing so drastically increases the complexity. Any line is of the form $y = mx + c$. To avoid infinite values of m and for easy operation, we assume the line to be represented by $p = x \cos \theta + y \cos \theta$, where p is the perpendicular distance of the line from origin and θ is the slope.

For each point in the graphic, we consider all the lines that go through the point at a particular angle θ to be chosen a priori. We calculate the distance d to the line through the point at angle θ. This is repeated for a number of angle θ that are discretized for computational purposes. The resulting distance d is also discretized. This gives us points of the form (θ, d), which are plotted in a space known as the Hough space.

Because we have already made the axis of the Hough space discrete, we now get boxes, or accumulators, in this space. These boxes accumulate the line possibilities and hence represent the density of the possibility of the lines at that angle and that distance. Ideally, for normally distributed data, the Hough space shows the Poisson ratio.

We then select the key points in the Hough space or lines in general. The Hough space for one of the inputs in the signature problem is shown in Figure 8.22. The bigger white boxes denote the region where the value was found to be greater than normal.

Some extracted matrix values from the Hough space are given in Table 8.8.

8.3.3 Artificial Neural Network

In the signature problem, we again used the ANN with BPA for the task of classification. This system takes as its input the extracted features from the previous step and assigns the output the degree of certainty associated with each input. This role of ANN with BPA has already been discussed numerous times in the text; therefore we avoid discussing it here.

8.3.4 Results

We collected three separate signature samples each from 50 users. The images taken were 255 × 105 pixels in size. These signatures were used for verification. The experiment was done in three cases.

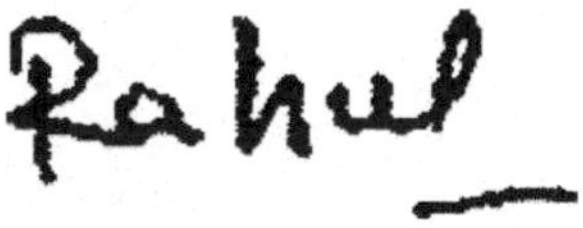

FIGURE 8.21c Signature preprocessing: Image binarization.

FIGURE 8.21d Signature preprocessing: Eroded and dilated image.

FIGURE 8.21e Signature preprocessing: Edge-detected image.

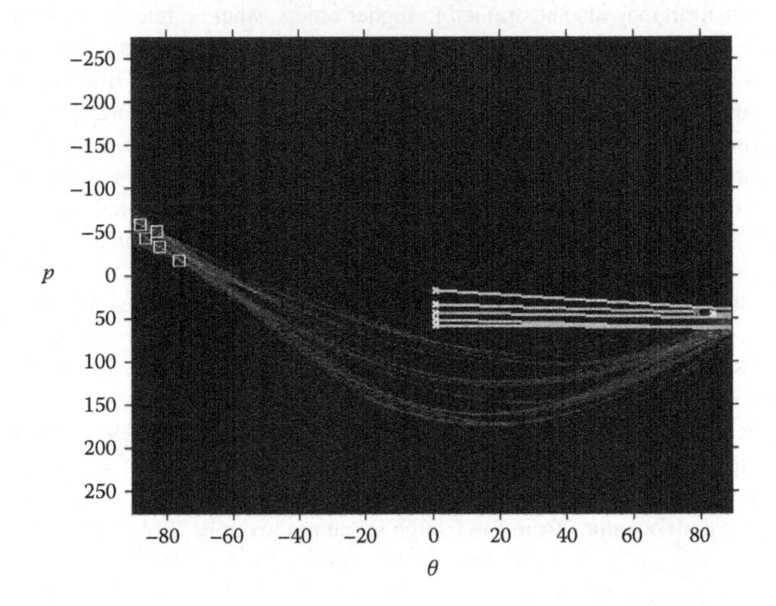

FIGURE 8.22 Hough space image of signature.

TABLE 8.8
Extracted Accumulator Array for Each Signature

Name	Sample 1	Sample 2	Sample 3
Person 1	1 36	1 37	1 41
	108 40	119 25	119 33
Person 2	6 1	1 6	5 1
	78 72	76 81	66 61
Person 3	1 56	2 72	1 29
	119 29	123 28	124 7
Person 4	1 41	1 47	1 1
	118 43	118 41	103 3

TABLE 8.9
Artificial Neural Network Architecture for the Three Cases

Parameter	Case 1	Case 2	Case 3
Momentum	0.90	0.85	0.78
Learning Rate	0.001	0.030	0.030
No. of Hidden Layers	2	2	2
Epochs	4,800	4,800	3,500
Nonlinear Function	Logsig	Logsig	Logsig
Training Fuction	Traingdm	Traingdm	Traingdm

In the first case, the same signatures were used for training and testing purposes. Each training and testing consisted of 150 images. In the second case, the data for training exceeded the data for testing, with 150 images for the former and only 120 for the latter. In the third case, we had fewer training data than testing data—100 images for training and 120 for testing.

The system was trained with BPA. The ANN architecture for the various cases is given in Table 8.9. Figures 8.23a, 8.23b, and 8.23c show the training curves of the three cases.

The system's FRR was found to be 0.1, while the FAR was 0.17. The system's efficiency for Case 1 was 99.87 percent; for Case 2, 97.56 percent; and for Case 3, 95.81 percent.

8.3.5 Concluding Remarks for the Signature Biometric

In this section, we used the Hough transform along with ANN and BPA for the purpose of user verification. In most cases, we were able to verify the correct identity of the person; hence the Hough

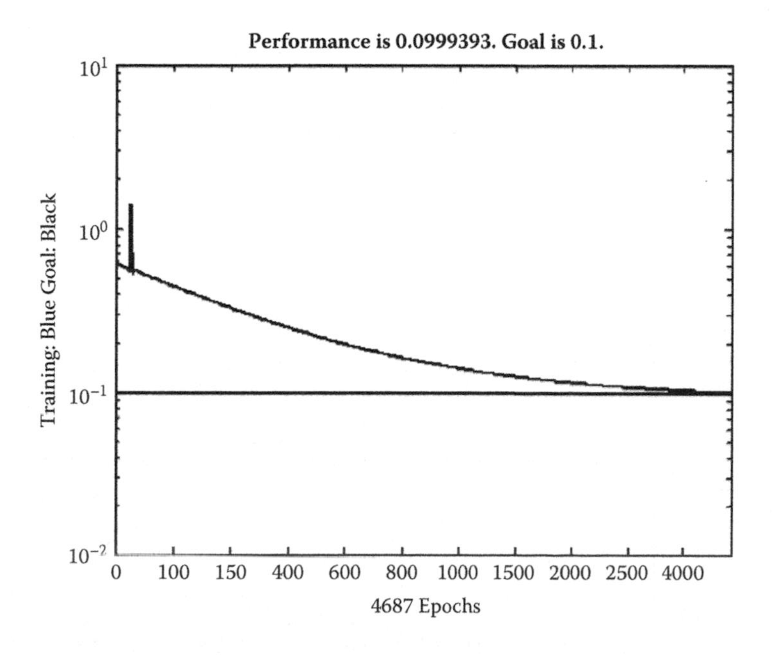

FIGURE 8.23a The training curves for the first cases in the signature recognition problem.

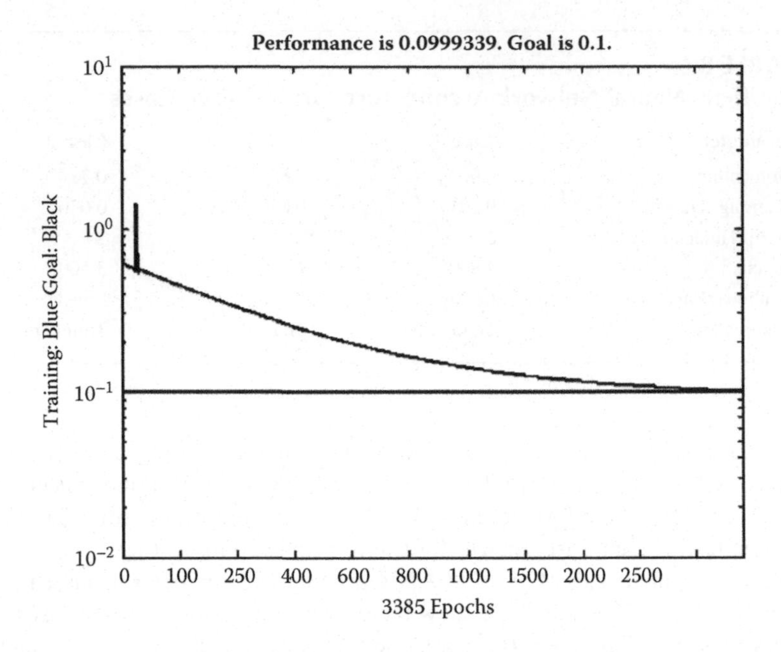

FIGURE 8.23b The training curves for the second cases in the signature recognition problem.

transform gives good features for the problem of offline signature verification. In addition, ANN with BPA acted as a good classifier to solve the problem and hence helped in the verification.

The other transforms, such as the Fourier or the wavelet transforms, may also be used for the same problem and may lead to better results. In addition, combining handwriting parameters with signature parameters can make a much better system.

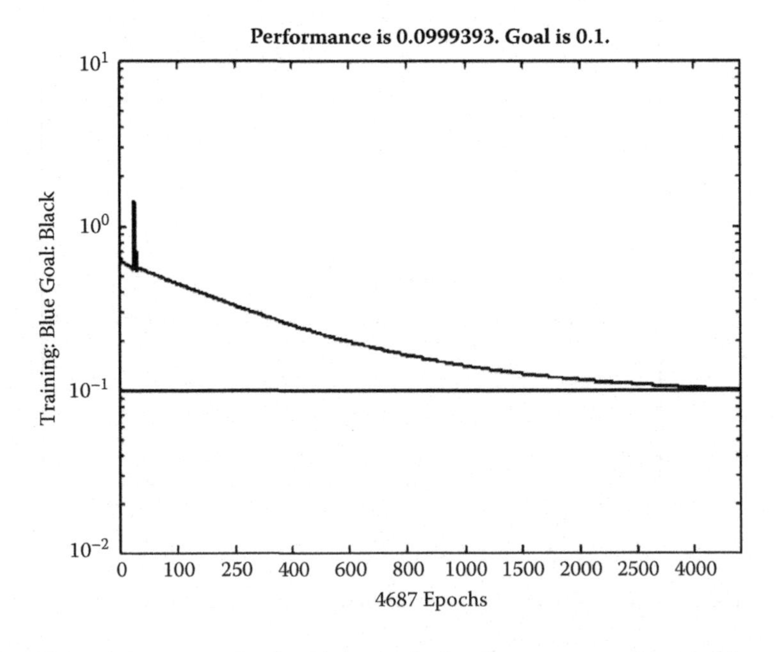

FIGURE 8.23c The training curves for the third cases in the signature recognition problem.

9 Fusion Methods in Biometrics

9.1 INTRODUCTION

In the past two chapters, we discussed identification and authentication systems that use various-soft computing techniques. We studied the different features that can be extracted from data and the various techniques that are applied to make real life applications that perform well under any circumstances. All of the experiments that we have performed throughout this book gave good results, and we have been able to prove the effectiveness of these systems under most conditions. We supported our statements by describing real life applications that use such systems and are performing highly well.

9.1.1 Problems with Unimodal Systems

Many times the systems described in this book will eventually face problems. For example, they may not be able to effectively identify a person. This could happen for various reasons. As we have mentioned, noise is one of the main reasons for a reduction in performance of any soft-computing technique. Noise often affects the biometrics to such a great extent that it becomes very difficult to train and test the systems at high accuracies. In addition, in many systems, the features are not good enough and change for the same person by large amounts. The possibility of a large number of classes or people also induces problems when using these systems in real life applications.

Another major problem with these systems occurs when the inputs are such that the intraclass distance is not very high. In such cases, the various classes or people lie very close to each other in the input space. The system is supposed to interpret the data and differentiate between the classes. But when the intraclass distance is not high, the various classes start intermingling, making it appear to the system that more than one class is at any given input. In such a case, the system has a hard time determining the different classes. This problem assumes great importance when the number of inputs is large or when the input space is very large.

9.1.2 Motivation for Fusion Methods

In the cases described, additional help is required for the system to start behaving normally again. For this help, we are looking for alternatives that might suggest the correct output class or might prevent the system from making the wrong decisions. This would give us the twin benefit of correcting mistakes and cross-checking results, which, in turn, would increase the system's efficiency and make the system more robust.

In classification problems, this help or assistance is usually done by adding attributes. These added attributes store a great deal of information and may be used for all of the identification purposes. This means that we can consult these added attributes for all decisions. Hence any decisions that we are not confident about can be resolved, and all decisions containing mistakes can be checked. All of this helps the system be more efficient.

In the context of biometrics, fusion means adding a few attributes that are then recorded. This may involve recording larger distances in face recognition or larger coefficients in wavelet analysis. But this solution faces problems. In most cases, the added attributes are prone to the same noise, errors, and conditions that made the recognition work difficult in the first place. These problems

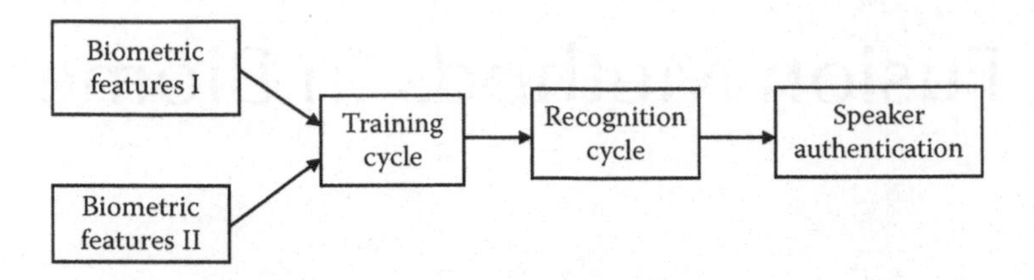

FIGURE 9.1 The general problem-solving methodology in fused biometric systems.

would affect the newly added attributes as well as the old attributes. In addition, these attributes often convey redundant information. Hence they do not provide any added information. In wavelet solutions, an increase in the number of coefficients might not help because, as we discussed, they contain much less information. Thus we may conclude that there is a need for new and different attributes for the benefit of system efficiency.

In real life systems, we get this functionality by mixing up the biometric identities. We solve the problem of identification by using two or more biometric identities in the place of one. The various attributes of the two systems are pooled and combined into the system, which can now more easily find the correct output. The features within the two modalities are as independent as possible, which is a desirable feature for classification systems or any other system in general. The general methodology of the algorithm is given in Figure 9.1.

9.1.3 Workings of Fusion Methods

The addition of attributes increases the ability of these systems to separate the regions of the various classes and thus to learn easily and perform well. The developed system can now use the best features of the individual systems and thus perform to its maximum. The increase in attributes increases the dimensions, and every dimension carries some classifying power. Hence the various powers are combined, and cumulative information is now available for the generation of real life systems.

Such a system, however, has disadvantage. The addition of attributes increases the system size, so that the system must now adjust more parameters. In the case of artificial neural networks (ANNs), these parameters are the modifiable weights and biases. In the case of fuzzy systems, these parameters may be the possibility of rules. The increase in the number of parameters requires a greater data set in order to adjust all the values in such a way that the system reaches its minima in terms of the performance error. In many systems, the data size is highly limited, which hinders the system's ability to train well. It is also possible that the system may be assuming arbitrary shapes, which would further cause problems in recognition. Furthermore the increase in complexity that results from adding attributes might prove bad for the system.

This difficulty may easily be seen in the input space, where the various inputs and outputs are plotted graphically. Consider the input space. Assume that instances of two classes lie very close to each other in this space. For a two-dimensional input space, this space may be considered a plane where the instances of the two classes lie close together. In this instance, classification would be difficult. But suppose we add a third attribute, converting the input space to three-dimensional space. Now the first class may be lying at one extreme of the new axis and the other may be lying at the other extreme, making classification very easy based on the newly added dimension or attribute.

9.1.4 What Can Be Fused?

Fused systems are very naturally observed in the natural world. For example, it is often difficult to recall a person by voice alone. However, the person may be easily recalled by seeing the person talk.

When we look at the person and hear him simultaneously, the features of both sound and speech simultaneously influence our decision. Likewise remembering someone sitting still may be much more difficult than remembering someone who is in motion. By the gestures, sight, voice, and so forth, we get a much better idea regarding the person and hence can recognize him more easily. This is true in the case of other biometrics as well.

The important decision to be made is which biometric identities or modalities can be mixed together. In practice, there is no restriction. We can fuse any two sets of modalities to make an effective identification system, and the resulting system would be able to reap the benefits of the fused biometric systems. However, in making this decision, we need to look at the system's practical aspects as well. Suppose, for example, that we mix the signature biometric with the speech biometric. The person would be asked to sign so that we can obtain the signature features to be extracted. Then the person would have to say something for speech features. This would mean a great deal of discomfort for the person and an unnatural manner of processing. This would also seem as if the person were passing through a series of security checks.

In fused biometric systems, we try to fuse together the systems that are most related so that there is no discomfort to the user. The recording procedure should also not be difficult. For these reasons, we fuse related features, such as speech and face. This example is very simple to achieve by placing a camera to record the face and a microphone to record speech. The person simply looks at the camera and says something. This completes the system, and identification can then be performed. Another similar fusion is speech with lip movements. This fusion enables us to clearly and simultaneously identify the words based on speech and lip movements. Other common examples include the fusion of signature and handwriting; ear and face; gait features and gestures; fingerprint and hand geometry; retina scan and iris; fingerprint, hand geometry, and palm vein; and DNA and odor. It may be easily observed the various modalities must integrate with each other to facilitate easy recording.

These systems of fused modalities do not act as cascade systems that apply security in layers, one after the other. The numerous differences between these two concepts play a major role in the system's performance. For example, cascading systems are able to model the dependencies among the attributes of the two modalities; hence they are not able to depict much of the behavior that fused methods can easily identify. This makes the fused methods much more resistant to noise and better performing than individual, cascading systems.

9.1.5 Unimodal or Bimodal?

Now that we have explained how fused systems perform much better than classical, single-modality systems, you may wonder why single-modality systems are used at all in real life applications. After all, when you have a system that performs better, why would you not use it for every instance? Recall that capturing the multiple modalities and making a system from them may not always be easy. Questions of integration, input devices, device placement, and so forth must be addressed. Furthermore the context in which the system is to be used is important. For example, if a face recognition system were used for identification in a crowded area, the speech may not be easy to record and identify.

In addition, even though these systems resist noise, they may not behave very well in all circumstances. If one of the modalities gives very bad results due to noise or similar limitations, the system's performance may decrease instead of increasing. Furthermore the modality with added noise might not contribute well to the system or it might negatively affect the decisions of the other modality. Thus, if one modality is having problems, it is best to use a single modality.

In this chapter, we highlight the various issues presented so far by developing two real life applications: fusion between speech and face and fusion between face and ear. We then study at the various means by which the problems can be solved.

9.1.6 Note for Functional Prediction Problems

So far we have discussed applying fused methods for the problem we want to solve. These have all been classificatory in nature, in which the task is to map an input to an output to which that input most probably belongs. The discussion regarding the use and methodology has concentrated on classification problems. But now we speak a bit about functional prediction counterparts, in which the aim is to find the output for some input so that the output is continuous in nature or in discrete incremental values.

For a classification problem, fusing or combining the attributes results in greater flexibility. However, for any functional prediction problem, it is assumed that all the major parameters that affect the output are known and present as an input to the system. This is a major point of difference between functional prediction and classification problems. In functional prediction problems, the inputs absent from the system are either so irrelevant that they do not largely affect the system or assumed to be constant among all the training and testing data. If these inputs do not remain constant for even a single test case, we would not get the correct output for that case. This means that fusion in the case of functional prediction might add irrelevant details to a system or might result in fine-tuning the system, depending on the relevance of the input parameters of the other fused modality. This is the case when the system worked fine in the unimodal case and we tried to add more attributes.

Many commercial biometric systems use fingerprint, face, or voice. Each modality has both advantages and disadvantages (e.g. discriminative power, complexity, robustness). User acceptability is an important criterion for commercial applications. Techniques based on iris or retina scan are very reliable but not well accepted by end users, whereas voice and face are natural and easily accepted by end users. Automatic face recognition has seen a great deal of activity in recent years. Speaker recognition is a natural way to solve identification and verification problems. Much work has been done in this field and has generated a certain number of applications of access control for telephone companies. Text-dependent and text-independent systems are the two major speaker-verification techniques.

Face is another of the more acceptable biometrics, in part because it is one of the most common methods of identification that humans use in their daily interactions. Human's ability to recognize faces is remarkable. We come across thousands of people in our lifetime and can identify them at a glance, even after years of separation. Even though we usually cannot recognize someone from ears alone, the human ear has both reliable and robust features that are extractable from a distance. Ears have several advantages over complete faces, including reduced spatial resolution, a more uniform distribution of color, and less variability in terms of expressions and orientation of the face. Compared with other biometrics, such as fingerprints, retina scans, iris scans, and hand geometry, the ear biometric does not require close contact of users and may not be considered invasive by users.

We discuss some of the major fusion modalities and techniques in the next sections of this chapter.

9.2 FUSION OF FACE AND SPEECH

The first system that we present here is the fusion of face with speech. We studied the two individual systems under physiological and behavioral biometrics, respectively. In this section, we make a combined system from the individual systems. These fused systems, which consist of a camera and microphone, may easily be used to capture and attain the data at the user's convenience. The camera records an image of the face, while the microphone records the speech. These systems are commonly found in real life applications, including video conferences, recording of an event, and so on. We have different devices for capturing the same event at the same time. This section discusses the work of Shukla and Tiwari (2008c, 2007c).

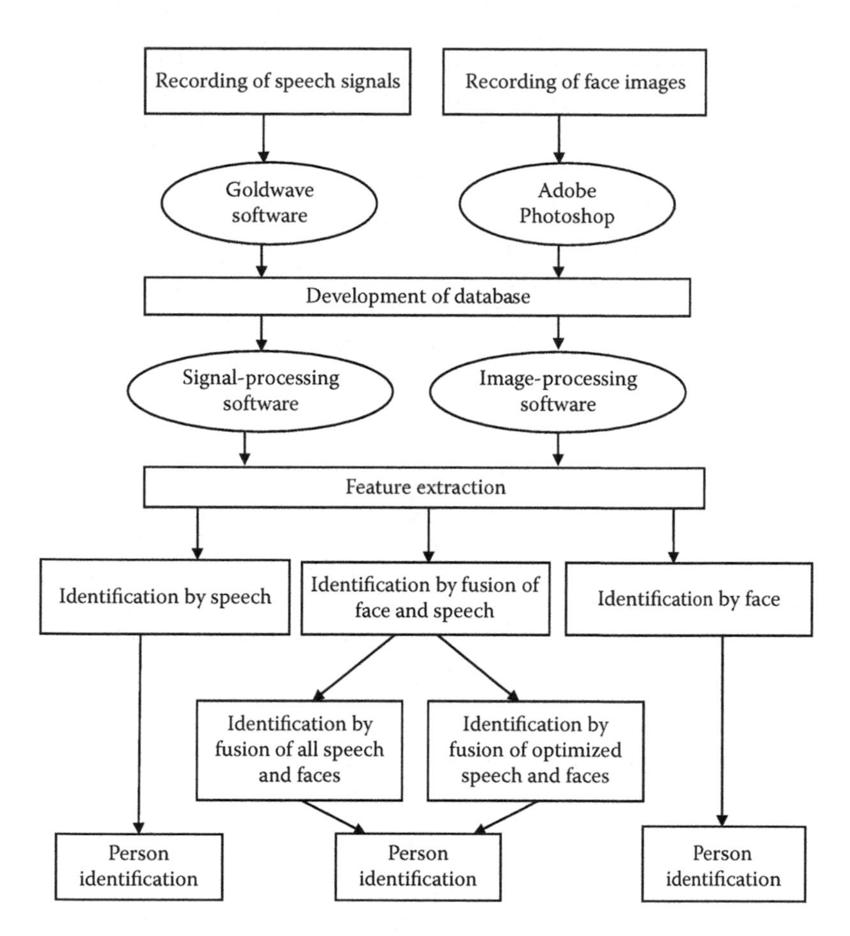

FIGURE 9.2 The basic methodology of the system for the fusion of speech and face.

9.2.1 Working

The basic methodology remains the same as for the individual systems. Data acquisition, preprocessing, and feature extraction are exactly the same as in the individual systems. The only difference is that they have been put together to make a combined system. The difference between the fused and individual systems is in the classification. The basic methodology is given in Figure 9.2.

The features we use from speech are maximum amplitude, cepstrum max, formant frequency (F1 to F4), peak and average power spectral density (PSD), time duration , number of zero crossing, pitch, and pitch amplitude, all of which have been discussed in earlier chapters.

The features that we use from the face are length of eye 1, width of eye 1, center dimension of eye 1, center dimension of eye 2, length of eye 2, width of eye 2, length of mouth, width of mouth, center dimension of mouth, distance between center points of eye 1 and eye 2, and distance between center points of eyes and mouth. We discussed the process for extracting these features in earlier chapters.

We first use ANN with back propagation algorithm (BPA) as a classifier. This classifier takes inputs from the individual systems for both speech and face. The inputs are applied simultaneously to the system for training and testing. The rest of the details are the same as those used in conventional ANN with BPA.

The entire experiment was done for two separate cases. In the first case, we want to find the total performance of the fused system. For this, we input the entire feature set for training and testing. The basic motive behind this approach is to make full use of the features for the purpose of classification. We ensure that we input any feature that might convey some information that is not conveyed by the other attributes. In this manner, we try to maximize the system's performance.

The second case is the optimized fused model, in which we try to input to the classifier the least number of features from the extracted features for recognition purposes. The system thus contains the features and advantages of the fused methodology, while also catering to the need for reduced dimensionality and, correspondingly, reduced system complexity. This system can easily identify people because of the diversity of the features used. It can also easily train itself and perform well, because we have restricted dimensionality to an optimal value.

The set of features that we select from face and speech are formant frequency (F1 to F4), peak and average PSD, length and width of the mouth, distance between the center points of eye 1 and eye 2, and the distance between the center points of the eyes and mouth.

9.2.2 Results

Data for both speech and face features were collected from 20 people. All the features for both the face and speech were extracted independently and made up the database used for experimental purposes.

We first solved the problem with Case 1, in which all the features were given as an input to the system, and the system was then asked to identify the person. The ANN architecture used for experimental purposes consisted of 2 hidden layers with 15 and 75 neurons each. The error goal was kept at 0.01; the learning rate, at 0.1; and momentum, at 0.5. The system's training curve is given in Figure 9.3.

The second experiment was for Case 2, in which we tried to restrict the number of features given to the system. The ANN architecture used for experimental purposes consisted of 2 hidden layers with 15 and 75 neurons each. The error goal was kept at 0.01; the learning rate, as 0.1; and momentum, at 0.5. The system's training curve is given in Figure 9.4.

9.2.3 Concluding Remarks for the Fusion of Face and Speech

In this section, we combined the features of speech and face to generate a fused model that uses both the speech and the face systems. The generated system was easily able to solve the problem with

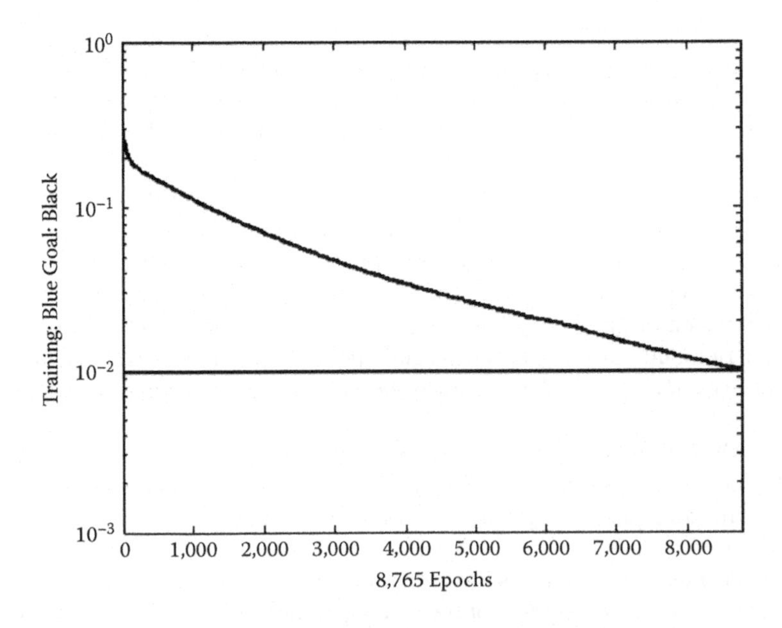

FIGURE 9.3 The training curve for Case 1 of the fusion of speech and face.

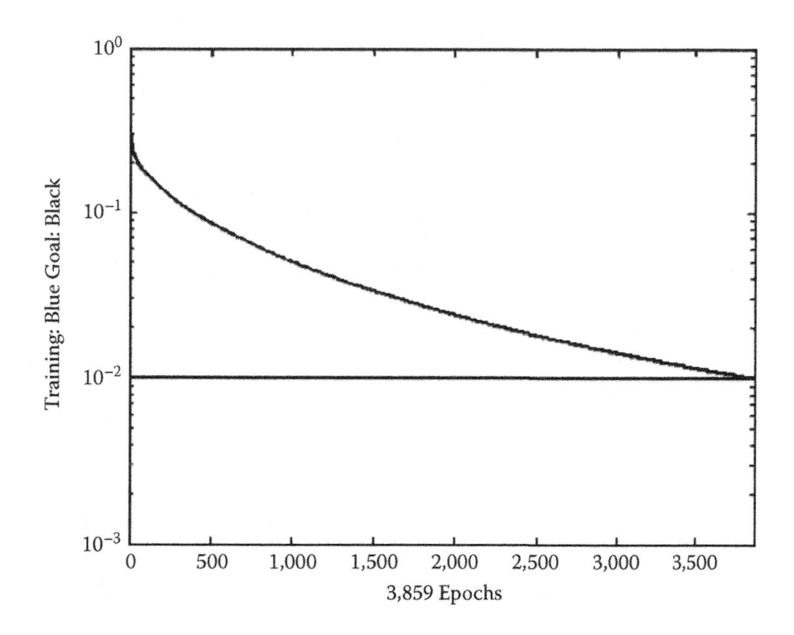

FIGURE 9.4 The training curve for Case 2 of the fusion of speech and face.

greater performances than the individual systems could do. The system also took much less training time, and the training efficiency was much better as compared with the two individual systems.

When we trained the ANN with only the face features on the same database, we found that the training could not be completed, even after a very large number of epochs at the specified goal. Similarly, the ANN model with speech features only was able to be trained but required a large number of epochs; hence the training time was very large. This limited the system's versatility by restricting the number of speakers to be authenticated. The fused ANN, on the other hand, was easily able to solve the problem and reached the required goal in a finite time and number of epochs.

We also tested the algorithm with a limited number of features in order to find the most optimal model of the system. In this model, we observed that the system was able to solve the problem in less time and fewer epochs. This enabled the model to include more speakers and thus increased its versatility.

The recognition ability of the fused model is very high as compared with the individual systems of face and speech. Whereas feature extraction requires a long time, the recognition process is much faster. Reducing the size of matrices can considerably reduce the time taken in the feature extraction process. This method was found to perform satisfactorily in adverse conditions of exposure, illumination and contrast variations, and face pose. This scheme is efficient because results have a smoother curve and a faster training cycle. These results further conclude that even a distorted image of the speaker, along with only few utterances of sound in a noisy environment, can be used to identify the person.

This work can be further extended for multiple-choice identification tasks, as are used for criminal investigation. Another extension is to include more biometric identities, such as retina, fingerprints, and so on, to identify people in security applications or for access control, telephone credit cards, and Internet banking. Further extensions can be used for talking faces, video sequences, and so on.

9.3 FUSION OF FACE AND EAR

In this section, we again fuse two biometric modalities that correlate very well to each other: the face and the ear. Both of these inputs are taken with the help of cameras. The only difference is that the face is shot from the front, while the ear photo is taken from the side. For practical deployment of this system, we may use either a single camera that takes two photographs or two cameras that take two separate images simultaneously. In the first instance, the user would have to change his angle in between the two photographs. The second option is more convenient and makes it much easier on the person being photographed and the user.

As we explained in earlier chapters, the face is a very exciting biometric identity that has various unique characteristics. The use of the face as a biometric identity is thus both natural and highly dimensional; as such, it necessitates the use of features that remain constant in multiple scans in order to reduce dimensionality. We discussed these issues in the chapter on face identification.

The use of the ear as a biometric modality has not been discussed. The ear comprises many characteristics that are unique to each person. However, ears have not traditionally been used for identification purposes. One reason for this may be that the ears are difficult for people to easily see and we rarely identify others by their ears.

In the face and ear identification that we do here, we use the Haar wavelet transform for the purpose of feature extraction. This transform also extracts the major features from the entire image and best represents the image with the least amount of information loss. The use of Haar for feature extraction for the face and ear gives us the flexibility to decide the dimensionality on our own with consideration to data and time constraints.

The classifier runs on a simple approach that calculates and compares for each class the distance of the unknown input with the distance of the templates that are already present in the database. The Hamming distance is used for measurement purposes. Ideally this distance must be as small as possible for members of the same class. The value of this distance is used to decide the class and the authenticity of a member in that class.

9.3.1 Haar Transform

Because the features of both the face and the ear have the application of Haar transform, we first study this transform and use it for feature extraction in this particular problem.

The Haar wavelet is the first-known wavelet. The Haar transform is used to transform any point of the form $[x(1), x(2)]^T$ into another space having the corresponding points $[y(1), y(2)]^T$. This is done with the help of a transformational matrix T. The relation between x and y is given by Equation 9.1:

$$\begin{bmatrix} y(1) \\ y(2) \end{bmatrix} = T \begin{bmatrix} x(1) \\ x(2) \end{bmatrix} \tag{9.1}$$

where the matrix T is given by Equation 9.2:

$$T = \frac{1}{\sqrt{2}} \begin{bmatrix} 1 & 1 \\ 1 & -1 \end{bmatrix} \tag{9.2}$$

This means that $y(1)$ and $y(2)$ are the addition and subtraction of $x(1)$ and $x(2)$, respectively, and are scaled by a factor of $\sqrt{2}$ to keep the energy constant. Note that T represents an orthogonal matrix whose rows have been normalized to have a magnitude of 1. In such a case, Equation 9.3 always holds true by the law of matrices. This also gives Equation 9.4:

$$T^{-1} = T^T \tag{9.3}$$

$$\begin{bmatrix} x(1) \\ x(2) \end{bmatrix} = T^T \begin{bmatrix} y(1) \\ y(2) \end{bmatrix} \tag{9.4}$$

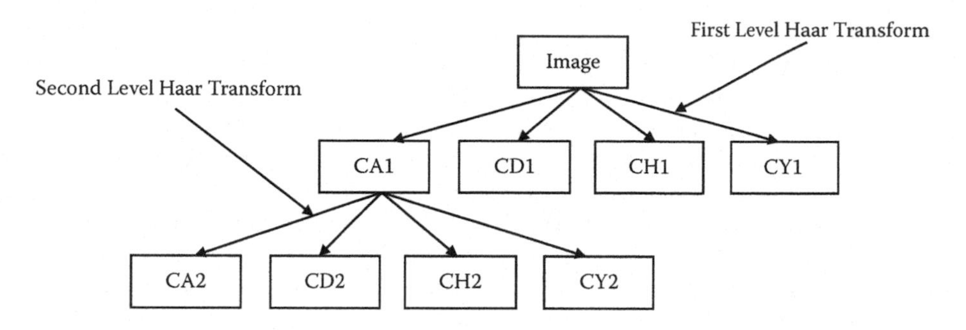

FIGURE 9.5 Graphical representation of wavelet decomposition.

If x and y denote 2×2 matrices, it may be derived that x and y are given by Equations 9.5 and 9.6:

$$x = \begin{bmatrix} a & b \\ c & d \end{bmatrix} \tag{9.5}$$

$$y = \begin{bmatrix} a+b+c+d & a-b+c-d \\ a+b-c-d & a-b-c+d \end{bmatrix} \tag{9.6}$$

The top-left $a + b + c + d$ in Equation 9.6 is a four-point average or two-dimensional low-pass filter, the top-right $a - b + c - d$ is an average horizontal gradient or horizontal high-pass and vertical low-pass filter, the lower-left $a + b - c - d$ is an average vertical gradient or horizontal low-pass and vertical high-pass filter, and the lower-right $a - b - c + d$ is a diagonal curvature or two-dimensional high-pass filter. Figure 9.5 shows the graphical representation of wavelet decomposition

This transformation is called the Haar transformation. Wavelet transform is capable of simultaneously providing the time and frequency information, hence giving a time-frequency representation of the signal. Higher frequencies are better resolved in time, and lower frequencies are better resolved in frequency. Extracted coefficients are approximation, vertical, horizontal, or diagonal. Approximate coefficients are decomposed further into the next level, and four-level decomposition uses the transformation discussed. The decomposition is shown in Figure 9.6.

9.3.2 Feature Extraction

We first study the feature extraction for faces. We then present the feature extraction for ears.

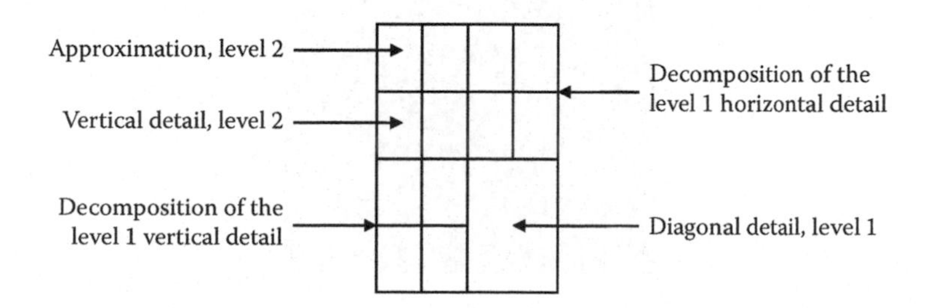

FIGURE 9.6 Wavelet decomposition.

9.3.2.1 Face

The face may be visibly different each time it is captured by the camera for various reasons, including changes in position, pose, and so forth. In addition, when taking a facial sample from the subject, the face must be extracted from the image, as it may include background or the person's neck, which are both unnecessary for facial feature extraction. To remove the background or extract the face from the image, we use triangularization. The face is selected manually by taking three points (left eyeball, right eyeball, and center of lower lip). The distance between each denotes the amount of scaling to be done.

Once this task of identifying and extracting the face is complete, a four-level wavelet transformation is applied to decompose the face and obtain the features. The feature template for the face is the fourth-level coefficient of the wavelet transformation. At this level, coefficient matrices are combined into a single feature matrix, or feature template $FV = [CD4\ CV4\ CH4]$. This is used to calculate the binarized code F_{Haar}, given by Equation 9.7:

$$\begin{aligned} F_{\text{Haar}}(i) = 1, \quad & FV(i) \geq 0 \\ 0, \quad & FV(i) < 0 \end{aligned} \tag{9.7}$$

where F_{Haar} is the binarized face code.

9.3.2.2 Ear Feature Extraction

Ears have several advantages over complete faces, including reduced spatial resolution, a more uniform distribution of color, and less variability in terms of expressions and orientation of the face. Profile images are acquired using a camera under the same lighting conditions with no illumination changes. All the images are taken from the right side of the face from a distance of approximately 15 to 20 centimeters. The images are carefully taken so that the outer ear shape is preserved. The less erroneous the outer shape, the more accurate the results.

The ear part of the image is manually cropped from the side face image and is then converted to a grayscale image. The edges of the ear are detected using the canny edge detection algorithm. Due to the noise in the image, the edge detector results in small noisy edges, which are of no use and moreover may reduce the algorithm's accuracy. Figures 9.7a and 9.7b show ear images with and without noisy edges, respectively.

The wavelet approach that was used for the face is also applied here. To get a binary template of the input image, the usual technique is as follows: the negative element of the coefficient matrix

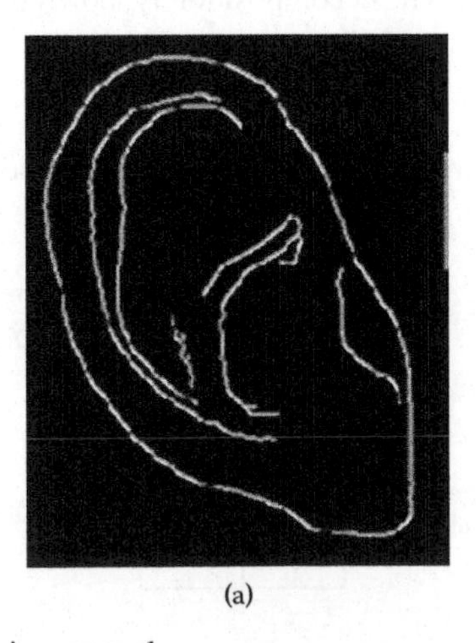

(a)

FIGURE 9.7a Ear image before noise removal.

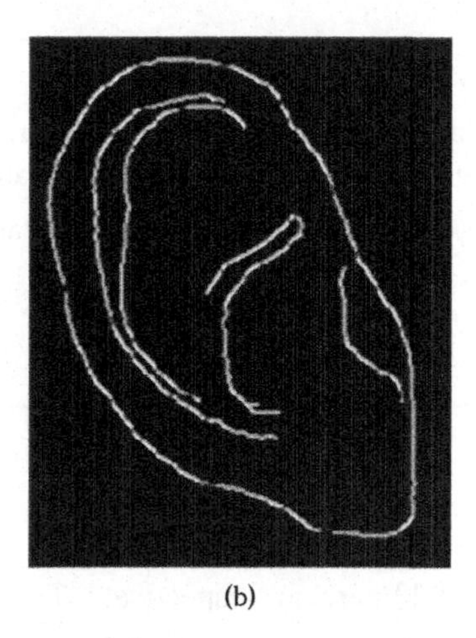

(b)

FIGURE 9.7b Ear image after noise removal.

is set to 0 and the positive is set to 1. This template is the binary template and is used as a feature matrix of the image. Ear images are decomposed in four levels; this decomposition compresses the image and extracts wavelet coefficients, which are clustered in a two-dimensional matrix.

9.3.3 Classification

Classification is done using the Hamming distance approach. This step is different from the previously discussed approaches that we seen throughout the text. The Hamming distance approach tries to find the deviation between the unknown input and the template of that same input that is already in the database. Hamming distance is used as a measure of this distance. The database consists of templates that are believed to be correct. This correctness is measured through the locations of the templates. In the training session, a database is created; the trained binary templates are then stored in that database using a unique index. The testing binary template (S) matches the database's query template (T) using Hamming distance. For the system to accept some claim of an identity, this distance from the unknown input to the template input must be less than some threshold value.

The formula for the Hamming distance (HD) is given by Equation 9.8:

$$HD = \frac{1}{n\ X\ m} \sum_{i=1}^{m} \sum_{j=1}^{n} T_{i,j} * S_{i,j} \tag{9.8}$$

where $T_{i,j} * S_{i,j}$ is the XOR operator. On iterations, this gives the total distance between T and S.

9.3.4 Results

The system was tested using a database built for 30 users. The training database contains a face and an ear image for each individual. The frontal-view face images were obtained under different orientations and lighting conditions.

In the first experiment, the individual systems for only the face and only the ear were developed and tested for FAR, FRR, and accuracy. Later the traits were combined at matching score levels using the sum of scores technique. The results were found to be very encouraging and promoted further research in this field. The system's overall accuracy is more than 97 percent, with an FAR

TABLE 9.1
False Acceptance Rate and False Rejection Rate of Individual Recognizers in Fusion of Face and Ear

Trait	FAR	FRR	Accuracy
Face	0.59	22	88.70%
Ear	11.00	6	91.05%
Fusion	2.46	1.23	97.00%

and an FRR of 2.46 percent and 1.23 percent, respectively. Table 9.1 shows the FAR, FRR, and accuracy of these systems.

9.3.5 Concluding Remarks for the Fusion of Face and Ear

In this section, we fused two very commonly used and similar biometric systems—the face and the ear. We saw that the fused system performed much better than the individual systems in terms of performance, training time, and training goal. We were easily able to solve the authentication problem using the fused system; this problem would have been very difficult to solve using only a single system. The performance table and accuracy curve show that the multimodal system performs better as compared with the unimodal biometrics, with an accuracy of more than 97 percent. However, it is worth studying the results by assigning different weights to different traits. At present, an equal weight is assigned to each trait. For the widespread use of biometrics to materialize, it is necessary to undertake systematic studies of the fundamental research issues underlying the design and evaluation of identification systems.

9.4 RECOGNITION WITH MODULAR ANN

In this section, we study another innovative way of fusing individual systems to make better systems. This technique is inspired by the modular neural network (MNN) architecture. Here we introduce into the system the concept of modularity. Then we try to solve the issues that the approaches discussed so far might face with certain inputs. We also solve the problem of fusing face and speech systems. The same features are used for this problem as were used in the preceding sections. We will see that the concept of modularity plays a key role in affecting the system's time and performance and makes it possible to accomplish tasks under the problem's demands.

9.4.1 Problem of High Dimensionality

We have seen that face and, to a large extent, speech both have a very high dimensionality. This high dimensionality introduces various restrictions into the systems. These restrictions include the need for a very large training data set for optimal training and the resulting long training time. As a result of the high training time, we must restrict the number of epochs and even the training output. This puts a very serious limitation on systems that have high dimensionality.

In fused systems, we mix two problems that are already of high dimensionality. This naturally results in a system with an even larger dimensionality. In an attempt to increase the recognition rate, we might choose to increase the dimensions, but this might actually lower the recognition rate due to the excessive dimensionality problems. Hence the solution might not always be this type of fusion of the system attributes.

An approach that we adopted earlier to deal with such scenarios was to limit the number of dimensions by taking up limited dimensions from each biometric modality. However, this might restrict the system's performance, as the feature we chose not to take might have conveyed valuable information for training and performance. This would impose further limitations to the systems.

This problem would gain an even higher concern if the inputs to the training data sets were very large. In this case, for every epoch, many calculations would need to be made. In addition, the inputs in such cases would generally be much more complex and scattered throughout the input space, which would make recognition very difficult. Any ANN has its scalability tested in large inputs, which rigorously checks the performance, time, and generalizing capability of that ANN. Any real life application would generally have a large data set size. This further necessitates the implementation of designs that result in systems that are scalable up to the demands of the real life applications.

9.4.2 Modular Neural Networks

The modular neural networks (MNNs) offer a good means for problem solving, in that they use the modularity of ANNs to divide the entire task into modules. Each module is solved by a separate unit of the system. Later the results of the various units are combined to yield the final system. This system thus requires an integrator that first divides the task into various modules and then assembles the results of each module to get the final output. Each module works independently of the others and supplies its final output to the integrator. The division of the task into modules is done during the design and is later implemented during training and testing. We studied about MNNs in detail in Chapter 6.

Here we follow the same principle of modularity. We try to break the entire task into a set of smaller tasks and then use those smaller tasks for the purpose of recognition. This procedure involves the creation of an integrator that divides the tasks and assembles the results. The basic aim is to reduce the dimensionality without losing information. This is achieved by introducing modularity at the attributes of the system.

A long list of attributes is a part of the system. These attributes come from both the face-recognition and the speech-recognition systems. Even in the absence of a large number of features, we can perform the task of verification with good accuracies. Recall that when there was no speech attribute and we had a single unimodal system, the accuracies were high. Even when the system was trained using a lesser number of attributes, we could receive sufficiently large accuracies.

This motivates the breaking up of the attribute set into various modules. Each module consists of a number of attributes of the system. The system is given only these attributes and no others; this restricts the dimensionality of every module. We then independently train the various modules by the selected attributes from the training data set. At the time of testing, each module is given its part of the attribute and calculates its own result.

Then we build an integrator that is responsible for integrating the individual outputs to yield the correct system output. The solution is built over a classification problem, in which the input must be mapped onto some class. Each ANN or module outputs the probabilities of the occurrence of the various classes as the final output. This probability lies between –1 to 1, where –1 means that the ANN is sure it is not the output and 1 denotes that the ANN is sure it is the output. Thus if there are c classes or people to be identified, each ANN gives c outputs with a probability associated with each class.

The integrator simply takes the average of the outputs of the various modules weighted by their performances. The performance is measured by the least error at the time of training. The integrator selects the class that averages the maximum, and this is regarded as the final output class.

In the next two sections, we discuss the division of the inputs and the integration of the outputs with reference to our problem of the fusion of face and speech recognition.

9.4.3 Modules

In this section, we discuss the breaking up of the entire problem into modules by dividing their attributes. In all, we have 11 attributes corresponding to speech: time duration, number of zero crossing, max cepstrum, average PSD, pitch amplitude, pitch frequency, peak PSD, and four formants (F1 to F4). We have 13 attributes for the face: length of eye 1, width of eye 1, center dimension x 1, center dimension y 1, length of eye 2, width of eye 2, center dimension x 2, center dimension y 2, length of the mouth, width of the mouth, center dimension x, center dimension y, distance from eye to eye, and distance from eye to mouth. Here x and y represent the image axes.

We begin by completely separating the face and speech into two different modules. The motivation for this is that the results obtained by the individual systems were highly encouraging in nature. We then divide the speech attributes into two parts and the face attributes into two parts. This results in four modules.

The first module covers some of the prominent speech attributes: time duration, max cepstrum, pitch amplitude, peak PSD, and F2 and F4. The second module contains the rest of the speech features: zero crossing, average PSD, pitch frequency, and F1 and F3. The same is true for the other two modules, which cover the facial features. The third module covers the features length of eye 1, width of eye 1, center dimension $i1$, center dimension $y1$, length of the mouth, and distance from eye to eye. The rest of the facial features belong to the fourth module: length of eye 2, width of eye 2, center dimension $x2$, center dimension $y2$, width of the mouth, center dimension x, center dimension y, and distance from eye to mouth.

The general mechanism is shown in Figure 9.8. The input comes to the system and is divided among these modules. Each module calculates its output and gives it to the integrator to decide the final output.

9.4.4 Artificial Neural Networks

Each module of the proposed MNN is an ANN. Each is trained using the BPA. We adopt a classification style of representation for the ANN, in which there are as many outputs as there are number of classes, or people. Each class denotes some person. Every ANN gives an output corresponding to

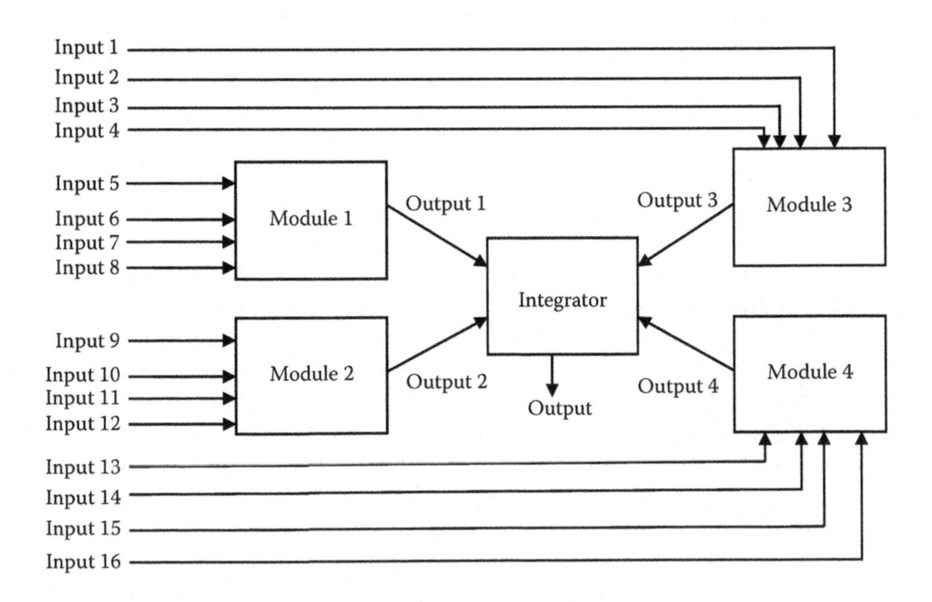

FIGURE 9.8 The division of inputs among modules.

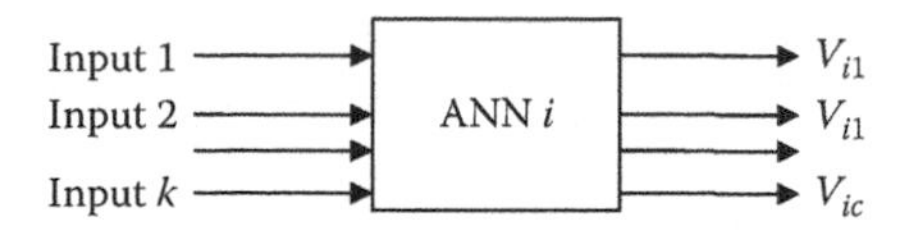

FIGURE 9.9 The inputs and outputs of the artificial neural network.

a class as the probability of that class being the final output: 1 denotes that according to the ANN, this class is definitely the output, while –1 denotes that this class is definitely not the answer. In an ideal case, the output of every ANN input should be 1 for just one class and –1 for all others. In practice, however, this does not happen. Instead the network outputs per class denotes the likeliness of the class being regarded as the system output. The use of ANN in this manner has already been discussed numerous times.

We assume that there are c classes, or people, to be identified. There are a total of n modules, or ANNs. Thus for every input I, ANN i gives c outputs in the form of vector $<v_{i1}, v_{i2}, v_{i3},\ldots, v_{ic}>$. This vector is represented by Figure 9.9.

9.4.5 Integrator

In this section, we study how to integrate the various solutions by the various ANNs in order to generate a final solution returned by the system.

Each ANN i gives c outputs $<v_{i1}, v_{i2}, v_{i3},\ldots, v_{ic}>$, where each v_{ik} is the probability of the occurrence of the class k measured on a scale of –1 to 1. The integrator simply averages the different probabilities classwise from the different ANNs. The weighted average is taken. We associate a weight for each ANN; this weight is its performance in the training data. Using this information, we take the weighted average over the inputs. Then the integrator simply takes the maximum of the various classes. The class that scores the maximum is returned as the final output, as shown in Figure 9.10.

9.4.6 Results

The experiments were done on a database of 20 people. The features mentioned earlier were used and divided as discussed earlier. All the inputs were normalized to lie in an interval between 0 and 1. Each ANN had one hidden layer. The activation functions of the layers were tansig and purelin. The training function used was traingd. The first ANN had 30 neurons in the hidden layer. The numbers of neurons in the hidden layers of the other ANNs were 28, 32, and 30. Training in each case was carried out for 3,000 epochs. The training curve for the first ANN is given in Figure 9.11.

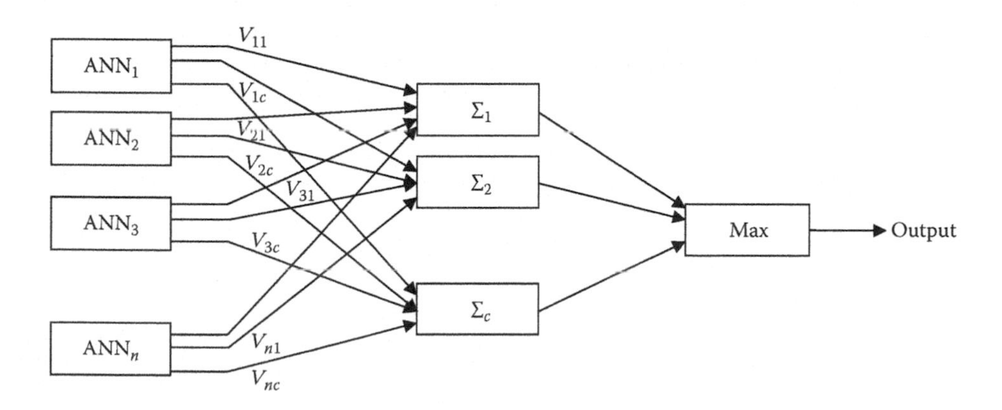

FIGURE 9.10 The integrator in the modular neural network approach.

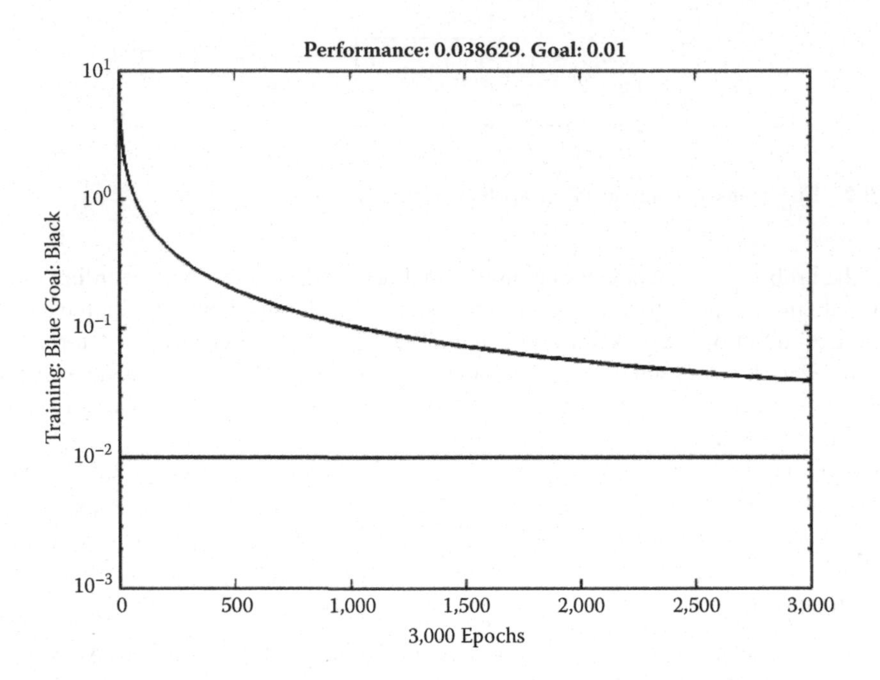

FIGURE 9.11 Training curve 1 of the artificial neural network for the fusion of face and speech.

Using this method, we achieved greater accuracies as compared with the previous approaches, in which the modular ANN also resulted in removing the problems of dimensionality. This greatly enhanced the systems. The accuracy after using this system came to 100 percent when the aforementioned configuration was used.

9.4.7 Concluding Remarks for Modular Neural Network Approach

In this section, we presented an approach that used modularity to solve the problem of high dimensionality faced by the fusion methods. We solved the problem by dividing the set of attributes into classes. Each class was given to a separate ANN, or module. These ANNs were trained independently of each other but in parallel. We combined the results using an integrator.

The resulting system solved the problem very effectively and hence can be used for fusion of the face and the speech. The accuracies received were very encouraging and proved that the system has a potential for use in real life applications. Additional work using the other multimodality systems may be done in future. Proper weighting strategies may be devised to calculate the weights of the different ANNs. The system may further be tested and verified using more diverse databases and databases of larger sizes to exploit the system's scalability factors.

10 Bioinformatics

10.1 ABOUT PROTEIN

The word protein is derived from the Greek word protas, which means of primary importance. Proteins are complex, high molecular mass organic compounds that consist of amino acids joined with the help of peptide bonds. Proteins play a vital role in the structure and functioning of all living cells as well as viruses. Different types of proteins perform various kinds of functions. These include the function as enzymes where proteins catalyze or accelerate the chemical reaction without self-consumption. Other proteins play structural or mechanical roles. Other functions of proteins include immune response, where proteins identify and neutralize foreign objects like bacteria and viruses. They are also used for storage and transport of various ligands (atom, ion, or molecule).

In chemistry, an amino acid is any molecule that contains both amine and carboxylic acid functional groups. Amines are organic compounds and a type of functional group that contain nitrogen as the key atom. Carboxylic acids are organic acids characterized by the presence of a carboxyl group that has the formula –C(=O)–OH, usually written as –COOH. In biochemistry, this shorter and more general term is frequently used to refer to alpha amino acids, which are those amino acids in which the amino and carboxylate functionalities are attached to the same carbon, also called the α-carbon. The structure of a general amino acid molecule is given in Figure 10.1. Here R depends upon the amino acid.

The relation between the amino acids and proteins is similar to the relation between letters and words. Letters combine in numerous ways to make different kinds of words. In a similar manner, the various amino acids may be linked together in various forms of sequences to make a variety of proteins. The unique shape of each protein determines its function in the body.

The general structure of an α-amino acid is given in Figure 10.2. In this figure R represents a side chain specific to each amino acid. Amino acids are usually classified by the properties of the side chain into four groups. The side chain can make them behave like a weak acid, a weak base, a hydrophile (if they are polar), or a hydrophobe (if they are nonpolar).

A peptide bond is a chemical bond formed between two molecules when the carboxyl group of one molecule reacts with the amino group of the other molecule, releasing a molecule of water (H_2O). The resulting CO-NH bond is called a peptide bond, and the resulting molecule is an amide. The reaction between the carboxyl group and the amino group resulting in the generation of a peptide bond is shown in Figure 10.3. The linking of two amino acids is shown in Figure 10.4.

Proteins are essentially polymers made up of a specific sequence of amino acids. The details of this sequence are stored in the code of a gene. It is very common for proteins to work together to achieve a particular function, and they often physically associate with one another to form a complex.

10.2 PROTEIN STRUCTURE

Proteins are assembled from amino acids using information present in genes. These are polymers built from 20 different L-alpha-amino acids (also known as residues) that fold into unique three-dimensional protein structures. The shape into which a protein naturally folds is known as its native state, which is determined by its sequence of amino acids. The native state of a protein is its operative or functional form. All protein molecules are simple unbranched chains of amino acids. The three-dimensional shape that they assume helps them to perform all their biological functions. In

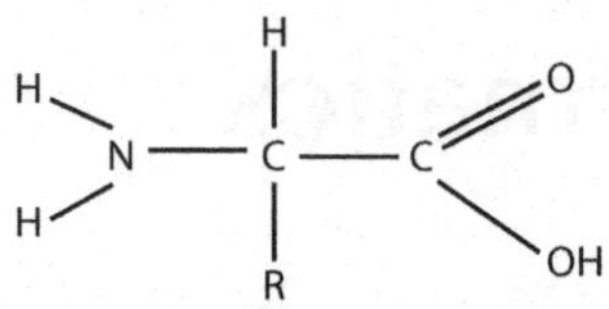

FIGURE 10.1 The general structure of an amino acid molecule.

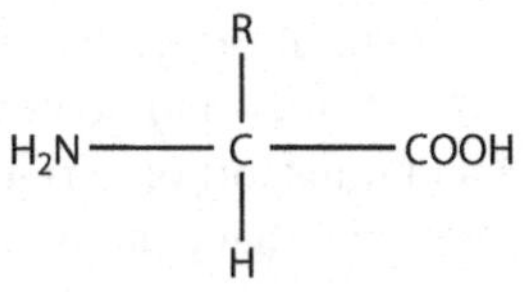

FIGURE 10.2 The general structure of an α-amino acid.

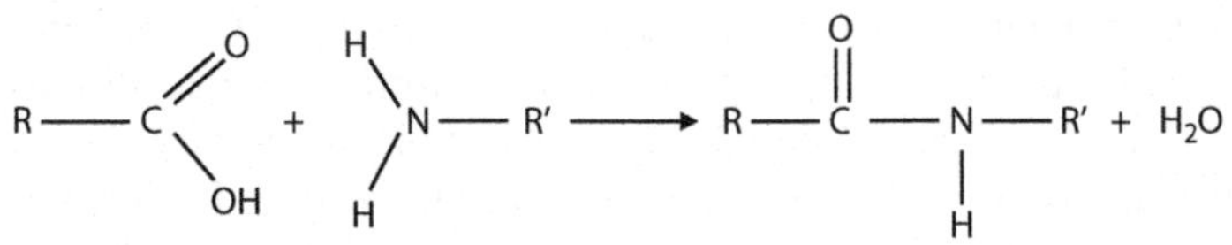

FIGURE 10.3 The formation of the peptide bond.

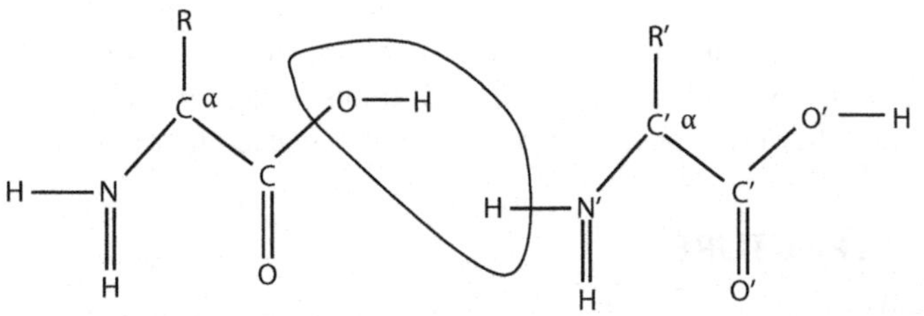

FIGURE 10.4 The linking of amino acids.

fact, shape changes in proteins are the primary cause of several neurodegenerative diseases. Protein folding is the process by which a protein structure assumes its functional shape or conformation. All protein molecules are heterogeneous unbranched chains of amino acids. By coiling and folding into a specific three-dimensional shape, they are able to perform their biological function. Heterogeneous means that they consists of a diverse range of different items

10.2.1 Four Distinct Aspects of a Protein's Structure

Proteins are generally studied in four distinct aspects or structures. These are primary structure, secondary structure, tertiary structure, and quaternary structure.

The primary structure is the sequence of residues in the polypeptide chain. The secondary structure is a local, regularly occurring structure in proteins and is mainly formed through hydrogen bonds between backbone atoms. So-called random coils, loops, or turns don't have a stable secondary structure. There are two types of stable secondary structures: alpha helices and beta sheets. Alpha helices and beta sheets are preferably located at the core of the protein, whereas loops prefer to reside in outer regions.

Tertiary structure describes the packing of alpha helices, beta sheets, and random coils with respect to each other on the level of one whole polypeptide chain. Quaternary structure only exists if there is more than one polypeptide chain present in a complex protein. Then quaternary structure describes the spatial organization of the chains.

10.2.2 Protein Folding

The process by which the higher structures form is called protein folding and is a consequence of the primary structure. It is not yet possible to predict the structures of proteins from basic physical principles alone.

Counting of residues always starts at the N-terminal end (NH_2-group). The rigid peptide bond angle is denoted by ω. This is the bond between C_1 and N and is always close to 180 degrees. The other important angle is the dihedral angle phi, φ. This is the bond between N and $C\alpha$. The other angle is denoted by psi, ψ. This is the bond between $C\alpha$ and C_1 and can have a certain range of possible values. These angles are the degrees of freedom of a protein; they control the protein's three-dimensional structure. They are restrained by geometry to allowed ranges typical for particular secondary structure elements. The various angles are shown in Figure 10.5.

10.2.3 Protein Secondary Structure Theory

The two most common secondary structure arrangements are the right-handed alpha helix and the beta sheet that can be connected into a larger tertiary structure (or fold) by turns and loops of a variety of types. These two secondary structure elements satisfy a strong hydrogen bond network within the geometric constraints of the bond angles ω, ψ, and φ. The β sheets can be formed by parallel or, most commonly, antiparallel arrangement of individual β strands. Only the atoms of the backbone are involved in secondary structure, not the amino acid side chains.

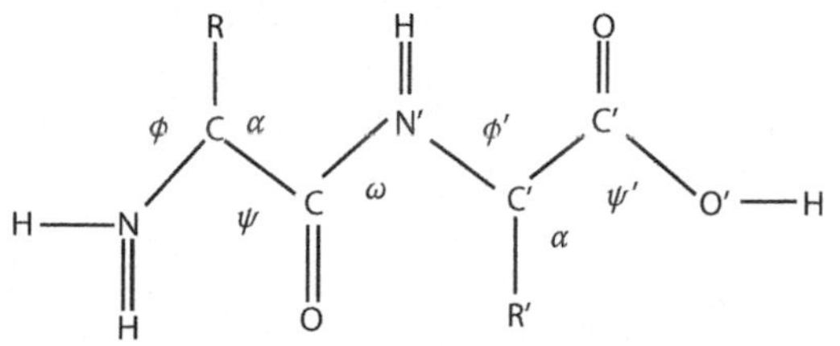

FIGURE 10.5 The various bond angles in proteins.

In detail, the secondary structure has three regular forms: alpha (α) helices, beta (β) sheets (combinations of beta strands), and loops (also called reverse turns or coils). In the problem of protein secondary structure prediction, the inputs are the amino acid sequences, while the output is the predicted structure (also called conformation, which is the combination of alpha helices, beta sheets, and loops). A typical protein contains about 32 percent alpha helices, 21 percent beta sheets, and 47 percent loops or nonregular structure.

A polypeptide is an unbranched structure of many amino acid sequence bonded with peptide bonds. An amino acid unit in the polypeptide chain is called a residue. The polypeptide chain starts at its amino terminus and ends at its carboxyl terminus.

Amino acids in the interior of the protein molecule come from the hydrophobic class, while amino acids from the polar class are at the surface of the molecule. Proteins evolved from a common ancestor are called homologous proteins; they usually have similar amino acid sequences and conformations, and hence similar properties and functions.

10.2.4 Characteristics of Alpha Helices

A common motif in the secondary structure of proteins is the alpha helix (α helix). This is a right-handed coiled conformation, resembling a spring, in which every backbone N-H group donates a hydrogen bond to the backbone C=O group of the amino acid four residues earlier ($i + 4 \rightarrow i$ hydrogen bonding). An alpha helix has a partial positive charge at the amino end and a partial negative charge at the carboxyl end. This in turn causes both ends of alpha helices to be polar, and therefore they are always at the surface of protein molecules. The lengths of alpha helices vary from 4 or 5 residues to over 40 residues. However, the average length is about 10 residues. The amino acids in an α helix are arranged in a right-handed helical (spring) structure, 5.4 Å (= 0.54 nm) wide. Each amino acid corresponds to a 100 degree turn in the helix (i.e., the helix has 3.6 residues per turn), and a translation of 1.5 Å (= 0.15 nm) along the helical axis. Most important, the N-H group of an amino acid forms a hydrogen bond with the C=O group of the amino acid four residues earlier; this repeated $i + 4 \rightarrow i$ hydrogen bonding defines an α helix. Similar structures include the 3_{10} helix ($i + 3 \rightarrow i$ hydrogen bonding) and the π helix ($i + 5 \rightarrow i$ hydrogen bonding). These alternative helices are relatively rare, although the 3_{10} helix is often found at the ends of α helices, "closing" them off. Transient $i + 2 \rightarrow i$ helices (sometimes called δ helices) have also been reported as intermediates in molecular dynamics simulations of α-helical folding. Residues in α helices typically adopt backbone (φ, ψ) dihedral angles around (–60°, –45°).

10.2.5 Characteristcs of Beta Sheets

A beta sheet is built from a combination of several polypeptide chains called beta strands. Beta strands are usually 5 to 10 residues long. They are aligned to each other such that hydrogen bonds can form between CO groups of one beta strand and NH groups on an adjacent beta strand and vice versa. Side chains point alternatively above and below the beta sheet. There are three ways to form a beta sheet: parallel (all beta strands are in same direction), antiparallel (beta strands alternate in direction), and mixed (a combination of parallel and antiparallel strands). However, mixed beta sheets occur rarely. All beta sheets have their strands twisted once.

The β sheet (also β-pleated sheet or β strand) is a commonly occurring form of regular secondary structure in proteins. It consists of a stretch of amino acids whose peptide backbones are almost fully extended, with dihedral angles (φ, ψ) = (–135°, –135°).

10.2.6 Characteristics of Loops

Loop regions occur at the surface of the protein molecule. The main chain CO and NH groups of the loop regions, which generally do not form hydrogen bonds with each other, are exposed to the solvent and can form hydrogen bonds with water molecules. Loop regions exposed to solvent have large quantities of charged and polar hydrophilic residues. It is possible to predict loop regions with higher accuracy than alpha helices or beta sheets. In homologous amino acid sequences, it is found that insertions or deletions of a few residues occur almost only in the loop regions. This is because during evolution protein cores are much more stable than loops (which are at the surface).

Each amino acid belongs to one of the eight following folding categories: H(α helix), G(310 helix), I(π helix), E(β strand), B(isolated β-bridge), T(turn), S(bend), and -(rest).

Knowledge of protein structure is generally considered a prerequisite to understanding protein function. Because months and sometimes years are involved in verifying protein structure through experimental methods, computational methods of modeling and predicting protein structure are currently viewed as the only viable means of quickly determining the structure of a newly discovered protein. The work presented in this chapter tries to make steps toward predicting protein secondary structure from sequence only.

10.3 PROBLEM OF PROTEIN STRUCTURE DETERMINATION

Proteins are very important to all living organisms. All the cellular chemical transformations are aided by proteins, and much of the structure of the cell is actually proteins. Proteins are chains of an "alphabet" of 20 amino acids, varying in length from 50 to 3,000, with an average length of 200. While it has been relatively inexpensive to sequence proteins, i.e., to discover the amino acids they are made of, it is expensive to discover the way the amino acids fold in three-dimensional space. In other words, the discovery of the protein structure, which is made through X-ray crystallography, nuclear magnetic resonance, etc., is time consuming and demands highly trained personnel. Thus, there is a great disparity between the number of proteins that have been sequenced and the number of proteins with known structure. A comprehensive understanding of the functional role of proteins depends on their structure and is extremely important in drug design. Because of the aforementioned situation, there is the need to develop algorithms for the prediction of the three-dimensional structure based on sequencing data.

Knowledge of protein structure is generally considered a prerequisite to understanding protein function. Because months and sometimes years are involved in verifying protein structure through experimental methods, computational methods of modeling and predicting protein structure are currently viewed as the only viable means of quickly determining the structure of a newly discovered protein.

The protein folding problem—i.e., how to predict a protein's three-dimensional structure from its one-dimensional amino-acid sequence—is often described as the most significant problem remaining in structural molecular biology; to solve the protein folding problem is to break the second half of the genetic code. On a practical level, solving the protein folding problem is the key to rapid progress in the fields of protein engineering and "rational" drug design. Moreover, as the number of protein sequences is growing much faster than our ability to solve their structures experimentally (e.g. using X-ray crystallography)—creating an ever-widening sequence-structure gap—the pressure to solve the protein folding problem is increasing.

At present, 100 percent accurate protein structures are determined experimentally using X-ray crystallographic or nuclear magnetic resonance (NMR) techniques. However, these methods are not feasible because they are tedious and time consuming, taking months or even years to complete. Around 90 percent of the protein structures available in the Protein Data Bank have been determined by X-ray crystallography. This method allows the exact three-dimensional coordinates of all the atoms in the protein to be determined to within a certain resolution. Roughly 9 percent of the known protein structures have been obtained by nuclear magnetic resonance techniques, which can also be used to determine secondary structure.

There are two main computational alternatives to experimental methods of determining or predicting protein structure from sequence data. The first approach is based on ab initio methods, which involve reasoning from basic physical principles. Ab initio methods rely on molecular physics and ignore any relationship of the molecule with other proteins. The second approach, often termed heuristic methods, is based on some form of pattern matching, using knowledge of existing protein structures.

The observation that each protein folds spontaneously into a unique three-dimensional native conformation implies that nature has an algorithm for predicting protein structure from amino acid sequence. Some attempts to understand these algorithms are based solely on general physical principles; others on observations of known amino acid sequences and protein structures. Here researchers are trying to reproduce the algorithm in a computer program that could predict protein structure from amino acid sequence. Most attempts (ab initio) to predict protein structure from basic physical principles alone try to reproduce the inter-atomic interactions in proteins, to define a computable energy associated with any conformation. Computationally, the problem of protein structure prediction then becomes a task of finding the global minimum of this conformational energy function. This approach did not succeed much, partly because of the inadequacy of the energy function and partly because the minimization algorithms tend to get trapped in local minima.

While ab initio methods of protein structure prediction can be used to identify novel structures from sequence data alone, they are too computationally intensive to work with all but the smallest proteins.

The alternatives to a priori methods are approaches based on assembling clues to the structure of a target sequence by finding similarities to known structures. These empirical or knowledge-based methods are becoming very powerful. Research toward predicting protein structures from sequence alone is a purely empirical one, based on the databases of known protein structures and sequences. These approaches hope to find common features in these databases, which can be generalized to provide structural models of other proteins.

The prediction of a protein's secondary structure—i.e., the formation of regular local structures such as alpha helices and beta strands within a single protein sequence—is an essential intermediate step on the way to predicting the full three-dimensional structure of a protein. If the secondary structure of a protein is known, it is possible to derive a comparatively small number of possible tertiary (three-dimensional) structures using knowledge about the ways that secondary structural elements pack.

The protein-folding problem deals with how to predict a protein's three-dimensional structure from its one-dimensional amino acid sequence. If this problem is solved, rapid progress can take place in the field of protein engineering and rational drug design. As the number of protein sequence is growing much faster than our ability to solve their structures, the pressure to solve the protein folding problem is increasing.

It seems obvious that it should be easier to predict secondary structure than tertiary structure, and to predict tertiary structure, a sensible way to proceed would be first to predict the helices and strands of sheets and then to assemble them.

10.3.1 Application of Artificial Neural Networks

Artificial neural networks appear well suited for the empirical approach to protein structure prediction. Similar to the process of protein folding, which is effectively finding the most stable structure given all the competing interactions within a polymer of amino acids, neural networks explore input information in parallel. Inside the neural network, many competing hypotheses are compared by networks of simple, nonlinear computational units. While many types of computational units exist, the most common sums its inputs and passes the result through some kind of nonlinearity. Three common types of nonlinearity are hard limiters, sigmoidal elements, and threshold logic elements.

The aim of secondary structure prediction (SSP) is to estimate the kind of conformation that the backbone of a residue adopts for a particular sequence. Here we consider the mainly helical (i.e., is part of an α helix) and extended (like a strand of a β sheet) elements. Everything else is "coil." These types are derived from the ϕ, φ angles measured in a protein's structure and so classify the residues of a protein structure by their local conformation class:

Protein sequence: ABABABABCCQQFFFAAAQQAQQA
Conformation class: HHHH EEEE HHHHHHHH

Here H means helical, E means extended, and blanks are the remaining coiled elements.

We know from chemistry that the kind of secondary structure formed depends on how the residue and its neighbors in the sequence are interacting. This is even true for β sheets; although there are rarely hydrogen bond interactions between the backbones or side chains in the same strand of a sheet, there are plenty between them. For the two types of secondary structure, it is also necessary for the side chains of residues to pack together correctly. These observations suggest that there must be some patterns of sequence that result in a particular type of secondary structure, and these patterns may not depend on the overall structure of the protein. In the context of prediction, the pattern is given in the "window" of neighboring amino acids around a residue:

Protein sequence: ABABAB**ABCCQ**QFFFAAAQQAQQA
Conformation class: HHHH EEEE HHHHHHHH

The problem of prediction, when viewed this way, should be solved by using a form of one-dimensional (1D) pattern classification. It is not straightforward, because there are many possible patterns, and the secondary structure may not always depend on just the local residue pattern; there is chance of exceptions. We use a classification method for secondary structure prediction in form of neural networks. They can learn which patterns of residues correspond to a particular type of secondary structure, based on the measurements from protein structure. The way this is done is by a mathematical computation, which can be viewed as a network (or directed graph) and is really just a set of simple equations with lots of variables.

The "learning" process just adjusts the values of the variables so that the output value is the correct one, or as close as possible to the one desired for each input pattern. After this learning, the network can be shown new patterns, and its output classifies them according to their resemblance to the training patterns; such ability is known as *generalization*. We rely on the network detecting the general properties of the secondary structure relationships so that it can assign the secondary structure for any sequence. To learn secondary structure types, we can train a network on a set of pattern pairs formed by the residue types for each sequence window and the secondary structure type for its central residue.

For protein secondary structure prediction, we use a feed-forward neural network. Feed-forward nets are the most well-known and widely used class of neural network. The popularity of feed-forward networks derives from the fact that they have been applied successfully to a wide range of information processing tasks in such diverse fields as speech recognition, financial prediction, image compression, medical diagnosis, and protein structure prediction; new applications are being discovered all the time. In common with all neural networks, feed-forward networks are trained, rather than programmed, to carry out the chosen information processing tasks. Training a feed-forward net involves adjusting the network so that it is able to produce a specific output for each of a given set of input patterns. Since the desired inputs are known in advance, training a feed-forward net is an example of supervised learning.

In this model, a window of residues is applied at the input side and the corresponding secondary structure (α helix, β sheet, or coil) of central residues is applied at target side for training. Also, nodes are connected in a feed-forward manner.

The back propagation training algorithm is used as the training algorithm. Here the weights are adjusted by a small amount to decrease errors. A window of training sequence is used as input to the network, and the predicted and expected structures of the central residue are compared. A set of small corrections is then made to the weights to improve an incorrect prediction, or the weights are left relatively unchanged for a correct prediction. This procedure is repeated using another training sequence until the number of errors cannot be reduced further.

10.3.2 System Analysis

System analysis, which is a management technique for designing a new system or improving an existing system, is implemented in order to analyze the system in an orderly grouping of interdependent components linked together according to a plan to achieve a specific goal. Each component is part of the total system and has to do its own scheme of work for the system to achieve the desired goal.

As our system protein secondary structure prediction is the task of a neural network, we apply system analysis for the following tasks:

1. Deciding the size of the input vector and the size of the output vector
2. Deciding which topology of neural network is suitable
3. Deciding which learning law is more suitable for our problem

The structure of a protein reveals important information, such as the location of probable functional or interaction sites, identification of distantly related proteins, discovery of important regions of the protein sequence that are involved in maintaining the structure, and so on. Along with a large number of protein sequences having been produced in high-throughput experiments, prediction of protein structure and function from amino acid sequences becomes one of the most important problems at the current time, and many researches have focused on the methods of protein structure determination. It is very difficult and time-consuming to experimentally determine protein structures (using techniques such as X-ray crystallography and nuclear magnetic resonance spectroscopy); therefore, computational methods to predict structures have been rigorously explored.

10.3.3 Approach

Outlined next is the typical methodology used to train a feed-forward network for secondary structure prediction, based on the approach adopted in (Qian & Sejnowski, 1988).

Each training pair is of the form

Pattern: LSADQISTVQASF

Target: H

That is, each pattern is a window onto a short segment of protein chain centered on the residue to be predicted (in this instance, the central Ser residue, which forms part of a helix).

Class of secondary structure: Each amino acid belongs to one of the eight following folding categories: H(α helix), G(310 helix), I(π helix), E(β strand), B(isolated β bridge), T(turn), S(bend) and -(rest).

This information has been used to train with supervised learning the three classifiers that we have described. Since the length of proteins, in terms of amino acids, varies, we have used a window of length n. In this implementation, we take three cases for $n = 13, 15, 17$. The window concept implies that the neighboring amino acids play a certain role in the current amino acid. For instance, given the following subsequence of a protein (upper line), we wish to train a classifier to predict the secondary structure of the middle amino acid (lower line).

Amino acid sequence: ISFAPQSGYTSGA

Secondary structure: T

TABLE 10.1
Mapping of Folding Categories

H, G	h (Helix)
E, B	e (strand)
Other	c (coil)

As is customary in the literature, we have mapped the eight folding categories into three, following the DSSP standard. These are given in Table 10.1. As we have used a window of length = n, in our implementation, we take three cases for n = 13, 15, 17. We will see all cases in detail. After using these eight folding categories in three classes, mapping that relation becomes

Amino acid sequence: ISFAPQSGYTSGA
Secondary structure: C

10.3.4 Encoding Scheme

One main problem is how to represent the amino acid sequence that is how to convert the letters for the amino acids and the secondary structure class into the values that are used in a neural network. That means how we are representing our training and testing data. Once we have decided how to do this, we can apply a learning algorithm, and then use the network to predict the secondary structure of other proteins.

We had adopted the unary coding for each amino acid, which produces an input vector of $21 * n$ dimensions. There are 20 amino acids plus another symbol to denote the left or right end of the protein.

Figure 10.6 shows a feed-forward neural network of the type used here for the prediction of secondary structure from a window of input amino acid sequences. Active nodes are shaded and arrows illustrate the connections between each node and all other nodes above it schematically. Only 5 input nodes are shown for each amino acid, although 21 were used.

For more detail, we model the primary structure of a protein as a finite string *str* of size 20. This is given by Equation 10.1.

$$Str = \{A, C, D, E, F, G, H, I, K, L, M, N, P, Q, R, S, T, V, W, Y\} \tag{10.1}$$

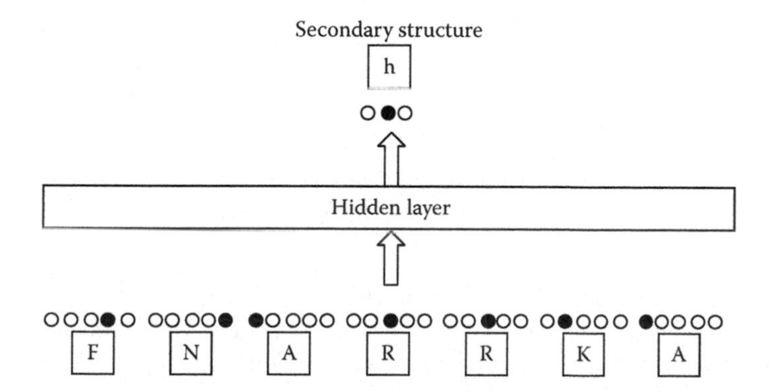

FIGURE 10.6 The general working of the artificial neural network in the problem.

TABLE 10.2
Coding of Amino Acids

Amino Aid	Code	Amino Acid	Code
Alanine	A	Leucine	L
Asparagine	R	Methionine	M
Aspartic acid (aspartate)	D	Phenylalanine	F
Cystine	C	Proline	P
Glutamine	Q	Serine	S
Glutamic acid (glutamate)		Theonine	T
Glycene	G	Tryptophan	W-
Histidine	H	Tyrosine	Y
Isoleucine	I	Valine	V

Here each letter represents one of the 20 amino acids. The codes for each of the 20 amino acids are given in Table 10.2. By including a gap (say the left/right end of the protein), string *Str* becomes as given by Equation 10.2. The size of *str* is 21.

$$Str = \{A, C, D, E, F, G, H, I, K, L, M, N, P, Q, R, S, T, V, W, Y, \text{-}\} \tag{10.2}$$

Here "-" is denoted by *X* (or *x*) in the implementation. Hence the general representation of *str* is as given by Equation 10.3.

$$Str = \{A, C, D, E, F, G, H, I, K, L, M, N, P, Q, R, S, T, V, W, Y, X\} \tag{10.3}$$

As mentioned earlier, we have applied unary coding for each amino acid. That means that each letter corresponds to the codes given in Equation 10.4.

$$\begin{aligned}
A &= [1\ 0\ 0\ 0\ 0\ 0\ 0\ 0\ 0\ 0\ 0\ 0\ 0\ 0\ 0\ 0\ 0\ 0\ 0\ 0\ 0] \\
C &= [0\ 1\ 0\ 0\ 0\ 0\ 0\ 0\ 0\ 0\ 0\ 0\ 0\ 0\ 0\ 0\ 0\ 0\ 0\ 0\ 0] \\
D &= [0\ 0\ 1\ 0\ 0\ 0\ 0\ 0\ 0\ 0\ 0\ 0\ 0\ 0\ 0\ 0\ 0\ 0\ 0\ 0\ 0] \\
E &= [0\ 0\ 0\ 1\ 0\ 0\ 0\ 0\ 0\ 0\ 0\ 0\ 0\ 0\ 0\ 0\ 0\ 0\ 0\ 0\ 0] \\
&\ \ \vdots \\
W &= [0\ 0\ 0\ 0\ 0\ 0\ 0\ 0\ 0\ 0\ 0\ 0\ 0\ 0\ 0\ 0\ 0\ 0\ 1\ 0\ 0] \\
Y &= [0\ 0\ 0\ 0\ 0\ 0\ 0\ 0\ 0\ 0\ 0\ 0\ 0\ 0\ 0\ 0\ 0\ 0\ 0\ 1\ 0] \\
X &= [0\ 0\ 0\ 0\ 0\ 0\ 0\ 0\ 0\ 0\ 0\ 0\ 0\ 0\ 0\ 0\ 0\ 0\ 0\ 0\ 1]
\end{aligned} \tag{10.4}$$

It may be seen that the size of input vector is $21n$. Here n is the window size.

We now discuss for the output side or the secondary structure. The secondary structure can be modeled by other finite strings over the alphabet given by Equation 10.5.

$$S = \{\alpha, \beta, c\} \tag{10.5}$$

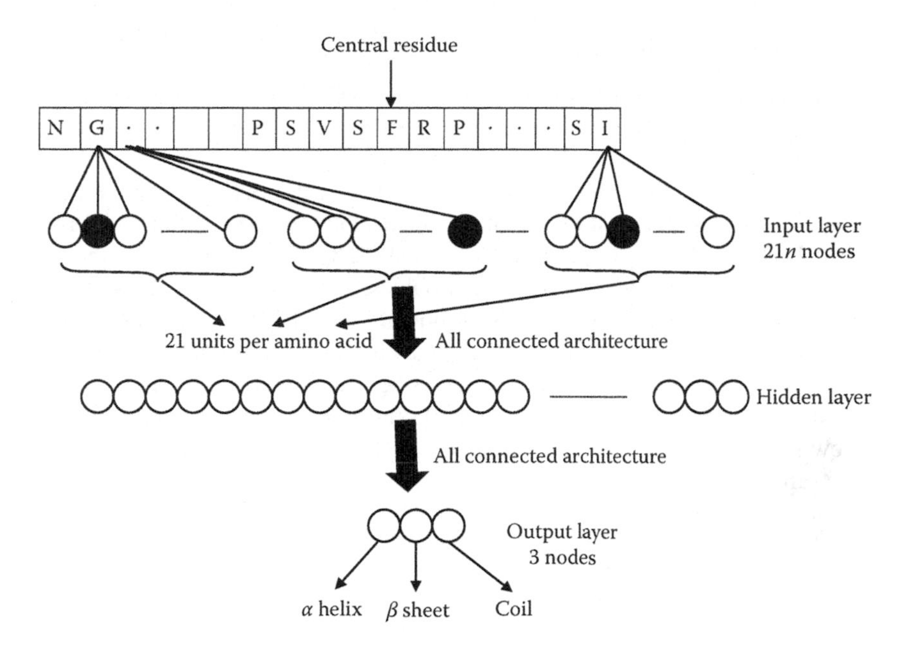

FIGURE 10.7 The architecture of the artificial neural network.

For implementation, in my work we take S = {h, e, c}, where each letter indicates that the amino acid is part of an α helix or a β sheet, or else it is classified as c-coil. The encoding is given by Equation 10.6.

$$
\begin{aligned}
h &= [1\ 0\ 0] \\
c &= [0\ 1\ 0] \\
e &= [0\ 0\ 1]
\end{aligned}
\tag{10.6}
$$

The advantage of this sparse encoding scheme is that it does not introduce an artificial ordering; each amino acid or secondary structure type is given equal weight. The main disadvantage is that it entails a large number of network parameters (especially at input vector size).

Slightly more concise encoding schemes are also possible, for example by treating a gap (for input patterns) or a coil residue (for output targets) as special "null" cases, i.e., a gap is represented by 20 zeros, a coil-target by 2 zeros. This approach leads to networks with only $n * 20$ inputs and/or two outputs and hence to networks with slightly fewer weights that may play a role in generalization.

10.3.5 Architecture of the Artificial Neural Network

A typical architecture is fully connected. The general architecture of the ANN is given by Figure 10.7. Here the input layer contains all the discussed inputs in their encoded form.

10.4 PROCEDURE

Using the concepts discussed in the previous steps, we now come to the step-by-step working procedure for solving the problem of protein structure prediction. Here we start with the manner in which the data is collected and processed. We then discuss the training and the testing of the ANN with BPA.

10.4.1 Data Source and Description

We use data for training and testing collected from the "RCSB Protein Data Bank." Each structure in the PDB is represented by a four-character identifier of the form [0–9][a–z, 0–9][a–z, 0–9][a–z, 0–9].

The training and testing data are both in MATLAB® .mat file format: "traindata.mat" and "data_for_testing.mat." In the traindata.mat file, we have 44 variables, corresponding to 44 proteins. The variable names always start with an "X" following the 4 PDB code of that protein in lowercase letters. In the data_for_testing.mat file, we have 12 variables, corresponding to 12 proteins.

The variables name always start with an "X" following the 4 PDB code of that protein in lowercase letters. The variables are formatted as a MATLAB character array. Each variable has two rows; the length of the column will depend on the length of the residue of the protein.

The first row is the one-letter amino acid code. Letter code "X" means unknown amino acid. The second row is the observed secondary structure code corresponding to the amino acid in the first row of the same column:

H = helix

E = sheet

C = coils

Example:

Protein (1DUR), with the corresponding variable name X1dur, has the following entries:

X1dur =

AYVINDSCIACGACKPECPVNCIQEGSIYAIDADSCIDCGSCASVCPVGAONOED
CEEECCCCCCCCCCHHHCCCCCEECCCCCEECCCCCCCCCHHHHHCCCCCEEECC

The array has the dimensions 2 × 55. The number of rows is 2, and the number of columns is 55, which means the protein has 55 amino acids. Starting from the first amino acid, with the corresponding index X1dur(1,1) = "A", the corresponding observed secondary structure is in X1dur(2,1) = "C". The second amino acid is in X1dur(1,2) = "Y", and its observed secondary structure is in X1dur(2,2) = "E", and so on. The last amino acid is in X1dur(1,55) = "D", and its corresponding observed secondary structure is X1dur(2,55) = "C". Some of the training data is given in Table 10.3.

TABLE 10.3
Training Data

155C	1EST	1TIM	2GCH	3CYT	7FAB
1ABE	1GCN	256B	2RNS	3DFR	7LYZ
1AZU	1HIP	2ACT	2SGA	3PGM	8TLN
1BP2	1MBN	2ALP	2SNA	4FXC	9PAP
1CA2	1PLC	2APR	351C	4INS	
1CYO	1PPT	2C2C	3APP	4RXN	
1DUR	1REI	2LHP	3DFR	5CPV	
1ECD	1SBT	2IZM	3CPA	6LDH	

10.4.2 Formation of Inputs and Outputs

In order to give the inputs and outputs to the ANN for the purpose of training and testing, we need to encode it using the strategy discussed. Consider the X1dur sequence. The size of the sequence as discussed is 2 * 55. Let the window size n be 17. Here the window is a shrinking window. Let p be the input and t be the output matrix. It may be seen that Size (p) = 17 * 41 and size (t) = 1 * 41.

To make p and t for each sequence with shrinking window size n, we follow the following steps:

```
Step 1: Initialization
        X = seq; %assign any protein sequence variable
        x = X(1,1:length(X));%assign primary sequence in x
        t1 = X(2,1:length(X));%assign secondary in t1
        x = ['x' x 'x' ] ; %appending gaps at end of sequence (x) ...denoted by x
        p1 = []; p = []; c = 0;
Step 2: Repeat for i = 1 to (length(x) – (n – 1))
Step 3: Repeat for j = i to (n + c)
Step 4:             x1 = (x(1,j)); %extracting jth position
                    p1 = [p1;x1]; %append columnwise
        [End of Step 3 loop]
Step 5:
        p = [p,p1]; p1 = []; c = c + 1;
        [End of Step 2 loop]
Step 6: t = t1(1,(n – 1) / 2:length(t1) – ((n – 1) / 2 – 1)); %target
Step 7: Display and store matrix p and t
Step 8: Exit
```

This algorithm of making input p and target t is for a single protein sequence (with using the general window size n). For a batch input px and target tx matrix we use these steps in a sequence. All the protein sequences are stored in a MATLAB variable in the following manner:

```
Xdata2 = [ ];
Xdata2{1} = X155c;Xdata2{2} = X1abe;Xdata2{3} = X1cyo,.,Xdata2{44} = X6ldh
```

For generating input px and target tx with window size n, we use the following steps:

```
Step 1 Initialization
        p = []; tx = []; px = []; k = 1; i = 1
Step 2 Repeat for k = 1 to length (Xdata2)
Step 3          X = Xdata2{k}
                f = X(1,1:length(X)); %f = primary seq. of protein
                s = X(2,1:length(X)); %s = secondary seq. of protein
                f = ['x' f 'x']; %appending gap at end of seq. "gap = x"
               p1 = []; p = []; c = 0;
Step 4          Repeat for i = 1 to (length(f) – (n – 1)
Step 5                  Repeat for j = 1 to (n + c)
Step 6                          x1 = (f(1,j));
                                p1 = [p1; x1];
                        [End of Step 5 loop]
Step 7                   p = [p, p1];
                        p1 = [];
                         c = c + 1;
                [End of Step 4 loop]
```

TABLE 10.4
Dimensions of Various Matrices for Varying Window Size

13 Window Size(n = 13)	15 Window Size (n = 15)	17 Window Size (n = 17)
$px = 13 * 6838$, $tx = 1 * 6838$	$px = 15 * 6750$ $tx = 1 * 6750$	$px = 17 * 6662$ $tx = 1 * 6662$
$Pb = 273 * 6838$, $Tb = 3 * 6838$	$Pb = 315 * 6750$ $Tb = 3 * 6750$	$Pb = 357 * 6662$ $Tb = 3 * 6662$

Step 8 p; $px = [px, p]$; $px3 = px$;
Step 9 $tx = s(1,(n - 1) / 2$:length(s) – ((n – 1) / 2) – 1));
Step 10 [End of step 1 loop]
Step 11 Exit

The final input matrix Pb is given by the following steps:

Step 1: load px and tx
Pb = []; %initialize
Step 2: Repeat for I = 1 to length (px)
Step 3: $S = px$ (1:n, I)';
Z = seq2bin (S);
Step 4: Pb = [Pb, Z'];
(End of Step 2 loop)
Step 5: Pb; save Pb
Step 6: Exit

The seq2bin function takes a string s as argument and gives output in binary using an encoding scheme as mentioned earlier. For each character there is a 21-bit code.

The final target matrix Tb is given by the following steps:

Step 1: load tx
Step 2: Tb = sec2bin (tx)
Step 3: save Tb
Step 4: Exit

Here the seq2bin function for each character returns a three-bit code.

In all these cases, the number of columns of the input matrix (p, px, or Pb) is same as the number of columns of the target matrix (t, tx, or Tb). The dimensions of the various input and output matrices discussed previously are given by Table 10.4 for various values of window size (n).

10.4.3 Training

We had selected a feed-forward network architecture. The network had an input layer, an output layer, and one hidden layer. Prior to training, the network weights and biases are initialized to small random values. Once the network weights and biases have been initialized, the network is ready for training. The training process requires a set of examples of proper network behavior. These are the inputs and target outputs as derived in the preceding section. During training, the weights and biases of the network are iteratively adjusted to maximize the network performance or minimize the error. The error of the ANN is measured by mean square error (mse). This is the average squared error between the network outputs and the target outputs. The back propagation algorithm (BPA) is used for training purposes. This algorithm uses the gradient of the performance function to determine how to adjust the weights to maximize performance. The errors are propagated backward from the output layer to the input layer.

TABLE 10.5
Training Parameters of Artificial Neural Network with Back Propagation Algorithm

Epochs	100
Goal	0
Maximum training time	Infinity
Minimum performance gradient	10^{-6}
Maximum validation failures	5
Learning rate	0.01
Delta increment (delt_inc)	1.2
Delta decrement (delt_dec)	0.5
Initial weight change (delta0)	0.07
Maximum weight change (delda_max)	50

These conventional training functions are often too slow for practical problems. There are many faster BPA invariants like variable learning rate BPA and resilient BPA. Resilient BPA is a heuristic-based technique developed from an analysis of the performance of the standard steepest descent algorithm. Here resilient BPA has been used to train the ANN.

Multilayer networks typically use sigmoid transfer functions in the hidden layers. These functions are often called "squashing" functions, since they compress an infinite input range into a finite output range (our requirement is in between 0 and 1). Sigmoid functions are characterized by the fact that their slope must approach zero as the input gets large. This causes a problem when using steepest descent to train a multilayer network with sigmoid functions, since the gradient can have a very small magnitude and, therefore, cause small changes in the weights and biases, even though the weights and biases are far from their optimal values. The purpose of the resilient back propagation (Rprop) training algorithm is to eliminate these harmful effects of the magnitudes of the partial derivatives. Only the sign of the derivative is used to determine the direction of the weight update; the magnitude of the derivative has no effect on the weight update. The size of the weight change is determined by a separate update value. The update value for each weight and bias is increased by a factor *delt_inc* whenever the derivative of the performance function with respect to that weight has the same sign for two successive iterations. The update value is decreased by a factor *delt_dec* whenever the derivative with respect to that weight changes sign from the previous iteration. If the derivative is zero, then the update value remains the same. Whenever the weights are oscillating, the weight change will be reduced. If the weight continues to change in the same direction for several iterations, then the magnitude of the weight change will be increased.

The ANN parameters used for the training are given in Table 10.5.

10.4.4 Testing

The next major task that decides the system performance is testing. Here we give unknown inputs to the system and see if the system is able to give the correct answers. The closeness of the outputs given by the system with the standard target output determines the system performance.

Consider the input X1fca. Here the primary sequence is as follows:

Primary Sequence =
AYVINEACISCGACEPECPVDAISQGGSRYVIDADTCIDCGACAGVCPVDAPVQA

By using the trained network, we get the result as follows:

CISCGACEPECPVDAISSQGGSRYVIDADTCIDCGACAGVCP
CCCCCCCHHHCCCCCEECCCCCCEECCCCCCCCCCHHHHCCC
CCCCCCCHHHCC-C----CCCCCEECCCCCCCCCCHHHH-CC

Here the first row is the primary sequence, the second row is the standard secondary structure, and the third row is the secondary structure predicted by the developed system; "-" means that the system was unable to predict the output at that location. The sequence length is 55, and number of residues predicted by the system for X1fca is 41. The number of correct predictions out of 41 was 35.

10.5 RESULTS

The system discussed was tested for numerous inputs to determine the net performance of the system. This is done by applying numerous unknown inputs and asking the system to predict the correct outputs. The system performance is measured by a term called quality. The quality (Q) is a measure of the closeness of the predicted sequence with the original sequence. This may be calculated by using Equation 10.7.

$$Q = \frac{Number\ of\ correct\ predicted\ residue}{N} \tag{10.7}$$

Here N is the total number of predicted residues.

Table 10.6 shows the prediction accuracies for various values of window size n. A neural network trained by a list of protein sequences gave an accuracy of about 80 percent for the testing sequence.

TABLE 10.6
Performance of the System for Various Values of Window Size

Sequence (length)	Q for $n = 13$	Q for $n = 15$	Q for $n = 17$
X1clf (55)	32/45 = 71	31/43 = 72	29/41 = 70.73
X1cto (185)	116/175 = 66	107/173 = 61.8	108/171 = 63
X1dox (96)	54/86 = 63	49/84 = 58	50/82 = 61
X1fca (55)	36/45 = 80	36/43 = 84	35/41 = 85
X1gbb (198)	100%	100%	100%
X1gdn (224)	130/214 = 60.6	127/212 = 60	134/210 = 64
X1jx6 (339)	185/329 = 56	185/327 = 56.57	193/325 = 59.38
X1ql3 (97)	60/87 = 69	56/85 = 66	64/83 = 77
X5cyt (104)	93/94 = 98.93	90/92 = 97.83	88/90 = 97.78
X8abp (305)	290/295 = 98.3	290/293 = 98.98	289/291 = 99.31
X2myc (153)	100%	100%	100%
X2frf (152)	114/142 = 80	113/140 = 81	113/138 = 82
Average	78.57	78.015	79.94

10.6 CONCLUSIONS

In this chapter, we took on the problem of predicting the protein structure. Here we first studied the basics of proteins. This included the protein structure, formation, amino acids, peptide bonds, etc. Various terms were introduced that gave an understanding of the way proteins function and perform. The three-dimensional structure of proteins was presented.

The latter part of the discussion concentrated on the problem of protein structure prediction. The encoding of the inputs and outputs was discussed to help us in the conversion of the inputs and outputs into a form that the ANN could handle. This made it possible to use ANN for this problem. The processing of data was a key step in the working of the algorithm that constituted the input and output encoding.

We used ANN to solve the problem. The resilient ANN model was followed, resulting in time-efficient neural network training. This further helped the ANN to learn or memorize the presented inputs and outputs and generalize this learning to the unknown inputs as well.

The system on testing showed a performance of almost 80 percent. This high accuracy of the ANN in this problem motivates its use for such problems of bioinformatics. Various other ANN models and training algorithms may be tried in the future, and their accuracies may be compared to produce a system that gives very high performance in this problem.

11 Biomedical Systems—I

11.1 INTRODUCTION

The long list of disciplines covered by soft computing includes the very exciting and prominent field of biomedical systems. Here soft computing collaborates with the discipline of medical science to lay the foundations of biomedical engineering, extending its fruits and solving many problems prevalent into the domain of medicine or medical technologies. In this chapter and the chapter to follow, we study the various diseases and develop systems that detect these diseases.

The use of soft computing in the medical domain might look very strange. If you ask a child the latest developments in this field, he might probably answer the discovery of new drugs, new surgical methods, etc. He might find it hard to visualize that the modern-day systems can do much of what doctors do: diagnosing a patient and finding out the presence of various diseases. He might even find it difficult to imagine that such systems are not a fantasy but in fact exist and have existed for quite a long time now.

Throughout the text we have talked about soft computing systems in terms of a design-oriented approach. Here the novelty lay in figuring out the correct or good parameters or features and using them in the correct soft computing design, considering all the practical and other constraints. All this requires a deep understanding of the system. The system understanding further helps in identifying all the inputs that the system may face, the output it is expected to return, the kind of conditions it would be used in, constraints, feasibility, etc. This gives a lot of help to the system designer to modify his system accordingly. Also finding out the correct and good parameters by understanding the way soft computing works is always a difficult or rather a tricky task. The better you understand the problem domain as well as the soft computing system, the better are the chances of good system design.

Unlike the other systems that we studied, this domain coves a very rich literature that has developed and grown extensively over the years. This was not the case with face recognition, where we just had to know a few basics of image processing, or speaker recognition, where some knowledge of signal processing was enough. This enormous literature acts as both a boon and a curse. On the positive side, it may be seen that the details of all the diseases, their causes, effects, symptoms, etc., are all documented and available. Hence we may not have to unnecessarily wonder what the good features are and experiment till good features are found. We just need to consult the literature to get a clear idea of the problem and the features that may be used for the system.

On the negative side, biomedical systems are usually not very easy to visualize or understand. This requires a sound knowledge of medical concepts and terminologies that are relatively difficult for engineers specializing in soft computing. In case of the other systems, we could easily visualize the system and understand and appreciate the various features. We could understand the reasons that they would be stable when recorded multiple times. But to do this with biomedical systems requires an understanding of the whole system in order to fully appreciate the features.

Explaining or understanding the complete system is naturally outside the scope of any person unless he has a medical background. It would not be possible to deeply study the problem and then try to work out the features and later the complete system. This further necessitates a proper collaborative effort among the people who fully understand the biomedical systems (the doctors) and the people who understand the soft computing systems (the engineers). This collaboration is the intent of biomedical engineering.

11.1.1 Need and Issues

Today doctors and engineers work together on every problem in order to try to devise systems that are able to automatically detect diseases. They try to find the best set of features and later develop classifiers for the purpose of the detection of diseases. Apart from the features, the system design is another step that requires people working together. The system behavior over the changing inputs is largely dependent on the inputs and thus the features used. An understanding of the features results in better systems that perform the task of detection of diseases much better.

The need for such systems is more than just a gift of soft computing. The rapid increase in the number of disease cases has resulted in the need of systems that can detect diseases to relieve doctors and very quickly analyze the possibilities of various diseases. Patients hence do not have to wait long for a verdict as to the presence of a disease. Further these systems aid doctors in diagnosing the patients and finding out the possibilities of various diseases. The time saved by these systems goes a long way toward the early detection of various diseases. This further may be very useful in effective prevention measures being taken at the correct time to stop and control the diseases. Hence the use of these systems is both essential and profitable.

Although soft computing measures have achieved great heights and have been able to solve numerous problems with very high efficiencies, yet the results are not 100 percent correct. This can prove fatal in the case of biomedical engineering. Imagine the system diagnoses a cancer patients and outputs the absence of cancer. This may lead to the loss of the patient's life due to taking the wrong decision. He may not receive any treatment at all and may suffer. Also imagine that the system diagnoses the presence of cancer in someone not suffering from cancer. He may wrongly receive a treatment for cancer, which may again be dangerous for his health. The kind of importance associated with a person's life prohibits the use of such systems even if the accuracies are very high. Here nothing less than 100 percent may do.

It may be noted that for this and various other reasons, these systems are not supposed to behave as autonomous stand-alone systems. The need for the physical presence and expertise of a doctor will always remain no matter how far the medical technologies and the clinical decision support systems grow. The doctor is usually able to give much better diagnostics than machines that just decide the presence or absence of some diseases. This happens because of the rich experience, vast knowledge, and understanding of the doctor that machines may not be able to imitate. But these systems do aid the doctor to make fast decisions and help in making timely, effective decisions.

11.1.2 Machine Learning Perspective

We have emphasized the importance of learning from the past data numerous time. This learning helped the systems to give correct output whenever that the same comes again. We further said that the soft computing systems are not only able to learn the historical data, rather they are also able to make out patterns in the data and generalize the data over unknown inputs. This made it possible for these systems to solve so many real life problems. Here we take the same principles as our motivations for solving the problem of identification of the presence of diseases.

From the historical database we have the data for the patients of some diseases. Over time the database will grow larger and carry reasonably good information. We assume here that the features used were good and stable. This means that in the near future if a patient comes having the characteristics that patients have in the database, he would most likely be suffering from the same disease. We can get this information by the concepts of machine learning. Hence the data that some patient is suffering from some disease is of great importance for future requirements of the system. This leads us to capture all such cases and later find trends in them using soft computing principles. Soft computing further helps in generalizing or finding the correct results to unknown inputs.

Imagine when you visit a doctor, how he diagnoses you. He may probably ask some questions. Further he may prescribe some tests. Drawing on all of these, he makes the final decision regarding the problem or the disease. Hence he is basically relying upon some attributes to make a final decision regarding the presence of the disease. This motivates the use of soft computing systems that work in almost the same manner by the application of attributes for detection. A doctor, however, first asks questions and performs basic tests to convince him that some disease may be present. Or in other words, he initially considers the various diseases that may be present and then measures the attributes or the features that would decide the presence or absence of the disease.

This chapter does not focus solely on the study of biomedical systems and diseases. We rather aim to understand soft computing techniques in relation to machine learning. The biomedical systems serve as ideal examples taken from real life applications in order to both experiment with and understand the effect of various parameters, concepts, etc., in determining the efficiencies of the resulting systems. We adopt a multidimensional approach here in which numerous diseases are approached in numerous ways. The inferences and results give us valuable data to gain a deep understanding of the ways in which the various techniques work for machine learning. We first lay down the foundations by studying the manner in which these systems work. Then we experiment with various real life biomedical applications. Finally, we use these inferences to validate or contradict the views we have presented on the working of these systems.

11.1.3 Diseases

The general term disease means dis-ease or not being at ease. It refers to the well-being and comfort of the person. A person may be said to be suffering from disease when he has some problems in the normal working of his body. The disease may happen in a person in numerous ways. Many are communicable and transferred from one person to another. Many others may develop inside the person. Various internal and external factors result in diseases, which are generally diagnosed by their symptoms. Each disease has certain symptoms that may be easily observed or felt by the patients. Based on these, the presence of diseases may be detected by a doctor. For this he may use various tests.

The diseases greatly vary in their impact and cure. Some diseases like headache, fever, etc., are very minor conditions that many of us face in daily lives. Others may be very serious, such as cancer or heart disease. Some others include mental illness, which has its own significance. Accordingly the cure varies. Some diseases need to be treated as soon as they are discovered. Others may not require immediate action as long as the patient is not uncomfortable. Yet others may disappear in the due course of time.

It is natural that one would not use any biomedical system for the purpose of detecting fever. A thermometer would definitely solve the purpose better. Similarly, a cough may be better detected by the patient himself. These are directly visible symptoms and hence do not require any additional systems, but they may be symptoms of various other conditions. The diseases we analyze here would be of much higher importance and value and would be much more difficult to analyze and work out without these systems.

It may be seen here that medical technology has made fabulous contributions in the past few decades. It has resulted in better solutions to diseases and better recognition systems that yield much higher accuracies. However, we still are not able to solve all diseases. For many diseases, soft computing systems still do not exist. For many others the systems do exist, but their performance is very poor. Hence better systems are needed here. This points to the need of positive developments in these fields for increased accuracies in identification for all these systems. These can change the health conditions of a country.

The problem of disease identification from the point of view of soft computing is primarily a classification problem. Here we have to classify the input among the two classes that stand for the presence of the inputs and the absence of the inputs. Models are designed on this basis similar to the verification and the authentication systems we developed in the previous chapters. They hence obey all the rules and design guidelines of the classificatory problems.

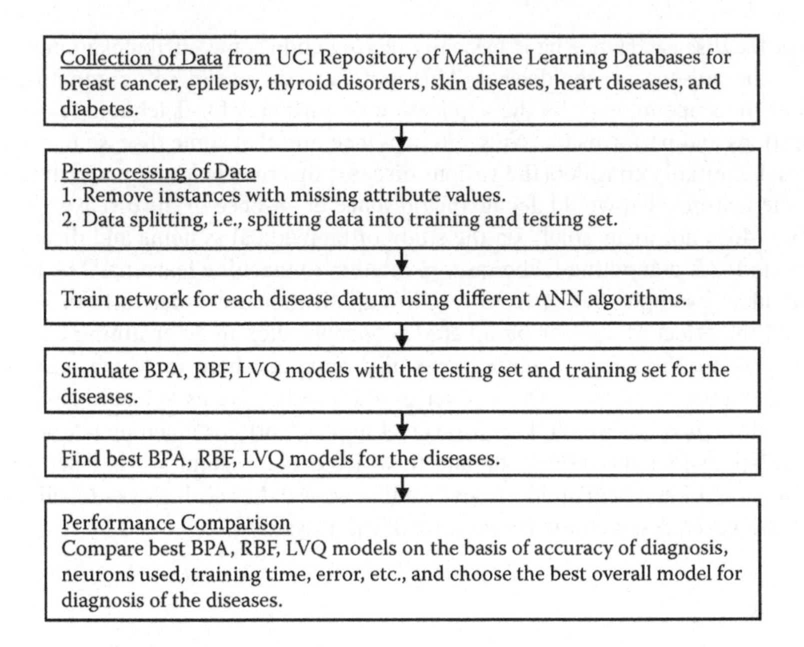

FIGURE 11.1 The overall methodology.

In this chapter we present various approaches of classification, including ANN with BPA, RBFN, and learning vector quantization (LVQ). This helps us understand the system behavior better and in turn helps in understanding the role and mechanism of the various classifiers.

11.1.4 Methodology

In this chapter we adopt an engineer's approach to solving the problem of disease identification. Here we do not go deep into the details of biology, medical systems, and diseases. Rather, we discuss the basic concepts and working of these diseases. We then discuss the features and their extraction. Then these features are used as inputs for the classifier that we explain.

Here we mainly discuss six different diseases and develop systems for them. We first consider breast cancer and then in turn epilepsy, thyroid disorders, skin diseases, heart diseases, and diabetes.

The general methodology adopted is given in Figure 11.1.

11.2 ANN CLASSIFIERS

Here we have used three ANNs for the purpose of classification: the back propagation algorithm (BPA), radial basis function network (RNFN), and learning vector quantization (LVQ). Although we have already discussed the role of each one of these in problem solving, we discuss these again with special reference to the problems in biomedical engineering. We basically try to compare and contrast these techniques from a theoretical point of view. This forms the basis of the experiments that we do later in the text. It may again be recalled that all these problems are classificatory in nature. Hence the entire discussion revolves around classification. To some extent readers may take this discussion as a continuation of our discussion in the use of various models of ANN, especially for classificatory problems. Readers may also take a quick look at our discussions there in order get the right context and background for our discussion in this section.

Biomedical engineering relies a lot on learning from past data. The fact remains that history repeats itself. If you have data recorded for someone, then this might prove useful for diagnosing some other patient. If he presents closely similar features to a patient, he is most probably suffering from the same disease. Hence the data for each and every patient, if learned, can supply lots and lots of information. Also, the larger and more diverse the training data set or the recorded data is, the better the system will perform. Here we are making a valid assumption. The disease and its attributes that we have taken in the system are ranked well according to the laws of soft computing literature.

A natural solution may sometimes arise in the mind of the readers. Why not store each and every datum into some storage and later use this storage primarily as a lookup table? Here there is one major issue. Two patients would never have the same set of attributes. They would always have almost the same set of attributes. Hence we have to keep room for the errors or noise or whatever may be called as per the situation and problem. The lookup table technique would seldom involve this ability.

The solution to this problem that may again arise in the minds of the readers is that by various ways the lookup table can be modified to incorporate this ability to withstand deviations up to a predefined level. Hence this is not necessarily a problem in the system.

We now discuss another problem using this approach. In any real life application, the number of cases may be too large. This would make the database massive in size. The number of test cases contained in it would be very large. Now it requires a large amount of storage to store all these data sets. Further even if we have stored these data sets, it again would require a lot of processing time for the system for every query made. It would have to iterate through all the various data sets in the database and then give the final output. These are both undesirable properties of the system.

This necessitates systems that may store the large volumes of data and also provide scope for errors or noise. The systems need to store this in a lesser volume of space. Also they must ensure that the processing time at the time of testing is as low as possible and as far as possible independent of the number of test cases in the database. This is an essential property that can provide a boost to the biomedical systems.

The ANNs do the same thing to a very large extent. They try to form associations in the inputs in the training data set and form rules that may be able to represent the data in shorter volumes. These rules are independent of the training input size to a good extent. Hence the testing time is finite and small. Here we would study three distinct ways to do this task of formation of rules or representation of the training data by a set of rules. These are discussed in next three subsections.

11.2.1 ANN WITH BPA

The ANN with BPA is one of the most common methods that we have been using throughout this text. The ease of use and design coupled with the great powers of machine learning and generalizing capability of the ANNs are the major reasons for their use in all these problems of machine learning.

The ANN with BPA looks at the data from a very generalized or global point of view. It is on....e of the most generalizing ANN models. It tries to form globally applicable rules. Each of these rules is framed by considering all the possible inputs that exist in the database. This makes the training very slow, requiring a considerable amount of time. But the ANN that comes out as a result of this is highly generalized and best suited to imitate the functionality of any commonly known function.

In case of the classificatory problems, the ANNs try to frame functions that have maxima around the regions that denote the presence of the disease and minima at the regions that denote the absence of the disease. These maxima and minima are spread around a very highly dimensional input space. The ability of the ANN to imitate this otherwise binary function decides the system performance. The performance indices here may point to the ability of the ANN to imitate the actual binary function, in other words, the closeness of its output to the desired output. At the same time, it would be important for the ANN to form as few rules as possible. In the ANN theory, this means that it must have as few neurons as possible. This means that the ANN must correctly identify the maximum possible training set inputs in the least amount of space. This further reduces the testing time requirements as well.

In short, the BPA approach says, as far as possible, form the global rules after considering all the training inputs and tries to form a function that covers the entire region that includes the input space where the class is present.

11.2.2 Radial Basis Function Networks

The RBFNs have quite a different manner of working and solving the same problem of learning. We discussed the mechanism in detail in previous chapters. From that discussion, it was concluded that the RBFNs have a highly localized approach, where the output at a point is decided by the nearby training data sets more than the whole global set of inputs. This reduces the training time to some extent and may sometimes result in higher memory complexities. We even discussed the contrasting nature of the two ANNs—BPA and RBFN.

Here we talk about the same issues and solutions with respect to the issues of machine learning that we introduced. The localized nature of RBFN works very well in the case of classificatory problems where every neuron covers a fixed set of regions that may even include the entire class. Each neuron has its effect reduced in the form of a radial basis function whose radius is a parameter that is adjusted at training time. Ideally the function to be imitated is a binary function that has maxima at all regions where class is found and minima at others. But in classificatory problems, this is not a hard and fast rule. This function may even be replaced by a function whose value is above a threshold at regions where the class is found and below the threshold where the class is not found. Here we advocate the fact that many times even a single neuron has the ability to represent the entire class region by using this property of the output function to be imitated. It would increase its radius to the order of the size of the class presence region. If not, then we can always put in more neurons to imitate the same functionality.

It must be noted that even the ANN with BPA actually is unable to imitate the binary function due to its characteristic nature where the outputs do not change at all over any change in inputs but then suddenly change. The same principle of thresholds applies there as well, except for the fact that it tries to predict the exact shape of this function as closely as possible using multiple neurons.

In short, the RBFN work in a localized manner to predict a function that correctly classifies inputs into outputs according to the threshold values. The localize approach results in faster training.

11.2.3 Learning Vector Quantization

The LVQs also represent a localized approach to problem solving by the use of ANNs. Given how they operate, LVQs may be considered even more localized than RBFNs. We say this because of their same fundamental working method, where every neuron represents a point in the input space. These ANNs further work in a strict classificatory style where the closeness of the input vector is seen to the neurons, making them more localized in nature. Due to their localized nature, LVQs function much like RBFNs with respect to the training and testing times and memory requirements. The major difference between these and the other ANNs is that these networks do not try to predict functions as the other two ANNs discussed do. This is chiefly because these are by design classificatory models that try to return a single class as their output.

In these networks as well, a neuron may itself be able to itself predict a large number of training data sets by actually enclosing a large region in the input space. This is particularly the case when no other neuron lies close to it in the competitive layer. This may be very useful when the intraclass separation is high. The correct placement of these neurons may hence result in a big boost where each neuron identifies its region well and make the correct prediction. For complex places, more neurons may be added as per the functional requirements.

For all these ANNs, we comment in general that they would be able to learn almost any training input data if we keep increasing the number of neurons. This might be very difficult, but still not impossible for the ANNs with BPA. For the RBFNs and the LVQs, this condition will be realized when there are as many neurons as there are training data sets. Hence the resulting system will be more or less a lookup table with a highly localized extent. In ANN with BPA, the function that the ANN tries to predict keeps getting sensitive to inputs and changes rapidly on increasing the number of neurons. A time will come when it will have enough freedom to change and assume very complex shapes and hence classify the problem.

11.2.4 Data Sets

Recall that any recognition system has the tasks of data acquisition, preprocessing, segmentation, feature extraction, and then recognition. Here we do not consider the first few steps. The data in the extracted form for various diseases is already present in the UCI Repository of Machine Learning Databases. This database provides the data collected along with features identified and extracted by various researchers and submitted to the database. From this database we straight away pick up the data for the six diseases that we present here: breast cancer, epilepsy, thyroid disorders, skin diseases, heart diseases, and diabetes.

It may be noted that the data collected in this database are not normalized and vary widely in range. Hence the normalization needs to be carried out before the application of any of the methods. Each data set was split into two parts: a training set and a testing set. It may again be noted that many attributes are missing for many inputs in the database. In this chapter the experiments do not consider this factor. The inputs with missing values are neglected and not used for training or testing purposes.

The engineers very frequently use the UCI Repository as a black box where they apply the methods of machine learning without considering the systems and attributes. To a large extent this approach is justifiable for two reasons: First, it would be very difficult for people to actually collect data, preprocess it, and then extract the features. This would require a lot of understanding of foreign systems, which soft computing engineers may seldom have. The cost and time for collecting the data are again a limitation. Second, this repository has been built and there is very little scope to change anything as per the developer requirements. You cannot re-extract the features. The simple way is to use them directly with as much understanding of the system as possible.

Here we face the same problems. Explaining the symptoms, causes, results, and dependencies of the various diseases would take volumes of text. Besides, proper understanding would require a very high level of understanding of the biological systems. This might not be true for all readers. In fact, this might not be true for any readers.

Next we discuss the various diseases along with the attributes and soft computing systems one by one. Every disease has been given a short description. It is highly likely that readers are unable to understand a large part of the text presented. It is even possible that readers find the text incomplete in terms of clarity of these diseases. These limitations should not greatly affect the usefulness of the presentation.

Again let us recall that the motivation behind this chapter is not only to learn about biomedical systems, but rather to learn about machine learning by using the biomedical systems as examples. In the preceding section we took a theoretical view of the concepts and operation of the various systems. We collect some valuable data from these systems and then continue our discussion in order to fully understand the working of the machine learning systems.

The rest of the chapter briefly discusses the work of Shukla et al. (2009c, 2009d, 2009i, 2009n).

11.3 BREAST CANCER

The first disease that we consider in this system is breast cancer. Cancer is a group of diseases in which cells are aggressive (grow and divide without respect to normal limits), invasive (invade and destroy adjacent tissues), and sometimes metastatic (spread to other locations in the body). Breast cancer is a cancer of the glandular breast tissue. Worldwide, breast cancer is the fifth most common cause of cancer death (after lung, stomach, liver, and colon cancers). In 2005, breast cancer caused 502,000 deaths (7 percent of cancer deaths; almost 1 percent of all deaths) worldwide. Early detection and the precise staging of therapies have contributed to higher success rates in the battle against cancer (Breast Cancer Statistics 2008; Cancer Research 2007; Wikipedia, "Breast Cancer"). It has become the most common health-related search topic among users of the Internet. Previous studies have evaluated use of the Internet by women with breast cancer and the quality of selected sites. Recent surveys show that 40 to 54 percent of patient's access medical information via the Internet and that this information affects their choice of treatment (Funda and Meric 2002). According to a survey among breast cancer survivors, a questionnaire assessing quality of life was sent to 325 breast cancer patients. A 66 percent valid response rate was obtained. Among these responses, 169 women were postmenopausal. More than 50 percent of postmenopausal women suffered from climacteric symptoms such as hot flushes, but few were taking a treatment to alleviate these symptoms (Antoine et al. 2008). The mainstay of breast cancer treatment is surgery when the tumor is localized, with possible adjuvant hormonal therapy (with tamoxifen or an aromatase inhibitor), chemotherapy, and/or radiotherapy.

11.3.1 Data Set of Breast Cancer

The data set consists of 699 instances. Out of these, values of one or more attributes are missing in 16 instances. These 16 instances have been left out while using this data set. So, 683 instances have been used, out of which 483 have been used for training the networks and 200 instances have been used for testing purposes. There are nine attributes per instance plus the class attribute. Each of the instances has to be categorized into either of the two categories: benign or malignant. The class distribution is: Benign: 444 (65 percent) Malignant: 239 (35 percent). Attributes of the data set are: (1) Clump Thickness; (2) Uniformity of Cell Size; (3) Uniformity of Cell Shape; (4) Marginal Adhesion; (5) Single Epithelial Cell Size; (6) Bare Nuclei; (7) Bland Chromatin; (8) Normal Nucleoli; (9) Mitoses. The output is classified into two classes. These are 1 for benign, 2 for malignant.

11.3.2 Results

11.3.2.1 ANN with BPA

The database as obtained for the problem was first trained by the use of ANN with BPA. The ANN consisted of two hidden layers. The learning rate of 0.07 and momentum factor of 0.8 was used. The number of maximum allowable epochs was 24,000. Table 11.1 shows the experimental results of diagnosis using ANN with BPA. From the table, it is clear that the ANN with BPA with 30 hidden

TABLE 11.1
Experimental Results for ANN with BPA for Diagnosis of Breast Cancer

No. of Hidden Neurons	No. of Epochs	Mean Sum-Squared Error (MSE)	Percentage Accuracy of Diagnosis	Remarks:
15 (10 + 5)	8406	0.010	98.5	No. of Hidden Layers = 2, Learning rate = 0.07, Momentum = 0.8, Function in Hidden Layer = Tan-Sigmoid Function in Output Layer = Purelin
20 (10 + 10)	24000	0.016	98.5	
25 (17 + 8)	20404	0.020	97	
30 (18 + 12)	6711	0.010	99.5	

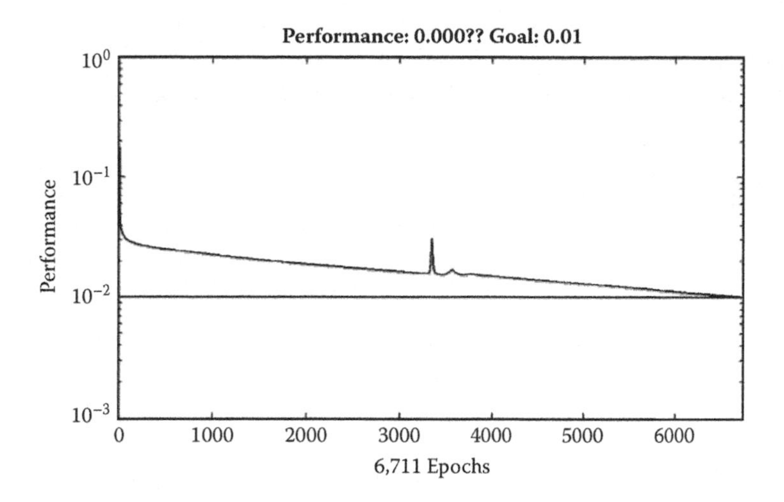

FIGURE 11.2a Training curve of best BPN for diagnosis of breast cancer.

neurons and having a percentage accuracy of diagnosis of 99.5 percent is the best BPA network for diagnosis of breast cancer. Figure 11.2a shows the training curve for the best ANN with BPA for the diagnosis of breast cancer, and Figure 11.2b shows the graphical representation of accuracy of the best ANN for diagnosis of breast cancer. This shows that the diagnosis is correct in 199 out of 200 cases in this case.

11.3.2.2 Radial Basis Function Networks (RBFN)

In Radial Basis Function (RBF) Networks, we need to specify the input values, target values, error goal, and spread value. The performance function named mean sum-squared error (MSE) was used. The value of spread was changed to get different models. This also resulted in changing of number

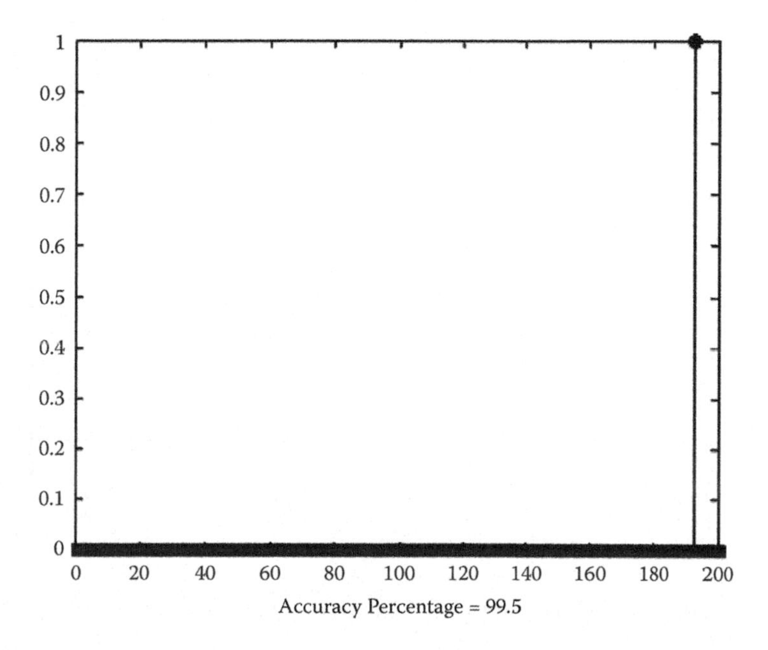

FIGURE 11.2b Percentage accuracy of best ANN with BPA for diagnosis of breast cancer.

TABLE 11.2
Experimental Results for RBFN for Diagnosis of Breast Cancer

Goal	Spread	Radial Basis Neurons	No. of Epochs	MSE	Percentage Accuracy of Diagnosis	Remarks:
0.5	6	200	200	0.660	96.5	Function Used = newrb
0.2	5	225	225	0.190	93	
0.01	4	275	275	0.018	87.5	
0.01	5	300	300	0.008	92.5	
0.01	**6**	**325**	**325**	**0.014**	**98.5**	

of epochs being used by the network. The experimental results of diagnosis using RBF networks are shown in Table 11.2. From this table, it is clear that the RBF network model with spread value of 6, 325 hidden neurons and accuracy of 98.5 percent is the best RBF model or diagnosis of breast cancer. Figure 11.3a shows the training curve for the best RBF network for the diagnosis of breast cancer, and Figure 11.3b shows the graphical representation of accuracy of the best RBF network for diagnosis of breast cancer. This shows that the diagnosis is correct in 197 out of 200 cases in this case.

11.3.2.3 LVQ Network

In our experiment, different LVQ models were obtained by changing the number of hidden neurons. The Kohonen's learning rate was set to 0.05. The performance function used was MSE. Table 11.3 shows the experimental results for LVQ networks, which differ from each other on the basis of the number of hidden neurons used. The network with 15 hidden neurons and an accuracy of 97 percent is the best LVQ network. Figure 11.4a shows the training curve for the best LVQ network for the diagnosis of breast cancer and Figure 11.4b shows the graphical representation of accuracy of the

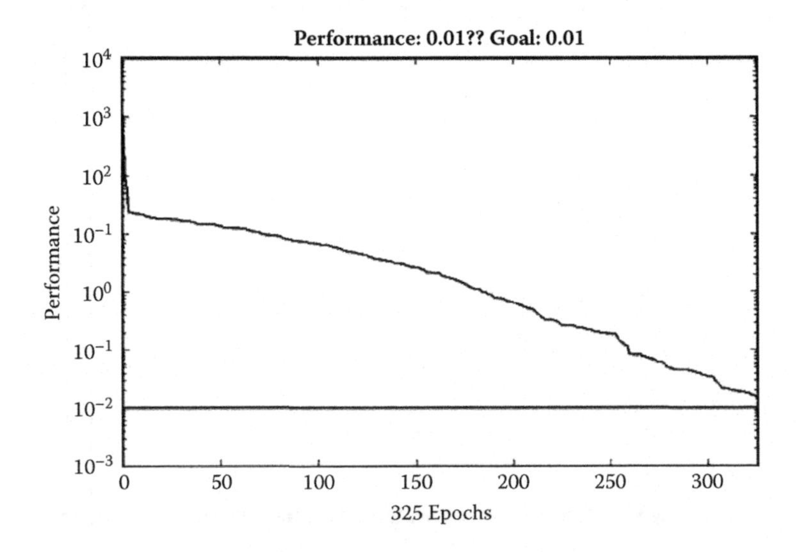

FIGURE 11.3a Training curve of the best RBFN for diagnosis of breast cancer.

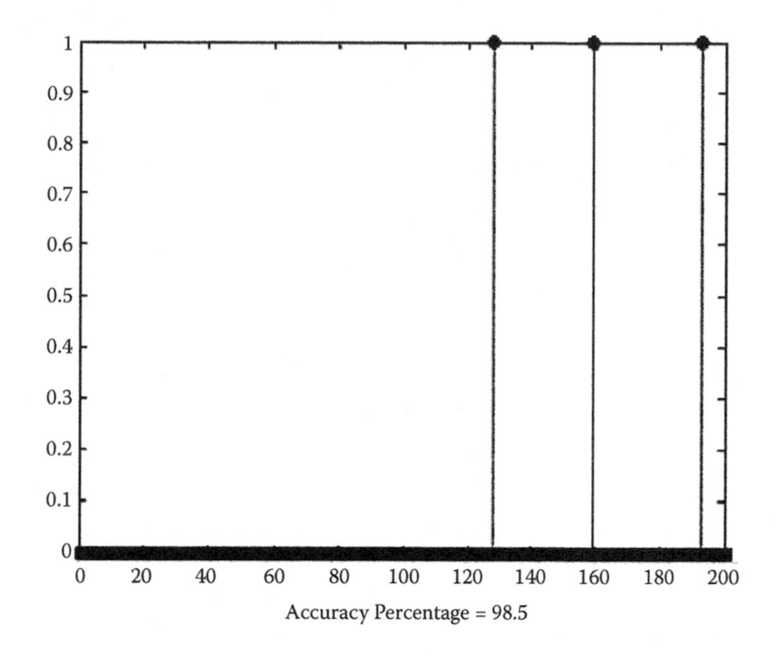

FIGURE 11.3b Percentage accuracy of the best RBFN for diagnosis of breast cancer.

best LVQ network for diagnosis of breast cancer. This shows that the diagnosis is correct in 191 out of 200 cases in this case.

11.3.2.4 Performance Comparison

Table 11.4 shows the performance comparison of ANN models to find the best diagnostic system for breast cancer. Networks trained with BPA were the best in classification accuracy when tested on a data set other than the one on which they were trained and produced an accuracy of 99.5 percent. Neither of the two network paradigms under consideration were superior in terms of classification accuracy to feed forward perceptrons trained with back propagation; however, some interesting characteristics were observed. RBFN produces the second best accuracy on testing data set. RBFN takes the least training time out of the three networks, which is 38.75 seconds. The RBFN came very close to matching the accuracy of BPA with a much shorter training time. The LVQ network was less accurate than ANN with BPA and RBFN and took the largest time for training.

TABLE 11.3
Experimental Results for LVQ Networks for Diagnosis of Breast Cancer

No. of Hidden Neurons	No. of epochs	MSE	Percentage Accuracy of Diagnosis	Remarks:
15	**300**	**0.030**	**97**	Learning rate = 0.05,
20	300	0.037	97	Learning Function =
25	300	0.032	96.5	learnlv1
30	300	0.031	96.5	

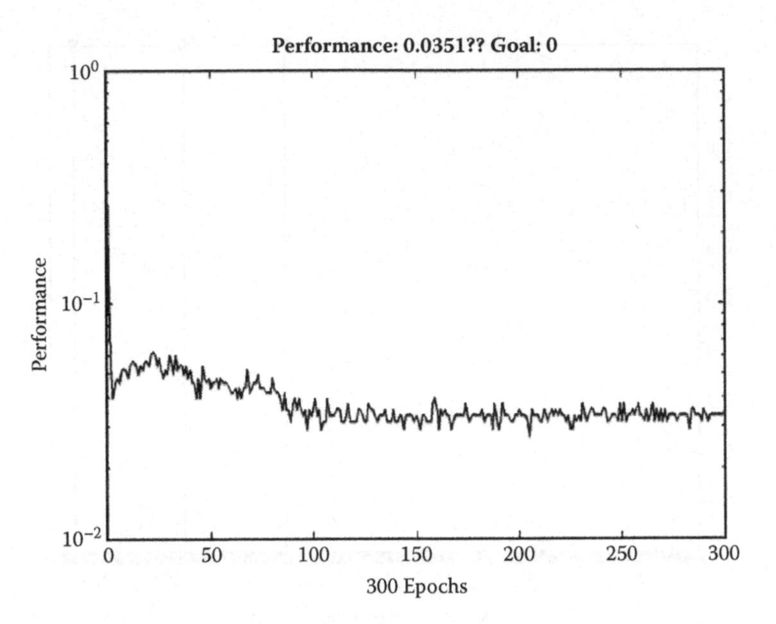

FIGURE 11.4a Training curve of the best LVQN for diagnosis of breast cancer.

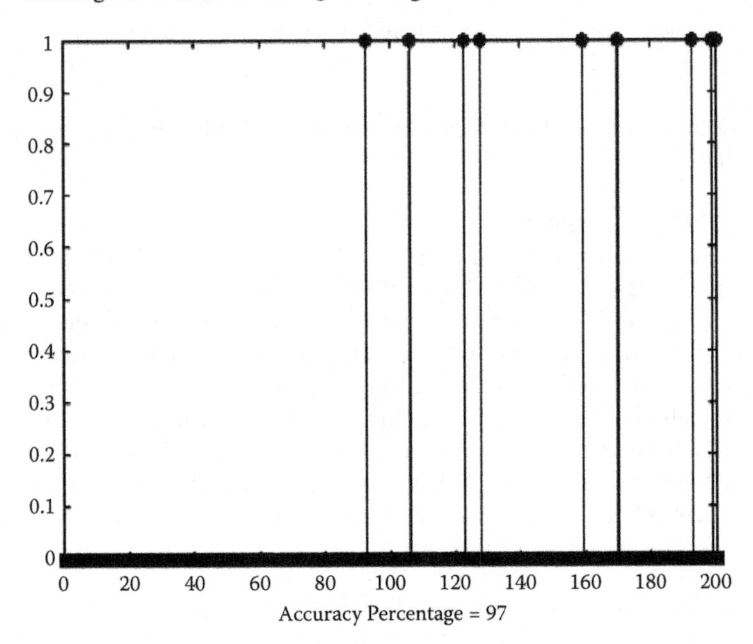

FIGURE 11.4b Percentage accuracy of the best LVQN for diagnosis of breast cancer.

TABLE 11.4
Performance Comparison to Find the Best Diagnostic System for Breast Cancer

Network	Percentage Accuracy of Diagnosis	Training Time (sec.)	MSE
ANN with BPA	**99.5**	**64.50**	**0.010**
RBFN	98.5	38.75	0.014
LVQ	97	155.78	0.030

11.4 EPILEPSY

The second problem that we take to build a soft computing system is epilepsy. Epilepsy is a common chronic neurological disorder that is characterized by recurrent unprovoked seizures. About 50 million people worldwide have epilepsy at any one time. Epilepsy is usually controlled, but not cured, with medication, although surgery may be considered in difficult cases. Epilepsy's approximate annual incidence rate is 40 to 70 per 100,000 in industrialized countries and 100 to 190 per 100,000 in resource-poor countries; socioeconomically deprived people are at higher risk. In industrialized countries the incidence rate decreased in children but increased among the elderly during the three decades prior to 2003 (Wikipedia, "Epilepsy"). Information about existing resources available within countries to tackle the huge medical, social, and economic burden caused by epilepsy is lacking. To fill this information gap, a survey of country resources available for epilepsy care was conducted within the framework of the ILAE/IBE/WHO Global Campaign Against Epilepsy. Data were collected from 160 countries representing 97.5 percent of the world population. The data reinforce the need for urgent, substantial, and systematic action to enhance resources for epilepsy care, especially in low-income countries (Dua and Tarun 2006). Epilepsy has serious social and economic consequences, too. People with epilepsy continually face social stigma and exclusion. The cost and burden of epilepsy varies between countries. In 1990, WHO identified that, on average, the cost of the anti-epileptic drug phenobarbitone could be as low as US$ 5 per person per annum (WHO 2009).

11.4.1 Data Set for Epilepsy

The nine input values are derived from a 1/4-second "SPIKE" event that is recorded on a single channel of an EEG monitor. A team of neurologists were asked to classify each spike as an epileptic event (True or False). The file contains 100 TRUE values and 165 FALSE values; total 265 patterns each having nine attributes. The patterns are in random order. First, all the data has been normalized so that the value of every attribute is between 0 and 1. Out of 265 instances, 200 instances have been used for training the system and 65 have been used for testing purposes. Class 1 of output signifies FALSE (not epileptic) and Class 2 signifies TRUE (epileptic).

11.4.2 Results

11.4.2.1 ANN with BPA

After the ANN with BPA was created, just as in the case of breast cancer, the values of momentum, learning rate, and number of hidden neurons were changed to get different BPA models and choose the best one. The number of maximum allowable epochs was 35,000. Table 11.5 shows the

TABLE 11.5
Experimental Results for ANN with BPA for Diagnosis of Epilepsy

No. of Hidden Neurons	Momentum	Learning Rate	No. of Epochs	Training Time (sec.)	Percentage Accuracy of Diagnosis	Remarks:
28 (16 + 12)	0.7	0.06	8869	39.47	86.2	No. of Hidden Layers = 2, Function in Hidden Layer = Tan-Sigmoid, Function in Output Layer = Purelin, MSE = 0.01
30 (20 + 10)	0.8	0.06	6546	37.05	93.85	
30 (18 + 12)	0.7	0.06	8044	37.86	93.85	
30 (20 + 10)	**0.8**	**0.07**	**16632**	**89.14**	**95.38**	
32 (22 + 10)	0.7	0.06	6372	35.36	90.77	
32 (22 + 10)	0.7	0.06	19215	105.89	90.31	

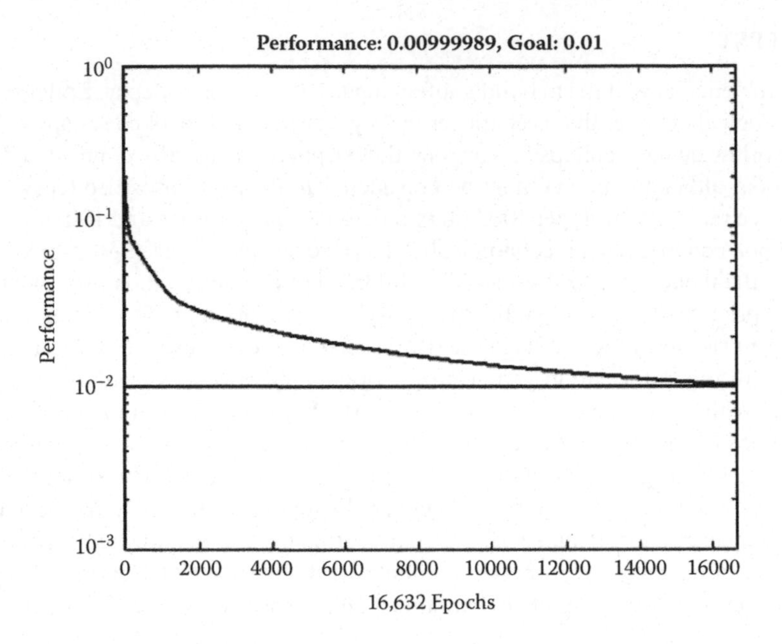

FIGURE 11.5a Training curve of the best ANN with BPA for diagnosis of epilepsy.

experimental results of diagnosis using ANN with BPA. From the table, it is clear that the ANN with 30 hidden neurons and having a percentage accuracy of diagnosis of 95.38 percent is the best ANN for diagnosis of epilepsy. Figure 11.5a shows the training curve for the best ANN for the diagnosis of epilepsy, and Figure 11.5b shows the graphical representation of accuracy of the best ANN network for diagnosis of epilepsy. This shows that the diagnosis is correct in 67 out of 70 cases in this case of epilepsy.

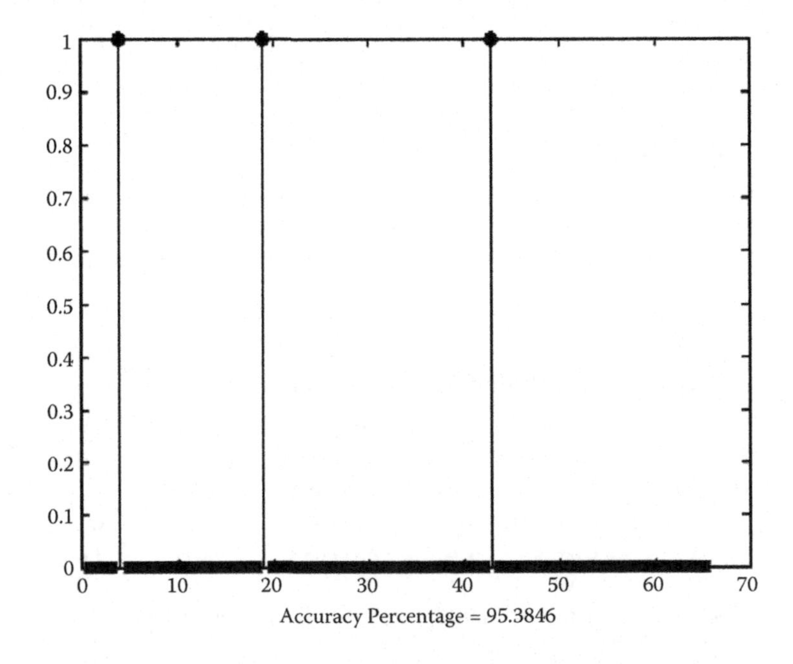

FIGURE 11.5b Percentage accuracy of best BPN for diagnosis of epilepsy.

TABLE 11.6
Experimental Results for RBF Networks for Diagnosis of Epilepsy

Spread	No. of Epochs	MSE	Training Time (sec.)	Percentage Accuracy of Diagnosis	Remarks:
2.5	150	0.03	4.2	86.20	Goal = 0.01, Radial
4.9	150	0.11	6.08	80.00	Basis Neurons = 150,
5.2	**150**	**0.02**	**5.97**	**87.69**	Function Used =
6	150	0.025	5.53	76.92	newrb

11.4.2.2 RBFN

After creating the Radial Basis Function Networks, just as in the preceding disease, the value of spread was changed to get different models. The experimental results of diagnosis using RBFN are shown in Table 11.6. From this table, it is clear that the RBFN model with spread value of 5.2, number of hidden neurons equal to 150 and accuracy of 87.69 percent is the best RBF model or diagnosis of epilepsy. Figure 11.6a shows the training curve for the best RBFN for the diagnosis of epilepsy, and Figure 11.6b shows the graphical representation of accuracy of the best RBFN for diagnosis of epilepsy. This shows that the diagnosis is correct in 62 out of 70 cases in case of epilepsy.

11.4.2.3 LVQ Network

The Kohonen's learning rate was set to 0.06, and the performance function used was MSE. Table 11.7 shows the experimental results for LVQ networks, which differ from each other on the basis of number

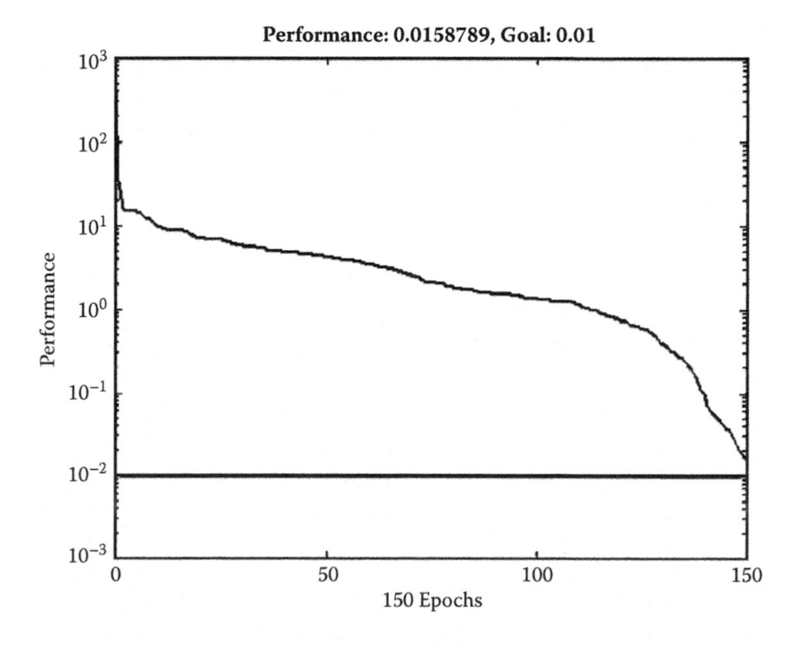

FIGURE 11.6a Training curve of the best RBFN for diagnosis of epilepsy.

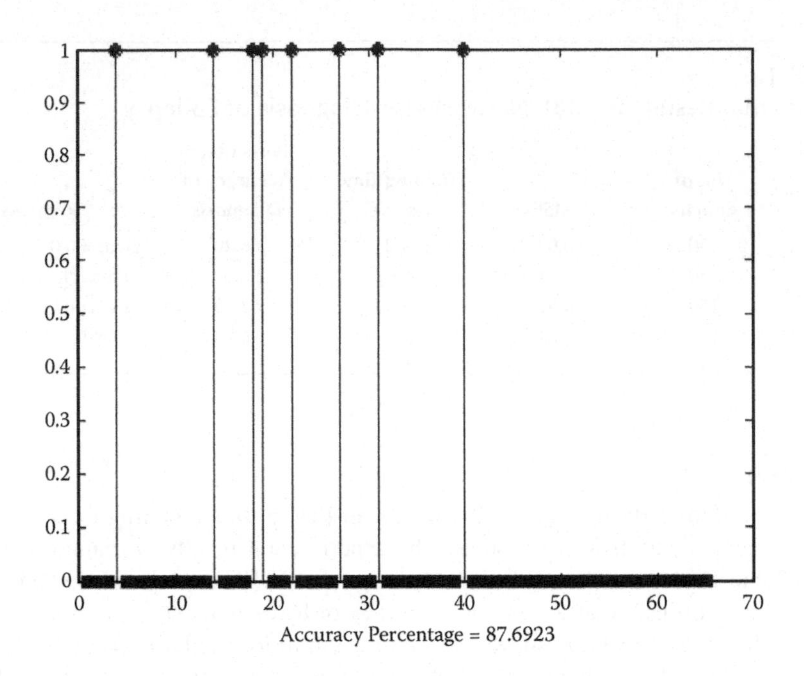

FIGURE 11.6b Percentage accuracy of best RBFN for diagnosis of epilepsy.

of hidden neurons used. The network with 10 hidden neurons and an accuracy of 95.38 percent is the best LVQ network. Figure 11.7a shows the training curve for the best LVQ network for the diagnosis of epilepsy, and Figure 11.7b shows the graphical representation of accuracy of the best LVQ network for diagnosis of epilepsy. This shows that the diagnosis is correct in 67 out of 70 instances in case of epilepsy.

11.4.2.4 Performance comparison

Table 11.8 shows the performance comparison of the three ANNs to find the best diagnostic system for epilepsy. For the diagnosis of epilepsy using the networks trained with the specified data, BPA and LVQ produce the same accuracy of diagnosis, but LVQ takes much less time in training than BPA to produce the same results. Hence is better in this case.

TABLE 11.7
Experimental Results for LVQ Networks for Diagnosis of Epilepsy

No. of Hidden Neurons	No. of Epochs	MSE	Training Time (sec.)	Percentage Accuracy of Diagnosis	Remarks:
10	**300**	**0.05**	**65.13**	**95.38**	Learning rate = 0.06, Learning Function = learnlv1
13	300	0.04	65.72	90.77	
15	300	0.04	65.70	90.77	
18	300	0.35	65.38	92.31	
20	300	0.04	68.56	92.31	

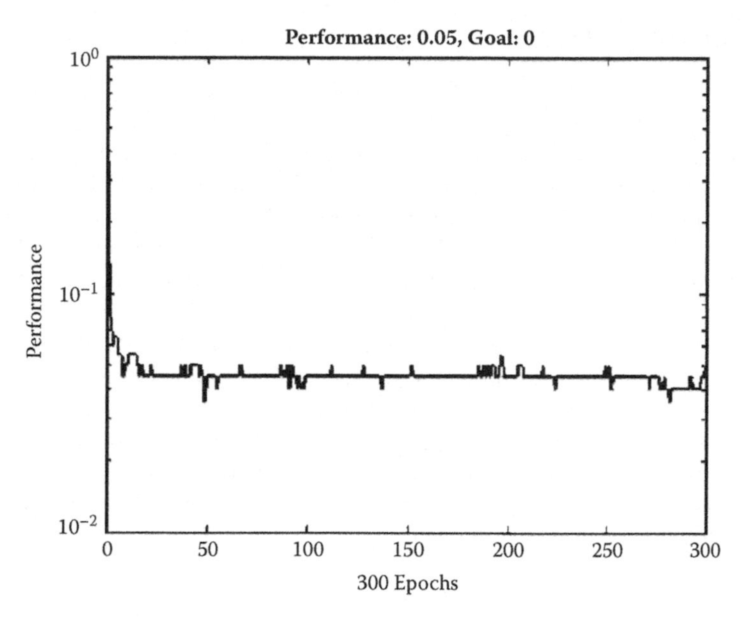

FIGURE 11.7a Training curve of the best LVQN for diagnosis of epilepsy.

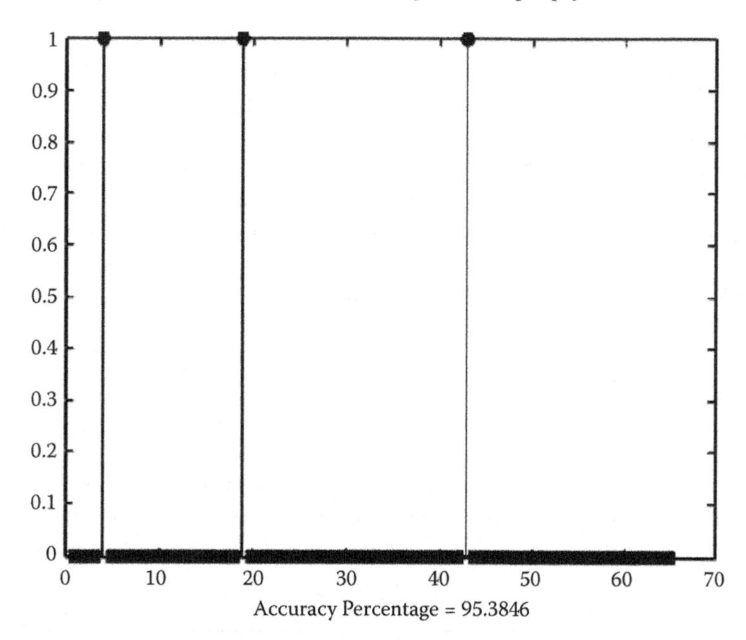

FIGURE 11.7b Percentage accuracy of best LVQN for diagnosis of epilepsy.

TABLE 11.8
Performance Comparison to Find the Best Diagnostic System for Epilepsy

Network	Percentage Accuracy of Diagnosis	Training Time (sec.)	MSE
ANN with BPA	95.38	89.14	0.01
RBF	87.69	5.97	0.02
LVQ	**95.38**	**65.13**	**0.05**

11.5 THYROID DISORDERS

The third disease under study is thyroid. The thyroid is one of the largest endocrine glands in the body. The thyroid controls how quickly the body burns energy and makes proteins, and how sensitive the body should be to other hormones. Because thyroid hormone affects growth, development, and many cellular processes, inadequate thyroid hormone has widespread consequences for the body. It is more common in women than men. Prevalence of thyroid disorders is more than 23 percent and undiagnosed prevalence rate is approximately 1 in 20 (Nussey and Whitehead 2001, Wikipedia, "Thyroid"). A survey of physicians' practice relating to radioiodine administration for hyperthyroidism was carried out in the UK over 15 years ago and showed wide variations in patient management. This led to the development of national guidelines for the use of radioiodine in hyperthyroidism (Vaidya el al. 2008). Researchers studied the relationship between thyroid disorders and glaucoma using a nationally representative sample of 12,376 adults in the U.S. who participated in the 2002 National Health Interview Survey. About 4.6 percent of respondents indicated that they had glaucoma, and 11.9 percent indicated that they had had a thyroid disorder (Cross 2008). Thyroid disorders could boost the risk of developing the eye disease glaucoma, according to research out today (Reinberg 2008).

11.5.1 Data Set for Thyroid Disorders

From the thyroid gland data set, 187 instances have been used for this work. Each instance has five attributes plus the class attribute. The five attributes are: (1) T3-resin uptake test (a percentage); (2) Total serum thyroxin as measured by the isotopic displacement method; (3) Total serum triiodothyronine as measured by radioimmuno assay; (4) Basal thyroid-stimulating hormone (TSH) measured by radioimmuno assay; (5) Maximal absolute difference of TSH value after injection of 200 micrograms of thyrotropin-releasing hormone as compared to the basal value. All attributes are continuous. Each of the instances has to be categorized into one of the three classes: (1) Class 1: normal; (2) Class 2: hyper; (3) Class 3: hypofunctioning. Out of 187 instances, 137 have been used for training and 50 have been used for testing purposes.

11.5.2 Results

11.5.2.1 ANN with BPA

The value of momentum, learning rate, and number of hidden neurons were changed to get different ANN models and choose the best one. The number of maximum allowable epochs was 35,000. Table 11.9 shows the experimental results of diagnosis using ANN with BPA. In this table,

TABLE 11.9
Experimental Results for ANN with BPA for Diagnosis of Thyroid Disorders

No. of Hidden Neurons	Momentum	Learning Rate	No. of Epochs	MSE	Training Time (sec.)	Percentage Accuracy of Diagnosis
25 (20 + 5)	0.8	0.07	13092	0.01	48.89	86
30 (25 + 5)	0.8	0.07	19437	0.01	84.91	88
34 (28 + 6)	0.8	0.08	20975	0.01	91.28	86
40 (35 + 5)	**0.95**	**0.08**	**7802**	**0.01**	**38.25**	**92**
46 (40 + 6)	0.95	0.08	33927	0.01	174.94	90

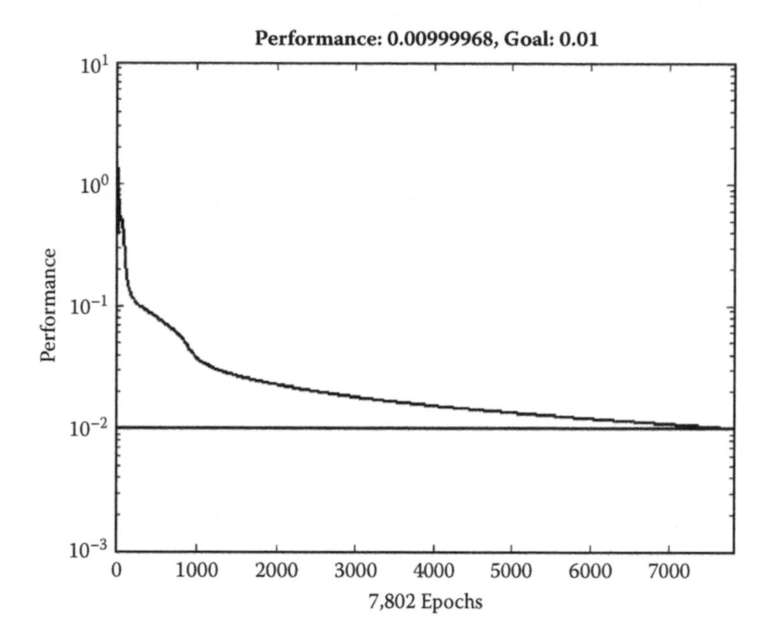

FIGURE 11.8a Training of best ANN with BPA for diagnosis of thyroid disorders.

experimental results for ANN are shown. From the table, it is clear that the ANN with 40 hidden neurons and having a percentage accuracy of diagnosis of 92 percent is the best ANN for diagnosis of thyroid disorders. Figure 11.8a shows the training curve for the best ANN for the diagnosis of thyroid disorders, and Figure 11.8b shows the graphical representation of accuracy of the best ANN for diagnosis of thyroid disorders. This shows that the diagnosis is correct in 46 out of 50 instances in this case of thyroid disorders.

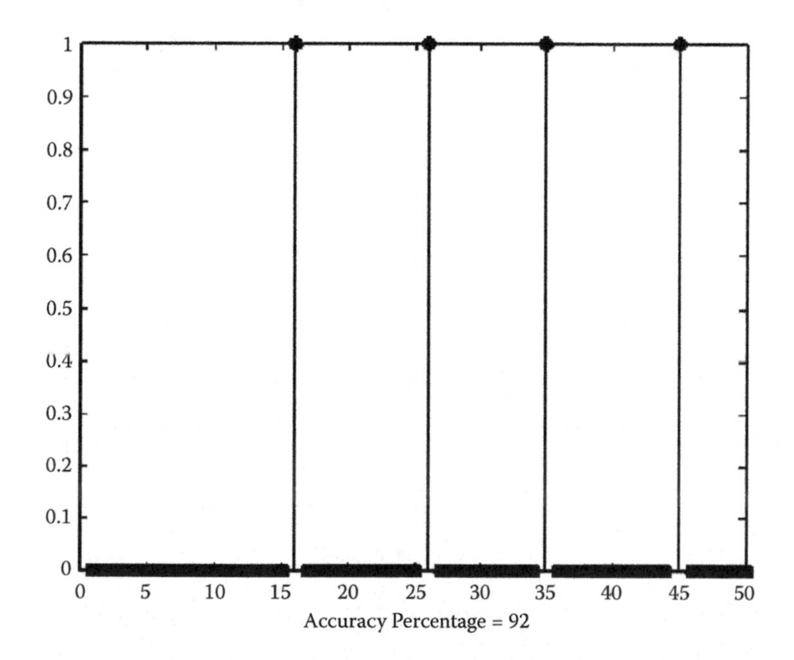

FIGURE 11.8b Percentage accuracy of the best BPN for diagnosis of thyroid disorders.

TABLE 11.10
Experimental Results for RBFN for Diagnosis of Thyroid Disorders

Spread	Radial Basis Neurons	No. of Epochs	MSE	Training Time (sec.)	Percentage Accuracy of Diagnosis	Remarks:
0.13	75	75	0.13	1.02	76	Goal = 0.01
0.15	**100**	**100**	**0.014**	**1.00**	**80**	
0.25	100	100	0.012	1.08	68	
0.3	100	100	0.03	1.00	70	

11.5.2.2 RBFN

The experimental results of diagnosis using RBFN are shown in Table 11.10. From this table, it is clear that the RBFN model with a spread value of 0.15, a number of hidden neurons equal to 150, and an accuracy of 80 percent is the best RBFN model or diagnosis of thyroid disorders. Figure 11.9a shows the training curve for the best RBFN for the diagnosis of thyroid disorders, and Figure 11.9b shows the graphical representation of accuracy of the best RBFN for diagnosis of thyroid disorders. This shows that the diagnosis is correct in 40 out of 50 instances in case of thyroid disorders.

11.5.2.3 LVQ Network

The Kohonen's learning rate was set to 0.09, and the performance function used was MSE. Table 11.11 shows the experimental results for LVQ networks. The network with 28 hidden neurons and an accuracy of 98 percent is the best LVQ network. Figure 11.10a shows the

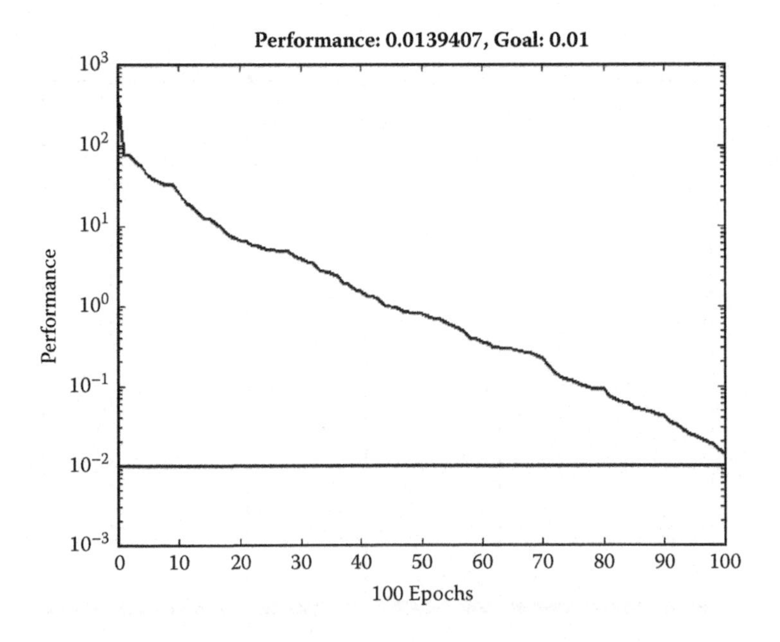

FIGURE 11.9a Training curve of the best RBFN for diagnosis of thyroid disorders.

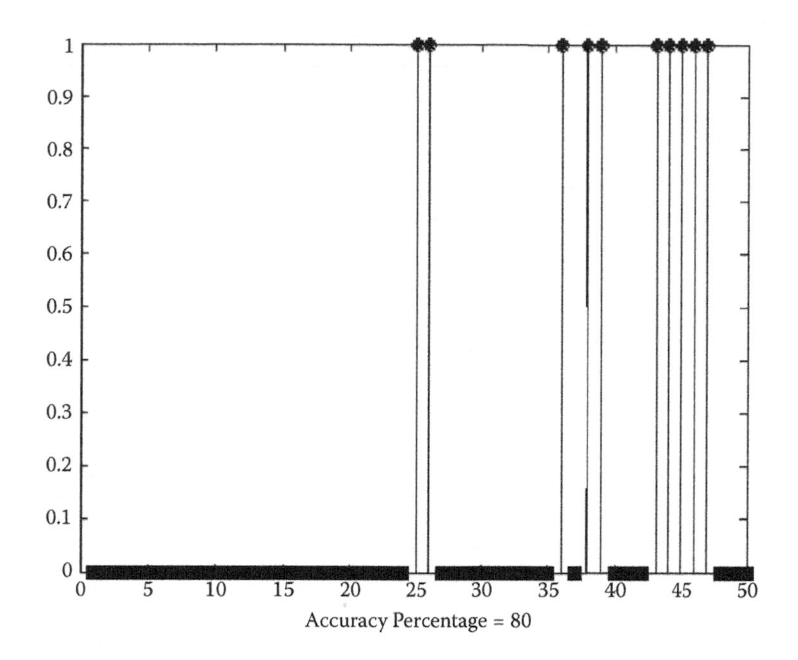

FIGURE 11.9b Percentage accuracy of the best RBFN for diagnosis of thyroid disorders.

training curve for the best LVQ network for the diagnosis of thyroid disorders, and Figure 11.10b shows the graphical representation of accuracy of the best LVQ network for diagnosis of thyroid disorders. This shows that the diagnosis is correct in 49 out of 50 instances in case of thyroid disorders.

11.5.2.4 Performance Comparison

Table 11.12 shows the performance comparison of the three ANNs to find the best diagnostic system for diagnosis of thyroid disorders. From the comparison of three networks, it is clear that for diagnosis of thyroid diseases using a system trained on the specified data, LVQ networks produce the best results in terms of accuracy. Hence LVQ neural networks are best in diagnosis of the thyroid problem in this case.

TABLE 11.11
Experimental Results for LVQ Networks for Diagnosis of Thyroid Disorders

No. of Hidden Neurons	No. of Epochs	MSE	Training Time (sec.)	Percentage Accuracy of Diagnosis	Remarks:
20	400	0.08	62.13	80	Learning rate =
25	400	0.07	61.88	82	0.09, Learning
30	400	0.07	61.11	82	Function =
34	400	0.07	64.45	82	learnlv1
28	**400**	**0.03**	**60.75**	**98**	

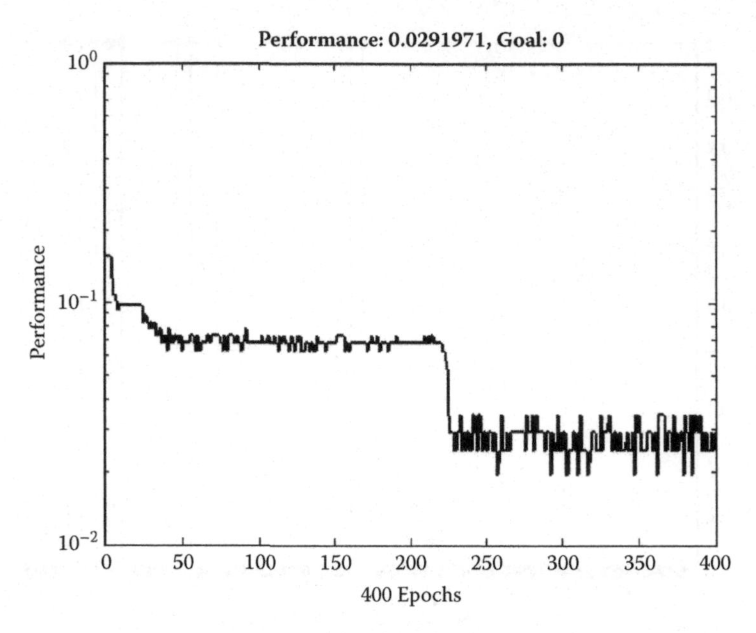

FIGURE 11.10a Training curve for the best LVQN for diagnosis of thyroid disorders.

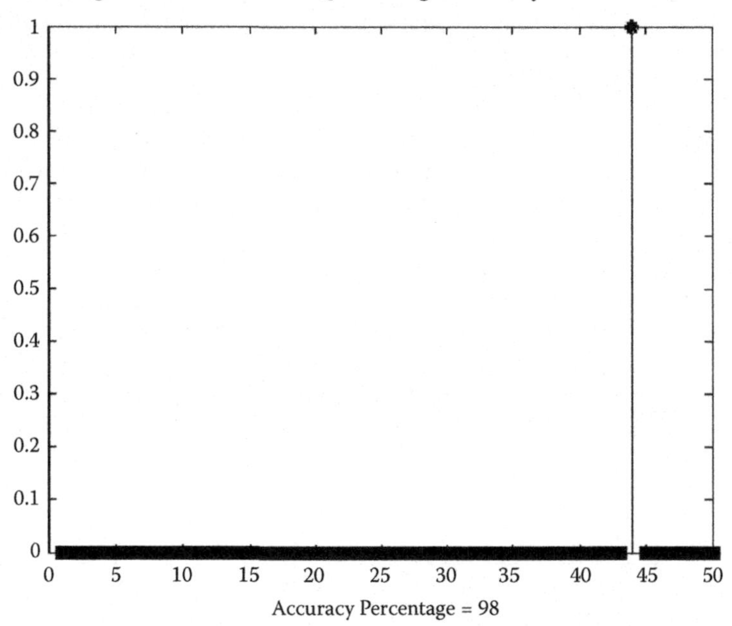

FIGURE 11.10b Percentage accuracy of best LVQN for diagnosis of thyroid disorders.

TABLE 11.12
Performance Comparison to Find the Best Diagnostic System for Thyroid Disorders

Network	Percentage Accuracy of Diagnosis	Training Time (sec.)	MSE
ANN with BPA	92	38.25	0.01
RBF	80	1.00	0.014
LVQ	**98**	**60.75**	**0.030**

11.6 SKIN DISEASES

Next we study the skin diseases. Skin diseases, or dermatoses, include skin cancer, eczema, psoriasis, acne, impetigo, scabies, sunburn, warts, skin tags, Fifth Disease, tinea, herpes, ulcers, and pruritus (Jobling 2007; Lofholm 2000; Schwartz, Janusz, and Janniger 2006). According to Bulletin of the World Health Organization, skin diseases are much more common in developing countries than in the developed world. Surveys have shown that up to 60 percent of people in both rural and urban areas in developing countries suffer from skin diseases. By contrast, one study in the United Kingdom estimated that 28 percent of people had a treatable skin disease. In contrast to the situation in developed countries, malignant melanoma and non-melanoma skin cancers are rare in the indigenous populations of most developing countries, and among those with pigmented skins in general. The exception is the very high risk of skin cancers in albinos (Kingman 2005). According to a survey of skin-related issues in Arab Americans, the most common self-reported skin conditions were acne, eczema/dermatitis, warts, fungal skin infections, and melasma. Significant associations exist between socioeconomic status and having seen a dermatologist. Attitudes surrounding skin perception were related to the number of years of residence in the U.S. Skin conditions and other related issues that affect Arab Americans are similar to those which affect other skin-of-color populations (EI-Essawi el al. 2007). According to a report on Occupational Skin Disorders by SHARP (Safety & Health Assessment & Research for Prevention), Washington, 13 percent of the population was suffering from skin disorders. Industries with high dermatitis claims reporting include Agriculture and Production Crops, Health Services, and Construction Special Trade Contractors (SHARP 1998).

11.6.1 Data Set for Skin Diseases

The data set contains 34 attributes, 33 of which are linear valued and one of which is nominal. In the data set, the number of instances is 366 and missing attribute values are 8 (in the Age attribute); the actual number of instances used is hence 358, and number of attributes is 34. Attribute information is given in Table 11.13. In the model, the number of instances in the training data set is 278, and number of instances in the testing set is 80.

11.6.2 Results

11.6.2.1 ANN with BPA

The experimental results of diagnosis using ANN with BPA are shown in Table 11.14. From this table, it is clear that the ANN model with 25 hidden neurons and accuracy of 96.25 percent is the

TABLE 11.13
Complete Attribute Documentation for Skin Diseases

Clinical Attributes	Histopathological Attributes
Erythema, scaling, definite borders, itching, koebner phenomenon, polygonal papules, follicular papules, oral mucosal involvement, knee and elbow involvement, scalp involvement, family history (0 or 1), Age (linear).	Melanin incontinence, eosinophils in the infiltrate, PNL infiltrate, fibrosis of the papillary dermis, exocytosis, acanthosis, hyperkeratosis, parakeratosis, clubbing of the rete ridges, elongation of the rete ridges, thinning of the suprapapillary epidermis, spongiform pustule, munro microabcess, focal hypergranulosis, disappearance of the granular layer, vacuolisation and damage of basal layer, spongiosis, saw-tooth appearance of retes, follicular horn plug, perifollicular parakeratosis, inflammatory monoluclear infiltrate, band-like infiltrate

TABLE 11.14
Training Parameters and Efficiency for Skin Diseases Using ANN with BPA

No. of Hidden Neurons	Learning Rate	No. of Epochs	MSE	Training Time (sec.)	Percentage Accuracy of Diagnosis	Remarks:
22 (20 + 2)	0.07	6936	0.01	41.11	91.25	Function in Hidden Layer 1 = Tan-Sigmoid, Function in Hidden Layer 2 = Log-Sigmoid, Function in Output Layer = Purelin, Momentum = 0.8.
25 (20 + 5)	0.06	8593	0.01	53.16	93.75	
25 (20 + 5)	0.08	3938	0.01	23.41	95.00	
25 (20 + 5)	**0.07**	**8540**	**0.01**	**52.02**	**96.25**	
30 (20 + 10)	0.07	6435	0.01	41.25	95.00	

best ANN model for diagnosis of skin disorders. Figure 11.11a shows the training curve for the best ANN. Figure 11.11b shows the percentage accuracy of diagnosis using the best ANN.

11.6.2.2 RBFN

In our experiment, the Radial Basis Function Network has an error goal of 0.01. The performance function MSE was used. The value of spread was changed to get different models. This also resulted in a change in the number of epochs being used by the network. The experimental results of diagnosis using RBFN are shown in Table 11.15. From this table, it is clear that the RBFN model with spread value 1.8 and accuracy of 82.50 percent is the best RBFN model for diagnosis of skin disorders. Figure 11.12a shows the training curve for the best RBFN, and Figure 11.12b shows the percentage accuracy of diagnosis using the best RBFN.

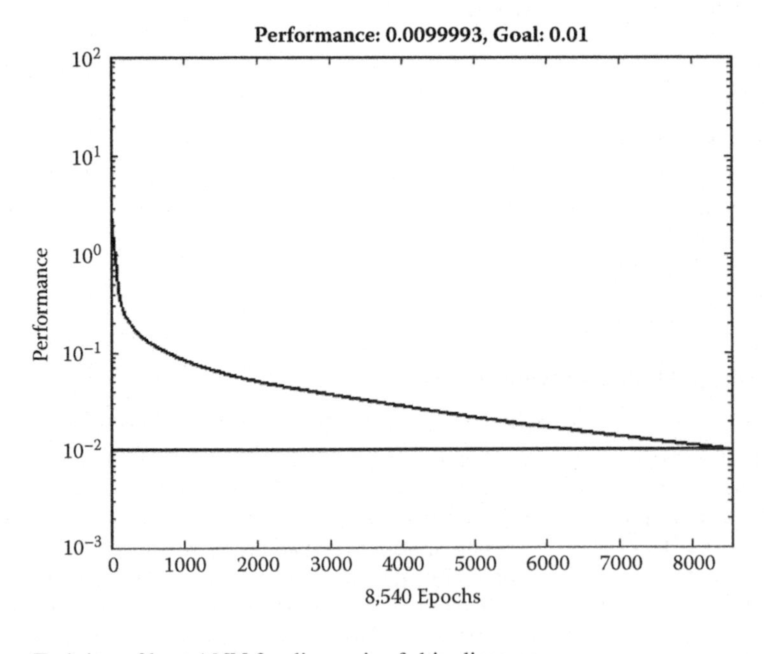

FIGURE 11.11a Training of best ANN for diagnosis of skin diseases.

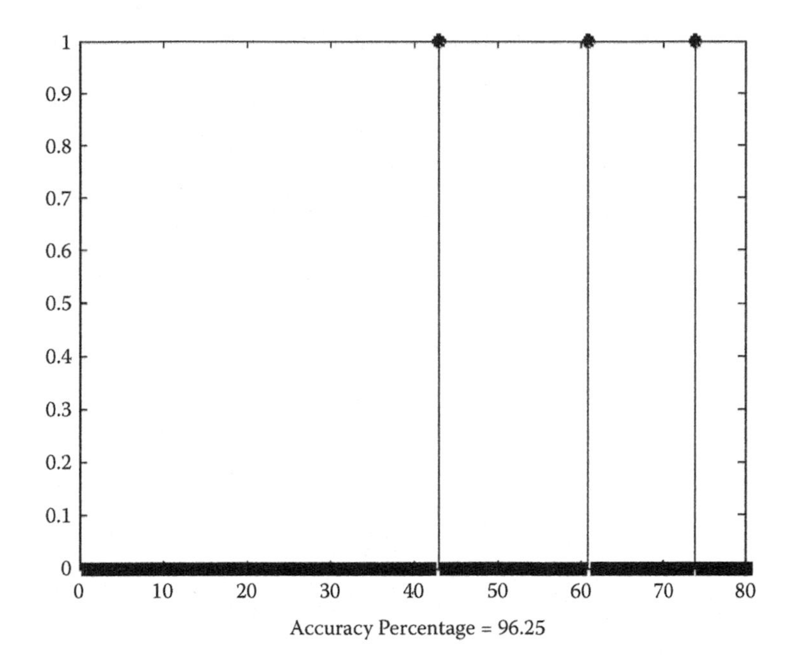

FIGURE 11.11b Percentage accuracy of the best ANN for diagnosis of skin diseases.

11.6.2.3 Diagnosis Using LVQ Network

Table 11.16 shows the experimental results for LVQ networks, which differ from each other on the basis of number of hidden neurons used. The network with 25 hidden neurons and an accuracy of 97.5 percent is the best LVQ network. Figure 11.13a shows the training curve for the best LVQ network. Figure 11.13b shows the percentage accuracy of diagnosis using the best LVQ network.

11.6.2.4 Performance Comparison

Table 11.17 shows the comparison of best results (for an optimized model) for three networks. On the basis of the experimental observations of three neural networks, LVQN gives the highest accuracy, 97.5 percent, for diagnosis of skin diseases.

TABLE 11.15
Training Parameters and Efficiency for Skin Diseases Using RBFN

Spread	No. of Epochs	MSE	Training Time (sec.)	Percentage Accuracy of Diagnosis	Remarks:
1.5	250	0.020	16.55	75.00	Radial Basis
1.7	250	0.030	16.88	81.25	Neurons = 250,
1.8	**250**	**0.025**	**17.59**	**82.50**	Goal = 0.01.
1.9	250	0.020	15.81	80.00	
2.0	250	0.030	17.95	76.25	
5.0	250	0.300	22.31	36.25	

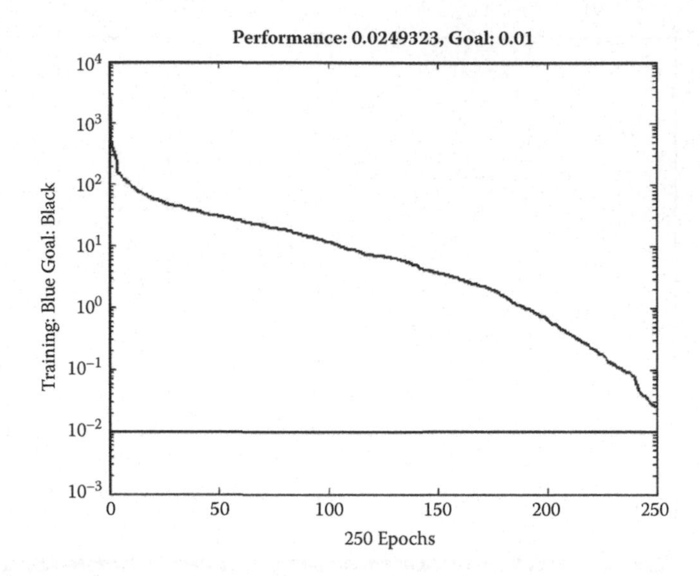

FIGURE 11.12a Training curve of the best RBFN for diagnosis of skin diseases.

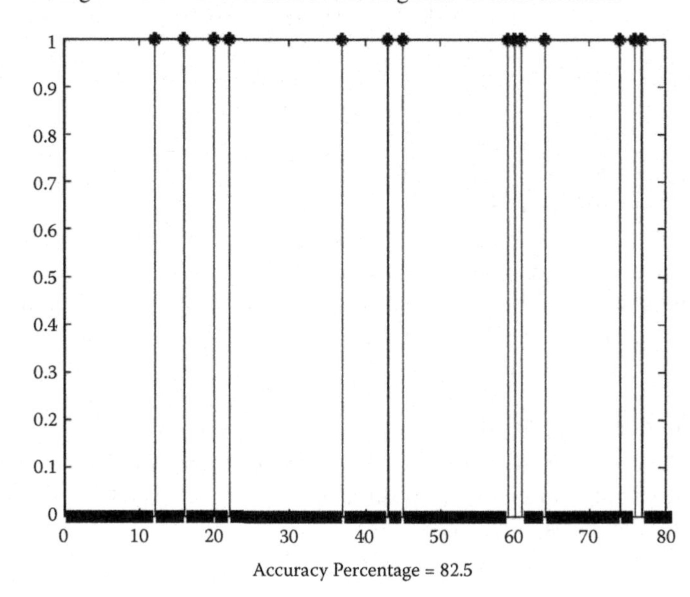

FIGURE 11.12b Percentage accuracy of the best RBFN for diagnosis of skin diseases.

TABLE 11.16
Training Parameters and Efficiency for Skin Diseases Using LVQN

No. of Hidden Neurons	Learning Rate	No. of Epochs	MSE	Training Time (sec.)	Percentage Accuracy of Diagnosis	Remarks:
10	0.06	400	0.041	174.41	82.5	Learning
10	0.08	400	0.041	172.27	82.5	Function =
15	0.06	400	0.041	179.38	82.5	learnlv1
20	0.08	400	0.042	191.52	82.5	
25	**0.09**	**400**	**0.009**	**194.05**	**97.5**	

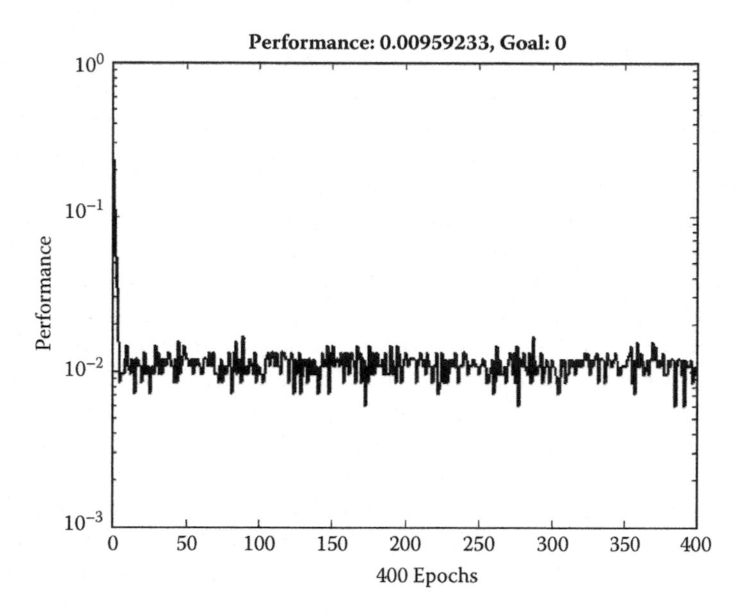

FIGURE 11.13a Training curve for the best LVQN for diagnosis of skin diseases.

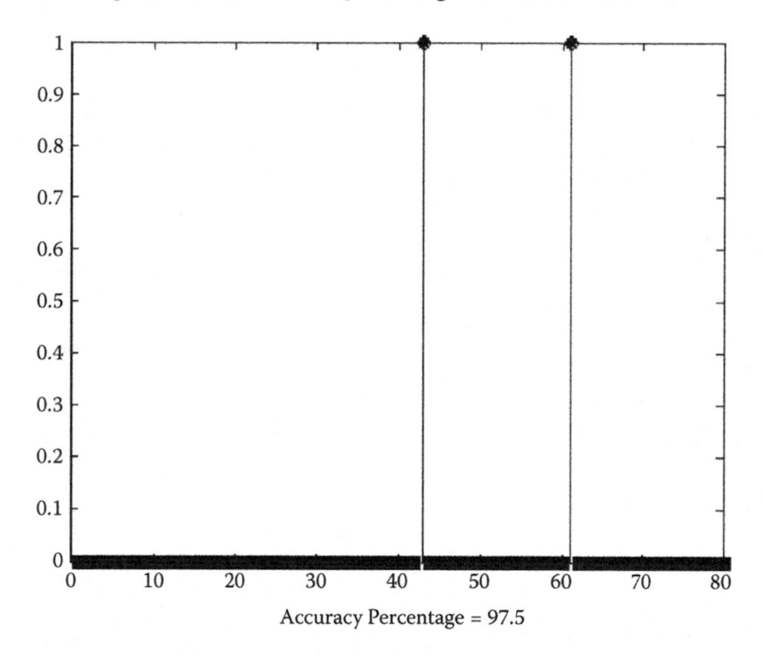

FIGURE 11.13b Percentage accuracy of best LVQN for diagnosis of skin diseases.

TABLE 11.17
Performance Comparison to Find the Best Diagnostic System for Skin Diseases

Network	Percentage Accuracy of Diagnosis	Training Time (sec.)	MSE
ANN with BPA	96.25	52.02	0.010
RBF	82.50	17.59	0.025
LVQ	**97.5**	**194.05**	**0.009**

11.7 DIABETES

Continuing our discussion in the study of diseases, we now discuss diabetes. The World Health Organization (WHO) recognizes three main forms of diabetes mellitus: type 1, type 2, and gestational diabetes (occurring during pregnancy), which have different causes and population distributions (WSEC 2004). According to the study by the Pan American Health Organization (PAHO), type 2 diabetes is increasing throughout the border area, along with risk factors for the disease. Some 1.1 million border residents aged 18 and older suffer from type 2 diabetes, and 836,000 are prediabetic (PAHO/WHO 2007). From the global Diabetes Impact Survey in conjunction with the 44th annual meeting of the European Association for the Study of Diabetes (EASD), less than 50 percent of people with type 2 diabetes are estimated to be at or below their blood sugar target. Results of the survey, conducted by IDF-Europe and the Lions Clubs International Foundation, showed that a large percentage of people with diabetes do not report being worried about the long-term diabetic microvascular complications associated with the condition, including diabetic retinopathy, which can lead to visual impairment and loss of vision. It was estimated that more than 2.5 million people worldwide experience vision loss due to diabetic retinopathy (Brussels 2002). According to the survey by American Diabetes Association (ADA), 49 percent of the U.S. adults polled said they most feared cancer as a potential health problem, while just 3 percent said they worried about diabetes. The survey results suggest that people need to assess their diabetes risk and take it more seriously (Doheny 2008).

11.7.1 Data Set for Diabetes

In the data set, the number of instances was 768, the number of attributes was 8, the number of instances in the training data set was 668, and the number of instances in the testing set was 100. For each attribute, the information taken was (all numeric-valued) as follows:

- Number of times pregnant
- Plasma glucose concentration at 2 hours in an oral glucose tolerance test
- Diastolic blood pressure (mm Hg)
- Triceps skin fold thickness (mm)
- Two-hour serum insulin (mu U/ml)
- Body mass index (weight in kg / (height in m)^2)
- Diabetes pedigree function
- Age (years)
- Class variable (1 or 2)

11.7.2 Results

11.7.2.1 ANN with BPA

The architecture used in this experiment consists of three hidden layers and one purelin output unit. Table 11.18 shows different models produced by changing the number of hidden neurons, momentum, learning rate, and transfer functions in hidden layers. From this table, it is clear that the ANN model with 54 hidden neurons and accuracy of 80 percent is the best ANN model for diagnosis of diabetes. Figure 11.14a shows the training curve for the best ANN. Figure 11.14b shows the percentage accuracy of diagnosis using the best ANN.

11.7.2.2 RBFN

From Table 11.19, it is clear that the RBFN model with spread value of 0.1 and accuracy of 68 percent is the best RBFN model for diagnosis of heart diseases. Figure 11.15a shows the training curve for the best RBFN, and Figure 11.15b shows the percentage accuracy of diagnosis using the best RBFN.

TABLE 11.18
Training Parameters and Efficiency for Diabetes Using ANN with BPA

No. of Hidden Neurons	Function in Hidden Layer 2	No. of Epochs	Training Time (sec.)	Percentage Accuracy of Diagnosis	Remarks:
49 (30,10,9)	Tan-Sigmoid	1228	1022.38	71	Function in Hidden Layer 1 = Tan-Sigmoid, Function in hidden layer 3 = Log-Sigmoid
51 (32,10,9)	Tan-Sigmoid	1196	1284.67	72	
53 (36,9,8)	Log-Sigmoid	1237	1133.95	68	
54 (37,9,8)	**Log-Sigmoid**	**1344**	**1259.36**	**80**	Function in Output Layer = Purelin, MSE = 0.01.

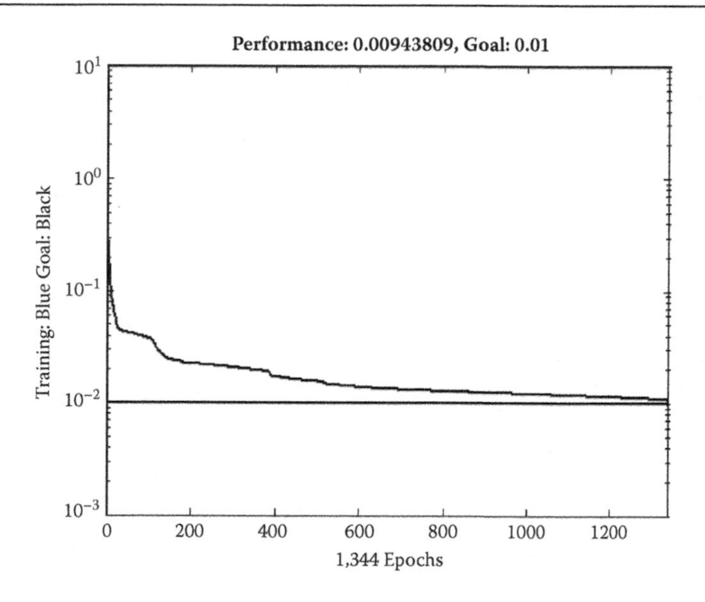

FIGURE 11.14a Training of best ANN for diagnosis of diabetes.

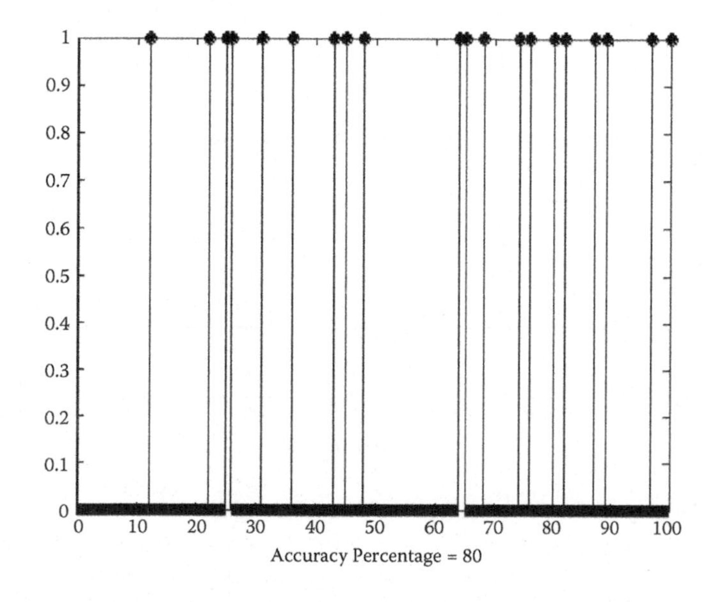

FIGURE 11.14b Percentage accuracy of the best ANN for diagnosis of diabetes.

TABLE 11.19
Training Parameters and Efficiency for Diabetes Using RBFN

Spread	Radial Basis Neurons	No. of Epochs	MSE	Training Time (in sec.)	Percentage Accuracy of Diagnosis	Remarks:
0.08	650	650	0.08	228.89	68	Goal = 0.01,
0.1	**650**	**650**	**0.08**	**222.16**	**68**	Function
0.3	650	650	0.02	230.91	59	Used = newrb
0.5	650	650	0.06	228.14	47	

11.7.2.3 LVQ Network

The parameters for the LVQ models are shown in Table 11.20. The network with 30 hidden neurons and an accuracy of 80 percent is the best LVQ network. Figure 11.16a shows the curve for training of a neural network with LVQ, and Figure 11.16b shows the percentage accuracy of optimized LVQ model on testing data set.

11.7.2.4 Performance Comparison

Table 11.21 shows the comparison of best results (for optimized model) for three networks. On the basis of the experimental observations of three neural networks, LVQN and BPA give the highest accuracy of 80 percent, but the training time for ANN with BPA is larger as compared to LVQN. Hence LVQN is the best network for diagnosis of diabetes.

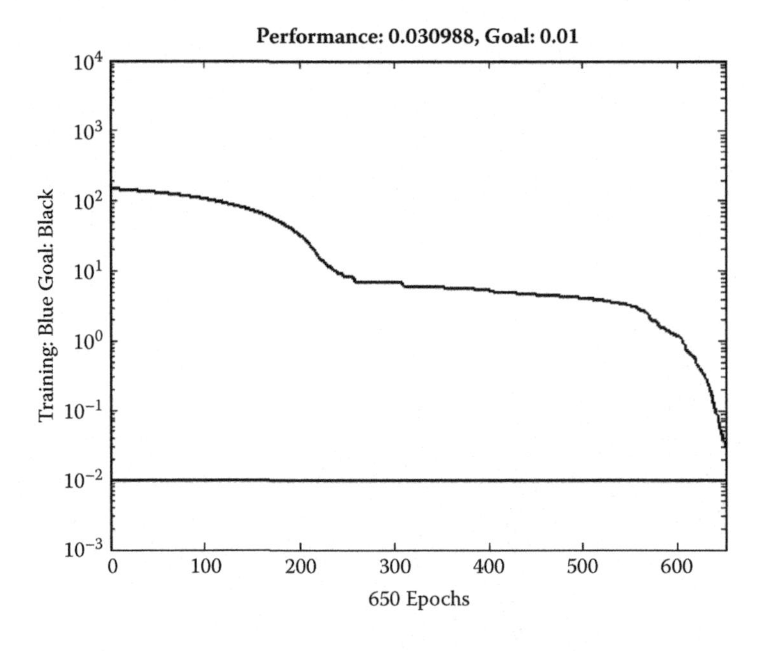

FIGURE 11.15a Training curve of the best RBFN for diagnosis of diabetes.

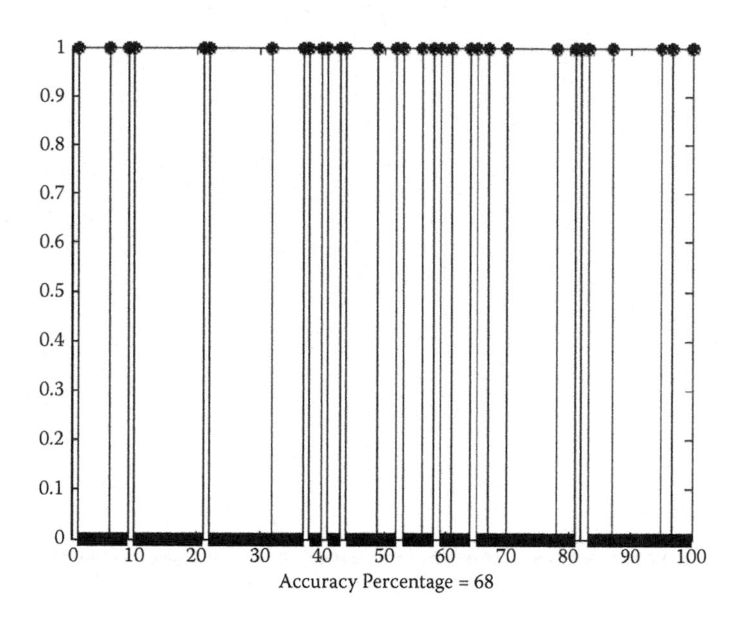

FIGURE 11.15b Percentage accuracy of the best RBFN for diagnosis of diabetes.

TABLE 11.20
Training Parameters and Efficiency for Diabetes Using LVQN

No. of Hidden Neurons	Learning Rate	No. of Epochs	MSE	Training Time (sec.)	Percentage Accuracy of Diagnosis	Remarks:
10	0.06	450	0.25	322.98	76	Learning Function = learnlv1
15	0.07	450	0.25	319.99	78	
20	0.07	450	0.24	327.69	77	
25	0.07	450	0.24	327.34	78	
30	**0.09**	**450**	**0.25**	**322.61**	**80**	

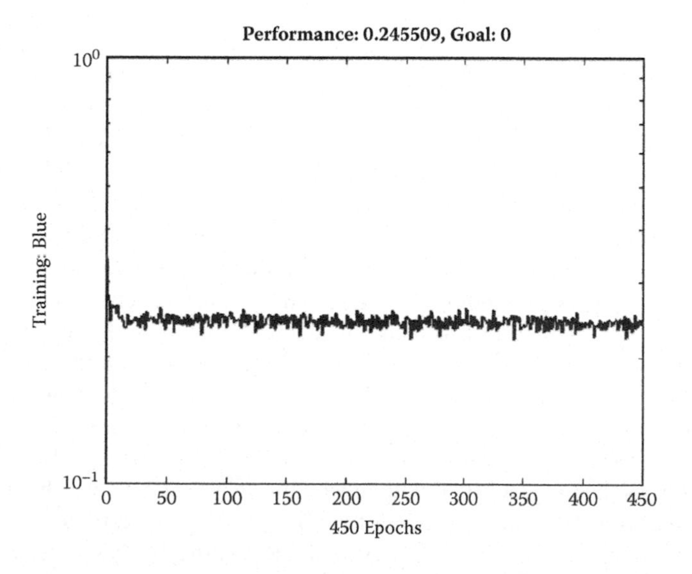

FIGURE 11.16a Training curve for the best LVQN for diagnosis of diabetes.

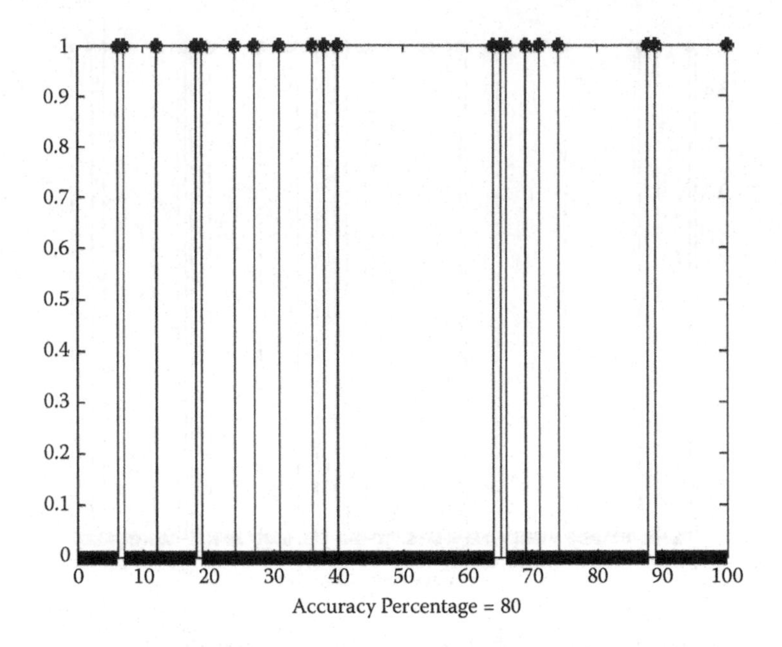

FIGURE 11.16b Percentage accuracy of best LVQN for diagnosis of diabetes.

TABLE 11.21
Performance Comparison to Find the Best Diagnostic System for Diabetes

Network	Percentage Accuracy of Diagnosis	Training Time (sec.)	MSE
ANN with BPA	80	1259.36	0.01
RBF	68	228.89	0.03
LVQ	**80**	**322.61**	**0.25**

11.8 HEART DISEASE

Heart disease is the last category of disease that we study in this chapter. Various forms of heart disease include aortic regurgitation, cardiogenic shock, congenital heart disease, coronary artery disease (CAD), hypertrophic cardiomyopathy, ischemic cardiomyopathy, peripartum cardiomyopathy, tricuspid regurgitation, etc. (Drugs.com). According to the World Heart Federation, rheumatic heart disease is the most commonly acquired cardiovascular disease in children and young adults and remains a major public health problem in developing countries. It kills some 350,000 people a year and leaves hundreds of thousands of people with debilitating heart disease (WHF 2007). An epidemiological survey of rheumatic heart disease in India, during 1996, found that the prevalence was significantly greater in rural schools (4.8 / 1,000) than in urban schools (1.98 / 1,000) (Thakur, Negi, Ahluwalia, Vaidya 1996). Most women mistakenly believe that breast cancer is the leading cause of death among females, underestimating heart disease as the nation's biggest killer. Results of a Newspoll survey released show awareness around heart disease and what causes it is "dangerously low" among a nationally representative sample of Australian women. Heart disease claims the lives of 30 Australian women every day, almost 11,000 a year, yet only 30 percent of those questioned named it as the biggest killer, with most naming breast cancer, which kills about 2,600 annually (McLean 2008). According to maps of world, the top 10 countries with the highest death rates are Ukraine, Bulgaria, Russia, Latvia, Belarus, Romania, Hungary, Estonia, Georgia, and Croatia (Maps of world 2008).

TABLE 11.22
Training Parameters and Efficiency for Heart Diseases Using ANN with BPA

No. of Hidden Neurons	Function in Hidden Layer 1	Function in Hidden Layer 2	Function in Hidden Layer 3	Momentum	Learning Rate	No. of epochs	Training Time (sec.)	Percentage Accuracy of Diagnosis
48 (30,10,8)	**Tan-Sig**	**Tan-Sig**	**Log-Sig**	**0.9**	**0.07**	**15380**	**129.28**	**83.33**
49 (32,9,8)	Tan-Sig	Tan-Sig	Log-Sig	0.9	0.07	6869	57.92	81.67
48 (30,10,8)	Tan-Sig	Log-Sig	Log-Sig	0.9	0.08	24637	209.89	71.67
55 (30,13,12)	Log-Sig	Log-Sig	Log-Sig	0.8	0.07	43659	400.64	75

Remarks: Function in Output Layer = Purelin, MSE = 0.01.

11.8.1 Data Set for Heart Diseases

From the data set the number of instances was 303, and the number of attributes was 13 with 1 class attribute. Attributes were sex, age, trestbps, chol, fbs, restecg, thalach, exang, oldpeak, slope, ca, thal, and class (the predicted attribute). The number of instances in the training data set was 243, and the number of instances in the testing set was 60. These are given in Table 11.21.

11.8.2 Results

11.8.2.1 ANN with BPA

The values of learning rate and momentum factor were varied to get the best results. The training function used was traingdm. Table 11.22 shows different models produced by changing the number of hidden neurons, momentum, learning rate, and transfer functions in hidden layers. From this table, it is clear that the BPA network model with 48 hidden neurons and accuracy of 83.33 percent is the best ANN model for diagnosis of heart diseases. Figure 11.17a shows the training curve for the best ANN. Figure 11.17b shows the percentage accuracy of diagnosis using the best ANN.

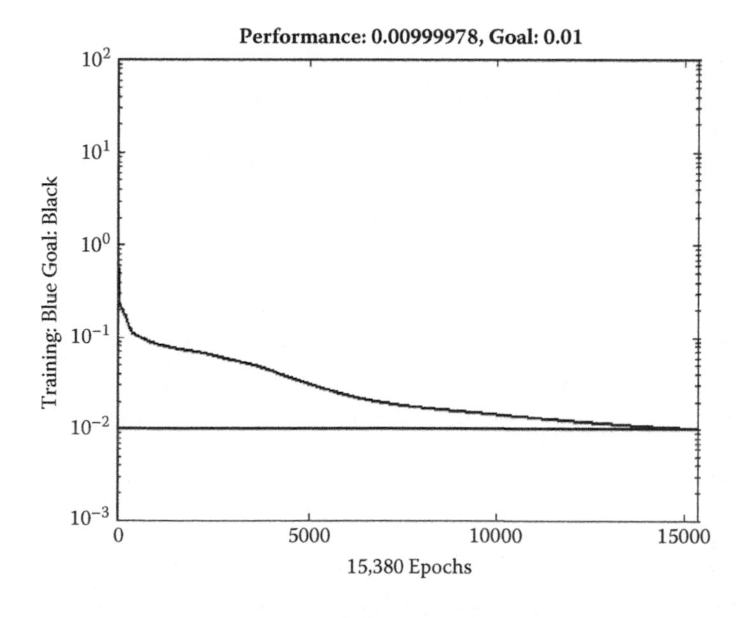

FIGURE 11.17a Training curve of best BPN for diagnosis of heart diseases.

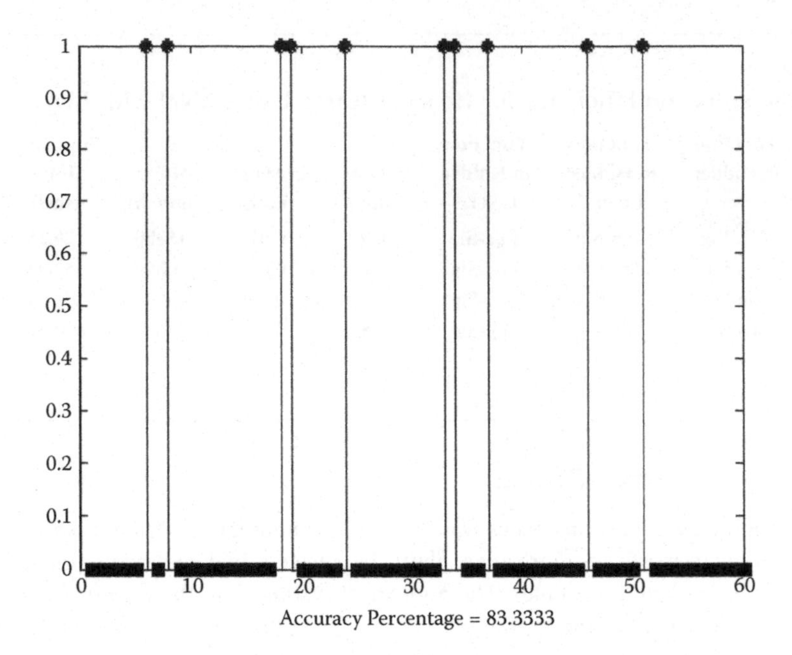

FIGURE 11.17b Percentage accuracy of the best BPN for diagnosis of heart diseases.

11.8.2.2 RBFN

Training parameters of different models based on RBFN are shown in Table 11.23. From this table, it is clear that the RBFN model with spread value of 1.05 and accuracy of 78.33 percent is the best RBFN model for diagnosis of heart diseases. Figure 11.18a shows the training curve for the best RBFN, and Figure 11.18b shows the percentage accuracy of diagnosis using the best RBFN.

11.8.2.3 LVQ Network

The network was created and the number of hidden neurons was changed to get the best results. The parameters for these LVQ models are shown in Table 11.24. The network with 25 hidden neurons and an accuracy of 86.70 percent is the best LVQ network. Figure 11.19a shows the curve for training

TABLE 11.23
Training Parameters and Efficiency for Heart Diseases Using RBFN

Spread	Radial Basis Neurons	No. of Epochs	MSE	Training Time (sec.)	Percentage Accuracy of Diagnosis	Remarks:
0.8	200	200	0.100	7.48	75	Goal = 0.01,
1.05	**225**	**225**	**0.040**	**8.13**	**78.33**	Function
1.2	225	225	0.036	7.56	73.33	Used = newrb,
1.5	225	225	0.025	8.22	63.33	
2.1	225	225	0.080	14.28	56.67	

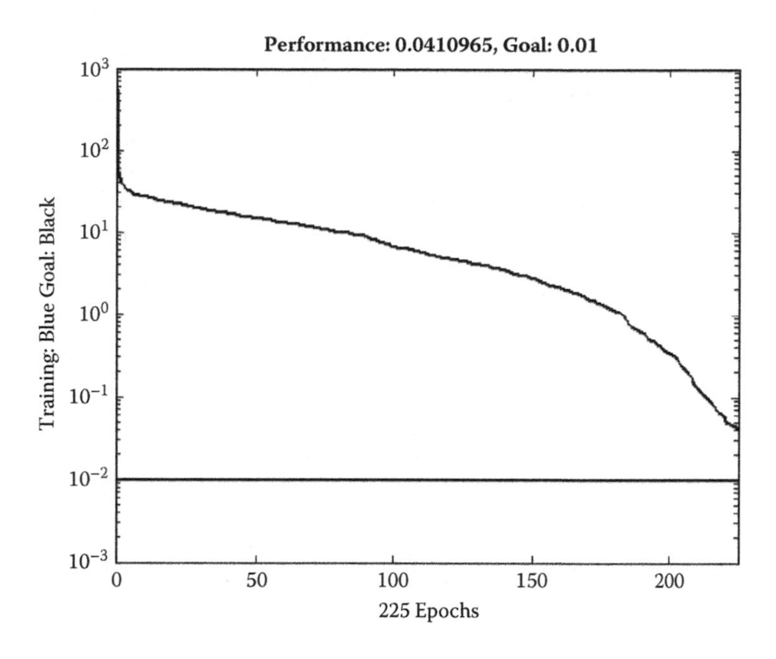

FIGURE 11.18a Training curve of the best RBFN for diagnosis of heart diseases.

of a neural network with LVQ, and Figure 11.19b shows the percentage accuracy of optimized LVQ model on testing data set.

11.8.2.4 Performance Comparison

Table 11.25 shows the comparison of best results (for optimized model) for three networks. On the basis of the experimental observations of three neural networks, LVQN gives the highest accuracy of 86.70 percent for diagnosis of heart diseases.

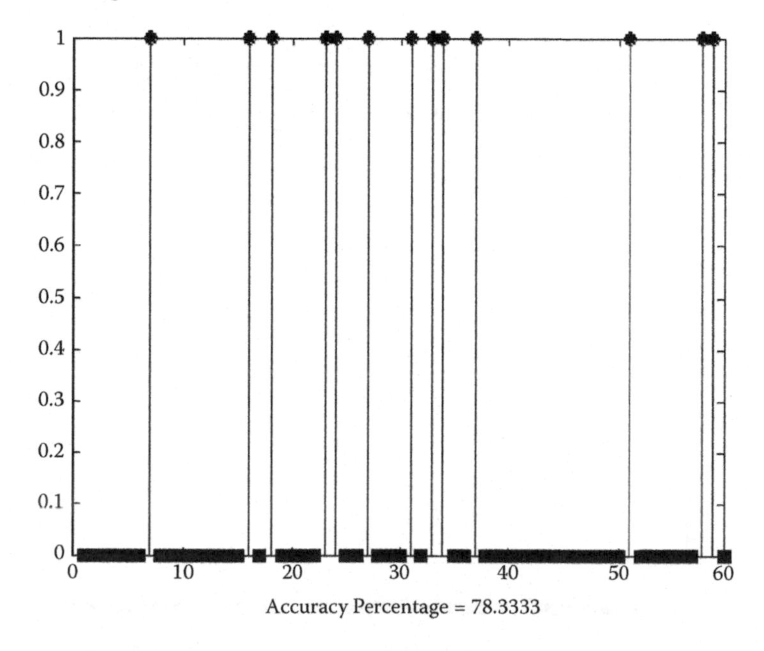

FIGURE 11.18b Percentage accuracy of the best RBFN for diagnosis of heart diseases.

TABLE 11.24
Training Parameters and Efficiency for Heart Diseases Using LVQN

No. of Hidden Neurons	Learning Rate	No. of Epochs	MSE	Training Time (sec.)	Percentage Accuracy of Diagnosis	Remarks:
25	**0.09**	**400**	**0.13**	**105.06**	**86.70**	Learning
32	0.08	400	0.15	106.63	85.00	Function =
27	0.09	400	0.14	109.36	86.70	learnlv1
20	0.09	400	0.14	104.24	83.33	
15	0.09	400	0.17	104.12	85.00	

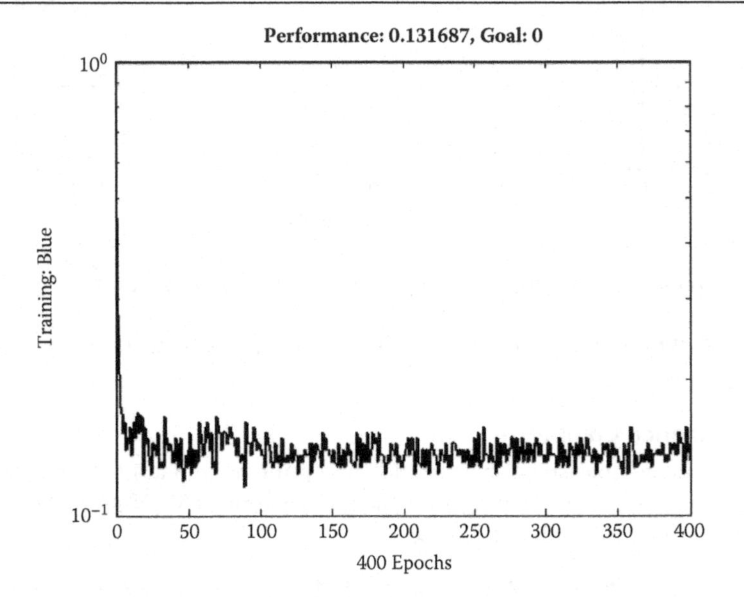

FIGURE 11.19a Training curve of the best LVQN for diagnosis of heart diseases.

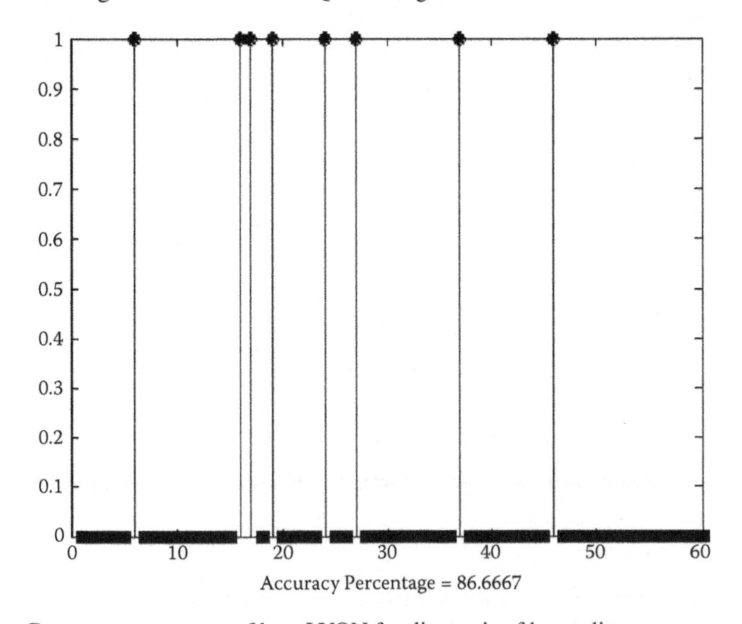

FIGURE 11.19b Percentage accuracy of best LVQN for diagnosis of heart diseases.

TABLE 11.25
Performance Comparison to Find the Best Diagnostic System for Heart Diseases

Network	Percentage Accuracy of Diagnosis	Training Time (sec.)	MSE
BPA	83.33	129.28	0.01
RBF	78.33	8.13	0.04
LVQ	**86.70**	**105.06**	**0.13**

11.9 CUMULATIVE RESULTS

On the basis of the experimental observations, the best accuracy of diagnosis in each of the six specified diseases is as summarized in Table 11.26.

In five out of six diseases, LVQ has produced the best accuracy of diagnosis, when trained on the specified data. Thus LVQ can be effectively used in medical diagnosis.

11.10 CONCLUSIONS

Now we have some useful data to continue our discussion related to machine learning that we started in the previous sections. It is known that classification is primarily a localized problem. Hence the localized models of ANN should be able to solve the problem much better than the generalizing models. This is again because of the fact that the localized ANNs see the possibility of the output by seeing the neighbors' output. Hence if a point in the input space is affected by some disease, it is highly likely that the points around also suffer from the same disease.

Although this must be generally true, yet breast cancer results show exactly the opposite. Here the most generalizing solutions are seen to return the best solution to the problem of breast cancer detection. Now the reader may question the theory we presented. It may, however, be possible that this shift happens due to experimental constraints and errors. Say the ANN with BPA was trained in the most optimal manner and the LVQ training was nonoptimal. These may be the reasons for poor performance most of the time.

However, we justify the results considering all experiments as optimal. The notion that a localized ANN performs better or outperforms a generalized ANN is usually true but cannot be generalized to all the possible input cases. Many times the data may be arranged in such a manner that the ANN with BPA is better able to model using its modeling capabilities. On the other hand, the other systems have a lot of problems in doing the same. The classificatory models are

TABLE 11.26
Best Accuracies and Methods to Attain These Accuracies for Various Diseases

Disease	Accuracy	Method
Breast Cancer	99.50%	ANN with BPA
Epilepsy	95.38%	LVQ
Thyroid Disorders	98.00%	LVQ
Skin Diseases	97.50%	LVQ
Diabetes	80.00%	LVQ
Heart Diseases	86.70%	LVQ

especially likely to be better in case the various models have high intraclass separation and low interclass separation.

The best way to understand this is by plotting the input space and seeing the distribution of the various classes there. However, we are able to easily see, infer, and interpret only graphs with a maximum dimensionality of 3. This puts heavy restrictions on our capability to understand these systems. It is not possible to get the complete idea of the inputs and their behavior. Had this been the case, we would have easily seen what the LVQ is trying to do and what the ANN with BPA is trying to do. For now, we can just visualize the problems and the possible solutions and infer why some system performs better in a given set of circumstances.

For the rest of the cases, it is natural that the LVQ performs better due to its localized nature. We only discuss these results here. The particular efficiencies, time requirements, and architecture of the systems may be easily verified and commented.

12 Biomedical Systems—II

12.1 INTRODUCTION

We talked a lot about biomedical systems in the preceding chapter. There we studied the general fundamentals of the biomedical engineering and even engineered systems using the concepts learned. We solved a variety of problems using these systems. We saw some of the most prominent diseases of the modern world and then made automated systems that could identify these diseases given some parameters. In all the diseases that we studied there, we found that the developed systems are easily able to solve the problems presented and the accuracy is quite high in majority of the cases.

Using biomedical problems as an example, we also studied how the various classifiers work. We had first developed theoretical foundations of these classifiers where we developed a base for the understanding of biomedical systems. Then we did a lot of experiments to get valuable results to validate our assumptions about these systems. At the end, we studied and discussed each of the results and offered comments.

This chapter continues and finally concludes our journey into biomedical systems. We study a couple more biomedical systems apart from the diseases. These systems so developed present another dimension to the infinite possibilities of biomedical systems. At the same time, these systems use the soft computing approach to solve very interesting real life problems. We first discuss a system that computes and tells us the nature of pregnancy. Here we solve the same problem using the artificial neural network (ANN) with back propagation algorithm (BPA), radial basis function network (RBFN), learning vector quantization (LVQ), and finally the adaptive neuro-fuzzy inference system (ANFIS). We will see the functioning of each one of these developed systems in deciding the final output of the systems. Here the ANN with BPA, RBFN, and LVQ will function just as they did in the preceding chapter. It will definitely be interesting to see if the results of this experiment follow the same rules as those of the previous experiment.

Moving further with the chapter, we would be discussing about the problem of Pima Indian diabetes, which is very common. Here we try to identify cases of diabetes from the various attributes. This problem may be solved in a variety of ways. Here we use ANN with BPA, ANFIS, and ensemble and evolutionary ANN. These solutions are applied to the problem, and we try to draw some conclusions from their results or training patterns. Again the manner in which the ANN with BPA would work can be imagined from our previous discussions, where we gave a multidimensional view of the use of ANN with BPA for classificatory problems.

The last section of the chapter is dedicated to the use of fetal heart rate monitoring. Here we develop adaptive filters to overcome the noise of the signals that are used for such monitoring. This would be ample to enable excellent analysis of the fetal heart rate even if monitoring is done in noisy conditions.

The preceding chapter was devoted to the use of soft computing systems and understanding how they work, especially for classificatory problems. We do the same thing here for hybrid systems. Here we try to understand from a very fundamental point of view how and why these systems work. We again develop analogies to predict which one is likely to work better for all problems. We also develop an understanding as to how these systems actually work and try to predict the outputs. Again let us see whether our analogies work out for the real life systems as they did in the preceding chapter.

The basics of classificatory problems remain the same as in the preceding chapter, and hence a large part of that discussion, where we studied the classificatory approach and fundamentals, may be recalled here. Also here we will restrict our discussion to hybrid systems, even though we use a variety of other methods to solve these problems. The other methods give us means to compare the hybrid systems as well as signify the commonly used solutions to these standard problems. Hence we experiment with them for the sake of completeness of the chapter as well as to provide a comparative metrics.

12.2 HYBRID SYSTEMS AS CLASSIFIERS

In this section, we explain the use and working of hybrid systems as classifiers. The hybrid systems are a collection of the basic systems of soft computing. These systems make use of a combination of systems in order to get the added advantages of both systems in use. The disadvantage of one is overcome by the advantage of the other and vice versa. This makes these systems excellent for use in soft computing. Classificatory systems are no exception; here hybrid systems have been a great boon and have efficiently solved many systems that could not have been solved by the use of single systems. By the use of combined systems, we can expect improved performance.

Before we continue discussing the various systems, it would be wise to analyze the problems associated with particular systems. It may be noted that the use of hybrid systems alone is no guarantee that the resultant system will work better. In fact in many cases, performance may be reduced. We need to clearly know the problems and whether the combination of systems would work better or not. The use of hybrid systems without understanding might result in magnifying the problems with the original system, and in turn we may end up with a poorer resulting system.

It may again be noted that the general method of problem solving with hybrid system involves first identifying the problems with the individual systems and then finding a reason for the occurrence of these problems. Finally, we must select the correct systems to make the hybrid systems or the correct model of the hybrid systems. We must analyze whether the resulting system would result in overcoming the shortcomings of the existing system even before experimenting.

In the rest of the discussion we take all these points into account. Further we use the basic concepts and philosophy of the hybrid systems that we studied in Chapter 6. Also we use the discussions we initiated in the preceding chapter in regard to classificatory problems and their solution by the various individual soft computing methods. Readers may recall all these before continuing reading the rest of the chapter.

One of the first systems that we discussed was ANN with BPA. Here we saw that these were the most generalizing networks that tried to generate solutions by the use of global rules. In the preceding chapter, we made a very brief comment on the number of neurons and the complexity of the problem. In any ANN trained with BPA, the complexity of the problem is said to be high if the curve to be imitated has a large number of turns. In classificatory problems, this may be interpreted as meaning the complexity of the problem is very high if the region of the presence of the class intermingles a lot with the region where the class is not found. Again if the interclass separation is not high, the complexity of the problem will be very high. A low interclass separation means that the curve to be imitated has its maxima immediately following its minima or vice versa. In other words, if we increase the value of some parameter even by a very small amount, the class will change. This makes the ANN very sensitive to inputs. As per our previous discussions, the ANN requires a comfortably large region in the input space to reach the minima from the maxima. If such a space is not available, the ANN with additional number of neurons gets an added ability to imitate very sensitive functions. This is why in systems where the intraclass separation is low the number of neurons required is high.

Now we discuss the problem with the ANN. We are given some training data to train the system with. But we do not know its complexity. Hence it would be very difficult for us to decide the number of neurons that we must ideally use. For this reason, we need to try to train the system again and again in order to get the most optimal ANN architecture. But the resulting system may still not be optimal. Hence there is definitely a problem in the use of ANN. Let us again remind ourselves of the notion that an ANN that takes the least number of neurons is always best if it gives the same performance. This limitation restricts the ANN from attaining various impossible structures. Further it limits the search space for the training algorithm, and this may drastically improve performance. Imagine the computational overhead, increase in dimensionality of error space, and memory requirements associated with the addition of a single neuron.

Another problem that may normally come with such systems is that the system may anytime get stuck at any time in local minima. This problem has already been discussed numerous times throughout the text. Also, being a generalized model, the ANN with BPA may not always bring good results. This again imposes a serious limitation in the use of these systems.

The next system that we used in the preceding chapter was RBFN. As compared to the ANN with BPA, the RBFNs represented much more localized ANNs, and hence we expected a little better performance, considering the localized nature of the RBFNs. These ANNs made use of multiple radial basis functions to imitate the actual function representing the region where the class lay. Another problem we've discussed is that the number of data centers or data guides and their region of impact may not be known initially. Also it may be visualized that it is very easy to get trapped in local minima. At the time of training, some configuration may look good enough. We may adjust it to find good solutions, but the actual global minima may lie somewhere at an entirely different region. It may be noted that most of ANN training algorithms have this problem. Here it is easy to get trapped in a local minimum. This is because of the inherent mechanism used by the training algorithms, which try to adjust the system parameters to give the correct output the next time. It is not ensured that traveling in this direction will really help us in reaching the global minima. Hence this is a serious limitation in almost all ANN models that we studied.

The last system that we studied and used in the preceding chapter was the best of all. This was the LVQ. These ANN models were the most localized in nature, where the output was calculated by measuring the distance from the closest known inputs. Due to their highly localized nature, these were advocated to be efficient for all general classificatory problems. Again the problems that we discussed in connection with the other ANNs are true for these as well.

So now we have some motivation behind the use of hybrid systems. We hope that the systems that we develop will cater to the needs of at least one of the problems that we mentioned. By removing this problem, we may further expect that if the output was really being affected by these problems, the hybrid systems may act as a good approach. These may solve the problem and hence result in better efficiencies.

Here we discuss three types of hybrid systems. These are the ANFIS, ensemble ANN, and evolutionary ANN. We study how good these models are for classificatory problems and how their use may result in better solutions. The input that we have is of very high dimensionality, and humans cannot perceive more than three dimensions. This restricts us from actually graphically seeing the training data or the system behavior in the input space. We can, to a large extent only, use the results and observations to predict the behavior and problem being faced by the system.

12.2.1 ANFIS

ANFIS is a hybrid system where a fuzzy model is trained by the use of ANN training. If we overlook the ANN that is used for training purposes, the ANFIS happens to be a fuzzy inference system (FIS). Since we have not yet formally discussed the working of the FIS in classificatory problems, we spend some time here discussing this topic.

The FIS has a great capability to map inputs to outputs using a set of rules that are cautiously framed considering the system behavior. These systems denote the natural way in which things work out. Here the output at any point depends upon the rules that drive the system and are made up from study of the system. In theory, FIS is able to form all kinds of rules, ranging from local to global. There are possibilities of making rules at all these levels whose combination gives the correct output or predicts the correct system behavior.

However, study of the generated fuzzy rules of any ANFIS would clearly point out the localized nature of these rules. Here all the problem inputs and outputs are broken down into membership functions (MFs), which lay consecutively one after the other, dividing the input and output axes in the input space. The whole problem may be supposed to be the calculation of the output with a set of inputs with reference to the points where the MF is centered. Hence the first task is that the output must be correctly calculated by the system at all combinations of inputs the MF is centered around. Say if an MF is centered at input I_1 and is called *low* and similarly the other sets of MF called low at other inputs are centered about points I_i, we must be able to get the correct output when the input I $<I_1, I_2,...I_n>$ is given to the system. It is not necessary that a training data set element be found near this input. If by any chance we are convinced that the output is correct for this input, then the task is relatively easy. The MF has a sudden or gradual decrease in its effect as we move away from its center. This helps in calculating the output for any input in the input space. The natural systems exhibit a similar phenomenon. The effect of MFs decreases sharply or gradually across some benchmark points where the output is well-known.

Now we discuss the working of fuzzy systems with relevance to the classificatory problems. We consider the problem to uncover the presence of disease, where the task is to classify the input into one of the two classes: whether disease exists or not. Other types of classificatory problems may be broken down to this type by considering each output as to whether this class is found or not. This is the same modeling of the output that we considered in the case of the ANNs for the classificatory problems.

In the case of the classificatory or the binary classificatory problem, the FIS must be able to imitate the binary function that gives maxima at regions where the class is found and minima at regions where the class is absent. Here we mainly study the rules, as they are the ones that drive the system outputs. The rules consist of the antecedents and the consequents. The antecedents contain all combinations of the various MFs of the various inputs. Imagine only the antecedents in the input space. These cover the entire input space, with the membership degree varying at different parts of the input space. At regions where there is some combination of points where each input has some MF of degree 1, the membership value happens to be 1. As we move away from this point on any of the input axes, the MF value keeps dropping. It may be noted that we assume the AND operator is used to join the conditions. Now first we make a big assumption here that may not always be true. This is that the outputs at these points with membership degrees are well known. This means that here the output would be a minimum or a maximum, depending upon the presence or absence of a class. This would also mean that the whole input space has fairly separated regions with known or assured outputs. It may easily be seen that now the MFs may be easily adjusted, as was the case with RBFN, to accommodate or not to accommodate the known outputs. The training algorithm so adjusts the graph of the concerned MF to easily accommodate the known input or to completely reject it. However, for evaluation purposes, a value above threshold is considered to belong to a class and vice versa.

It may be visualized here that changing the shape of an MF has a global impact. This is in the sense that each MF was used with all combinations of MFs of other inputs. If we change this MF seeing some particular inputs, the membership values of all the other combinations get changed accordingly. This is the globalized nature of these rules. Hence even though the FIS seems localized, it is not localized to the degree of RBFN or LVQ.

Here, readers may sometimes argue that because of this property, it may not be possible to adjust the functions according to problem demands. It is as if you were asked to fix the width and height of an image, at the same time keeping the aspect ratio constant. We know that changing width would naturally mean a change in height and vice versa. But remember that the globalized networks aim at extracting the global rules that are believed and experimented to hold good. There is a lot of hope that these MFs follow globalized patterns. In other words, we will be able to adjust these functions such that the outputs are above and below threshold wherever required. This is possible and is achieved by the training algorithm.

But it may not even be possible to adjust the MFs such that this behavior is achieved. This is clearly the case when we are trying to achieve more generalization than may be possible. Here we need to localize the FIS more by addition of the MFs or rules. This divides the axis on whose input the MF is added into more parts and naturally contributes toward the localization of the FIS.

Our entire discussion so far has been centered on the presumption that the outputs are clearly known at the points where the membership value is 1 or the input corresponds to all points where MFs are centered. This would very rarely be the case in real life applications. Even in RBFNs, we stated that the inputs are far apart from each other and not uniformly distributed. Even though we are dividing the input axis into fuzzy regions, it may be recalled that this division is not uniform. The characteristic of a good division technique or what was previously referred to as a good MF selection criterion is that the MF width or coverage is as per the trends of the output. If the output is suddenly changing over some short region and then changing gradually, we may wish to keep the width or the coverage of the corresponding MF smaller first and larger next.

To ensure or try to ensure that the outputs are correctly predicted at these points, the system uses the training algorithm. Here it tries the various alterations to maximize the final output or performance of the system. This ensures that the overall performance is optimal, which would only be possible in the case of the correct mapping of the outputs to the discussed set of points. The rules play a major role in getting the desired behavior in these systems.

The inputs form only one part of the rule; the other part or the consequent is composed of the output MFs. It may be naturally visualized that these function in the same way or rather support the work of the inputs. The clear mapping of the inputs to the outputs may be accomplished by placing the MFs of the outputs at desired locations. Again these MFs may be added to make the system more localized. The role of the output part is to map the so-far discussed membership value, considering both input and output, into the corresponding hard-coded outputs. This transformation is done using the fuzzy arithmetic that we discussed earlier. It may be verified that our arguments hold whether we study the variations in the MFs of inputs or the final outputs.

The FIS is trained by use of ANNs. We have already seen the working of ANNs as classifiers and their learning in the preceding chapter.

12.2.2 Ensemble

The ensemble is a modular neural network (MNN) that forms a very promising solution to most of the problems. Here we divide the whole problem into modules. Each module represents an ANN that is trained using its own means independent of all other ANNs or modules. An integrator collects the results per ANN and uses them to calculate the final system output. Each time an input is given to the system, it is distributed among all modules or ANNs. Each decides its own output. Then polling is done between ANNs by the integrator. The class that gets the most votes is declared the winner. In this section we try to appreciate the novelty of the systems and the problem that they solve that was prevailing in the ANNs.

As discussed, the ANNs are always likely to be stuck at some local minima. The reason behind this may be the nature of the BPA that tries to calculate the minima by the present location of the network. Even though the momentum tries to pull the ANN out of being trapped from the local minima, yet chances exist that the ANN will again gets trapped at some minima or fail to get out at all. In any case the global minima may lie at some very different place in the configuration space of the ANN. This would be a location that the network never even got close to.

The characteristic of the local minima especially in the case of classificatory problems is that even though they give correct output for most of the problems, yet there may be some inputs for which they fail to give correct outputs. Since the total error is appreciably less, it may be assumed that the answer is correct for most problems. This is the reason we achieved good efficiencies in our systems with the use of ANN with BPA. Hence even though most of the inputs do not create a problem, some inputs do lead to wrong results.

It may again be noted that the entire configuration or error space may be filled with numerous local minima. Each would be at some distinct region of the configuration space. Hence the ANN while training may get trapped in any one of them. This even depends on the initial weights and biases that are kept random. Hence the algorithm starts from a random location in the configuration space and may decide the local minima in which it may possibly get trapped. The momentum and epochs naturally play another important role. Of course if we add or delete some neurons, the configuration space changes with the decrease or addition of dimensions, but the issues remain the same.

If we train an ANN numerous numbers of times, it may again and again get trapped at some different minima. This minimum will give the correct results to most of the inputs in the input space and wrong answers to some of them. In other words, this ANN can see a large part of the complete picture clearly, but another part is incorrectly visible to it. At the same time, if it were at some different minima, another part would have been clearly visible to it with some other part unclear. This forms the basis of the ensemble.

Here, we try to form a globalized view by taking or considering the localized view of the various ANNs or modules. Whenever an input is given, all the ANNs depict it using their own means. They all correctly or incorrectly calculate the final output and return the class. Now since most of the ANNs had a correct view of a large part of the picture or gave correct output to most of the inputs, the output returned by most of the ANNs is correct. This outperforms the wrong answers being returned by the ANNs that were incorrectly giving outputs to those inputs. The number of correct votes is large enough to make the genuine class win the poll.

It may sometimes be argued, why don't these ANNs return a 100 percent correct output? The reason is that many times the training data has noise due to which the inputs are present at wrong locations. This makes the systems impossible to train unless they are localized to a very low level. Also many times the outputs have characteristic changes such that it is very difficult to make out the class to which they may belong. This happens due to the small interclass separations for some inputs that makes these systems very difficult to realize. All this poses a limitation on accuracy. Many times all the ANNs or modules give the wrong prediction of a class. This may be due to its characteristic location in the input space or the presence of large amount of noise.

Hence, it is clear that these approaches are primarily for the classificatory problems where they try to use the localized or the suboptimal performances of the individual ANNs to form a system with optimal performance. This may have a very positive deep impact on the working of the algorithm.

12.2.3 Evolutionary ANN

These ANNs evolve with the help of evolutionary algorithms (EAs) or genetic algorithms (GAs). Here we can keep the entire architecture variable. Also these methods solve many problems that we face in training the ANNs where they get trapped in local minima. Hence these approaches many times result in better solutions. The GA decides the correct values not only of the weights and the biases, but also of the entire ANN architecture. As the generations increase, the system grows more nearly optimal and the fitness value or the performance improves.

The evolutionary ANNs are primarily ANNs that are formed and trained using the GA. We have already discussed the role and functionality of the ANNs numerous times. We saw how they imitated the behavior of the function and how the training over them was performed. We even discussed the problems that are associated with these ANNs and the causes of these problems.

Here, we would mainly study and visualize how the training is done in the ANNs. Before we move on with the discussion, let us clarify two very distinct words that we have been using throughout the text. We might use these words simultaneously in this section. These are input space and configuration space. The *input space* is a high-dimensional space with each axis representing an input. The output is usually visualized as the vertically top axis. The other term is the *configuration space*. This is the space where each axis denotes some weight or bias that is used in the ANN. This is highly dimensional in nature and multiplies its dimensionality on the addition of every neuron in the system. The error (or negative performance) of the system for the training or the validation data set is usually marked as the vertical axis and plays the same role in our discussions as the output in the input space. In fixed architecture–type systems this space is constant. On the other hand, when the architecture of the ANN changes, this space also changes.

Let us see what happens at the time of training of the ANN. Here there is a transformation of the status or the position of the system in the configuration space. In any training algorithm, the intention is to reach a region in this configuration space that gives us the least error or the lowermost point in the output axis. This makes the training algorithm walk, wander, or jump in search of such points. We discussed in the chapter 2. how the BPA uses the steepest descend approach in this search. We further studied in considering how the GA invests various individuals in the configuration space and makes them move and jump in search of the point with the minimum error.

Let us move from the configuration space to the input space. Here a position in the configuration space corresponds to an entire function being imitated in the input space. The point corresponding to the least error in the configuration space means the most ideal imitation of the function serving as a solution to the problem in the input space. As we move about the configuration space, the function changes its shape in the input space.

We have basically two phenomena that the evolutionary GA incorporates. First is the change of configuration with same architecture, and the other is the change in the architecture itself. We know the kind of results the former phenomenon produces in both the configuration and input spaces.

The change in the architecture is a complexity or the sensitivity control of the ANN. Architectures with limited connectivity and a limited number of weights prohibit the ANN from imitating complex functions that are prone to too many changes and sudden changes. The configuration space changes completely if we change the ANN architecture. Hence a jump from one particular architecture to the other, or in other words the change in the configuration space, makes the resulting function being imitated simpler or more complex in the input space. A simpler structure due to limited dimensionality of configuration space is much easier to train and find the most optimal solution, if the optimal error is up to toleration limits.

Now, let us combine the two and see the working of the GA. As the configuration space is changing, the discussion requires a much broader insight into the functioning of the ANNs. Of course the actual space that incorporates the GA encoding of the chromosome is much easier to discuss as it obeys all laws of GA and is equally difficult to visualize and comprehend, especially with respect to the ultimate ANN and its performance.

The GA at any time consists of numerous individuals placed at multiple configuration spaces and multiple points at each configuration space. Here the convergence may be considered at two levels. One is that the GA at some configuration space tries to pull as many individuals as possible toward it that belong to the same configuration space. This is much like the trivial way of functioning of the GA. The other possibility is that the solution at any configuration space with decent performance tries to pull individuals from the other configuration spaces. These may end up at a configuration space that is in between the two spaces. Hence there is jumping of configuration spaces in order to reach the configuration space where the optimal solution resides. At the same time all the individuals of a configuration space try to converge to a point. At any time during the journey of any individual at any generation, a more nearly optimal solution is likely to be found. The necessary changes happen the next generation onward that attract the individuals toward this new optimal solution. This was the crossover operation. Mutation plays a similar role of adding to randomness.

We present a brief note concerning the classificatory problems where we tried to classify inputs or imitate a binary function. Imagine the condition of the input space for all the individuals as the GA runs across generations. Initially there are random functions ranging in complexity that claim to be imitating the ideal function. A convergence here toward the minima in the same configuration space would mean that the corresponding function in the input space approaches the idealistic one. This means that the function that had a wider range of areas being covered gets its area reduced and vice versa. Further, the jumping of neurons from higher configuration space toward lower configuration space would mean that the undue number of turns and sharp turns all get filtered out and the resulting function is retained. A transformation to the higher space means that the function now adds more turns and makes more turns to be capable of imitating the actual function. Readers may visualize that as we proceed with the algorithm, the individual representing the best solution keeps adding/loosing complexity and getting more realistic closer to the ideal curve.

Now that we have a good insight into how the various hybrid systems must work and how they must react to the inputs, we come to the real life applications, real life data, and real life systems. Let us see how these systems behave given these real life inputs and whether they show the desired traits.

12.3 FETAL DELIVERY

The first problem that we discuss here is the fetal delivery. Here we discuss the nature of pregnancy and classify the pregnancy into either of the two classes, i.e., normal or operational. The inputs consist of the pathological attributes that would help us in our prediction of whether fetal delivery should be done normally or by surgical procedure. These consist of the pathology tests like blood sugar, blood pressure, resistivity index, and systolic/diastolic ratio. In practice these tests are recorded at the time of delivery, and based on them, the decision on the nature of the pregnancy may be made. All attributes lie within a specific range for a normal patient. The database for this problem is collected from a local hospital, which consists of the data for cases, i.e., normal deliveries and surgical procedures. Here we would be using ANN with BPA, RBFN, LVQ, and neuro-fuzzy systems to build these systems. We try to judge the performance of the hybrid system, i.e., the neuro-fuzzy system against the other systems. This section discusses the experimentation and results of Shukla et al. (2009c).

These days, expert systems are widely used in health to predict and diagnose a particular disease. Medical expert systems are also useful in certain situations where either the case is quite complex or there are no medical experts readily available for patients. The field of medicine is such a complex and sophisticated one that safety is always a major issue.

The motivation behind the development of such systems is that these systems can be used to assist doctors in deciding whether the pregnancy should end normally or by the means operation. This would contribute a lot toward taking effective decisions at the correct time. The doctors can also compare their diagnoses with the results computed by these systems and thus cross-verify their results or rather reconsider their decisions in case an anomaly arises in the results of the doctors and those of these systems.

The basic functionality or the working of the system is shown in Figure 12.1. The first task is to acquire the input, which is in the form of blood samples. The various tests are performed on these

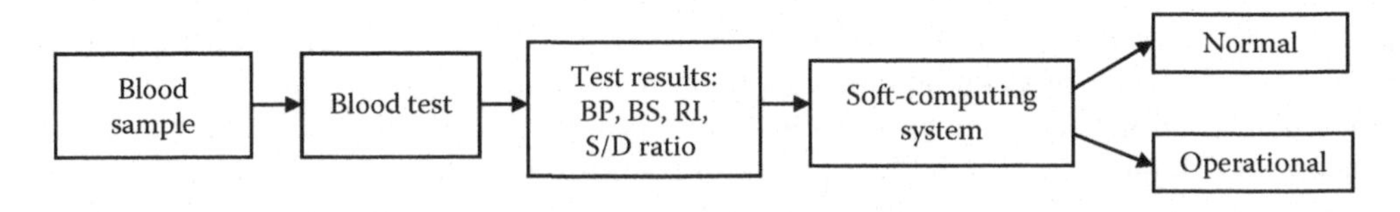

FIGURE 12.1 The fetal delivery detection system.

TABLE 12.1
Complete Attribute Set of the Fetal Heart Rate Monitoring Problem

Test Attributes	Range
Blood Pressure (BP)	Normal upper 120, Lower 80
Blood Sugar (BS)	120 Normal
Resistivity Index (RI)	0.7 Normal
Systolic/Diastolic (S/D) ratio	3 Normal

blood samples that results in the blood report that gives us the data for our system. This includes blood pressure, blood sugar, resistivity index, and systolic / diastolic ratio. This is used by the prediction system to find out whether the pregnancy must end normally or by operation.

12.3.1 Description of Data Set

The data set for women, pregnancy, and delivery was collected from Pt. Jawaharlal Nehru Government Medical College, Raipur, Chhattisgarh, India. This data is recorded for 640 instances at the time of delivery. This consists of four attributes: blood pressure (BP), blood sugar (BS), resistivity index (RI), and systolic/diastolic ratio (S/D Ratio). The attribute ranges have been given in Table 12.1. For all experiments that we perform here, the number of instances in the training data set is 480 and in the testing set is 160.

12.3.2 ANN with BPA

The first experiment was with the use of ANN trained with BPA. The ANN consisted of one hidden layer along with input and output layers. There were a total of 50 neurons in the hidden layer. The ANN had a goal of 0.01. The transfer functions in hidden layer neurons and output layer neurons are sigmoid and purelin respectively. The performance function used was mean sum-squared error (MSE). Figure 12.2 shows the training curve of the function. The system attained an accuracy of 95.83 percent in the training data. The efficiency of the testing data was 93.57 percent.

12.3.3 RBFN

The second experiment was done with the use of RBFN. The error goal of 0.01 was used. The performance function was MSE. The value of spread was modified to get an optimized model using RBF. The RBFN had spread value 0.05 and 50 neurons in the hidden layer. The system gave an accuracy of 99 percent for this problem. The accuracy was same for both the training and testing data. Figure 12.3 shows the training curve for the best RBFN.

12.3.4 LVQ Networks

LVQ models are simulated for different neurons in hidden layer. The performance function used is MSE. The network used had 50 neurons in the hidden layer. The network attained an accuracy of 97.5 percent for the training data. The performance was 99 percent for the testing data. Figure 12.4 shows the training curve for the best LVQ network.

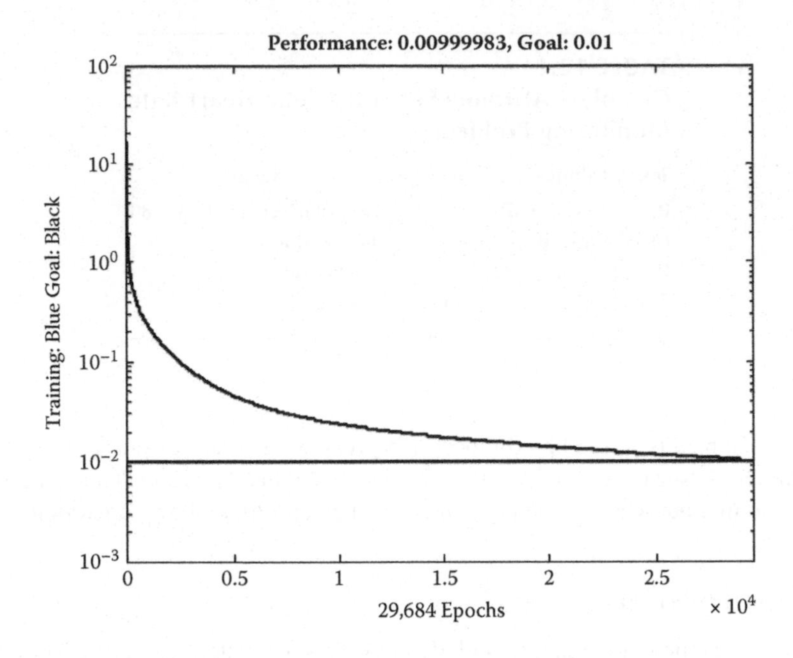

FIGURE 12.2 Training curve of the artificial neural network for the fetal delivery problem.

12.3.5 ANFIS

The last system implemented was with the help of ANFIS. For this the MATLAB® ANFIS editor was used. Sugeno FIS was used. The membership functions for one of the input variables are shown in Figure 12.5. Training result is shown in Figure 12.6. The training data consisted of 480 instances

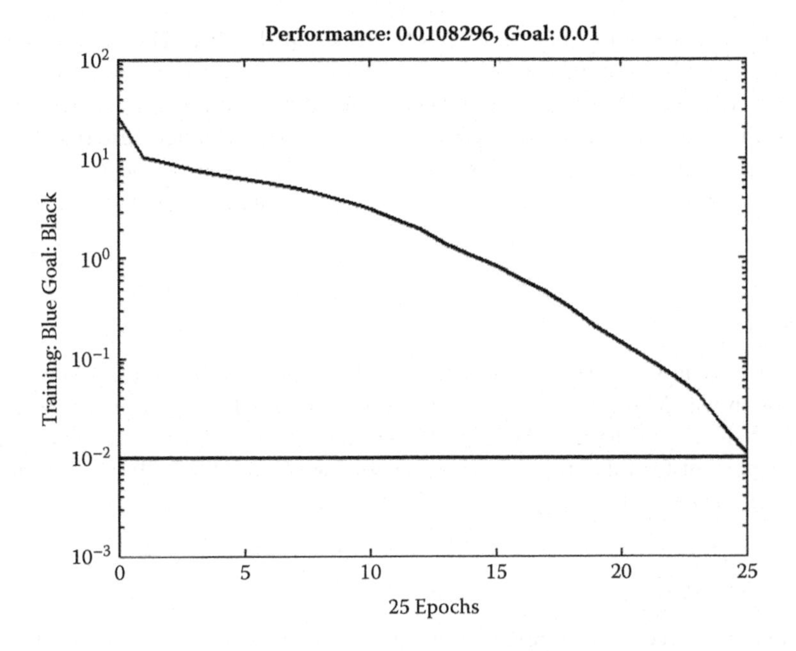

FIGURE 12.3 Training curve of the best RBFN model for the fetal delivery problem.

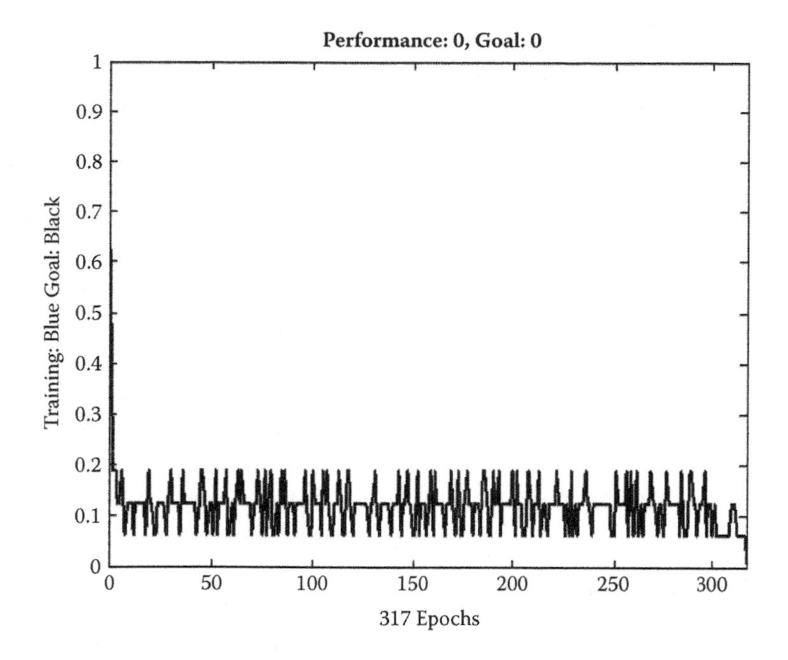

FIGURE 12.4 Percentage accuracy of diagnosis for the best training LVQ network for the fetal delivery problem.

and the testing data consisted of 160 instances. The system achieved 100% accuracy in both the training as well as the testing data.

12.3.6 Comparison of Results

The results have been summarized in Table 12.2.

On the basis of the experimental observations of the previously discussed systems, it is clear that ANFIS gives the highest accuracy, 100 percent, for the diagnosis of fetal delivery system. Further the results obtained in case of RBFN were very high, having an efficiency of 99 percent. In this case ANFIS is the best method to solve the problem. This work can be extended by the use of various other models and techniques of soft computing. These techniques may be used for other problems such as blood pressure identification and heart attack prediction.

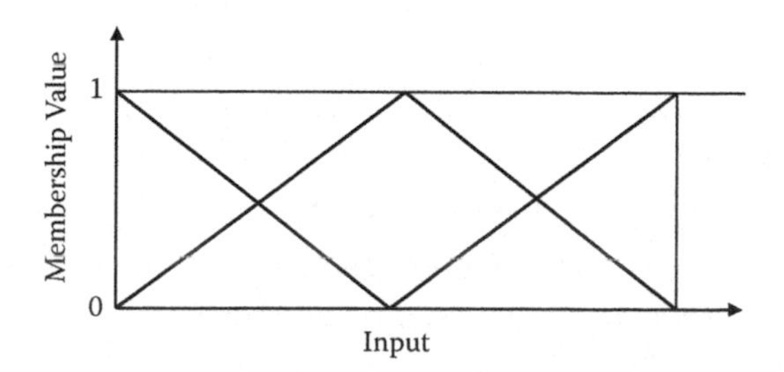

FIGURE 12.5 Input membership functions for the fetal delivery problem.

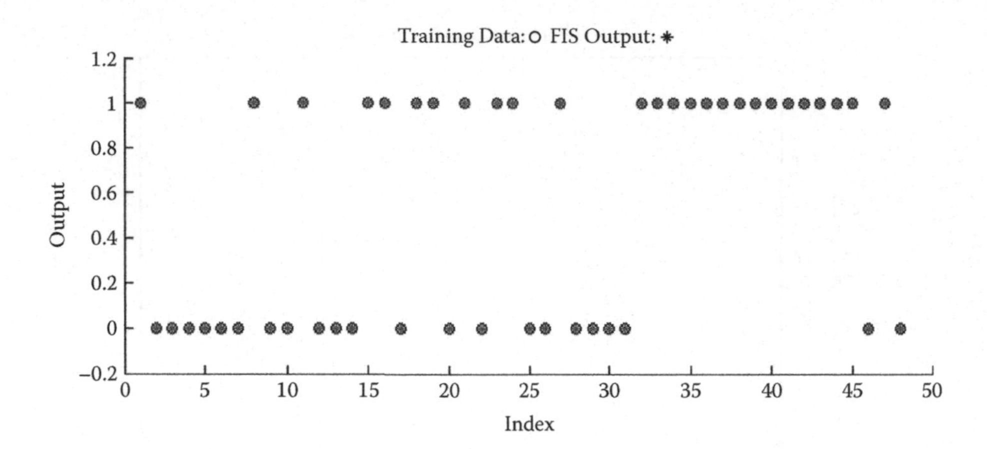

FIGURE 12.6 Outcome of testing the training data sets for the fetal delivery problem.

12.4 PIMA INDIAN DIABETES

In this section, we try to make system that tells us about the presence or the absence of Pima Indian Diabetes by looking at the records of a patient. Diabetes is another problem that calls for great concern. This requires the development of automated systems that may themselves detect the diabetes in a patient. This would help doctors in numerous ways, such as in making decisions regarding the presence of diabetes, and would hence contribute toward control of this disease.

12.4.1 Data Set Description

The data set of this problem comes from the UCI Machine Learning repository. The Pima Indian Diabetes data set consists of a total of eight attributes. These decide the presence of diabetes in a person. This database places several constraints on the selection of these instances from a larger database. In particular, all patients here are females at least 21 years old of Pima Indian heritage.

The first attribute is the number of times the women were pregnant. The next attribute is plasma glucose concentration at 2 hours in an oral glucose tolerance test. We further have the attributes diastolic blood pressure (mm Hg), triceps skin fold thickness (mm), 2-hour serum insulin (mu U/ml), body mass index (weight in kg/(height in m)^2), diabetes pedigree function, and age (years).

The database contains a total of 768 instances of data; 535 (~70 percent) instances of the data were used for training purposes and the rest 233 (~30 percent) of the data was used for testing purposes.

TABLE 12.2
Results for Fetal Delivery Problem

S. No.	Neural Network Algorithm	Efficiency %
1	Neural Network with Back Propagation Algorithm (ANN and BPA)	93.75
2	Radial Basis Function Network (RBFN)	99.00
3	Learning Vector Quantization (LVQ)	87.5
4	ANFIS	100

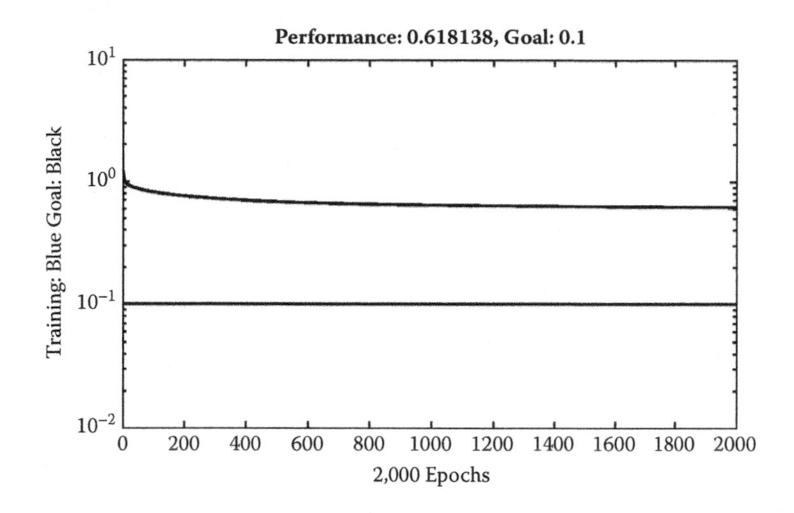

FIGURE 12.7 The training curve of ANN for the Pima Indian diabetes problem.

12.4.2 ANN with BPA

The first method used for this was ANN with BPA. Here we used a single hidden layer, which consisted of 12 neurons. The activation functions for the hidden layer was tansig and purelin. The training function used was traingd. The other parameters were a learning rate of 0.05 and a goal of 10^{-1}. Training was done for 2,000 epochs.

After the network was trained and tested, the performance of the system was found out to be 77.336 percent for the training data set and 77.7358 for the testing data set. The training curve of the ANN is given in Figure 12.7.

12.4.3 Ensemble

The second experiment was done on ensembles. Here we had used four modules or ANNs. Each one of them was trained separately using the same training data set. The four ANNs were more or less similar with small changes to the ANN used in the preceding section. The first was exactly the same as discussed previously. The other ANNs only had changes made in the number of neurons in the hidden layer and the number of epochs. These had 14, 10, and 12 neurons respectively. The numbers of epochs were 2,500, 200, and 4,000. The four ANNs were trained separately.

Here, we used a probabilistic polling in place of the normal polling. Each module or ANN returned the probability of the input belonging to some class. This means that there are as many outputs as there are classes. These probabilities for similar classes get added up. The final total is calculated for all the classes. The class with the highest total is declared as the winner.

The resulting system had a total performance of 78.7276 percent for the training data and 76.9811 percent for the testing data. It may be noted that the performance was 78.33 percent for the training data and 76.2264 percent for the testing data in the use of conventional ensemble where each module votes for some class.

12.4.4 ANFIS

The next experiment was done using ANFIS as a classifier. Here we used the same training as well as testing data sets. The FIS was generated using a grid partitioning method. Each of the attributes had two MFs with it. The system was allowed to be trained for a total of 100 epochs. The resulting training graph is shown in Figure 12.8.

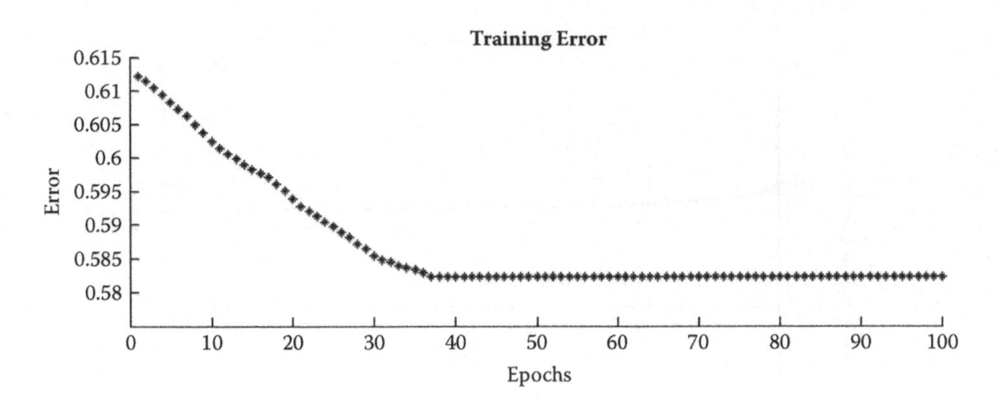

FIGURE 12.8 The training curve for ANFIS for the Pima Indian diabetes problem.

The final system so obtained had a performance of 88.9720 percent for the training data and 66.5236 percent for the testing data.

12.4.5 Evolutionary ANN

The last hybrid system that was applied to the same problem was the evolutionary ANN. Here we tried to evolve the ANN weights as well as the correct connectionist architecture. The connectionist approach is something that we had not discussed clearly earlier.

The ANN that we generally take for problem solving in the problems is a fully connected model where every neuron of every layer is connected to all the neuron of the next layer. The same thing, however, does not necessarily exist in the brain, where connections are made and destroyed automatically every second. The fully connected architecture has numerous problems associated with it that restrict performance in real life applications. Imagine an ANN used for some real life application. If it fails to train, the general approach is to add a neuron. But imagine the immense increase in dimensionality as a result of this addition of a single neuron. This expands the dimensionality by a big amount. This may make it very difficult for the training algorithm to train the network as a result. But then we have no option other than to add the neuron because the system is not getting trained otherwise.

Hence, we make use of the concept of a connectionist approach, which tries to inhibit or stop certain connections. This is done so that the complexity of the problem may be limited to a great extent. Hence in the resulting model, all the neurons are not allowed to connect to the neurons of the next layer; rather, they are allowed to connect to only some of the ANNs of the next layer. This makes the calculations very fast, and even the further training can take place easily as the connections are limited.

In this problem, we first use GA to evolve connectionist architecture. The other role of the GA is to fix the weights of the ANN. We take the maximum complexity that the ANN can span as a single hidden layer model with 25 neurons. Hence the problem is to decide the connections between the input layer and the hidden layer and then the hidden layer and the output layer that are allowed or that exist. All the other possible connections are inhibited. These may be taken to be connections with a weight of 0.

The first section of the chromosome consists of a 0 or 1, depending upon the existence of connection between the input layer and the hidden layer neurons. A 1 depicts the existence of a connection, and a 0 depicts an absence of a connection. Similarly the next section depicts the existence of a weight between a hidden layer neuron and an output layer neuron. The bits have their same meaning.

The other half of the chromosome contains the actual values of weights between the various connections. Here the first section of the second half contains weights between input and hidden

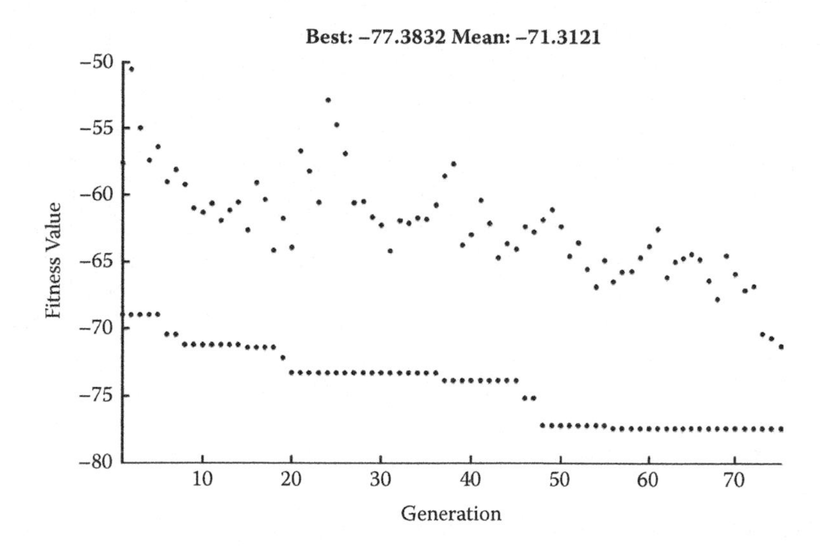

FIGURE 12.9 The training curve for evolutionary artificial neural network for the Pima Indian diabetes problem.

layer neurons, and the second section between hidden and output layer neurons. These are the actual values of connections that hold whether the connection exists or not. If the physical connection is not there, the corresponding weight value may be ignored. The last part contains the values of the biases.

The crossover and mutation were modified to match the requirement of the problem. The crossover used was a two-point crossover, which was applied in such a way that the first point always lay in the sections that constitute the architecture and consisted of the bits. The second part always lay on the section containing the values of weights and biases. These were the real values. Similarly the mutation was modified to ensure that the section reserved for bits always contains only one of the two valid inputs, i.e., 0 and 1.

In order to determine the extreme values or ranges of weights and bias, we saw some trained ANNs of fully connected types and then saw the extreme values of the weights and biases that they take. We selected a range that comfortably covered the observed range while at the same time not resulting in the search space or the configuration space becoming very vast.

The other parameters of the GA were 25 as the population size with an elite count of 2. The creation function was uniform, and double vector representation was chosen. Rank-based fitness selection was used. A stochastic uniform selection method was used. The crossover ratio was 0.8. The algorithm was run for 75 generations. The training curve is shown in Figure 12.9.

The final system had a performance of 77.38 percent for the training data set and 73.819 percent in the testing data set.

12.4.6 Concluding Remarks for Pima Indian Diabetes

We saw four different methods to solve the problem of the Pima Indian diabetes in this section. Here we first presented the different theoretical foundations of these systems, and then we studied these systems in terms of their application to the database of the Pima Indian diabetes. We observed the performance of the systems in both the categories of the training as well as testing database.

The first experiment done was of the use of ANN with BPA. We saw that the ANN was able to solve the problem appreciably well, and hence we got a good recognition that was capable of solving the problem. Then we tried to use an ensemble to remove some of the shortcomings of these systems. We observed that this gave a better training data set performance as well as a comparable testing data set performance. The next method discussed was the ANFIS. Here we observed that the system was able to give a good performance. The results reveal that the system performed very well on the training data set but did not do that well on the testing data set. This may be attributed to the failure of the system to generate rules that could be generalized over the network. At the training phase, the system must have adjusted the MFs so as to meet the requirements, but the rules so generated failed to generalize. This may be even because of the selection of the wrong data that essentially belonged to some parts of the network. Then we discussed the evolutionary ANNs. These systems also gave a good performance where they could decide their own architecture as well as the values for the weights and biases. We saw a good performance of these systems in the presented problem.

There are numerous possible ways to solve some problem, but a merely theoretical knowledge of the systems might not be helpful in deciding or predicting the system or model that must be used. Many times, data behaves in a characteristic manner that makes it difficult for some systems to perform while aiding in the performance of other systems. Hence all this requires a lot of practice, experience, and ability to depict the results.

12.5 FETAL HEART SOUND DE-NOISING TECHNIQUES

In this section, we use soft computing techniques for another very exciting use. So far we had been using soft computing systems as classifiers, where these systems calculated the existence of the diseases. Here we will use the soft computing system for a functional, performance-related task rather than for a classificatory task. Here we try to engineer a system that is able to remove or reduce the level of noise in the fetal heart sounds that are recorded in the experiments.

Fetal heart rate (FHR) monitoring refers to the monitoring of the heart rate of the fetus. This enables doctors to analyze the activity and hence the health of the child in the womb of a pregnant woman. This further helps in querying the health of the child from time to time so that measures may be taken at the early stage of any health hazard.

Numerous techniques are actively used for the FHR. One commonly used technique is ultrasound Doppler-based cardiotocography (CTG). In this technique, an ultrasound transducer sends ultrasound waves through the maternal abdominal surface toward the fetus. The ultrasound wave gets echoed as it is modulated by the movements of the fetal heart. One of the greatest disadvantages of these systems is the harm caused to the child from prolonged ultrasonic exposure. Also the cost factor is very high in terms of machinery and expertise, creating further difficulties especially when such a system is needed for prolonged periods in order to help anticipate fetal distress.

Fetal electrocardiography (FECG) is another tool that is used for monitoring. This uses electrodes on the maternal abdominal surface. However, the signal processing requirement is very complex in these cases. This is because of the difficulties faced in the separation of the material EEG signals from the needed EEG signals in the FECG signals. Although numerous methods have been proposed for the rejection of the maternal signal, the automated evaluation of FECGs is less accurate than the Doppler-based ultrasound method. Also in these systems the signal quality largely depends on proper electrode placement, which is another problem. Hence the prolonged recording is difficult to carry out and requires electrode readjustments due to fetal movements. The main benefits of FECG compared to the CTG are the lower cost and its noninvasive nature.

To further overcome these difficulties, we use a technique called phonocardiography. In this technique, instead of sensing and measuring fetal heart movement by an active ultrasound, natural heart sound is perceived and assessed from the subject abdomen. This technique applied is fully passive and noninvasive, thus it does not cause a problem on exposure. However, phonocardiography is susceptible to ambient noise. Unfortunately, the fetal heart activity produces very little acoustic energy, and, moreover, it is surrounded by a very noisy environment. The major factors contributing toward the noise in the system are acoustic noise produced by the fetal movement, maternal heart sound, maternal respiratory sound, maternal digestive sound, movement of the measuring head during recording—sheer noise, external noise originating from the environment—ambient noise.

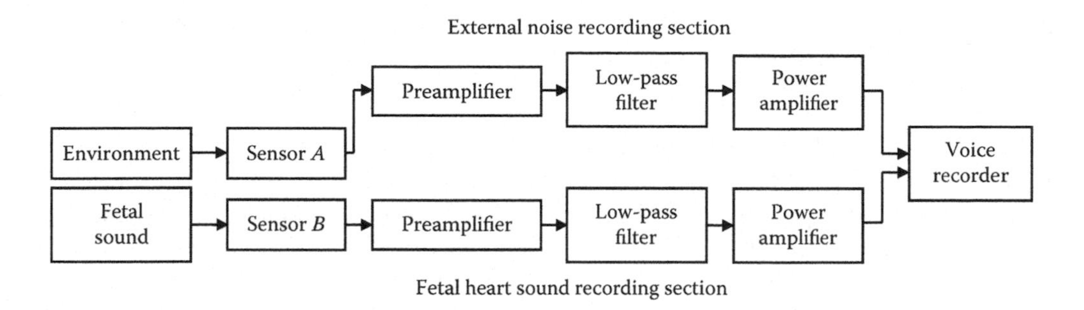

FIGURE 12.10 Block diagram of DRM module.

These factors are dynamic in nature and keep changing with time and situation. The objective here is to separate these factors from the useful parts of the sound signal or the fetal heart sound. This further requires very complex signal processing for the separation of sounds.

In this section we first present the signal detection and recording techniques. This is the first step in the de-noising that we are trying to achieve here. Then the various de-noising techniques are explained in general. Then we move on to the development of the adaptive filter that would adaptively remove the noise from the fetal phonocardiography signals.

12.5.1 Signal Detection and Recording

The basic motive of this problem is to remove the ambient noise that creates a lot of problems at the time of signal processing. In order to remove noise, we develop specialized techniques that require recording of the noise that is present in the signals. Hence we record ambient noise and fetal heart sound on two separate channels on a portable memory device. The first channel is used to record acoustic signals from the abdominal surface. The second one is used exclusively for recording external noise. This section discusses the experimentation and findings of Mitra et al. (2007).

The block diagram of the developed signal detection and recording module (DRM) is illustrated in Figure 12.10. The two sensors record their signals, which go to separate preamplifiers. This amplifies the signals and makes them capable of better noise rejection. The IC LM 381 is used here for this particular purpose, which raises the signal from the transducer level to the line level. The ambient noise is controlled with the help of an active low-pass filter having a cutoff frequency of 70 Hz. This value of cutoff frequency is selected because most of the fetal heart sound spectrum lies within the frequency range of 20 to 70 Hz. The commonly available operational amplifier IC 741, with a suitable resistance capacitor network, is used to implement the active low-pass filter. The output that the filter generates is further strengthened with the help of a power amplifier. This gives us the de-noised fetal heart sound. This sound is then fed to the recording device and headphone.

12.5.2 Signal De-Noising

As must be clear from Figure 12.10, we use two different sensors to record the signal. The first, or the sensor '*A*', records only the noise and is placed at a region near the abdomen to record nearly the kind of sound that would be present at the abdomen. The second sensor, or the sensor '*B*', is placed very near to the abdomen to record the fetal heart rate. This fetal heart rate gets mixed with a damped version of the external noise when it is recorded. This noise badly affects the original signal, which drastically changes from one beat to the next. The signal-to-noise ratio (SNR) hence becomes very poor. These unwanted, disturbing signals contribute to difficulty in identification of the principal heart sounds (*S*1 and *S*2). Here we study and discuss the various methods that help in the removal of these noises from the environment and ultimately lead to a better SNR. All these aim at cancellation of the noise by de-noise or anti-noise. These techniques for the removal of noise are called de-noising techniques.

12.5.2.1 Band-Pass Filtering Method

This method only carries forward the signals that lie within a comfortable range of 40 Hz $< f <$ 80 Hz. All the other frequencies are deleted. The selection of the lower limit is based on the experimental fact that a significant part of the maternal heart and digestive sound lies below this border, while the fetal heart sounds are not dominantly present there anymore. The band-pass filter is realized by designing two separate eighth-order Butterworth-type IIR digital filters, one low-pass filter and one high-pass filter. However, this method does not provide satisfactory results, as a significant part of the external noise falls in the pass band frequency spectrum, and there is no provision to limit these unwanted signals.

12.5.2.2 Signal Difference Method

We recorded the signals from two sensors, *A* and *B*, that recorded the noise and the noisy signal. Here, we make use of both of these to calculate the final system output. The external channel amplitude is adjusted and then filtered through a low-pass filter to accommodate the effect of sound insulation medium transfer characteristics, which are around the abdominal microphone. Subsequently, the filtered external noise signals are subtracted from the signals of the abdominal channel. One of the major limitations of this method is that the applied stationary model on the external channel is only an approximation of a nonstationary abdominal channel. Hence performance is limited.

12.5.2.3 Blind Separation of Source Method

Blind source separation (BSS) refers to the problem of recovering signals from several observed linear mixtures. The strength of the BSS model is that only mutual statistical independence between the source signals is assumed, and no prior information about the characteristics of the source signals, the mixing matrix, or the arrangement of the sensors is needed. The cancellation of external noise from abdominal signals can be stated as a two-source, two-drain problem for BSS analysis. The separation algorithms, when applied to nonstationary external noise, will require frequent recalculation of the separation matrix. This requirement of the BSS method eventually takes a lot of computational time and hence makes it unsuitable for the present application.

12.5.2.4 Adaptive Noise Cancellation Method

This is a very innovative technique of removal of the noise from the signal that uses an adaptive architecture for the purpose of the removal of the noise based on the noise in the background and the recorded fetal phonographic signal. This method is detailed in the following section.

12.5.3 Adaptive Noise Cancellation

We know that the actual system is very dynamic, where the fetal sound as well as the environment noise keeps changing from time to time. This necessitates the need of adaptive filters that can fulfill the requirements for the identification as well as the cancellation of noise. These systems calculate the noise that is most probably in the system and hence separate the noise from the rest of the signal. The goal is to make the output almost the same as the original signal not containing any instances of noise. The filter coefficients are changed for these purposes. Alternatively, it is possible to say that adaptive filter coefficient or weight changes over time to adapt to the statistics of the signal. These filters learn themselves, as the signals are being fed into the filters, and adjust their weight to achieve the desired result. There are numerous applications where adaptive filters are used. Some common applications are system identification, inverse system identification, noise cancellation, and signal

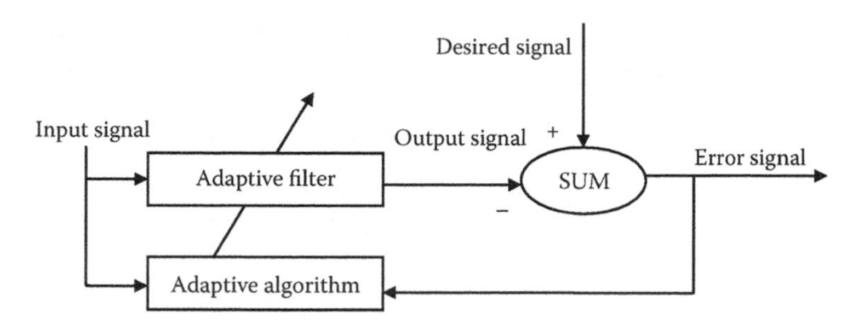

FIGURE 12.11 General adaptive filter.

predictions. Figure 12.11 shows the generalized adaptive filter block diagram for noise cancellation application.

The system that we develop here takes as its input both the recorded signal of the fetal heart as well as the sample noise. As the noise on the input port remains correlated to that on the desired port, the adaptive filter adjusts its coefficients. This change in coefficients reduces the difference between output signal and desired signal, which results in a clean signal on the error port. The generalized mathematical relations between desired signal $d(k)$, input signal $x(k)$, output signal $y(k)$, error signal $e(k)$, and filter weight $w(k)$ are given by Equations 12.1, 12.2, and 12.3.

$$e(k) = d(k) - y(k) \tag{12.1}$$

$$y(k) = \text{filter}\ \{x(k),\ w(k)\} \tag{12.2}$$

$$w(k+1) = w(k) + e(k)x(k) \tag{12.3}$$

This filter is used for removal of noise that gets mixed with the genuine signal of the fetal heart. The block diagram of the system or the adaptive filter used in this section is given in Figure 12.12.

Signal $d(k)$, which is to be cleaned, is applied on the desired port. It comes from the abdominal microphone, which contains the noise $n(k)$ and the desired signal $s(k)$. The external microphone carries only background noise $n_2(k)$ and is applied as the input signal $x(k)$ to the filter. As long as the input noise to the filter $n_2(k)$ remains correlated to the unwanted noise $n(k)$, the adaptive filter adjusts its weights $w(k)$, reducing the difference between $y(k)$ and $d(k)$. This results in removal of the external noise, and the de-noised fetal heart signal $s(k)$ will appear on the error port. The generalized mathematical relationship between various signals and filter weights is given in Equations 12.4,

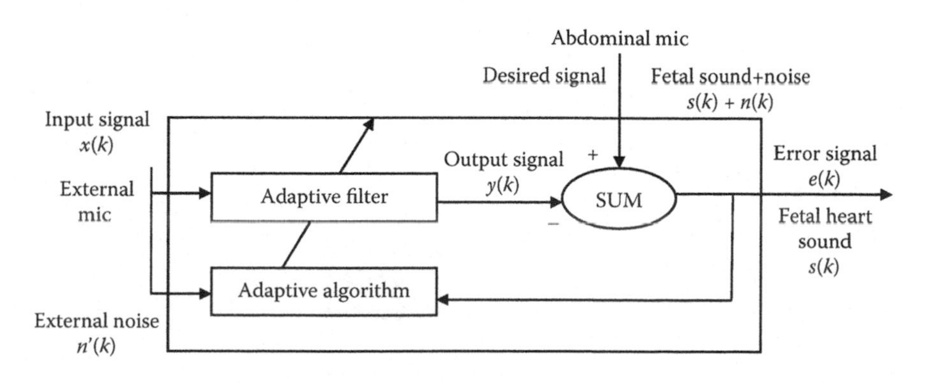

FIGURE 12.12 Adaptive filter used for de-noising.

12.5, 12.6, and 12.7. It is important to note that in this implementation, the error signal converges to the input data signal, rather than converging to zero.

$$d(k) = s(k) + n(k) \tag{12.4}$$

$$y(k) = \text{filter}\{x(k), w(k)\} \tag{12.5}$$
$$\text{or } y(k) = w(k)\,.\,n'(k)$$

$$e(k) = d(k) - y(k) \tag{12.6}$$
$$= s(k) + n(k) - w(k)\,.\,n'(k)$$
$$\approx \text{s}(k)$$

$$w(k+1) = w(k) + e(k)\,.\,x(k) \tag{12.7}$$

12.5.4 System Simulation

Various algorithms have been simulated in order to study the effect of adaptive filtering. Algorithms took, as their input, the fetal heart sound signals of widely distinctive nature and quality. MATLAB Simulink is used to model and simulate these algorithms, which are then analyzed for their output performance in the time and frequency domain. These models are built by connecting the required blocks, which are available in the software library, and their parameters are entered while designing them for simulation. One such test model is illustrated in Figure 12.13.

The fetal heart sound signal detected from the mother's abdomen is fetched from the corresponding *.wav file and applied to the desired port of the filter block. This signal carries the fetal heart sound and a damped version of the external noise. The unwanted external noise is available in another *.wav file and is applied to the input port of the filter. The de-noised signal comes out through the error port, which is connected to the time scope block, where original signals and de-noised signals are displayed and compared.

With many adaptive filters to choose from, selection of one that best meets the specific requirement needs very careful consideration. After making elaborate and extensive tests using various

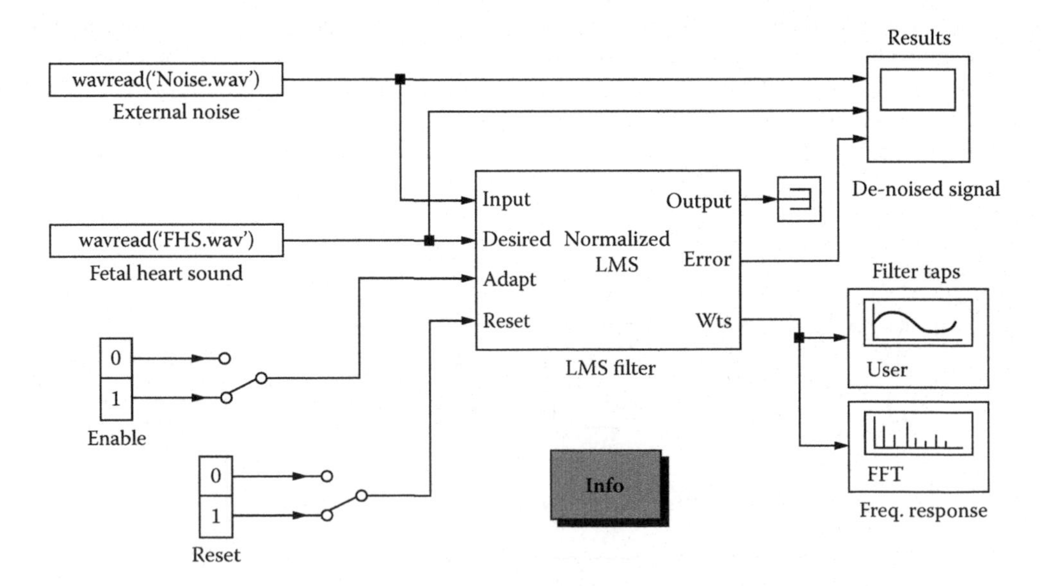

FIGURE 12.13 Computer simulation for adaptive noise cancellation.

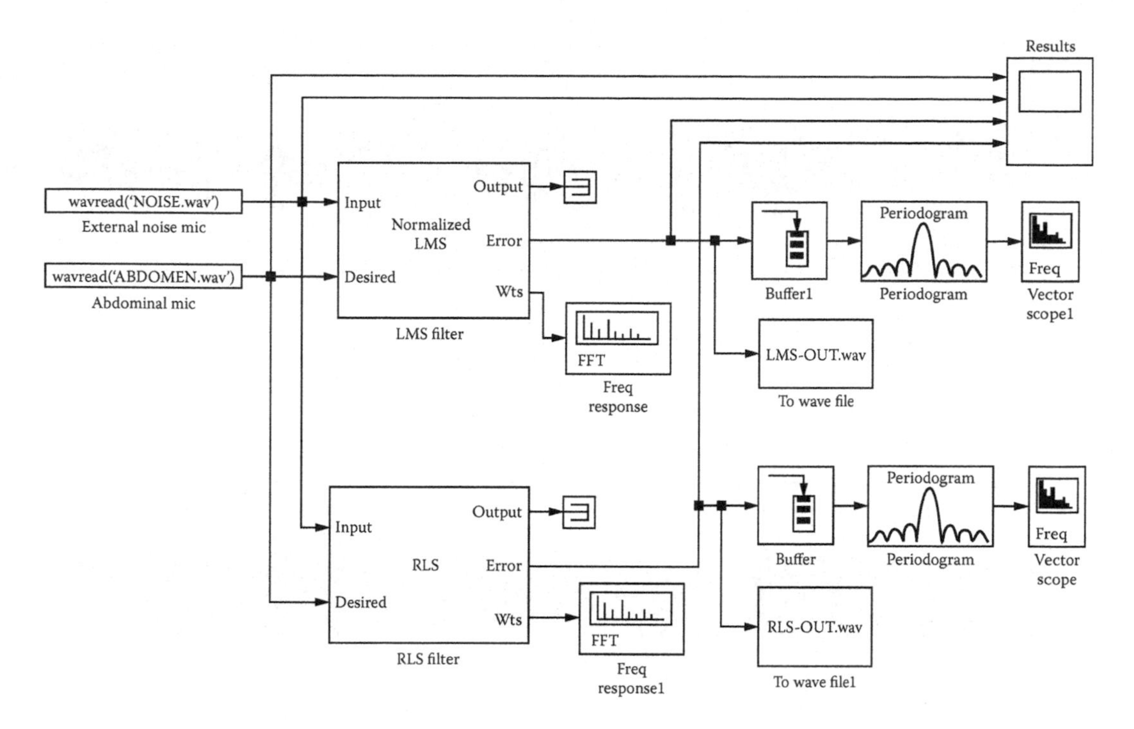

FIGURE 12.14 Computer simulation for comparative analysis between two adaptive filters.

simulation models as already outlined, the recursive least squares (RLS) and normalized least mean square (NLMS) adaptive filter algorithms were found to be more suitable, with superior performance and consistent results. The NLMS adaptive filter is straightforward to implement and sufficiently effective in use. It provides a solid baseline against which other, more complex adaptive filters can be compared and investigated. For making a fine comparison, another computer simulation is developed and employed to compare different adaptive algorithms. Figure 12.14 shows such a simulation for comparison between the normalized LMS and RLS filters. In this model, signals identical in magnitude and phase are applied to the input and desired ports of the filters under comparison. De-noised signals come out from the error ports, which are subsequently analyzed and compared in the time and frequency domains. These signals are further stored in separate *.wav files for further quantified analysis through the wavelet transform method.

12.5.5 Results

A prototype of the phonocardiography recording device designed and developed has been used for fetal heart sound and external noise recording. More than 20 fetal heart sound recordings were taken from different women, who were between 36th and 40th week of singleton pregnancy. Recorded data were transferred to a personal computer as *.wav files through a multimedia card. The MATLAB simulation considered previously was used to process and display the recorded sound from the wave files. Figure 12.15 shows the waveform of one such record, in which the x-axis represents the time in seconds and the y-axis represents the amplitude of the signal in volts. Figure 12.15a represents a 2-second time span of the representative sample heart sound of a fetus after the 38th week of gestation. Figure 12.15b is the correlated external noise, which is principally from electrical gadgets and other commotions present in the maternity room. It can be noted that the external noise is too high, which also pollutes the fetal heart sound to a larger extent. This representative sample of noise is used as the reference input of adaptive filters in the signal processing stage of simulation. Figures 12.15c and 12.15d represent the results of normalized LMS and RLS

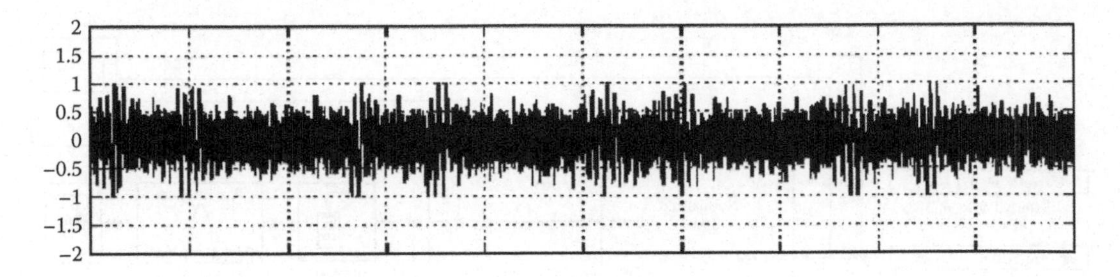

FIGURE 12.15a Simulation results: Signal from abdominal channel.

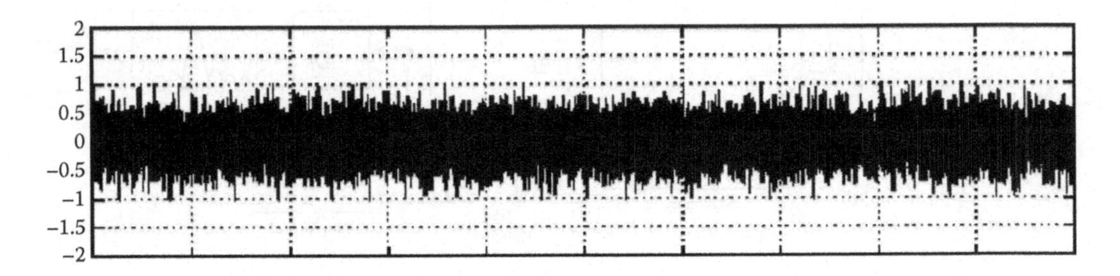

FIGURE 12.15b Simulation results: Signal of external channel.

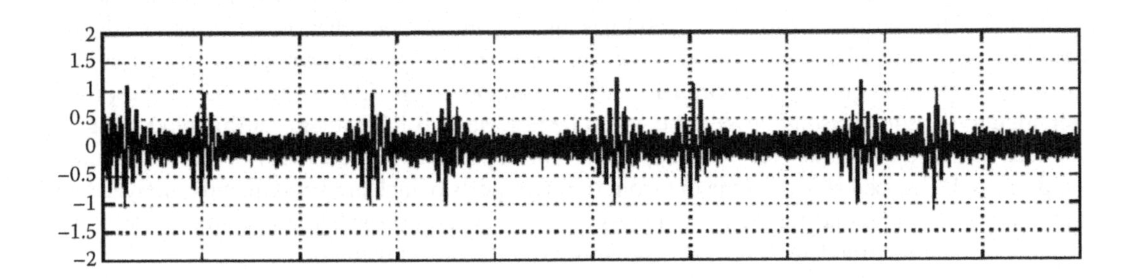

FIGURE 12.15c Simulation results: De-noised signal by NLMS filter.

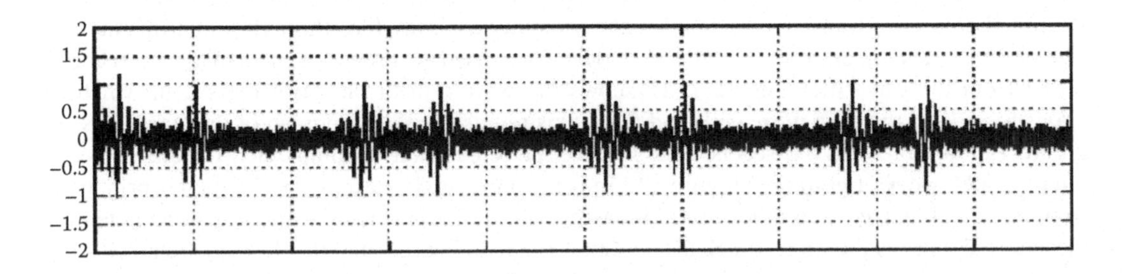

FIGURE 12.15d Simulation results: De-noised signal by RLS filter.

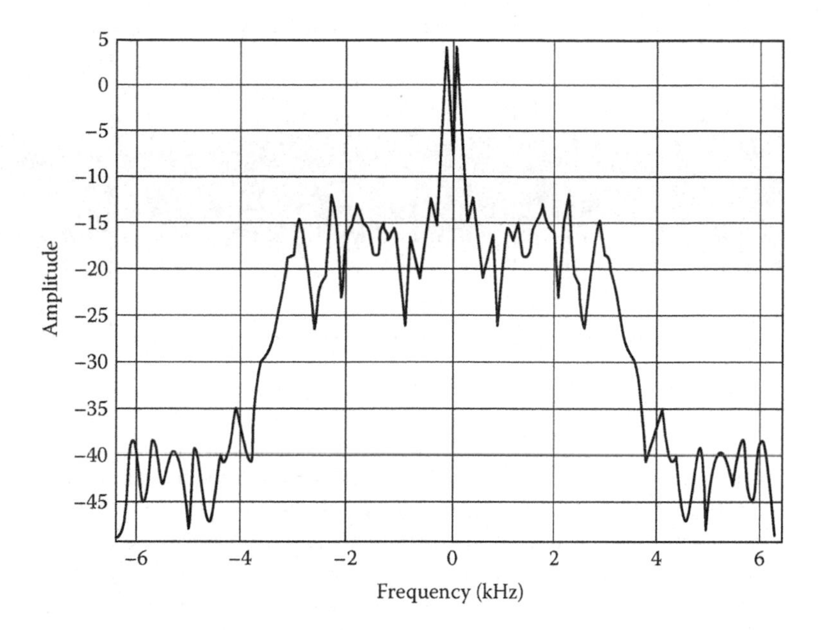

FIGURE 12.16a Frequency spectrum of signal from NLMS.

adaptive algorithms, respectively. It can be clearly deduced that the fetal heart sound is meticulously extracted from the abdominal signal and the signal-to-noise ratio is considerably increased by the use of these methods. Although the output of two algorithms looks very similar, making detailed close observations one can see the minute differences and it can be deduced that the RLS algorithm performs better than the normalized LMS algorithm.

Along with time domain response, the simulation also shows the frequency spectrum of the filter's output signals. The spectral display for fetal heart sounds around the 0.55 second instant of recorded signal is shown in Figures 12.16a and 12.16b.

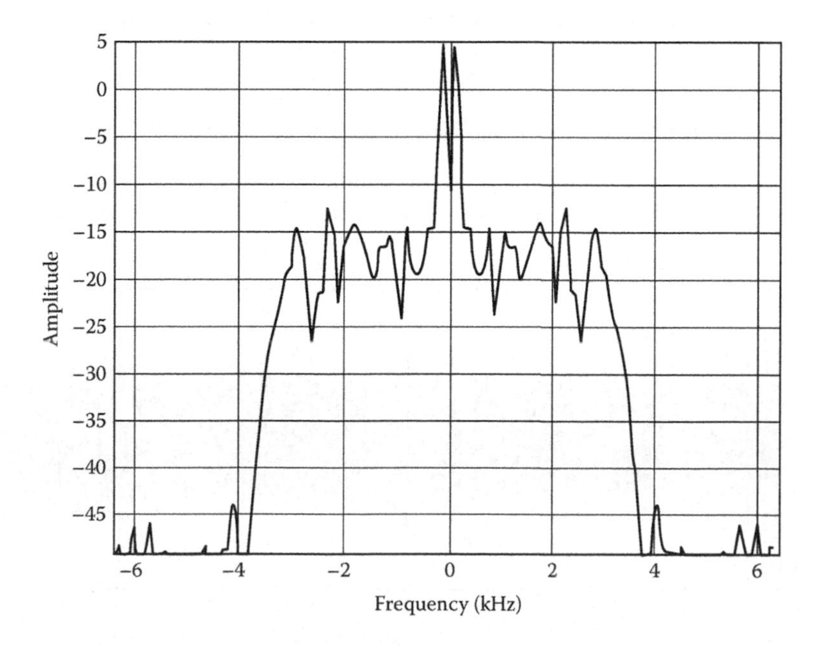

FIGURE 12.16b Frequency spectrum of signal from RLS adaptive filters.

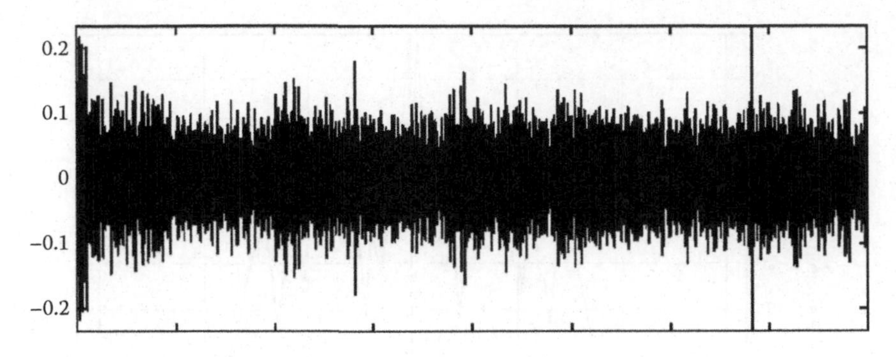

FIGURE 12.17a Detail signal component of NLMS.

It is evident from the spectrogram that the high-frequency components are notably more prominent in the NLMS filter output in comparison with RLS filter output. In order to quantify these results, filtered phonocardiograph data were further analyzed by the wavelet transform method. Using the Daubechies–1 wavelet, signals from NLMS and RLS adaptive filters are decomposed to level one. The decomposition gives both high- and low-frequency elements, called detailed and approximation components, respectively. By means of short-time Fourier analysis, it is determined that the dominant part of the acoustic energy produced by the fetal heart activity lies in the frequency band 20 Hz $< f <$ 80 Hz. Hence, approximation components of the wavelet transform will mainly contain fetal heart sounds, whereas detailed components will include only noise and unwanted signals. Figures 12.17a and 12.17b display reconstructed detailed signal components from NLMS and RLS filters. It is important to note that these signals do not carry fetal heart sounds and are mainly composed of high-frequency noise.

The statistical analysis for the detailed component of recorded de-noised signals is carried out by the wavelet transform toolbox of MATLAB. Table 12.3 summarizes these results in which the first record is the analysis of the waveform shown in Figure 12.17.

It is clearly observed from these results that the difference in the standard deviation by two methods is very small; however, it is invariably lower with the RLS algorithm than with the NLMS algorithm.

12.5.6 Concluding Remarks for Fetal Heart Sound De-Noising Techniques

Cardiography and cardiotocography are the most commonly used noninvasive clinical diagnostic tools that provide accurate determination of FHR. These devices have a few inherent limitations

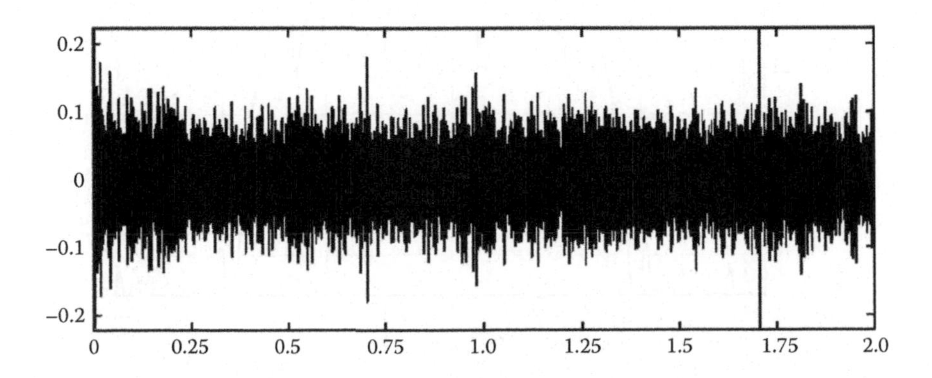

FIGURE 12.17b Detail signal component of RLS adaptive filters.

TABLE 12.3
Comparison Between Statistical Results for Fetal Heart Sound De-Noising Techniques

NLMS Algorithm			RLS Algorithm		
Maximum	Minimum	Standard Deviation	Maximum	Minimum	Standard Deviation
0.2345	–0.2345	0.0433	0.2227	–0.2227	0.0415
0.305	–0.305	0.035	0.285	–0.285	0.0312
0.401	–0.401	0.041	0.399	–0.399	0.04
0.2157	–0.2157	0.0311	0.201	–0.201	0.0301
0.421	–0.421	0.0251	0.4011	–0.4011	0.0231
0.354	–0.354	0.0711	0.3211	–0.3211	0.0701
0.2591	–0.2591	0.0401	0.2231	–0.2231	0.0399
0.315	–0.315	0.0931	0.209	–0.209	0.2511
0.2541	–0.2541	0.0451	0.2211	–0.2211	0.0441
0.351	–0.351	0.0314	0.0331	–0.0331	0.0295

due to the use of ultrasound. To overcome this, an alternative technique, phonocardiography, was used in which acoustic vibrations produced by mechanical activity of heart are detected. This technique is fully passive and therefore provides absolutely noninvasive fetal diagnostic measurement. A portable lead-free fetal heart sound recording system was designed and developed, which can be used for nonstress tests and for long-time monitoring of fetal heart sounds in high-risk pregnancies. It was exhibited that fetal heart sounds can be recorded using acoustic signals of fetal heartbeat, and this can be done even in highly disturbing external noise conditions. Recorded signals were transferred to a standard PC and digitally processed by different filtering and noise cancellation methods. MATLAB simulation and wavelet transform were used to compare different adaptive algorithms, and the most suitable one was identified. It can be safely concluded that the RLS adaptive noise cancellation technique definitely gives most the stable and satisfactory results, which can be used in the designing of the preprocessing stage of phonocardiography-based fetal monitoring instruments.

Section III

Soft Computing in Other Application Areas

13 Legal Threat Assessment

13.1 INTRODUCTION

Chapter 4 introduced you to fuzzy systems, including the ways in which fuzzy systems could be used as an effective means for determining the vulnerability of any system. In this chapter, we develop a complete expert system for legal threats using the same concepts. First we introduce the basic terminologies used in the text as well as discuss the current works in this area, along with the objectives for the system. Then we deal with the architecture of the expert system and model to carry out threat management. We also discuss the input modeling and fuzzification of the input. Next we analyze the rule and the expert system in terms of the relationship between the linguistic variables and output. Then we will evaluate the expert system as a whole with given test cases. We conclude the chapter by considering future work that may be carried out in this field. We first define a few terms used in the text.

13.1.1 THREAT, JUDICIARY, AND JUSTICE

Threat may refer to: an act of coercion wherein a negative consequence is proposed to elicit response, such an act of coercion between nations, a threat of force (public international law), and an expression of intention to inflict evil, injury, or damage. (Source: Merriam Webster Dictionary)

We are all used to threatening others or being threatened by others in our daily lives. Threats may take numerous forms and have numerous causes. Consider, for example, the case of a factory, where the employees are constantly threatening the employers with strikes aimed at deriving additional benefits. Similarly the employers threaten the workers with expulsions and unfavorable actions. Likewise, a parent may threaten a child, or a friend may threaten another friend. People who are threatened may often feel compelled to act in accordance with the desires of the threatening person or situation. Usually this is to avoid the negative consequences of the threat. This largely depends upon the negative impact that the threat may cause.

In this chapter, we specifically deal with the legal threats. Every society or government follows laws for the maintenance of law and order. These are legal rules that are binding on all. The defiance of these laws leads to punishments as per the laws. These laws ensure the smooth functioning of the society without any worry or "threat." The threat arises in the system when we fear some element of the society acting in an unlawful manner and hence making us suffer loss. The level or risk associated with the threat depends heavily on the effort of the person threatening to execute his or her dangerous plans and leading to a loss on the part of the threatened personnel. Naturally, if someone can pose a threat and hope to derive huge profits with little effort and risk, he or she may most probably carry forward his or her plans and finish them successfully. Hence the associated risk is very high. Consider how revealing some piece of information may be very easy by using post, forums, blogs, and the like. A person threatening may easily be able to do it without much risk. But stealing some protected information may be very difficult with high risk of being caught in the process. A person may not do this very easily.

In mathematical terms: "In each period, t, the victim chooses a payment, pt is greater than or equal to 0, to make to the threatener, after which the threatener must decide whether to carry out the threat. If the threat is carried out, the game ends; the victim suffers a loss, v, and the threatener gains b (in addition to the stream of past payments, ps, where s is less than or equal to t). If the threatener does not exercise the threat, the game proceeds to period $t + 1$, and the stage game is repeated" (Source: Steven Shavell and Kathryn E. Spier, "Threats Without Binding Commitment," *Topics in Economic Analysis & Policy*: Vol. 2: Iss. 1, Article 2, 2002). In the case of a blackmail that can be played over an infinite time horizon, "The present value of the payment stream is bounded between the value of immediately carrying out the threat to the threatener and the harm to the victim were it carried out."

The legal rules in any state or society are implemented through a fixed judiciary system that runs on the basis of rules framed during the making of the constitution of the state. These become the guiding principles for making all kinds of decisions. Decisions may take the form of deciding whether the person is guilty or not and, if guilty, the seriousness of the crime that he or she is convicted of. The punishment depends upon this. The judiciary is framed keeping in mind the ethics, values, and culture of the people as well as the state. The key point here is that the judiciary or the law-implementing body makes decisions based on the known evidence. Some amount of fuzziness is always associated with the decisions. The fuzziness may depend on numerous factors, such as the availability and precision of the evidence, the guiding judicial principles, and the time taken. Milton Friedman has this to say about the role of government in capitalism and free societies: There is no formula that can tell us where to stop. We must put our faith, here as elsewhere, in a consensus reached by imperfect and biased men through free discussion and trial and error.

Another commonly used term in this context is *common law*. The common law describes the body of legal principles and concepts that were evolved over many centuries by judges in the English courts of law. These principles include such values as honesty, the pursuit of truth, responsibility, fairness in interpersonal relations, concern for one's immediate neighbors, respect for property, loyalty and duty to one's spouse and children, the work ethic, and keeping one's word. The common law decides the fate of any case when it comes to be solved under any judiciary system. The two conflicting parties are dealt with according to the norms of the common law. An understanding of the context of the problem or dispute and the associated norms of the common law leads to a solution of the problem that respects the clauses of the parties and leads to the most fit solution. The common law is flexible enough to adapt to the specific problem and its associated complexities.

A law enforcement agency (LEA) is any agency that enforces the law. This may be a local or state police department, a federal agency such as the Federal Bureau of Investigation (FBI) or the Drug Enforcement Administration (DEA) in the U.S., or else the Chief Bureau of Investigation (CBI). It could also refer to a national police force such as the Indian Police Services (IPS), and it can be used to describe an international organization such as Europol or Interpol as well.

Justice, or precisely a liberal justice system, has three major factors associated with itself. The first factor deals with the resolution of conflicts and the need to reach a proper verdict to each and every case based on the principles of common law. This factor tries to resolve all the conflicts between the two conflicting parties and to make proper decisions based on the social, interpersonal, and related factors. The second factor that is associated with justice is that no one should be punished or disadvantaged except for fault. This prohibits the punishment in any sense or degree of a person who is likely to be innocent. The justice system makes every attempt to ensure a person is actually guilty before reaching any verdict and punishing him or her for the offense. The third factor associated with the judiciary stresses the procedures. Procedures limit power by providing for wide consultation among interested parties.

13.1.2 Role of Time in Threat

In this section, we discuss one of the major reasons for the development of threat assessment systems, which is the factor of time, something that plays a huge role in any system. The threat assessment system is no exception. The burning demand of time heightens the need for threat assessment that can be used as decision support systems to support the verdict by using analysis.

Today the crime rate has increased, and hence the number of cases in the courts has increased. Justice has no meaning if it cannot be achieved on time. The growing number of cases has already resulted in cases remaining under consideration for many years altogether and the verdicts taking a painfully long time. By the time a final decision is made, all the effectiveness of the case is lost. The verdict hardly compensates for any loss to the threatened or sufferer. Also the large amount of time spent results in heavy losses to both parties. It may hence be realized that decisions lose their value if they take a long time.

Over three million cases are pending in India's 21 high courts, and an astounding 26.3 million cases are pending in subordinate courts across the country. At the same time, there are almost a quarter million defendants languishing in jails across the country. Of these, some 2,069 have been in jail for more than five years, even as their guilt or innocence is yet to be ascertained.

If we look at India and study its stability and susceptibilities to threat, we would identify two major sources or classes of threat. The first is the threat from the internal sources. There are over 1.1 billion people residing in the country. There are huge possibilities of internal stability being lost due to the conflict between individuals or social groups. Further, there is always a risk of external conflicts. These are the conflicts that may arise between Indian and any other country on any issue. These two sources determine the threat to the country.

13.1.3 Key Outcomes

The main objective of this chapter is to develop a system that can be used for effective threat assessment. As an outcome of this system, we should be able to get the following benefits. In the last part of the chapter, after we have made an in-depth analysis of the system, we will see how the developed system is able to achieve these functionalities.

- *Manage risk of a particular threat*: The main objective of the system being discussed is to prioritize the threats and then evaluate the risk associated with them. By this means, we should be able to understand the various kinds of threats that exist in the wider system. We should further be able to understand which of them are alarming and need serious attention. Then we need to develop some mechanisms by which the threat can be reduced to manageable levels. All this takes place through the use of proper mechanisms to understand the relative threat at any point.
- *Devise a suitable model for threat categorization and input modeling*: Numerous kinds of threat may be potentially exist. The actual concept of threat is vast and can be generalized to a very large level. It is not possible to study and experiment with all the various kinds of threats. For this, we need to place threats in various categories. Also we need to identify the common and most alarming threats at any place and time. Further each threat uses its own set of inputs that play a major role in deciding the threat value. Choosing and representing the correct parameters and measuring them properly is yet another aspect of study of the system.
- *Appreciate the validity of the fuzzy inference engine in threat management*: The fuzzy theory is a good fit for this problem because of its flexibility and wide applicability. Fuzzy applications are much closer to natural applications and hence have a great capability to model the system much as the natural systems model does. In the system that we are developing, we mainly try to adapt the FIS to the problem.

13.1.4 Motivation

The current socioeconomic condition of the country urgently calls for such an expert system for automatic threat assessment based on some parameters that affect the threat. Automation must play a vital role as an expert system or decision support system for the aid of the authorities.

The process of threat modeling helps system architects assess and document the security risks associated with a system. There are various issues and factors in the process of modeling. All the factors need to be very carefully engineered and then used and depicted with some metrics. The system that evolves is a great boon in various ways. The system helps in determining the threat in any given scenario. Based on this threat, if the threat level requires concern, proper actions may be taken to curb and control the threat. This may be done by identifying and installing the necessary controls at the site of the threat. Security arrangements may be one such kind of control that is extensively used for the security of all public places.

In this chapter, a fuzzy logic–based technique is designed for effective threat modeling. Thus, a fuzzy rule–based inference engine has to be developed to aid in modeling the threats and evaluating the risks associated with them given the threat prioritization. The higher the threat, the higher the vulnerability from that threat, and hence proper actions and measures have to be taken for such threats.

13.1.5 Literature Review

In this section, we present a brief overview of the various kinds of approaches and systems that have been developed for the purpose of vulnerability assessment. The systems depend upon the kind of threat that we are trying to identify. This may be a legal threat, a software threat, a systems threat, or any other type of threat. Based on these threats, various statistical and other models have been developed that try to rate or categorize threats at a place or system.

Threat or risk is prevalent in software and creeps in at the design stage of the software. Based on the decomposition of applications, analysis evaluates the threats and risks to a system and chooses techniques to mitigate the threats. Another common way of dealing with threats is by modeling them with the attack trees, which describe the decision-making process attackers would go through to compromise the system. Casteele presented a paper on threat modeling for web applications. This work focused on the important OWASP (Open Web Application Security Portal) top 10 web application security vulnerabilities/problems. An architecture that is based on the STRIDE model was used for the analysis. Klein presented a similar technique of carrying out threat modeling on a voting system. Several academic teams jointly wrote a paper on analysis of threats that occurred when smart cards are used in web applications. Their analysis was a part of the Designing Secure Applications (DeSecA) project, funded by Microsoft. The aim of their project was to provide an application developer with a tool that allows him or her to prevent the exploitation of a broad range of threats. They investigated common threats in five areas, each focusing on one particular technological building block for web applications. One of these was the smart card, and in particular the electronic identity card.

Sodiya presented an architecture for producing secure software, and threat modeling was the foundation of this architecture. A comprehensive Internet browser threat model was discussed in Griggs. He first listed various well-known threat models on the Internet and then identified the ones specific to browsers. The main types of attacks discussed are phishing, eavesdropping, man-in-the-middle, and denial of service (DoS).

Most of the papers discussed threat modeling for applications and how to incorporate it into development process of application products (in quantitative terms). This means applying these threat modeling techniques to real life threats so that they can be applied and risk can be prioritized using the threat mitigation process.

13.1.6 Approach and Objectives

The primary objective of this chapter is to develop an expert system to calculate the prioritized threat and risk. The objectives at the ground level are:

- Identification of the attributes that affect the vulnerability to threats
- Input modeling of the attributes by giving them appropriate weights and quantification
- Development of a rule-based inference engine and expert system

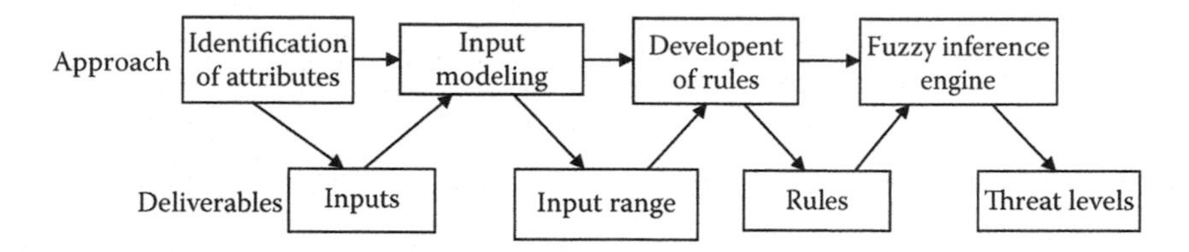

FIGURE 13.1 Approach for expert system.

We now state the basic approach used in the development of the threat assessment system. This involves all the various steps we named earlier as making up any expert system. The basic approach starts with the identification of the input attributes. Here we list all the factors that can contribute toward the threat. Each and every input then needs to be engineered. We need a proper metrics using which we can represent it numerically. Further we need minimum and maximum values of all the inputs under which our system performs. For the identification of attributes, proper surveys of both courts and police cases have to be done.

Following this, rules have to be devised from the pattern of the cases. These rules are more or less based on common sense, the way we expect the system to work with the given inputs. Then we integrate all these into the FIS. The FIS takes in the used entered input values and returns the threat in the scenario we engineered it for. The approach is represented in Figure 13.1.

13.2 EXPERT SYSTEM

We have already discussed the architecture, properties, and working of the expert systems in Chapter 1. In this section we give a very brief overview and then proceed with making the expert system. Expert systems are designed to behave just like experts in the field in which they have been trained. For the customer or the end user, it makes no difference whether he or she is front of a human expert or a designed expert system. These systems consist of a knowledge base that is developed by acquiring knowledge over a long period of time and stored in a permanent knowledge database. The expert systems also include a rule base, which stores and summarizes the knowledge and is used to give expert output to any of the presented inputs. The expert systems are very flexible in terms of knowledge capture, which means that they can learn a lot over time. The rules, however, may need to be refreshed with time. The output of the expert systems may be of great value when either directly implemented in the system or acting as a guide for the human expert who makes the decisions. The expert systems provide a sophisticated mechanism of knowledge acquisition and use to make effective decisions for the decision support systems (DSS).

The expert system that we develop here also follows the same principles. We develop a knowledge acquisition system that uses a Java interface to acquire new knowledge and store it in a database or knowledge base. The fuzzy rules are the rules that map the inputs to the outputs. The outputs are validated against all known knowledge or cases.

13.2.1 Expert System Architecture

The proposed expert system for the threat management is categorized in three sections.

- **Threat capture system**: The threat capture system takes the input from the user, target, or third party (courts or law enforcing agencies) to develop the knowledge base. Java Swing is used to develop the user interface, and the data is stored in an MS Access database as a knowledge base.

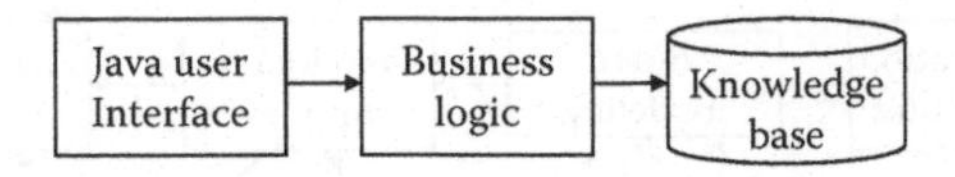

FIGURE 13.2 Threat capture system.

- **Input modeling**: The input modeling is the key part in the expert system. It transforms the input knowledge to quantize it into some range using a logic described in following section. The input modeling also includes fuzzification of the input as a linguistic variable.
- **Inference engine**: The third and most important part is the rule-based inference engine. This part does the job of mapping the inputs to the outputs on the basis of the rules that are worked over. This will be discussed in next section.

13.2.2 Threat Capture System

This system captures the threat from the source, target, or third party, such as police personnel or courts. The main objective of this system is to develop a knowledge base from the collection of legal threats and their outcomes. Prior of this system, an intensive study of the possible attributes was made; this identified knowledge was then taken from the source, target, or third party and stored in a database for the knowledge creation. The system is shown in Figure 13.2.

Another major aspect of the system developed is the inputs. The specific domain of threat assessment that we are discussing is the legal threat. We need to clearly identify the inputs that can affect the threat. This requires a very clear understanding of the crime, criminal psychology, proceedings of the law, and risk factors that play a role in the entire process. Once we develop a good insight into this entire operation of crime, we are able to work out the various inputs that affect any big or small criminal operation. In this manner, we jot down all the major inputs of the system. In order to keep the number of inputs manageable and finite, various similar inputs may be clustered together. Further if any two inputs are likely to behave in a similar manner or likely to have a very high correlation factor, they may be clustered together or any one of them may be taken. In this manner, we would be able to figure out and come up with a set of good inputs for the effective system design.

In order to completely understand the system and identify the inputs, we need to study the existing cases that may be obtained from public legal databases, court proceedings, police records, etc. The police themselves are a good source of information for an insight into the criminal psychology, the kinds of commonly occurring cases, and threats, and the like. All this gives us a valuable source of information from which we may be able to mine the input factors that affect the threat.

Drawing on all this analysis, work, surveys, and study, we identified the following inputs for the knowledge base of the system. These all cover a wide variety of threats and possible risks. Using the information entered here for the knowledge base, we would be able to derive the identified system inputs that we study in the next section.

1. ***Threat type***: The threats can be of different types. As per surveys and the possible number of cases in courts, we have identified seven types of threat in a court case:
   a. **Criminal**: This covers criminal charges or threats involving law enforcement. All terrorist threats such as bomb attacks, attacks to a building, robbery, smuggling, or kidnapping should be treated as criminal threats.
   b. **Denial of access**: This is when a right of access has been denied (e.g., denials of access to documents, public places, or court hearings).
   c. **Disciplinary action**: This covers actions by employers, schools, and other organizations that have adverse consequences (e.g., suspensions, demotions).

   d. **Correspondence**: This is for informal threats such as email or letters.
   e. **Lawsuit**: This is for threats involving a civil court proceeding.
   f. **Subpoena**: This applies if a subpoena has been used as a legal threat (e.g., to discover someone's identity or to request news gathering materials).
   g. **Other**: This is for any other type of threat.

2. ***Date***: The date of legal threat (e.g., date lawsuit was filed; subpoena issued, letter sent, or disciplinary action taken) is given in the DD/MM/YEAR format.
3. ***Party issuing threat***: This is the full name of the person or organization that initiated the legal threat.
4. ***Party receiving threat***: This is the full name of the person or organization that received or responded to the legal threat.
5. ***Source and target locations***: This covers the location or state of the source and target.
6. ***Source and target type***: The source and target types can be:
   a. **Individual**: If a person is acting in his or her individual capacity
   b. **Organization**: If the entity is a business or group of individuals acting as an organization
   c. **Large organization**: If the entity is well known or has more than 500 employees (e.g., General Motors, Viacom, RIAA)
   d. **Government**: If a local, state, or central government is involved
   e. **Intermediary**: If the party's role in the case or controversy involves hosting or transporting content for others (e.g., Facebook, YouTube, Blogger)
   f. **Media company**: If the party is a large or well-known content creator (e.g., Viacom, CNN, New York Times)
   g. **School**: If the party is a public or private educational institution

7. ***Source income***: This is the income of the source: i the income of an individual or the revenue of any institution.
8. ***References***: The are possible references the source can produce.
9. ***Threat content type***: The threat content type may be audio, video, plain text, etc.
10. ***Publication medium***: This input covers the means/form of communication at issue in the dispute. This input may be "blog," "broadcast," "email," "forum," "podcast," "print," etc., where all words have their usual meanings. The input is "social networking site" for content on social sites such as MySpace and Facebook; "user comment" for blog comments; and "website" for web content that does not fit into another category.
11. ***Court type***: This is the court where the case has been filed.
12. ***Status***: This input decides whether the legal threat remains pending. Lawsuits and criminal matters that are on appeal should be marked "pending."
13. ***Disposition***: This input is "dismissed (partial)" if a party or claim has been dismissed but some portion of the case remains; "dismissed (total)" if the entire case has been dismissed; "lawsuit filed" if a letter or other informal threat has resulted in a lawsuit; "settled" if the case or controversy has settled; "verdict" if the case has resulted in liability for a party; and "withdrawn" if the entire case has been withdrawn.

13.2.3 Input Modeling

The input modeling takes the knowledge of the knowledge base to model the input for the fuzzy inference engine. The nature of the attribute cannot be a crisp number because the threat depends on all attributes and these attributes are psychological and behavioral in nature. We cannot just give any weights and then use some mathematical model to find the vulnerability and associated risk of the threat. Hence we apply the fuzzy logic and use the rule-based inference engine to calculate it.

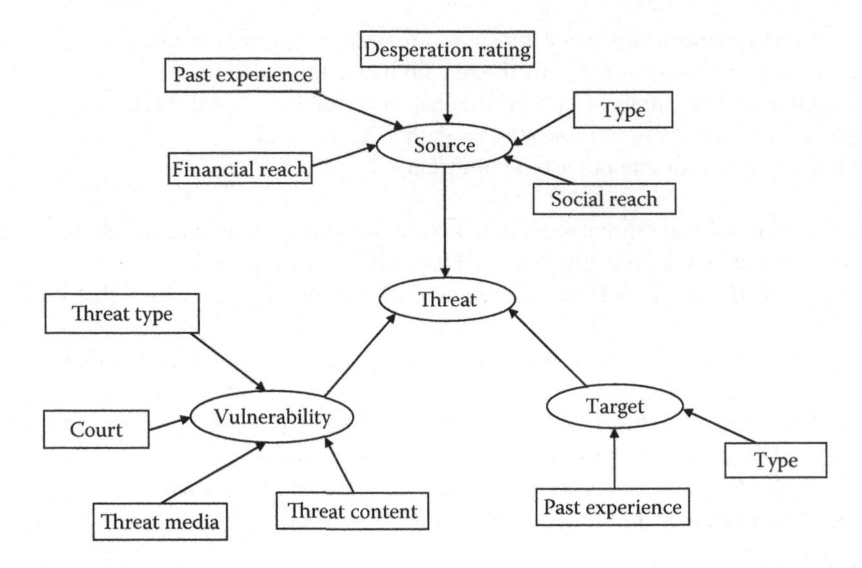

FIGURE 13.3 Input categorization.

The fuzzification of the input has to be done. The first step in input modeling is to identify the inputs from the knowledge base and their quantization.

13.2.3.1 Input Quantization

It has been observed that vulnerability to any threat depends upon the three major characteristics. Hence, the input for the fuzzy inference engine can be divided into three parts: source inputs, target inputs, and threat inputs. This is shown in Figure 13.3.

13.2.4 Source Input

All the input related to source is taken into consideration. The source plays a very important role in identifying the risk of the threat. The following input is taken:

13.2.4.1 Source Type

This input is the type of the source. Sources are broadly categorized into seven categories, with category 6 as "Others" if the source type does not fall in any of the other categories. The source type is 0 for school, 1 for intermediary, 2 for government, 3 for individual, 4 for organization, and 5 for large organization. The range for this input is 0 to 6. The value of source type is summarized in Table 13.1a.

TABLE 13.1A
Source Type Value for Various Kinds of Sources

	Source Type
School	0
Intermediary	1
Government	2
Individual	3
Organization	4
Large organization	5
Others	6

13.2.4.2 Past Experience

Source past experience is the behavior of the source. If we get any cases of source from the knowledge base, we can be able to derive the *source_past_experience* as the positive cases from the source. This is given in Equation 13.1. This input lies in the range of 0 to 1.

$$\text{source\_past\_experience} = \frac{\text{number\_of\_positive\_dispositions(source)}}{\text{total\_number\_of\_cases}} \tag{13.1}$$

13.2.4.3 Desperation Rating

This input is measured from the number of threats. The range of this input is between 0 and 4. The desperation rating is taken as the number of threats, up to a maximum of 4. Initially the rating is 0. This is given by Equation 13.2.

$$\begin{aligned} \text{Desperation\_rating} &= 4, \text{ if number\_of\_threats} >= 4 \\ &= \text{number\_of\_threats, otherwise} \end{aligned} \tag{13.2}$$

13.2.4.4 Financial Capability

This input depends upon the source type and source income. It is a real input between 0 and 1.

13.2.4.5 Social Capability

This input measures the source's social reach and behavior in society. It is very unlikely that if the source has good social capability, it can produce an effective threat. Hence an empirical method of number of references can be used for the social capability. Social capability lies in the range of 0 to 4. Source capability is equal to the number of references, up to a maximum of 4. This is given in Equation 13.3.

$$\begin{aligned} \text{Social\_reach} &= 0, && \text{if number\_of\_references} = \text{“none”} \\ &= 4, && \text{if number\_of\_references} >= 4 \\ &= \text{number\_of\_references}, && \text{otherwise} \end{aligned} \tag{13.3}$$

13.2.5 Target Input

Like the source, the target also greatly determines the vulnerability. The following factors are considered in this model.

13.2.5.1 Target Type

This type is the sum of seven constituent units. Each unit is assigned a weight. If that unit is present, its weight is added to the target type inputs. In this manner, target type adds the weights of present constituent units. The final weight is the computed input value.

This weight of school is 0, intermediary is 1, media company is 2, government is 3, individual is 4, organization is 5, and large organization is 6. It may be calculated such that the total range lies between 0 and 21. The weights are summarized in Table 13.1b.

13.2.5.2 Target Past Experience

This type is the number of cases where a verdict has been reached and the verdict is against. The range of this input is between 0 and 10. Any input larger than 10 is restricted to 10.

13.2.6 Vulnerability Input

The threat also depends on the way the threat is conveyed or filed, for example. Hence the following input is taken:

TABLE 13.1B
Target Type Weights for Various Kinds of Targets

	Target Type Weight
School	0
Intermediary	1
Media company	2
Government	3
Individual	4
Organization	5
Large organization	6

13.2.6.1 Threat Type

This input lies in the range of 0 to 6. The threat type is 1 for denial of access, 2 for correspondence, 3 for disciplinary action, 4 for subpoena, 5 for lawsuit, 6 for criminal, and 0 for others. This is summarized in Table 13.1c.

13.2.6.2 Content Type

This input is the sum of six constituents that collectively form the content. Each has a weight associated with it. If the constituent is present, its weight is added to the content type. These are virtual with 0 weight, text with a weight of 1, graphics with a weight of 2, audio with a weight of 3, photo with a weight of 4, and video with a weight of 5. The range lies between 0 and 15. The weights are summarized in Table 13.1d.

13.2.6.3 Court Type

This input lies between 0 and 4. It is 0 for panchayat, 1 for tribunal court, 2 for district court, 3 for High Court, and 4 for Supreme Court. These values are summarized in Table 13.1e.

TABLE 13.1C
Threat Type Input for Various Threats

	Threat Type
Others	0
Denial of access	1
Correspondence	2
Disciplinary action	3
Subpoena	4
Lawsuit	5
Criminal	6

TABLE 13.1D
Content Type Weights for Various Kinds of Content

	Source Type
Virtual	0
Text	1
Graphics	2
Audio	3
Photo	4
Video	5

13.2.6.4 Publication Media

Like the content type, this input is also the sum of individual constituents. The weights are 1 for user comment, 2 for Website, 3 for social networking sites, 4 for forum, 5 for blog, 6 for broadcast, 7 for email, 8 for phone, 9 for podcast, 10 for print, and 0 for others. The net range is between 0 and 66. The weights are summarized in Table 13.1f.

13.2.7 Fuzzification of Input

This is the process of generating membership values for a fuzzy variable using membership functions. The first step is to take the crisp inputs and determine the degree to which these inputs belong to each appropriate fuzzy set. This crisp input is always a numeric value limited to the range of values that the input can take. Once the crisp inputs are obtained, they are fuzzified using the designated fuzzy membership functions.

A membership function is designed for each potential threat; it is a curve that defines how each point in the input space is mapped to a membership value (or degree of membership). Linguistic values are assigned for each threat level as Very Low, Low, Rather Low, Medium, Rather High, High, and Very High.

The MFs for the inputs and outputs are shown in Figure 13.4 for the source system, Figure 13.5 for the target system, Figure 13.6 for the vulnerability system, and Figure 13.7 for the threat management system.

TABLE 13.1E
Court Type Value for Various Kinds of Courts

	Source Type
Panchayat	0
Tribunal	1
District court	2
High Court	3
Supreme Court	4

TABLE 13.1F
Publication Media Type Weights for Various Kinds of Sources

	Source Type
Others	0
Comment	1
Websites	2
Social networking sites	3
Forum	4
Blog	5
Broadcast	6
Email	7
Phone	8
Podcast	9
Print	10

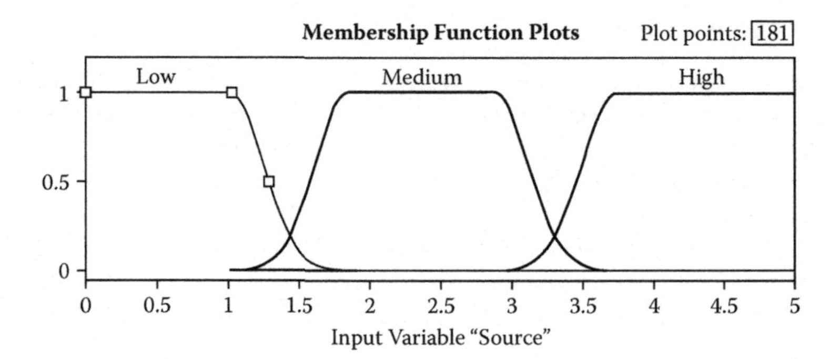

FIGURE 13.4a Membership functions for source: source.

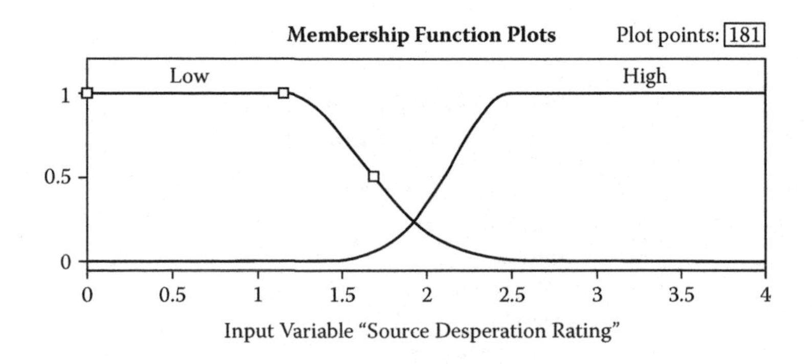

FIGURE 13.4b Membership functions for source: desperation rating.

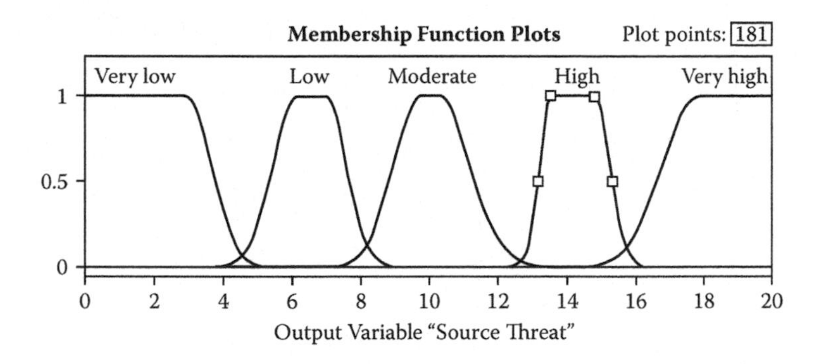

FIGURE 13.4c Membership functions for source: financial reach.

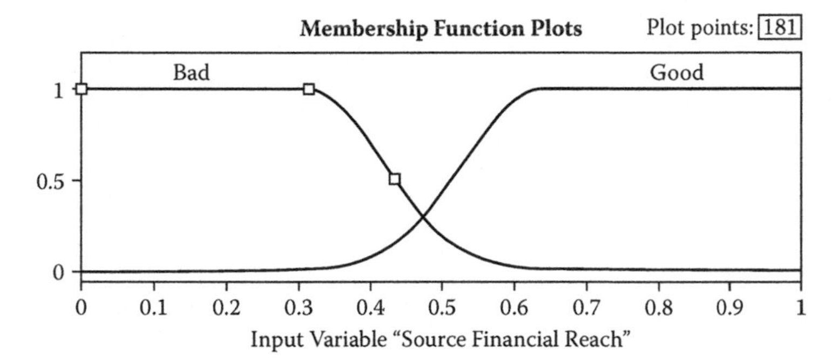

FIGURE 13.4d Membership functions for source: social reach.

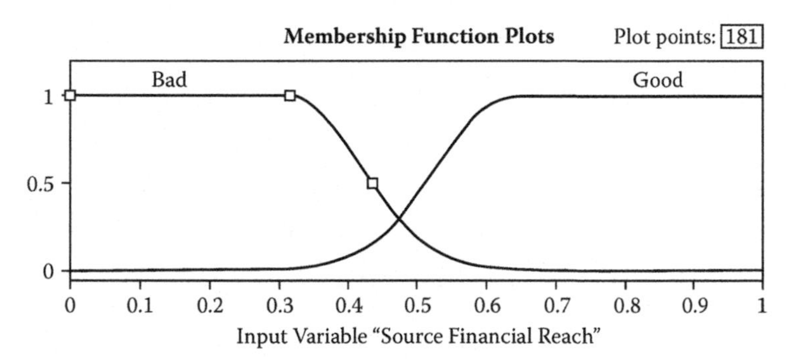

FIGURE 13.4e Membership functions for source: past experience.

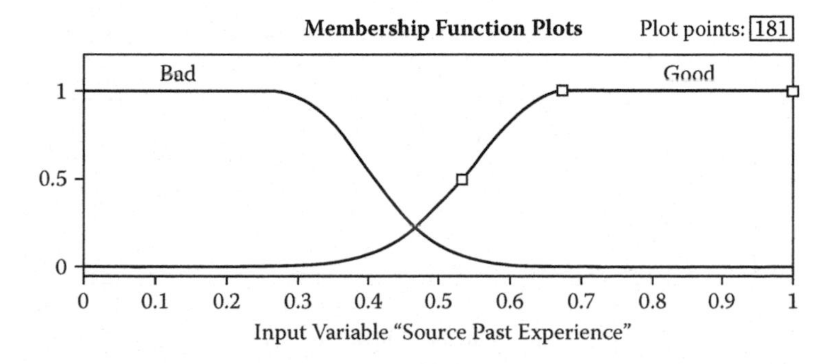

FIGURE 13.4f Membership functions for source: output source.

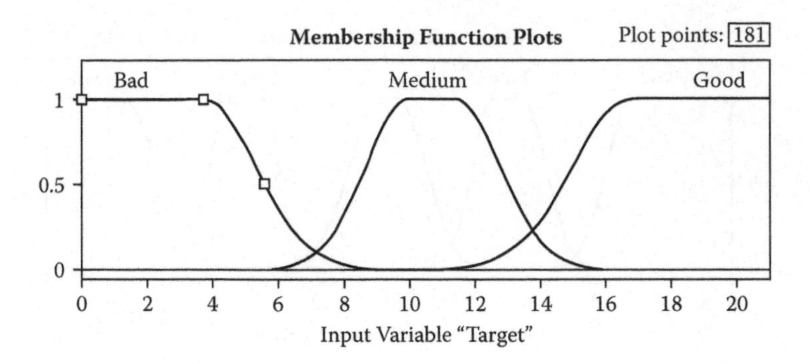

FIGURE 13.5a Membership functions for target: target.

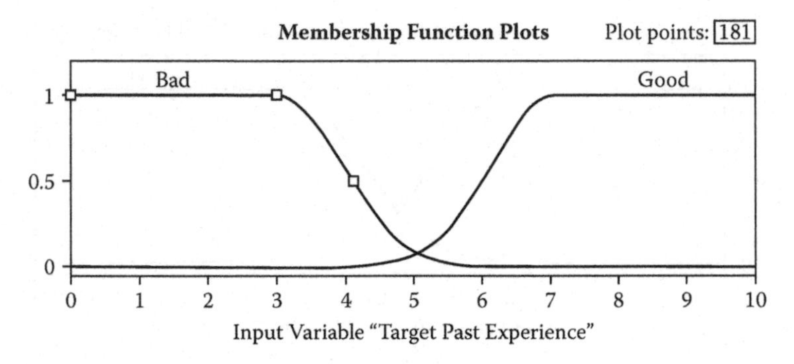

FIGURE 13.5b Membership functions for target: target past experience.

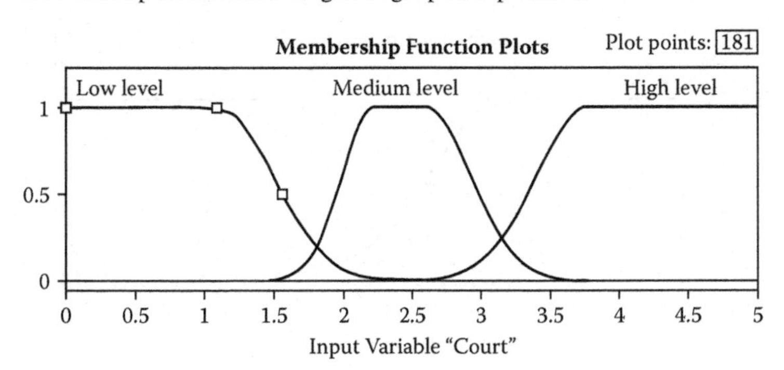

FIGURE 13.6a Membership functions for vulnerability: court.

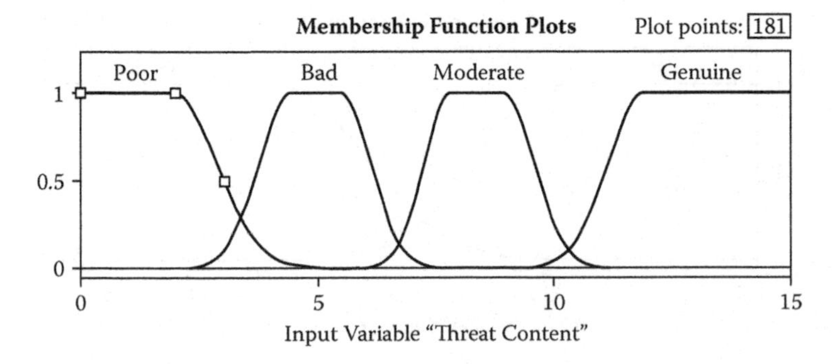

FIGURE 13.6b Membership functions for vulnerability: threat content.

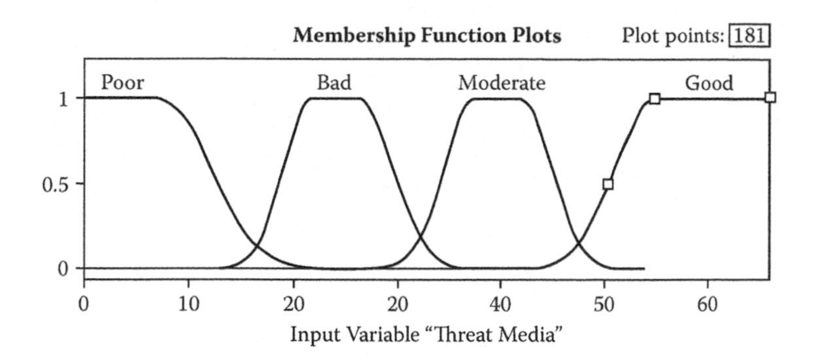

FIGURE 13.6c Membership functions for vulnerability: threat media.

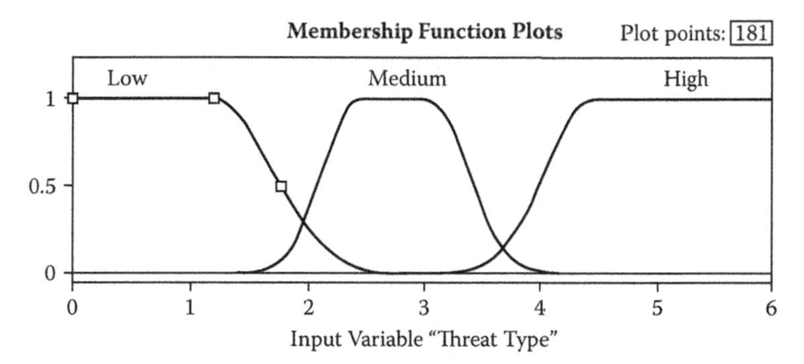

FIGURE 13.6d Membership functions for vulnerability: threat type.

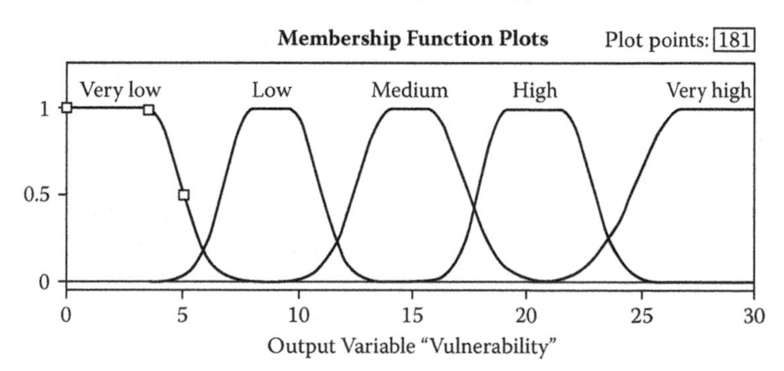

FIGURE 13.6e Membership functions for vulnerability: vulnerability.

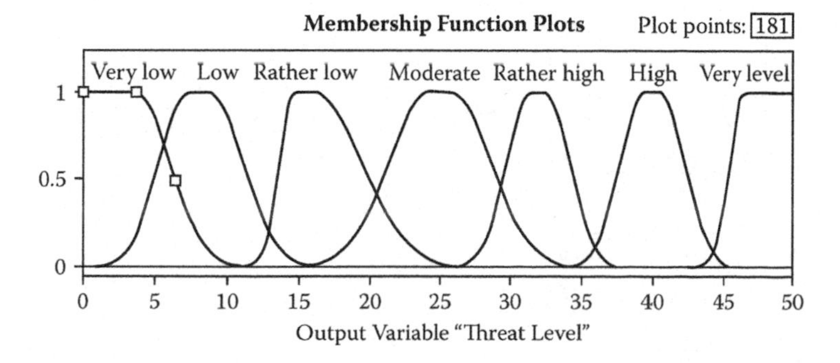

FIGURE 13.7 Membership functions for threat level.

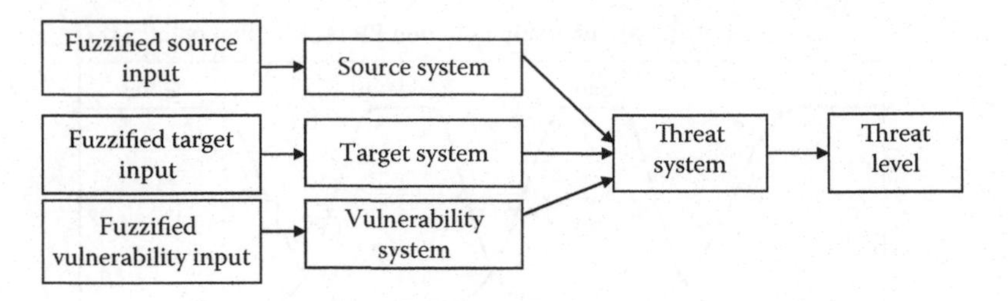

FIGURE 13.8 The multilevel design of inference engine.

13.3 FUZZY INFERENCE SYSTEM

The task of FIS is to use the inputs and map them to the corresponding outputs using the fuzzy rules. The fuzzy rules depict the general rules for the mapping using simple rules that are built. The FIS mapping of the inputs to the outputs is on the basis of fuzzy operations and fuzzy relations that we have already studied.

In order to develop the FIS for the threat management, four inference systems have been developed. In the preceding section, we have seen that a threat depends upon three input types: the source, target, and vulnerability of the threat. So, different rule-based inference systems are developed for each input type and are later combined to give a threat level. The system is shown in Figure 13.8. This will be elaborated later in the chapter. The inference engine used is the Mamdani type for all the rule-based engines using the centroid algorithm for defuzzification.

13.3.1 Source System

The source system plays very important role in determining the threat level. We had already identified the fuzzified inputs for the source in previous section. In a nutshell, the source engine can be described as illustrated in Figure 13.9.

The fuzzified source inputs go into the Mamdani-type rule-based engine, and then the output will be a source threat of range [0 20]. The rules are designed using the MATLAB® rule editor. Rules are developed entirely in unison with the survey and observed pattern of the threat and disposition from the Citizen Media Legal Database. For example, in the case of source analysis, the desperation rating plays a very important role in identifying the source threat level. If the source_desperation_rating is low, then the source_threat will be less than or equal to moderate,

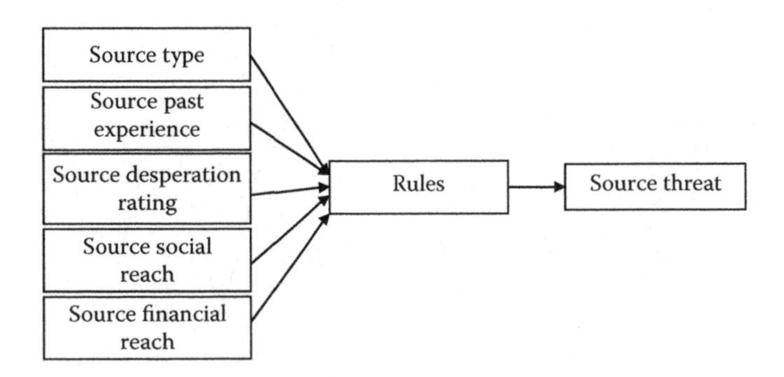

FIGURE 13.9 Source inference engine.

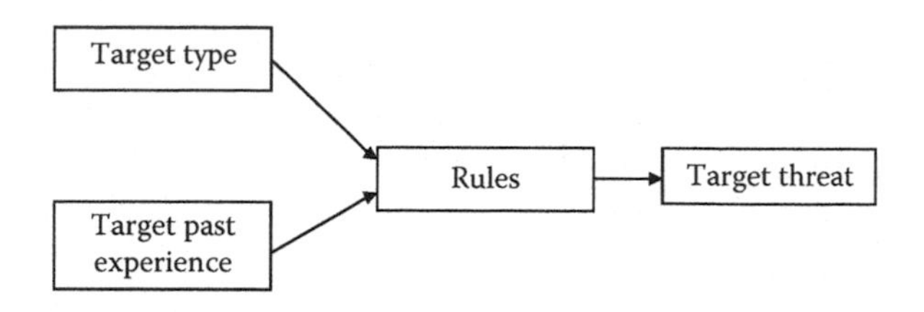

FIGURE 13.10 Target inference system.

and if the source_desperation_rating is high, then the source_threat will be greater than or equal to moderate, i.e.:

If (source_desperation_rating is Low) then (Source_threat ε {Very_low, Low, Rather_low, Moderate })
If (source_desperation_rating is High) then (Source_threat ε { Moderate, Rather_high, High, Very_high})

With the rules and the fuzzified inputs, the output can be determined at different levels. After the input to the fuzzy inference engine, we get numerical data within a range that describes the output. With defuzzification using the centroid algorithm, we interpret the results, which are again input to another inference engine. This inference engine calculates the final output, i.e., the threat level.

13.3.2 Target System

Similar to the source system, the target system is also a fuzzy inference engine with a Mamdani-type engine. The rules were made. The target system with the inputs and outputs is shown in Figure 13.10. The input variables for the target system are target and target_past_experience.

Unlike the source, the target can be any one or more of the identified target types. The threat given of field can affect an individual (for instance, a threat to murder/kidnap/subject to criminal activity), an organization (a threat of defamation or denial of access), or both; for instance, the threat of a bomb attack is a vulnerability to individual and organizational property (such as a building). Therefore, unlike the source, the target can have a multiple selection of inputs. Hence the range of the target is [0 21].

The target_past_experience is the variable that determines the past experience of the target and the vulnerability of target itself. If the target was more vulnerable in the past, it may be more vulnerable to threat this time also. Its past experience is modeled using the database. We had the status and disposition of the threat filed as a case in the court. If we find an entry in a database and the result of the case is against the target, then the target is more vulnerable; else the target is less vulnerable. In algorithmic form,

```
        Initially, target_past_experience = 0; (signifies low)
        If (disposition(target) = favorable)
        target_past_experience ← target_past_experience + 1
    Else
        If(target_past_experience > 0)
                target_past_experience ← target_past_experience – 1;
```

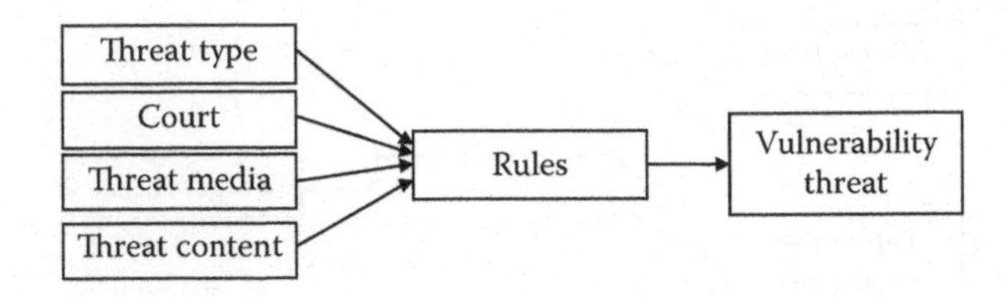

FIGURE 13.11 Vulnerability inference system.

13.3.3 Vulnerability System

Sometimes, the source and target demographics alone are not able to identify the true nature of the threat. The way of conveying the threat or the challenges it creates also determines the nature of the threat. Accordingly, the vulnerability to the threat plays an equal role to the demographics of the threat. The linguistic variables of the threat identified for the vulnerability system are threat_type, court (where the threat is challenged), threat_media (the media of the threat), and threat_content. The basic system is shown in Figure 13.11.

The important variables for the vulnerability of the threat identified are threat_type and threat_media both. The threat is broadly categorized in six categories and fuzzified in three levels [Low, Medium, High], and threat_media is the media of the threat, such as video, audio, or print. The level of the threat_media is [Poor, Bad, Moderate, and Good]. So

If (threat_type ε [Low]) and (threat_media ε [Poor, Bad]) then (Vulnerability ε [Very_low, Low])
If(threat_type ε [Medium]) and (threat_media ε [Poor, Bad]) then (Vulnerability ε [Low, Moderate])
If(threat_type ε [High]) and (threat_media ε [Poor, Bad]) then (Vulnerability ε [Moderate, High])
If(threat_type ε [High]) and (threat_media ε [Good]) then (Vulnerability ε [Very_High])

And so on.

The other key variable is the threat_content. The rules devised for the vulnerability system are based on calculating the vulnerability by the mode of threat. The threat content type also shows the desperation of the source. The case of the threat and the level of the court are also important to help us recognize the vulnerability of threat.

The output range of the vulnerability is [0 30].

13.3.4 Threat Management System

The threat management system is the final inference engine of the expert system. The motivation of the chapter was to prioritize the threat and identify the levels of the threat using the environmental factors of the source, target, and threat. Since all these attributes are different from each other to some extent, a different inference engine is designed for each of these to calculate the threat from the source perspective, the target perspective, and the threat perspective. The threat management system is designed to integrate the threats obtained from all the different sources of threat and combine them to produce a integrated threat level that is the ultimate threat level. The motivation for the different threat engines lies, in fact, with the availability of data. If only source data is available, then the source threat can be identified, and hence the threat_level in that case would be source_threat alone. Similarly, if we have target variables or threat variables, then the target_threat or vulnerability threat can be obtained. But the system will work well with all the inputs and variables. The system is shown in Figure 13.12.

The source_threat and target_threat with the vulnerability determines the threat_level. The most important variable for the threat_level is source_threat, with the range [0 20] and levels [Very_low, Low, Moderate, High, Very High].

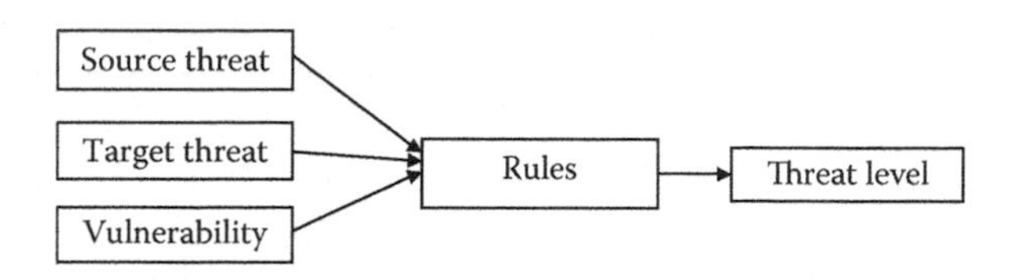

FIGURE 13.12 Threat management system.

If (source_threat is Very_low) then (Threat_level ε [Very_low])
If (source_threat is Low) then (Threat_level ε [Very_low, Low, Rather_low])
If (source_threat is Moderate) then (Threat_level ε [Low, Rather_low, Moderate])
If (source_threat is High) then (Threat_level ε [Rather_high, High, Very_low])

And so on.

13.4 ANALYSIS OF RULES AND INFERENCE ENGINE

This section will discuss the rules, including their effect on the output and input for each inference engine.

13.4.1 Source System

The source system is the inference engine for the source_threat. The linguistic variables and fuzzification have already been discussed in the preceding section. In this section, we discuss the relationship of an individual variable to the output variable as per the rule base. For the source inference engine, the output is source_threat with range [0 20]. The relation between different variables with the source_threat is shown in Figure 13.13. Figure 13.13a shows that as we increase the source, the source_threat increases. This increase is much in the form of steps as seen in the figure. Similarly Figure 13.13b shows that the source_threat increases with the increase in financial reach. The increase is very sudden at the middle of the graph, as compared to the other areas, where the increase is relatively constant. The curve of source_threat with source desperation shown in Figure 13.13c shows another interesting trend, where the increase in desperation causes an increase of source_threat that is slow at the start and rapid at intermediate values. The past experience in Figure 13.13d shows a negative trend with vulnerability with a decrease in vulnerability with the increase in past experience of the source.

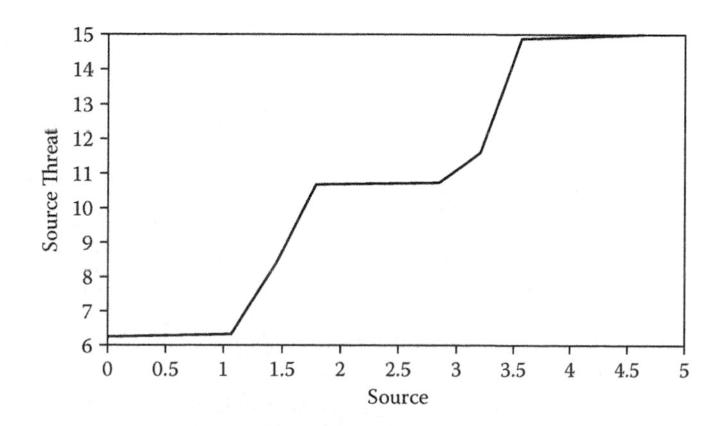

FIGURE 13.13a The relation between source and source_threat.

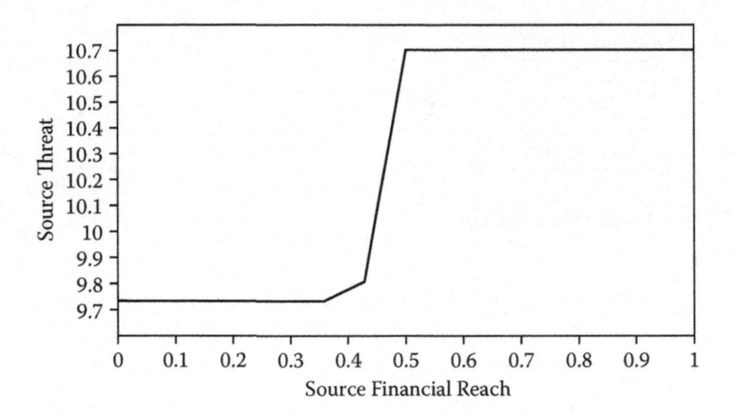

FIGURE 13.13b The relation between source_financial_reach and source_threat.

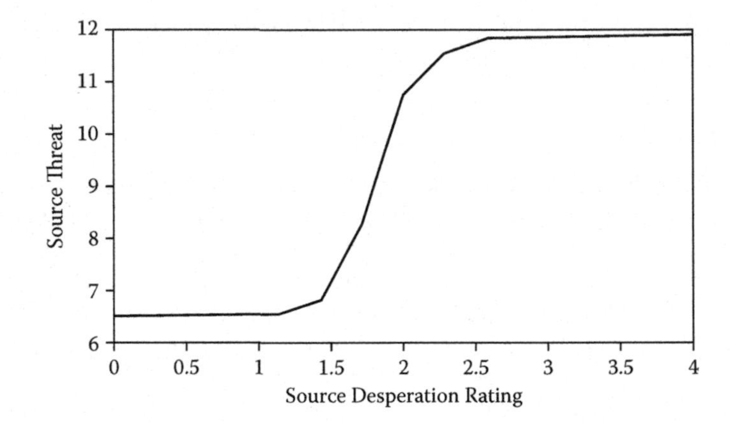

FIGURE 13.13c The relation between source_desperation_rating and source_threat.

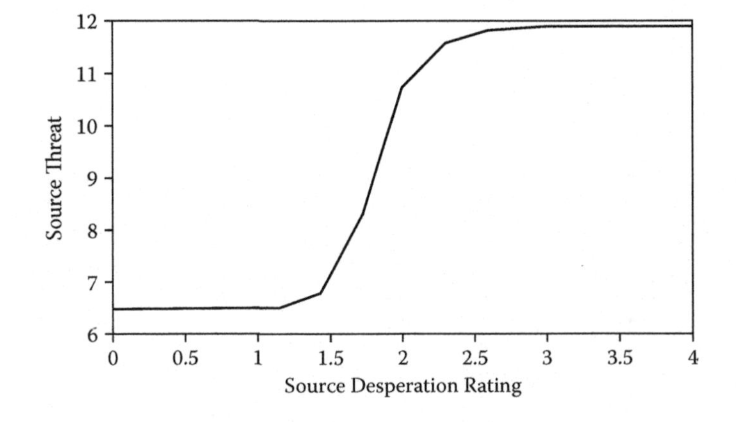

FIGURE 13.13d The relation between source_past_experience and source_threat.

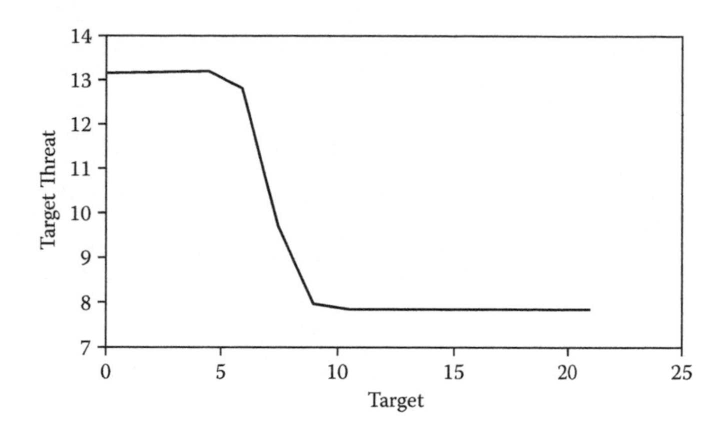

FIGURE 13.14a The relation between target and target_threat: target_past_experience.

13.4.2 Target System

The target system is the inference engine for the target_threat. The linguistic variables for the target system are target and target_past_experience. The fuzzification and input modeling of these variables have already been discussed in the preceding section. The output range of the target_threat is [0 20]. The relation between different variables with the target_threat is shown in Figure 13.14. Figure 13.14a makes it clear that there is a decrease of target_threat with the increase in the value of target. Figure 13.14b shows almost a similar trend with past experience and target_threat.

13.4.3 Vulnerability System

The vulnerability system is the inference engine for the finding the threat vulnerability and is the last key ingredient for the threat management system. The range of the output vulnerability is [0 30]. The relation between different variables with the vulnerability is shown in Figure 13.15. Figure 13.15a depicts the relation between the vulnerability and threat_content. Here it may be easily observed that the vulnerability remains almost zero initially with very little increase. Then it suddenly increases toward the end. The increase in the value of the variable court causes a uniform

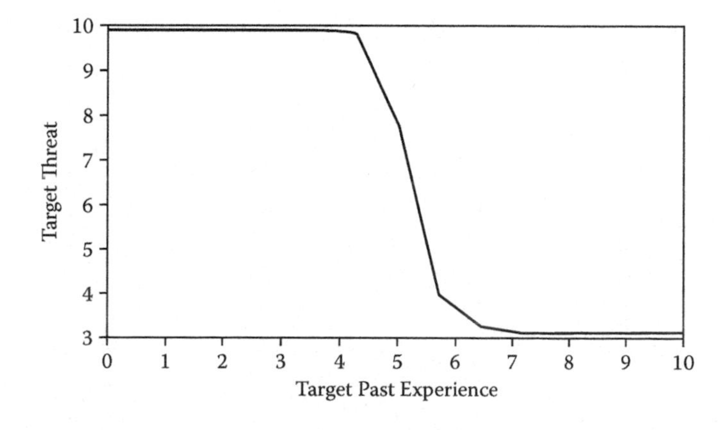

FIGURE 13.14b The relation between target_past_experience and target_threat.

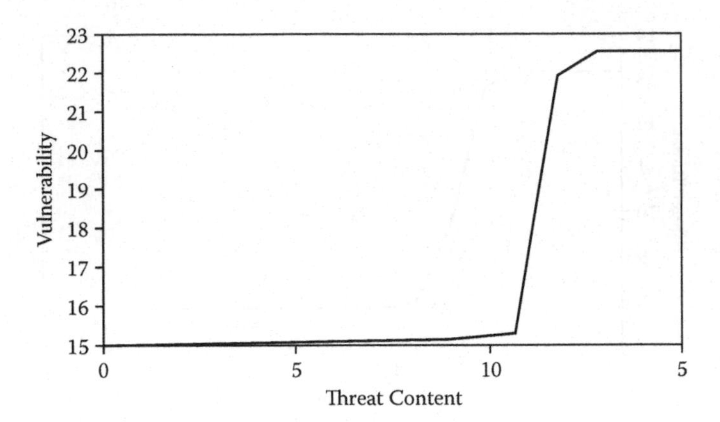

FIGURE 13.15a The relation between threat_content and vulnerability.

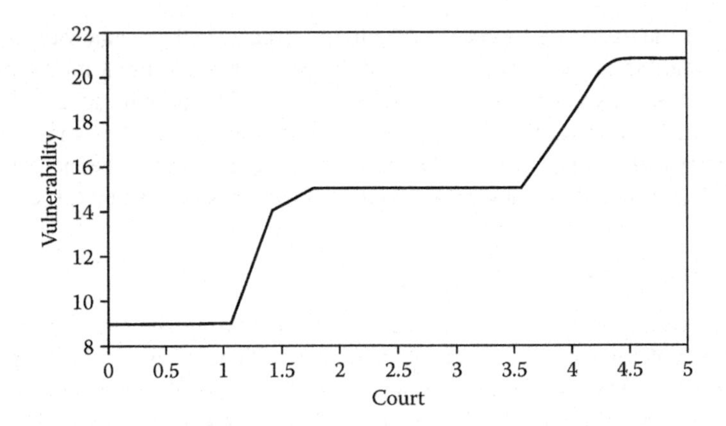

FIGURE 13.15b The relation between court and vulnerability.

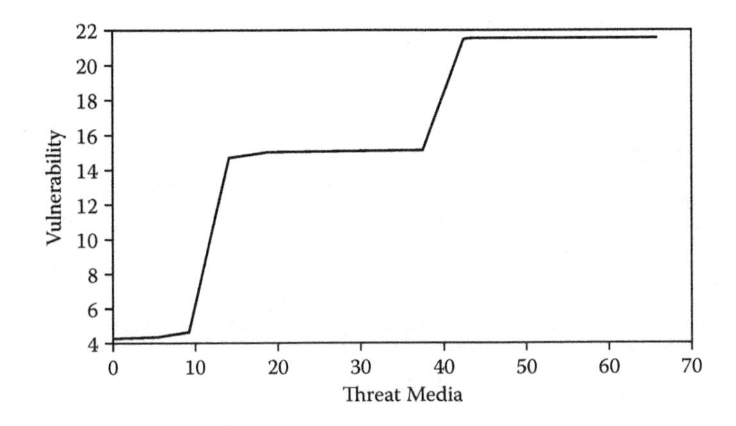

FIGURE 13.15c The relation between threat_media and vulnerability.

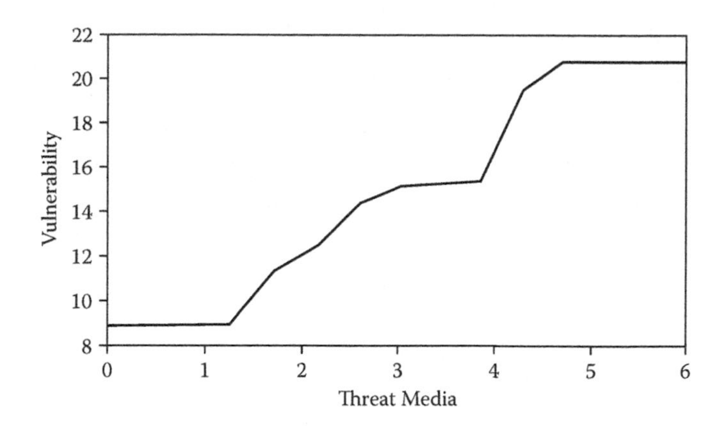

FIGURE 13.15d The relation between threat_type and vulnerability.

increase in vulnerability as seen in Figure 13.15b. Threat content and threat type show similar curves shown in Figures 13.15c and 13.15d.

13.4.4 Threat Management System

The threat management system is the final inference engine used to determine the threat_level from the variables source_threat, target_threat, and vulnerability of the threat. The output range of the threat_level is [0 50]. The threat_level would be again helpful in prioritizing the threats among the given set of threats. The relation between different variables with the threat_level is shown in Figure 13.16. The curve given in Figure 13.16a shows a very characteristic relation between the threat_level and target threat. Here the threat level first decreases and then in the middle increases with the increase of target threat. The other two variables, source_threat and vulnerability, have a similar curve, which shows a uniform increase in threat level with their increase. These are given in Figures 13.16b and 13.16c.

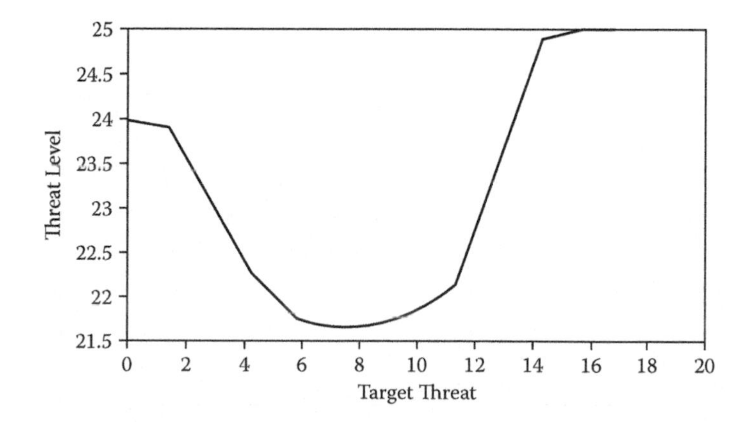

FIGURE 13.16a The relation between target_threat and threat_level.

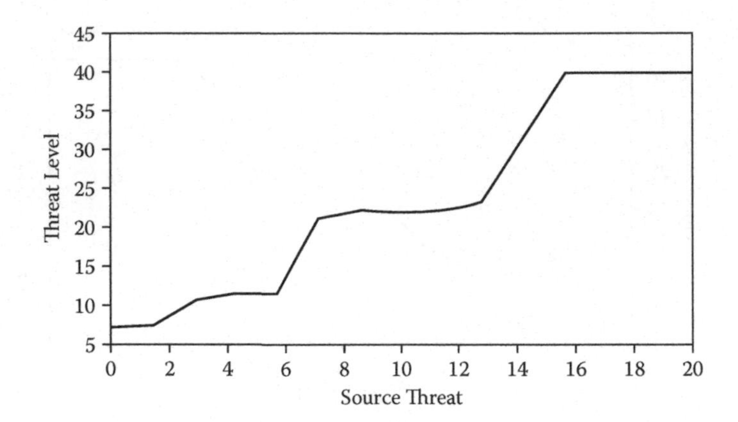

FIGURE 13.16b The relation between various variables with threat_level: source_threat.

13.5 EVALUATION OF THE EXPERT SYSTEM

The expert system was evaluated with a focus upon the risk and level of threat. The first two areas of testing solely focused upon input modeling and level of each inference engine for a given test case. The final and most important area is the evaluation of how threat inference engine validates the correct rules. To determine the appropriateness of the approach taken in the chapter, the threat level of the court cases is assessed. To do this, we compare the threats register in court and their disposition and status from Citizen Media Legal Database with the threat levels of the expert system.

13.5.1 Aim of Testing

The overarching aim of the testing was, of course, to determine if the approach taken in creating the expert system to determine the threat level was appropriate. That is, should the domain of rule-based reasoning be used exclusively by courts and law enforcement agency to prioritize threats? To assess the appropriateness of the approach taken, different areas of testing were determined: validity, correctness, input contribution, and threat prioritization.

13.5.2 Performance Benchmark

In this section we shall present theorems that show that the knowledge representation and inference engine of the threat management system is *valid*, *correct,* and *optimal.*

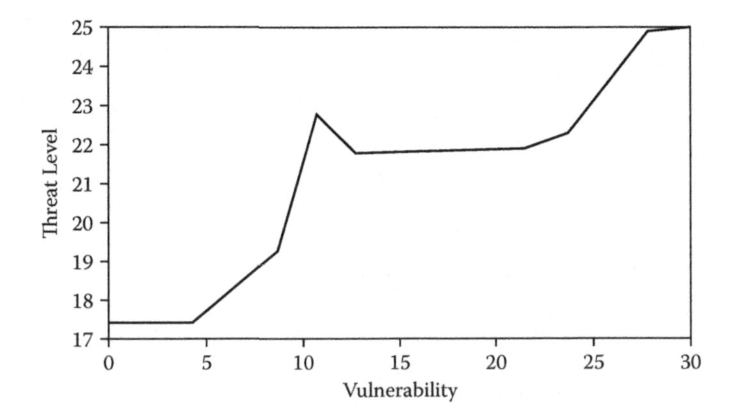

FIGURE 13.16c The relation between vulnerability and threat_level.

13.5.2.1 Validity

The legal threat depends on the socioeconomic factors of the target and the source. The expert system uses the knowledge base to model the linguistic variables and then apply the rule-based inference engine. The system will be valid if it is able to model the threats and their related factors mathematically and give the correct result. Hence, the system is valid if it is able to give results at each and every condition of the input. Mostly, systems fail or throw exceptions in extreme boundary conditions. So in a nutshell, the *validity* of the system is the proper function of the system at the extreme condition.

13.5.2.2 Correctness

Input to a system is a sequence of values to be assigned to attributes, and output consists of a sequence of actions that are executed. So, *correctness* is defined as the ability of the system to give the same result as is given by the courts and other experts under the same scenario. The mapping of different dispositions to the threat levels is given in Table 13.2.

13.5.2.3 Effectiveness

The input contribution for any inference engine is depicted by the chart that shows how the inference engine is affected by the inputs. The basic definition of effectiveness is the extent to which planned activities are realized and planned results achieved. In our case, the effectiveness could be defined as the correspondence of inputs to the output using rules formed to get the prioritized threats.

13.5.2.4 Competence

The basic aim of expert system is to prioritize the number of threats/cases received over a period. The prioritization of threats should be done on the basis of risk and vulnerability associated with them. The competence of the system is, then, the ability of the system to prioritize the task.

13.5.3 Test Pack

The test cases are taken from the Citizen Media Law Project. The database contains lawsuits, cease and desist letters, subpoenas, and other legal threats directed at those who engage in online speech.

Test Case: *State v. King*

Description: Oklahoma has a criminal libel statute, making libel "punishable by imprisonment in the county jail not more than one (1) year, or by fine not exceeding One Thousand Dollars

TABLE 13.2
Threat_Level and Disposition Match

Disposition	Threat_level
Withdrawn	[Rather_low, Low]
Verdict(Plantiff)	[Moderate, Rather_low]
Verdict(Defendant)	[Rather_high, High]
Subpoena quashed	[High, Rather_high]
Settled	[Very_low, Low]
Lawsuit filed	[Very_high, High, Rather_high]
Injunction (issued)	[High, Rather_high]
Injunction (denied)	[Moderate, Rather_low]
Dismissed (partial)	[Moderate, Rather_low, Low]
Dismissed (total)	[Low, Very_low]

($1,000.00), or both." 21 Okla. Stat. § 773. "Libel" is defined as "a false or malicious unprivileged publication by writing, printing, picture, or effigy or other fixed representation to the eye, which exposes any person to public hatred, contempt, ridicule or obloquy, or which tends to deprive him of public confidence, confidence, or to injure him in his occupation, or any malicious publication as aforesaid, designed to blacken or vilify the memory of one who is dead, and tending to scandalize his surviving relatives or friends." 21 Okla. Stat. § 771.

Harold King runs the forum site "McAlester Watercooler," which he describes as a forum for citizens of McAlester, Oklahoma, to "voice their views about the on-going City events." In August 2005, former state senator Gene Stipe complained to the local police that King had published false information about Stipe and his family on the forum (the precise nature of the statements was not disclosed). The police passed Stipe's complaint and evidence on to the local district attorney, but the district attorney did not pursue criminal charges.

Threat type: Criminal
Date: 08/16/2005
Party Issuing Threat: State of Oklahoma
Party Receiving Threat: Harold King
Type of Threatening Party: Government
Type of Threatened Party: Individual
Location of Source: Oklahoma
Location of Target: Oklahoma
Status: Concluded Disposition: Withdrawn
Court Type: District
Publication Medium: Forum, Website
Content Type(s): Text

The linguistic variables from input modeling are given in Table 13.3.

The threat_level is 19.572589677217, which corresponds to ***rather_low***. Hence, the disposition is withdrawn from the source (State of Oklahoma), as the chances of winning the case is rather_low.

TABLE 13.3
Values of Inputs for the Problem

Source	2
Desperation_rating	2
Source_pastexperience	0
Source_financial_reach	0 (because it is government)
Source_social_reach	0
Source_threat	**3.045132792420**
Target	5
Target_past_experience	0
Target_threat	**3.314935190260**
Threattype	6
Courttype	2
Threatmedia	6
Threatcontent	1
Vulnerability	**15.532028269963**
Threat_level	***19.572589677217***

13.5.4 Performance of the Expert System

Now, after the development of test packs with a lot of test cases from the Citizen Media Law Database, extensive testing is being done to check the performance of the expert system. The performance benchmark as already seen in the preceding section was *correctness*, *optimality*, and *validity*. A basic design for a test is shown in preceding section. The similar kind of evaluation is done in various other cases from the Citizen Media Law Database. The experiment was first performed on a test pack of 30 cases. Then the test was performed on the complete database.

13.5.5 Results

On the basis of the previous experiment and result, we can check our expert system on the performance benchmark discussed in the preceding section.

13.5.5.1 Correctness

The correctness is the measure of the system's capability to produce a correct result. In the 30 test cases, the system correctly identified all of them. In the database of 2,000 cases from Citizen Media Law Database, the correctness of the expert system was 98.20 percent, with 1964 correct results out of a total of 2,000 test cases. Although 98.20 percent seems to fall short of a 100 percent result, we should not deny the fact that we are dealing with human psychology and behavior, which is very difficult to model and hence predict. Therefore, a 98.20 percent success rate would be considered good.

13.5.5.2 Validity

The system will be valid if it is able to produce results in any dire situation. Even if there are no inputs, the system should be able to give results. The system's extreme conditions would be the low and the high.

From the Tables 13.4a through d, it is clear that the system is valid at the extreme boundary conditions and that a similar test is being done at all the possible values of the input. The system was able to give all the results, and hence the system is valid. This implies that rules were devised properly, and the input modeling module is correctly derived from the knowledge base.

13.5.5.3 Effectiveness

The dependency of the outputs or the rules is checked with the help of 30 test cases of the test pack. The contribution of input variables to the output variables was studied here.

13.5.5.4 Competence

Suppose the court or law enforcing agency received *n* cases/threats. The question is prioritizing the threats according to their potential for damage and risk of occurrence. The threat management system gives the output threat_level for the threats. This table is sorted, and the threat with the highest value will be the first one to tackle. This will help the courts and law enforcements agencies treat the risky threats in time. For example, consider the test pack of 30 test cases. The threats with priority for first three cases come out as shown in Table 13.5.

TABLE 13.4A
Validity Check for Source

Source Type	Past Experience	Desperation Rating	Financial Reach	Social Reach	Source Threat (Output)
0	0	0	0	0	1.899177827385
5	1	4	1	4	18.21232342424

TABLE 13.4B
Validity Check for Target

Target Type	Target Past Experience	Target Threat (Output)
0	0	3.109164997941
21	10	18.21234643125

TABLE 13.4C
Validity Check for Vulnerability

Threat Type	Court Type	Threat Media	Threat Content	Threat Vulnerability (Output)
0	0	0	0	2.598176752581
6	4	66	15	27.21234643125

TABLE 13.4D
Validity Check for Threat

Source Threat	Target Threat	Vulnerability	Threat Level (Output)
1.899177827385	3.109164997941	2.598176752581	3.421918805959
18.21232342424	18.21234643125	27.21234643125	47.234

TABLE 13.5
Competence of Three Threats

S. No.	Threat Number	Threat_Level
1	13	44.29308
2	23	40.00911
3	25	40.00006

13.6 CONCLUSIONS

Here we again make an overview of the system, considering the objectives and trying to analyze how far we have been able to achieve these objectives. We also look at the system applicability that we had specified in the introduction of the chapter.

13.6.1 Implication of the Results

The main objective of the system was to develop a system that would model the threats and prioritize them by risk and vulnerability. A rule-based fuzzy inference system has been used to find the threat level. The results of the system imply that system works well as predicted; 98.20 percent correctness for a system that deals with human psychology is good and proves that we were able to find the input and attributes affecting the threat correctly. Furthermore, the validity of the system implies that system will be able to take any inputs and perform and give desired results by prioritizing those results. The system will stand up even in extreme boundary cases. This shows that fuzzy rules are properly designed and so is the input modeling from the knowledge base.

13.6.2 Key Outcomes

The following outcomes were the motive of the chapter. It is expected for the developed systems to meet these outcomes in order to be able to be used in real life.

- *Managing risk of a particular threat*: The main objective of the chapter is to prioritize the threats and then evaluate the risk associated with them. The risk or vulnerability of any threat depends upon the attributes of the source, target, and threat. A threat is managed using prioritization of the risk. The threats having higher value are more risky and hence have been given a higher priority than other threats.
- *To devise a suitable model for threat categorization and input modeling*: Threat categorization and input modeling are done on the basis of inputs, and different levels of the inference engine are built so that their inputs could be modeled independent of each other. A multilevel inference engine is the key to threat categorization. For input modeling range of the knowledge base is formulated. A mathematical logic is developed to isolate the linguistic variables from the knowledge base.
- *Validity of fuzzy inference engine in threat management*: The threat and input attributes that directly or indirectly influence the threat cannot be a crisp input. The socioeconomic and psychological factors of the source and target determine the vulnerability to the threat. Hence, fuzzy logic is helpful to model the threat and calculate the risk. The validity of the inference engine is checked by testing the system regressively and at all combinations of rules and variables. A total of 43,030 combinations were tested. The system proved to be valid across those combinations. Boundary conditions were presented in the preceding section. This proves the validity of the inference engine.

13.6.3 Future Work

There is still a lot more that can be added to the current system and developed even more to be better applicable in real life. Following the same trend and observing the same principle, we can explore some further directions to work to produce a better system:

- We can further explore the input identified for the knowledge base. A number of other parameter may be helpful to properly formulate the system.

- A few parameters, like source financial reach, source social reach, and past experience, were calculated according to a naïve approach. A comprehensive study is needed in this direction to properly model those parameters.
- The membership functions of the variables can also improved to give more options while developing the rules.
- Risk mitigation and other risk management concepts can also be applied to the threat level to manage the risk properly.

14 Robotic Path Planning and Navigation Control

14.1 INTRODUCTION

With the rapid growth in technology coupled with the increasing demands from the people and industry, one field that has emerged magnificently is robotics. The growth and development in this field has been deeply studied for many years. The robots have now started replacing humans, especially in tasks that the humans were not comfortable at or found very difficult and laborious to do. The security of humans further restricts them from performing certain actions or tasks that can be easily undertaken by the robots. Housecleaning, carrying heavy loads in factories, and going to unknown lands for surveys are common examples. The field of robotics is vast and interdisciplinary in nature. It has numerous issues, complexities, and applications in various domains, and the combined efforts of people from various disciplines results in the effective development and deployment of robots. In this chapter, apart from discussing general robotics, we mainly deal with the problem of planning the paths of robot and moving a robot using the planned path.

In practical world, the robot needs to move in an environment that is filled with static and dynamic obstacles. There are paths where it may move, and regions where it may not be safe to move. Under all these cases, we need to ensure a collision-free navigation plan for each of the robots. Further, after we have decided the navigation path, we need to physically move the robot using control mechanisms. In this chapter we first use the AI and soft computing algorithms to find out the path of the robot, which turns out to be almost the optimal path possible. We make the use of three algorithms for the problem. These are genetic algorithm (GA), artificial neural network (ANN) with back propagation algorithm (BPA), and A* algorithm.

After deciding the path, we also design a robotic controller using a neuro-fuzzy system to physically move the robot in a way that avoids the obstacles. The controller, as we shall see later, is completely dynamic and able to survive rapid changes in the environment if they happen to occur, for instance, if someone suddenly gets in the way of the robot, or some object suddenly lands on the robot's path. The controller is modeled using a behavior-based approach that is implemented by using neuro-fuzzy systems. The robot can be moved wherever is useful without fear of collision.

The first problem that we deal with is the problem of path planning. Here we are given a robotic map. The map marks the presence and absence of obstacles along with the accessible and inaccessible regions. We need to devise a strategy to find a collision-free path for the robot from a predefined point (or source) to another predefined point (called a goal). The quality of the path generated is measured by the optimality of its length. The map is made by the robot using information acquired from the cameras, sensors, etc. Numerous representations of the map are possible, and numerous algorithms aid in this process of map formulation.

The chapter first makes use of GA, ANN with BPA, and the A* algorithm to find out the optimal path for the robot. The robot physically moves according to these results using the designed controller that runs on the neuro-fuzzy system. It is not ensured that every move will bring the robot closer to the goal. If a path is possible that leads to the goal starting from the source, then that path must be returned by the path planning algorithm. This is known as the completeness of the algorithm. In case it is not possible to reach the goal at all, the robot becomes stationary.

The robotic procedure of working is built upon the cognition model that tells us about the various tasks and the order in which they need to be performed in order to successfully execute a robotic task. This model of cognition includes a “sensing-action” cycle that is a prime area of study of the chapter. Here we get or sense the location of the obstacles and need to generate the necessary control actions using which these obstacles may be avoided. This is the first cycle of the cognitive model. The second cycle passes through perception and planning states of cognition, and the third includes all possible states, including sensing, acquisition, perception, planning, and action.

The real-world environment during mobile robot navigation has the following problems: (a) Knowledge of the environment is partial, uncertain, imprecise, and approximate; (b) the environment is vast and dynamic and the obstacles can move, appear, or disappear; and (c) due to the quality of the ground, sensor data received are not completely reliable.

Many methods for robot control have been developed, but they can be generally grouped into two categories: deliberative and reactive. In the *deliberative* approach the entire details of the environment are believed to be known beforehand and using this known information, the task of planning is carried out. These methods build the paths for reaching the target without any collision. A global optimal solution can be achieved with this approach. However, this scheme has a major drawback: It is highly unlikely to get all the details of the environment. Further, any change in the environment in dynamic time causes a problem for these algorithms. In the *reactive* approach, no model of the world is needed; actions are determined according to information gathered from sensors. The robot has to react to its sensor data by a set of stimulus-response mechanisms. The drawback of these systems is that the sensor data is available for a very short range. This limits the planning that can be done beforehand. The final solution generated may not be globally optimal.

The A*, ANN, and genetic algorithms we use here are primarily deliberative in nature. They do the higher-end planning. The ANFIS (adaptive neuro-fuzzy inference system) used to move the robot is reactive in nature ; it physically moves the robot and accustoms itself to most of the dynamic changes that in the robotic environment. Hence the solution tries to integrate the best practices available in literature.

We first talk about the motivation of the problem. Then we discuss the simulation model which includes modeling of the problem and the algorithms. Here we also introduce the readers to the world of robotics, where we spent a fair amount of time discussing the robotics basics and AI robotics. We then discuss the three algorithms one by one. These are the GA, ANN, and A* algorithms. Then we design the robotic controller that would be used to physically move the robot. Then we test all these algorithms and present the results. At the end we offer conclusions.

The problem of robot navigation control, due to its applicability, is of a great interest. We have already seen good research in various modules. A lot of work exists to model the entire problem. There exist good algorithms to scan the environment and represent all the obstacles in form of a grid. Also various algorithms have been proposed to plan the movement of the robot using various conditions.

The whole problem till now has been seen under separate heads of planning navigation control of a static environment and planning navigation control of a dynamic environment. If we come to a static environment, many algorithms have been implemented and results verified. In planning a dynamic environment, the steps are a little different as the environment continuously changes.

We also have encountered various works of research in which people have tried to solve the navigation problem using genetic algorithms. The basic principles in all these have been to take a fixed solution length and find the solutions by using genetic operators. In this paper, we propose a graphical node representation that will work for all sorts of highly chaotic conditions where the number of obstacles is very large.

Also similar work exists in neural networks, which have been applied mainly to static data. Here we have adopted a representation that minimizes the errors that might come about due to neural noise or an incomplete database. Some work also exists on the A* algorithm. Mainly people use Manhattan distance, or the simple distance between current position and goal position. In this chapter, we have used a heuristic function that optimizes the path and at the same time resolves the conflicts when two paths may have the same heuristic values by considering the rotational factor as well.

The neural-integrated fuzzy controller (NIF-T) approach has been applied to the problem. This approach was deliberative and not reactive, which was disadvantageous in such an environment. Similarly, work exists in the use of robot navigation and control under uncertainty through neuro-fuzzy. Other major work includes behavioral modeling, input modeling, and controller optimization using ANFIS. We describe the mobile robot navigation techniques using a mathematical model for behavior selection; they do not deal with a natural way of describing the behavior rules and behavior selection during behavior conflicts. This chapter discusses a detailed approach to solve the problem through ANFIS, which is a fuzzy inference system implemented in the framework of adaptive networks. By using a hybrid learning procedure, ANFIS constructs an input-output mapping based on both human knowledge (in the form of fuzzy if-then rules) and input-output data pairs.

14.2 ROBOTICS AND SIMULATION MODEL

In this section we first discuss some of the basics of robotics. Here we open the world of robotics, which has infinite possibilities to readers. We then discuss the way we model the whole problem and also the way we apply all the three algorithms in this problem. The whole model includes modeling of the robot and environment, modeling of the algorithms, and modeling of the results, so that they may be implemented.

14.2.1 Robotic Hardware

The *robot* is a much generalized concept that may be applied in numerous applications and domains. Each has its own set of hardware specifications and design. The good design of a robot plays a very important role in fuel economy, ability of the robot to pass uneven, rough terrain, or to survive such extreme conditions.

The robotic hardware essentially consists of wheels, shafts that connect wheels to motors, and motors that are responsible to turn the wheels of the robot. The speed of the motor decides the speed with which the robot moves. The speed may be programmatically controllable via a motor hardware interface. In addition, the wheels may be made to turn about in both directions in order to steer the robot. This again may have a programmatically controllable interface. All the robot controls are interfaced with a computer that can send signals to the robot in order to control it. The computer uses these signals as devices in order to communicate with the robots. In addition, the processing chip may be integrated in the robotic body. This is what happens with autonomous robots that do not have a computer interface. Wireless controlled robots, in addition, have wireless receivers and processors integrated in them.

The wheels and body of the robot and the various motor specifications depend upon the condition of use, terrain of movement, friction, and other mechanical parameters.

One of most extensively used robots resembles a car (Figure 14.1). This robot has four wheels, two at the back and two in front. Normally, the two front wheels are to drive and steer the robot, and

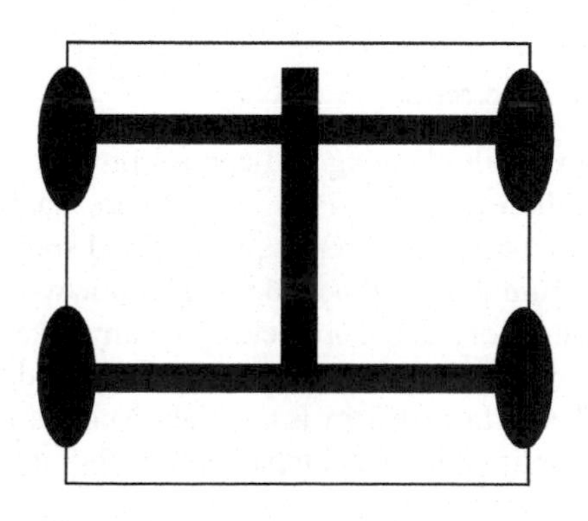

FIGURE 14.1 The car-like robot.

the back two are to drive the robot. The robot can be turned about through the forward two wheels. To move the robot left, the wheels are made to rotate leftward and vice versa. The robot can be made to move both forward and backward. This depends upon whether the wheels rotate in a forward direction or a backward direction.

Numerous other types of robots are also commonly used. One such type is used in sea and naval operations either to float on the sea or to be submerged. The ones that float may be designed in form of boats or ships. Many others have a human-like structure with narrow legs and wheels attached at the base. These are very useful to glide through obstacles in crowded places. Many may even be provided with a leg-like structure to enable walking like humans. Many robots, on the other hand, are big and bulky, requiring more than four wheels to balance and move. Robots are used in space, where they are designed and handled differently, as different many physical laws apply. Robots are also used as robotic hands for various operations. Surgical robots need to be able to handle surgical instruments and to be able to operate them with the utmost precision. In many cases, these operations may be at a nano-scale, where again different physical laws apply.

14.2.2 Robotic Sensors

The "autonomous" robots make extensive use of sensors to guide them in and enable them to understand their environment. Numerous kinds of sensors are extensively used, including infrared and Bluetooth sensors. Apart from these, GPS devices may also be used as sensors for planning purposes. Often video and still cameras are used to get a picture of the environment. This is then depicted using image processing techniques. The goal, obstacles, paths, etc., are identified by using image processing and AI algorithms. In this manner, the robot comes to know its environment.

Often robots may need to receive and understand speech signals, especially if they are speech-controlled robots. Also many robots use infrared cameras and again use image processing techniques to figure out the paths and obstacles from the images captured.

14.2.3 Robotic Map

A *map* is a representation of the robotic world that is useful for numerous purposes, especially for the purpose of path planning that we deal in this chapter. The robot gathers information on paths, obstacles, and accessible and inaccessible areas in numerous ways. It may even be guided for these operations. This information is then converted into a map. Numerous types of maps are used for the purpose of planning. The most basic is a geometric map that has the whole information in form of a grid of size $M \times N$. This map is very simple to form and work. Unfortunately, it leads to slower processing by many planning algorithms. Many like areas may be clustered to produce even better maps that are compact with little loss of information. Other types of maps include Vornoi maps, topographical maps, hybrid maps, etc. We do not discuss these maps in this text.

14.2.4 Path Planning and Control

We have already discussed both path planning and control in the introduction of the chapter. We once again formally deal with these concepts. *Path planning* in robotics is the problem of devising a path or strategy to enable a robot to move from a prespecified source to a prespecified goal. The map thus generated needs to ensure that the robot does not collide with obstacles. The path planning will depend upon the size of the robot and the obstacles. In many cases the robot may be assumed to be of point size. This happens when we are sure that the robot would be in most cases able to glide through obstacles, assuming the obstacle density is not high. Another important characteristic of the path planning algorithm is its optimality and completeness. Optimality refers to the quality of the path that the algorithm generates. Completeness refers to the property that the algorithm is able to find a solution to the problem and return the solution in finite time, provided a solution exists.

The path planning returns a layout that helps in navigating the robot. The actual movement may be made using a robotic controller. The controller is responsible for operating the motors and shaft of the robot to make it travel in such a way that it reaches the goal without colliding with obstacles. The controller again needs to avoid very sharp turns, as it is very unrealistic for any car-like robot to take sharp turns at moderate speed. These are known as the non-holmic constraints of the robot.

14.2.5 AI Robotics and Applications

Here we give a brief picture of the exciting world of robotics. AI robotics makes use of AI techniques to enable robot to do tasks autonomously, meaning that the robot does not require any human guidance to carry out the task. One of the prominent areas of application is space exploration, where the robots go to explore planets like Mars and its moons. Again robots may be used for many rescue operations that may be risky for humans to carry out. Robots are commonly used to clean the house or to transport materials in factories. Robots may also be used to chase thieves or for robotic surgery, surveying operations, underwater surveys, automatic driving or sailing, military operations, and more.

14.2.6 General Assumptions

The robot we consider here is a general car-like robot. The robot makes use of some steering mechanism to steer itself at some angle. Further, the robot can move backward or forward using the controlling mechanism that moves the robot. Here we consider the speed of the robot to be constant. This is done to limit the complexity of the problem.

The path planning algorithms further are very sensitive to the complexities. Here we have to further limit the robotic moves and place suitable conditions on the map so that the problem can be solved. The following assumptions are further being made for the problem.

The robotic map that we consider for the problem is divided in grids of $M \times N$ size. The size of the grid plays an important role in the path planning algorithms. Grids that are too small may result in a lot of time being taken by the path planning algorithms. Many times the algorithms may not return a solution for the same reason. Large grid sizes may lead to a loss of optimality of the generated solutions. Accordingly, we decide and implement the solution. Further it is assumed that each obstacle as well as the robot can make only a unit move in a unit time called the threshold time (η). This means that the path planning algorithm may assume the environment to be completely static for the η amount of time. It must return its solution within this time to avoid any collisions. If it does not return the solution within this time, it is possible that the algorithm was working on an updated map and the moves decided according to the updated map may result in collisions.

Alternatively, the dynamic obstacles may be handled by the neuro-fuzzy controller that can resist heavy changes in environment with the loss of path optimality.

For this problem, we take the following movements as valid ones that can be performed by the robot. The movements have been quantized to restrict the complexity of the algorithm. If we add more moves, the algorithmic complexities would increase and again the algorithms may take too long or may not return a solution at all.

- Move forward (unit step).
- Move at an angle of 45 degrees forward (unit step) from the current direction.
- Move clockwise/anticlockwise (45 degrees).
- Move clockwise/anticlockwise (90 degrees).

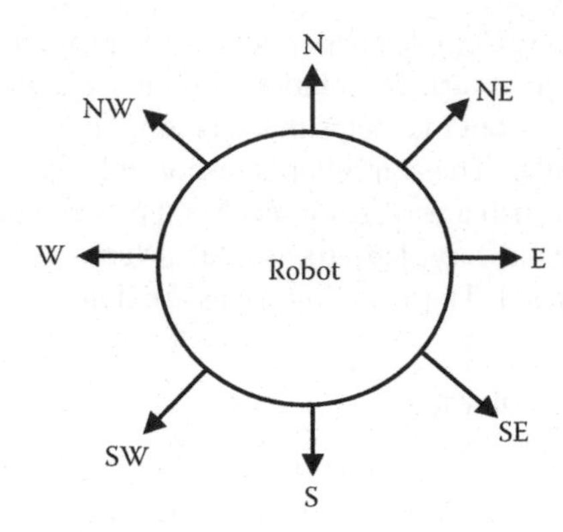

FIGURE 14.2 The various directions the robot can move in.

It is assumed that the robot only rotates/moves in angles of a multiple of 45 degrees. Hence we can only move the robot in the directions (by rotations and forward move) north, northeast, northwest, south, southeast, and southwest as given in Figure 14.2. This quantization of the movement of the robot saves a lot of time for the planning algorithms. They restrict the whole search space of the problem and hence reduce the algorithmic complexity.

For algorithmic purposes the directions have been denoted by numerals. This helps in computational purposes. The numeral depiction of the angles is given in Table 14.1.

We briefly discuss the movements of the robot and their representations now. Imagine that the robot is facing northward, i.e., at an angle of 0. Now there are eight kinds of moves that it can perform. (1) The first move possible would be to move forward a unit step. The next couple of moves possible would be to turn left or right by 45 degrees. This would change the direction of the robot to (2) 7 or (3) 1 respectively. The next couple of moves do the same thing at an angle of 90 degrees. This changes the direction of the robot to (4) 6 or (5) 2 respectively. It may even change its direction as it did in moves two and three and take a unit step forward. This means turning to the direction (6) 7 and (7) 1 and taking a unit step forward. (8) The last move is no motion. The robot remains stationary in this step.

Here we discuss the simulation model and results of Kala et al. (2009c) and Shukla et al. (2008d).

TABLE 14.1
Representation of Various Directions

N	0
NE	1
E	2
SE	3
S	4
SW	5
W	6
NW	7

14.3 GENETIC ALGORITHM

The first algorithm that we develop and implement for this problem of path planning is the genetic algorithm (GA). As already discussed, GA is inspired by the process of natural evolution. This algorithm works by the principle of trying to find the best solution from some combination of the existing solutions. The whole algorithm proceeds in generations. Each generation is attempted to be better than the preceding one. This keeps improving the generations. The children or solutions of every generation are generated from the preceding generation. This happens with the use of genetic operators. The main operators used are selection, crossover, and mutation. Selection selects the individuals from the population. These are fused to make new individuals from the parents. The child takes some characteristics from the first parent and some from the other. The new characteristics are introduced into the individual by the genetic operator of mutation. The algorithm uses a fitness function to measure the fitness of various individuals. The role of the GA is to optimize the fitness of the individual. The final solution returned is the globally best available solution.

The GA in this problem is used to find the optimal path that leads to the goal (final position) starting from the source (initial position). GA as applied by us in this problem differs from the other genetic algorithms in that we have considered a graphical node as a chromosome rather than as a predetermined-length bit sequence. The graphical manner of the application of GA helps us in the easy generation of individuals in the solution pool that are feasible in accordance with the problem constraints. Now we study the various aspects of the GA that we have used.

14.3.1 REPRESENTATION

In this problem, we have applied GA graphically. Thus the solution is represented in the same manner, where the individuals are a collection of nodes or vertices of the graph that make the path. We know that according to the problem constraints, each solution must start from the source node and must end at the goal node. The path consists of a series of vertices or nodes in between the source and the goal. Hence the solution or the individual representation in this case is of the form $<P_1, P_2, P_3 \ldots P_n>$. Here each P_i denotes some node in the graph. P_1 is the source, and P_n is the goal node; n is a variable number here. The path traced by the robot is $P_1 \rightarrow P_2 \rightarrow P_3 \rightarrow \ldots \rightarrow P_n$. It may be noted that in order for the solution to be feasible, each set of points (P_{i-1}, P_i) must be connected to all others in some or the other manner for all $i = 2$ to n. This means that there must be some move out of the eight allotted moves of the robot using which the robot comes to state P_i from a state of P_{i-1}.

Each point P_i has three attributes associated with it. These are the x coordinate, the y coordinate, and the direction in which the robot is facing. Here the direction is any one of the eight possible directions that we had discussed earlier. Any point can thus be represented by

$$(x,y,d)$$

where:

x = x coordinate
y = y coordinate
d = direction in which the robot is facing (between 0 to 7)

Consider the node (10, 15, 3). We can connect to the node represented by (10, 15, 5) adjacent to this node, as this can be achieved by a valid move of rotating clockwise by 90 degrees. But we cannot connect to the node (11, 16, 3), as no move would result in this position being reachable by the robot.

It may, however, not be possible that every individual in the population pool represents a feasible solution that starts from the source and ends at the goal. In this algorithm, we consider the partial solutions as well that either do not start from the source or do not reach the goal. We shall see later in the algorithm that these solutions can be of a lot of use as well, where different incomplete solutions can collectively result in the generation of better solutions that are feasible and complete. Hence we categorize the solution or the genetic individuals in three categories: (1) Full solutions that represent the class of solutions that are complete. These start from the source and reach the goal position. (2) Left solutions: These solutions reach the goal position but do not start from the source position. (3) Right solutions: These solutions start from the source but do not reach the goal node. All the solutions in other respects generated at any time by any genetic operator are fully feasible. They all obey the problem constraints of valid moves, and no node in these algorithms leads to a collision.

14.3.2 Evaluation of Fitness

The fitness function does the task of rating and comparing of the individuals of the GA. The goal of the GA is to generate solutions with an optimal fitness value. In this problem the objective is to minimize the fitness value of the individuals. The fitness function here is the number of nodes in the chromosomal representation of the individual. It may be seen that it represents the total distance traversed by the robot when implementing the solution represented by this individual. Hence we are optimizing the total path that needs to be traversed by the robot.

In case the individual represents a partial solution, i.e., a left solution or a right solution, there is an extra penalty that we add. In any of the cases, we add the square of the physical distance that the solution did not cover to the fitness value. For the left solution, we take the distance between the last traversed point and the goal. For the right solution, we take the distance between the source and the first traversed point. The square generates enough of a penalty that the full-solution individuals are better than their partial-solution counterparts.

Along with the penalty of the physical distance, another penalty is added to the partial solutions. This penalty is useful while comparing two individual that have the same physical distance measure as the penalty. This penalty is the cost needed to join the available solution with the solution that we assumed would be generated at the maximal cost of the physical distance penalty. This penalty is the minimum number of moves that the robot would take to rotate in its left journey from the direction it is currently pointing at, assuming no obstacles in the left part. This is also called as the rotational parameter and is denoted by $R(n)$.

Consider Figure 14.3. If the direction is 2 (east) and the final position is just in front, $R(n)$ has the value of 0. This means that the robot need not make any turn. It may simply march toward the goal. Keeping the positions of the goal and the robot as same, we get different values of $R(n)$ for the different directions of the robot. If the robot is facing at any point in the region depicted in Figure 14.3 as region 2, then the value of $R(n)$ is 2. This necessitates two turns, as the total angle to be turned is more than 90 degrees. In the other regions denoted in the figure as region 3 and region 4, the value of $R(n)$ is 1.

The $R(n)$ term ensures that the number of rotations are minimal. As in the practical scenario, we need to avoid rotations as much as possible. Good solutions travel as straight as possible with as few turns as possible.

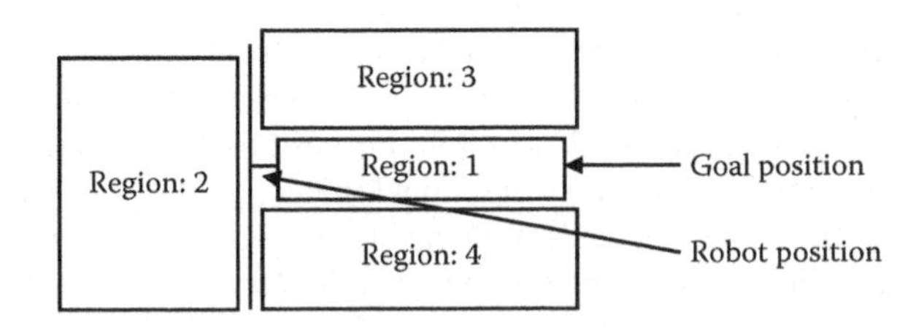

FIGURE 14.3 The values of $R(n)$ at various places.

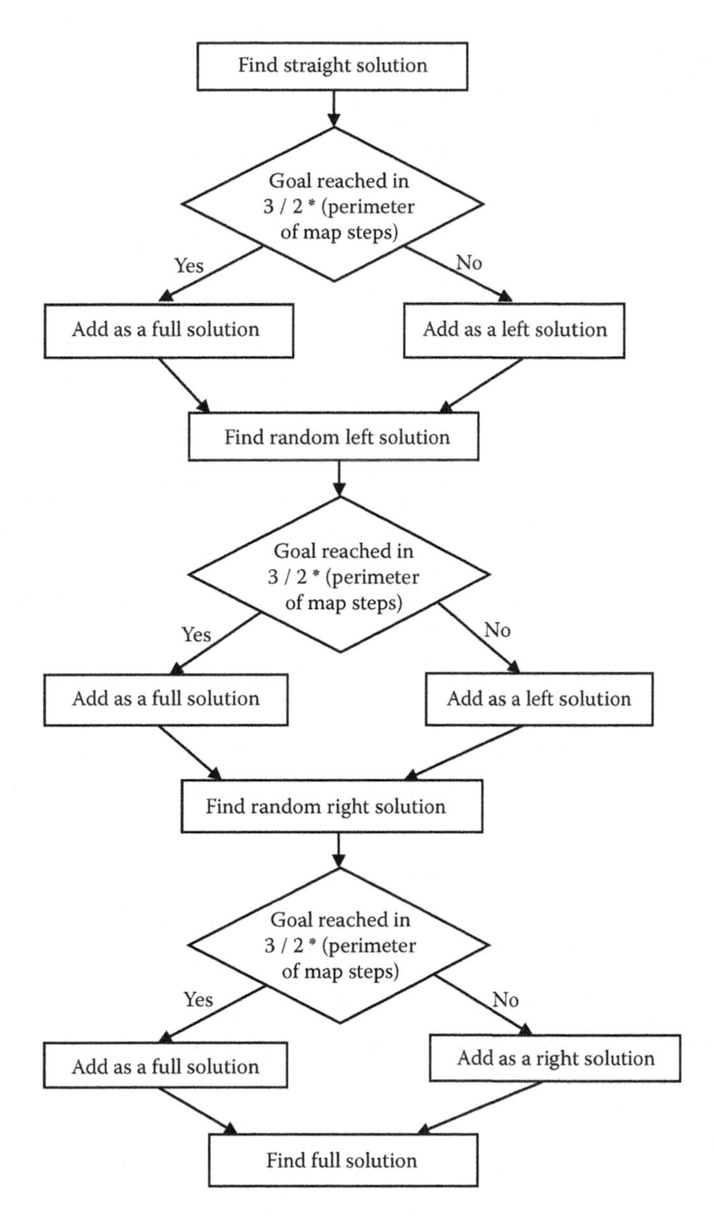

FIGURE 14.4 The steps for generating an initial solution for the robot.

14.3.3 Initial Solutions

The initial solutions play a big role in the GA. They more or less define the scope of the solutions. The GA in the course of the generations tries to find more and more solutions based on the initial solutions generated. The layout of the initial solution is explained in Figure 14.4. The initial solution is found by applying the following strategies one by one. We likewise discuss these one by one.

14.3.3.1 Find a Straight Path Solution Between Source and Destination

From the source, march straight ahead toward the destination. Whenever traveling at any point, there is a possibility of collision, take *randomMoves* (a number of random steps, turning in random directions in between). Now repeat this process of marching toward the destination.

FindStraightSolution(currentPoint,destinationPoint)

Step 1: for $i \leftarrow 1$ to (no of initial solutions intended)
Begin
Step 2: for $j \leftarrow 1$ to (3/2)*perimeter of map (Maximum moves per solution being tried)
Begin
Step 3: if $p \leftarrow$ destinationPoint
Step 4: add p to current solution set and break
Step 5: find the move m in moves that brings the robot closest to destination
Step 6: if point p we reach after m is already in the current solution set, then delete earlier points and add this point
Step 7: if no move takes the robot closer to target
Step 8: make randomMoves number of random moves in any direction
Step 9: for each move m in these moves, if point p we reach after m is already in the current solution set, then delete earlier points and add this point
Step 10: if destinationPoint is reached, then add this solution set to fullSolution
Step 11: else add this solution set to leftSolution

14.3.3.2 Find Random Left Solution Between Source and Destination

At any point take random moves at any direction. Repeat doing this. Use *direction* = *left* in the algorithm *FindRandomSolution.*

14.3.3.3 Find Random Right Solution Between Source and Destination

At any point take random moves from the destination in any direction. Add every solution at the head of the final solutions sequence; so that it looks as if we approached the destination from a point. Repeat doing this. Use *direction* = *right* in the algorithm *FindRandomSolution.*

FindRandomSolution(currentPoint,destinationPoint,direction)

Step 1: $p \leftarrow$ currentPoint
Step 2: for $i \leftarrow 1$ to noIterationsInitial
Begin
Step 3: for $j \leftarrow 1$ to (3/2)*perimeter of map
Begin
Step 4: if $p \leftarrow$ destinationPoint
Step 5: add p to current solution set and break
Step 6: moves $\leftarrow$ make a random move
Step 7: if point p we reach after move m is already in the current solution set then
Step 8: delete all points from the previous point p to end
Step 9: add p to the current solution set
Step 10: if destinationPoint is reached, then add this solution set to fullSolution
Step 11: else add it to the left or right solution, whichever is applicable

14.3.3.4 Find Full Solutions

Once we have sufficient entries for the left and right solutions, there are chances of getting more full solutions from these data. Search all the left and right solutions obtained so far. If you find any point in common between any solution of the left and right solutions, mix them to get a full solution. Hence if any of the left solutions is

$$(x_{11},y_{11},d_{11})(x_{12},y_{12},d_{12})\ldots\ldots\ldots(x_i,y_i,d_i)\ (x_{1i+1},y_{1i+1},d_{1i+1})\ldots\ldots\ldots(x_{1n},y_{1n},d_{1n})$$

and right solution is

$$(x_{21},y_{21},d_{21})(x_{22},y_{22},d_{22})\ldots\ldots\ldots(x_i,y_i,d_i)\ (x_{2i+1},y_{2i+1},d_{2i+1})\ldots\ldots\ldots(x_{2n},y_{2n},d_{2n})$$

then the integrated full solution will be

$$(x_{11},y_{11},d_{11})(x_{12},y_{12},d_{12})\ldots\ldots\ldots(x_i,y_i,d_i)\ (x_{2i+1},y_{2i+1},d_{2i+1})\ldots\ldots\ldots(x_{2n},y_{2n},d_{2n})$$

We take the left part from the left solution and the right part from the right solution.

FindSolutionFromLeftAndRight(currentPoint,destinationPoint)

Step 1: for each solution set s_1 in leftSolution
Begin
Step 2: for each solution set s_2 in rightSolution
Begin
Step 3: for each common point p in points
Begin
Step 4: $sol_1 \leftarrow$ all points in parent1 till p + rotations necessary to transform direction of p to the one of point p of parent2 + all points in parent2 after p
Step 5: Add sol_1 to the fullSolution set

14.3.4 Crossover

Crossover is the genetic operator that generates a new individual by combining the parents. The new individual takes on characteristics from both the parents. In this problem, we use crossover to fuse existing solutions and to generate the new solutions.

Here we have a constraint for the parents of the GA. Not all parents can genetically reproduce. The crossover operation can only take place between the following pairs of parents. Here M stands for the male and F stands for the female counterpart. The M or F property is not the personal characteristic associated with any individual. Rather it is the way or the role that the parent is playing in the crossover operation.

- Full Solution (M) and Full Solution (F)
- Left Solution (M) and Full Solution (F)
- Full Solution (M) and Right Solution (F)
- Left Solution (M) and Right solution (F)

The crossover always takes place between the male and the female. The male is the one who represents the initial part of the newly generated solution. The female contributes to the latter parts. The new individual has the initial or the left characteristics contributed by the male and the right ones contributed by the female.

If the left sequence selected is

$$(x_{11},y_{11},d_{11})(x_{12},y_{12},d_{12})\ldots\ldots\ldots(x_i,y_i,d_i)\ (x_{1i+1},y_{1i+1},d_{1i+1})\ldots\ldots\ldots(x_{1n},y_{1n},d_{1n})$$

and right sequence selected is

$$(x_{21},y_{21},d_{21})(x_{22},y_{22},d_{22})\ldots\ldots\ldots(x_i,y_i,d_i)\ (x_{2i+1},y_{2i+1},d_{2i+1})\ldots\ldots\ldots(x_{2n},y_{2n},d_{2n})$$

The integrated full solution will be

$$(x_{11},y_{11},d_{11})(x_{12},y_{12},d_{12})\ldots\ldots\ldots(x_i,y_i,d_i)\ (x_{2i+1},y_{2i+1},d_{2i+1})\ldots\ldots\ldots(x_{2n},y_{2n},d_{2n})$$

We take the left part from the male participating in the crossover and right part from the female. The final solution is added to the solution set, if crossover is possible. If there are more than one common points, the crossover takes place for each of these points. All these are added to the solution set.

CrossOver(parent1,parent2)

Step 1: points $\leftarrow$ find common points in parent_1 and parent_2 that have same x and y coordinate
Step 2: if points $\neq$ null
Step 3: for each common point p in points
Begin
Step 4: $sol_1 \leftarrow$ all points in parent_1 till p + rotations necessary to transform direction of p to the one of point p of parent_2 + all points in parent_2 after p
Step 5: Add sol_1 to fullSolution set

14.3.5 Mutation

The *mutation* is the method by which we try to add newer characteristics to the system. This is important to ensure that the algorithm does not get converged at local minima due to the absence of characteristics. The newer characteristics are generated by the current individuals in the population.

This process occurs with a very small probability. In this process any one full solution is chosen. We take any two arbitrary points on this solution. Then we try to find out full solutions between these two points. Hence we try to find out a path between these points using an algorithm similar to the algorithm used to find the initial solution. If such a path is found, we replace the path between these two points with this new path. If more than one solution are found, we replace the old path by the one with the least value of fitness function. This is added to the solution set.

Mutate()

Step 1: if number of solutions in full solution set > 0 then
Step 2: $r_1,r_2 \leftarrow$ any two random points on a random solution from fullSolution set with r_1 on the left of r_2
Step 3: find solutions as found in initial solution with start point as r_1 and goal point as r_2
Step 4: if full results generated > 0
Step 5: newSolution $\leftarrow$ most fit solution in this generated solution set
Step 7: $sol_1 \leftarrow$ all points in solution till r_1 + rotations necessary to transform direction of r_1 to the one of first point of newSolution + all points in newSolution + rotations necessary to transform direction of last point in newSolution to r_2 + all points in solution after r_2
Step 8: add sol_1 to solutionFull

14.4 ARTIFICIAL NEURAL NETWORK WITH BACK PROPAGATION ALGORITHM

The second algorithm we implement is the artificial neural network (ANN) with back propagation algorithm (BPA). The ANN is used for the purpose of machine learning, i.e., the learning of the historical data. The ANN is able to reproduce the output whenever the input is again given. The ANNs are robust against the noise that may creep into the input data. Another important aspect of the ANNs is their power to generalize and form rules out of the presented data. They are able to give correct results to unknown inputs as well.

It is known that the various conditions that affect the movement of a robot would repeat themselves over and over again in a given period of time. This means that if we learn the conditions and the calculated output or move of the robot, we will be able to repeat the move each and every time whenever the same conditions arise. The job now is to make an extensive database of situations that the robot may face and the associated moves that are best for the robot to make. We would then make the robot learn this database. Then the robot would be easily able to make its way out of any situation or around any obstacle.

Another motivation for the use of ANNs is the natural way in which humans and other animals plan their way. We know that there is no major algorithm or big computation going on to make a move to reach the goal. We simply walk down step by step and achieve the target in a way that is fairly optimal in most of the cases. This ease of decision making is mainly due to our past understanding of the situations and the associated actions that we learn initially as well as keep learning in our lifetimes. The use of ANN with BPA largely follows the same concepts and fundamentals.

The ANN takes as its input the robot's current situation. The output of the ANN is the optimal next move. If we keep on inserting the conditions and executing the moves like this, we will get the optimal path that the robot must travel.

14.4.1 INPUTS

There are total 26 inputs to the neural network (I_0 to I_{25}). These are as follows:

The I_0 and I_1 denote the rotations needed for the robot to rotate from its present direction to the direction in which the goal is present. We know that the rotation can be –2 or +2 (180 degrees left or 180 degrees right), –1 (90 degrees left), 0 (no rotation), or else +1 (90 degrees right). These are represented by these two inputs as follows:

$(I_0, I_1) = (0,0)$ *represents* $+2$ *or* -2
$(I_0, I_1) = (0,1)$ *represents* $+1$
$(I_0, I_1) = (1,1)$ *represents* 0
$(I_0, I_1) = (1,0)$ *represents* -1

It should be noted that these are numbered according to gray codes so that even if one of the bits is corrupted by noise, the effect on the neural network is minimal.

The other bits (I_2 to I_{25}) represent the condition of the map. Considering the whole map as a grid of $(M \times N)$, we take a small portion of it (5×5), with the robot at the center of the map. Each coordinate of this map is marked as 1 if the robot can move to this point in the next step, or 0 if the point does not exist or the robot cannot move to this point because of some obstacle. So we have taken a small part of the graph. If (x,y) is the coordinate of the robot, we take $(x-2,y-2)$......$(x+2,y+2)$ as the graph. These are 25 points. These points, excluding the center of the graph (where our robot stands), are fed into the neural network.

The layout of the 24 inputs (I_2 *to* I_{25}) is given in Figure 14.5. Here (m, n) is the coordinates of our robot. I_2–I_{25} denotes the inputs. These are 1 or 0, depending on the obstacles.

I_2 (m−2,n−2)	I_3 (m−2,n−1)	I_4 (m−2,n)	I_5 (m−2,n+1)	I_6 (m−2,n+2)
I_7 (m−1,n−2)	I_8 (m−1,n−2)	I_9 (m−1,n)	I_{10} (m−1,n+1)	I_{11} (m−1,n+2)
I_{12} (m,n−2)	I_{13} (m,n−1)	**Robot (mXn)**	I_{14} (m,n+1)	I_{15} (m,n+2)
I_{16} (m+1, n−2)	I_{17} (m+1,n−1)	I_{18} (m+1,n)	I_{19} (m+1,n+1)	I_{20} (m+1,n+2)
I_{21} (m+2,n−2)	I_{22} (m+2,n−1)	I_{23} (m+2,n)	I_{24} (m+2,n+1)	I_{25} (m+2,n+2)

FIGURE 14.5 The inputs for the artificial neural network.

14.4.2 Explanation of the Outputs

There are total of three outputs (O_0,O_1,O_2), each of which can be either 0 or 1. We can make eight combinations of outputs using these. The meanings of these outputs are as follows:

$(O_0, O_1, O_2) = (0,0,0)$ *Rotate left* 90 *degrees.*
$(O_0, O_1, O_2) = (0,0,1)$ *Rotate left* 45 *degrees.*
$(O_0, O_1, O_2) = (0,1,0)$ *Move forward.*
$(O_0, O_1, O_2) = (0,1,1)$ *Rotate left 45 degrees and move forward.*
$(O_0, O_1, O_2) = (1,0,0)$ *Do not move.*
$(O_0, O_1, O_2) = (1,0,1)$ *Rotate right 90 degrees.*
$(O_0, O_1, O_2) = (1,1,0)$ *Rotate right 45 degrees and move forward.*
$(O_0, O_1, O_2) = (1,1,1)$ *Move right* 45 *degrees.*

It should be noted that this sequence is in gray code as well; if we arrange it in gray code sequence, the order of the moves will be 90 degrees rotate left, 45 degrees rotate left, 45 degrees rotate left and move forward, move forward, 45 degrees rotate right and move forward, 45 degrees rotate right, 90 degrees rotate right.

Hence there is a transition in series. This ensures that the effect of noise is minimal.

14.4.3 Special Constraints Put in the Algorithm

There were a few limitations of the algorithm, but these were removed by applying some constraints to the algorithm, which ensured the smooth working of the algorithm:

1. It may happen that the robot does not move or keeps turning in place continuously. Since there is no feedback in test mode, this is likely to continue indefinitely. For this, we apply the constraint that if the robot sits idle for more than five time stamps and does not make any move (excluding rotations), we take a random move.
2. It may happen that the robot develops an affinity to walk straight in a direction; hence if it is situated in the opposite direction, it may go farther and farther from the destination. For this we limit that if the robot makes five consecutive moves (excluding rotations) that increase its distance from the destination, we make the next one move only one that decreases its distance (excluding rotations).

14.4.4 Procedure

The algorithm first generated test cases, so that the established neural network can be trained. These are done by generating input conditions and solving them using A* algorithm and getting the inputs and outputs. The ANN is established and trained using these test cases. We lay more stress on the rotational parameters of the robot, i.e., the capability of the robot to rotate from its present direction to the goal direction. After the training is over, we enter into the test phase. Looking at the condition of the map, the input cases are generated and are fed into the ANN. The output is fetched, decoded, and followed, and the robot is moved.

The following is the algorithm

GenerateTestCases()

Step 1: initialize map and generate N obstacles at positions (x_i, y_i, d_i)
Step 2: while(noTestCaseGenerated < noTestCasesDesired)
Begin
Step 3: Generate the robot at random position (x_i, y_i, d_i)
Step 4: $I \leftarrow$ Input sequence
Step 5: $m \leftarrow$ getNextRobotMoveUsingAStar(CurrentPosition)
Step 6: moveRobot(m)
Step 7: $O \leftarrow$ Output Sequence
Step 8: Add (I,O) to training data

NavigationPlan()

Step 1: $n \leftarrow$ Generate Neural Network
Step 2: $n \leftarrow$ TrainNeuralNetwork()
Step 3: while(CurrentPosition $\neq$ FinalPosition)
Begin
Step 4: $I \leftarrow$ generateInputSequence()
Step 5: $O \leftarrow$ generateOutputSequence(I)
Step 6: if O gives valid move
Step 7: moveRobot(O)
Step 8: if robot not changed position in last five steps
Step 9: makeRandomMove()
Step 10: if distance of robot decreasing in last five steps (excluding rotations)
Step 11: move next move only if the next move decreases distance or is of equal distance

14.5 A* ALGORITHM

We have also implemented the A* algorithm for the same problem. The A* algorithm tries to minimize the path traveled and the path that is left to be traveled. It hence tries to optimize the total path to be traveled by the robot. We take a heuristic function to predict the number of moves that are left in a particular position. The heuristic function is very important, as it decides the closeness of the position to the goal. We select the closest position, at every step, so that the travel route is minimized.

This algorithm uses a heuristic function to optimize the path traversed from the start to the end. The algorithm used is the conventional A* algorithm to find out the goal (final position) starting from the initial position. In this algorithm, we use the following:

$g(n)$ = depth from the initial node, increases by 1 in every step
$h(n)$ = square of the distance of the current position and the final position + $R(n)$
$r(n)$ = minimum time required for the robot to rotate in its entire journey assuming no obstacles
$f(n) = g(n) + h(n)$

where n is any node.

The function $R(n)$ is exactly similar to what we had taken in genetic algorithm discussed earlier.

The following is the algorithm:

gePath(CurrentPosition,GoalPosition)

```
Step 1: closed ← empty list
Step 2: add a node n in open such that position(n) = CurrentPosition
Step 3: while open is not empty
            Begin
Step 4:         extract the node n from open with the least priority
Step 5:         if n = final position then break
Step 6:         else
Step 7:             moves ← all possible moves from the position n
Step 8:             for each move m in moves
                    Begin
Step 9:                 if m leads us to a point that can be the point of any obstacle
                        in the next unit time, then discard move m and continue
Step 10:                if m is already in the open list and is equally good or bet-
                        ter, then discard this move
Step 11:                if m is already in closed list and is equally good or better,
                        then discard this move
Step 12:                delete m from open and closed lists
Step 13:                make m as new node with parent n
Step 14:                Add node m to open
Step 15:            Add n to closed
Step 16:            Remove n from open
```

14.6 COMPARISONS

We studied three different algorithms for robotic path planning, i.e., GA, ANN with BPA, and the A* algorithm. All three of these algorithms solved the problem of path planning and gave good results to the maps that were presented. In this section we look at some of the differences between the three algorithms in regard to the time taken for execution, computational and input constraints, and optimality of the path generated. The following are the key points of comparison between these algorithms:

- GA takes a long time for generation of good results as compared to ANN and GA. However, it is an iterative algorithm. In case the map is much too complex for the ANN to solve, optimality is a necessary constraint, and the A* version fails because of the large input size, then GA can be used as a very effective algorithm. The algorithm would specially attract attention because of its iterative nature where the solutions are generated and keep improving. We can terminate the algorithm at any time and accept the solution that is the best so far.
- The A* is the best algorithm that gives us the optimal solutions. The optimality is better than for the ANN or the GA. The algorithm, however, may not work in case the map is very large and complex. The time required by the A* would be very high in such conditions.
- The ANN gives very fast results and leads to a very quick generation of a path. The path, however, may not be optimal. Also, the ANN will fail badly if used in highly chaotic conditions or given a maze-like problem to solve.

- The ANN has the lowest memory requirements. The memory requirements of the GA are high but constant and normally do not depend upon the size of the input. For the same reasons, the GA is able to perform even when large input maps are given. The A* has memory requirements that directly depend upon the input map size and the complexity of the map.

In general, the various algorithms studied could solve the problem of path planning quite efficiently and result in good paths in finite amount of time.

14.7 ROBOTIC CONTROLLER

Neuro-fuzzy systems, as we saw in Chapter 6, integrate the two well-used systems of ANN and FIS. The resulting system uses the best components of both these individual systems and hence results in a system that is scalable to meet the requirements of real life applications. The neuro-fuzzy system uses fuzzy rules to map inputs to outputs. This provides us the power to use linguistic rules to map the inputs to the outputs. Further, the ANN is used to tune the rules. These are excellent agents for learning historical data and hence benefit from such data. The neuro-fuzzy system thus generated behaves very realistically and is able to solve the given problem.

The robotic controller is responsible for moving the robot in a way that it reaches its goal without colliding with any obstacles. The controller uses sensors to sense its immediate surroundings. Using this sensed information, it is able to understand the surroundings. It then uses the fuzzy rules to physically move in such a way that it reaches a point near the goal and at the same time avoids collision with obstacles. In this way, the robot keeps moving until it reaches its destination. It is ensured that any move does not end in a collision.

14.7.1 Inputs and Outputs

Here, we have identified four inputs of the system. These are Left Obstacle Distance (LOD), Right Obstacle Distance (ROD), Forward Obstacle Distance (FOD), and Target Angle (TA). The LOD measures the distance from the current position of the robot to an obstacle that lies exactly to the left of the robot. This sensor is placed to the left side of the robot and uses the sensing mechanism to find the physical distance at which the nearest obstacle is placed. The ROD and FOD work similarly. The TA is a measure of the angle that leads us to the target. The LOD, ROD, and FOD are given in Figure 14.6.

The system outputs the steering angle. The robot needs to turn using this steering angle in order to reach the goal without collision with the obstacles.

The membership functions of the various inputs are given in Figure 14.7.

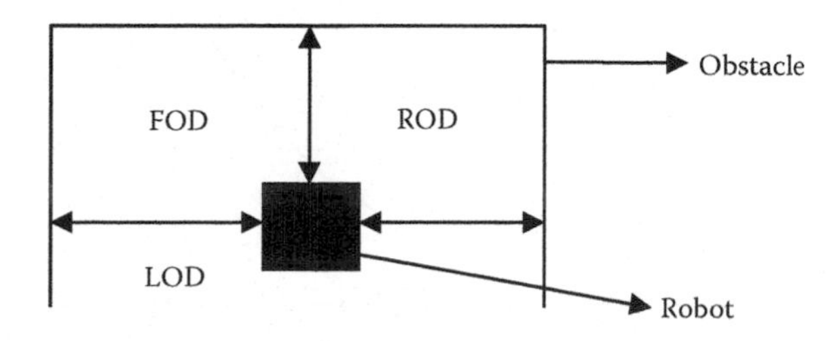

FIGURE 14.6 The inputs to the adaptive neuro-fuzzy inference system.

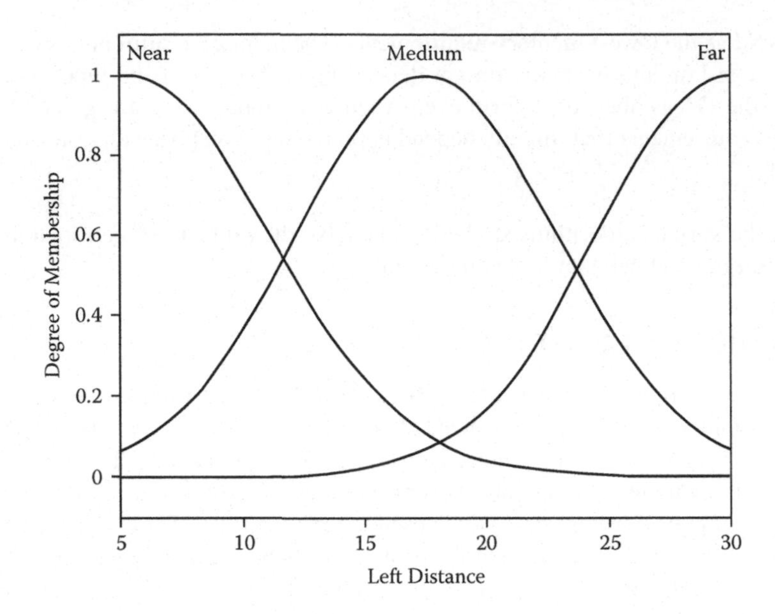

FIGURE 14.7a Membership functions for left obstacle distance.

14.7.2 Rules

A total of 81 rules were generated in order to effectively control the robot using the designed controller. The rules help in effective mapping of the inputs to the outputs and are hence sources of concern in the algorithm. Here are some of these rules:

Rule 1:
If (left_distance is near) and (right_distance is near) and (front_distance is near) and (target angle is negative), then (steering angle is positive).

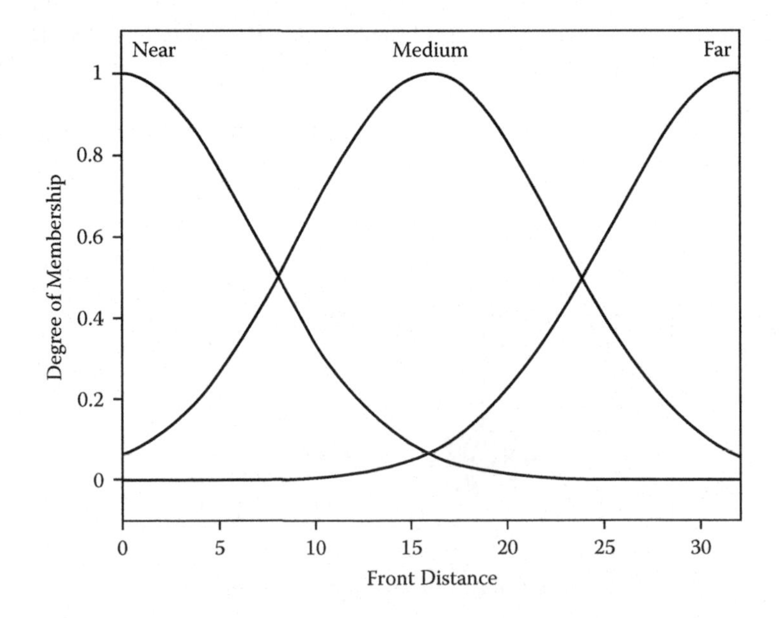

FIGURE 14.7b Membership functions for front obstacle distance.

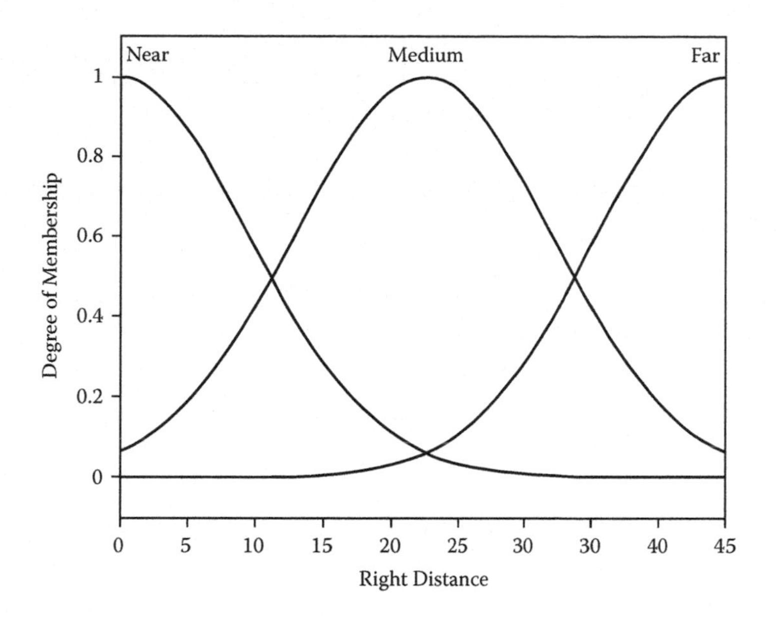

FIGURE 14.7c Membership functions for right obstacle distance.

Rule 2:

If (*left_distance is near*) *and* (*right_distance is near*) *and* (*front_distance is near*) *and* (*target angle is zero*), *then* (*steering angle is positive*).

Rule 3:

If (*left_distance is near*) *and* (*right_distance is near*) *and* (*front_distance is near*) *and* (*target angle is positive*), *then* (*steering angle is negative*).

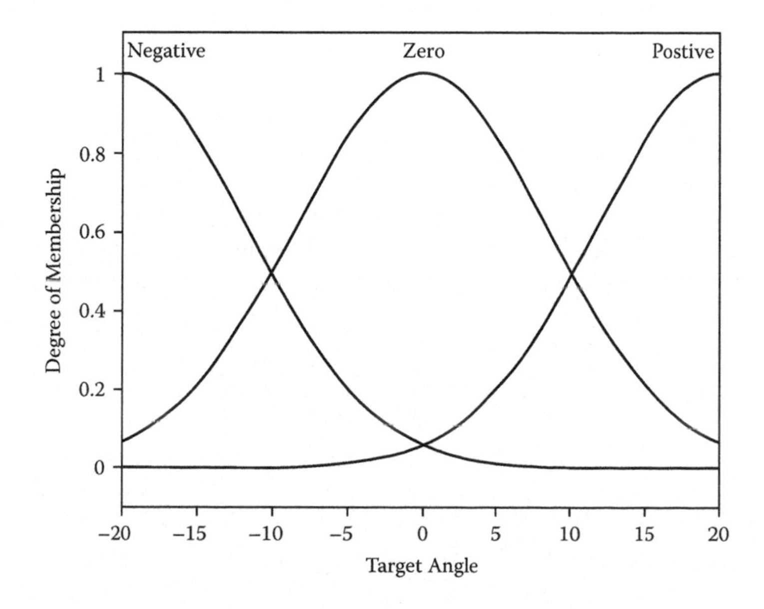

FIGURE 14.7d Membership functions for target angle.

Rule 4:
If (*left_distance is near*) *and* (*right_distance is near*) *and* (*front_distance is medium*) *and* (*target angle is negative*), *then* (*steering angle is positive*).

Rule 5:
If (*left_distance is near*) *and* (*right_distance is near*) *and* (*front_distance is medium*) *and* (*target angle is zero*), *then* (*steering angle is zero*).

Rule 6:
If (*left_distance is near*) *and* (*right_distance is near*) *and* (*front_distance is medium*) *and* (*target angle is positive*), *then* (*steering angle is negative*).

Rule 7:
If (*left_distance is near*) *and* (*right_distance is near*) *and* (*front_distance is far*) *and* (*target angle is negative*), *then* (*steering angle is positive*).

Rule 8:
If (*left_distance is near*) *and* (*right_distance is near*) *and* (*front_distance is far*) *and* (*target angle is zero*), *then* (*steering angle is zero*).

Rule 9:
If (*left_distance is near*) *and* (*right_distance is near*) *and* (*front_distance is far*) *and* (*target angle is positive*), *then* (*steering angle is negative*).

Rule 10:
If (*left_distance is far*) *and* (*right_distance is near*) *and* (*front_distance is near*) *and* (*target angle is negative*), *then* (*steering angle is positive*).

Rule 11:
If (*left_distance is far*) *and* (*right_distance is near*) *and* (*front_distance is near*) *and* (*target angle is zero*), *then* (*steering angle is positive*).

Rule 12:
If (*left_distance is far*) *and* (*right_distance is near*) *and* (*front_distance is near*) *and* (*target angle is positive*), *then* (*steering angle is negative*).

14.8 RESULTS

The testing of the system was done using Java as a platform. Here we randomly generated various obstacles. The robot was controlled by the path planning algorithm. The algorithm moved the robot from the start position and made it reach the goal position. All this was shown using Java applets.

All the obstacles and our robot were displayed in the applets. The directions were displayed with the head pointing toward the direction as shown in Figure 14.8. Here the first figure shows any robot/obstacle pointing out in the upward (north) direction. The direction number of this is 0

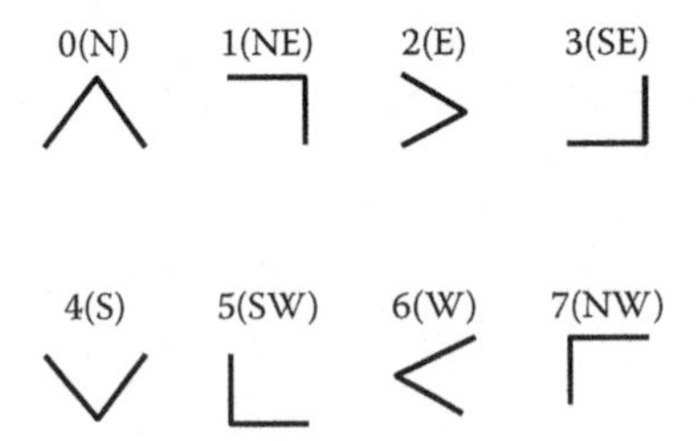

FIGURE 14.8 The representation of obstacles in various directions.

(as explained earlier). Similarly the subsequent figures show the robot/obstacle pointing northeast (1), east (2), southeast (3), south (4), southwest (5), west (6), and northwest (7) respectively. An open figure denotes an obstacle, and a closed one denotes a robot. The applet also collected the complete path followed by the robot to give a path trace of the robot.

The algorithms were tested using the technique described earlier. In all cases, the movement of the robot at each step was recorded using all the three algorithms. The results are as follows.

14.8.1 Genetic Algorithm

The first algorithm used was the genetic algorithm. Here we had used a sample grid space on which the robot is to be moved that was [100 × 100] in dimension. The coordinates could vary from (0, 0) to (99, 99). We used total 750 obstacles, which all could move a unit step at any unit time. The threshold time was fixed to be 9 seconds. We moved the robot from position (0, 0) to the goal position (99, 99).

Some of the constants as mentioned in the algorithm were fixed as follows. The path traced by the robot is given in Figure 14.9 (for four runs).

14.8.2 Artificial Neural Networks

A similar run was performed for the artificial neural networks as well. The parameters were quite similar to the ones used in the previous algorithm. The sample grid space on which the robot is to be moved was [100 × 100] in dimension, where the coordinates could vary from (1, 1) to (100,100). MATLAB® was used for the simulation. We used a total of 500 obstacles, which all could move a unit step at any unit time. The threshold time was fixed to be 1 second. We moved the robot from position (1, 1) to the goal position (99, 99).

The path traced by the robot is given in Figure 14.10 (for four runs).

14.8.3 A* Algorithm

The third algorithm used for the problem was A* Algorithm. Here also the grid size was [100 × 100]. The coordinates could vary from (0, 0) to (99, 99). We used a total of 1,000 obstacles, which

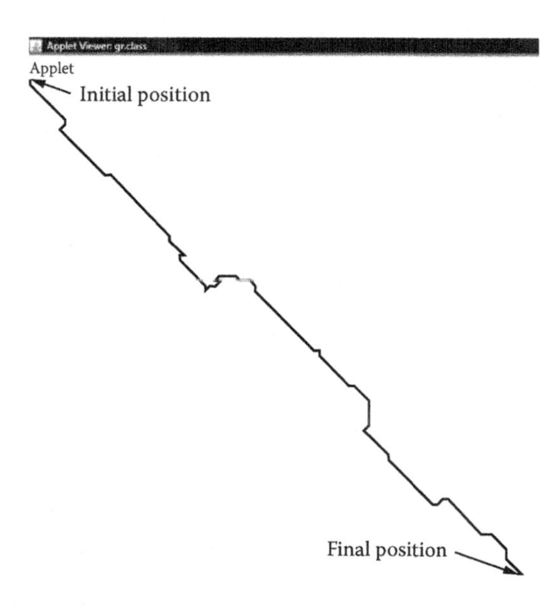

FIGURE 14.9a The result of the first run of the genetic algorithm.

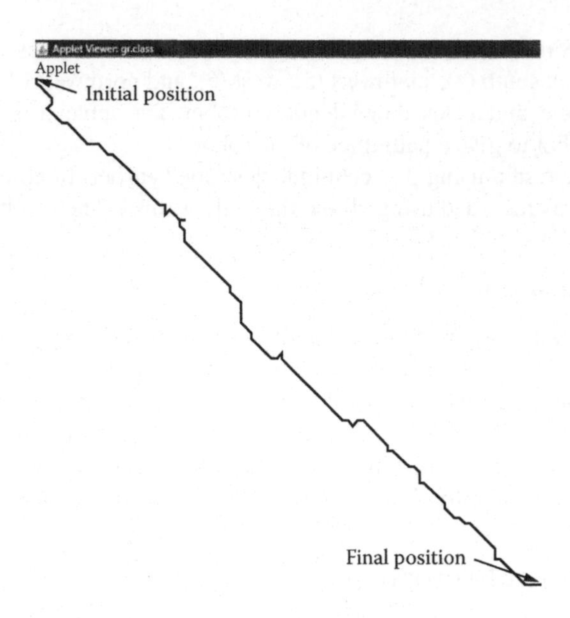

FIGURE 14.9b The result of the second run of the genetic algorithm.

all could move a unit step at any unit time. The threshold time was fixed to be 1 second. We moved the robot from position (0, 0) to the goal position (99, 99).

The path traced by the robot is given in Figure 14.11 (for four runs)

The condition of the board at a random time is given in Figure 14.12.

Closely watching the robot go toward its goal, we see that there is no collision on its way. Hence we have been successful in avoiding collisions. Also, we find looking at the path traced, that the path is optimal with respect to the conditions given.

Looking at the path traced by the robot in its journey from the source to the goal, we find that the whole path is a collection of various kinds of points. These points depict the state of the

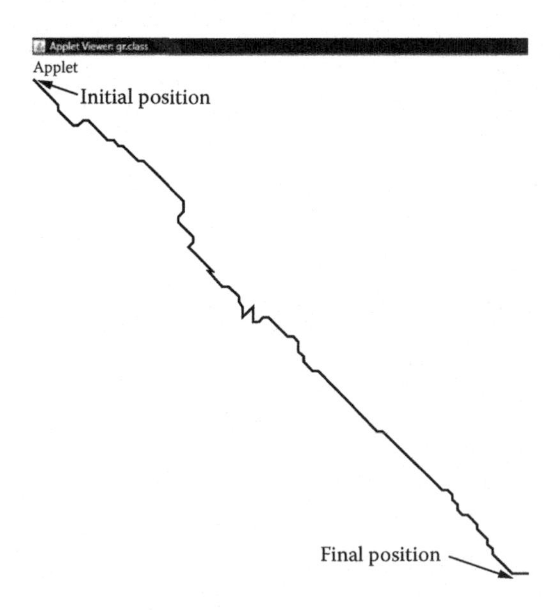

FIGURE 14.9c The result of the third run of the genetic algorithm.

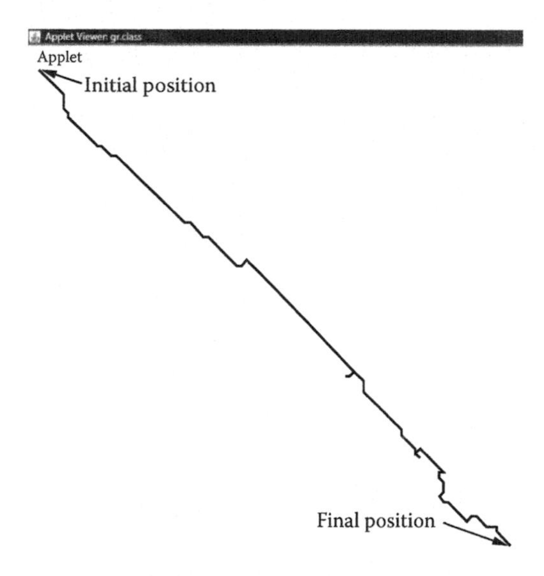

FIGURE 14.9d The result of the fourth run of the genetic algorithm.

surroundings and the obstacles. Accordingly, we classify these points into four major categories. These are the following:

1. **Stationary phase:** In this phase, the robot is completely stationary. This happens because there is no way that the robot can reach the goal. Additionally, when the algorithmic complexity is high and the algorithm is unable to return a solution, the robot may remain stationery.
2. **Straight phase:** This is the most desirable kind of phase, where the robot marches ahead in almost a straight-line path. This is when there are no obstacles at all or there are very few obstacles in the path of the robot and the goal.
3. **Collision avoidance phase:** If an obstacle is very close to the robot, it takes a sharp turn to avoid collision and keep going. This phase is known as the collision avoidance phase.

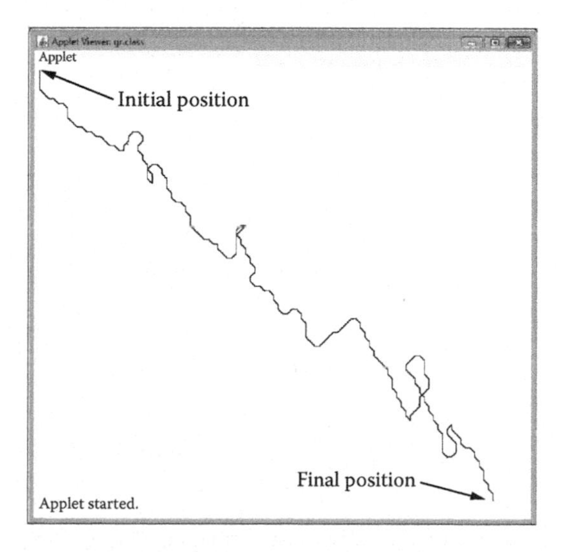

FIGURE 14.10a The result of the first run of the artificial neural network.

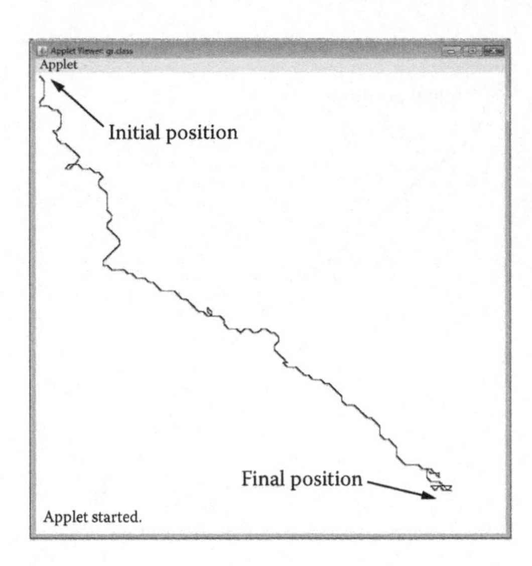

FIGURE 14.10b The result of the second run of the artificial neural network.

4. **Backtracking phase:** If the robot happens to deviate greatly from its path due to an excessive number of obstacles in its close vicinity, it backtracks to its path and then continues to move further.

These are shown in Figure 14.13.

14.8.4 Robotic Controller

This section presents the experimental results that show the performance of the neuro-fuzzy system. The results were simulated on MATLAB. Figure 14.14 shows the training curve for ANFIS. The

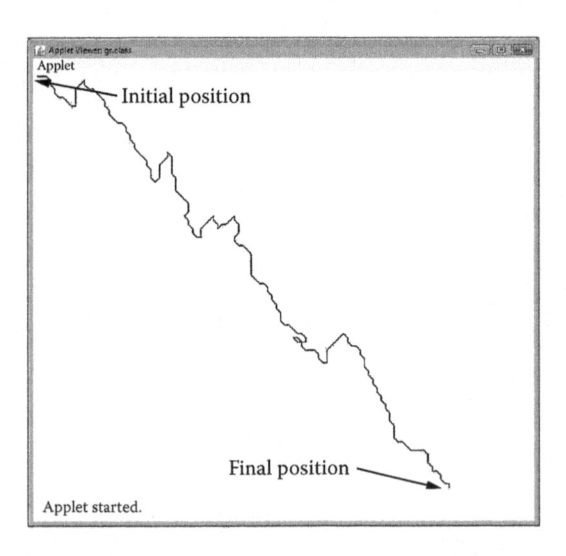

FIGURE 14.10c The result of the third run of the artificial neural network.

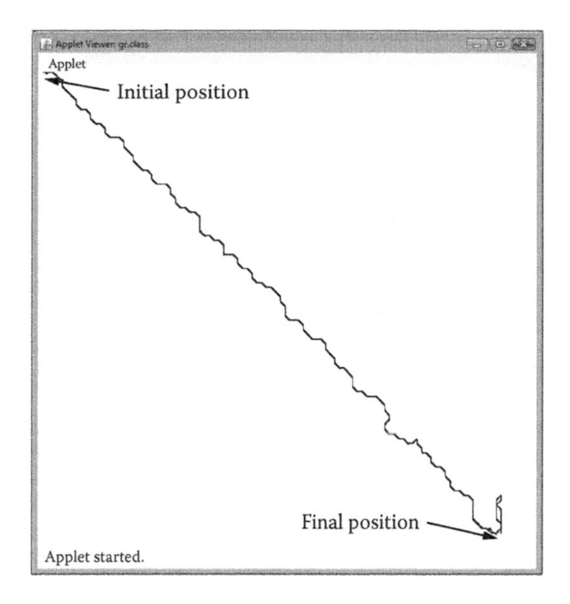

FIGURE 14.10d The result of the fourth run of the artificial neural network.

result of the algorithm for robotic runs is given in Figure 14.15. Looking at the path traced by the robot using the constructed neuro-fuzzy system, we can easily see that the robot is able to avoid all the obstacles and reach the goal position starting from the initial position. The strategy behind the robot movement is to steer the robot in such a way that it avoids all obstacles that happen to come its way. Along with doing that, the robot marches toward the goal and finally stops at the goal position.

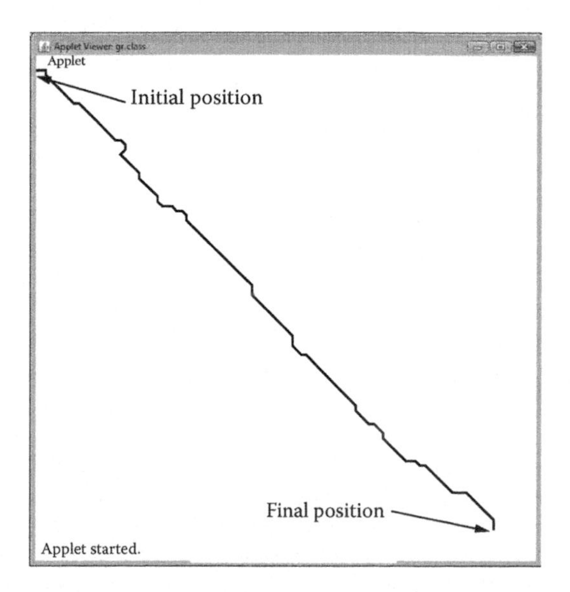

FIGURE 14.11a The result of the first run of the A* algorithm.

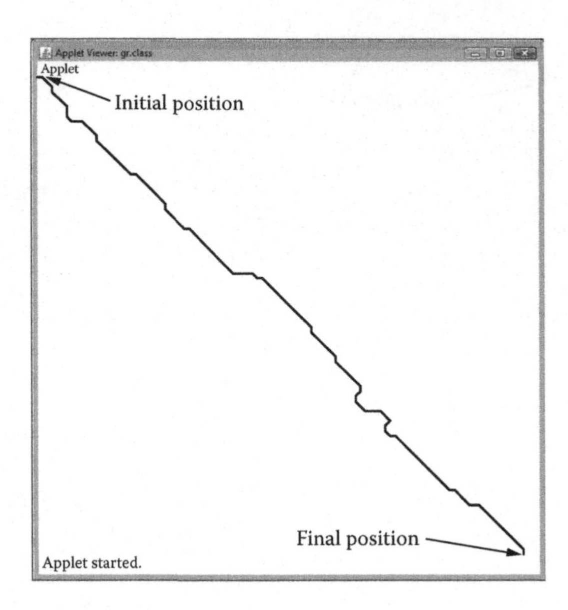

FIGURE 14.11b The result of the second run of the A* algorithm.

14.9 CONCLUSIONS

In this chapter, we first introduced path planning and solved the problem of path planning using three algorithms: GA, ANN with BPA, and the A* algorithm. We studied the theoretical basis of each of these algorithms and later applied these algorithms to solve the problem. In all cases, we saw that the algorithms were able to solve the problem and gave reasonable and optimal solutions in the map that was given to the algorithm as input. We also compared the three algorithms over numerous issues and studied which one would perform better under what circumstances. We even studied the algorithmic performance over randomly generated obstacles.

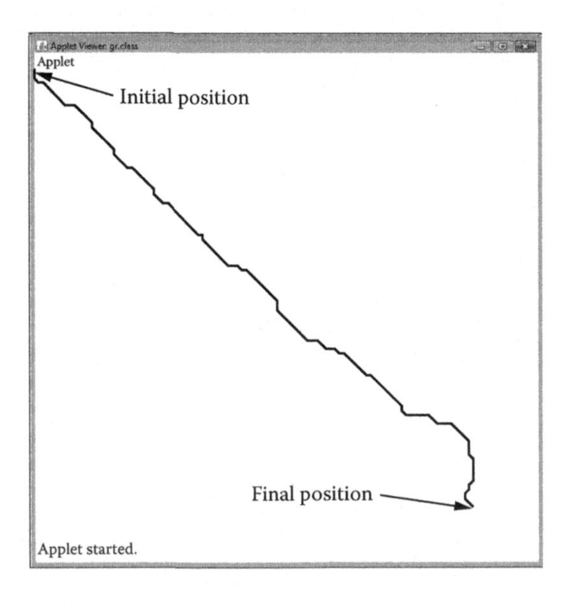

FIGURE 14.11c The result of the third run of the A* algorithm.

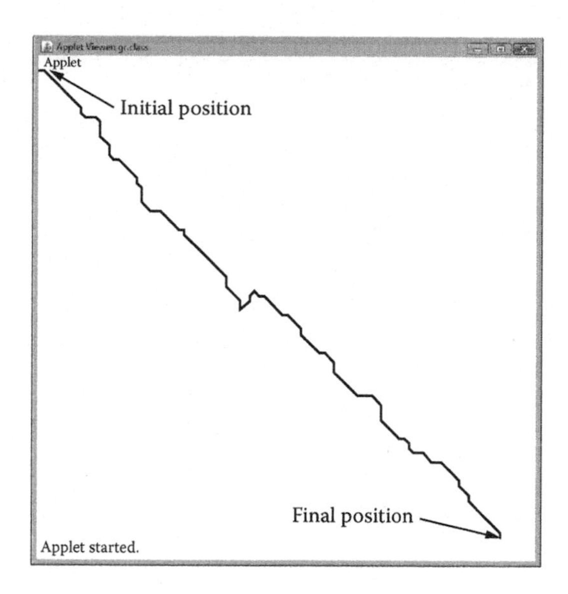

FIGURE 14.11d The result of the fourth run of the A* algorithm.

The first algorithm studied was GA. Here we observed that the algorithm takes a time for the generation of a really optimal solution. The algorithm has an added advantage because of its iterative nature. We even discussed that this algorithm would give a good solution to a very large map while the other algorithms might fail to generate a solution in finite time.

The second algorithm that we studied was the ANN. Here we saw that the algorithm learns from the past data that for implementation purposes we had generated using A* algorithm results. We said that these conditions would repeat themselves over and over again in the course of time. The

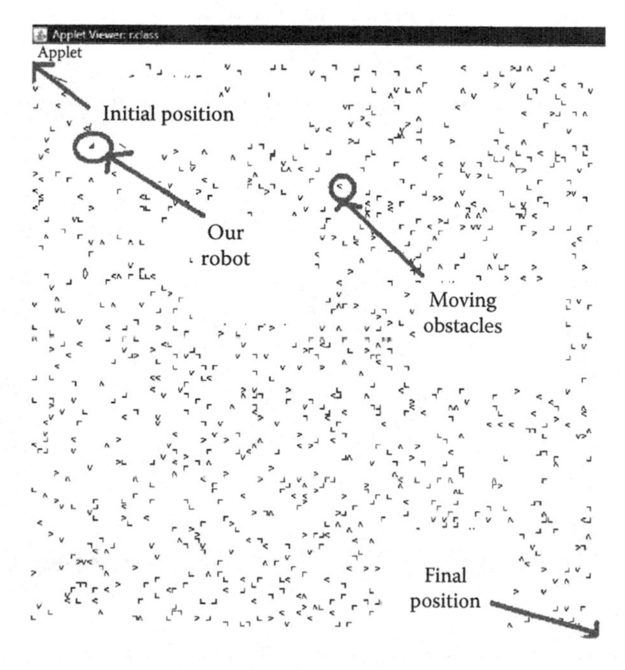

FIGURE 14.12 Obstacles and robot in motion.

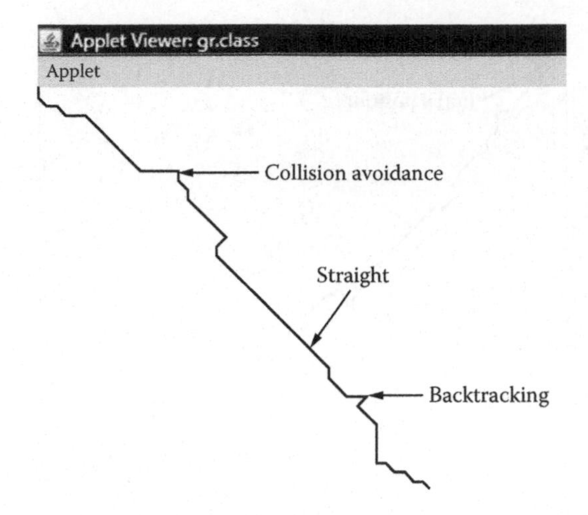

FIGURE 14.13 The various points on the path traced by the robot.

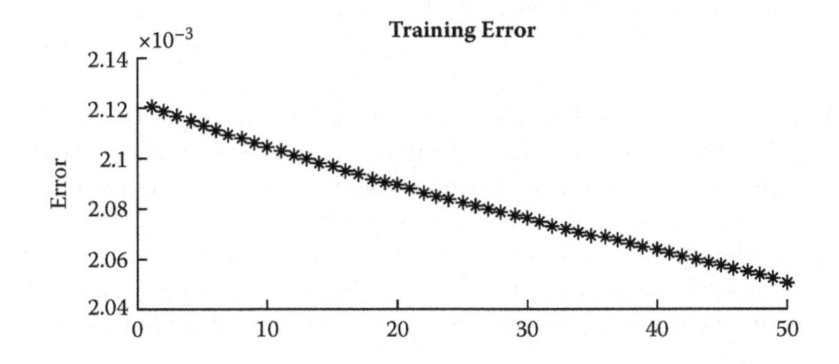

FIGURE 14.14 Plot showing errors during the training of ANFIS structure.

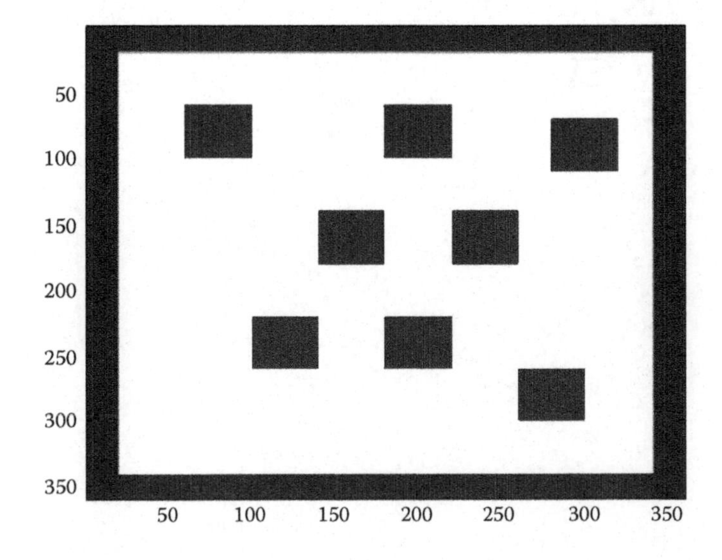

FIGURE 14.15a Snapshot of the environment with eight obstacles.

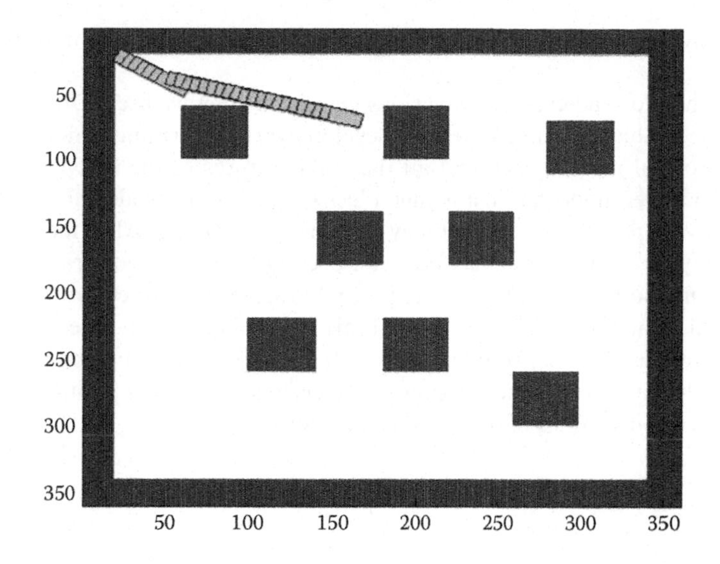

FIGURE 14.15b Motion of the robot in the environment with eight obstacles at the middle of the journey.

algorithm required a long training time. But the resting phase took very little time for the execution purpose. We, however, argued that this algorithm would fail when given a complex map to solve.

The last algorithm used was the A* algorithm. The algorithm was very efficient in terms of both time and the optimality of the path generated. We, however, further argued that the algorithm may face problem in the case of large and complex maps.

This chapter further discussed the implementation and the design of neuro-fuzzy robotic controller. This was used to move the robot on the map by using a rule-based fuzzy approach optimized by ANN learning. The controller guided the robot from the source to the destination. The robot moved well and very optimally covered the entire path. We stated the fact that this system is adaptive and can accommodate sudden changes in the environment. The changes would be reflected or sensed by

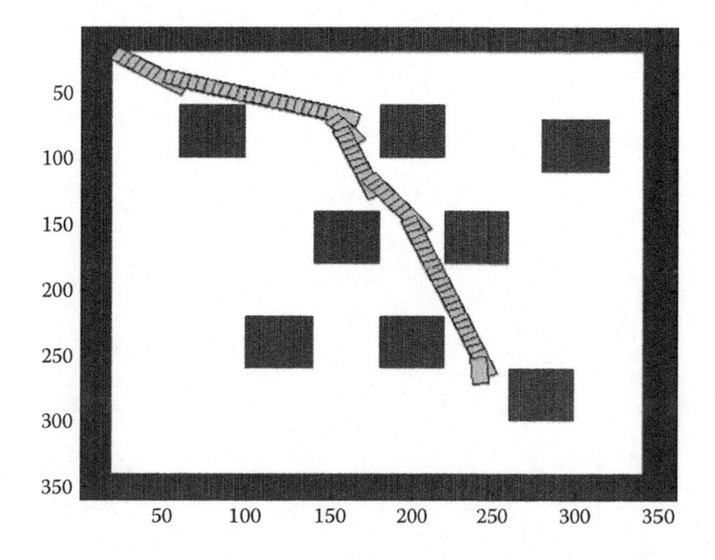

FIGURE 14.15c Motion of the robot in the environment with eight obstacles toward the end of the journey.

the sensors used by the robot. This would then be used for calculating and implementing the next robotic move.

We discussed the pros and cons of the various algorithms that we used. Drawing on these algorithms and other available algorithms, we may develop hybrid algorithms or implement these algorithms in a hierarchical manner such that all the good features of the individual algorithms join together to yield a better algorithm that is much less limited over a wide variety of map and constraints. This is a very promising task that may be done in the future. Also, the various algorithms discussed here may be modified to adapt them for specific maps and constraints according to the place of application and the type of robot. This would further help in controlling the algorithmic complexity as well as maximizing of the algorithmic optimality over the specified constraints and demands of the problem. The capabilities of the robot control system can be extended by adding new behaviors. A behavior remembering previously visited goal object positions and encouraging the robot to go unvisited sections of the world can be added.

15 Character Recognition

15.1 INTRODUCTION

The traditional way of working in any organization or department involved a lot of paperwork where documents were typed, printed, and circulated. The forms were manually filled and processed. Then the automated systems came and took on the bulk of the job. Even though automation has covered most fields, paper still continues to survive and intermingles with the automated counterparts at various fronts. This necessitates and motivates the use of systems that can automatically read and extract characters from any picture, scanned document, web-acquired image with text, etc. Character recognition is a field that deals with the recognition of characters embedded in pictures that may be acquired in numerous ways, depending upon the system under study. Optical character recognition (OCR) applies to the printed documents that have optical characters placed on them. A related field is handwriting recognition, where we try to extract text scribbled by a person.

The problem of character recognition has been applied to various systems. Extracting text from images, converting scanned documents to text, product identification, and vehicle identification are few of the systems that are extensively used in daily life. The problem of character recognition especially becomes very difficult when there is noise, which occurs naturally in most real life applications. Another common problem is that there are numerous ways in which a character can be written. We need to identify all of these. The presence of multiple languages is yet another problem. In this chapter, we study optical character recognition as well as handwriting recognition separately using different techniques.

OCR refers to the problem of identifying the character that is written or drawn. The recognition gives us in text form the character that is likely to be present in the given picture. The problem comes under the category of classification. In these types of problems, we are to classify the input into a set of output classes. This classification problem has as many output classes as there are characters in the language or languages being considered. It is natural that the complexity of the problem increases as we increase this number of classes that the input can map to. If the language contains more characters, identification becomes much more difficult than if the language contains fewer characters. Similarly we need to consider how the various characters are written and the differences between the various characters. They always have an effect on the performance of the handwriting recognition system.

Just like any recognition system, character recognition is usually studied under two separate heads: offline character recognition and online character recognition. The *offline* technique has the final character or picture given directly. There is no information on what makes up this picture. This happens in case of scanned images, images acquired by cameras, etc. The other type of character recognition is the *online* type. Here the character is written in the presence of the system. Hence we may sample the character at various times to learn about its formation. Examples include the light pen or other pointers that are used to write text on a digital screen.

We first discuss a technique for multilingual character identification. This technique uses a mixture of a rule-based approach along with the ANN to identify characters. The rule-based approach is used to identify the language to which the input belongs. Every language has its own ANN. Once the language is known, we use the language-specific ANN to identify the exact character. In this manner, we first identify the language and then the character. This algorithm, when tested for all

English capital letters, English small letters, and Hindi characters, yielded an appreciable level of accuracy using test cases of all languages. This proves the efficiency of the algorithm.

Later we study the problem of handwriting recognition and develop a genetic algorithm solution to the problem. The issues with handwriting are little different. There are numerous styles that may be used to write the various alphabets. These styles change as the person grows and gets exposure to newer styles. Hence many people develop hybrid styles for writing the characters. This further makes the process of learning difficult in the case of handwriting. The task, however, seems to be relatively simple for most humans, who can easily identify the character written no matter how it is written. This fact motivates the use of genetic algorithms for the problem of handwriting recognition. Here we try to use genetic operators to evolve hybrid styles that may more closely match the given character. At any time, we try to mix existing styles to make newer styles of the character we are trying to match the given character with. In this way, the matching process is optimized and so we get the correct answer to the problem of recognition of the character.

In a rule-based approach, we keep firing the rules, one after the other, to identify the final class to which the output belongs. Every rule tries to find an association of the input to its class. The first input that succeeds is associated with its class. This approach may be taken as a sequence of "*if... else if...else*" statements. The general structure of this kind of approach is

If condition$_1$ → result$_1$
Else if condition$_2$ → result$_2$
Else if condition$_3$ → result$_3$
...
...
Else result$_n$

This type of approach is very useful when it is well known how the inputs are mapped to the outputs. If we can write down all the relevant rules, then the system will work without any error. The major problem that people usually face with this system is that the rules are difficult to imagine and find out and sometimes even too numerous to write down. Another important point of concern with this approach is that there should not be any error in the input. Even a very small error may cause the outputs to change.

ANN is used here for its power to learn historical data and to generalize and store those data. In this system, we map the inputs to the outputs using a network of processing units called *neurons*. The neurons are connected to each other by weights that can vary. The output that any input would produce is decided on the basis of these weights between any two neurons. The neurons are arranged into horizontal layers, and every neuron of a particular layer is joined to a few neurons of its next and previous layer. This network is first trained by using known inputs and outputs. During training, the weights between the various neurons are modified. The system is tested. Here unknown input is applied to find out the final projected output.

Here we have made different ANNs for the various languages. These ANNs are trained from the training data in the respective languages. We take an input and apply a rule-based approach first. The rule-based approach finds out the language that the given character belongs to. This is then fed into the neural network of that particular language. In this way, we are able to recognize the character.

Similarly in the handwriting recognition system, for every character in a language, the existing forms of styles mix to form hybrid styles of writing characters. This goes on for numerous generations. At the end, the character that has the maximum mapping to the unknown input is the final solution computed by the algorithm.

We give a brief introduction to the approaches to identification used in this chapter. Then we present the algorithm and discuss its various aspects. We discuss two approaches here. The first is the multilingual optical character recognition system that uses a combination of a rule-based approach along with ANN. The second is handwriting recognition, which uses GA. These are one after the other. We then present the results of the algorithm. At the end, we offer our conclusions.

Due to the large application and scope of character recognition, this field attracts the attention of many people in related fields of research. This has resulted in various models being devised in order to give better results. A lot of work exists in this field, but most of it has been done using a single language. The complexity of the algorithm increases with the increase in languages dealt with. The major technique used to recognize characters is the artificial neural network. Another major approach is using the Hidden Markov Model (HMM). Various genetic optimizations have been applied to this problem with reference to feature extraction, matching, etc.

Neural networks are a good means of learning from the historical data and then applying the algorithm to a new input. The basic approach in this method is to break the sample given into a grid. Each member of this grid is either 0 or 1, depending on whether something was written in that position. The neural networks are applied straight to the input to give us the output.

The Hidden Markov Model, on the other hand, is a completely statistical model. This model is used to find out the probability of the occurrence of any character, based on the given input. This method is used extensively for these kinds of problems where a classification needs to be made. The HMMs are used to map the input to the output based on the statistical model formed.

A rule-based approach has been applied to various problems. It is a good approach and works very well when it is possible to write down such rules. The rules may be written if they are simple to write and few in number. Usually these rules are written when they are distinctly visible. Many classification problems have been solved using this approach.

15.2 GENERAL ALGORITHM ARCHITECTURE FOR CHARACTER RECOGNITION

The problem that we are discussing in this chapter involves numerous steps. Consider any OCR system, say, scanning a document to get the embedded text. The task is quite sophisticated, involving numerous algorithms and various stages of processing. This chapter primarily deals only with the recognition of a character. It is assumed that all other steps have already been performed.

In this section, we briefly discuss the big picture of the whole process of conversion. Each and every stage we discuss has its own assumptions, limitations, problems, and solutions. It is not necessary that all these steps have equal weight ages in the algorithm. The importance largely depends upon the kind of inputs. e.g., a very clearly taken image might not need noise removal. The general steps involved in the algorithm are as follows:

- **Image acquisition**: In this step we shoot the image or acquire it using any standard means or device. Scanners and cameras are common ways to acquire the image.
- **Binarization**: The acquired image may be in the grayscale or color format. Usually it becomes very computationally expensive to use these representations. Also, the workspace becomes too large to be handled. Hence the algorithms face problems. In this step, we convert the image to a binary format where each pixel is either black or white.
- **Layout analysis**: We need to understand the layout of the image acquired. We need to explicitly cite the text and understand the way it is oriented in the entire document. This helps in the uniform representation of the image containing text, after which the image can be used as a standard input for the identification purposes.
- **Preprocessing**: This phase is involved with the processing of the input character and converting it into a form that can be used as an input to the recognition system. This step does minor processing like noise removal. In our algorithm, we assume that this step has already been done.
- **Segmentation**: In this phase, the entire input is broken down to a character. We may have a whole page as an input, but our algorithm is only for a single character. Hence we need to first break it down into lines and then into characters. This step gives as its output a

character that can be used as input for the next step. In our algorithm, we assume that the segmentation has already been done and we already have a segmented image of a single character with us.
- **Character recognition**: In this phase, the character is given as an input to the algorithm. This character is processed by all the stages of the algorithm. Finally the algorithm classifies the character and so recognizes it.

We discuss some of these steps in the subsequent sections, one by one.

15.2.1 Binarization

Here we are converting the image into a binary format of black and white. The basic objective is to set a threshold. All pixels above this threshold will be marked as white, and all pixels below this threshold will be marked as black. This way, the whole image gets converted into a black-and-white image. The major task here is to decide the threshold for the black and white colors. If the image is very clear, as in a printed word document, the two may easily be separated, as they would lie at the two extremes. But the same may not be true in case the character to be recognized is on a background image, dirty paper, or the like.

Two commonly used techniques include global and local feature thresholding. In global technique, the gray level of the text remains almost the same throughout the document. Hence we plot a histogram of the image. This gives us two extremes corresponding to the background and the text. The threshold may be the mean of the two. In the local technique, however, there might be large variations. Hence we handle each small segment of the image separately and locally. We may calculate the local extremes, and the mean of two may serve as the threshold.

Another technique is feature-based thresholding. This uses the features of a region to detect the edges. Special operators are used for the purpose. Common edges detected are double edges, step edges, corner edges, and edge boxes.

15.2.2 Preprocessing

Here we discuss the various preprocessing techniques that must be performed on the image in order to effectively extract and recognize the text. The procedure involves image noise removal, correction, regularization, etc. The preprocessed image is then ready to be used to extract the features or be segmentated. Noise may be introduced in the image in any time during the process of acquiring and handling the image. This has an ill effect on the recognition system and so must be reduced beforehand.

15.2.2.1 Filters

One commonly used method of noise removal is the application of *noise filters*. Here it is assumed that the noise is linearly added to the image to give the corrupt image. The use of low-pass filters is very common to correct the variations in the higher frequencies where the noise is present. Gaussian filters are also used for noise removal. Multiscale representation may also be used for regularization from a finer scale to a courser scale.

15.2.2.2 Smoothing

Smoothing is another major operation that needs to be applied over the image. Smoothening causes an effect of blurring, which further helps in the reduction of the noise effects. Blurring causes the removal of small parts that may have been left over in the image. If the image has small trails left over in the curve, this operation will suppress the unwanted trails and so help in improving the image. Similarly the disconnected curves may be connected by this operation. Filtering is the common way to apply these effects. In filtering, we use a mask around a pixel to fix the values of the

	1	1	1
1/9	1	1	1
	1	1	1

FIGURE 15.1a The 3 × 3 averaging filter mask.

pixel, depending upon the values of the pixels around. The mask is rotated in the entire image to correct the image. The filtering may be linear or nonlinear. In *linear* filtering, the output at a pixel is the linear combination of the pixels around and vice versa. An example of this is the averaging filter that averages the value of a pixel based on the values of surrounding pixels. One such mask is given in Figure 15.1. The averaging filter is applied a predefined number of times; otherwise, all the important features, such as details and edges, may get completely removed.

Another filter used is shown in Figure 15.2. This is used in case of binary images. These masks are passed over the entire image to smooth it, and this process is repeated until there is no change in the image.

These masks begin scanning in the lower-right corner of the image and process each row, moving upward row by row. The pixel in the center of the mask is the target. If all pixels that lie at positions marked by '=' have the same binary value, then the pixel at the center is given the same value, else no change is made to the pixel. The other pixels marked by *X* are ignored.

15.2.2.3 Skew Detection and Correction

Skew is the rotation that may have occurred in the image due to improper placing of paper while scanning or improper shooting of the image or any other reason. The image segmentation and detection algorithms normally assume that the text is oriented normal to the image. The undue rotation may cause an adverse effect in the performance of the system. The skew hence needs to be detected and corrected. An example of skew is given in Figure 15.3.

A very common way of detecting skew is to figure out the individual characters in the text. These are called the connected components. The average angle of the line between the centroids of two consecutive characters gives a measure of the skew. Another common technique used is the projective analysis, where the projections are measured at several angles. The skew angle is one that gives the minimum variance in the number of black pixels per projected line. The document may be rotated along the skew angle using coordinate transformation to correct the skew angle.

15.2.2.4 Slant Correction

Many times the character may be inclined at a certain angle over the text in the document. This is what happens in case of cursive handwriting, where the curves are not normal to the baseline. There is a certain slant. This slant further induces an ill effect in the recognition of the characters in text. The slant may be detected by enclosing the character in a rectangle and measuring the inclination of this rectangle with the coordinate axis. After the inclination has been detected, a coordinate transformation is applied to balance the slant of the character. The slant is shown in Figure 15.4.

15.2.2.5 Character Normalization

In this step, the image is normalized and is made to lie in some standard coordinates with some standard size. This fixes the area covered by the character. Based on this, the recognition algorithm

	1	2	1
1/16	2	4	2
	1	2	1

FIGURE 15.1b The averaging filter mask.

=	=	=
=	T	=
X	X	X

FIGURE 15.2a The filter for binary images: Type 1.

=	=	X
=	T	X
=	=	X

FIGURE 15.2b The filter for binary images: Type 2.

X	=	=
X	T	=
X	=	=

FIGURE 15.2c The filter for binary images: Type 3.

X	X	X
=	T	=
=	=	=

FIGURE 15.2d The filter for binary images: Type 4.

FIGURE 15.3 The skew effect.

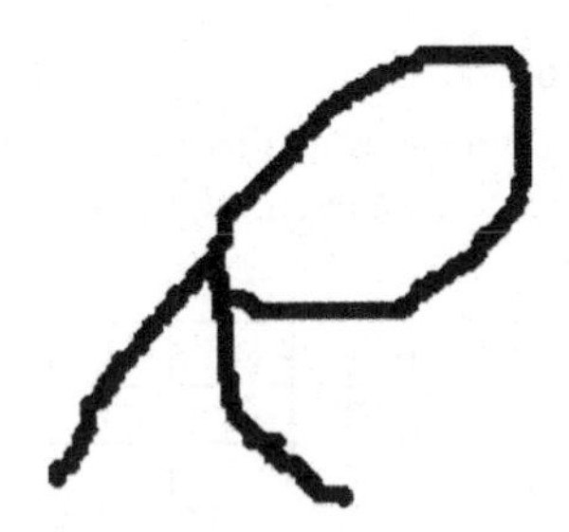

FIGURE 15.4 The slant.

FIGURE 15.5a The original characters.

can use this as a standard input in the standard size for the purpose of recognition. This even reduces the dimensionality of the image, further helping in the process of recognition by reducing computational time as well as in limiting the input space. The normalization may be done by linear mechanisms, moment-based mechanisms, or nonlinear mechanisms.

15.2.2.6 Thinning

Thinning is the process by which the character's lines are narrowed to an extent that all strokes in it are of a unit thickness. This again plays a great role in dimensionality reduction, which is of great use in the recognition process. A good thinning algorithm maintains the connectivity among the components as well as places the thinned lines and curves at the centers of the entire curves. These are the essential properties for the thinning algorithms. The thinning is shown in Figure 15.5.

Two commonly used thinning algorithms are Hilditch's Thinning Algorithm and Zhang-Suen's Thinning Algorithm.

15.2.3 Segmentation

Segmentation is normally done using edge detection mechanisms. Here we try to detect edges present in the entire image. The edges are the boundaries of the objects that separate the object from the rest of the surroundings. This segments the entire images into segments. Each segment can be used separately for recognition purposes. One of the commonly used methods is the Gaussian kernel. This mechanism can withstand noise to some extent. A multiscale approach is used for regions with large-intensity variations to give a good quality of data. The operator used is called as the Laplacian of Gaussian, or LoG, operator. The KCS filter is another commonly used technique for this purpose.

FIGURE 15.5b The character after thinning.

15.3 MULTILINGUAL OCR BY RULE-BASED APPROACH AND ANN

In this section, we study the way in which we solve the problem of OCR using a hierarchical approach involving a rule-based approach and ANNs. This model comes from the work of Shukla et al. (2009a). The basic principle we use is that first the rule-based approach analyzes the given character. We assume here that the difference between the individual languages is so easy that we can make very simple rules to distinguish the languages from one another. Using these simple rules, the rule-based approach is able to identify the language. The identified language has an ANN of its own. This ANN was trained using the characters of this language exclusively. We select this ANN and give the character as an input to it. The output returned by this ANN is the final identified output. In this manner we have used a combination of a rule-based approach along with ANN to solve the problem. The fusion of these two methods results in a good, efficient solution that could have otherwise not been possible.

We have used English capital letters, English small letters, and Hindi letters for this algorithm. The general structure of the algorithm is given in Figure 15.6. In this algorithm, the input is a matrix of 14×14. Each cell in the matrix represents a pixel. The pixel is marked as 1 or –1, depending on whether something was written in the area of that pixel or not. If something was written, the pixel is marked as –1; else, the pixel is marked as 1. The input of the algorithm, when the input was *A*, is given in Figure 15.7.

The language identified by the rule base may be any of Capital English letters, Small English letters, or Hindi letters. We have differentiated between the capital and the small English letters because they were distinctly different from each other and could be easily separated with the help of rules.

If the given input is found to be an English capital letter, the original input matrix is scaled down to a scale of 7×7. This may be done by deleting alternative rows and columns. If there is a vertical or horizontal line over the column/row being deleted, then the other row may be deleted. This is given as the input to the ANN of the English capital letters. If the given input is of English small letters, a similar procedure is followed. The whole process is of breaking the matrix into smaller size of 7×7 and giving it to the neural network of the English characters.

If the given input is of Hindi characters, we do not scale down the original image. Rather we break it into sections. These sections are made in accordance with the philosophy of the Hindi characters. There is a top section that is used to accommodate the overhead projection of the vowels. Similarly there are left, right, and bottom sides. These are used to accommodate their respective projections. The various projections collectively determine the final character. The leftover space at the center is for the main character. This is shown in Figures 15.8a, 15.8b, 15.8c, 15.8d, and 15.8e.

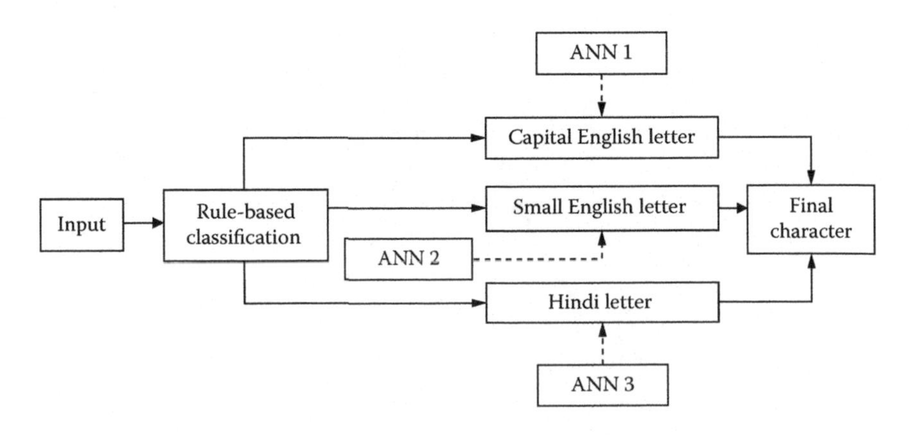

FIGURE 15.6 The general recognition algorithm.

						*							
					*		*						
				*				*					
				*				*					
			*						*				
			*						*				
		*								*			
		*	*	*	*	*	*	*	*	*			
	*										*		
	*										*		
	*										*		
*												*	
*													*
*													*

FIGURE 15.7a The logical algorithm input.

−1	−1	−1	−1	−1	−1	1	−1	−1	−1	−1	−1	−1	−1
−1	−1	−1	−1	−1	1	−1	1	−1	−1	−1	−1	−1	−1
−1	−1	−1	−1	1	−1	−1	−1	1	−1	−1	−1	−1	−1
−1	−1	−1	−1	1	−1	−1	−1	1	−1	−1	−1	−1	−1
−1	−1	−1	1	−1	−1	−1	−1	−1	1	−1	−1	−1	−1
−1	−1	−1	1	−1	−1	−1	−1	−1	1	−1	−1	−1	−1
−1	−1	1	−1	−1	−1	−1	−1	−1	−1	1	−1	−1	−1
−1	−1	1	1	1	1	1	1	1	1	1	−1	−1	−1
−1	1	−1	−1	−1	−1	−1	−1	−1	−1	−1	1	−1	−1
−1	1	−1	−1	−1	−1	−1	−1	−1	−1	−1	1	−1	−1
−1	1	−1	−1	−1	−1	−1	−1	−1	−1	−1	1	−1	−1
1	−1	−1	−1	−1	−1	−1	−1	−1	−1	−1	−1	1	−1
1	−1	−1	−1	−1	−1	−1	−1	−1	−1	−1	−1	−1	1
1	−1	−1	−1	−1	−1	−1	−1	−1	−1	−1	−1	−1	1

FIGURE 15.7b The converted algorithm input.

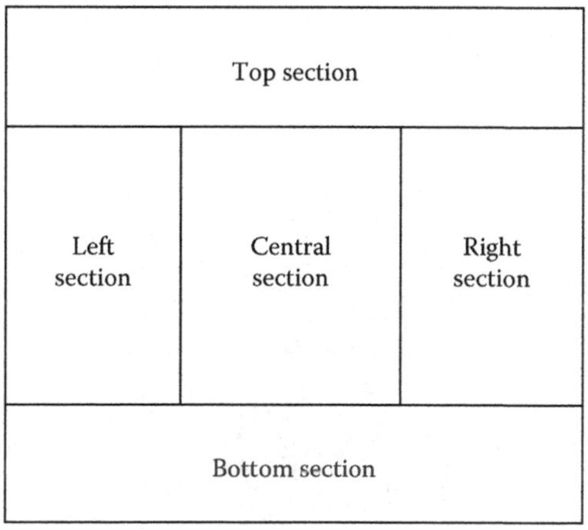

Top section

Left section

Central section

Right section

Bottom section

FIGURE 15.8a Segmentation of the Hindi characters.

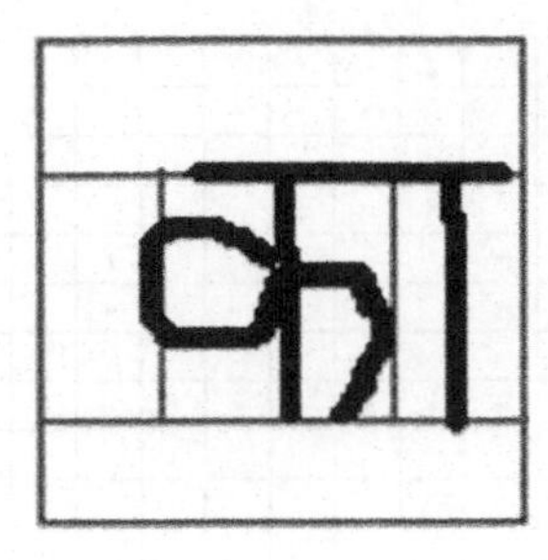

FIGURE 15.8b Use of various segments: Sample 1.

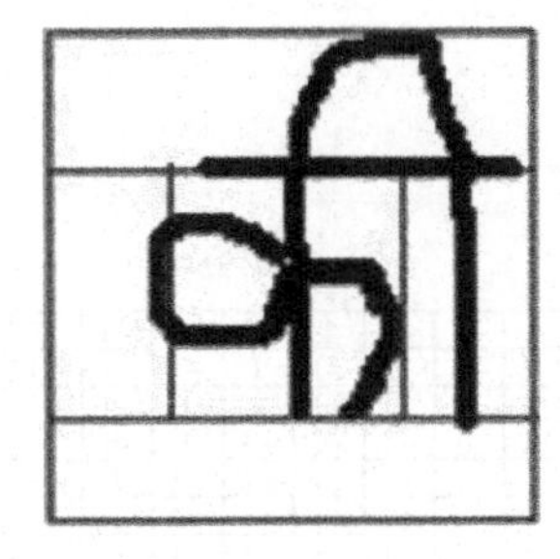

FIGURE 15.8c Use of various segments: Sample 2.

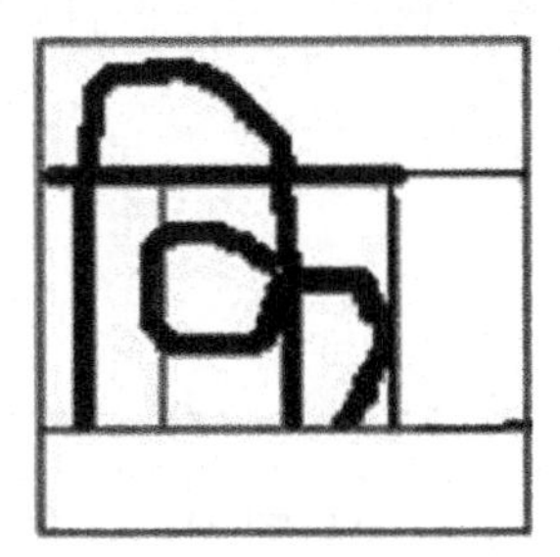

FIGURE 15.8d Use of various segments: Sample 3.

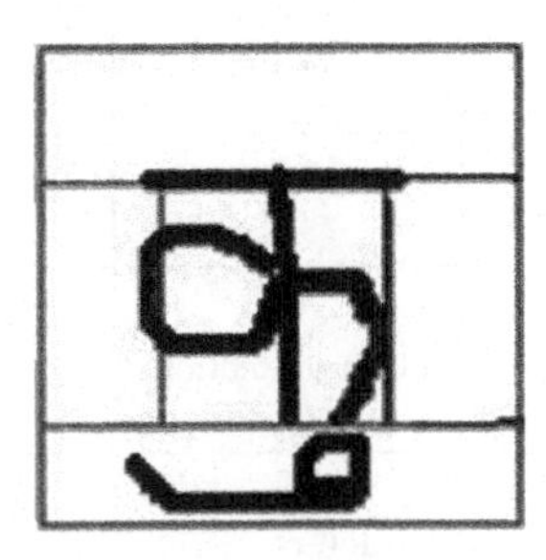

FIGURE 15.8e Use of various segments: Sample 4.

15.4 RULE-BASED APPROACH

This is the first step that is applied when the input is given for recognition to the system. The algorithm tries to find out the class to which the input belongs. In this problem for identification of English/Hindi characters, we have taken three classes. The first class corresponds to the capital English characters (A–Z). The second class corresponds to the small English characters (a–z). The third class corresponds to the Hindi characters. Hence after the algorithm, we know clearly the class to which the given input belongs to.

The input is present in the form of a grid of 1 and –1 as explained already. We apply a series of inputs one after the other that would give us the correct cluster to which the input belongs. This is determined by the rules that are fired.

15.4.1 CLASSIFICATION

In this problem, we have three major classes that the rule-based approach is basically supposed to classify the inputs into. The three classes are Class A comprising the capital English letters, Class B comprising the small English letters, and Class C comprising the Hindi characters.

It would still be very difficult to write rules to differentiate the various inputs and classify them into these three classes. Hence we further cluster the various inputs according to their similarities. The subclasses for the capital English letters follow. We will see the reasons behind this clustering later in this section.

Class A-1: A, B, D, E, F, I, J, K, M, N, P, R, T, Y
Class A-2: C, O, S, U, V, W, X, Z
Class A-3: E, F, I, T, Z
Class A-4: H, L
Class A-5: G, Q
Class A-6: J

Similarly the classes for the small English letters are

Class B-1: b, d, h, i, j, k, l, p, t
Class B-2: a, e, f, g, m, n, q, r, x, y
Class B-3: c, o, s, u, v, w, x, z

For the Hindi characters, we have just one class. This class comprises all the characters found in Hindi and also incorporates all the combinations allowed by the language.

Now we study the common characteristics between the different subclasses. These characteristics are important, as we will primarily exploit these characteristics to differentiate between the various classes. This will help in classification using the rule-based approach.

- All the characters of class C have a line at the top. In Hindi language, this line is always present and may be regarded as the base line. This line may not be present at the start but is always present at the end. This is shown in Figures 15.9a and 15.9b. Further, it is not possible to make any character in Hindi with the use of lines alone. We would be required to use curves for the purpose of writing the character.
- Classes A-3 and A-6 are the only characters of the Class A that have a horizontal line at the top. In rest of the characters, there is no such horizontal line at the top.
- More specifically, A-3 is the class all of whose members have a horizontal line at the top and consist of a set of lines only with no curve. On the other hand, class A-6 has a horizontal line with a curve at the bottom.
- Classes A-2 and B-3 are the similar classes. Here the character is written in the same way in English capital and English small letters. As a result, we may not identify whether the

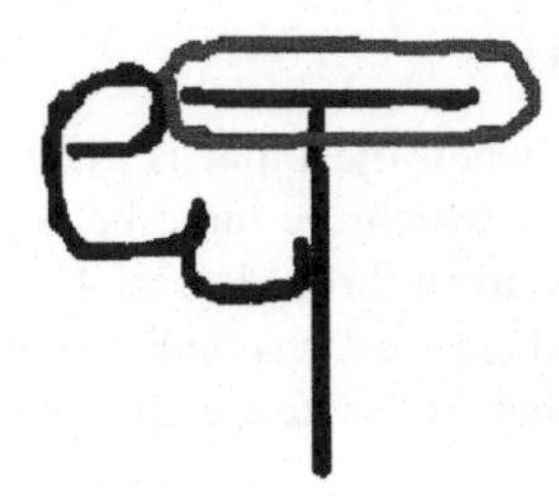

FIGURE 15.9a The top line (Half).

character belongs to the A-2 class or the B-3 class. Both the answers would be permissible to the further algorithms.

- Class B-1 is the set of characters in small English characters, where there is a small vertical line at the top. This is a very small line of more than 1/3 the length of the total character. This line is shown in Figure 15.10.
- Class A-4 is the only set of characters in English capital characters that have a vertical line at the top. This line is shown in Figures 15.11a and 15.11b.
- Class A-1 is the set of characters in capital English characters where there is always a straight line that joins the topmost and leftmost point to any other point. This line is shown in Figures 15.12a and 15.12b.
- Class A-5 is the class that does not contain a line from the topmost and leftmost point. Rather it is predominantly a curve.

15.4.2 Tests

We need some mechanism to exploit the characteristic differences in the various classes that we discussed in the preceding section. Before we develop formal rules to distinguish the various classes, we first define some tests. These tests will enable us to check for the presence of various characters and subclasses. Once we are through with these tests, then we define the formal rules that help us in the process of classifying the given input into the output language.

Before we proceed with the tests, we first need to model the given input in the form of a graph. This will enable us to carry out the tests that we discuss next. We specifically need to figure out edges, lines, and curves present in the given figure. In our modeling of the graph, all the points where edges are present act as vertices. Edges exist between any two vertices that are adjacent to each other on the graph. All this happens when we represent the input as a graph.

The first major concept is *traversal*. Here we travel the points where ink is present. The traversal starts from a predefined location on the given input. If we are on a line or curve, we get only one option to proceed with our traversal. But instead, if we are at some point of intersection, we get multiple options for traversal, depending upon the number of lines or curves that intersect at the point. Each point is traversed only once in the entire algorithm. The general algorithm for this traversal is given in Figure 15.13.

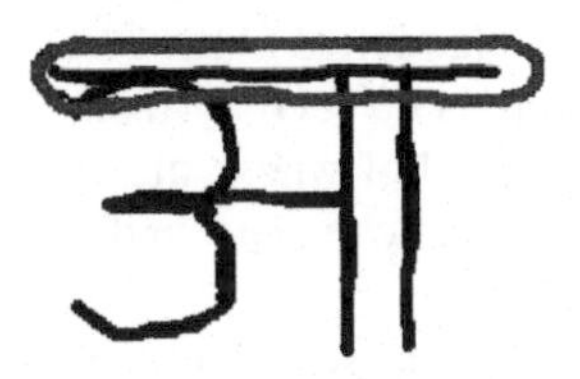

FIGURE 15.9b The top line (Full).

FIGURE 15.10 The top line in B-1.

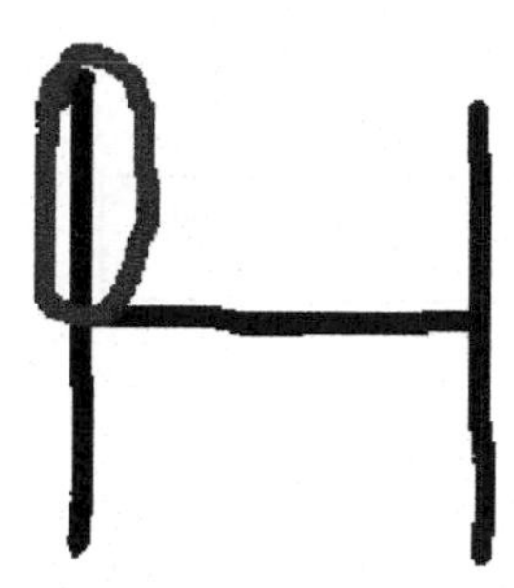

FIGURE 15.11a The vertical line at Class A-4 (Half).

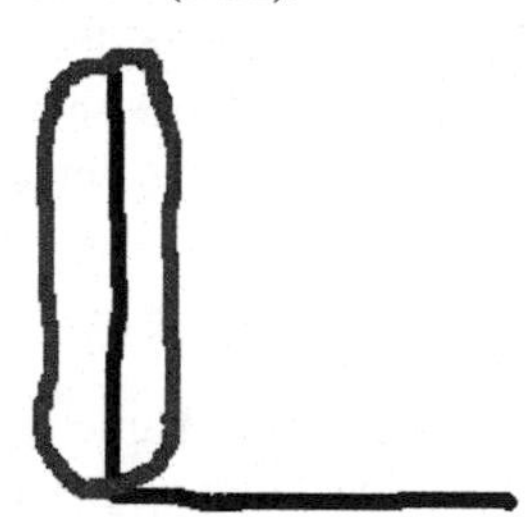

FIGURE 15.11b The vertical line in Class A-4 (Full).

FIGURE 15.12a The straight line in A-1, Case-1.

FIGURE 15.12b The straight line in A-1, Case-2.

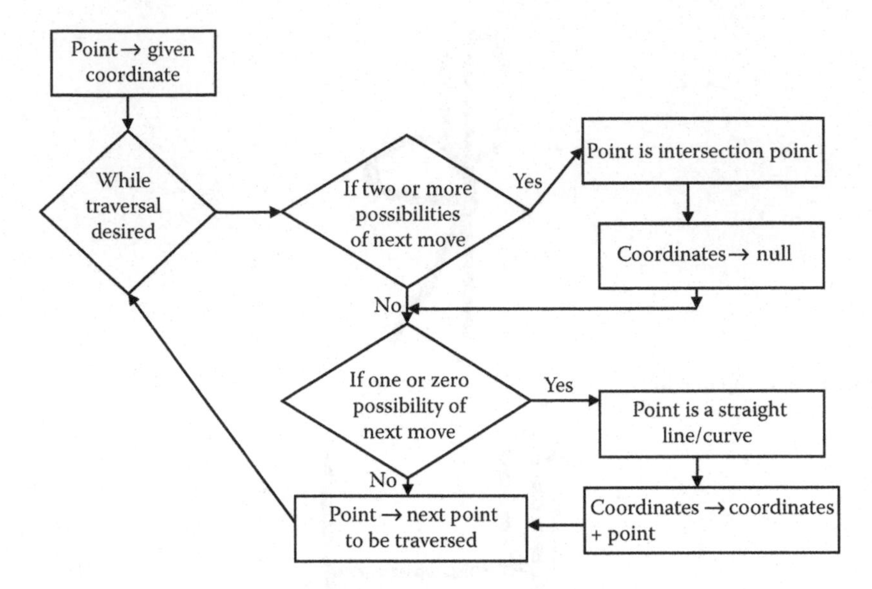

FIGURE 15.13 The traversal algorithm.

As we traverse the graph, we need to figure out the lines and the curves present in the given input. For this, we use the cosine formula of triangles. At any point of intersection, we know that we have traversed a line or a curve. The distinction between these is made on the basis of the angle between the start for the line/curve, the end, and the midpoint. If this angle is approximately 180 degrees, we assume that the object traversed was a line; else we assume that what was traversed was a curve. The angle is given in Figure 15.14. For this line detection, it is desired that the inputs have high dimensionality.

15.4.2.1 Class C Test

In this test, we check for the top vertical line in the character. The general way to do this test is to find out the rightmost point in the top 1/3 of the input grid. This can be taken as the rightmost point in the horizontal line that we wish to find. Starting from this point, we keep traversing till a point of intersection is met. Using this set of coordinates, we decide whether it is a line. If this set of coordinates comes out to be a line, we say that the test was positive. If, on the other hand, the set of points does not lie on a line, we say that the test failed.

15.4.2.2 Class A-3 Test

In this test, we check whether the figure contains only lines, or it contains both lines and curves. We start from the topmost and leftmost point. Using this point, we start traversing. At every intersection, we traverse all the points possible. Each set of paths traversed is recorded as the set of coordinates. At the end of this procedure, we clearly know all the sets of paths that may be lines or curves. We apply the algorithm to each and every path to find out whether it is a line. If all the paths are

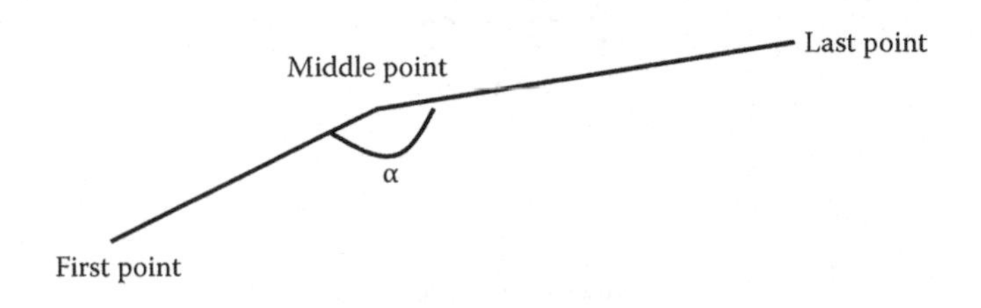

FIGURE 15.14 The angle for line detection.

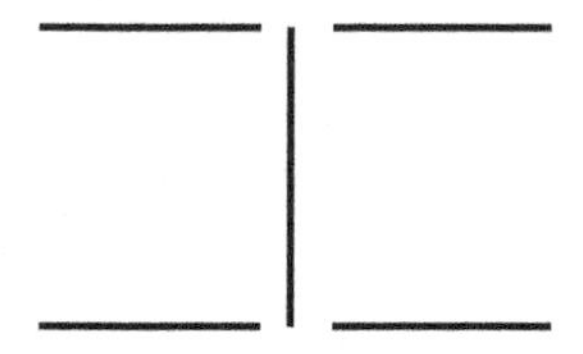

FIGURE 15.15 The five different paths for letter *I*.

found to be lines, we say that the Class A-3 test has been passed; else we say that the class A-3 test has failed. The five paths generated when *I* was the input are given in Figure 15.15.

15.4.2.3 Class A-6 Test

The purpose for this test is to test whether the given input is *J*. This test checks for the presence of one top line and one curve at the bottom. Both these may be easily done using information provided in the Class C test and Class A-3 test.

15.4.2.4 Class B-1 Test

This test checks for the presence of a small vertical line of at least 1/3 the length of the whole character at the top of the character. We find out the topmost and the leftmost point in the given input. We start traversing from this point. We need to make sure that we travel almost straight down. Once the intersection point is reached, we break. We also check the total length of the path traveled and whether it was a line or not. If both the conditions are satisfied, we say that Class B-1 test succeeded.

15.4.2.5 Class A-4 Test

The purpose of this test is to check for the availability of *H* and *L*. These are the two characters with a vertical line and without curves. The test is similar to the B-1 test. Here we also apply a test similar to A-3 test to find five straight lines for *H* and two straight lines, one vertical and one horizontal (or one curve) for *L*. When these conditions are met, we say that the Class A-4 test has passed.

15.4.2.6 Class A-1 Test

In this test, we check for the presence of a straight line traversing from the top-left point. This test is similar in nature to the Class C test.

15.4.2.7 Class A-5 Test

In this test, we check for the presence of *G* and *Q*. These are predominantly curves in nature. If we traverse leftward from the top-left position, we will find a curve in both of these figures. This curve would be intercepted by a straight line. This may be used to distinguish the class.

15.4.3 Rules

In this section, we write down rules to distinguish between the various languages that we have considered. In this problem, we have considered three classes A, B, and C that we need to classify the input into. While framing these rules, we need to ensure that they must be exhaustive to distinguish between any pair of classes. In other words, we should be able to distinguish between members of class A and class B, class B and class C, and class A and class C.

The following are the rules that we use to distinguish among the various classes. Whenever any input is given, we apply these rules to get the corresponding language to which the input belongs.

If Class C test is positive and Class A-3 test fails → Class C
Else if Class C test is positive and Class A-3 test is positive → Class A

Else if Class C test is positive and Class A-6 test is positive → Class A
Else if B-1 test is positive and A-4 test is negative → Class B
Else if B-1 test is positive and A-4 test is positive → Class A
Else if A-1 test is positive → Class A
Else if A-5 test is positive → Class A
Else Class B

15.5 ARTIFICIAL NEURAL NETWORK

After we know the language or the class, the problem is simple. Here we have to simply solve the problem of character recognition using ANN. In this problem we use ANN with BPA for the process of learning. The database of every ANN consists only of the limited number of characters that belong to the same class or language. As a result, the problem is much simpler and becomes easy to solve with the help of the ANN.

It may be seen that the total algorithm that we develop is very much along the lines of modular neural networks (MNNs) that also use ANNs in a hierarchy for the purpose of solving the problem by reducing its complexity and making use of the available modularity.

In this problem, the ANNs of the English capital and English small letters are identical. The ANN of Hindi characters differs in some amount, where different ANNs were applied to different segments of the image that we discussed in the earlier sections. The segments into which we broke a Hindi character were top segment, bottom segment, center segment, left segment, and right segment. For the left and the right segments, there are only two possibilities. These are the presence or absence of a line. These can easily be identified without the use of ANNs.

15.5.1 Inputs

The input of any neural network is a grid. The grid is a collection of 1 and –1. The 1s denote that there was something written in that position. On the other hand, the –1s denote the absence of anything written in that position. The input of the English capital characters and English small characters consist of the scaled input from the previous step. On the other hand, the Hindi image is internally segmented and every segment is handled independently.

15.5.2 Outputs

In this problem, we have used a special kind of neural network, specially designed for the purposes of the classification problems. We know that the main aim is to classify the inputs into a set of output classes. Both capital and small English has 26 such output classes. The number of outputs in our neural network is actually the number of such output classes. For any input *i*, there are the same number of outputs as there are classifying classes. Each of these outputs shows the possibility of the final class of which the input is a member. This value may lie between –1 and 1. The highest probable is 1 and the least is –1. Hence after the application of any input *i*, we find out the output with the maximum possibility. This is regarded as the final output.

Let's say we have the neural network of the capital English characters. It takes a grid of 7×7 as input. Hence there are 49 inputs, $<I_1, I_2, I_3, I_4 \ldots I_{49}>$. Corresponding to this there can be 26 classes possible as outputs (A–Z). The output array would be in form $<O_1, O_2, O_3, O_4 \ldots O_{26}>$. Here O_1 corresponds to *A*, O_2 to *B*, O_3 to *C*, and so on. If we are giving the training data for *B*, we know that the possibility of O_2 will be 1 and the possibility of any other output will be –1. Hence the training data output for *B* will be $<-1, 1, -1, -1 \ldots -1>$. While testing we find the highest output, $max(O_i)$, for all *i* from 1 to 26. The character corresponding to this is the final answer.

FIGURE 15.16a Various styles of writing *A*: Style 1.

15.5.3 Identification

In capital English characters and small English characters, we straightaway get the final answer. But in Hindi character recognition, we get the presence of various signs at the top, bottom, right, left, and center positions. Once it is known what lays at all the positions, we may easily come to know the final character (character and the vowel associated).

15.6 RESULTS OF MULTILINGUAL OCR

The discussed system was implemented in MATLAB®. The specific code that deals with graphs was implemented in Java. The training data sets were made to train the individual ANNs of the system. Then we made the main logic that differentiated languages. We developed test cases for various languages and characters that the system may take. For the Hindi characters, different permutations for the different segments were tried. The system so developed yielded very good performance. We set up a total of 218 test cases, out of which the system could correctly identify 193 test cases. This returned an efficiency of 88.53 percent in the presence of numerous characters and languages.

15.7 ALGORITHM FOR HANDWRITING RECOGNITION USING GA

Every soft computing system requires a good and diverse training database that must cover all diversities that may be present in the data for best performance. The case of character recognition is no different, where we need to ensure that the training database is filled with almost all styles in which a character may be written. This, however, is practically very difficult, especially when it comes to handwriting. There are numerous styles in which a character may potentially be written. All these vary from person to person. People many times mix two styles to form a new style of writing altogether. The three different styles in which people usually write the letter *A* are given in Figure 15.16.

FIGURE 15.16b Various styles of writing *A*: Style 2.

FIGURE 15.16c Various styles of writing *A*: Style 3.

In this section as well, we look at the character in the form of a graph. The graph is a collection of vertices and edges. In this section, all the points of intersection and sharp turns denote the vertices, and the edges are the lines or curves or both that connect the vertices.

Here we use GA to solve the problem of handwriting recognition. We have numerous ways in which a character can be written already available in the database. These ways are combined to form newer ways of writing down the same character. This forms a diverse set of training data sets that combine the best matching characters to generate even better characters. This is done for all the characters in the training database. The unknown input is matched to these characters, and the best match is found. In this way, the GA is used to optimize the match between the unknown character and the hybrid characters generated from the characters available in the training data set.

Now we discuss the general procedure of the working of the algorithm. The algorithm does the task of matching an unknown given input character to the ones present in the training dataset. Based on the input character, the training data set characters are used to form hybrid characters that more optimally match the unknown input character.

HandwritingRecognition(Language,TrainingData,InputImage)

Step 1: For every character *c* in language
  Do
Step 2: For every input *i* for the character *c* in test data
  Do
Step 3: Generate graph g_{ci} of *i*
Step 4: Generate graph *t* of input image
Step 5: For every character *c* in language
  Do
Step 6: Use the genetic algorithm to generate hybrid graphs
Step 7: Return the character corresponding to the graph with the minimum fitness value (out of the graphs generated in any genetic operation)

Seeing the previous algorithm, it is clear that we have first generated graphs and then used genetic algorithms to mix these graphs and find the optimal solution. The two algorithms are explained in the next section.

15.7.1 Generation of Graph

The whole algorithm treats the input as well as the training data as graphs. The image hence needs to be first converted into a graph so that the same may be used for the purpose of recognition. The algorithm is given in Figure 15.17. We use a similar concept of traversal and detection of intersections, lines, and curves that we had done in the earlier section.

A graph here represents the vertices and the edges. The edges are the lines or curves connecting any two points. Every point where an edge ends/starts is regarded as a vertex. We are also interested

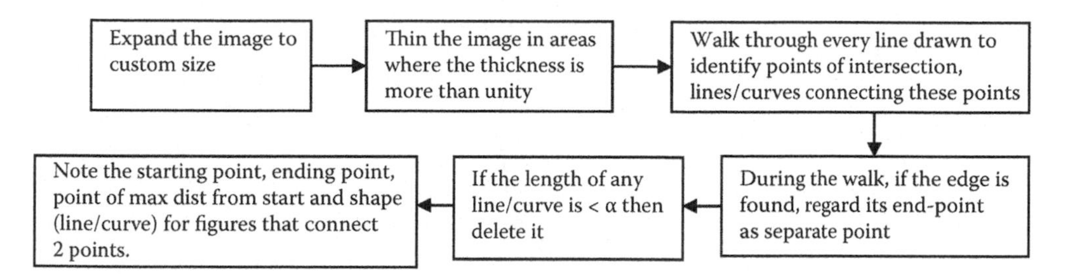

FIGURE 15.17 The graph generation algorithm.

in knowing the point for every line/curve that is at the maximum distance from the start point. This is useful when the graph may contain a closed curve, e.g., *O* would be regarded as a curve from a vertex that ends at the same vertex. Each edge must hence represent the start vertex, the end vertex, the shape (line/curve), and the point of maximum separation from the start vertex.

The preprocessing part of the algorithms that include the image expansion and thinning has already been discussed in the previous sections. The same fundamentals are used for the detection or separation of lines and curves that we discussed in the preceding section. Figure 15.18 shows the graph generated when the input was *J*.

15.7.2 Fitness Function of GA

GA plays the role of optimizing the match between the unknown input graph and those generated by the training data sets. The GA generates newer and newer graphs from the existing graphs in the training data set, which match better with the unknown input. The fitness function here represents the match between the unknown input and the graph whose fitness is to be found out. The match is the deviation between the two graphs. In the rest of the section, we derive the formula to measure this deviation.

15.7.2.1 Deviation Between Two Edges

We first define a function $D(e_1, e2)$ that finds the deviation between any edges of a graph (e_1) with any edge of the other graph (e_2). Here an edge may represent a line or a curve. This may be calculated by measuring the distance between the start and end points of the line or curve being considered.

The start point of the edge e_1 may match with the start point of the edge of e_2, and the end point of e_1 may match with the end point of e_2. Or the start point of e_1 may match with the end point of e_2, and the end point of e_1 may match with the start point of e_2. The match is shown in Figures 15.19a, 15.19b, and 15.19c.

If, however, e_1 is a line and e_2 is a curve or vice versa, an overhead cost of β is added.

If both e_1 and e_2 are curves and start and end points of e_1 or e_2 are less than α units apart (it is almost a circle), then we take point of maximum distance in place of end points in the preceding formula. Here the point of maximum distance is the point in the curve that is at a maximum distance from the start point of the curve.

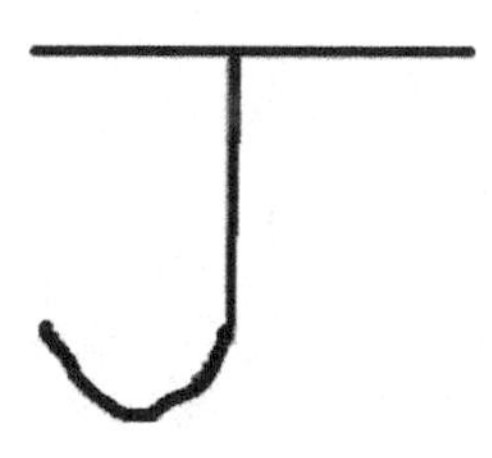

FIGURE 15.18a The letter *J* (Style 1).

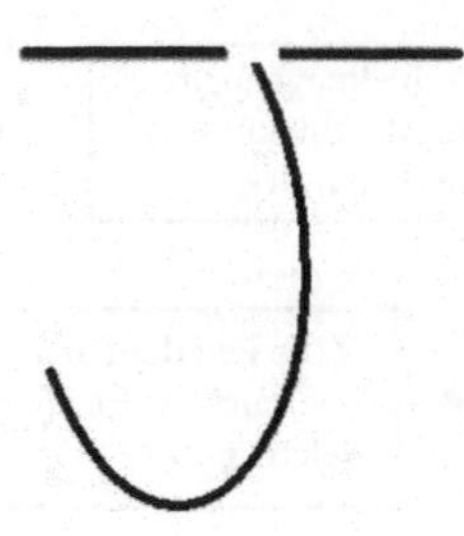

FIGURE 15.18b Generated graph for the letter *J* (Style 1).

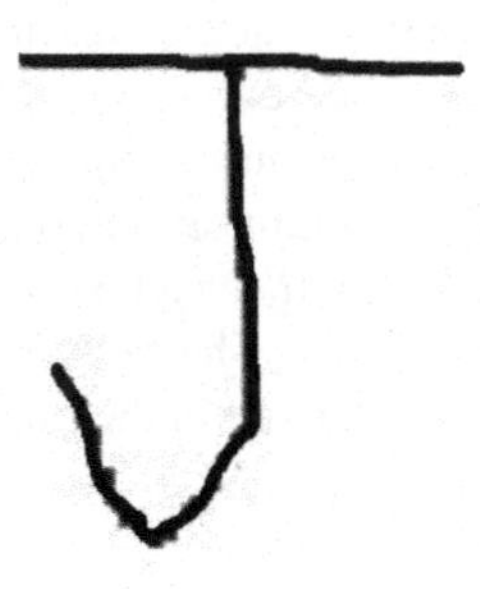

FIGURE 15.18c The letter *J* (Style 2).

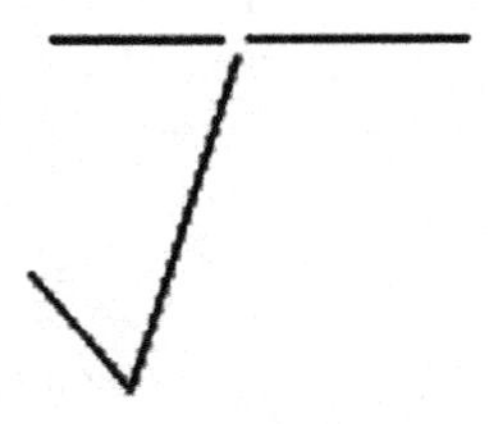

FIGURE 15.18d Generated graph for the letter *J* (Style 2).

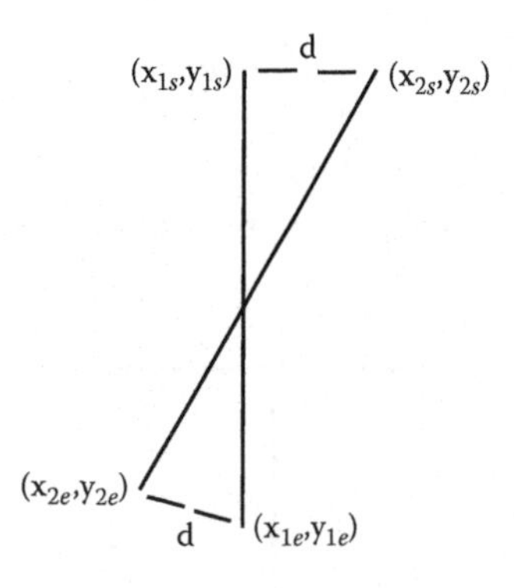

FIGURE 15.19a Calculating D_l with two lines.

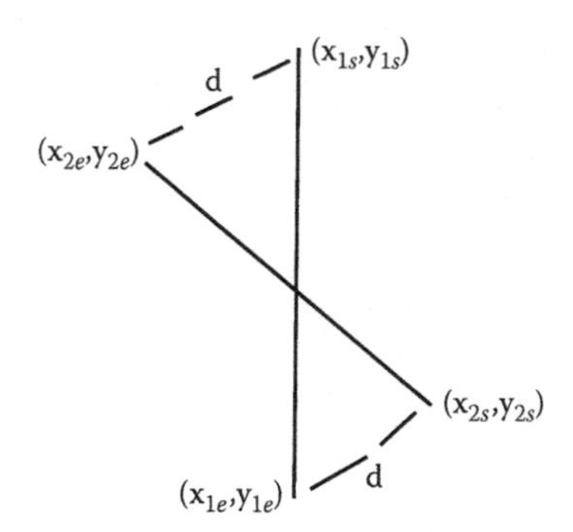

FIGURE 15.19b Calculating D_2 with two lines.

We even generalize the formula to the condition when either e_1 or e_2 is null. This means that we can find the deviation of a line or a curve with nothing. This is a feature useful in finding graph deviation when there are an unequal number of edges in two graphs.

The combined formula is given by Equations 15.1, 15.2, and 15.3.

$D_1(e_1,e_2)$ = square of distance of start points of e_1 and e_2 + square of distance of end points of e_1 and e_2, if $e_1 \neq$ null, $e_2 \neq$ null, e_1 and e_2 are both lines or curves with length $> \alpha$

= square of distance of start points of e_1 and e_2 + square of distance of end points of e_1 and $e_2 + \beta$, if $e_1 \neq$ null, $e_2 \neq$ null, e_1 is line and e_2 is curve or e_1 is curve and e_2 is line

= square of distance of start points of e_1 and e_2 + square of distance of point of maximum separation of e_1 and e_2, if $e_1 \neq$ null, $e_2 \neq$ null, e_1 is curve and e_2 is curve and length of e_1 or $e_2 < \alpha$

= Distance between the starting point and end point of e_1, if e_2 is null

= Distance between the starting point and end point of e_2, if e_1 is null (15.1)

$D_2(e_1,e_2)$ = square of distance of start point of e_1 and end point of e_2 + square of distance of end point of e_1 and start point of e_2, if $e_1 \neq$ null, $e_2 \neq$ null, e_1 and e_2 are both lines or curves with length $> \alpha$

= square of distance of start point of e_1 and end point of e_2 + square of distance of end point of e_1 and start point of $e_2 + \beta$, if $e_1 \neq$ null, $e_2 \neq$ null, e_1 is line and e_2 is curve or e_1 is curve and e_2 is line

= square of distance of start point of e_1 and point of maximum separation of e_2 + square of distance of point of maximum separation of e_1 and start point of e_2, if $e_1 \neq$ null, $e_2 \neq$ null, e_1 is curve and e_2 is curve and length of e_1 or $e_2 < \alpha$

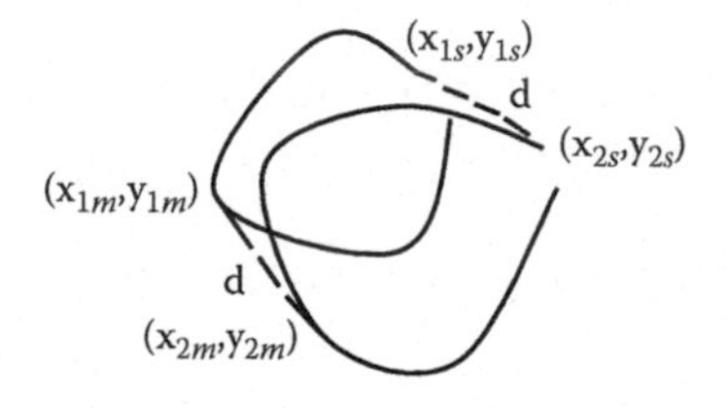

FIGURE 15.19c Calculating D_1 with two curves.

= Distance between the starting point and end point of e_1, if e_2 is null

= Distance between the starting point and end point of e_2, if e_1 is null (15.2)

$$D = \min \{D_1(e_1, e_2), D_2(e_1, e_2)\} \quad (15.3)$$

Suppose that the edge e_1 in the first graph has start points as (x_{1s}, y_{1s}) and end points as (x_{1e}, y_{1e}). Similarly suppose that the edge e_2 in the second graph has start points as (x_{2s}, y_{2s}) and end points as (x_{2e}, y_{2e}). The point of maximum distance of e_1 is (x_{1m}, y_{1m}) and e_2 is (x_{2m}, y_{2m}). The formula just explained can also be stated as given by (15.4), (15.5), and (15.6):

$D_1(e_1,e_2) = (x_{1s}-x_{2s})^2 + (y_{1s}-y_{2s})^2 + (x_{1e}-x_{2e})^2 + (y_{1e}-y_{2e})^2$, if $e_1 \neq$ null, $e_2 \neq$ null, e_1 and e_2 are both lines or curves with length $> \alpha$

$= (x_{1s}-x_{2s})^2 + (y_{1s}-y_{2s})^2 + (x_{1e}-x_{2e})^2 + (y_{1e}-y_{2e})^2, + \beta$ if $e_1 \neq$ null, $e_2 \neq$ null, e_1 is line and e_2 is curve or e_1 is curve and e_2 is line

$= (x_{1s}-x_{2s})^2 + (y_{1s}-y_{2s})^2 + (x_{1m}-x_{2m})^2 + (y_{1m}-y_{2m})^2$, if $e_1 \neq$ null, $e_2 \neq$ null, e_1 is curve and e_2 is curve and length of e_1 or $e_2 < \alpha$

$= (x_{1s}-x_{1e})^2 + (y_{1s}-y_{1e})^2$, if e_2 is null

$= (x_{2s}-x_{2e})^2 + (y_{2s}-y_{2e})^2$, if e_1 is null (15.4)

$D_2(e_1,e_2) = (x_{1s}-x_{2e})^2 + (y_{1s}-y_{2e})^2 + (x_{1e}-x_{2s})^2 + (y_{1e}-y_{2s})^2$, if $e_1 \neq$ null, $e_2 \neq$ null, e_1 and e_2 are both lines or curves with length $> \alpha$

$= (x_{1s}-x_{2e})^2 + (y_{1s}-y_{2e})^2 + (x_{1e}-x_{2s})^2 + (y_{1e}-y_{2s})^2 + \beta$, if $e_1 \neq$ null, $e_2 \neq$ null, e_1 is line and e_2 is curve or e_1 is curve and e_2 is line

$= (x_{1s}-x_{2m})^2 + (y_{1s}-y_{2m})^2 + (x_{1m}-x_{2s})^2 + (y_{1m}-y_{2s})^2$ if $e_1 \neq$ null, $e_2 \neq$ null, e_1 is curve and e_2 is curve and length of e_1 or $e_2 < \alpha$

$= (x_{1s}-x_{1e})^2 + (y_{1s}-y_{1e})^2$, if e_2 is null

$= (x_{2s}-x_{2e})^2 + (y_{2s}-y_{2e})^2$ if e_1 is null (15.5)

$$D = \min\{D_1(e_1,e_2), D_2(e_1,e_2)\} \quad (15.6)$$

15.7.2.2 Deviation of a Graph

The deviation of the graph uses the derived formula of the deviation of edges to find out the total deviation between any pair of edges. In order to restrict the time complexity, we start picking up the pairs from the pool of available edges and keep adding their deviation to the total deviation. The algorithm is as given here:

Deviation(G1,G2)

Step 1: dev ← 0
Step 2: While first graph has no edges or second graph has no edges
  Do
Step 3: Find the edges e_1 from first graph and e_2 from second graph such that deviation between e_1 and e_2 is the minimum for any pair of e_1 and e_2
Step 4: Add its deviation to dev
Step 5: Remove e_1 from first graph and e_2 from second graph
Step 6: For all edges e_1 left in first graph
  Do
Step 7: Add deviation of e_1 and null to dev

Step 8: For all edges e_2 left in second graph
 Do
Step 9: Add deviation of null and e_2 to dev
Step 10: Return dev

The goal here is to minimize the total deviation that is the objective of the GA. We select the pair of edges, one from each graph, such that their deviation is minimal. We keep selecting such pairs until one graph gets empty. Hence we keep proceeding by keeping the total deviation at a minimum. In the end, we add the deviation of all the left edges. In this way, we find the minimum deviation between the two graphs.

15.7.3 Crossover

In crossover, we try to mix the existing graphs or solutions in order to generate hybrid solutions or graphs. These graphs form the next level of generation from the previous generation. In this manner, we keep moving from one generation to the other in search for the optimal match. In this algorithm, we try to create graphs that take some edges and vertices from the first parent and some from the other. The aim is to generate a graph that may better match the input.

The basic motive of using this operation is to mix styles. If the two graphs have characters in different styles, we would be able to mix them and form a style that is intermediate between the parent styles. The crossover operation makes sure that the style of writing a particular section of the character is taken from one of the graphs. This section is removed from the other graph, and the new section is added. Hence using the crossover operation, we may be able to mix the styles to form unique new solutions. Many solutions are possible for every combination of parents. In this algorithm, we generate all the forms and add them as a solution. Hence one crossover operation results in many solutions being generated.

We select two vertices in the graph as the crossover points along which the crossover operation will take place. The first parent graph provides the base over which the child will emerge. We try to find paths that connect these two crossover points. We lay down a condition that any point can be visited at most once while finding the path. Also the total path length must be less than four edges. This means we must be able to reach the second vertex from the first, using a maximum of three intermediate distinct vertices. Once such a path is found in the first graph, we carry out a similar task using the same points in the second graph as well. If the path is found, we are ready for crossover. The final generated solution is same as the first graph. The only difference is that we delete the entire path that was found between the two chosen points from first graph. We then insert the path that was chosen from the second graph between the same chosen points. Hence a region of style between the chosen points is changed from the old style of graph 1 to the new style of graph 2. The style of the remaining character remains same as graph 1.

However, a point in the first graph may not be the same in the other graph as well. For this, there happens to be a need to match the points. A point of the first graph is tried to match to the closest possible point in the second graph. The points that fail to match to any other point are added in the end to the other graph as well. This means that if a point in graph 1 failed to match any other point in graph 2, it would be simply added as a vertex in graph 2.

The following is the algorithm used for the crossover of the two graphs:

CrossOver(G_1,G_2)
Step 1: match ← MatchPoints(G_1,G_2) (Find points in G_1 corresponding to G_2 and vice versa)
Step 2: W_1 ← GenerateAdjacency(G_1) (Generates the Adjacency Matrix of the Graph)
Step 3: W_2 ← GenerateAdjacency(G_2)
Step 4: if $W_1 \neq W_2$
Step 5: FindPaths(W_1) (Finds the paths of all lengths between any pair of points)

Step 6: FindPaths(W_2)
Step 7: $p_1 \leftarrow$ path between vertices v_1 and v_2 in G_1 of length 1, for all v_1, v_2 in W_1 and length $1 = 1, 2, 3$, or 4
Step 8: $p_2 \leftarrow$ path between vertices match[v_1] and match[v_2] of length l_2 in G_2 of length $l_2 = 1, 2, 3$, or 4
Step 9: If p_1 exists and p_2 exists and $p_1 \neq p_2$
Step 10: Remove all edges from W_1 that are found in p_1
Step 11: Add all edges in W_1 that are found in p_2
Step 12: $g \leftarrow$ MakeGraph(W_1,W_2) (Generate graph out of Adjacency matrix of W_1)
Step 13: Add g to solution set
Step 14 else $g \leftarrow$ MakeGraph(W_1,W_2)
Step 15: Add g to solution set

Each of the steps is discussed in the further sections.

15.7.3.1 Matching of Points

This is the algorithm that we use to match or get the corresponding points between the two graphs. If a point in graph 1 is said to match a point in graph 2, it may be assumed that a reference to the point in graph 1 would be taken as analogous to the reference to the matching point in graph 2. In other words, suppose both graph 1 and graph 2 are the character *I* written in two different styles. Further suppose the point s_1 in graph 1 is the leftmost point of the topmost line. If we say that this point s_1 matches the point s_2 in graph 2, then we may assume that point s_2 in graph 2 is the leftmost point of the topmost line. This is shown in Figure 15.20.

The algorithm is as given here:

MatchPoints(G_1,G_2)

Step 1: $ver_1 \leftarrow$ all unique points in first graph which are at least α units distance apart from each other
Step 2: $ver_2 \leftarrow$ all unique points in second graph which are at least α units distance apart from each other
Step 3: match $\leftarrow$ null
Step 3: While ver_1 is not null or ver_2 is not null
Do
Step 4: match[s_2] $\leftarrow s_1$ such that s_1 is a vertex in ver_1 and s_2 is a vertex in ver_2 and the distance between s_1 and s_2 is least for any combination of vertices in ver_1 and ver_2 and distance between s_1 and s_2 is less than α units
Step 5: Remove s_1 from ver_1
Step 6: Remove s_2 from ver_2
Step 7: For all vertex v in ver_1
Do
Step 8: match[v] $\leftarrow$ null

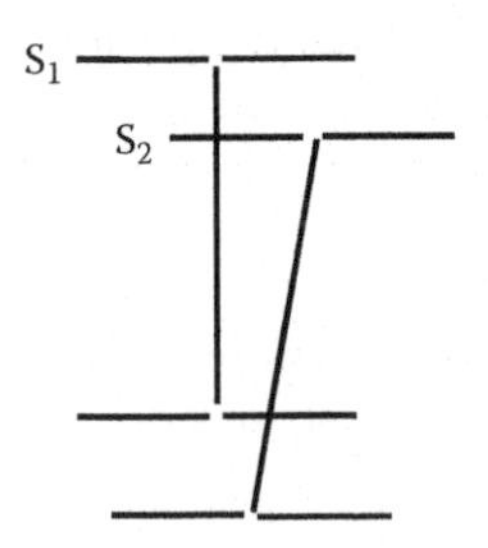

FIGURE 15.20 Matching points in two graphs.

Step 9: For all vertex v in ver_2
Do
Step 10: match[v] ← null

15.7.3.2 Generate Adjacency Matrix

Once the points have been identified and matched, the next step is to generate the graph. We use the adjacency matrix way of graph representation for this problem. We know that since this is a nondirectional graph, the path from a node x to a node y would also imply the path from the node y to the node x. This means the final adjacency matrix (W) would be symmetric. Further, we know that any two points can be simultaneously connected by a curve and/or a line. Hence we use the convention for the representation of a cell W_{ij} in the adjacency matrix for any pair of vertices i and j as given by (15.7). All other states are invalid.

$W_{ij} = 0$ (If the node i is not directly connected by the node j) (15.7)

$W_{ij} = 1$ (If the node i is directly connected by the node j by only 1 line)

$W_{ij} = 2$ (If the node i is directly connected by the node j by only 1 curve)

$W_{ij} = 3$ (If the node i is directly connected by the node j by only 1 line and 1 curve)

The algorithm for the generation of the adjacency matrix is given here:

Generate Adjacency(G)

Step 1: $W \leftarrow 0$
Step 2: For all edges e in graph joining points i and j
Do
Step 3: $W_{ij} \leftarrow W_{ij} + 1$ (if e is a line)
Step 4: $W_{ij} \leftarrow W_{ij} + 2$ (if e is a curve)
Step 5: $W_{ji} \leftarrow W_{ij}$
Step 6: Return W

15.7.3.3 Find Paths

Looking at the algorithm for the crossover, the next job is to find the paths between the crossover points. We use the concept of dynamic programming to solve this problem. We know that we have a graph. This consists of a number of vertices. For this problem, we restrict ourselves to paths that are of length maximum 4. Hence the problem is to find the existence of a path p, from a vertex i to a vertex j of a distance of exactly l. Here i and j can take any values and l is 1, 2, 3, or 4. We define a variable $F(i,j,l)$ that stores the existence of the path. It stores the last vertex traversed while traveling from i to j using path of length l. If such a path is not possible, it stores null.

We also know that for single length ($l = 1$), a path is only possible if there is a direct edge between vertex i and vertex j. Hence the formula (15.8) is valid.

$$F(i,j,0) = i, \text{ if edge } (i,j) \text{ exists or edge } (j,i) \text{ exists} \quad (15.8)$$
$$= \text{null, otherwise}$$

$F(i,j,l)$ may be calculated as given in Equation 15.9.

$$F(i,j,l) = k, \text{ if there is a path possible of length (1–1) between vertex } i \text{ and vertex } k \text{ and } j \text{ is directly connected to } j \ (l \neq 0)$$
$$= \text{null, otherwise} \quad (15.9)$$

This can also be written as given in Equation 15.10.

$$F(i,j,l) = k, \text{ if } F(i,k,l-1) \text{ exists and (vertex } (k,j) \text{ exists or vertex } (j,k) \text{ exists)}, \ l \neq 0 \quad (15.10)$$
$$= \text{null, otherwise}$$

Hence using Equations 15.9 and 15.10, we can find whether the path of length l exists between nodes i and j. If it exists, the value of $F(i,j,l)$ is not null. If the path exists, we can find the path by reverse-iterating the data structure F.

However, while executing the algorithm, we need to take care of the constraint that any vertex can occur a maximum of once in any path. Hence before making any selection of k in the preceding formula of $F(i,j,l)$, we need to make sure that we do not include a point twice in the entire path from i to j. Also it is possible for various values of k to satisfy the preceding formula of $F(i,j,l)$. In such a situation, we first find out all such possible values of k and then select any one randomly.

15.7.3.4 Removing and Adding Edges

As mentioned in the preceding algorithm of crossover, we apply the preceding data to find paths in the first and second graphs between any two points. If both paths are found, we remove the edges of the first path and add the edges of the second path from G_1. In such a manner, we are able to mix two distinct graphs and generate a new graph.

However, there may be various possibilities in the addition or removal. While removing an edge from vertex i to a vertex j in the graph, we take care of the following conditions to maintain the consistency of the graph:

- If $W_{ij} = 0$, or in other words there exists no edge from i to j, we break the operation.
- If $W_{ij} = 1$, or in other words there exists a line from i to j, we remove the line.
- If $W_{ij} = 2$, or in other words there exists a curve from i to j, we remove the curve.
- If $W_{ij} = 3$, or in other words there exists a curve and a line from i to j, we remove any one of them randomly chosen.
- $W_{ij} > 3$ is an illegal state and the operation is broken.
- After the operation, W_{ij} is made equal to W_{ji}.

Here W_{ij} refers to the adjacency matrix of graph 1 on which the removal operation is applied. Here we have taken care that after the operation is over, W_{ij} should be between 0 and 3, which are the only legal values it can take.

Similarly while adding an edge from vertex i to a vertex j in the graph, we take care that after the operation, W_{ij} should be between 0 and 3, and W_{ij} should be equal to W_{ji}.

Intermixing of graphs may generate many impossible graphs. By proper checking and breaking operations, we save the solution set from getting wrong data.

15.7.3.5 Generation of Graph

The last step is to convert the graphs in the form of an adjacency matrix into graphs in the form of an edges list. This graph contains the list of vertices along with their locations and details of the edges. Here we need to take care of another kind of optimization. This is the distance optimization of the algorithm. Earlier we introduced the concept of matching points. This concept found out pairs of matching points between the two graphs. A pair here consisted of a point s_1 in the first graph and s_2 in the second graph. We said that the role of s_1 in the first graph is the same as the role of s_2 in the second graph.

The points from the graph were matched with those of the unknown input, and the deviation or the distance between these points was used in the fitness function. Moving the two ends of the lines or the curve might thus prove to be useful for optimization of the fitness value. This optimization is called the distance optimization. We optimize the distance by moving the two ends of the line or curve in the new individual formed by crossover. This is done by placing the points or vertices of the new individual at a location that is the arithmetic mean of the locations of the parents. In this manner, the child takes the characteristics of the parents in regard to the placement of vertices as well.

In place of taking vertices from any one of the graph, we take the mean of the matching vertices of the two graphs. This optimizes the distance as well, along with the style. It may be noted that in case the two graphs match in style, which is the case most often for higher-fitness individuals, the distance optimization proves very useful in optimizations.

The algorithm for the formation of graph is given here:

GenerateGraph(W)

Step 1: for all vertices in W
    Do
Step 2: for all vertices j in W that come after or at i
    Do
Step 3: if $W_{ij} \geq 2$
Step 4: add a curve from x to y
Step 5: $W_{ij} \leftarrow W_{ij} - 2$
Step 6: if $W_{ij} = 1$
Step 7: add a line from x to y

Here x = mean of vertices i in the first graph and the corresponding matching vertex k in the second graph, such that match[k] $\leftarrow i$), y = mean of vertices j in the first graph and the corresponding matching vertex k in the second graph (such that match[k] $\leftarrow i$)

15.8 RESULTS OF HANDWRITING RECOGNITION

In order to check the working of the algorithm, we coded the algorithm. Java was used as a language. Input was given in the form of images. Test data was also stored in the form of images. We applied the algorithm over the capital letters of English language. The language contains 26 characters.

In order to make the database, we first wrote each letter twice. The first time the letter was written perfectly, like the way we are taught. The second time we used a raw hand to introduce some imperfection. Letters like *A* had some types of style associated with them. Thus we wrote these letters more times as per requirements. The input data set was made by writing each character some enumerable number of times, using different ways. The motion of the hand was shaken while writing, to introduce some unknown imperfections in the test data. The algorithm was made to first process the training data. Then it was made to run on the test data by iterating through the test data one item after the other.

We used a total of 69 characters as the training data. The test data was a collection of 385 inputs. When the algorithm was made to run, it correctly identified 379 characters. It wrongly identified 6 test cases. This gave the efficiency of 98.44 percent.

15.8.1 Effect of Genetic Algorithms

This problem could have been solved with the absence of genetic algorithms as well. In order to see the importance of genetic algorithms in the problem, we tested the data in the absence of the application of genetic algorithms. We found that the genetic algorithms were useful in the following manner.

15.8.1.1 Distance Optimization

The character that is input for identification may have the vertices of its graph at a point quite far from that of the training data. Actually language allows us to end the various lines over a large distance. This depends on the writer and his writing style. But when we would compare the closeness of such an input, it is clear that the differences would be large. On the other hand, identification by a human of the same character would be very easy.

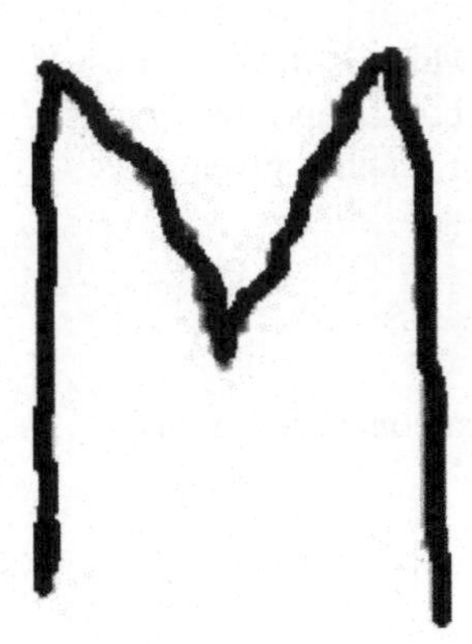

FIGURE 15.21a Input identified wrong without genetic algorithm.

Such a problem was highlighted when we tried to recognize the character *M* without the genetic algorithm. The input given is shown in Figure 15.21a. As it can be seen in this figure, the two slanting lines of *M* are quite medium in size and relatively near to one another. This was not true in any of the training data, where one example contained the entire *M* distributed normally. This made the slanting lines quite big. Another example was written in a quick jerk to the middle section, making the section quite short. This is shown in Figures 15.21b and 15.21c. When the algorithm was made to run, it happened to match with the character *X*. But on applying genetic algorithms, the distance got optimized, and the character was identified correctly.

15.8.1.2 Style Optimization

We have already discussed how the algorithm is very efficient in mixing two styles and generating a mixture of styles. This property of the algorithm to generate newer and newer styles proves very useful in the working of the algorithm. In the training data set, we had given two distinct *B*s in the training data of *B*. When operated without genetic algorithms, these two failed to match with the input of the data. On the contrary, the best match was found out to be with the letter *S* with a high deviation. But when the same was done with genetic algorithms, the *B* matched correctly. Figure 15.22 shows graphs of the input, the training data, and one of the genetically produced results. Here the top curve of first training data was added to the second training data's top set of lines. The genetic algorithm was applied across the vertical top line.

FIGURE 15.21b Closest training data1 in the wrongly identified input problem.

FIGURE 15.21c Closest training data2 in the wrongly identified input problem.

FIGURE 15.22a Input identified wrong without genetic algorithm.

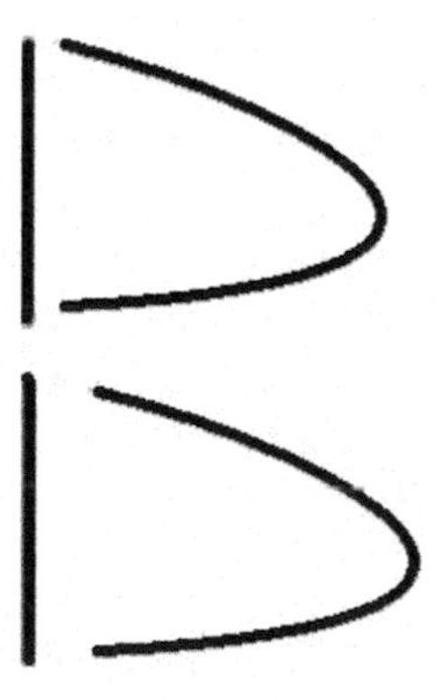

FIGURE 15.22b Graphical representation of closest training data1 in the wrongly identified input problem.

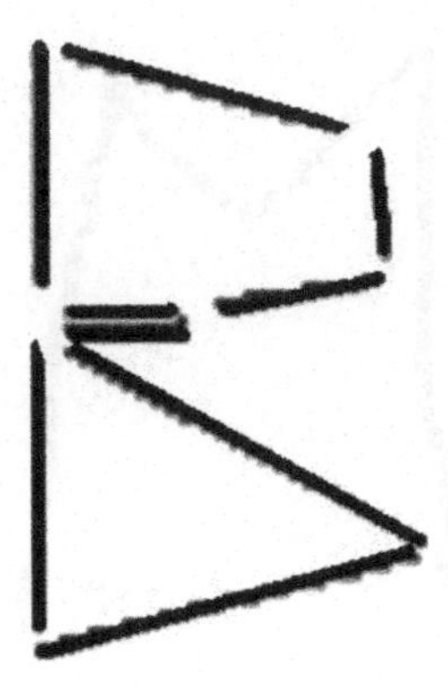

FIGURE 15.22c Graphical representation of closest training data2 in the wrongly identified input problem.

FIGURE 15.22d Genetic mixture of training datal and training data2 in the wrongly identified input problem.

15.9 CONCLUSIONS

In this chapter, we first presented an approach to solve the problem of OCR. We devised a hierarchical model that first identified the language using a rule-based approach. Then the language-specific ANN was used to map the detection of the actual character that was identified. Using this method, we were able to easily identify the English capital letters, English small letters, and Hindi letters. Specific rules were devised that could identify or classify the given input into these classes.

Later in the chapter, we introduced another concept that used GA for the purpose of handwriting recognition. This algorithm used the available training data set to generate hybrid characters formed by the available characters in the data set. Using this approach, we were able to easily classify the English uppercase characters. This algorithm treated the individual characters as graphs, and the entire operation was done on the generated graphs. The GA helped in generating newer styles that better matched with the unknown input. Further the GA helped in optimizing or minimizing the deviation between the unknown input and the characters available in the training data set.

In this chapter, we proposed some methods that helped us identify lines, curves, or edges. These means can be improved and may be done in the future. Also exhaustive testing needs to be done using various languages, styles, and representations of the characters using both the presented methods.

16 Picture Learning

16.1 INTRODUCTION

Picture learning is a widely studied problem that deals with the ability of the machine to learn and reproduce a picture. In other words, the machine parameters may become the replica of the picture which may be used to replicate or reproduce the image. The main motivation behind this is its application in image compression. Images being compressed may be mainly divided into three categories. There are color images, grayscale images, and black-and-white images. In this chapter, we only consider black-and-white images.

The learning of black-and-white images may be visualized as a classificatory problem where the x and y coordinates are given and the task is to classify that point into one of the two classes, black or white. Classification is the problem where we are given some input and asked to map the input into some predefined classes. There are numerous examples of classification that include face recognition, character recognition, speaker verification, and so on. Due to the immense application of these problems, there is a constant rise in the study of methods and applications. Classification is one of the fields where the developed systems lag behind the biological counterparts. Humans can easily identify faces, speech, words, and such. The systems that we develop, however, still struggle to gain high efficiency.

The first part of this chapter is devoted to the development of hybrid systems for solving real life problems. We study the special case of classificatory problems and model a hybrid system to solve the problem. This system is inspired by the neuro-fuzzy architecture or methodology of problem solving. We study each and every step of the neuro-fuzzy system and look at the means by which we make a similar system that is more suited for the classificatory problems. Using this architecture, we then solve the problem of picture learning. This enables us to use the proposed mechanism in a real life application.

The algorithm that we develop and implement here uses clustering of the training data set. This limits the training data to a workable amount. The clustering takes place with reference to the output class. We use a kind of rule-based fuzzy approach that uses the summarized information from the training data set to map any given input to its corresponding class. Each representative of a cluster forms a rule of its own where it tries to select the inputs like itself and tries to map them to its class. The rules use two parameters: the center and power of the cluster representatives. The parameters are optimized using a training mechanism that uses ANN and is further optimized by using GA.

This algorithm is applied over the problem of picture learning. Here the algorithm is presented with a picture to learn. The algorithm is only shown a part of the picture and not the entire picture. The algorithm is supposed to learn the picture and reproduce it whenever demanded. We take the case of a black-and-white picture, where the problem can be generalized to a classificatory problem of mapping some coordinates into either of two classes, black or white.

In the latter part of the chapter, we solve the same problem with the help of a new class of ANNs. These ANNs are the instantaneously trained ANNs. They take virtually no time to train themselves. The training can be simply done by an inspection of the inputs. They, however, require a lot of storage capacity to store the various training data sets, which can be a disadvantage. This is comparable to the phenomenon of short-term memory in humans. If we are to learn something for a very small duration, we usually do not take time. However, the memory is soon lost and lacks generalization.

Here we talk about the problem of picture learning and the various soft-computing approaches to solve this problem. Then we present the modified algorithm for the classificatory problems. Here we look at the theory of the algorithm and the various steps involved. We then use this algorithm for the purpose of picture learning. At the end we give the conclusions.

As discussed, there are so many problems that are classificatory in nature. A lot of work has been done in each of the individual problems. These problems mainly use artificial neural networks to approximate the output. The neural networks with back propagation have been extensively used in almost all the problems. Genetic algorithms have also been applied to the neural networks for optimized training. A lot of work is being done in the field of neural network for validating, generalizing, and better training of the neural network.

The problems also employ the concept of Hidden Markov Models. These are statistical models that can be used to predict the consequence of the unknown input. These models have found a variety of use in handwriting recognition, speech recognition, and related fields. Instantaneously trained neural networks are a good approach for faster training with a smaller generalization capability. These networks require very little training time, as the weights are decided just by seeing the inputs. Neuro-fuzzy networks are relatively new. These systems emerged as a motivation for the combination of the powers of neural networks and fuzzy logic. These have been widely applied to a variety of problems in various domains. These networks offer very optimal solutions to the problem. Much is being done with various aspects of these algorithms, in such areas as clustering, genetic optimizations, rule forming, and parameter updating.

Classification is a major problem of study that attracts immense interest because of its applicability. A lot of work is being done to adapt the neuro-fuzzy systems to the classification problems. Researchers in these algorithms try to optimize the performance in clustering by designing various models based on the architecture of neuro-fuzzy systems. Self-organizing maps have also been used extensively for problem solving. Various other mathematical models have been proposed. These employ mathematical techniques like point-to-point matching to solve problems.

16.2 PICTURE LEARNING

In this section we talk about the problem of picture learning and discuss some of its methods. This will give some background to the motivation behind our development of the algorithm and its use.

In the problem of picture learning, the machine is first presented with a picture. Only some parts of the picture are shown to the machine. The machine is supposed to learn the picture. The machine makes use of learning techniques for this. Later, the machine is asked to reproduce the picture that it learned. The efficiency of the system is measured by the closeness of the reproduced image to the original image.

The part of the image or sometimes the entire image that is shown to the image forms the training data. Here each data set that is presented as training data to the system used by the machine has the x and y coordinates of the pixel as its two inputs. The output is the color of the pixel. The color may be a set of red, blue, and green components in case of color images. For grayscale images, it may be the gray level of the pixel being considered. A black-and-white image has an output of 0 or 1 to denote the presence of black or white at the particular location.

Once the system is trained, it becomes synonymous with the image. Now we may set aside the image, as it can be reproduced whenever required using the trained system. We may additionally communicate with this system whenever required in place of the image itself. The way to represent this system is in the form of network architecture and weights in the case of ANN and rules and MFs in the case of FIS. The representation of this system usually requires much less space than the complete image. Hence this method of picture learning can be used for image compression purposes.

The compression of the image may be lossy or lossless. This depends upon whether we have allowed the errors to remain to some extent in the training step. If the training is done such that the error level is completely zero, the picture compression will be lossless. In other cases, the compression will be lossy. The amount of loss depends upon the error in the trained system. Usually the addition of memory storage (weights, number of MFs, number of neurons, number of rules, etc.) reduces the errors. But at the same time, this results in a decrease of the compression factor.

At the time of testing or reproducing the image, the system is given all the points that lie on the picture as input one by one. Each input is the x and y coordinates. The algorithm returns the color at that location. The color may be physically plotted by the testing algorithm at the canvas. This will complete and reform the entire image. Usually the plotted image is not same as the original image, as we allow for errors for better compression. The difference in the two images determines the effectiveness of the algorithm. An effective or efficient algorithm redraws the best possible picture using the least amount of storage space. This, however, largely depends upon the picture being taken and the kind of quality or compression desired.

Traditionally various statistical approaches were used for the problem of picture learning. These calculated the differences between the consecutive pixels and stored the nonzero differences. This greatly reduced the storage capacity that was required to store the image. Various factors controlled the quality of the image and the level of compression. In recent years, the soft-computing techniques have evolved and have shown great performance in terms of image quality, compression, and the time required for the compression and decompression. The soft-computing techniques, because of their close resemblance with the natural world and great modeling capabilities, are easily able to learn the image, even when the entire image is not shown but only a part of it. Hence there has been a great research and use of soft-computing techniques in this field as well.

The image learning depends upon the type of image. Here we classify the images into three categories, depending upon the color that is the output of whatever system we are building. These are color images, grayscale images, and black-and-white images. In the case of *color* images, the output may be the red, green, and blue parameters of the color. Hence the system returns these three values as its output. The *grayscale* images have the entire image represented in shades of gray. This is one of the most studied formats in all image-related research. Here each pixel is denoted by a single number that denotes the gray level. This number lies in between 0 and 1. Here 1 represents white, 0 represents black, and any number in between represents gray. The closer the number to 1, the lighter its shade. The third form is the *black-and-white* image. Here all pixels are either black or white. Hence the image may be represented or stored in the form of bits, with 0 representing black and 1 representing white. In these types of images, the output of the system is 1 or 0. In this way, the problem becomes classificatory in nature, where we have to classify each input in one of the two classes: 0 or 1.

Any black-and-white (b/w) picture may be converted into grayscale and vice versa. Both these formats are able to depict the actual image. The viewer would be easily able to recognize the object being displayed, the surroundings, and so on. Hence the choice of representation entirely lies in the hands of the user. Normally a grayscale image gives a smoother look to the landscapes and such images that mark a transition of colors. However, in images like scanned word documents, black-and-white images look better because the natural possibility of limited colors. Black-and-white images further require much less storage space than grayscale images. The grayscale image and its black-and-white equivalent are shown in Figure 16.1.

Now we will study some of the techniques. Later in the chapter we build an algorithm for the classificatory problem. We then use this algorithm over binary images and see how the machine is able to learn and redraw the image on being shown only some points in the image.

16.2.1 Coding Techniques

In this section we look at some of the fundamental techniques that are used for the coding purposes. Using these fundamental techniques, we later on discuss how these techniques are employed in real life applications.

FIGURE 16.1a The grayscale image.

One of the techniques used is *predictive coding*. In this technique, we try to predict the value of the color of any pixel, based on the values of the neighboring pixels. This technique exploits the property in images that the value of any pixel largely depends upon the surrounding pixels and can hence be predicted by knowing its surroundings. A common example is differential pulse code modulation (DPCM).

The other technique is the transformational technique of coding. Here we break the picture into blocks of pixels. Then we transform the blocks using some transformational technique. These can be used for compression and storage purposes. Later the inverse of the transformation is used to get the original image back. The principal component analysis (PCA) is one such method of coding.

The third type of coding is seen in vector quantization–based systems, where the various levels that any pixel can have are quantized up to some levels. These levels are decided using a codebook that is defined externally. The Euclidian norm is commonly used to decide the levels.

FIGURE 16.1b The black-and-white image.

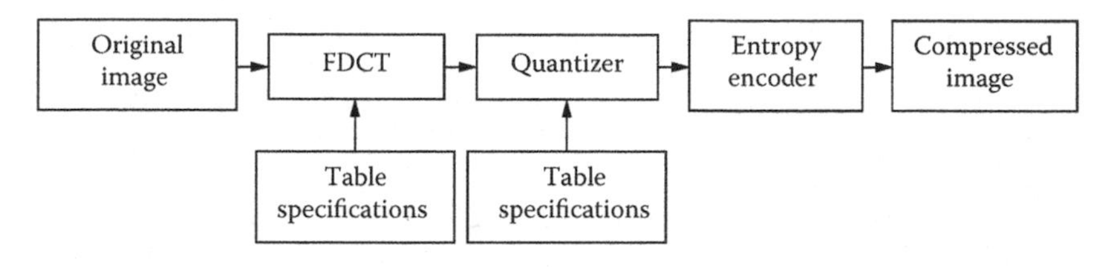

FIGURE 16.2 The DCT-based JPEG encoder.

16.2.2 JPEG Compression

JPEG stands for Joint Photographic Experts Group. It is a joint standard for image compression developed by the ISO and CCITT. It is an extensively used image compression and storage format. We are all used to the .jpg and .jpeg images stored on our systems and over the Internet. All these use JPEG compression. The JPEG software compresses the picture and stores it on the system. This picture can be later reproduced by the image viewing software by performing a decompression. The format has specifications for the way the image is compressed, stored, and later decompressed. In this section, we take a brief look at the compression and decompression done by JPEG.

JPEG compression uses DCT (direct cosine transformation) for the compression, which is a lossy compression. A predictive method is applied for lossless compression. A baseline method, a subset of the other DCT-based modes of operation, is an extensively used compression technique. The goals of JPEG compression are to accommodate various qualities of images, to be applicable to almost all sources of images, and so on. Further the JPEG operates in numerous modes. This includes sequential coding that takes place pixel by pixel from left to right and top to bottom. Another common mode is the progressive encoding mode, where the image is encoded in multiple resolutions, from courser to finer. Lossless encoding and hierarchical encoding are the other modes of operation.

The DCT-based encoder and decoder of the JPEG system are shown in Figures 16.2 and 16.3 respectively.

In the encoder, the image is broken up into blocks of pixels. These pixels are then applied through FDCT (forward discrete cosine transformation). The information after the transformation is quantized by the use of a quantizer. This restricts the values up to some predefined levels and is helpful for the lossy compression purposes. The 64 DCT coefficients are mapped to the prespecified values in the specification table. This table must be available with the JPEG encoder beforehand and is specified by the user or software being used. The quantized information is encoded using the entropy encoder. This encoding is a lossless encoding technique that compresses the transformation coefficients using a statistical analysis. The most commonly used techniques are Huffman Coding and Arithmetic Coding. The baseline method of JPEG compression only uses the Huffman Coding technique. This gives us the compressed image.

The decoder follows a procedure that is the reverse of that of the encoder. Here first decoding takes place, followed by dequantization and inverse DCT operation. This finally gives us the original image back, with a little loss of information.

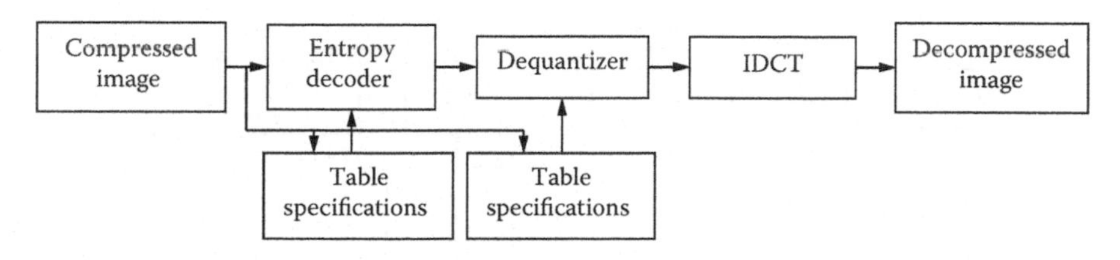

FIGURE 16.3 The DCT-based JPEG decoder.

16.2.3 Soft-Computing Approaches

From the previous discussions, you can easily see the usefulness of soft-computing techniques for image compression and learning. They can be used for all sorts of coding: predictive coding, transformational coding, or vector quantization.

The ANNs are commonly used for all sorts of coding techniques. Their use in picture compression may be easily visualized and appreciated. The self-organizing maps (SOMs) have extensive use. Similar is the use of recurrent ANNs and learning vector quantization (LVQ). The BPA is used as well for the coding purposes. Hence we see that the ANNs have an extensive role to play in the coding and compression of the image. All these techniques give great results when applied to the compression of images.

Fuzzy logic and FIS are also extensively used in image compression. These techniques are commonly used because of their ease of modeling and representation. These also give good results when applied to most images. Various hybrid techniques are also applied to image compression.

16.3 HYBRID CLASSIFIER BASED ON NEURO-FUZZY SYSTEM

In this section, we develop a hybrid algorithm that solves the classificatory problem. The model is derived from the work of Kala et al. (2009b). The whole architecture and problem-solving methodology of this system is based on the neuro-fuzzy architecture. In this section, we study each and every step that we discussed in Chapter 6 in describing neuro-fuzzy systems and we modify these steps to be better applicable to the classificatory problems. Every step that we take in this algorithm will be handled separately for each and every class. This offers the flexibility of equal interclass treatment that is not present in conventional neuro-fuzzy systems. The evolved system using the modified algorithms will thus work better in situations especially where there are a large number of classes involved that repeat themselves in various clusters in the input space.

In this algorithm, the first task is to cluster the input data set into clusters. The clusters, as we shall see later in the text, are class specific. For the same reasons, we develop a clustering algorithm of our own. This does not take the number of clusters as input; rather, it decides the optimal number of clusters on its own.

Then we use a fuzzy logic system that maps every input to one of the classes that the system is asked to classify to. This step too involves a competitive approach between the various classes. The class that gets the maximum score is declared the winner. The mapping rules are decided by the clusters that were initially formed, where one member per cluster tries to influence the decision of the class in its favor. Likewise the sum of influence of all the clusters is done for every class and on that basis, the final winner is declared.

In order to optimize the system, we use a training algorithm that uses gradient descent approach just like the ones used by ANNs. The training algorithm optimizes the parameters of the fuzzy system so developed. The parameters are additionally introduced to fine-tune the system that evolves. The GA is further used to ensure that the training does not get stuck at some local optima but reaches the global optima. The neural network and genetic algorithms are applied over the validating data, which should be set aside for testing the algorithm. The whole process of the algorithm is shown in Figure 16.4. We discuss the various steps of the algorithm in detail one by one.

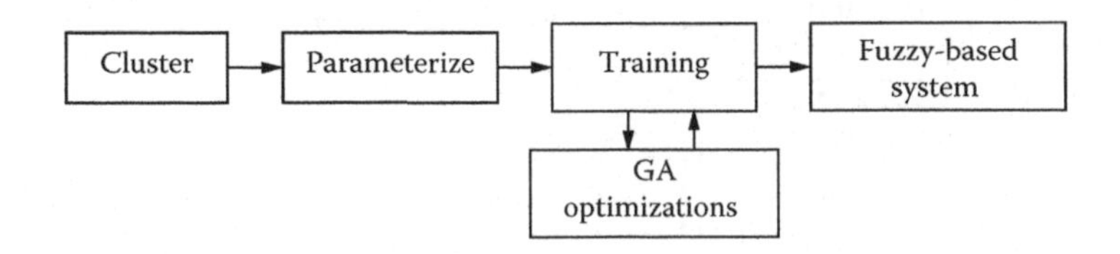

FIGURE 16.4 The general classification algorithm.

16.3.1 Clustering

Clustering is the very first step that is performed in the algorithm. The clustering is useful for limiting the amount of data. This saves large computational and memory requirements in the following steps. The clustering selects the best data items from the training database that would be useful in the subsequent steps. Clustering does the task of grouping data items from the same class that lie close to each other.

We introduce a major constraint in the clustering algorithm. Usually the clusters are made on the basis of the input data only. Here in addition we place a constraint that all data members of a cluster need to belong to one class only. This means that we will have output class–specific clusters where some clusters may belong to the first output class, some to the second output class, and so on.

Clustering in this algorithm encloses a region in the input space where only the members of one particular class are found. This region is circular in nature. The clusters thus formed represent circular regions that exclusively belong to a particular class. The clusters further do not entertain any member of the foreign class in the circular region enclosed by them. One such cluster for a two-input problem is given in Figure 16.5.

The basic philosophy of the clustering is to enable us to replace the whole cluster by one representative that may effectively replace the entire cluster. Likewise all clusters may be replaced by the class-specific clusters. At the time of testing, as we shall see, these representatives fire their influence in order to decide the final output. These try to influence the decision of the algorithm in favor of the class to which they belong. In this step we ensure that the cluster representatives easily dominate within the region enclosed by these classes. In other words, the sphere of influence of the clusters has high influence in the region enclosed by their clusters.

The representative of the cluster is placed centrally at the cluster. Suppose the input space was an n-dimensional space in which the training data in a particular cluster are placed at $<I_{11},I_{12},I_{13},I_{14}\ldots I_{1n}>$, $<I_{21},I_{22},I_{23},I_{24}\ldots I_{2n}>$, $<I_{31},I_{32},I_{33},I_{34}\ldots I_{3n}>\ldots<I_{p1},I_{p2},I_{p3},I_{p4}\ldots I_{pn}>$. The center for this data is given by Equation 16.1.

$$<C_1,C_2, C_3, \ldots, C_n> = <(\max(z_1)+\min(z_1))/2, (\max(z_2)+\min(z_2))/2, (\max(z_3)+\min(z_3))/2\ldots(\max(z_n)+\min(z_n))/2> \quad (16.1)$$

Here $z_1 = \{I_{11},I_{21},I_{31}\ldots I_{p1}\}$
$z_2 = \{I_{12},I_{22},I_{32}\ldots I_{p2}\}$
$z_n = \{I_{1n},I_{2n},I_{3n}\ldots I_{pn}\}$
C is the cluster center

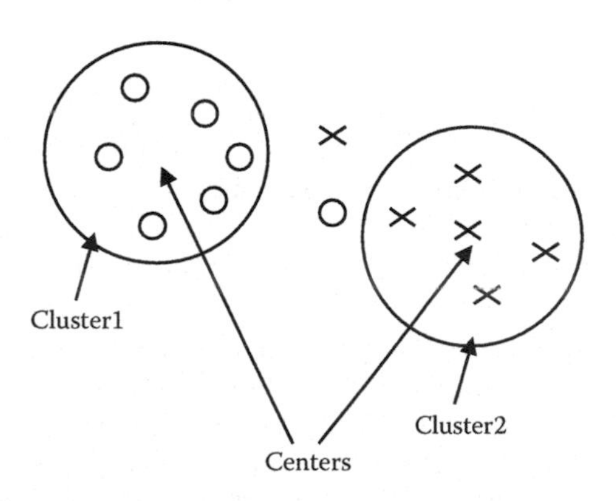

FIGURE 16.5 The clusters of the given input.

We define the radius of the cluster as the square of distance between the center of the cluster and the most distant point in the cluster. Hence the radius of the cluster can be written as

$$R = \text{Max}\{(C_1 - I_{i1})^2 + (C_2 - I_{i2})^2 + (C_3 - I_{i3})^2 + \cdots\cdots + (C_n - I_{in})^2\} \quad (16.2)$$

For all i in cluster

Here $C = <C_1, C_2, C_3, C_4 \ldots C_n>$ is the cluster center.

Further we placed an additional constraint on the cluster that all members must belong to the same class. In other words, there should be no foreign member in the cluster. Hence for any input $<I_{x1}, I_{x2}, I_{x3}, I_{x4} \ldots I_{xn}>$ in the system that belongs to a different class, Equation 16.3 must always be false.

$$(C_1 - I_{x1})^2 + (C_2 - I_{x2})^2 + (C_3 - I_{x3})^2 + \cdots\cdots + (C_n - I_{xn})^2 <= R \quad (16.3)$$

This concept in a two-input system is given in Figure 16.5.

The second parameter that we introduce for every cluster is called the *firing power* parameter. Each cluster is assumed to influence the output in its favor by a certain factor. This factor depends upon the radius R calculated by Equation 16.2. The net influence of the class is calculated using this constant parameter called firing power and denoted by P. The formula is given by Equation 16.4.

$$P = \alpha R \quad (16.4)$$

Here α is const

R is radius of the graph

P is the power

The algorithm for the clustering is given by the following algorithm:

Cluster()

Step 1: While all points are not clustered

Do

Step 2: Add any random point to a new cluster

Step 3: While it is possible to add new points in this new cluster

Do

Step 4: Search for the point p that is closest to any point in the cluster

Step 5: If cluster formed by adding p to this cluster is possible, then add p to new cluster

Step 6: Find center and initial power of this cluster

16.3.2 Fuzzy Logic

The major base of the algorithm is a fuzzy-inspired approach that maps the inputs to the outputs. Here we make use of the summarized data items from the training data that we receive from clustering algorithm. Each of the representatives of a cluster plays the role of a rule in this approach and tries to influence the decision in its favor. Like the fuzzy approach, this influence to a certain degree depends upon the membership degree of the input to this cluster center or rule. The fuzzy procedure involves parameterization in the form of parameters for every cluster or rule. There are two parameters per rule that we discussed in the preceding section. These are the firing power and location.

As per the basic motive of the algorithm, this process of application of fuzzy logic is class specific. The whole operation is done separately for each and every class. In this manner each class gets an equal opportunity to be selected as the final answer that the input classifies to. The model adopted is a competitive model where the various output classes compete with each other in order to be selected as the final output. The decision is made on the basis of final scores of each class. The scores of each class are contributed by the clusters belonging to the specific class. Here the influence parameters as well as location parameters both play a prominent role.

We draw an analogy between the developed system and the conventional fuzzy system. A conventional fuzzy system works by the mechanism of application of inputs. Various rules work over the input, and each contributes toward mapping the input to the output. Similarly in this system the inputs are applied. Each cluster representative acts as a rule that tries to contribute toward the output or affect the output. The conventional fuzzy systems measure fuzziness in input by the closeness of the input to the region catered to by the specific membership function. Here we do the same thing by measuring the membership in the form of degree of closeness of the input to any cluster representative. The fuzzy system rules generate a fuzzy output with a certain degree of confidence or membership value. Similarly in this system, each cluster representative output is fuzzy that has a certain confidence or influence or membership value. The fuzzy aggregation is performed by simple addition of the like classes, and the fuzzy defuzzification is performed by the competition among the classes. Further the fuzzy rules have weights that denote their importance. Here we introduce weights for each and every cluster representative or rule.

Now we take a step-by-step approach to all the points discussed, starting from the rules. In this system, every cluster representative represents a rule of its own that tries to find inputs like it and influences the system to generate outputs similar to one passed by this cluster. This way, the rules are class specific. The output of the rule is given by Equation 16.5.

$$O_i = P_i / D_i \tag{16.5}$$

Here O_i is the influence of the individual rule for cluster i in favor of its class c
P_i is the firing power of the cluster i
D_i is the square of distance between the center of cluster i and the given input

Here P may be assumed as the weight of the rule, which is an adjustable parameter.

The next major task is aggregation, which is simply addition in our case. This operation is also made class specific where only influences of the same class are together. In this way, we get as many outputs as the number of classes. The final influence for any specific class is given by Equation 16.6.

$$C_j = \Sigma\, O_i \tag{16.6}$$

Where C_j is the output for any class j
O_i is the output for individual rule i

The defuzzification is a competitive model where the different classes compete with each other. Every class has an influence factor calculated by Equation 16.6. The class with the maximum influence is declared the final winner. The final output is the class corresponding to the maximum influence. This is given by Equation 16.7.

$$\text{Class} = i \text{ such that } C_i > C_j \text{ for all } j \neq i \tag{16.7}$$

16.3.3 Network Training

In order to optimize the system, we had introduced two parameters to every cluster or rule. One was the firing power, which played the role of weights in a conventional fuzzy system. The other was the location, which played the role of membership function in the conventional fuzzy system. We need to modify these parameters and fix them to correct values in order to get an optimal system. For this purpose, we use a training mechanism that uses an algorithm similar to the back propagation algorithm. This algorithm also runs on the steepest descent approach that we studied earlier in Chapter 2.

The training is performed in a batch processing style where the various training data sets are fed one after the other. This takes place in multiple epochs during the training period. In each epoch, the difference between the actual and desired outputs is calculated and used to update the parameters. If the difference is large, the change is large, and vice versa following the steepest descent approach.

We take another concept from the ANN: the validation data set, which is used by the conventional ANN to measure the generalizing capabilities. The ANN is trained only until the performance against the validation data set improves. The validation data is especially reserved data out of the available data set. Here we use the validation data set for training and try to train the network for optimal performance against the validation data set.

The other concept we take from the conventional ANN is that of the learning rate. The learning rate decides the rate at which the network will train itself. In a conventional ANN, a very small learning rate usually results in the system taking a long time to train itself. A very high learning rate makes the system oscillate in between high error contours.

In this system, the training data set is used by the clustering algorithm to cluster and give the cluster representatives that serve as rules. The training is performed on the validation data set over a predefined number of epochs or any other stopping criterion. The training in this algorithm might not always result in lowering of the training error rate. Hence we train the system and extract the globally minimum error that occurred during the training phase.

The general algorithm for training the system is given in Figure 16.6.

The parameter modification modifies the center and the firing power of every cluster. The modification does not use a steepest descent approach. The modification algorithm iterates through every

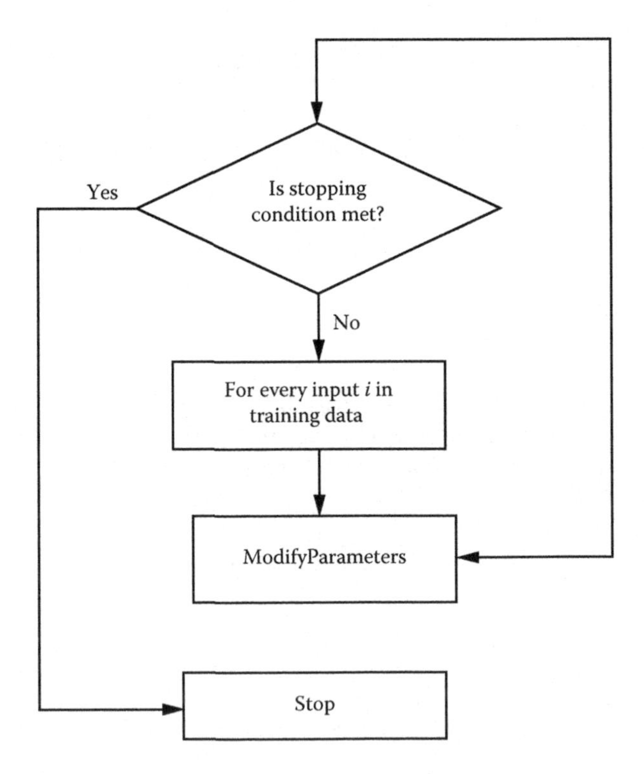

FIGURE 16.6 The training algorithm.

cluster. It sees whether the class to which the cluster being considered belongs is same as the class of the training data being considered. If the two classes are same, the algorithm tries to increase the contribution of this cluster. On the other hand, if the classes do not match, the contribution is tried to be minimized.

The rate of increase or decrease of the contribution depends on the distance between the training data and the cluster being considered. The greater the distance, the lesser would be the change made. Hence an input only affects the regions near it in the n-dimensional space.

The change or modification in weight is class specific, where every change happens by considering its class. We consider the difference between the influences of the desired class with the rest of the classes and try to maximize the difference in favor of the desired class. By this we assume that when the same input is applied the next time, the difference will be large and we will be able to clearly identify the actual output class. The modification in weight is directly proportional to the difference in the influence.

We take the difference between the influence of the desired class and the class that wins the competition. If the class that wins the competition is the same as the desired class, the difference with the next highest class is taken. If the difference is large, the change is small and vice versa. However if the condition is that the difference is negative, the converse is true. This means that whenever the contribution of the class being considered is more than that of the desired class, we try to make a large change, so that this difference can be reduced and eventually reversed.

The change is multiplied by learning rate that does the same job as it did in the ANNs. The learning rate may be changed as needed. It may be fixed or made variable as per the training strategy.

The algorithm for the modification of the parameters is given as follows:

ModifyParameters

Step 1: For every cluster c in sample space
Do
Step 2: $a_1 \leftarrow$ contribution of desired output class in deciding final answer
Step 3: $a_2 \leftarrow$ contribution of class c if different from desired output class, else contribution of second highest contributing class for the final output
Step 4: if desired class = c
Step 5: With a probability of 0.5 increase its power by K percent
Step 6: With a probability of 0.5 move center closer to the given input by C percent
else
Step 7: With a probability of 0.5 decrease its power by K percent
Step 8: With a probability of 0.5 move center away from given input by C percent

The increase/decrease in percent in power is given by Equation 16.8.

$$K = \frac{\eta_1}{(a_1 - a_2 + c_1) * D} \quad (\text{If } a_1 > a_2) \tag{16.8}$$

$$= \frac{\eta_2 * (a_2 - a_1 + c_2)}{MD} \quad (\text{in all other cases})$$

Where K is the percent change in firing power
D is the distance between input and cluster being considered (c)
M is the maximum distance between any two points in the n-dimensional space
η_1, η_2 is the learning rate
c_1, c_2 are constants

The percent increase/decrease in center is given by Equation 16.9.

$$C = \frac{\eta_3}{(a_1 - a_2 + c_3) * D} \quad (\text{If } a_1 > a_2) \tag{16.9}$$

$$= \frac{\eta_4 * (a_2 - a_1 + c_4)}{MD} \quad (\text{in all other cases})$$

Where K is the percent change in firing power
D is the distance between input and cluster being considered (c)
M is the maximum distance between any two points in the n-dimensional space
η_3, η_4 is the learning rate
c_3, c_4 are constants

16.3.4 Genetic Algorithms

In order to save the training algorithm from being trapped in some local minima, we apply GA as well. The role of GA is to generate diversities and help avoid the local minima that would have occurred if only ANN training was used. The GA combined various semitrained networks to generate more networks. This further helps the algorithm explore the entire search space and converge at the local minima.

16.4 PICTURE LEARNING USING THE ALGORITHM

So far in this chapter, we have studied picture learning and the hybrid algorithm for classification. Now we use this algorithm for picture learning. The algorithm was given a part of small black-and-white picture. The algorithm as supposed to memorize it. Here the picture contained black pixels and white pixels that were given the values of 0 and 1 for the algorithmic implementation. The pixels shown to the algorithm were from all sections of the picture. These pixels were randomly selected initially from the picture. The selected pixels were later divided into training data sets and validating data sets. Those present as the training data sets were used by the clustering algorithm for the purpose of learning and making initial rules. The others were used by the training algorithm for the purpose of learning and modification of the network parameters.

This even tested the generalizing capability of the system, where the system had to predict the output at the completely unknown inputs. The output predicted at these unknown points should be precise as per the requirement of the algorithm. The efficiency of the algorithm was measured as its capability to correctly classify as many points as possible. The more the algorithm correctly classifies the points, the more realistic will be the regenerated image after reconstruction by the algorithm.

The efficiency of the algorithm in various phases is given in Table 16.1. It can be easily seen that the efficiency improves as a result of each and every step that was discussed in the algorithm. Since the classificatory problems are highly localized in nature, the efficiency may further improve with an increase of the training data points. As we keep increasing these points, the efficiency goes on increasing. The same problem was also tried to be solved by a neural network and a conventional neuro-fuzzy system. The same training data was given to the two systems as was given to our neuro-fuzzy system. The results are shown in Table 16.2. The results clearly show that the neural network failed to give results to the problem with much computation.

Hence it can be easily inferred from the results that the neuro-fuzzy system that we have proposed here gives the best results as compared to the other systems. The system gave better results

TABLE 16.1
Results of the Algorithm in the Problem of Picture Learning

Original Image

```
. . . . . . . . . . . . . . . .
. . . . . . . . . . . . . . . .
. . . . . . . # # # . . . . . .
. . . . # # # # # # # # # . . .
. . . # # # # # # # # # # # . .
. . # # # # # # # # # # # # # .
. . # # # # # . . . # # # # # #
. . # # # # . . . . . # # # # #
. . # # # # . # . . . # # # # #
. . . # # # # # . . . # # # # #
. . . . # # # . . . . # # # # #
. . . . . . . . . . . # # # # #
. . . . . . . . . . . # # # # #
. . . . . . . . . . # # # # # #
. . . . . . . . . # # # # # # #
. . . . . . . . # # # # # # # #
```

No. of points: 256
No. of #: 115
No. of .: 141

Training Data (I: Training Data of Fuzzy Clustering V: Training Data for Neural Network)

```
. . I . . . . . I . . . . I . .
. . . . . . . . . . . . . . . .
. I . . . . . # I # . V . I . .
. . . . V # # # # # # # # . . .
. . . # # # # I # # # # # V . .
. . # # I # # # V # I # # # # .
. . I # # # # . . V # # I # # #
. . # # # V . . I . . # # I # #
. . I # # # . # . . V # # # # #
V . . # I # # # . . . # # I # #
. . . . V # # V . I . # # # # #
. . I . . . I . . . . # # # # #
. . . . . . . . I . . # # I # #
. I . . . . . . . . V # # # # #
. . . . I . . . . # # # # # # #
. I . . . . . . # # # I # # I #
```

No. of training data: 37
No. for clustering: 26
No. for neural: 11

Image after Only Application of Fuzzy System

```
. . . . . . . . . . . . . . . .
. . . . . . . . . . . . . . . .
. . . . . . . # # # # . . . . .
. . . . . . # # # # # # . . . #
. . # # # # # # # # # # # # # #
# # # # # # # # # # # # # # # #
# # # # # # . . . # # # # # # #
# # # # # # . . . . # # # # # #
# # # # # . . . . . # # # # # #
. . # . . . . . . . # # # # # #
. . . . . . . . . . # # # # # #
. . . . . . . . . . # # # # # #
. . . . . . . . . . # # # # # #
. . . . . . . . . . # # # # # #
. . . . . . . . . . # # # # # #
. . . . . . . . . # # # # # # #
```

Total: 256
Correct: 218
Incorrect: 38
Training size: 26
No. of Clusters: 11
Efficiency: 85.15625

Image after Application of Neural and Fuzzy

```
. . . . . . . . . . . . . . . .
. . . . . . . . . . . . . . . .
. . . . . . # # # # . . . . . .
. . . . # # # # # # # # . . . #
. # # # # # # # # # # # # # # #
# # # # # # # # # # # # # # # #
# # # # # # # . . . # # # # # #
# # # # # # . . . . # # # # # #
# # # # # # . . . . . # # # # #
# # # # # . . . . . . # # # # #
. . . . . . . . . . # # # # # #
. . . . . . . . . . # # # # # #
. . . . . . . . . . # # # # # #
. . . . . . . . . . # # # # # #
. . . . . . . . . . # # # # # #
. . . . . . . . . . # # # # # #
```

Total: 256
Correct: 223
Incorrect: 33
Training size: 11
Efficiency: 87.109375

Image after Application of Neural and Fuzzy with Genetic

```
. . . . . . . . . . . . . . . .
. . . . . . . . . . . . . . . .
. . . . . . # # # # . . . . . .
. . . . # # # # # # # # . . . #
. # # # # # # # # # # # # # # #
# # # # # # # # # # # # # # # #
# # # # # # # . . . # # # # # #
# # # # # # . . . . # # # # # #
# # # # # # . . . . . # # # # #
# # # # # . . . . . . # # # # #
. . . . . . . . . . # # # # # #
. . . . . . . . . . # # # # # #
. . . . . . . . . . # # # # # #
. . . . . . . . . . # # # # # #
. . . . . . . . . . # # # # # #
. . . . . . . . . . # # # # # #
```

Total: 256
Correct: 223
Incorrect: 33
Efficiency: 87.109375

TABLE 16.2
Results of the Conventional Classifiers in the Problem of Picture Learning

Original Image	Neural Network	Conventional Neuro-Fuzzy System
....	....	...#####......##
....	...###..........	..######.......#
.......###......	...#######......	..########......
....#########...	...#########....	..#############.
...###########..	..##############	..###.########..
..#############.	..##############	..#############.
..#####...######	..######....####	..######...####.
..####.....#####	..#####.....####	..######....###.
..####.#...#####	..####......####	..#####.....##..
...#####...#####	..###......#####	..####......###.
....###....#####	...##......#####	...###......####
...........#####	....##....######	...##........###
...........#####	....##....######	...#......#..###
..........######	.....##..#######	..##.....##...##
.........#######	.....##.########	..##....###...##
........########	......##########	..##....####.###
No. of points: 256	Total: 256	Total: 256
No. of #: 115	Correct: 215	Correct: 188
No. of .: 141	Incorrect: 41	Incorrect: 68
	Training size: 37	Training size: 37
	Efficiency: 83.984375	Efficiency: 73.4375

than both the conventional neuro-fuzzy system and the artificial neural networks. Our designed architecture, as expected, gave very efficient results. The same problem and the computed results may be generalized for any classificatory problem based on the results and increase in efficiency obtained in this algorithm.

16.5 INSTANTANEOUSLY LEARNING NEURAL NETWORK

We have discussed over and over again in the text the power of ANN to learn the past data and to generalize the learned data to effectively apply to the unknown inputs. This gave great powers to the ANNs, which was the reason behind their use in various systems that are used in the real life applications. The ANNs have been combined with various other systems for the generation of hybrid systems. Here the ANNs play their own role in contributing to the effective use and application of the system.

However, it may be seen that the ANNs require a lot of time for learning. Any learning algorithm continues for a large number of epochs, where it takes a lot of time to constantly apply the inputs in a batch processing mode, get the answers, calculate the errors, and later back-propagate them to the previous layers for the modification of the weights and the biases. This whole procedure is very time-consuming and may result in large training times.

In this section we develop a system that is much faster in training of the system. This is the special class of ANNs that are called the instantaneously trained ANNs. The ANNs have some restrictions on the inputs and the outputs. Based on these restrictions, they may be easily trained in short training times. In fact, the training can be done by an inspection of the inputs themselves. This network may then be used for the testing purposes where it may be given both the known and unknown inputs. In this manner, the ANN gives the needed performance.

These ANNs train themselves very fast with a loss in generality. They are inspired by human short-term memory. People are very easily able to remember some things when they are exposed to them. They may easily be able to recall these whenever required in a short span of time. However, the same may not be true for a large span of time. Also the amount of data that people memorize is limited. They can memorize just a few things that they retain for some time and later forget. The learning does not require much time. To learn some things permanently, on the contrary, may require a lot of time and may require a constant exposure to them over and over again.

The case of these ANNs is similar. They are able to learn the training data very well, but the training data has to be small in size. This is the short-term memory equivalent in terms of the ANNs.

In this chapter, we first present the networks that can only take binary inputs in the form of 0 or 1. These can further only generate binary outputs that are also 0 or 1. We use the previously studied architecture of the conventional ANN and modify the weights and the activation function as per the demands. We see that these networks so formed do not require training. The weights can be decided merely by looking at the inputs and their corresponding outputs.

However, this form of inputs may give rise to a large number of inputs to the system. This is because when we convert the input from the present form into a binary format, the number of inputs increases rapidly. This results in associated problems because of the high number of inputs. Later we modify the algorithm in order to take complex inputs as well. In this manner, we get a larger variety of inputs.

The basic objective in these algorithms is to train the network by isolating the corner in the n-dimensional cube of the input space. These algorithms include the CC1 algorithm. In this the weights were obtained by the use of the perceptron algorithm. The next were the CC2 algorithms. In this the weights were obtained by inspecting the data. These algorithms, however, did not account for the generalization. The next were the CC3 algorithms, which were like CC2 algorithms. In these, however, the generalization was provided. This was a result of the randomization that sometimes even led to misclassification of the inputs. As a result, further checking and adjustment of the weights was necessary. Next were the CC4 algorithms, which are the most advanced; they place some conditions on the inputs but give good generalizations. We will mainly study the CC4 algorithm in this section.

We first present the CC4 algorithm. Then we study the complex input coding that includes the binary and quaternary encoding. After that, we present the instantaneous training of ANNs. Finally we discuss the ACC algorithm.

16.5.1 CC4 Algorithm

The CC4 is an advanced algorithm that is used for its high classificatory powers and fast training. Here we fix the weights of the various inputs by just inspecting the training inputs. Drawing on these, we apply the weights to the various connections. This gives a very fast learning to these networks. Further the architecture is made in such a way that these networks have a very high classifying power.

The CC4 algorithm takes the same ANN architecture as used by the other corners algorithms. This is the same architecture that we studied in Chapter 2. We, however, modify the working principle in order to adapt them as per the requirements. The network can take only binary inputs and can give only binary outputs. This means that all the inputs and outputs are either 0 or 1 in order to use this system. In many problems, even though the inputs and outputs are not in the binary format, it may be possible to convert them into the same format.

The CC4 algorithm uses a concept of radius of generalization that is essentially a generalization control. This is an additional input that is added to the network apart from the custom inputs. The ANN architecture consists of an input layer, a hidden layer, and an output layer.

16.5.1.1 Corners Algorithm

The basic principle used behind these algorithms is the Corners Theorem, which states that the minimum number of hidden neurons H required to separate M number of regions in a d-dimensional space is given by Equation 16.10.

$$M(H,d) = \sum_{k=0}^{d} \binom{H}{k} \tag{16.10}$$

Here $\binom{H}{k} = 0, H < k$

Let the number of regions that the hidden neurons separate be equal to C, where $C \leq M$. Since number of classes at output is 2, these C regions coalesce into the two classes at output. Let the input space dimension be d and let each dimension be quantized so that the total number of binary variables is n. Not every dimension may require the same precision. If the average number of bits per input dimension is q, then n is given by Equation 16.11.

$$N = q * d \tag{16.11}$$

If the number of training samples is T, then $M \leq T$, for $d - 1$, H is given by Equation 16.12.

$$H = M - 1 \tag{16.12}$$

16.5.1.2 Learning

Learning in these ANNs is just by inspection of the weights. We consider the case when there is only one output class for simplicity. The cases when there are multiple output classes can be handled in similar ways. The ANN architecture consists of one input layer, one hidden layer, and one output layer. The number of neurons in the hidden layer is always equal to the number of training data elements. This means that in these networks, each hidden layer neuron represents a training data element.

The activation function of the neurons is such that it outputs 1 only when the sum of the inputs exceeds 0. This means that as per the working of the neurons, the various inputs that are presented to the output neuron are added with multiplication to their weights. If the number so generated is greater than 1, the output neuron outputs 1. In all other cases, its output is 0.

We add an extra input to the system. This is similar to adding bias in the conventional ANN. This input is supposed to supply the threshold that would be needed to control the output. Remember that the output of every neuron is high only when the inputs sum to a number greater than 0. The value supplied to this neuron is always 1. We control the output by modifying the weight of this input.

Now we start the process of setting up the weights. We first set the weights of the neurons from the hidden layer to the output layer. We know that each hidden layer neuron i corresponds to some data set that has an input $I_i <I_{i1}, I_{i2}, I_{i3},...I_{in}>$ and an output O_i. The output O_i can be either 0 or 1. If O_i is 1, we set the corresponding weight from the hidden neuron to the output as 1, and if O_i is 0, we set this to –1.

Similarly the weight from j^{th} input neuron to the ith hidden neuron is kept to be h if the corresponding j^{th} input in the ith training data set (I_{ij}) is 0; else if I_{ij} is 1, the weight is set to be 1. Here h is a parameter that is usually kept as –1.

We added another additional input to the system. At the end we set its weight to $r - s + 1$. Here r is a parameter that is called as the radius of generalization, and s is the number of 1s in the corresponding input (I_i). The radius of generalization affects the inputs that are at a region r units away from it. This may be easily verified by taking all 0 input ($s = 0$) that gives the weight to be $r - 1$.

In this manner by inspection of the training data set, the weights can be set. This ends the learning.

TABLE 16.3
XOR Problem

Inputs		Outputs
X	Y	Z
0	0	0
0	1	1
1	0	1
1	1	0

In order to further understand the problem, we solve the XOR problem using the discussed procedure. The truth table of XOR showing the inputs and outputs is shown in Table 16.3.

Here we fix the value of the radius of generalization, *r*, to be 0. The final structure of the ANN is shown in Figure 16.7. It may be verified that it gives correct output for all the inputs when presented during testing.

16.5.2 Input Encoding

We know that in these systems we have the restriction that the input has to be binary always. Hence we need to convert the inputs into binary before giving them to this system. One of the easiest ways to do this is naturally to convert the number into the binary format directly. But this does not give good generalization because of the large disparity in the inputs even when they would otherwise be very close. Consider the inputs 7 and 8 that would be represented by 0111 and 1000. It can be seen that the decimal numbers are very close to each other but their binary counterparts are very different.

Hence we adopt a different strategy. We fix the length of the inputs to the maximum input possible. The input now contains a series of 1s. Here the number of 1s is the input that needs to be given to the system minus 1. The rest of the space is filled with 0s. Using this system, 7 and 8 may be represented as (for a system that takes maximum 16 as an input) 000000000111111 and 000000001111111. Here there are six and seven 1s respectively. It may even be possible to reverse the 1s and 0s to form a new style of coding. This manner of encoding the input is called unary encoding.

This other way of encoding is called the quaternary encoding. Unlike binary encoding where the base values were 0 and 1 only, quaternary encoding has four base values. These are 0, 1, *i*, and 1 + *i*. Here *i* is the axis of the complex domain ($i = \sqrt{-1}$). This is thus a much more compact form of representation. In this representation, we again need to look at how close the inputs are to each other

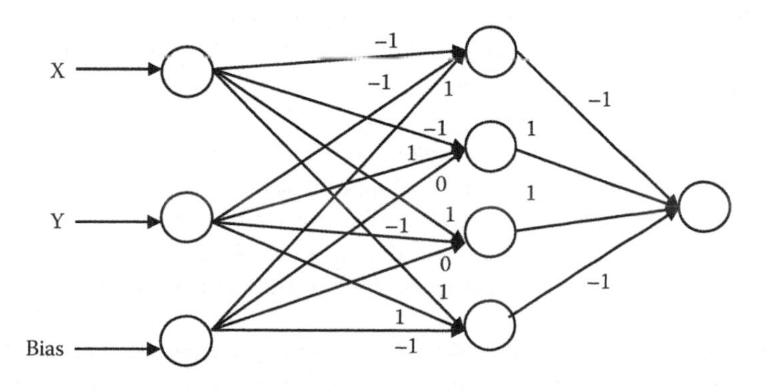

FIGURE 16.7 The network for the XOR problem.

TABLE 16.4
Quaternary Encoding of Inputs

Number	Quaternary Code				
1	0	0	0	0	0
2	0	0	0	0	1
3	0	0	0	1	1
4	0	0	1	1	1
5	0	1	1	1	1
6	1	1	1	1	1
7	1	1	1	1	i
8	1	1	1	i	i
9	1	1	i	i	i
10	1	i	i	i	i
11	i	i	i	i	i
12	i	i	i	i	i+1
13	i	i	i	i+1	i+1
14	i	i	i+!	i+1	i+1
15	i	i+1	i+!	i+1	i+1
16	i+1	i+1	i+!	i+1	i+1

in the numeric domain. We hence start filling the numbers with 1s. Later when all inputs are filled with 1s, we start filling the digits with *i*. When all digits are filled with *i*, we do the same for the $1 + i$. In this way we are able to represent all numbers in a compact form using the quaternary coding. The representations for the case when the maximum input can be 16 are given in Table 16.4.

Let N be the largest number that needs to be represented (in our example case 16). Let L be the length of the codebook. Using quaternary codes, we can clearly divide the entire set of inputs into three regions. The first $L + 1$ set of inputs ranges from 1 to $L + 1$ and consists of only 1s and 0s. The next set of inputs is L in number and ranges from $L + 2$ to $2L + 1$. These consist of only 1 and *i*. Similarly the next set of inputs is L in number and ranges from $2L + 2$ to $3L + 1$. It may be seen that L can be calculated by using Equation 16.13.

$$L = ceil\left(\frac{N-1}{3}\right) \tag{16.13}$$

16.5.3 Modifications for Complex Inputs

We introduced the concept of the complex inputs in the preceding section. In this section we mention the modifications that need to be carried out in the CC4 algorithm to enable it to work with complex inputs. We only discuss the modifications here; the rest of the algorithm remains exactly the same as per our previous discussion.

Just like 0 and 1, the inputs i and $i + 1$ work in pairs. Here i is considered to be a low input (analogous to 0) and $i + 1$ is considered to be a high input (analogous to 1). The first modification is in the weights from the inputs to the hidden neurons. Here if the corresponding input is i, the corresponding weight is fixed to be $-i$. In case the corresponding input is $1 + i$, the corresponding input is fixed to be $1 - i$. It may hence be observed that the weights are complex conjugates of the corresponding inputs.

Recall that the bias was given a value of $r - s + 1$. In this case the value of s is calculated differently. Here the value of $\underline{s}$ is the total number of 1s, total number of *i*s, and twice the total number of $(1 + i)$s. Hence consider that the input is <*1, 0, 0, i, 1 + i*>. For this the value of s would be 4.

The last change is applied to the activation function of the hidden layer neurons. The activation function works over the complex input that is given to it by the weighted input summation. The activation function outputs 1 high (1) only when the imaginary part of the input is 0 and the real part is greater than 0. In all other cases, it outputs low (0). The activation function is given by Equation 16.14.

$$f(Z) = 1, \quad \text{if } Im(Z) = 0 \text{ and } Re(Z) > 0 \qquad (16.14)$$
$$= 0, \quad \text{otherwise}$$

It may easily be seen that whenever any of the input is applied from the training data set, only one neuron receives noncomplex inputs. This is the reason we fix the activation function only to respond to noncomplex inputs. This gives the correct answers to the training data set inputs. The activation function of the output neurons is kept to be same as in the CC3 algorithm.

Again suppose that we apply an input from the training data set. Here each input is multiplied with its corresponding weight; 0 multiplies with –1 to give 0. Similarly 1 multiplies with 1 to give 1, i with $-i$ to give 1, and $1 + i$ with $1 - i$ to give 2. For this reason we take s to be the sum of 1s, is, and twice the sum of $(1 + i)$. The bias with a value of $r - s + 1$ would be added as an additional input to the applied input, and this helps both in getting correct answers to the unknown input and in generalization. Consider the value of r to be fixed at 0. In this case, the total input would be s from the weighted summation of inputs and $-s + 1$ from the bias, summing to a total of 1. Hence the corresponding neuron receives a real high input for all training data set inputs. For some inputs it might be possible for two neurons to be simultaneously active, depending upon the inputs and weights. The summation at the output neuron in that case decides the final output.

Again consider the XOR problem we solved previously with the real inputs. Here we solve the same problem by encoding the inputs in the complex domain. The resulting truth table is given in Table 16.5. The network architecture of the resulting ANN is shown in Figure 16.8. Here r is kept to be 0, since no generalization is needed. The inputs and outputs at the various layers are given in Table 16.6.

16.5.4 Restrictions on Inputs

It was discussed earlier that some inputs may activate more than one hidden layer of neurons. This is an undesirable property for the network. This happens because more than one vector may result in generating a high real output for any hidden neuron vector. For these reasons, we restrict two kinds of inputs from being used in the system.

- If an m element vector with n wholly real high and n wholly real low elements is given as an input, then another m element vector with n wholly complex high and n wholly complex low elements cannot be given as an input, if these high and low elements of both the

TABLE 16.5
XOR Problem with Quaternary Inputs

Inputs		Outputs
X	Y	Z
I	I	0
$i+1$	1	1
$i+1$	$i+1$	1
$i+1$	$i+1$	0

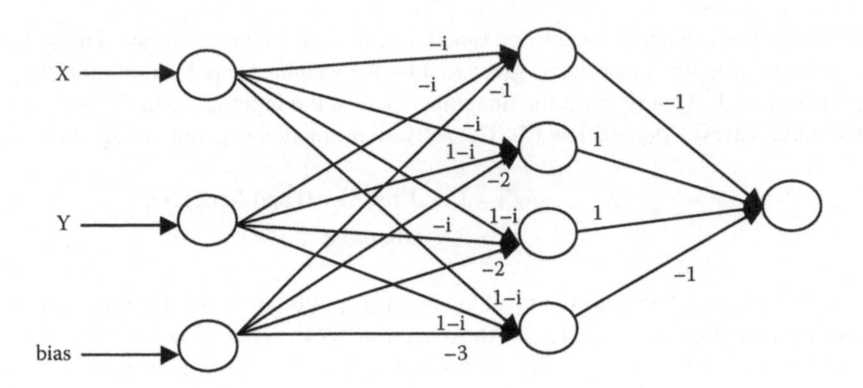

FIGURE 16.8 The network architecture for XOR problem.

vectors appear in the same positions, while all other $m - 2n$ elements of both the vectors are identical. For example, the following pairs are illegal as input: ...1...0... and ....1 + *i*...*i*... , ...*i*...1 + *i*... and ...0...1....

- If an m element vector with n complex high elements is given as an input, n being even, then another m element vector with $n / 2$ real high elements and $n / 2$ complex low elements in the same position as the n complex high elements of the first vector cannot be given as an input, if all the other $m - n$ elements of both the vectors are identical. For example, the following pairs are illegal as input: ...1 *i* 1 *i*... and ...1 + *i* 1 + *i* 1 + *i* 1 + *i* ..., ...1 + *i* 1 + *i* 1 + *i* 1 + *i*... and ...1 *i* 1 *i* ...

16.5.5 ACC Algorithm

The advanced corner classification (ACC) algorithm removes the restrictions on inputs that were placed by the CC4 algorithm. The basic algorithm is same as for CC4. Here we discuss the modifications made to the CC4 algorithm to enable it to work in the restricted inputs as well. The rest of the architecture and working of the ACC is similar to that of the CC4 algorithm.

In this algorithm all inputs are treated as complex. We break every input into a real part and a complex part to decide the corresponding weight. If the real part is 0, then the real part of the corresponding weight is fixed to be –1. If the real part is 1, then the real part for the corresponding weight is also 1. Similar is the procedure with the imaginary part of the input. If the imaginary part of the input is 0, then the imaginary part of the corresponding weight is fixed to be –1, and if the imaginary part of the input is 1, then the imaginary part of the corresponding weight is fixed to be 1.

TABLE 16.6
Inputs and Outputs of Various Layers (Here $H_i = i^{th}$ Hidden Neuron)

Inputs			s	Weights			Input to				Output of				Input to Y	Output of Y
							H1	H2	H3	H4	H1	H2	H3	H4		
i	*i*	1	2	–*i*	–*i*	–1	1	*i*	*i*	–1+2*i*	1	0	0	0	–1	0
i	1+*i*	1	3	–*i*	–*i*1 – *i*	–2	–*i*1 – *i*	1	0	*i*	0	1	0	0	1	1
1+*i*	*i*	1	3	–*i*1 – *i*	–*i*	–2	–*i*1 – *i*	0	1	*i*	0	0	1	0	1	1
1+*i*	1+*i*	1	4	–*i*1 – *i*	–*i*1 – *i*	–3	1-2*i*	–*i*1 – *i*	–*i*1 – *i*	1	0	0	0	1	–1	0

TABLE 16.7
Results of the CC4 Algorithm in the Problem of Picture Learning

Original Image	Training Data (Denoted by I)	Reconstructed Image by CC4
. . . .	. . I . . . . . I . . . . I . .	. . . .
. . . .	. . . .	. . . . . # . # . . . .
. . . . # # #	. I . . . . . # I # . I . I . .	. . . . # # # # # . . . .
. . . . # # # # # # # # # . . .	. . . . I # # # # # # # # . . .	. . . # # # # # # # # . . . . .
. . . # # # # # # # # # # # . .	. . . # # # # I # # # # # I . .	. . # # # # # # # # # # # # . #
. . # # # # # # # # # # # # # .	. . # # I # # # I # I # # # # .	# # # # # # # # # # # # # # # #
. . # # # # # . . . # # # # # #	. . I # # # # . . I # # I # # #	. # # # # # # # # . . # # # # #
. . # # # # . . . . . # # # # #	. . # # # I . . I . . # # I # #	# # # # # # . . . . . . # # # #
. . # # # # . # . . . # # # # #	. . I # # # . # . . I # # # # #	. # # # # # . . . . . # # # # #
. . . # # # # # . . . # # # # #	I . . # I # # # . . . # # I # #	. . # # # # # . . . . . # # # #
. . . . # # # . . . . # # # # #	. . . . I # # I . I . # # # # #	. . # # . # . . . . . . # # # #
. . . . # # # # #	. . I . . . I . . . . # # # # #	. . . . # # # # #
. . . . # # # # #	. . . . I . . # # I # #	. . . . # # # # #
. . . . # # # # # #	. I . . . . I # # # # #	. . . . # # # # # #
. . . . # # # # # # #	. . . . I . . . . # # # # # # #	. . . . # # # # # # #
. . . . # # # # # # # #	. I . . . . . . # # # I # # I #	. . . . # # # # # # #
No. of points: 256 No. of #: 115 No. of .: 141	No. of training data: 37	Total: 256 Correct: 224 Incorrect: 32 Efficiency: 87.5

This scheme of weights and inputs always results in real numbers being generated that go to the hidden neurons. As a result the activation function of the hidden neurons may be simply a binary step function that outputs a high whenever it gets an input greater than 0.

16.6 PICTURE LEARNING

In this section, we use the CC4 algorithm for solving the same problem of picture learning as we had done in the preceding section with the built hybrid algorithm using a neuro-fuzzy architecture. We took some training data elements in the picture and used them as hidden neurons in the network. Then the algorithm was supposed to solve or reproduce the entire picture. The original picture, the training points, and the reconstructed picture by the algorithm are given in Table 16.7. Since we are not showing the entire picture to the algorithm, generalization is needed. We fix the radius of generalization of the algorithm to 3.

Recall that since these are instantaneously trained ANNs, they require much smaller training time but much larger memory for solving the problem as compared to the previously discussed hybrid method or any of the generally known mechanisms. The efficiency of 87.5 percent shows the generalizing and the learning capabilities of these ANNs.

16.7 CONCLUSIONS

In this chapter, we first introduced picture learning. We saw the various phases and mechanisms of picture learning. We even learned about the standard JPEG compression. Also we learned about the various types of soft-computing approaches that can be used for the purpose of picture compression and learning.

Late in the chapter, we developed a model that could solve the classificatory problems. We made an algorithm inspired by the neuro-fuzzy architecture. The motivation was to adapt the neuro-fuzzy architecture in such a way that it can better handle the classificatory problems. We studied each and every step and modified it according to the problem. In this way we were able to evolve a good system for dealing with classificatory problems.

We then applied this developed architecture to the problem of picture learning. A black-and-white picture was taken that could be easily generalized to a classificatory problem where the task was to classify a point in any of the two classes: black or white. We got a good efficiency, 87 percent, which shows the impact of the algorithm. We also compared this efficiency with the efficiency of neural network and conventional neuro-fuzzy system. The difference between the efficiencies clearly shows that our algorithm outperformed the other algorithms.

Later, we also studied another class of algorithms, the corners algorithms. We studied the CC4 algorithm, which was a good means to solve the classificatory problem where all inputs and outputs could be represented in binary numbers. We saw the quaternary manner of encoding the inputs, which was much more compact than its binary counterpart. We studied the manner training is performed in these ANNs by just inspecting the inputs. We even studied the ACC algorithm.

At the end of the chapter, we solved the same problem of picture learning using the CC4 algorithm. We saw that again we were able to solve the problem and we received a good efficiency of 87.5 percent. In this problem we had to fix the radius of generalization to 3 in order to make the algorithm achieve generalization.

We mainly presented a solution to the classificatory problem using neuro-fuzzy architecture. The model can be improved by adapting the mathematical model more according to the problem. Keeping the learning rate variable might also give better results. All this may be seen as an opportunity for the future. Besides, the adaptation of this algorithm to the other classificatory problem needs to be done.

We also studied the corners algorithm later in the chapter. This algorithm needs to be modified to handle the nonbinary inputs and outputs. This would help a lot in increasing the applicability of the algorithm to other areas and applications as well. Also, there must be mechanisms to limit the memory requirement of these algorithms, as in many real life applications the training data set size might be too large.

17 Other Real Life Applications

17.1 INTRODUCTION

Soft computing can never be restricted into a small set of applications or systems. It is an immensely vast field that spans numerous areas and spheres. It covers almost all sections of industry and society that you can possibly imagine. Soft computing has already dominated major application areas and is growing tremendously over all areas. Character recognition, picture learning, and robotics form a small part of the numerous possibilities that soft computing holds. The areas are numerous; the applicability is immense. The ways to apply soft-computing techniques are likewise numerous. This unfolds huge opportunities and potentials in this field that bring people from all domains to contribute in their own way to soft computing.

At the same time, after reading about the theory and applications of soft computing, you would certainly realize that the applications keep changing. The ways to apply the soft-computing methods largely remain the same. They just need to be designed in the right way as per the problem demands. This presents an entirely different view of soft computing, which means primarily achieving an in-depth understanding of the problem and the issues and then realizing a correct design perspective. This presents the soft-computing practitioners and researchers as designers who are schooled and skilled for evolving effective designs for the systems. This is independent of whatever domain the problem comes from and whatsoever the background of the problem may be.

It would not be possible for the authors to cover all the application areas of this vast domain. In this book we have picked up diverse areas of applications. The intention was to introduce the readers to the different applications of soft computing. In this chapter, we give a glimpse of some of the other areas where soft computing has cast its impact.

In this chapter, we mainly explore four other applications that are chosen from real life. The first application is *automatic document classification*. This takes us to the domain of web search, where document classification finds immense applications. Here we would apply Bayes' theorem and ANN for the purpose of classifying documents. The second application that we have chosen is *negative association rule mining*. This comes from the domain of data mining, which is again a very vibrant area of work. Here we use genetic algorithms (GA) for this problem. Later we go into the musical world. Here we study ways of classifying genres automatically using modular neural networks. Then we go to the financial domain. Here we solve the problem of assigning credit ratings by the use of evolutionary ANNs.

17.2 AUTOMATIC DOCUMENT CLASSIFICATION

With the ever-expanding web, the importance and need of effective information retrieval has gained an importance of its own. The expansion has resulted in the search space becoming very vast, needing proper management in order to give the best results in the least possible time. Currently search engines use keywords for retrieving the relevant documents to meet the user demands. In this section, we present means to classify these documents so that they may be grouped together in different categories or genres. The search now may be performed in the closest category, which may result in a tremendous reduction in search time and improvement of the search results. The classes are prespecified into which the documents may map to.

We discuss two methods to solve this problem of automatically classifying documents. The first approach is Bayes' theorem. This is a mathematical model that calculates the probabilities of the input belonging to a particular class. We then study the same problem and solve it using artificial neural networks (ANN). The section briefly explores the work of Kumar and Shukla (2008).

17.2.1 About the Problem

17.2.1.1 Information Access and Retrieval

The bulk of the information available over the Internet has demanded that effective information retrieval systems be developed. This bulk of information is highly dynamic, and we need effective management in order to enable people to get the most relevant information to meet their needs. Another important task in effective information retrieval is drawing relevance from external sources. The increase in web traffic has not been in text only. The web now comprises graphics, multimedia, and documents that also need to be searched and retrieved. A good search engine effectively returns the results present over the web that meet the search query as well as the needs of the user. Spam filtering is one of the common examples, where we try to identify and filter out the unwanted content. There are various standard and specialized applications of machine learning such as information extraction, latent semantic indexing, information filtering, and spelling correction that are extensively used in such applications. The language barriers to information have also been removed by focusing on cross-linguistic information retrieval.

17.2.1.2 Document Classification

The Web is filled with numerous documents that are uploaded and downloaded numerous times by numerous people every day. These documents are in numerous forms and convey a wide variety of information. Document classification can be defined as a document analysis methodology that assigns a document to one (or many) of the predefined classes. It is natural that it would be impossible to find out all possible classes that are enough to accommodate all the available documents. This is because of the wide diversity in the documents. We can still have some classes, and we can classify the document in the closest possible class.

Document classification is formally defined as follows: Given a set of documents D and a set of classes C, a classifier for the class c_I is a function $f_I: D \rightarrow \{0,1\}$ that approximates an unknown function $f'_I: D \rightarrow \{0,1\}$, which expresses the relevance of the documents for the class c_I. The document classification task can be a simple text categorization task, for instance, whether a document belongs to a particular class or not, or it can be a multiclassification task where a document can have membership in one or more classes.

In the search approach with document classification that we discuss, the document classifier becomes a part of the web-based search. The classifier module in the architecture includes a data preprocessing module, a learning module, and the classification module. The user gives the search query, which is processed to give the search results. The search engine makes use of a web crawler to crawl the web, and this returns the relevant web pages or other results. In the suggested approach, the search is class based, where each query is mapped to one or more classes. The search is only performed on the web pages or documents belonging to the selected classes. The general architecture of the system is given in Figure 17.1. This limits the search space, and as a result the search is faster.

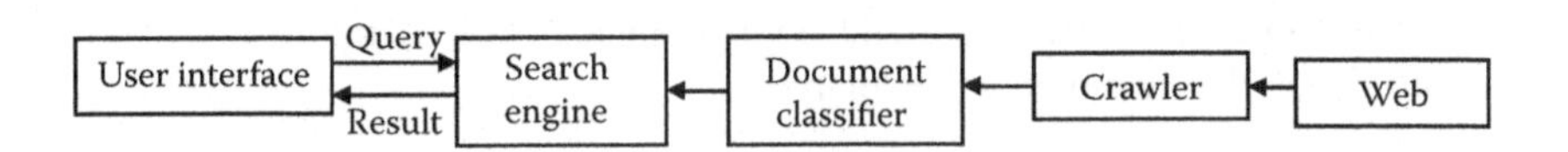

FIGURE 17.1 Document classification supported search architecture.

17.2.1.3 Automatic Document Classification

Document classification is similar in nature to any classificatory problem. As discussed, any classifier mainly uses numerical information or features to classify the various inputs into classes. The documents on the contrary contain textual information in the form of words. We use feature extraction to extract numerical information from the available document. We mainly do this by measuring the word frequencies.

There are two kinds of document classifiers that are used. These are *supervised* classifiers where some external help is available in form of human feedback or any other means and *unsupervised* classifiers where no external information is available. Many classifiers assume that each document can map to only a single class. These are mainly implemented using the supervised approach. Common single-label discrete supervised algorithms used in automatic genre classification studies are discriminant analysis, logistic and multiple regression, and machine-learning-based classifiers, such as C4.5, naïve Bayes classifiers, neural networks, or support vector machines (SVMs).

The identification of features is one of the most important steps in document identification as in any soft-computing system. Then we may use any of the classificatory models for the task of classification. Many classifiers use a content-dominated classification, where word frequencies are measured for classification. This also leads to the development of probabilistic inference models. There are mainly three types of classifiers:

- **Rule-based classifiers:** These classifiers work on rules. The rules may be derived after the training data set or from logic. Some common examples of these classifiers are decision trees, theorem proving, propositional logic, or first- and second-order logic.
- **Linear classifiers:** In these algorithms, we compute a profile for each class. We maintain weights that are used to map inputs to classes. For each class and document, a score is obtained by taking an in-product of class profile and document profile. The common examples of this type are Bayesian classification and heuristical learning algorithms, including ANN.
- **Example-based classifiers:** These classifiers classify a new document by finding the k documents nearest to it in the training data set and doing some form of majority voting on the classes of these nearest neighbors.

17.2.2 Architecture of a Classifier

The document *classifier* consists of both the training and the testing phases. In the training or the learning phase, the system is presented with inputs and known outputs. The system is supposed to fix its parameters or learn. In the testing phase, the system is given unknown inputs and is asked to classify these inputs into classes. Any data needs to be preprocessed before being given to the system. The general architecture is given in Figure 17.2.

17.2.2.1 Preprocessing

This is one of the major steps involved in the system that removes noise and extracts the correct context out of the data. The classification context here refers to the features used for classification. The features would depend upon the context in which the classifier works. This includes (1) a content-based classification that classifies on the basis of the content present in the web page or document, (2) a link-based classification that uses the incoming and outgoing links as a clue to the kind of page or for classification, or (3) using other external evidence such as anchor text. The preprocessing is used to filter out

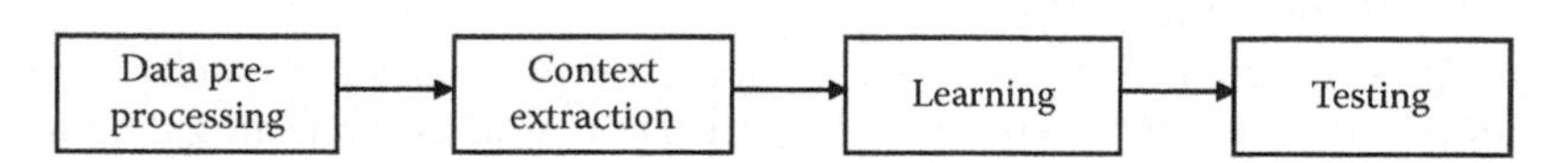

FIGURE 17.2 The general architecture of a classifier.

the irrelevant content; e.g., if we are using only a content-based classification, then we have to filter out the tags and extract the content in HTML web pages. Similarly, in case of content- and link-based classification, the incoming or outgoing links are to be extracted along with the content.

17.2.2.2 Learning

The learning in such systems varies from implementation to implementation. It may be done in supervised or unsupervised mode. Every implementation may have its own learning algorithm. The Bayes algorithm uses probabilistic learning. ANNs use the back propagation algorithm (BPA).

17.2.2.3 Outputs

The classifier system so developed may allow the document to only map to a single class (in case of mono classification systems), or it may allow documents to map in more than one class (in case of multiclassification systems). In the former case, the output of the classifier is just the class to which the document maps. In the latter case, the output of the class is a set of probabilities. Here each probability denotes the belongingness of the input to that class. Hence there are as many probabilities as there are classes. One of the ways of implementing a single-output classifier is to find the maximum probability of the outputs and return the corresponding class.

17.2.3 Naïve Bayes Model

This section aims to approach the classification problem with a probabilistic model approach. Here we develop a naïve Bayes model to classify the documents. The model goes through the training and testing phases. A probabilistic approach ensures that the document as a whole (content) is taken into consideration during classification, which reduces bias.

The model makes use of the Bayes' theorem with strong (naïve) independence assumptions for calculating the probabilities. The training phase adjusts the various parameters of the Bayes model to maximize efficiency. In spite of their naïve design and apparently oversimplified assumptions, naïve Bayes classifiers often work much better in many complex real-world situations than one might expect. An advantage of the naïve Bayes classifier is that it requires a small amount of training data to estimate the parameters necessary for classification.

17.2.3.1 Bayes' Theorem

Bayes' theorem is a way of inverting a conditional probability. It is given by Equation 17.1.

$$P(x|y) = P(y/x)\frac{P(x)}{P(y)} \tag{17.1}$$

This defines the probability of occurrence of x given the occurrence of y.

Let $P(c)$ be the probability of any document belonging to class c and $P(X)$, the probability of the vector X occurring. Let $X = (a_1, a_2, ..., a_n)$. Let us also assume $P(X)$ to be the same for all label values. It may easily be observed that this probability is proportional to $P(a_1, a_2, ..., a_n/c).P(c)$.

Assuming independence of attributes, we may take the probability proportional to Equation 17.2.

$$\prod_{j=1}^{n} P(a_j/c).P(c) \tag{17.2}$$

17.2.3.2 Classification Algorithm

Here the algorithm is presented that does the complete task of learning as well as classification. The algorithm takes as its input the training data set (D) as well as the testing data set (D_T). The algorithm first trains the system. Then this system is used for the purpose of classification. The final result is the classified documents as per the classification algorithm.

The first task is that of the training. Here the algorithm first parses the training data set D. Then the various words are identified and added in the word frequency database. The next task the algorithm demands is to calculate the probabilities associated with each word for belonging to any category c. We need to calculate another probability here. This is the absolute probability of the category over the entire document. Both these probabilities may be easily computed by using simple conditional probability laws. These probabilities denote the system or the classifier. These are saved for the reference and working of the classifier.

The next task is the classification itself by using the classifier that we obtained in the preceding step. This classifier works over the testing data set (D_T). Here for all the documents in the testing data set, we calculate the conditional probability given by $\Pi_{j=1}^{n} P(a_j/c).P(c)$. Here a_j represents a word, and we calculate the product of probability of occurrence of each word with the probability of that particular word belonging to a category. The higher the probability value for different categories, the more probable it is that the document belongs to that genre. Using the same probabilities, we then generate the ordered list of categories that it might belong to for each document d_i. This is the output of the classifier.

In text classification, our goal is to find the best class for the document. The best class in naïve Bayes' classification is the most likely or maximum a posteriori (MAP) class c_{map}. This is given by Equation 17.3.

$$c_{map} = \text{argmax } \hat{}P(c|d) = \text{argmax } \hat{}P(c) \, \Pi \hat{}P(a_j|c) \tag{17.3}$$

We write ^P for P because we do not know the true values of the parameters $P(c)$ and $P(t_k|c)$ but estimate them from the training set, as we will see in a moment.

Equation 17.3 may even be represented in another form given by Equation 17.4. This avoids any possibility of floating point underflow due to the multiplication of a large number of probabilities that all lie in the interval of 0 to 1.

$$\text{cmap} = \text{argmax } [\log \hat{}P(c) + \Sigma \log \hat{}P(a_j|c)] \tag{17.4}$$

The preceding equation has a simple interpretation. Each conditional parameter $log(\hat{}P(a_j|c))$ is a weight that indicates how good an indicator a_j is for c. Similarly, the prior $log \hat{}P(c)$ is a weight that indicates the relative frequency of c. More frequent classes are more likely to be the correct class than infrequent classes. The sum of log prior and term weights is then a measure of how much evidence there is for the document being in the class, and the equation selects the class for which we have the most evidence.

17.2.3.3 Implementation Procedure

Step 1: Extract the content from the HTML document by stripping the HTML tags.
Step 2: Parse the text to calculate the word frequencies in a document.
Step 3: Calculate the total number of words in the document.
Step 4: Calculate the probability of the occurrence of word in the document, $P(x/y)$.
Step 5: Also calculate the probability of the document in the class c (genre).
Step 6: Use the training set documents to train the naïve Bayes model.
Step 7: Feed the new document to the model, to know the ordered list of most probable category for that document.

17.2.4 Artificial Neural Network Model

The second method discussed in this section is the artificial neural networks (ANN) with back propagation algorithm (BPA). Here the identification of the inputs is a little more difficult, and that makes the ANN more complex. The ANNs are very sensitive to the number of inputs, as the rise in the number of inputs rapidly increases the system complexity. Hence we cannot pass all the extracted words to the ANN as input. We need to do some intelligent feature extraction. The approach with the neural network model is to extract major concepts from the corpus that are representatives of the prospective classes in the corpus. Besides, ANNs do not understand text; all they need is a numeric input. Here we have taken a concept extraction–based approach to neural classification.

17.2.4.1 Concept Extraction and Classification

Concept extraction is a method by which we try to identify and extract only some major words or topics out of the entire document. The whole document should be able to generalize to these set of words. If the corpus only contains documents about information technology, then database, soft computing, networks, and communication systems can be considered as concepts representing the data set. Each concept or extracted word may be closely related to many words, which can be collected from any thesaurus or by similar lookup, for example, *<Soft Computing> = {Artificial Neural Networks, Genetic Algorithms, Fuzzy Logic …}*. Such a thing, however, needs to be done automatically. In this experiment, we extracted the concepts from the Reuters-21578 corpus. Reuters-21578 is a set of news articles from Reuters, and the data set consists of five major classes: *<Topics, Places, People, Exchanges, Organizations>*. These classes also contain a lot of subclasses, for example, *<Topics> = {grains, wheat, ..}, <Places> = {USA, Australia, UK ..}*.

These subclasses can most likely be used as representatives of the whole corpus. However, it may not always be true that if a word occurs, then the document is related to the associated class. Consider that the <Topics> field contains <D>USA</D>; it cannot be assumed, however, that the topic is associated with the USA. Thus, we extracted all the subclasses present in all the superclass tags and considered them as the representative concepts. These concepts act as input to the neural network; the concept values can be quantified by word counts of the subclass in the corresponding documents. For example, if USA occurs 12 times in D1, wheat occurs 4 times, and NSE occurs 1 time, the inputs to neurons in the first layer would be 12/17, 4/17, and 1/17. The inputs are normalized before feeding them to the network.

17.2.5 Experiments and Results

The discussed Bayes and ANN approaches were validated by testing against a standard data set. The data used for experimentation is a standard Reuters-21578 collection. The documents in the Reuters-21578 collection appeared on the Reuters newswire in 1987. The documents were assembled and indexed with categories by personnel from Reuters Ltd. and Carnegie Group, Inc., in 1987. There are multiple categories, the categories are overlapping and nonexhaustive, and there are relationships among the categories. The five categories are: *<Topics, Places, People, Exchanges, Organizations>*.

The Reuters-21578 collection is distributed in 22 files. Each of the first 21 files (reut2-000.sgm through reut2-020.sgm) contains 1,000 documents, while the last (reut2-021.sgm) contains 578 documents. The files are in SGML format.

Each article starts with an "open tag" of the form *<REUTERS TOPICS=?? LEWISSPLIT=?? CGISPLIT=?? OLDID=?? NEWID=??>*, where *??* are filled in an appropriate fashion. Each article ends with a "close tag" of the form *</REUTERS>*. In all cases the *<REUTERS>* and *</REUTERS>* tags are the only items on their line.

The *TOPICS* categories are economic subject categories. Examples include *"coconut," "gold," "inventories," and "money-supply."* The *EXCHANGES, ORGS, PEOPLE,* and *PLACES* categories correspond to named entities of the specified type. Examples include *"nasdaq"* (*EXCHANGES*), *"gatt"* (*ORGS*), *"perez-de-cuellar"* (*PEOPLE*), *and "australia"* (*PLACES*). For the *LEWISSPLIT* tag, the possible values are *TRAIN, TEST,* and *NOT-USED. TRAIN* indicates it was used in the training set in the experiments conducted by Lewis. *TEST* indicates it was used in the test set

TABLE 17.1
Results of the Two Classification Techniques

Classification Technique	Accuracy
Naïve Bayes	75%
Back propagation	84%

for those experiments, and *NOT-USED* means it was not used in those experiments. Similarly for the *CGISPLIT* tag the possible values are *TRAINING-SET* and *PUBLISHED-TESTSET*, indicating whether the document was in the training set or the test set for the experiments conducted by Hayes.

After all the test documents have been classified, they can be compared with their class tags to see if they have been classified correctly or not. We classified the documents using both naïve Bayes and back propagation neural network models. The classification results are shown in Table 17.1.

17.3 NEGATIVE ASSOCIATION RULE MINING

Association rule mining is a very active area in the field of databases. Here we try to extract the rules or patterns out of the given database. The motivation behind these systems is to predict the future trends and maximize the future profits of a company by making the most effective decisions. This is done by a study of the historical trends already stored into the database, which emphasizes on the importance of data. The rules that we extract in the case of association rule mining are of the form $X \rightarrow Y$. This states that if a person buys the items or item set X, then he is very likely to buy the item or item set Y. This information or rule mined from the database is of value to the company or store. This may result in the company keeping the item set Y near the item set X. It may also result in various special offers or discounts being decided by the company.

In this section, however, we deal with the negative rules. Here the rules are of the form $X \rightarrow \neg Y$. This means that if a person purchases the item set X, then he is likely not to purchase the item set Y. This negative rule mining also has a special relevance to the associative rule mining and leads to a better understanding of the historical trends.

The solution that we build in this section is over genetic algorithms (GAs). The GAs can be very effective tools for solving such problems because of their great optimization powers as well as their capability to find the global optima without getting stuck at some local optima. This further means that we would be able to get quality results in a finite span of time. Most of the other methods would take days altogether to mine any real life databases. The iterative approach followed by the GA is a great boon in such situations.

The use and role of such mining algorithms is especially important because of the explosion of data that has occurred in the past few years. Imagine any supermarket or store these days. The amount of data that they store within one year is immense. As a result, the mining algorithms have to be very scalable to work on massive data that is collected over the years. The high volume of data is useful, as it leads to better results that are very noise resistant. This, however, results in the traditionally used algorithms taking a painfully long time. In most cases, the algorithms fail to give results even when executed for days on end. The field of databases that deal with the extraction of knowledge is known as KDD, or knowledge discovery in databases. Knowledge in this context stands for the summarization of the trends, rules, or other facts that are present in the database in usable form. This knowledge discovery can be done in various ways, such as decision trees, association rule mining, or Bayesian classifiers.

17.3.1 Association Rules

As already stated, the association rules take the form of $X \rightarrow Y$. Suppose that there is a database (say D) containing a total of N transactions. The rule states that if a transaction contains X, then it is most likely to contain Y as well. To define how likely is it for the rule to hold, we define a term called a *support*. The support for the rule is the probability that X and Y both hold together among all the possible transactions. We define another related term to understand the rule. This is called *confidence*. Confidence is defined as the percentage of transactions containing Y in addition to X with regard to the overall number of transactions containing X. This probability can be represented as conditional probability $P\ (Y\ \varepsilon\ T\,/\,X\ \varepsilon\ T)$.

Normally the most studied rules are positive rules that express relations stating the likehood of purchase of products given some conditions. This is essentially useful if the store manager or other authority wants to know or predict the products likely to be purchased. The positive rules are widely studied and easy to visualize and implement. The other type of rules that we study here consists of negative rules. These express relations such as that if a customer buys X, then he is not likely to buy Y. Such relations further help in devising marketing strategies and other related applications. These rules are slightly more difficult to visualize, implement, or discover. As a result, negative rules are still not very widely studied.

Consider Table 17.2, which shows synthetic data for some store. We assume that a minimum support of 30 percent and a minimum confidence of 70 percent is needed for every rule. Here the age has been classified in two categories. Category 1 comprises young people with age less than or equal to 30. Category 2 comprises old people with age greater than 30.

Item sets that satisfy the criterion of minimum support are shown in Table 17.3. The rule that satisfies both the minimum support and minimum confidence criteria is $age < 30 \rightarrow item = pen$. The confidence of this rule is 75 percent.

In case the negative rules are mined, then another negative rule may be worked out. This rule is age > 30 $\rightarrow \neg$pen, which has a confidence of 83.3 percent.

Hence we see from our previous discussion that the negative rules convey meaningful and useful information. However, negative rules are much difficult to mine than positive ones. The major problems come in setting the values of the minimum support and the minimum confidence that would yield good and effective results. The other major problem that usually comes is the fact that the databases usually contain numerous items. Of these, many items either are not present in the transactions or present very insignificant numbers of times. An absent item is defined to be the one that occurs an insignificant number of times in the transaction set. The negative rules containing a mixture of absent and nonabsent data items would be numerous in number and would not give any

TABLE 17.2
Store Purchase Information

S. No	Name	Age	Item
001	A	15	Pen
002	B	35	Paper
003	C	48	Magazine
004	D	24	Pencil
005	E	75	Pencil
006	F	24	Pen
007	G	33	Paper
008	H	10	Pen
009	I	80	Pen
010	J	72	Pencil

TABLE 17.3
Large Item Sets with Minimum Support of 30 Percent

Item Set	Support
Age ≤ 30	40%
Age > 30	60%
Pen	40%
Pencil	30%
Age ≤ 30, Pen	30%

additional information. Hence the mining algorithm must avoid these items and try to find negative rules from the effective items.

17.3.1.1 Formal Definition of Negative Association Rule

A negative association rule is an implication of the form given in Equation 17.5.

$$X \rightarrow \neg Y \tag{17.5}$$

The same rule may even be represented by the form $\neg X \rightarrow Y$, which also conveys the negative association rule between X and Y. It may be seen, however, that the expression $\neg X \rightarrow \neg Y$, which contains negation at both sides of the rule, is a positive rule, as it may be converted to a general positive rule. We define the term support for the negative rules much as we did with the positive rules. A rule is said to have a support of s if s% of transactions contain item set X but do not contain item set Y. Similarly the confidence is defined for these rules. Let U be the set of transactions that contain all items in X. The rule is said to have a confidence of c%, if c% of transactions in U do not contain item set Y.

17.3.2 Application of Genetic Algorithms

The problem of negative rule mining has been solved by the application of GA. GAs have been used because of their potential to solve such problems in finite time while giving the best possible results. They search the entire search space for a solution. The convergence happens at the search space where the best fitness or the best solution is found. This is mostly the global minima. The GA is known to escape the local minima, which are the problem with most of the algorithms. Also the iterative nature of the GA enables it to generate quality solutions very early that may be used for implementation purposes.

We study the major steps of the GA one by one.

17.3.2.1 Individual Representation

Representation of the individual is one of the major steps in GA. The individual representations denote the encoding of the rules for individuals in the population. There are two major kinds of representations of the individual: Michigan and Pittsburgh. In the Michigan approach, all individuals in one population represent one solution. Here each individual is a collection of several blocks that together denote a condition that is the left-hand side of the rule. This also has a conclusion that is the right-hand side of the rule. In this method the individual is represented as follows:

<110|0001|1001> → <011>

Here each separator stands for some condition that is joined with the use of the AND operator. The various conditions within their region may be encoded as per their requirements; e.g., the condition *age > 30* may be simply encoded by the binary equivalent of the number 30. The presence of a characteristic may be shown by the binary digit 1, and the absence may be shown by the binary digit 0. In this way, all the building blocks of the rules may be encoded in the genetic individual. These are joined by concatenation. The same process may be used for the concluding part.

The preceding example may be depicted as "it can walk, it can jump, but it cannot fly AND it barks AND it is 90 cm long—it is a dog." Each digit in first expression, 110, stands for the presence (denoted by 1) or absence (denoted by 0) of the characters walk, jump, and fly taken in the same order. Similarly the next set denotes other characteristics. The last set is the binary equivalent of the length of tail. The right-hand side of the rule is the numeral depiction or class of the animal.

The other commonly used approach is the Pittsburgh approach. This approach closely follows the standard functions of the GA. Here the solutions are represented by individuals. The whole algorithm operates in generations where the best and the average fitness of the individuals in a population keep improving. Always the fittest solutions pass from one generation to the other. The passage of individuals from one generation to the next is through reproduction, carried by the crossover operator that tries to generate good solutions. Here an expression "the head is round and the jacket is red, or the head is square and it is holding a balloon" can be encoded as

(<S=R> & <J=R>) (<S=S> & <H=B>).

Using this approach, however, longer individuals are formed that make computation very difficult in terms of search space as well as the calculation of the fitness function. In addition, it may require some modifications to standard genetic operators to cope with relatively complex individuals. In the Michigan approach, the individuals are simpler and syntactically shorter. This simplifies the whole process and makes it time effective. However, in the Michigan approach as the individuals represent individual rules, it is not possible to judge the quality of the entire solution set.

Here we use Michigan's approach for the representation of the individual. In order to understand this approach, consider the general rule "If a customer buys pen and pencil, then he will also buy paper." We can simply write this rule in the following format:

If pen and pencil then paper

We encode this rule by means of the Michigan approach. The rule can be represented as **00** 01 **01** 01 **10** 01, where the bold d_i-digits are used as product id, like 00 for pen, 01 for pencil, and 10 for paper, and the normal d_i-digits are 00 or 01, which shows absence or presence, respectively.

17.3.2.2 Genetic Operators

The second important task in the GA is the genetic operators. Here we use all conventional genetic operators for the sake of finding the optimal solution of the problem. These are selection, crossover, and mutation. Roulette wheel sampling is used for selection. The single-point crossover is applied.

17.3.2.3 Fitness Function

The *fitness function* decides the goodness of a solution or a rule. The GA primarily does the task of optimization of the fitness function. The requirement of a good rule is that it must have a high predictive accuracy; it must be comprehensible; and it must be interesting—thus choice of this function is very important to get the desired results.

Consider that the rule is of the form "*If A Then C*," where *A* is the antecedent and *C* is the consequent preceded class. In order to study the effectiveness of this rule, we make the use of a simple 2×2 matrix that is called the confusion matrix. This matrix for the rule is given in Figure 17.3.

		Actual Class	
		C	Not *C*
Predicted Class	*C*	TP	FP
	Not *C*	FN	TN

FIGURE 17.3 Confusion matrix for a classification rule.

The labels in each quadrant of the matrix have the following meaning:

TP = True Positives = Number of examples satisfying *A* and *C*
FP = False Positives = Number of examples satisfying *A* but not *C*
FN = False Negatives = Number of examples satisfying *C* but not *A*
TN = True Negatives = Number of examples neither satisfying *A* nor *C*

A good rule has higher values of TP and TN and lower values of FP and FN. The fitness of a rule is measured by the confidence factor. This is given by Equation 17.6.

$$\mathrm{CF} = \mathrm{TP}/(\mathrm{TP} + \mathrm{FP}) \tag{17.6}$$

Completeness is the next important thing in the fitness function. This measures the proportion of examples having the predicted class *C* that is actually covered by the rule antecedent. The rule completeness measure, denoted Comp, is computed by Equation 17.7.

$$\mathrm{Comp} = \mathrm{TP}/(\mathrm{TP} + \mathrm{FN}) \tag{17.7}$$

The net fitness function takes into account both the factors that are obtained by Equations 17.6 and 17.7. This is given by Equation 17.8.

$$\mathrm{Fitness} = \mathrm{CF} \times \mathrm{Comp} \tag{17.8}$$

We can even add a comprehensibility measure to this fitness function. A simple approach is to define a fitness function as given by Equation 17.9.

$$\mathrm{Fitness} = w_1 * (\mathrm{CF} * \mathrm{Comp}) + w_2 * \mathrm{Simp} \tag{17.9}$$

Here, *Simp* is a measure of rule simplicity (normalized to take on values in the range 0 to 1) and w_1 and w_2 are user-defined weights. The *Simp* measure can be defined in many different ways, depending on the application domain and on the user. In general, its value is inversely proportional to the number of conditions in the rule antecedent—i.e., the shorter the rule, the simpler it is.

17.4 GENRE CLASSIFICATION USING MODULAR ARTIFICIAL NEURAL NETWORKS

Music lovers can well explain why music is so captivating and such a growing area of interest. The volume of music generated over the years has now grown to a size that it would not be possible for any person to arrange all the volumes or classify them into genres, artists, and so on. This is true even though this classification of music is always carried out in music stores, music channels, or even personal collections. People prefer to keep music in categories so that they may easily access the desired music according to their mood and interests. The massive and ever-increasing burst of music emphasizes the need for systems that can automatically classify it.

One of the most important issues in music is that it comes from various lands; in various tastes, styles, and languages; and from different artists. This makes any collection of available music not only massive in size but also very diverse in nature. The large volume of music can be easily observed in any prominent music store or music company. The immense number of types and varieties of music leads to problems with most classifiers, which are not scalable enough to handle such large volumes of data.

Music, just like any signal processing application, represents a very high dimensionality of data. Good-quality music is recorded in a way that fits numerous frames per second, with each frame consisting of high-dimensional data. This greatly multiplies the volume of data and results in large volumes of data being available for the system to handle. The better the system handles this volume of data, the better is the scalability of the algorithm.

This large dimensionality motivates the use of good feature extraction algorithms that can be representative of the data without occupying much space or consuming much computational time. These play a vital role in deciding the scalability and hence the applicability of the algorithm. The ability of the system to handle large volumes of data is yet another capability that any algorithm needs to have. The algorithm must be able to learn as early as possible, as in this case it may not have enough computational time for the training to go in numerous epochs. The learning has to produce good results in the fewest possible cycles.

The application of these systems is immense. They may be used to automatically classify the music by genre. This eliminates the need for humans to do so and hence makes classification possible. People can hence listen to or search for the music they want to listen to without the need of long searches. The automatic classification also results in the desired music being automatically delivered or played to the interested parties without the need of specifying a playlist or selecting songs manually.

In this section, we propose the use of ensemble that is a modular ANN for the task of classification. The ensemble contains a number of classifiers that are all presented with the input. These are used to solve the problem of classification and return the result. The results then get collected and a majority vote is performed. The class that receives a majority of votes in this step is regarded as the final output.

The ensemble is good way of dealing with such problems of huge training data set size. Ensembles are easily able to gain from the parallel architecture that they have. Many tasks can be performed in parallel in a modular manner. This modularity that is built into the architecture of the MNN enables it to handle large volumes of data and learn in a finite time. The ensemble deploys various ANNs for this purpose. In other words, it may be visualized that the whole task can be divided between the different ANNs in a modular fashion.

We mainly discuss the work of Bergstra et al. (2006), where they used ensembles as the classifier along with numerous features for the inputs for the system. They had given a system that was scalable with the inputs of MIREX 2005 international contests in music information extraction and won first prize for genre identification and second prize for artist recognition there.

In this section we also first discuss some of the concepts of genre identification. Then we discuss the various approaches used for feature extraction and classification in order to solve the problem in literature. Then, we discuss some of the features that need to be extracted in this problem. At the end, we discuss the ensemble that does the work of classification.

17.4.1 Genre Identification

The genre identification is the task of identifying the genre or the style to which the music belongs. Here the genre refers to the various classes or styles that any person may be easily able to distinguish on listening. This may be folk, jazz, rock, etc. The identification seems to be very simple for human beings. This is because of their ability to easily distinguish one category from the other. For this, people may generally not require long training sessions or learning effort. However, the case of the machines is different, where the identification of the genre is quite a difficult task, especially when the amount of data is large.

The whole task of genre identification may be easily divided into three steps. The first step is the extraction of the features from the audio frames. Here we do an analysis of the sound to identify the various features. The next step is the aggregation of the extracted features to express them in more compact form, at the segment level. These make the algorithm scalable for the input size requirements. The third and the last stage is classification. Here the algorithm does the actual classification to map every input to a genre or a corresponding class. All these are discussed one after the other.

17.4.1.1 Feature Extraction

In *feature extraction*, we take an audio frame and extract the features from it. The audio frame represents a highly dimensional input that would amount to a large number of bits if used directly. It is natural that any ANN would not be able to take as input such a high amount of data. This necessities the need of intelligent feature extraction mechanisms that can reduce the dimensionality of the input, at the same time not resulting in much loss of information that may be useful for classification purposes.

Numerous techniques of feature extraction have been proposed and used in many real life applications. Here we list a few major ones. Fourier analysis is one of the most widely used techniques. This technique breaks up the signal into its corresponding frequencies and gives the frequency coefficients. We chose only a few higher coefficients. The other, lower coefficients are not very important and may be neglected. The original signal can almost be constructed with the help of these extracted coefficients. The higher the coefficients we extract, the better the information content of the extracted features.

A similar class of feature extraction is the cepstral analysis. This includes the mel-frequency cepstral coefficients (MFCC), which are similar in concept to the Fourier analysis. Another class of feature extraction may be done with wavelet coefficients. Autocorrelation is another statistical method of feature extraction out of the sound signal. The other features include zero-crossing rate, spectral centroid, spectral rolloff, and spectral spread.

The frames of data from which these features are extracted are usually restricted to short frames of about 50 milliseconds in order to avoid loss of information, which may occur in case of fine timescale structures like the timbre of specific instruments.

17.4.1.2 Aggregation

Aggregation follows feature extraction and does the same task of limiting the dimensionality of the inputs that would lead to better scalability as well as learning. This happens by the compression of the features that were extracted from the preceding step. We hence limit the size of the features for the classifier.

One of the approaches is to directly compress the features by representing them at the song level. The mixture of Gaussians is an approach that is commonly followed. The other way in which this task is done is to leave the features at the frame level itself and to implement a classifier that classifies the frame-level features. A majority vote might be implemented in order to get the final classified result. This adds the modularity feature to the system. The other option is to aggregate the features of the frame over a segment that is smaller than the entire song but larger than the frame. This is called as a windowing technique where the size of the window is the length over which the aggregation is being carried out. The variation in this window size would vary the speed as well as the efficiency of the algorithm. The window size may hence be fixed accordingly.

17.4.1.3 Classification

The last job in the problem of genre classification is to map the inputs to some output class or genre. The classification is done by a classifier that analyzes the input and, using the information stored in its database, decides the class to which the given input would map. Here again a variety of classifiers may be used. A few commonly used methods are k-nearest neighbors, linear discernment analysis (LDA), mixture of Gaussians, and regression trees. We have more or less covered each one of these methods in some of the other chapters. The other classifiers include the support vector machine (SVM), ANNs, and Bayesian networks.

17.4.2 Extracted Features

Here, we mainly discuss the features extracted for the automatic classification of the songs into their genres. The inputs or the features extracted are given and explained here:

- **Fast Fourier transform coefficients (FFTCs):** As discussed in the preceding section, the Fourier analysis does the frequency breakup of the signal, and the most prominent frequencies out of all the decomposed frequencies are selected.
- **Real cepstral coefficients (RCEPS):** These coefficients try to separate the vocal excitation from the effects of the vocal tract. RCEPS is defined as *real(F_(log(|F(s)|)))*, where *F* is the Fourier transform and *F_* is the inverse Fourier transform.
- **Mel-frequency cepstral coefficients (MFCC):** These features are similar to RCEPS, except that the input x is first projected according to the mel scale, which is a psycho-acoustic frequency scale on which a change of one unit carries the same perceptual significance, regardless of the position on the scale.
- **Zero-crossing rate (ZCR):** This is the number of times or the rate at which the signal crosses the x-axis or the time axis.
- **Spectral spread, spectral centroid, spectral rolloff:** Spectral spread and spectral centroid are measures of how power is distributed over frequency. Spectral spread is the variance of this distribution. The spectral centroid is the center of mass of this distribution and is thus positively correlated with brightness. Spectral rolloff is the a-quantile of the total energy in |Fs|.
- **Autoregression coefficients (LPC):** The k linear predictive coefficients (LPC) of a signal are the product of an autoregressive compression of the spectral envelope of a signal.

Once the features have been extracted, the next job is to aggregate these features over the segment. For this, nonoverlapping blocks of m consecutive frames are grouped up into segments. Each segment is summarized by fitting independent Gaussians to the features (ignoring covariance between different features). This forms the input to our classifier system, which we study next.

17.4.3 Classifier

The next important aspect of the system is the *classifier*. This takes as its input the compressed inputs that we obtain from the previous steps. The output of the classifier is the precise class to which the input belongs. Here an ensemble is implemented for the task of classification.

The inputs are given to the individual ANNs that act as weak learners. These weak learners get a part of the database to learn. The training is carried out in supervised mode where the input and outputs are supplied by the database. After the training with the supplied data, any ANN changes its state to the weakly learned state. Here it can be used for testing purposes. The ANN has as many inputs as there are classes. The output for any class is ideally –1 if, according to the ANN, the input does not belong to that class. If, according to the ANN, the input belongs to that class, the output is 1. In practical terms, the class with the maximum output is selected and declared as the winner.

The final output is calculated using a majority rule. All the classes vote their output to a central counter that uses the fundamental voting technique to find the final output class. In this way if some ANN was buggy, the errors can be rectified by the other ANNs. Using this system, the problem of classification can be easily solved by using weak learners as the ANNs.

17.5 CREDIT RATINGS

We now move into the world of financial analysis, where we will develop models for credit ratings. In order to do so, let us first briefly understand how the financial world moves. We then develop an evolutionary neural model for solving the problem of credit ratings.

Firms require capital in order to start projects, or for any activity within their jurisdiction. It is likely that a firm does not have the needed capital on hand at the outset of a project. For this reason, the firm will typically raise its shares, which people purchase for cash. The firm may even raise debt capital for the same purposes. The bond is defined as a debt security, which acts as a promise from the side of the issuing firm that it will pay the standard rate of interest based on the face value of the bond and will redeem the bond at its face value on maturity. In order to enter into shares or debts, firms usually acquire a credit rating, which represents to what extent the firm is worthy enough to acquire credit. This factor is high if the firm is believed to be very trustworthy and the credit is likely to result in a profit.

This rating is very useful for the firm as well as the investors. Any person investing in a firm will look at its credit rating. The credit rating denotes the probability of the company making the interest payments and maturity of the bonds. The credit rating goes a long way in fixing the prices of the shares. Further any bank or firm making any investment in a project will consult the credit ratings to know the possible returns. These shares are traded in the stock markets.

Here, evolutionary ANNs are used to solve this problem. The credit information is easily available for numerous companies. This provides plenty of training data for the system to be built. Further since the system is much like any natural system that depends in a characteristic way on its inputs, the use of ANNs is natural under such circumstances. The ANNs are used because of their vast modeling powers and ability to generalize the historical data over unknown inputs.

In this system, however, numerous factors may affect the output. A fully connected ANN might take too long to train itself. Also the ANN may not be able to reach the global minima and get trapped in some local minima. For these reasons, the ANN is used in combination with the GA. The GA not only builds the correct architecture for the system but also searches for the best combinations of weights that give an optimal solution. The optimization of the GA in terms of architecture and weights aids in the evolution of an ANN that reaches global minima. The GA further restricts the number of connections, saving the ANN from being excessively complicated or having a large number of inputs. The large number of inputs and their abstract nature can only be handled with an effective use of GA along with the ANNs.

This section uses the work of Perkins (2006), where he uses GA with MLP to develop the credit rating system. Here we first present the fundamentals of credit risk ratings or bond ratings. We then present the system using evolutionary ANNs for solving this problem.

17.5.1 CREDIT RATING

The *credit rating* is the rating that is given to a company or a firm and that denotes how worthy the company is of getting credits from the market. This denotes the probability that the company will be able to make best use of and generate income from the money that it takes from the market. The credit rating hence decides the nature and the condition of the company to sustain in the market and achieve its goals. The credit rating as discussed previously carries a lot of meaning and importance to the company as well as to the investors. This plays a key role in the regulation of the market of the company.

The credit rating is usually denoted by discrete and nonoverlapping grades. Here each grade has a distinct score attached to it. The grades are usually denoted alphabetically and may be converted to their decimal equivalents whenever required. An S&P credit rating consists of ten distinct grades. The grade AAA is the strongest grade, and the grade D is the weakest grade. Ratings between AAA and BBB are considered to represent investment-grade ratings. A C grade stands for bankruptcy, and a D grade represents a case where the borrower is currently in default on its financial obligations.

The credit rating is provided by the credit rating agencies on demand by the firms. The credit rating agencies assign a rating or grade to the company in return for a fee after analyzing the various documents and condition of the company. The rating company deeply studies all the financial documents, financial statements, previous debts, and so forth, in order to assign a grade to a company. The nonfinancial statements and documents are also considered in the process of grade assignment. The rating agency considers numerous factors while assigning a grade, which is done in coordination with the company's management. Based on all the analysis of the company, a report is made, which is then used by the rating committee for the assignment of the grade. The various factors and their weights by which the firm decides these grades are usually not made public. The firm may again apply for the grading if there is a change in its environment. Here there may be no change in the grade, the grade may be upgraded, or the grade may be downgraded, depending upon the current scenario of the firm.

Numerous methods are commonly used for the purpose of rating. These include statistical methodologies such as linear regression (OLS), multiple discriminant analysis, the multinominal logic model, and ordered-profit analysis. The AI and soft-computing methods consist of ANNs, case-based approaches, support vector machines, fuzzy systems, and other classifiers as well.

17.5.2 Methodology

In this section, we mainly discuss the role and use of evolutionary ANNs for assigning credit ratings to firms. The assigned grades may be matched against the standard grades in the database. This validates the model and gives us the system efficiency. For this problem the main task is the identification of the input. There are numerous inputs that would be possible and can play a role. We need to select the best inputs that can play a major role in the system. Then we develop the GA-based ANN to solve the problem over the available database. The attempt here is to limit the connections in order to save the ANN from becoming too complex. This would help in attaining good efficiencies and ensuring that the system reaches the global minima. Here the output has been restricted to only four classes or grades: A, BBB, BB, and B.

17.5.2.1 Inputs

The inputs play a key role in any system and hence must be figured out with great care and concern. Both the number and type of inputs are important. Too many inputs may make the system fail to train. Too few may again result in the poor performance of the system. The effective design of a soft-computing system involves the selection of the best possible inputs.

Here the scenario is quite the same, where there are numerous inputs that are possible that affect the system directly or indirectly. But we must restrict our system to a limited number of inputs that have a significant impact. The inputs of the system are sought by literature survey and previous experiments. The different variables must be collectively able to convey tall needed information without much overlap in individual variables. Out of the 200 inputs that were present in literature, 14 were selected based on these conditions. All selected ratios demonstrated a statistically different mean at the 1 percent level between the investment grade and junk bond categories.

17.5.2.2 Evolutionary ANN

The evolutionary ANN performed the task of optimizing the rating against that of the standard database. The ANN tried to maximize the efficiency. For this, the GA was used. The GA performed the role of selecting the best inputs out of the selected inputs, making and breaking connections to maintain algorithmic simplicity along with efficiency, and deciding the correct activation of the hidden-layer neurons. The restrictions placed on the GA were that the number of inputs could be a maximum of eight, the activation function at each processing node could vary between linear or logistic, and the number of hidden-layer nodes could vary up to a maximum of four.

The representation of the individuals in the GA consisted of the bits representing the absence or presence of each of the input neurons or hidden-layer neurons. Further bits were used for the representation of the activation function, which could be linear or logistic. This resulted in the complete architecture of the ANN. For training purposes, though, the authors used the back propagation algorithm (BPA). As per the standards of the soft-computing principles, the whole database was partitioned into training and the testing data sets.

17.6 CONCLUSIONS

Throughout this book, we have seen numerous applications, their working, fundamentals, concepts, and so on. We've studied these applications in depth in regard to their characteristics, behavior, and needs. All these applications worked using soft-computing solutions. In all these, we saw diverse ways to use these soft-computing solutions and the ways and manners by which the different soft-computing methods are adapted to the problem requirements. The applications presented cover almost every aspect of the soft computing.

But as we browse the Web or glance through any journal or proceeding, we find that we have discussed only a small part of the entire picture. The actual applications covered by soft computing are beyond the scope of any standard text. This fact makes it impossible for any person to master the complete domain of soft computing. One is only able to master the methods or the tools and techniques, along with some applications. Further it is impossible for any book to cover the complete domain of soft computing. After all, the number of pages is always limited and finite. So too is the capacity of the readers and the authors.

In almost every system we studied, we spent a lot of time understanding the system that belonged to numerous domains. Even within this book, we discussed to some extent the fundamentals of image and signal processing, robotics, and more. This is probably not what readers might have expected before taking up this book. After studying these systems, we got a clear idea of the problem requirements. Then using our understanding of soft-computing techniques, we were able to develop some method to solve the problem. In reality, to design an effective soft-computing system, one needs an understanding of both the problem and soft-computing techniques. This makes the field multidisciplinary. Either people skilled with soft computing have to learn about the domain where they are going to apply the techniques, or the people skilled in their domain must learn and apply soft computing. An amalgamation of both these ways leads to the generation of effective systems that make a real life application.

This means that this conclusion does not end the chapter. It rather marks the start of the chapter where we have just begun to explore the space of soft computing. We have already discussed some of the major applications of soft computing. We included applications from diverse domains. There are numerous applications that may still be explored in each of the discussed domains. Further every application may be solved by many methods. The choice of method lies in the hands of the designer. Every method has some limitations, advantages, and disadvantages. Every solution may further be adapted to the various contexts a problem or application is used in. This is a continuous work of research that is actively going on in each and every application or application domain. Besides, there is a constant need of new algorithms and systems using existing or modified soft-computing methods so that they can be generalized to a variety of problems. This again is an active area of research where better optimization, classification, and other models are being built. Readers may contribute and join the research force in any of these domains.

Section IV

Soft Computing *Implementation Issues*

18 Parallel Implementation of Artificial Neural Networks

18.1 INTRODUCTION

We have seen over the past chapters that the artificial neural networks (ANNs) are extremely useful for machine learning in real life applications. The ease with which these systems are designed, implemented, and put into use lends itself to their wide application in speech recognition, facial recognition, character recognition, financial analysis, bioinformatics, credit analysis, and more. The ANNs form great tools not only for learning the historical data but also for generalizing this learning to unknown inputs.

The basic motivation for the ANN is the human brain, which possesses immense capabilities because of the very large number of neurons the brain holds and the large number of connections that exist in the entire brain. Connections are likewise of great importance in the ANN, since the entire memory of the ANN takes the form of the connections of varying weights. The connectionist architecture of the human brain is one of the major reasons behind its potential and capabilities.

Parallel processing is a technology that is on a constant rise. We see many algorithms now being made with a parallel implementation to aid the simultaneous working of many units. This naturally gives great performance, with the computation distributed among many nodes. This is done by the introduction of parallelism in the algorithm, where different processing units perform different parts of the algorithm simultaneously. The back propagation algorithm (BPA) used for ANN training is an algorithm that is parallel by its very nature: the various neurons work in parallel for the generation of the final result. The BPA consumes a lot of time in training, and hence we need to take the advantage of this parallelism to divide the whole computation among many elements.

The basic philosophy behind the parallel implementation of the ANN is partitioning. Here we try to partition the entire problem into a set of smaller computational problems that can be executed in parallel by the various ANNs. The major kinds of partitioning that happen in neural networks are data set partitioning, node partitioning, and layer partitioning. These all divide the computation among various nodes and thus help in parallel execution of the ANN, resulting in better performance of the whole system.

Many times in this text we have tried to mix systems in order to achieve greater performances. We did this in the case of the hybrid systems, where we saw that mixing existing systems gave better performance and thus resulted in a more optimized solution. Here we try to learn from the same fundamentals as we mix the various partitioning techniques to generate a technique that must give better resulting performance. This is the hierarchical partitioning. In order to fully exploit the potential of these partitioning techniques, we apply all of them one after the other. This takes the form of data set partitioning followed by layer or node partitioning. This type of system ensures that we have an optimal structure for the network.

A basic fact that is seldom considered in case of parallel systems is that the various processing nodes are of different processing powers. This difference in processing powers may be due to the difference in configuration of the two systems. At many other times, the speed or performance may get low when the system suddenly becomes slow or temporarily loses its performance. This underscores the need for unequal distribution of computation between the various processing nodes so that all nodes finish their assigned tasks at almost the same time.

Further it is necessary for the system to be self-adaptive in nature. This would work across all the partitioning to ensure a distribution of work among all the nodes in such a way that no node sits idle for long. All the small changes in performances of the constituent nodes are monitored, and necessary actions are taken for the optimal performance of the system at every instant of time.

In this chapter, we talk about the parallel implementations of the ANN. We first discuss the basic three partitioning techniques one by one. Later we discuss the hierarchical partitioning technique. Here we mix the partitioning techniques and come up with a model of partitioning that will be able to give greater performance. We try out these techniques and finally offer some conclusions.

A lot of work has been done on neural networks to adapt them to various problems. They are finding growing application in every field. This intense demand is a key reason for the growing consciousness of the need for scalability and speed. The parallel algorithms that have been introduced to various problems become even more prominent when it comes to neural networks. A lot of work is being done to make systems more parallel. Back propagation is a major area where parallel algorithms find application. Various types of node, data, layer, hybrid, and other partitioning have found use and offer solid speed and performance benefits as well. Newer areas being explored include genetic algorithms, hardware support exploitation, and more. Research is going on to implement the back propagation algorithm in various situations, as well as to adapt parallel techniques to various related technologies such as self-organizing maps and perceptrons.

Efforts are under way to depict the human brain more closely. Researchers are trying to understand more about the functioning of the human brain and incorporate these insights into the neural network. The study of the human brain and computational neurosciences has led to new dimensions of thought. An increase in automation also is emphasizing on the need for systems to be self-controlled rather than under human control. There is a persistent need of automation at all levels. This has given rise to self-adaptive systems that can make autonomous decisions, which inspires us to use such systems in the neural networks as well.

This chapter discuss the work of Kala, Shukla and Tiwari (2009f).

18.2 BACK PROPAGATION ALGORITHM

One of the most commonly used algorithms for the purpose of machine learning is the BPA. This algorithm operates in a supervised mode, where the outputs are presented to the system at the time of training along with the inputs. The BPA adjusts the system bias and weights to optimize performance to the entire system, where the trained ANN better imitates the function desired. You already learned about the BPA in previous chapters. In this section, we give a brief overview of this algorithm. This may be taken as an attempt to first learn the serial algorithm so that we are able to convert it effectively into a parallel algorithm.

The whole algorithm works in epochs, which are the iterations of the algorithm. At each epoch, the ANN is fed all the inputs and outputs one by one. The ANN operates in a batch processing mode, and this operation is repeated in ever epoch. At every epoch, generally, the ANN adapts itself well in order to better imitate the function and result in lower errors when presented with the training data set.

The ANN operates in two stages: feed forward and feedback. First comes the feed forward stage, where the given input is applied. All the inputs are given to the system one after the other, and the result is calculated. This happens by means of a flow of information from each layer to the next. This goes on until the output layer is reached, where finally the answer is evaluated. The second stage is the feedback stage. Here we use the results of the feed forward stage. These results are used to calculate the differences between the actual results and the desired results.

These differences or errors are back-propagated to the previous layers and used to adjust the ANN weights and biases.

In the forward phase, each neuron acts on the input that is given to it. The output of this neuron is propagated to the next-layer neuron of the network. Each neuron consists of a number of connections from the previous layer. Let these connections be numbered *1, 2, 3...n*. Each connection has a weight W_i associated with it. Each neuron performs two steps. The first is the weighted summation, shown in Equation 18.1.

$$O_1 = \sum W_i x_i \tag{18.1}$$

Here, O_1 is the first level output of the neuron
W_i is the weight of ith connection
x_i is the ith input of this neuron

The second is the addition of nonlinearity to the output. This is done by passing the input to a non-linear function $f(x)$ as is given in Equation 18.2.

$$O_2 = f(O_1) \tag{18.2}$$

Usually we use the sigmoid function as the nonlinear function. This function is given in Equation 18.3.

$$f(x) = \frac{1}{1+e^{-x}} \tag{18.3}$$

In the backward phase, the error is calculated and is propagated backward to all the previous layers. This error is used to make adjustments in the weights of the neurons. The error of a given output node is denoted by δ. This is given in Equation 18.4.

$$\delta = O_2 * (1 - O_2) * (O_d - O_2) \tag{18.4}$$

Here O_2 is the output received in forward phase
O_d = Desired output

The weight of this neuron is changed by a small fraction ΔW. Here ΔW is given by Equation 18.5.

$$\Delta W = \acute{\eta} * \delta * O_2 \tag{18.5}$$

Here, $\acute{\eta}$ is the learning rate.

The final updated weight is given by Equation 18.6.

$$W_{new} = W_{old} + \Delta W \tag{18.6}$$

This formula is used in the backward phase for all neurons from end to the start. Hence all the neurons have their weights updated. In a similar manner, the biases are updated.

18.3 DATA SET PARTITIONING

For most of the applications for which the parallel implementation is applied, the training data set is usually high. *Data set* partitioning tries to distribute the training data among the processing elements (PEs). Each of the PEs receives some part of the training data and is supposed to train itself

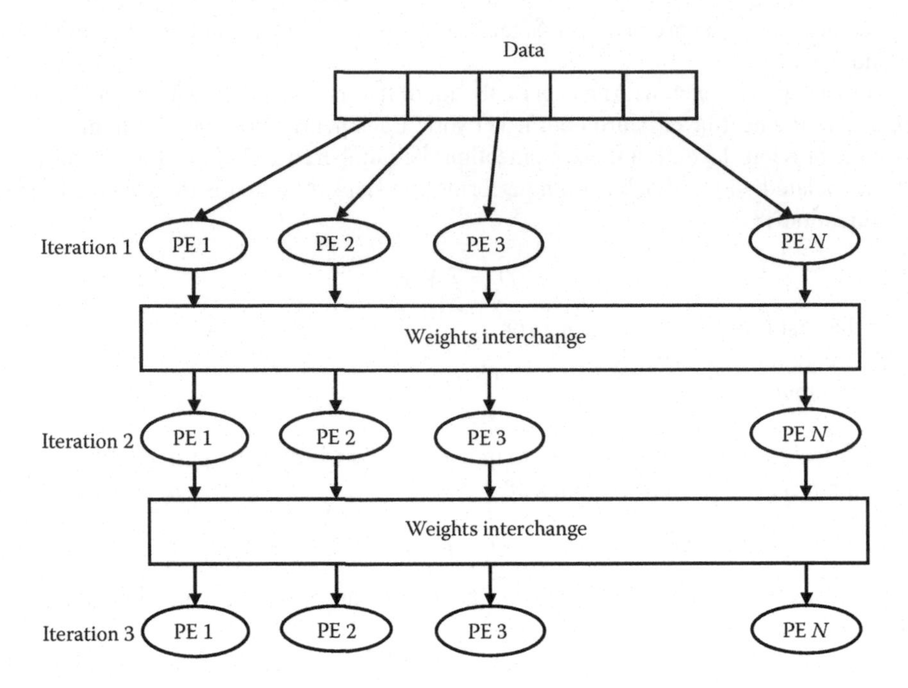

FIGURE 18.1 Data set partitioning.

by this data. The various PEs are trained simultaneously. When all the PEs have trained themselves sufficiently well, the results are combined and are then communicated to all the PEs. This helps all of them update the weights and the biases. It may be noted here that each PE keeps its own copy of the ANN. All the copies of the ANN are similar to each other.

In order to update the weights and the biases, each PE takes the mean of all the weights calculated by the different PEs. This is given in Equation 18.7. The final resulting weights and biases are reflected in all the copies of the ANNs. The update is done by the PE itself in its copy of the ANN once it knows the calculated weights of all the PEs.

$$W_{ij}^{k} = \Sigma\,(W_{ij}^{k})_l \,/\, N \tag{18.7}$$

Here, W_{ij}^{k} is any weight from any node i to any node j in layer k
l is any of the data set partitions
N is the number of data set partitions

The basic idea of this type of partitioning is shown in Figure 18.1.

A major point of concern in this type of partitioning is the communication between the nodes. The communication needs to be kept at a minimum in order to minimize the communication overhead. If there is a lot of communication, making and breaking of connections, and so on, it is possible that these times spent on communicating data may turn out to be much more than the total time saved in partitioning. In order to minimize this time, we use the concept of peer-to-peer connections. The connections are preestablished, where every node is connected to the next node to form a circle. Each node communicates its final answer to the next node. It also communicates the next answers it gets to the next node. This cycle of communication of final answer is carried out until $n-1$ cycles. At the end of these $n-1$ cycles, every node is aware of all n answers calculated by the n nodes. They then update their weights. This communication is shown in Figure 18.2.

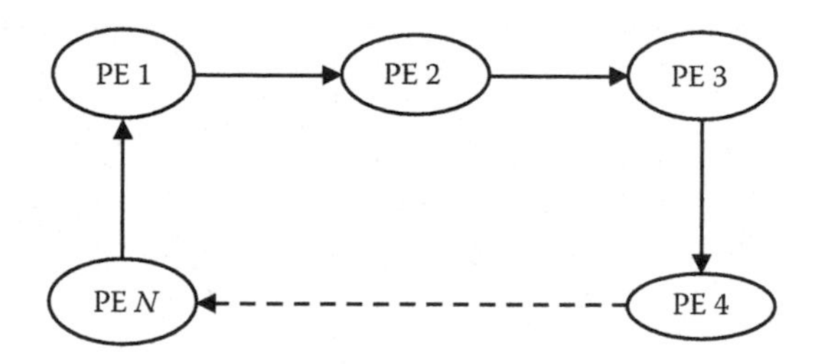

FIGURE 18.2 Communication in data set partitioning.

18.4 LAYER PARTITIONING

Layer partitioning is another type of partitioning. Here we partition or divide the layers of the ANN. Each layer or set of layers is allotted to a separate PE that carries forward all the computation of these layers. It needs to be seen here that a layer will be able to work for some input only after the preceding layer has finished working with it. The results are supplied from the preceding layers to the next layers, and in this manner the transfer of information is carried out. This happens in both the feed forward and feedback stages.

In order to maintain parallelism, the different layers work in a pipelined architecture. The different layers work on different data items. Once the first layer makes its calculations for the first data item, it communicates the results to the second layer. It itself starts working on the next data item while the second layer is working on the first. A similar process goes on the reverse side as well. This type of pipelined architecture allows the different layers to work on different data at the same time. The weights are updated at the backward propagation cycle and need to be stored temporarily when the forward and backward cycles finish. This process is shown in Figure 18.3.

The general architecture of the layer partitioning is given in Figure 18.4. This partitioning is applied over the different layers. The communication follows a similar approach, where the computations are transferred from a backward layer to a forward layer (in the forward pass) and a forward layer to a backward layer (in the backward pass). Hence every node (layer) is connected to the next node (layer) and the previous node (layer). The communication is shown in Figure 18.5. The mathematical model for this partitioning is given in Equation 18.8.

For forward phase we have

$$O_l = \sum W_i I_{ij} \tag{18.8}$$

$$\text{Here } I = \begin{bmatrix} I_1 \\ I_2 \\ I_3 \\ I_4 \\ I_5 \\ \vdots \\ I_n \end{bmatrix}$$

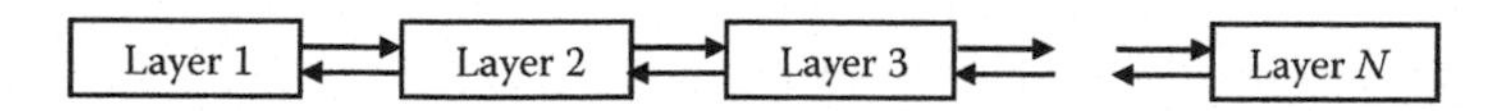

FIGURE 18.3 How layer partitioning works.

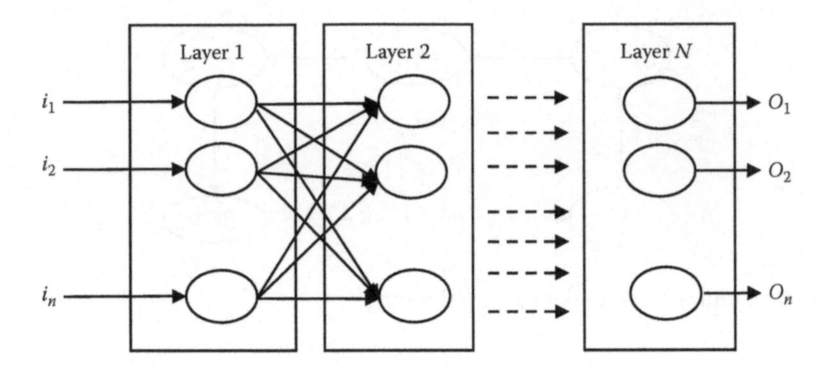

FIGURE 18.4 The layer breakup of a neural network.

Similarly in the backward phase we propagate the array of desired valued backward. The error E that is propagated from one layer to another is given in equation (18.9).

$$E = \begin{bmatrix} E_1 \\ E_2 \\ E_3 \\ E_4 \\ E_5 \\ \vdots \\ E_n \end{bmatrix} \tag{18.9}$$

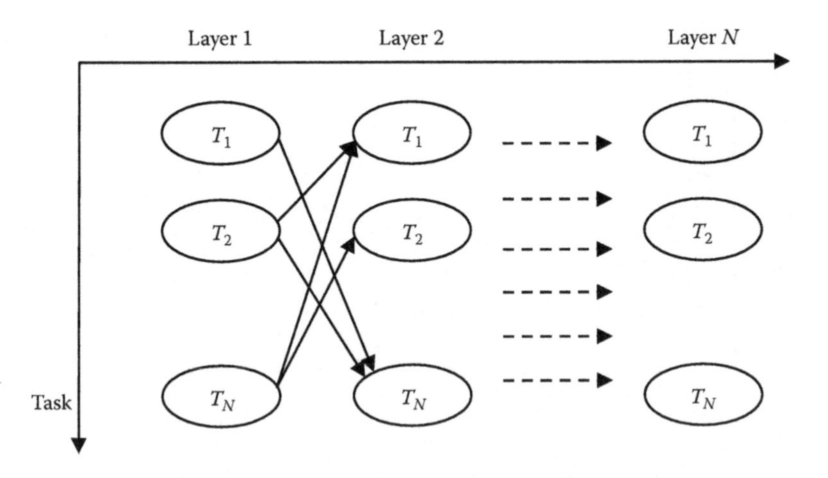

FIGURE 18.5 The working of node partitioning.

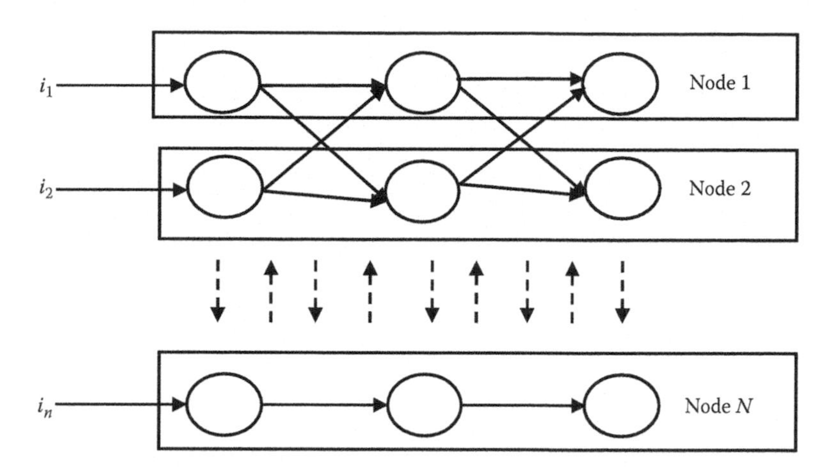

FIGURE 18.6 The structure of node partitioning.

18.5 NODE PARTITIONING

In this type of partitioning, we try to partition or divide the nodes of the various layers horizontally. Each partition is handled by a separate PE. The partitioning consists of nodes from all layers.

Here, the computation is slightly more complex. The computation has to precede layer by layer. Hence at the start, all first nodes of various PE may start their functioning. Once this is done, then only the second nodes will get a chance to operate. This way, the processing propagates simultaneously among all processing elements, from one layer element to the other. The computation in the backward phase also follows the same principles in the reverse direction. The working of this computation is given in Figure 18.5.

The general structure of this partitioning is given in Figure 18.6. The computation of the weights in this type of partitioning is a little complex. The weights are usually stored reputedly to save time. Due to the number of connections, the total time is usually greater. The communication is from one processing element to all the processing elements as the connections may exist that require such a data transfer. The communication is shown in Figure 18.7. The mathematical model of this type of partitioning is given in Equation 18.10.

For forward phase we have

$$O_l = \Sigma\, W_i I_i \tag{18.10}$$

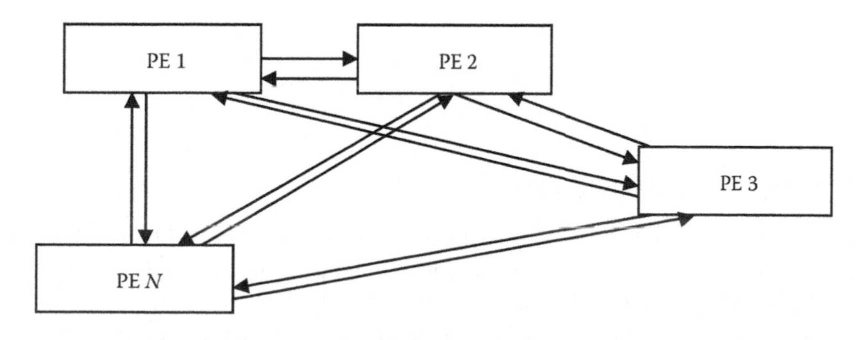

FIGURE 18.7 Communication in node partitioning.

$$\text{Here } I = \begin{bmatrix} I_{11} \\ I_{12} \\ \vdots \\ I_{1n} \\ I_{21} \\ I_{22} \\ \vdots \\ I_{2n} \\ \vdots \\ I_{m1} \\ I_{m2} \\ \vdots \\ I_{mn} \end{bmatrix}$$

This is the straight input as arrays from the various nodes that are working. The equation shows m processing units. Any processing unit gets $m - 1$ arrays of inputs from other units and 1 from its own unit. Here, we have assumed that any PE carries n neurons in a single layer in itself. The rest of the operation is the same as discussed for the layer partitioning.

Similarly in the backward phase we propagate the array of the desired valued backward. The error E that is propagated from one layer to another is given in Equation 18.11.

$$E = \begin{bmatrix} E_{11} \\ E_{12} \\ \vdots \\ E_{1n} \\ E_{21} \\ E_{22} \\ \vdots \\ E_{2n} \\ \vdots \\ E_{m1} \\ E_{m2} \\ \vdots \\ E_{mn} \end{bmatrix} \tag{18.11}$$

18.6 HIERARCHICAL PARTITIONING

It is natural that the individual systems or partitioning techniques face problems that restrict their performance. This necessitates more intelligent partitioning techniques that might further give a performance boost. This is done with what we refer to as the *hierarchical* partitioning technique. Here the aim is the judicious mixture of the existing partitioning techniques so that the resulting partitioning technique performs better.

First, we partition the data as per the data set partitioning norms. After this partitioning has been applied, the various PEs get a copy of the data that they need to train. This is further broken down into node partitioning or layer partitioning. A general method for applying this partitioning technique is to use half of the PEs for the node partitioning and the other half for layer partitioning. Each layer partition uses its own set of PEs to work. Each PE represents a layer, and the number of such elements is the number of layers. Similarly, each node partition has its own set of PEs. Each partition has its own share of the ANN. This results in the generation of unique architectures of the resulting systems where the different architectures mix together. By varying the role and the degree of partitioning of the various techniques, we may control the system performance or the contribution of any single partitioning technique to the entire system.

When the system is asked to train itself, the first job is the data set partitioning. The data set is divided and distributed among all the PEs of the data set partition. Each is given its own share of the data, along with a copy of the ANN with initial weights and biases. These data set partition or PE sets train the system independent of each other. When they all finish a prespecified epoch, the circular communication starts and the result is communicated everywhere. Each PE set of the data set partition uses data set or layer partitioning to distribute its computation among the PEs. This communication does not affect the communications between the data set nodes. The result is computed internally by these PEs. The final answer or the weight vector is calculated. This is then communicated to all the nodes in the data set partition.

The general structure of this system is given in Figure 18.8. It may be noted here that there is no physical element that acts as the data set partition processing element. Rather the extra functionalities of the data set partition are added to the first member of the nod/layer partition. Hence it performs both the tasks. We use two different sockets for both these communications.

18.6.1 Self-Adaptive Approach

We have already mentioned how important it is that the systems be self-adaptive. This is because the various PEs may have different processing powers. Some may be able to perform very fast, while others may take time for the same computation. This difference in the processing powers needs to be taken care of. Also the performance of the PEs may change over time. This may be due to external events at the PEs or the external computation that the PEs perform. It may happen that the performance of any PE might suddenly or temporarily reduce and later become normal. All these temporal variations in the speed of the systems also have an impact on the performance of the entire system. Hence there is a need for performance optimization in such conditions in order to reduce the idle time of every PE.

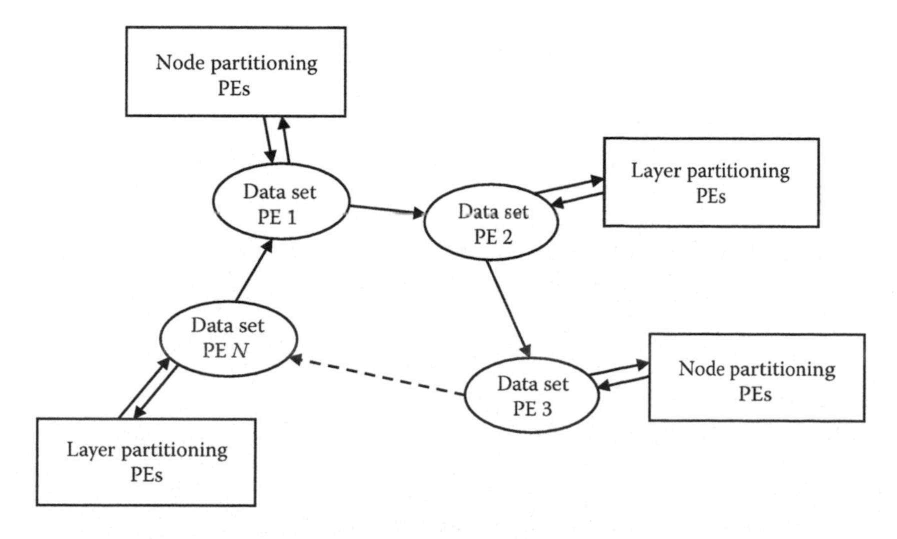

FIGURE 18.8 The general structure of hierarchical partitioning.

The general algorithms would not give optimal performance in such cases. Here we have to make a system that is able to adapt itself to such conditions. The adaptation will enable us to save valuable time. It is carried out by varying the load of the workstations. If we find that the workstation is performing very well, we increase its workload. If a PE is not able to perform that well, we reduce its workload. The performance here is seen in relative terms. It is the performance of a PE as compared to all other PEs.

For the sake of implementing this concept, we go with the idea of a server and a client. The server here does the job of load balancing. The load balancing is done on the data set partitioning workstations. The server is responsible for the performance evaluation of all the nodes and then division of workload among them.

A major question here is the frequency with which the server monitors and reallocates the load. If the frequency is too low, it is possible that the system may recover completely from the low performance peaks and the problem may go unnoticed. If the frequency is very high, it may result in a huge loss of time in communication between the nodes. This communication may increase unwanted overhead, and the total time may shoot up. Hence we need to take care when deciding the operating frequency.

The communication required for this system is carried out in peer-to-peer mode. Each workstation informs the next workstation of its performance characteristics. This cycle goes for $n-1$ cycles. While implementing this step, we kept it common with the step where the data partitioning nodes communicate their weights to peers. The information on both weights and performance goes simultaneously. This is communicated by the first node to the server at the desired frequency. The server studies the performances and then redistributes the work load among all.

In order to facilitate this distribution, every node is provided with a number that denotes the number of test cases that it will be given. If this number is changed, the next time the workstation will accept only that many test cases. The circulation of test cases also takes place in the circular queue. The test cases are passed from one node to another. Each node picks up the mentioned number of test cases and passes on the others to the next node.

Workload may be easily distributed according to the ratio of the performances as reported by the various nodes. The performance here is the actual time taken by the nodes in execution, neglecting the times it was waiting for the response of another node. The number of test cases n_i to be given to any node i may be calculated as given in Equation 18.12.

$$ni = \frac{ni * p}{T} \tag{18.12}$$

Here n_i = Number of test cases to be given to node i

N = Total number of test cases

p_i = Performance of node i

T = Total sum of performances $(\sum p_i)$

Here p_i is calculated as the difference between the processing time of that node and the time taken by serial execution of the algorithm. This is given in Equation 18.13.

$$p_i = t_i - S \tag{18.13}$$

Here t_i = Processing time of node i

S = Time taken by serial execution of the algorithm

The last node gets the remaining number of test cases $(T - \sum n_i)$

The general architecture of this system is shown in Figure 18.9.

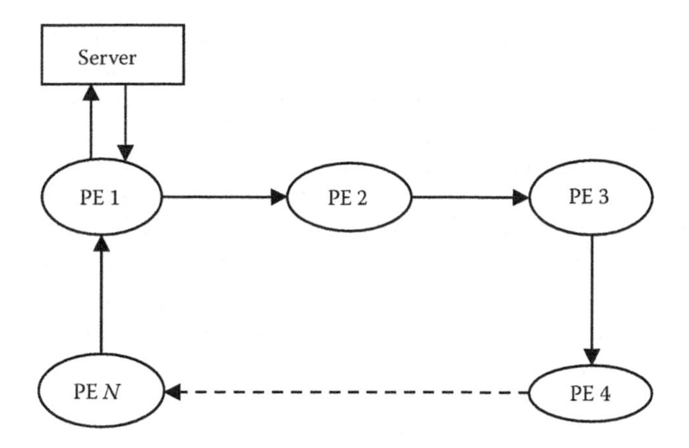

FIGURE 18.9 The server/client self-adaptive model.

18.6.2 Connections

The connection referred to here involves the sockets used between the PEs for the transfer of information from one PE to the other. These connections are established as per the decided architecture design at the start of the algorithmic run itself. The basic connectionist model of the algorithm is shown in Figure 18.10. Here we observe that the whole architecture of the algorithm is based on a single client-server relationship. This is to carry on the load balancing by the server in order to

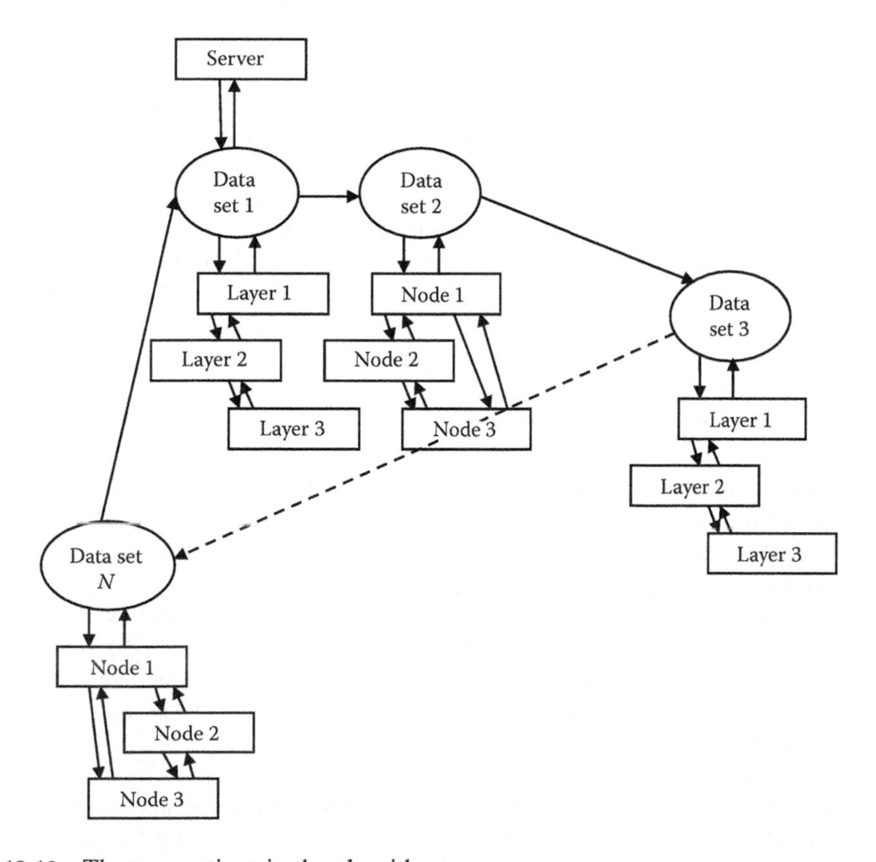

FIGURE 18.10 The connections in the algorithm.

distribute workload equitably among the various PEs. All the transactions are carried out from this server and are communicated to all the clients in a circular manner. The server takes in the performance of the various nodes and then calculates the workload that every node should have. This is propagated to all the nodes.

Then we come to the outer ring between the PEs. This ring belongs to the data set partitioning PEs. Each ring connects one member of a PE set to the entire ring. This is the ring through which the data set partition data are floated. The various PEs or PE sets take their share of the data set. In a similar manner, the final calculated weight needs to be carried over all the PE sets. For this, also, the same ring is used. The results are propagated in a peer-to-peer manner.

Each PE set of the data partitioning is internally composed of various PEs that need to be connected to each other for the proper flow of data or information. This happens for both node and layer partitioning. In the case of layer partitioning, the various layers or PEs are connected in a linear manner one after the other. Each layer or PE connects to the next as well as the previous layer or PE. The node partitioning is somewhat different. Here the PEs represent horizontal nodes. This means a PE now needs to transmit and take information from all the available nodes in the entire PE set. These connections are provided from all PEs to all PEs.

18.6.3 Working

The basic purpose of the algorithm is to train the ANN using the BPA. This happens in a parallel manner by using the various kinds of partitioning that we studied. In this section we come to grips with the basic working of the resulting system that is able to use the parallel concepts well and result in a performance boost. The parallelism may even be applied for testing purposes. The first task of the algorithm is to establish all the connections that we discussed in the preceding section. These connections are formed once at the time the program starts and are maintained throughout the algorithm run.

Then the data set partitioning is made active. Here the data are partitioned and divided among the various data set partitions or PE sets. All the data are circulated on the circular ring of the data set partition. The data set partition PEs take the required number of test cases and pass the rest to the next nodes. After this step, all the nodes involved in data set partitioning have a data set with them. Also each one of them gets a copy of the ANN comprising the architecture, weight, and biases.

Once this communication is over, each ANN starts the process of training itself with its copy of the data. After the training is over for all these ANNs for a prespecified number of epochs, the results consisting of the final weights and the biases are communicated to all the members of the data set partition using the same ring. This entire process of a fixed number of epochs is carried a specific number of times for the desired duration.

Each member of the PE set uses a layer or a node partitioning to carry out the task of computation or training. The input that the leading member of the partitioning technique receives on behalf of the PE set is shared among the other members for the algorithm to run as per the demands. These two approaches of layer partitioning and data set partitioning work exactly as their concepts and use the established connections for the transfer of the information from one PE to the other. This happens for all epochs.

From time to time the server checks the performance of the various nodes to carry out the correct node balancing. The server interrupts these process (in synchrony with the completion of all pending computations) and then distributes the computation. Based on the individual performances, the revised workload for the nodes is calculated and the same is communicated. This process is repeated at a predefined frequency.

The training finishes whenever the specified stopping criterion is met. The final list of weights and biases as computed by the data set partition nodes is regarded as the final answer and is used for testing purposes. After the training is completed, we observe that the system reached good levels of speed as a result of the robust technique we had developed for training the artificial neural network.

The most critical part of the algorithm is to main a proper synchronization among the different steps to ensure that the activities take place in the correct order and no activity starts before the previous one has ended. This is very important to keep the consistency of the program.

18.7 RESULTS

The hierarchical algorithm was tested against a synthetic problem to study its working and effect on the performance of the system. Java was used as a platform for the implementation, and socket programming was used for the transfer of the information across the discussed connections. The whole network architecture was logically formed before the simulation and was reflected in the code so that the systems connected to the correct systems and the data were sent over to the correct system and correct port based on the logical architecture decided beforehand. Then the algorithm was simulated. The master server initiated the training by the mechanisms discussed earlier.

Here, a synthetic data set was generated for the training of the ANN. The data set generated was by taking random inputs. The outputs for every input set were calculated using Equation 18.13. Hence the task of the ANN was to imitate the function given by Equation 18.14.

$$f(a,b,c,d,e) = (a * 100 + b * c + \mathrm{Sin}(d)) / e \tag{18.14}$$

The performance of this algorithm was matched against the performance of a single-processor singe system where the simple BPA was made to run. The times of execution for both these systems comprising multiple and single workstations were compared to know the amount by which the parallel systems prove to be better.

The two systems are compared by the speedup. This is a ratio between the times taken by the single workstation algorithm (a sequential algorithm) and a multisystem algorithm or a parallel algorithm. The higher the ratio, the better is the parallelism in the algorithm, and this means that the algorithm makes the best possible use of the various workstations. This ratio is given by Equation 18.15.

$$\text{Speedup} = \frac{\text{Sequential execution time}}{\text{Parallel execution time}} \tag{18.15}$$

We also make a study here of the self-adaptive nature of the algorithm. We try to analyze the performance of the algorithm with and without this feature. This helps us in knowing the difference that can arise in the use of such a system and the performance boost by the use of this feature.

The speedup for various input sizes is given in Figures 18.11a through 18.11d. The various times taken by the system and the architecture are given in Tables 18.1 through 18.4.

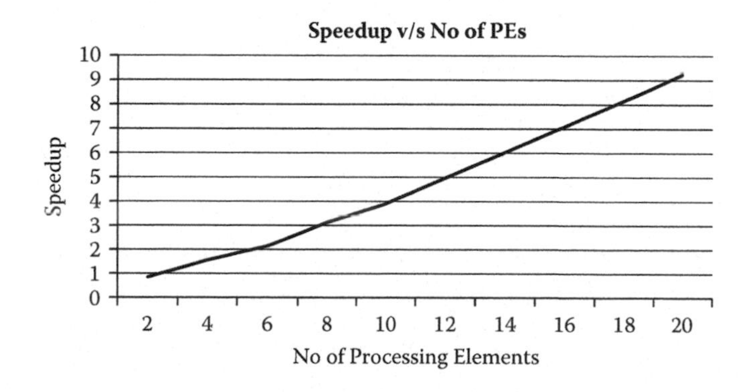

FIGURE 18.11a Speedup vs. number of processing elements: Input 1.

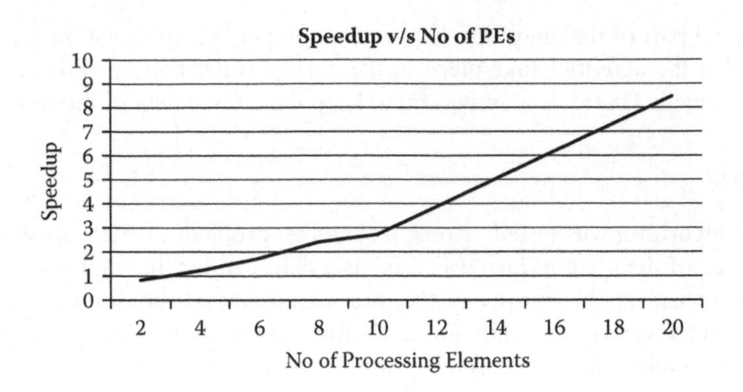

FIGURE 18.11b Speedup vs. number of processing elements: Input 1 without self-adaptive approach.

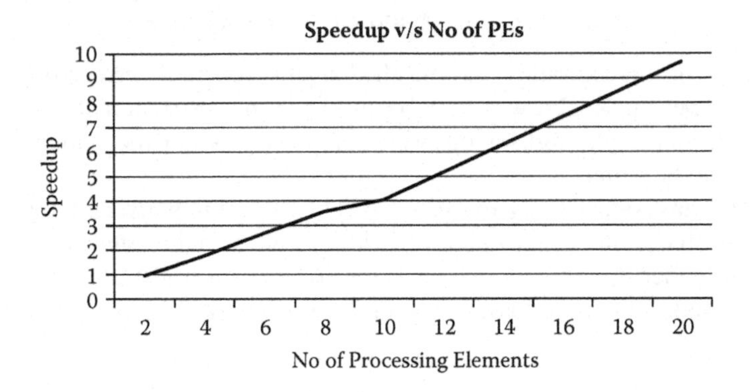

FIGURE 18.11c Speedup vs. number of processing elements: Input 2.

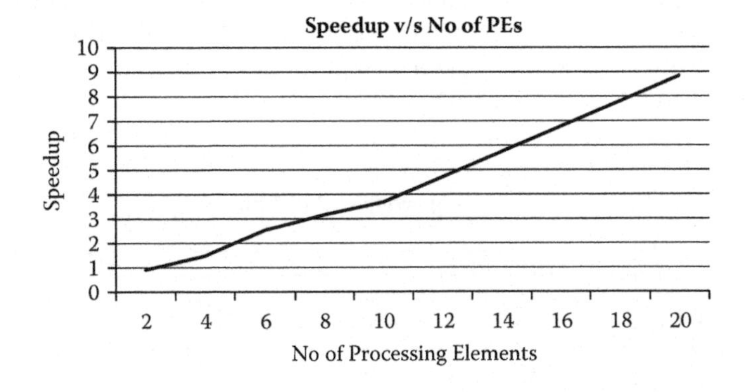

FIGURE 18.11d Speedup vs. number of processing elements: Input 2 without self-adaptive approach.

TABLE 18.1
Speedups for Input 1

PE	No. of Inputs	Network Architecture	Iterations	Server Sync after Iterations	Time Serial	Time Parallel	Speedup
2	500	8-15-1	150000	50000	823737	982521	0.838391
4	500	8-15-1	150000	50000	823737	530306	1.553324
6	500	8-15-1	150000	50000	823737	384294	2.143507
8	500	8-15-1	150000	50000	823737	263511	3.126006
10	500	8-15-1	150000	50000	823737	211249	3.899365
20	500	8-15-1	150000	50000	823737	89225	9.232132

TABLE 18.2
Speedups for Input 1 Without Self-Adaptive Approach

PE	No. of Inputs	Network Architecture	Iterations	Server Sync after Iterations	Time Serial	Time Parallel	Speedup
2	500	8-15-1	150000	NA	823737	993483	0.829141
4	500	8-15-1	150000	NA	823737	674358	1.221513
6	500	8-15-1	150000	NA	823737	473899	1.738212
8	500	8-15-1	150000	NA	823737	342174	2.407363
10	500	8-15-1	150000	NA	823737	303214	2.716685
20	500	8-15-1	150000	NA	823737	97180	8.476405

TABLE 18.3
Speedups for Input 2

PE	No. of Inputs	Network Architecture	Iterations	Server Sync after Iterations	Time Serial	Time Parallel	Speedup
2	500	11-25-1	150000	50000	1733690	1829312	0.947728
4	500	11-25-1	150000	50000	1733690	984654	1.76071
6	500	11-25-1	150000	50000	1733690	644070	2.691773
8	500	11-25-1	150000	50000	1733690	486311	3.564982
10	500	11-25-1	150000	50000	1733690	429221	4.039155
20	500	11-25-1	150000	50000	1733690	179378	9.665009

TABLE 18.4
Speedups for Input 2 Without Self-Adaptive Approach

PE	No of Inputs	Network Architecture	Iterations	Server Sync after Iterations	Time Serial	Time Parallel	Speedup
2	500	11-25-1	150000	NA	1733690	1911289	0.907079
4	500	11-25-1	150000	NA	1733690	1175223	1.475201
6	500	11-25-1	150000	NA	1733690	683619	2.536047
8	500	11-25-1	150000	NA	1733690	548465	3.160986
10	500	11-25-1	150000	NA	1733690	471293	3.678582
20	500	11-25-1	150000	NA	1733690	196229	8.835035

18.8 CONCLUSIONS

In this chapter, we solved one of the major problems that restricted the functioning of the ANNs, which was the training time. We made the whole system applicable in multiple systems, and the training was completed. It may be seen that there was no loss of generality in the process, which is one of the major achievements of the algorithm. This opens the way to choose very complex structures for the ANN and very large data sets without caring about the training difficulties.

We studied three basic partitioning techniques: node partitioning, data set partitioning, and layer partitioning. We saw how these three partitioning techniques helped in the parallel implementation of the ANN. The first technique was data set partitioning. Here we divided the whole training data set into parts, and each of the PE was given a part of it. In this way, the parallelism was exploited through the division of the training data set. We further proceeded with layer partitioning. Here each PE was given a layer to handle. This way, the parallelism was carried forward by the division of computation between the layers. The last technique was node partitioning. Here the nodes were distributed horizontally among the PEs. Each PE got a horizontal strip of the ANN. In this manner, the work was carried out and the parallelism was met.

We later discussed the hierarchical partitioning technique, which is an amalgamation of the simple partitioning techniques. This resulted in better performance for the individual systems. The speedup was used as a factor to measure the performance, which improved as a result of the use of different techniques one after the other. The self-adaptation strategy was used to even improve the system, and the experimental results revealed that such a strategy worked well and resulted in increasing the system performance.

Finally, we developed a robust method of implementing BPA across various kinds of nodes. The algorithm so far has no capability to change the architecture of the system. This is something that is believed to happen in the human brain, where new connections are made and older ones are destroyed. Such a capability would make the system completely dynamic. Such a system needs to be designed in the future.

19 A Guide to Problem Solving Using Soft Computing

19.1 INTRODUCTION

So far we have studied numerous aspects of soft computing. We saw the principles and theories of each of the systems of soft-computing. Here we studied the foundations of these systems and how and why they perform. We also studied ways and means to model the entire problem according to the requirements of soft-computing systems. Each of the systems that we discussed had a different way of functioning and different types of problems that it addressed. We discussed numerous times in the text how all this converts the domain of soft computing into a design-oriented approach where the role of the engineer is to use the best of his knowledge and experience to design systems. He is to ensure that these designed systems are scalable up to the mark that the real life application requires. He needs to ensure system performance, efficiency, resistance to noise, and so on. We made numerous mentions in the text that this is a piece of the art that we have tried to cover as much as possible.

Much of the content of this text is devoted to discussing real life systems. We studied numerous systems that had their own ways of solving problems, modeling problems, and dealing with factors that influence the performance. We saw that all these systems used soft-computing approaches in various ways and came out with effective solutions that could be deployed in real life applications. Much of the text was spent in learning and discussing topics that primarily belong to other domains. In order to discuss and apply the soft-computing approaches, we were exposed to image processing, signal processing, financial management, and other areas. We explored all these fields before coming out with systems that used soft-computing approaches to solve the problems associated with these domains.

We discussed various tools and techniques for the problem solving in this text. All these tools and techniques were a result of both the theoretical understanding of the systems and the practical experience that we gained by working with real life problems and the real life databases. Here we learned that many times the problems could be solved or the solution could be improved by the application of some trick or technique. We even advocated certain systems over others by presenting supporting views after noting the characteristics of the problem. All this led to the creation of good and effective systems for almost all the problems that we studied, be they real life or synthetic. It looked as if problem solving is a really easy job in the domain of soft computing.

But in reality people face problems of various kinds when they themselves apply the same techniques that other people easily succeeded in. This makes it very difficult for them to appreciate soft-computing systems, since they themselves are unable to get good answers even to standard problems. It is natural that this is due to the mistakes that they commit in the process. These may be common mistakes made by people: misunderstood literature and procedures, wrong visualization and conception of the problem and the solution used, and so on. All in all, there are numerous problems that people may commonly face. We discussed much of the content as professionals trying to explain the systems. But this approach often leaves behind unclear issues that contaminate our understanding of these problems. So it would be wise to invest some time and space in clearing up many things that the readers may have got wrong.

In this chapter, we discuss some of the very common mistakes that people make, along with some of the common issues and problems that the people face. We try to address all these problems so that readers may learn what not to do or what to avoid, not just what is to be done. This chapter is devoted to the study the systems that we have been using and to point out ways of using them for the creation of systems without bugs and that perform very well in real life scenarios. This chapter acts as a working guide to the readers to consult for all the general queries, problems, and issues related to soft computing.

Many of the points discussed in this chapter have already been under discussion. Right from the theoretical chapters to the real life applications, we kept giving tips and tricks for the use of the soft-computing systems. Vigilant readers might feel this chapter to be a repetition of content. But it is always better to keep such advice in a compiled format somewhere that may be referred to and benefited from. Many other readers may not have noticed much of the advice that we present here. They may find this chapter to present a good summation of all that the book talked about. In short, this chapter is a small, informal discussion of the myths, beliefs, misconceptions, and real picture of soft computing.

Before we carry forward our journey in discussing these tools and techniques, let us discuss the major systems that we have studied. We first of all studied artificial neural networks (ANNs). Here we first discussed the ANN with the back propagation algorithm (BPA). We saw the great generalizing powers of this algorithm that led us to use it for various real life problems. We then studied various other models of the ANN. These mainly constituted of learning vector quantization (LVQ), radial basis function networks (RBFNs), adaptive resonance theory (ART), Hopfield neural networks, and self-organizing maps (SOMs). We studied the similarities and differences between all of these. We further studied the impact of these algorithms on numerous types of problems. They all in general proved to be very comprehensive problem-solving tools. They were used primarily for learning the historical data and to generalize this learning into the unknown data as well.

Later we even studied fuzzy inference systems. These systems made use of the concept of fuzziness or impreciseness to model many real life applications. These systems were driven by rules, which were more or less driven by lingual expression. Here we presented the fuzzy algebra that aided in mathematical modeling of the problem based on lingual rules. These systems were mainly used to find the output to unknown problems after the correct modeling of the problem based on the associated rules.

We also studied evolutionary algorithms and systems inspired by natural evolution. We discussed the ways the genetic algorithm (GA) is able to come up with good solutions to problems and the methodology and reasons behind the ability of the GA to result in great convergence and effective exploration of global minima. We mainly used the GA for optimization purposes.

Then we studied the hybrid algorithms. Here the motivation was to mix two or more of these basic systems to result in the generation of systems that complement each other. The advantages get added up, and the disadvantages get removed of each of the corresponding systems. The resulting systems are able to much better solve the problems. Here we studied adaptive neuro-fuzzy inference systems (ANFISs), evolutionary ANN, evolutionary GA, modular neural networks (MNNs), fuzzy neural networks, etc. All of them were made to solve real life problems. By their own ways and means, they all resulted in the generation of the optimized solutions to the problems presented. The resulting solutions were shown to be better than each of the constituent systems for the purpose of problem solving using soft-computing techniques.

We also studied a lot many methods of problem solving in the form of real life applications. These were integrated into various applications and various parts of the text. An example is the use of fusion techniques in biometrics that can be generalized to other systems as well. Another example is the Bayesian classifier, which was also discussed in the text.

So this book is filled with numerous ways, tools, and techniques. This also increases the scope of mistakes in the use of these systems or the difficulty in understanding of the actual process or the system at the back end. We suppose that by now readers must have tried their hands at some data sets, either standard or self-generated. Readers might have tried to apply the soft-computing approach to one problem or another. They are likely to have been able to achieve good accuracy through the use of these systems. It is also possible that the resulting system was not efficient enough. In this chapter, we analyze the problems and try to find out what might have hampered the performance of the system.

Here we take the various major systems under discussion one by one for the purpose of in-depth analysis of their working and the common problems associated with them.

19.2 ANN WITH BPA

This is one of the most common algorithms that we use in this text. This algorithm always performed well and achieved decent accuracies for whatsoever problems it was applied to. In practice, however, when you train an ANN, the results may not always be good. Many times the system shows good performance in the training data set but fails in the testing data set. We highlight some major reasons or mistakes that designers commit one by one.

19.2.1 Number of Hidden Layers

The drive to achieve better performance very commonly leads to an increase in the number of hidden layers. It is common for newbies to increase the number of hidden layers without realizing the effect the addition of each hidden layer has.

Now being mature enough, we know that the ANN starts depicting very complex functions with each increase in hidden layers. This might increase the performance in the training data set, but it will result in very poor performance in the testing data set. The whole ANN becomes capable of depicting the most sensitive data that is characterized by many sharp changes in outputs even on very minute changes in the input. This is because of the great degree of freedom that the increase in layers gives. Now the ANN can imitate almost anything that one may imagine. The sad part is that this appearance is deceiving. The obtained ANN with the larger number of layers may actually perform better than the ANN with smaller number of layers in the training data set. This is again attributed to the high sensitivity that the ANN can handle. Many times this may even lead the researcher to conclude that the ANN has been trained, whereas in reality it might not be true. If this ANN is verified against a completely new testing data set, the results might reveal that the errors are high, to that extent that the ANN might be completely worthless. Many times not using a testing data set and giving results only on the basis of the training data set might lead to very wrong conclusions.

Further every hidden layer leads to a tremendous increase in the dimensionality of the problem in the error space or the configuration space, where the training algorithm is now supposed to find the global optima in a larger section of more dimensions. This makes it very hard for the ANN to find global optima. Also as the function predicted is very sensitive, the configuration space might be filled with too many mountains and valleys. Hence the ANN is very likely to keep getting trapped in these valleys and have no idea how to reach the final global optima.

In order to better understand this, let us train a data set with outputs corresponding to some known function by two ANNs. One will have lesser dimensionality, and the other will have a high dimensionality. The results are shown in Figure 19.1.

In general the number of hidden layers must always be one. If the ANN does not work out and completely fails to get trained, then the number of layers may be increased to two, but the increase must always be controlled, as it might hit the generalization aspects of the ANN very hard. More than two hidden layers are never used and must always be avoided. If the ANN is failing to get trained, finding better attributes might help more than working over the number of hidden layers.

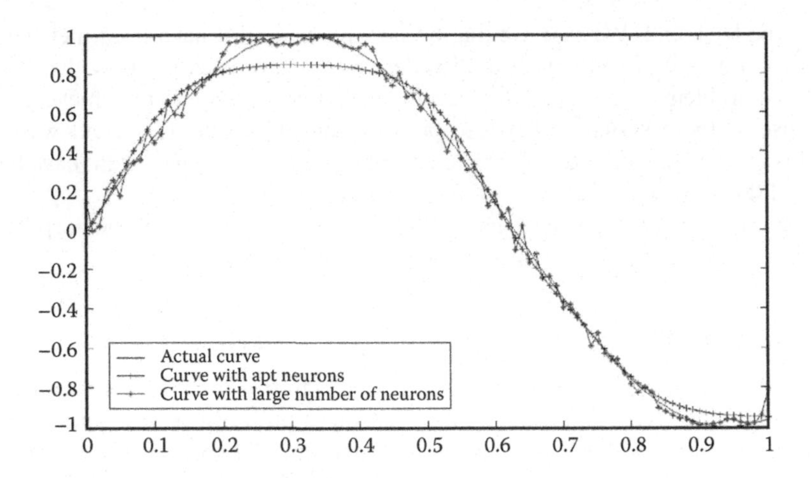

FIGURE 19.1 The artificial neural network with less and more of layers.

19.2.2 Number of Neurons

Many people are aware that adding hidden layers is a bad thing. Many somehow get accustomed to the use of only one hidden layer. But they again develop another very bad habit, which is to increase the number of neurons in the hidden layer. Again it is the same drive to increase performance or, in other words, to see the training curve show fewer errors that makes people adopt this habit.

The results of the increase in the number of neurons are more or less same as the results of increasing the number of layers. The difference is that the former is less suicidal than the latter. The effect of increasing the number of neurons is not as big as that of increasing the number of layers. Effectively, the number of neurons denotes the turns that the function depicted by the ANN is allowed to take. A function involving more turns takes a larger number of neurons than a function that involves less turns. The addition of every neuron adds to the degrees of freedom that ultimately empower the ANN to make a nonlinear movement in the function depicted. It may be recalled that the neurons used in the ANN had a nonlinear activation function that aided in the use of these networks for nonlinear problems. In classificatory terms, this was to separate the classes with nonlinear functional depictions, and in functional approximation, it was to depict a general function that might not be linear.

In order to prove this, we do another experiment. We use the simple *sin* function that has been scaled in the x-axis. We solve this for three functions that have increasing number of curves. We see that the number of neurons keeps increasing as we increase the number of turns. The plots for three versions of the functions along with the imitated functions are given in Figures 19.2a through 19.2c. The graph shown in Figure 19.2a used three neurons. Similarly the neurons used for Figures 19.2b and 19.2c are 5 and 7, respectively. Below this number of neurons, training was not possible.

19.2.3 Learning Rate

Learning is another factor that plays a major role in the training of the ANN. The large learning rate results in the system failing to converge to minima. The system keeps oscillating between high error values. On the other hand, a very low learning rate results in a very slow movement toward the minima. Hence the learning rate at one end determines the amplitude of the oscillation of the error values in the error minima valley. At the other end, it determines the speed at which the system moves toward the minima.

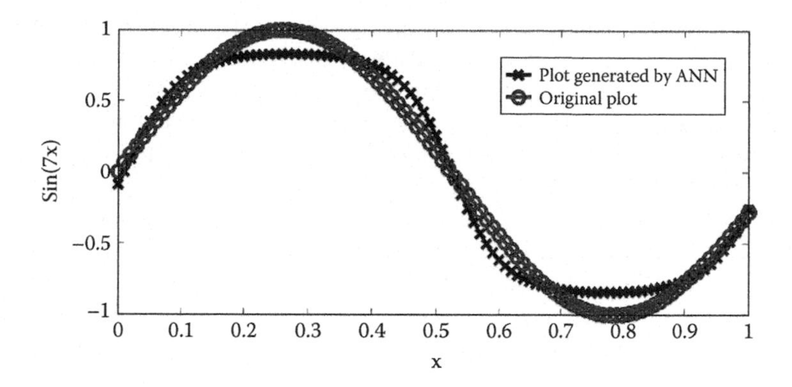

FIGURE 19.2a Increasing turns need more neurons: curve with 3 neurons.

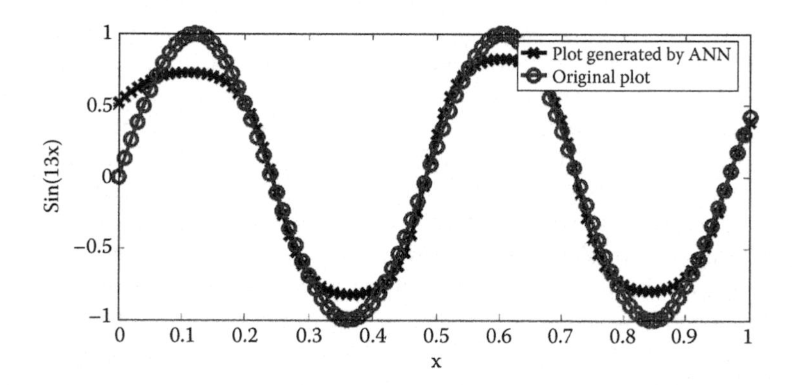

FIGURE 19.2b Increasing turns need more neurons: curve with 5 neurons.

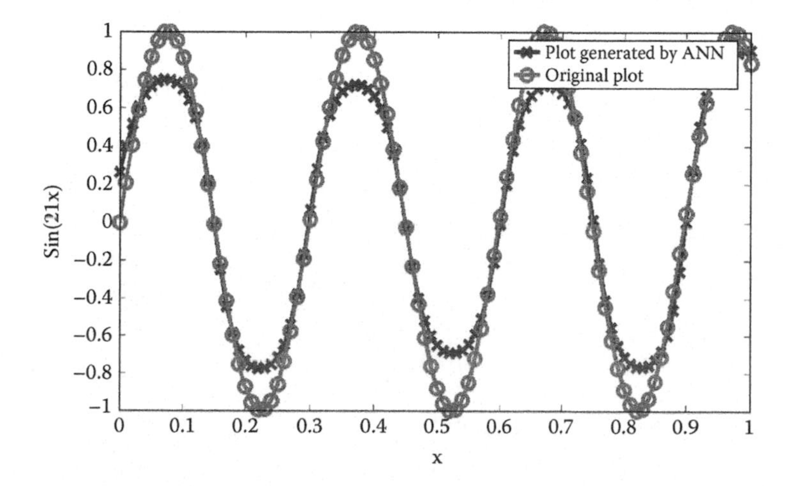

FIGURE 19.2c Increasing turns need more neurons: curve with 7 neurons.

The effect of minima is often difficult to see and observe in the training of the ANN. This usually results in people not experimenting much with the learning rate. But many times people may see the training curve converge at some point much before the goal. People may be tempted to react by increasing the learning rate. On the other hand, knowing about this problem, people may keep the rate too low to get the best final system and may advocate keeping the training on for a long time. It may, however, be observed that both these approaches may fail. The solution to the convergence problem might not lie with the learning rate at all. Increasing the learning rate might result in putting the system at stake. At the same time, training the system for a very long time with a very low learning rate might actually not benefit much. The final system may still be undertrained. It may be recalled that we follow a steepest descent rule where the bulk of the performance improvement happens at a first few iterations. The last ones are more or less passive without a large performance boost.

For the graph depicted by Figure 19.2a, we try to study the same effect of the learning rate. We try to see the effect produced in the final output if we keep increasing the learning rate. We saw that when very small number of epochs was allowed (300), very low and very high learning rates gave a very poor performance. The performance was best for the medium learning rates. Figures 19.3a through 19.3c show the graphs when the system was trained with learning rates of 0.001, 0.1, and 3, respectively. Figures 19.4a through 19.4c show the corresponding training curves.

Observe Figure 19.4c. It shows numerous ups and downs in its training plot. This is a very interesting behavior and deserves a separate discussion. We spoke of oscillations, and these are clearly visible in the training curve so obtained. Let us understand what in the configuration space ultimately leads to these oscillations. Consider Figure 19.5. It shows the error when plotted against any one parameter that can be adjusted by the training algorithm. This means that in order to facilitate better understanding, we are taking the error or the configuration space to consist of a single dimension rather than being n-dimensional. Suppose at any time in the training, the algorithm is situated at some location A. Now we understand that the steepest descent approach would try to make the system reach the point O, which is at the minima. This would mean a rapid change in position toward O. The learning rate being high further results in a big movement. Suppose as a combined effort the system reaches a point B. Now again the same principles hold. This time the movement will lead in crossing over the point O and the system reaching a point C.

Now two things may happen as the BPA tries to place the system toward point O. The first possibility is that the system, in its attempt to reach O, overshoots O and reaches the point B backward. This would mean that the next time the system would reach point C, then B, then C, and so on.

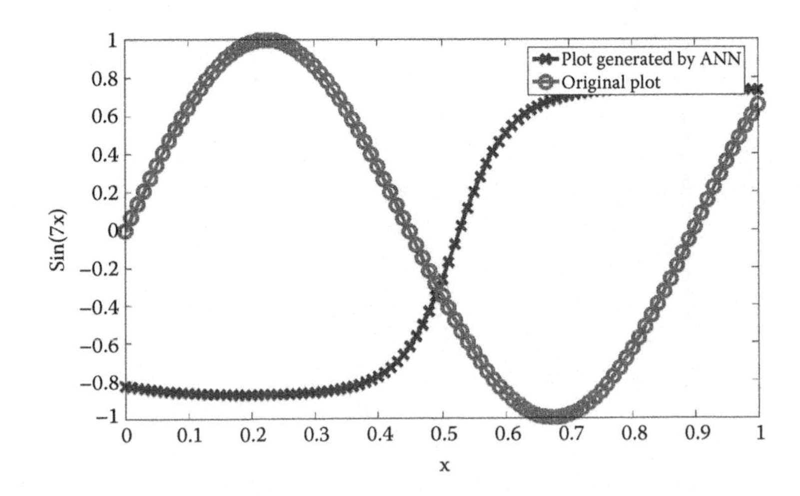

FIGURE 19.3a The effect of learning rate in system performance: very low learning rate.

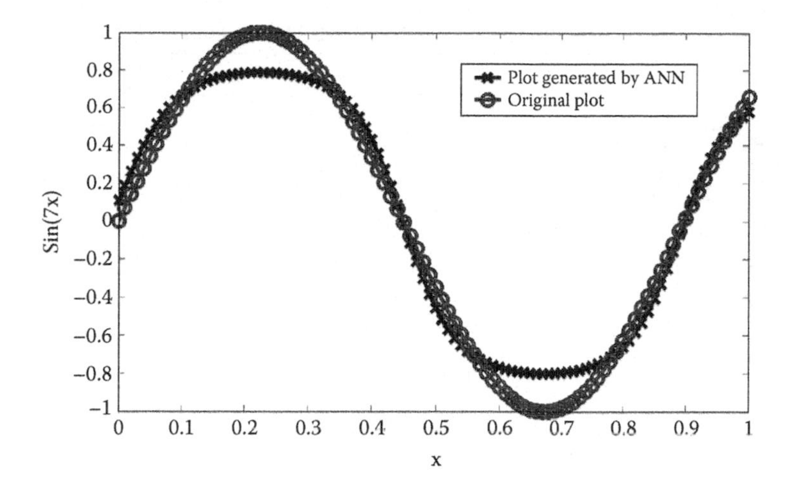

FIGURE 19.3b The effect of learning rate in system performance: moderate learning rate.

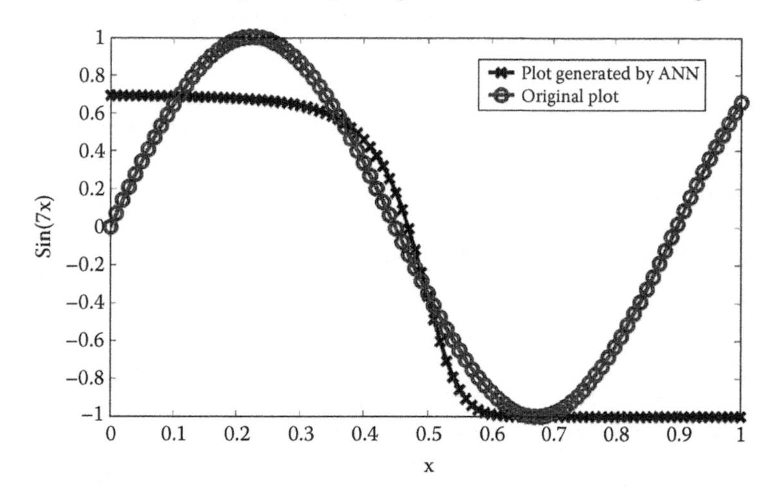

FIGURE 19.3c The effect of learning rate in system performance: very high learning rate.

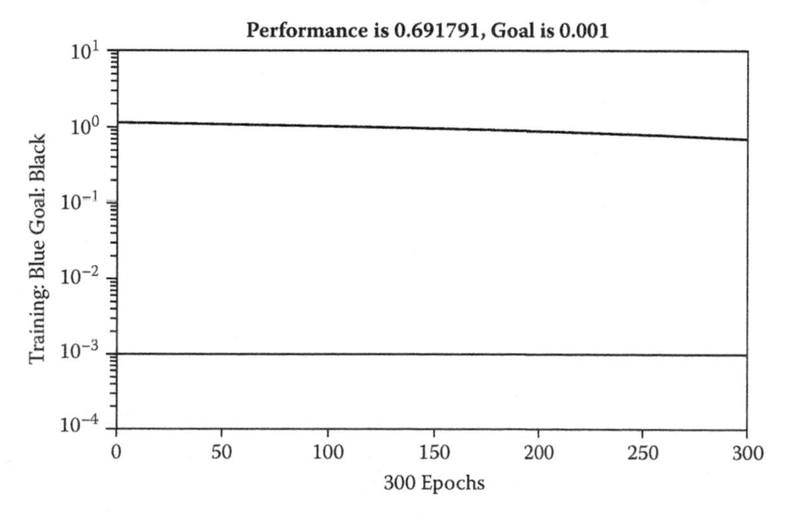

FIGURE 19.4a The effect of learning rate in training curve: very low learning rate.

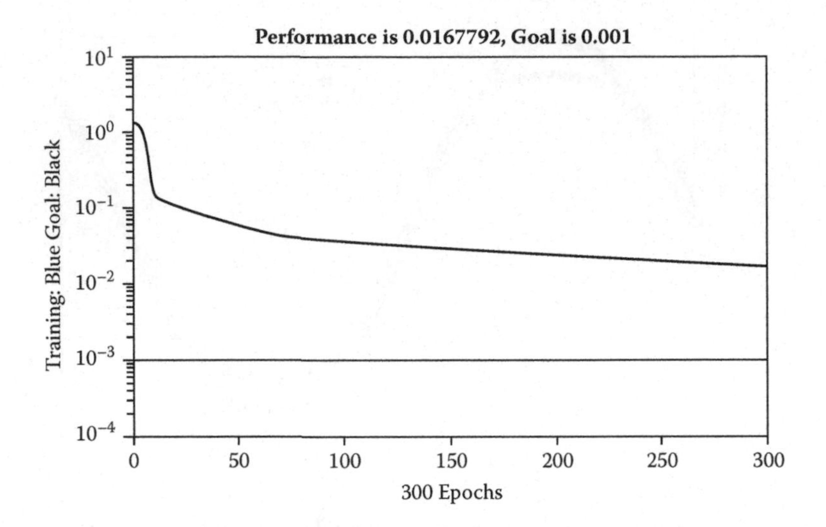

FIGURE 19.4b The effect of learning rate in training curve: moderate learning rate.

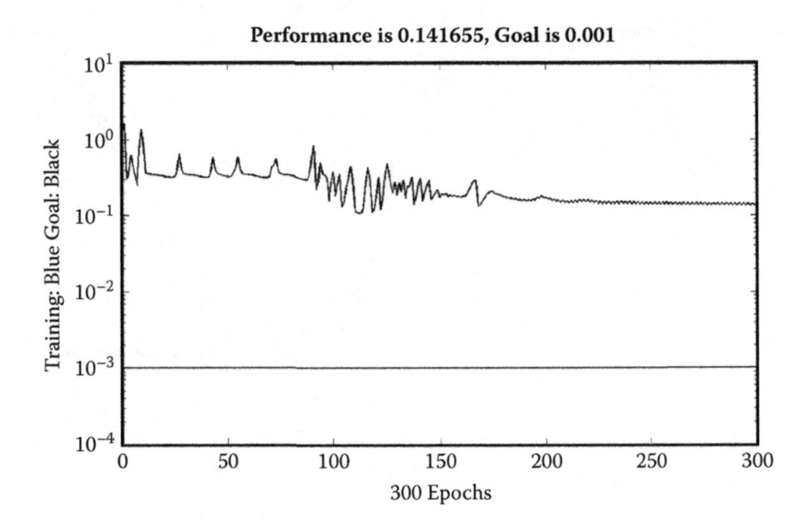

FIGURE 19.4c The effect of learning rate in training curve: very high learning rate.

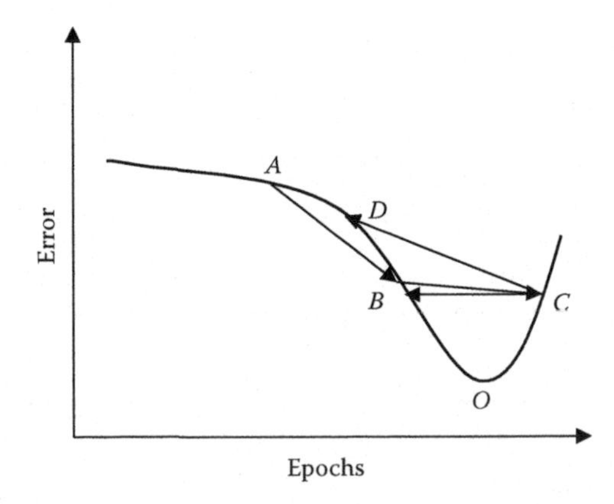

FIGURE 19.5 The error plot in the error space.

This is what we refer to as oscillation. Now as we see, the errors at points *B* and *C* are exactly the same. Therefore the training curve shows no change in error and there is a line parallel to the *x*-axis in the training curve.

The other option is even more exciting. After the system reaches point *B*, it may happen that it lands at point *D* at the next jump. Here the error is larger than that of points *B* and *C*. Depending upon the location, the next jump might take the system at some point of higher or lower error. This would go on and on. Since the error is decreasing and increasing, the change will be visible in the training curve over some region. This is something that we see as visible oscillations in the training curve.

19.2.4 Variable Learning Rate

Now we have seen the concept of learning rate in ANN with BPA. Here we discuss another, related concept, that of variable learning rate. It is clear from our previous discussion that the learning rate plays a key role and must be chosen judiciously for better performance of the algorithm. The algorithmic performance depends upon the learning rate to a very large extent. Choosing a constant learning rate is disadvantageous, as advantages of the larger and slower learning rates are not regarded. The faster and slower learning rates, on the other hand, are useful at different amounts of times.

Initially, the system is at a point where the minimum is quite far off. Hence the learning rate must be high in order for the system to approximate the minimum as soon as possible. Hence the learning rate is usually kept high at the start of the training. Now as the system approaches the minima, there are chances of oscillation due to the high learning rate. Hence at this stage we need to slow down the learning rate. Also the lower learning rates are preferable, since the system is quite close to the minima and hence will be easily able to reach them without much loss of time. Hence a lower learning rate is desirable. In other words, we start with larger learning rates and in the course of time we keep decreasing the rate.

But here we are making a big assumption. This is about the absence of momentum. Ideally the momentum will pull out the system the moment it spends some considerable amount of time near some minima. This is to save the system from being trapped in local minima. Hence we need to incorporate this in the algorithm as well. Whenever the momentum pulls the system out of the local minima, the learning rate should be kept high. This will enable the ANN to reach the next minima as soon as possible.

In short with the use of a variable learning rate, the basic strategy is simple. We see if some movement is resulting in an increase in error or a decrease in error. If the error is decreasing, it means that the ANN is marching toward the minima. We hence increase the learning rate. This is done with the belief that the ANN will not be in an oscillatory position till the error starts increasing. If, on the other hand, the error happens to increase, we interpret this as the start of possible oscillations. We hence decrease the learning rate in such conditions. Similar is the contribution of the momentum, where the increase in error means a decrease in the learning rate.

Figure 19.6 shows one of the graphs of the system. Here both the training curve as well as the learning rate have been shown. Here it may be seen that the training error keeps reducing uniformly while the learning rate changes a lot. Learning rate may be seen to oscillate between values. This is a combined result of the momentum and the position with respect to the minima.

19.2.5 Momentum

The next important factor for the system is the momentum. The momentum is a tool to pull out the ANN from local minima. This ensures that the ANN does not get trapped in any local minima. This increases the chances of the system to reach the global minima.

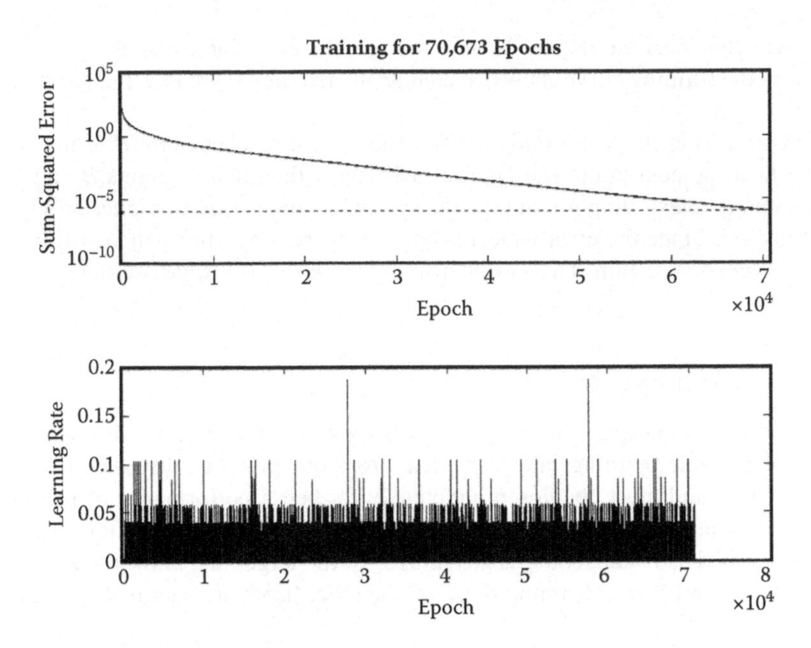

FIGURE 19.6 The variable learning rate.

Again the effect of the momentum may not be very easily seen, and that often results in not a lot of experimentation with this parameter as compared to the other parameters in the ANN. People may often try to suboptimize their problem with the hope of getting lower errors in a way that may not be globally best but still within desired limits. This may make them fix a low value of the momentum. However, this is not the correct move. The local minima might give a very high error. On the other hand, some good minima might be somewhere near. Lowering momentum unnecessarily suboptimizes the system.

Another set of people may believe in wandering and trying to cover the entire configuration space with the hope of finding the optimal point. This results in keeping very large values of momentum. However, this again has limitations. The high momentum will restrict the ANN from going into any level of depth to search for the minima. Even if a global minimum is in the vicinity, it might actually get missed as a result of the wilderness.

We had kept the momentum zero when we studied the learning rate. This made it possible for us to cleanly study the learning rate without the effect of the momentum. But unfortunately the converse is not true. We naturally can't keep the learning rate zero to study the effect of the momentum. Keeping this in mind, Figures 19.7a and 19.7b show the training curve with momentum as 0 and 0.98, respectively. Figure 19.7a shows the system eager to converge at some point. On the contrary, Figure 19.7b is interested in exploring newer area and moving away from the minima. It may be noted that sometimes the zero momentum graph got badly struck at a minimum where the error was very high. In these times we had to repeat the experiment to find better answers where the errors are low. The graph presented in one such case where the ANN was initially in a region of low error. Figure 19.7c shows such a graph where the ANN is trapped in local minima and can naturally not come out. The corresponding training curves are given in Figures 19.8a, 19.8b, and 19.8c. It may be seen that when the ANN was initially in a region of low error, it readily got converged, giving good performance even though at a local minimum. On the other hand when it was at local minima with a large error, it gave very bad performance. With large momentum, it was always trying to go out in search of newer areas, and thus the performance was not that good.

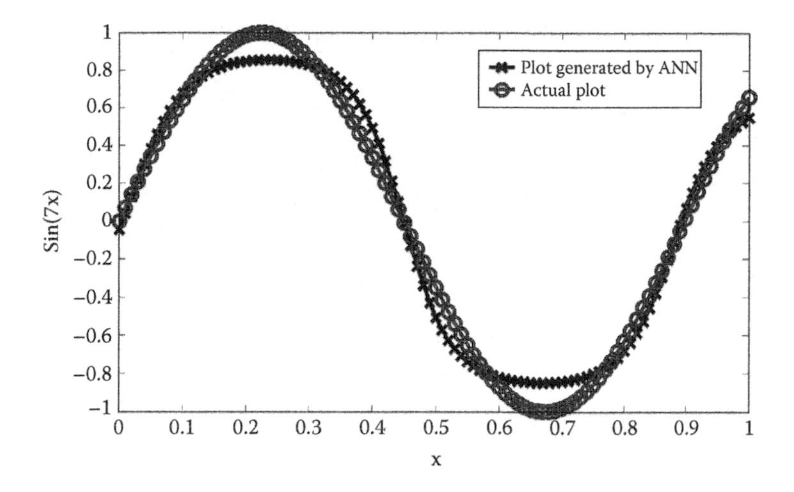

FIGURE 19.7a Output curve with zero momentum in global minima.

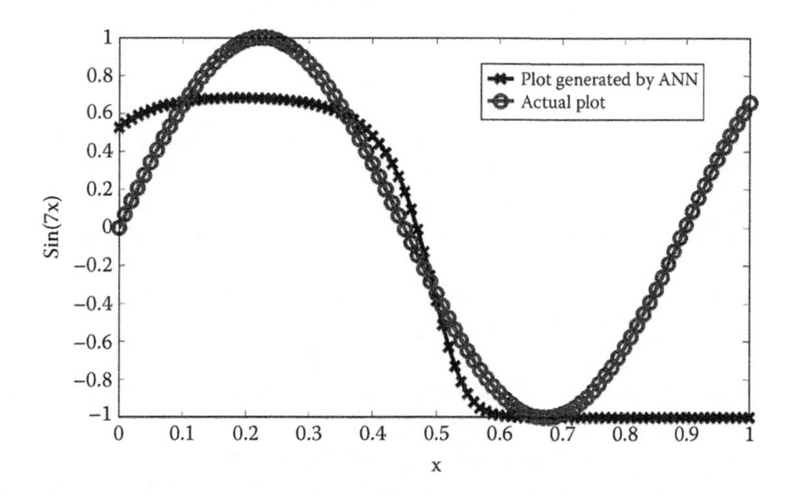

FIGURE 19.7b Output curve with high momentum in global minima.

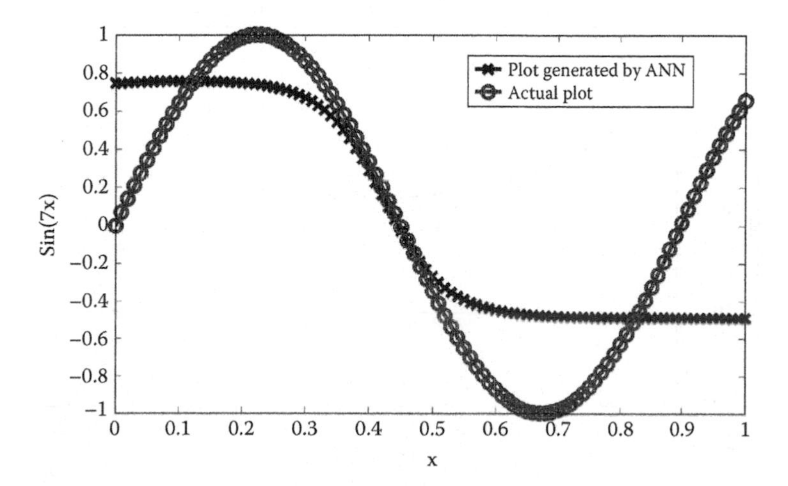

FIGURE 19.7c Output curve with low momentum in global minima.

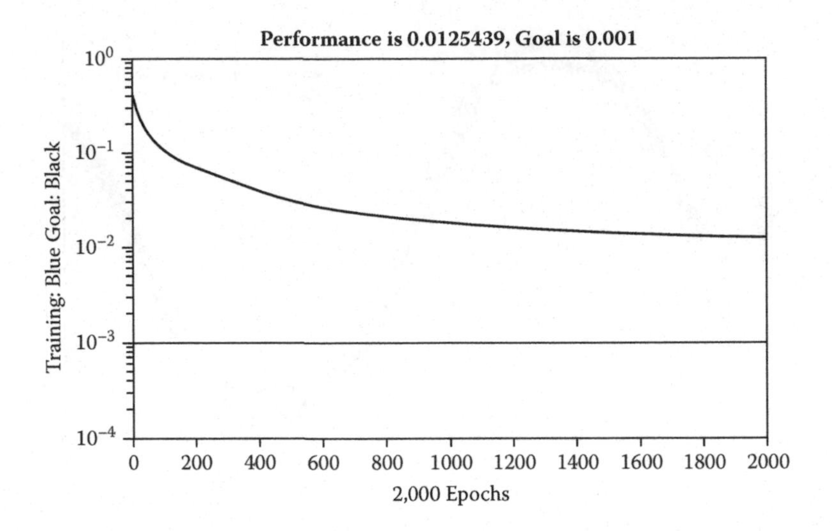

FIGURE 19.8a Training curve with zero momentum in global minima.

19.2.6 Training Time, Epochs, and Overlearning

The ANN curve is many times viewed as the curve where the error keeps reducing with the number of epochs. Even though the change is very small, people like to invest time. The natural solution that might click in their mind is to let the ANN get trained for a day; we'll at least have the error reduced. The next day, you actually see the training curve in much the same state with almost the same performance.

It is natural that misconceptions lead to waste of time. An ANN trains itself to the maximum in the first little iteration. This is due to the steepest descent approach that is followed by BPA for the training purposes. Afterward the training makes much less difference. Now readers may again make the mistake of supposing that the number of iterations may be appreciably small, especially when they only care for a major improvement in the system. But this again will not help, because the ANN needs time to get out of the local minima, get retrapped in others, and so on till it finds global minima. One such situation is given in Figure 19.9. Hence many times the ANN training curve may

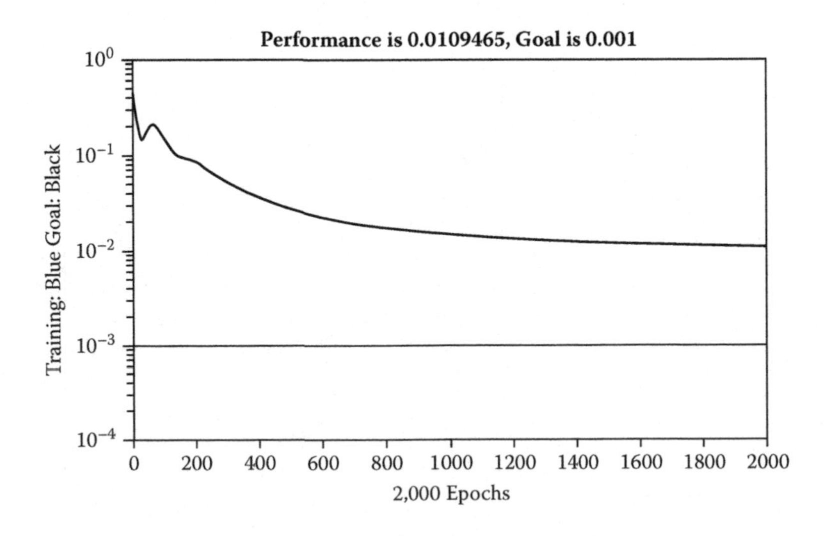

FIGURE 19.8b Training curve with high momentum in global minima.

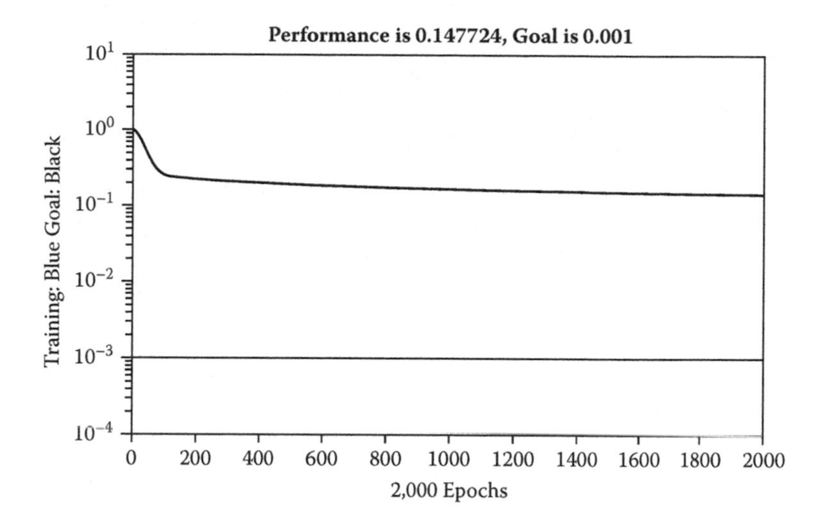

FIGURE 19.8c Training curve with low momentum in local minima.

seem to have trained as much as it was possible, and then suddenly it gets out of a local minimum and the training error suddenly reduces. Of course it is possible that leaving the ANN for one day might result in the exploration of some global minima, but chances are practically rare.

Many times the ANN carries neurons more than usual. This results in the ANN being able to imitate more complex functions than it usually may. Training for longer times or larger epochs may actually result in the ANN having a very long time to complicate the actual function. This basically happens as the ANN tries to keep the errors as small as possible and in the process ends up in overfitting the training data points. Hence the training may reduce the system performance if tested on the testing data. For this the training must be carried out for a limited time and number of epochs. The unnecessary complication of the imitated function by increasing the number of epochs is termed overlearning. Looking at the complex nature of curves of most real life systems, it is natural that the ANN will have extra neurons, which increase the possibility of overlearning.

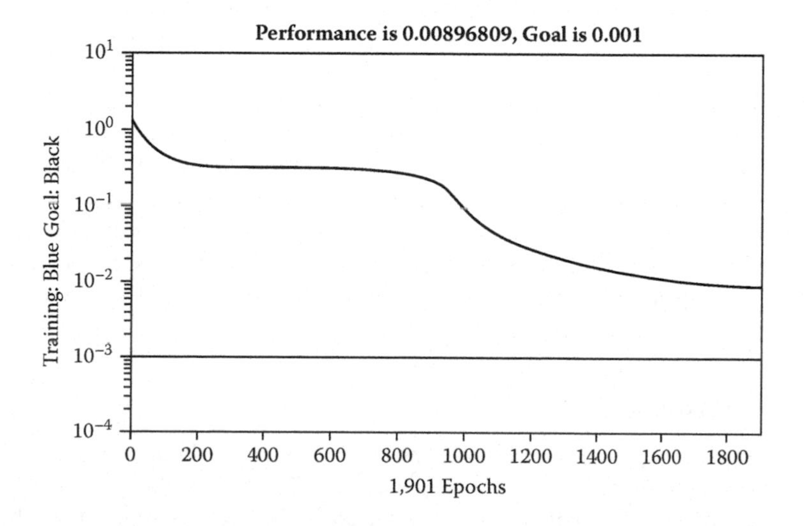

FIGURE 19.9 Effect of epochs and time on training.

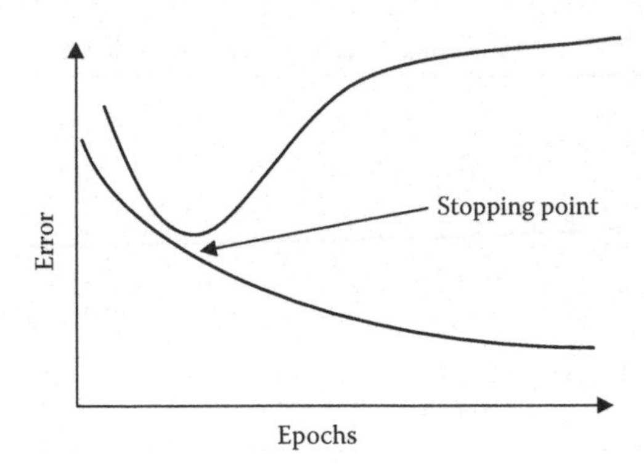

FIGURE 19.10 The early stopping.

19.2.7 Validation Data and Early Stopping

Here we continue our discussion from the preceding section. So we know that the ANN must be trained for a limited number of epochs. But the question is how to decide the actual number of epochs. This may be done by using the concept of validation data. These are special data items that are reserved to carry forward validation. As we start training the ANN, first the error on both the training and the validation data will decrease. However, as we continue, the error on the training data will decrease and that of the validation data will increase. This is the region when the ANN is complicating itself more than required degree. This is the time when it must be stopped, as further training will result in reduced performance when applied to the testing data set. This stopping of the ANN training is referred as the early stopping. This is given in Figure 19.10.

19.2.8 Goal

The goal plays a role only as a member of the stopping criteria of the algorithm. The training stops immediately when the goal is met. This is more or less correctly applied by people when they first try to experiment and find the minimal goal that the system easily reaches. The actual goal is set based on these observations. It may be noticed that the goal may scale the error function such that many times the same training curve may be interpreted differently for different goals. We may look at a training curve and say that the training did not reduce the errors. But this might not be true. The illusion of low training might be because the goal might be too low. This contracts the entire graph.

One of the misconceptions that people might develop is that it is mandatory for the ANN to meet its goal for the testing to be carried out. People suppose that an ANN with its goal not met is worthless. Even if the ANN almost reached the goal, they will not test the system. Rather they work with the goal or other parameters in order to make the goal meet. This, however, is not a valid approach. At every stage of an ANN, the system is ready for testing with the general performance and the error level prevalent in that part of the training curve. Even if an ANN fails to train, it will continue to work with a performance or error level at the training data set as it was at the time the training finished and depicted as the last point in the training curve.

19.2.9 Input Distribution

We have talked numerous times, especially during the introduction, about the input distribution and the characteristics of good inputs of the training data sets. We again summarize some of the major points here for the purpose of completeness of this chapter. The inputs must be preferably with some finite range. Also the inputs must be normally distributed in this range, covering as many areas of the input space as possible. Another major feature is that noise must be as low as possible. The testing data best performs if it has enough training data available, preferably in the vicinity. For almost all real life problems, patterns exist in the training data that the ANN tries to exploit. For classificatory problems, the interclass distance must be as large as possible and the intraclass distance must be as small as possible for the best results.

19.2.10 RANDOMNESS

You design an ANN to solve some problem with BPA. You try again and again, and you finally come up with a solution that gives very good performance and less errors. You train and test the data and get very good results. Then you are asked to reproduce the results. You confidently run the code again only to find that the results are very bad this time.

This is a very common experience and may happen to many people: that the same code gives a very bad result which was earlier giving good results. The problem does not lie with the code. It's not that the system has become nasty. The issue is that while training and then testing, we usually forget a very important fact. This is that the ANN may be using random weights and biases as the initial parameters. Again let us visualize the entire problem in the error or configuration space, which is a highly multidimensional space consisting of the weights and biases as its axis. The ANN even before a single epoch assumes random values for all of these. This may be visualized as the ANN randomly places the system at some point in this highly dimensional configuration space. Then starts the journey aimed at lowering the error. There the training algorithm tries to find the local optima, get out of the local optima, and try to attain global optima.

Now if the random placement of the ANN places it at a location that is somewhere in a valley comprising the global minima, the system will readily converge to attain the global minima. If, on the other hand, the ANN is initially placed in a local minimum surrounded by lots of local minima, it will get trapped again and again at the minima. The performance will naturally be very poor no matter how long you train.

19.2.11 GENERALIZATION

Generalization is the property of the ANN that helps it to give correct results to the unknown inputs. We have discussed various issues and factors affecting generalization. These are the training parameters, input characteristics, and function and problem characteristics. To add to all this, we present another point of view of generalization in this section: the globalized learning strategy of the ANN. Here we discuss the role of ANNs as functional regressors where they fit any curve over the training points by varying their weights.

The ANN with BPA always tries to find global patterns, global rules, and global characteristics from the data presented. It tries to find a function (whose complexity is decided by the number of neurons and layers) that best fits the training data points. Each and every neuron has a contribution to make here. The change in weight of every neuron affects the output of all the training data items. The whole function may change as a result of the change of one particular weight. This is the globalized nature of the use of ANN with BPA, where all the training data set items are seen as a global function that the system is asked to imitate.

There is nothing localized here. This means that we can never say that the output is not good for this particular input. So we may do something so that the output of only this particular input gets improved without disturbing the rest of the system. Here each input is evaluated against the global trends, and the modifications carried out in the weights are dependent on the contribution of this input in the total set of inputs given to the ANN as the training data set. Hence the ANN may completely reject some input or class of inputs and disregard them, as they might fail to fit in what the global understanding of the ANN reveals. It is also possible that some set of inputs affect the ANN to the extent that the trends present exclusively in this part of the input are generalized as global

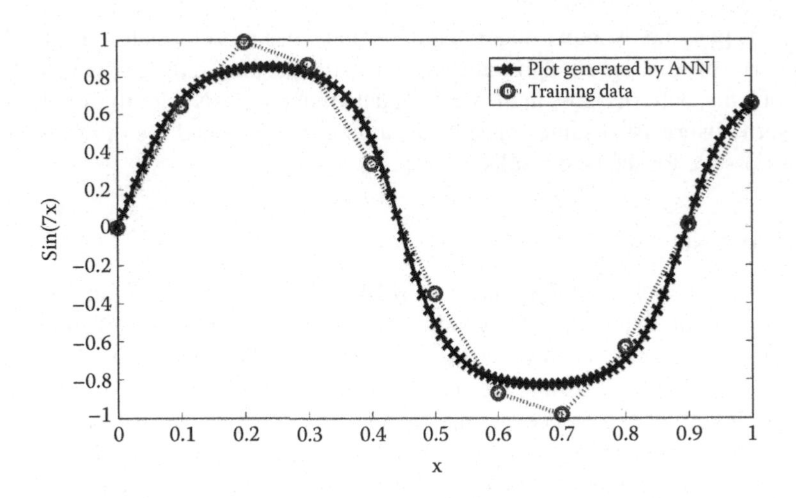

FIGURE 19.11 The artificial neural networks as regressor.

trends and the system implements them at the globalized level. Figure 19.11 shows the training data set and the predicted function by the ANN. It may clearly be seen that the ANN is basically trying to apply some sort of regression on the training inputs.

Now the natural question that would arise in one's mind is how to try to make the system behave better with some particular inputs for which it is giving bad outputs. Even though we advocate and experimentally presented in the text that most real life applications follow global trends and generalizations that make the ANN with BPA capable of solving these problems, this problem exists in numerous real life applications. These applications have some characteristic data that may not necessarily be noisy, but the system always gives bad performance, especially with these inputs. Since this problem exists in real life applications, it is necessary to devise methods to solve these problems.

There is no general method to solve these problems. The only suggestion may be to either change the model of ANN being used or try collecting some more data and adding it to the database. The former is an achievable method. One may exploit the use of LVQ, RBFN, or any other method for the purpose of problem solving. The latter is a very different approach that may not always be feasible. Many times costs and operational feasibility restrict the addition of new data in the data set. The problem is even more pronounced because the data set may usually be present with some data that is very rare and hard to find. This includes unique combinations of attribute values that are not normally found. Hoping to find more of this type of data may naturally mean being too optimistic.

19.2.12 Global and Local Optima

We have discussed these terms of global optima and local optima many times in the chapter so far. So let us clearly study these in this section. The global optima refer to the position where the error is minimum using any combination of the weights and biases. Usually the entire configuration space will have a single global optimum. On the other hand, the local optima are many in number. These are the places where the error is lowest in the surrounding region.

Let us again visualize the error or configuration space where the error is plotted against all the weights and biases in a highly multidimensional space. This space will be filled with valleys and hills, where each point plots the error corresponding to the combination of weights and biases as the other axis. These hills and valleys are traversed by the training algorithm. Local minima are the bottommost points of a valley. The valley with the least value of error corresponds to the global minima. This is shown in Figure 19.12.

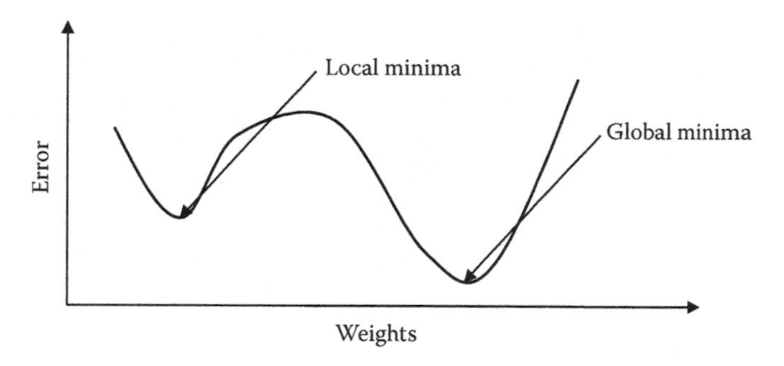

FIGURE 19.12 Global and local minima.

Now another interesting question is, whenever you are at a minimum, how to decide whether it is a local minimum or a global minimum. This is one of the most difficult or rather impossible questions to answer. The only thing you can do is to explore as much as possible. Try to find out all the valleys and their minima. At the end, the one with the lowest value is regarded as the global minima and the rest as local minima. Two points may now be noted. First, we may never reach the bottommost point in any minima; we may rather only reach a point very close to it. Second, it is only our belief that the point we reached was the global minima. There is no proof or guarantee to that. It is possible that global minima actually lie somewhere very far apart. The truth is that the whole configuration space can never be constructed in totality. It can never be explored fully. We can only explore a very minute part of it. We are asked to find the minima in such a situation where only a minute part can be explored given the finiteness of the time. The beauty of the training algorithm lies in how well it is able to use this constraint and come up with an effective solution that the real-life application uses.

19.2.13 Noise

The ANN with BPA is very strong in handling noise. Here the system continues to perform well even if there is some noise in the training data set. In other words, the noise has a very little effect in the system performance. The system is capable of handling noise at a very good level. This may again be attributed to the globalized nature of the ANN. The ANN with BPA basically tries to form global rules that can imitate the actual function. This means that the system is trying to find the functions that the training data match to. Now suppose that the system has noise. This will make the outputs deviate from the actual outputs. This means that now the ANN will have problems in understanding the actual function that it is trying to build. Since the training data set has errors, it gets difficult to trace the global patterns or rules out of the system. But still the ANN is able to appreciably link everything and come up with good functions that are able to imitate the desired functions despite the noise levels. This is again due to the global nature of the ANN, where the global patterns are mined, and this results in selection of good patterns despite noise. The ANNs try to make patterns or rules that give the least errors or fit in the outputs to the greatest extent possible. Many times this results in lowering the error rate. The rules so mined are the actual rules, and the rest may be the errors. Whatever the case may be, it may be seen that however generalized or globalized the system is, the effect of noise will be reduced. This is because the output is not disturbed much by the temporal noise in the vicinity of the inputs. This is the case with most localized networks where the output largely depends upon the surrounding outputs.

The errors in the ANN may be of two types. The first are general errors that are present in the entire data set. This forms a very practical situation. Say the training data set is recorded by the use of microphones. Now background sound will act as noise for all the recorded data sets. We assume here that the sound is low enough for the system to perform. This sound will again have an impact

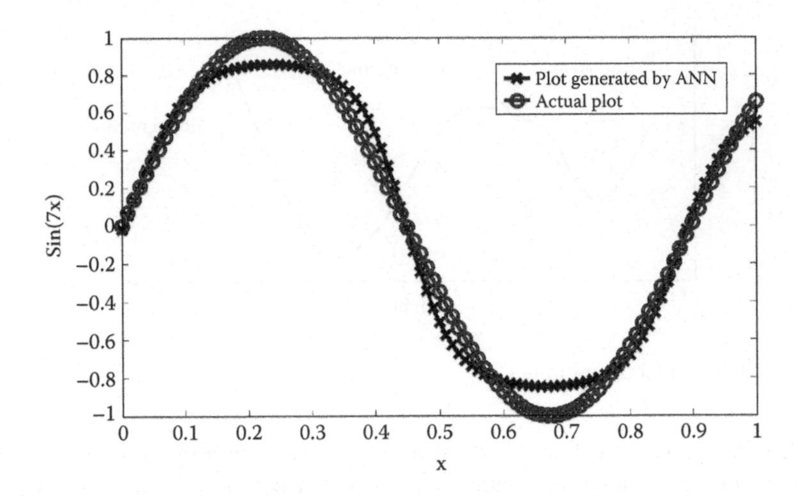

FIGURE 19.13a Original training and testing plot.

on the extracted features. All the data set will be affected as a result of this noise. The ANN is able to handle this noise as the global trends are sought.

In order to study this, we induce some random error into the entire training data set. This makes it difficult for the system to predict the correct output. Figure 19.13a shows the actual training data and the function obtained while testing. Figures 19.13b through 19.13f show the same with the addition of 10 percent, 30 percent, 50 percent, 70 percent, and 90 percent noise in the training data set. It may be noted that the number of neurons for this test was restricted to three, which means that to a large extent the ANN was forbidden to imitate very complex functions. This may not necessarily be true for most real life applications. Also the function used for experiment was rather simple and could be easily imitated by the ANN.

19.2.14 Classificatory Inputs

We have talked about the manner in which we handle the classificatory outputs. We solved all the identification and authentication problem using this. However, classificatory attributes may be

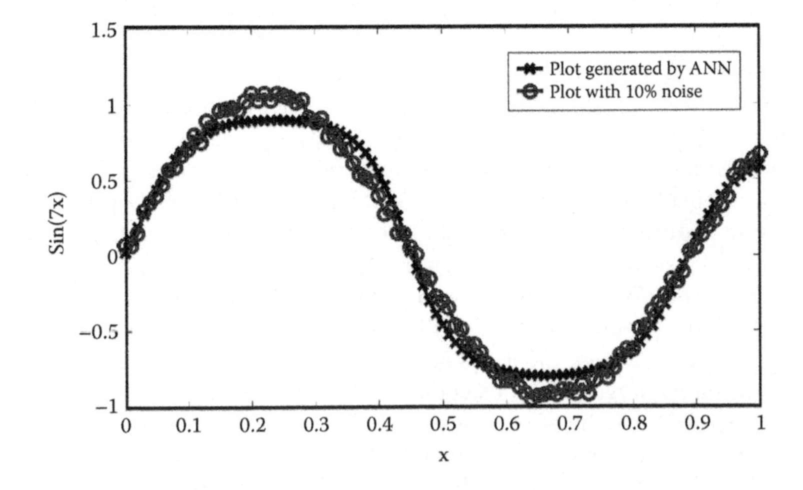

FIGURE 19.13b Plot with 10 percent noise.

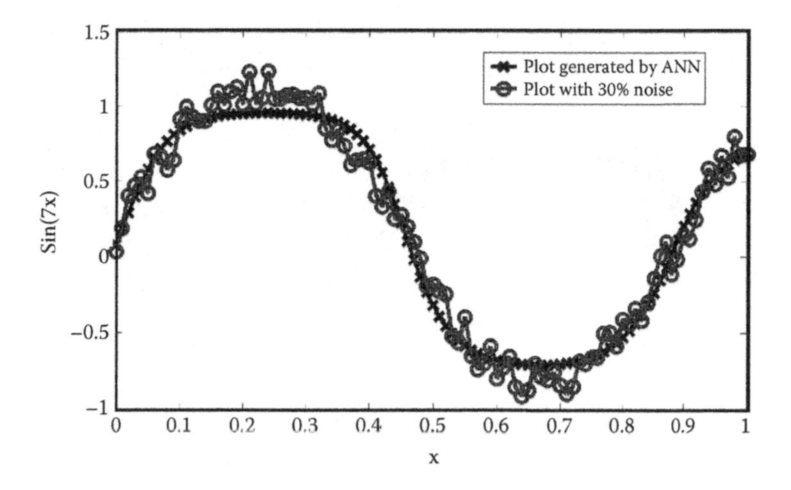

FIGURE 19.13c Plot with 30 percent noise.

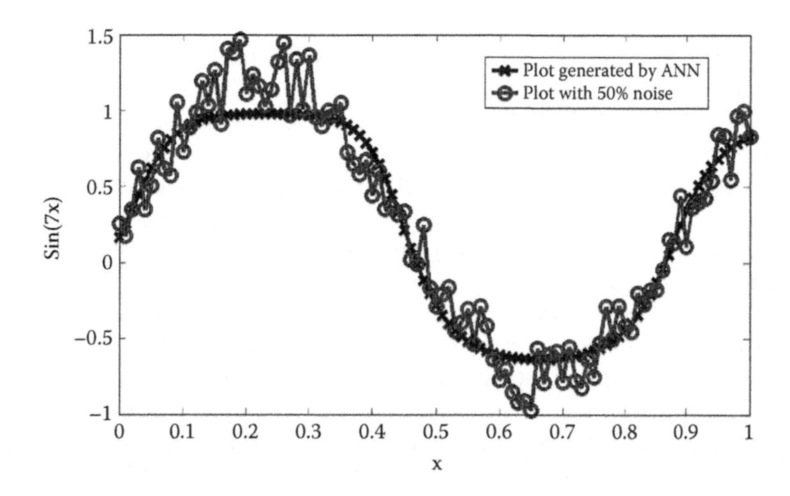

FIGURE 19.13d Plot with 50 percent noise.

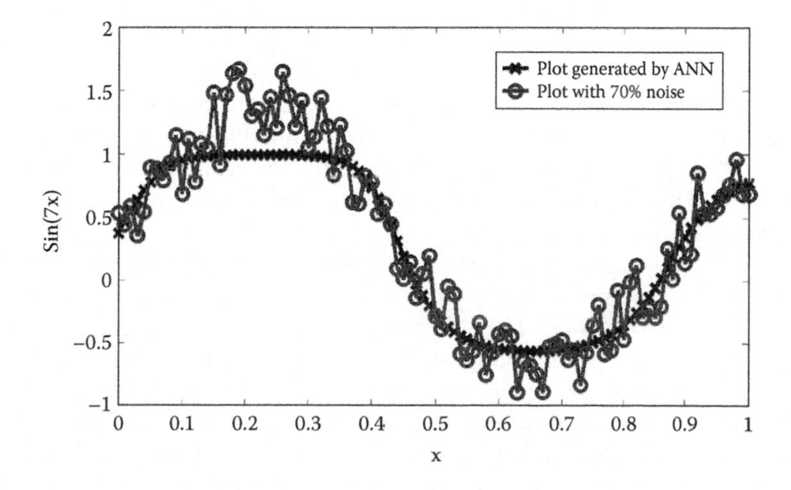

FIGURE 19.13e Plot with 70 percent noise.

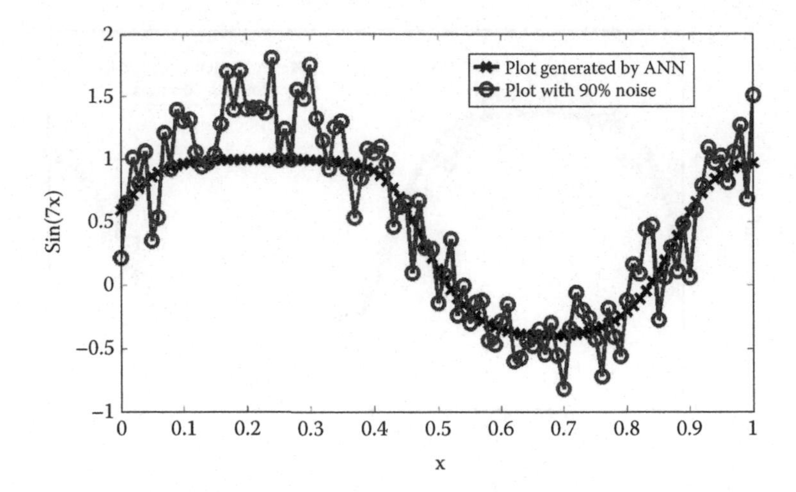

FIGURE 19.13f Plot with 90 percent noise.

present in the input as well. Many of the problems we solved had classificatory inputs which we substituted with some numeric counterparts. This is in fact the usual technique that is followed. In this section we study the problems associated with this method along with the solution and the associated issues. It may again be seen that one of the easy ways of solving the problems in the case of classificatory outputs is to assign them numeral equivalents. We did not opt for that approach for the reasons we have mentioned.

We first consider the case in which the input attribute is classificatory. Let the attribute be such that it takes values out of n classes associated with it. Now suppose that we assign numeral equivalents to it. The BPA mainly tries to use norms or distances for the calculation of the errors and modification of weights and biases. It further uses the norms or distances for the prediction of the correct output. Suppose that some input has an output with this attribute set to 1. Now suppose that the output has the same inputs to other attributes and the input to this attribute set of 2 is missing. Now the ANN is supposed to use its generalizing powers to calculate the correct output. Since the output with the parameter set to 1 is available and 1 is quite close to 2, the ANN may err in assuming that the output should be similar to when class was 1. This may, however, not be generally true. If the same experiment was repeated with class taken as n, it is likely that the output might be completely different. The ANN may assume class n to be very far away from class 1. Hence the ANN is taking the distance between classes n and 1 to be very large as compared to classes 1 and 2, which is wrong. Both these are equidistant in practice. This may hence lead to bad answers. Even if the training data set is available for the input corresponding to the attribute set as 2, it is likely that the output might be very different even though the inputs are interpreted to be almost the same. This will force the ANN to predict a very delicate function that changes very sharply on a very small change in the input. This is naturally wrong, and such sensitive systems have been shown not to perform well, especially on the testing data.

The effect of the classificatory inputs is more profound when the percentage of classificatory inputs is rather high as compared to the other inputs in the system. Also it depends upon the number of classes in the system (n). The system has bad performance if the number of classes is high. Classificatory data is a natural problem. It may be seen that this type of input in common understanding exists at either of the two extremes of the axis. This if seen with regard to the nonclassificatory inputs is a very bad thing, where we keep talking about a small change in input corresponding to a small change in output. Also the generalization is extremely difficult to carry out, as no information apart from the extremes at the two axes is available. The trends cannot be visualized.

It may be seen here that classificatory inputs like age group are by definition classificatory in nature. On the other hand, they do not face the problem discussed. Here the difference between the first class (say 0–30) and the second class (say 30–50) is actually smaller than that between the first and the third classes (say 50–70). Hence the norms or the numeric differences give a clear idea of the separation of these classes.

The classificatory outputs follow the same problem with the numerical replacements of the classificatory outputs. Here the difference between the actual and the desired outputs is calculated, and this decides the change in the output. The nearness or non-nearness of two classes with respect to their numeric representation hence plays a key role in deciding the system output. Imagine the input space. Here the various classes are scattered in the entire space. We discussed the role of high interclass separation and low intraclass separation in this space with respect to the classificatory problem. Suppose that the intraclass separation is not very high. As an extreme case, consider that the classes actually intermingle in the input space. Now the change in some input by small amount changes the class to which the input belongs or classifies to. Suppose this is a change from a class x to a class y given a small change in input at some region of the input space. Now if the x and y numeric representations are very far apart, we may clearly see a very steep change in imitated function value by the ANN. This makes the ANN very sensitive and necessitates a large number of neurons. This is, however, relaxed to some extent if the numeric representations of x and y are close. We do not state here that if the numeric representations are close, the system will be good, because the interclass separation may be low. The system still has a large change in output for a small change in input.

We have already discussed the solution to the problem in case of the classificatory outputs, where we took as many outputs as the number of classes, with each output denoting the probability of occurrence of a class. The same may be done in case of the input as well.

19.2.15 Not All ANNs Get Trained

It is very natural that you are given a database and you put everything behind its training. You let it train for days on end, train it numerous numbers of times, change the training parameters, and do everything that you may think of. But at the end, it may be possible that the system still does not give you the desired results. This naturally would leave you frustrated, because you were so keen to have a good system that you could not attain. Here it is important to realize that not all ANNs get trained. The BPA can only make an attempt to train the ANN. It does not have any guarantee that the ANN will get trained with good performance in the training as well as the testing data sets. However much you wish, you cannot force the ANN to train itself.

The lack of training of an ANN might happen for various reasons, which need to be worked on if you seriously want the system up and running. For many problems, other systems may perform much better than ANN with BPA, such as ANFIS or RBFN. Also the best possible option will be to work over the attributes. It may happen that some very important attributes are missing in the system, and that is causing problems in the training. This is especially important for problems where functional approximation is required. Here the attributes play a major role in deciding the system output. If any of them is missing, the ANN might not get trained. For the classificatory problems, many times the attributes may be redundant and may generally not be required. Many other times, the existing attributes may be enough to separate the classes. The other thing that may be responsible for the poor training is noise. The ANN can withstand noise to a large degree, but this always has disadvantages. The extra noise will reduce system performance by a very large degree, and the system might ultimately fail to train itself.

In the next three sections we repeat our discussion of three issues concerning ANNs with BPA. These are their use as classifiers and functional predictors. Having the background information presented in this chapter along with the experimental results of the other chapters and experience of the other systems, we are now in a better position to understand and appreciate this discussion.

Further, this completes all the aspects that the chapter was meant to cover. Here we do not present the machine learning perspective that we presented in Chapter 11, where we showed ANNs as effective tools to learn large volumes of historical data with much lower computational and memory requirements at the testing time. Readers may consider giving that chapter a second look.

19.2.16 ANN with BPA as Classifier

In classification we are supposed to find the correct class to which the given input belongs. The ANN we discuss here is based upon the classificatory model. Here there are as many outputs as there are classes. Each output here denotes a class. The output numerically stands for the probability or chance for the given input to belong to the class being considered. Suppose that the system has a record of n different users. Here there will be n different outputs for every input I of the form $<o_1, o_2, o_3..., o_n>$, where any general output o_i denotes the probability that the input I belongs to the class i. The probability can be a maximum of 1 and a minimum of –1. A probability of 1 denotes that the input for sure belongs to the same class, and a probability of –1 means that the input for sure does not belong to the same class. Hence whenever we apply any input, the system gives us n number of outputs. We now select the maximum of the outputs, and the class corresponding to it is returned as the answer. Ideally one on the n outputs must be 1 and the rest must be –1. But in reality this will not be possible for obvious reasons. Hence we select for the maximal output.

In case of the classificatory problems, the ANNs try to frame functions that have maxima around the regions that denote the presence of the class and minima at the regions that denote the absence of the class. These maxima and minima are spread around a very highly dimensional input space. The ability of the ANN to imitate this otherwise binary function decides the system performance. The performance indices here may be the ability of the ANN to imitate the actual binary function, in other words, the closeness of its output to the desired output.

The ANN with BPA classifier believes in performing the task of classification by generalizing the class and the class boundaries. It tries to predict the regions in the input space where each of the concerned classes is present or is influential. As a result of the training, it is successful in making some function that separates the region in the input space where the class is present from the rest of the region. The ANN with BPA is known for its generalizing capability. The generalization is the concept used in the training as well. Here the ANN tries to look at the global input space, and the decision at any point is the result of the globally available data in the data set. This makes these systems more resistant to noise in general, as the effect of any data in the data set is limited in nature.

These ANNs use BPA as the training algorithm, which attempts to set the values of the unknown parameters in such a way that the total error for the training data set is reduced. The BPA fixes the values of the parameters of the system or the weights in the same respect. It tries to keep the boundary separating the class regions and the nonclass regions as optimal as possible. If any data set belongs to the class, it tries to modify the parameters such that this training data set lies in the class region as comfortably as possible. This means that the surrounding areas are also likely to lie in the same class region. The converse is true if the data set does not belong to the class. Here the BPA tries to completely remove the data set from the region accommodated by the class. This will result in the parameters being adjusted in a manner that this point is nowhere near the region accommodated by the class. This may again mean that the surrounding points are not in the region occupied by the class.

As we increase the number of neurons in the hidden layer, the number of parameters in the system keeps increasing. This makes the system even more flexible. In the preceding paragraph, we discussed that if an input belongs to some class, then the surrounding inputs may also belong to the same class. This is because the ANN will occupy a larger region in the transition between the input space that belongs to a class and the input space that does not belong to the class. This transition is not sudden but requires some space.

If the number of neurons is far too many, the ANN will be very sensitive. This will enable it to make very rapid or sharp transitions in very little space. This is because of the flexibility being introduced by the large number of system parameters, which give the ANN the capability to imitate any function that may be as complex or unrealistic as possible.

The converse is true if the number of neurons or parameters is less. In this case, the ANN is restricted in attaining a complex architecture or imitating complex functions. It can only make a gradual transition from the class-occupying region to the non-class-occupying region. This inhibits a very sensitive structure. It may be noted that a very sensitive structure is highly unlikely due to the fact that the ANN can now imitate anything one imagines and hence the performance with the unknown inputs is very bad.

19.2.17 ANN WITH BPA AS FUNCTIONAL PREDICTOR

Here we discuss the role of ANN with BPA in functional prediction problems. For a functional prediction problem, the data here must be diverse, carrying good information on trends. In easy terms, the values of output at various locations of the input space must be known. This requires data to be normally distributed over the input space.

Any real life problem has a high dimensionality that stands for millions of possible options for the various inputs. This puts a serious demand on the training data set to be diverse and large. But this requirement is very difficult to meet when you practically design a system. Most of the data in possession might not cover the diversity, and getting good data might be too expensive in terms of time and money. As a result, we have to work over the system with the help of available inputs in terms of both number and quality. Out of the entire range of possibilities, we may observe that for any given problem only a few values actually occur. This limits the input space by a reasonable amount. Out of all these values, still fewer may occur commonly. We may sometimes observe the input lying at places scattered over a few areas of the entire input space possible.

In such a situation, the ANN with BPA does its best to find patterns and generalize the system globally. It is natural that if a datum comes to the system that is very distinct to those data sets that were used for training purposes, it might not lead to the correct output. On the other hand, if an input comes whose nearby locations are associated with known output, it is highly likely that the output will be correctly computed. This is because the training over that region must have fine-tuned the network well for inputs belonging to those regions.

The best part in the use of ANN with BPA is that they can act as regressors and can hence try to evolve functions that best cover the entire input space based on all the points available, even if those points lie scattered at some prominent regions in the input space. This power of regression is of a lot of use; it makes these the most generalized networks for better performance with functional prediction problems than any other network or model. The various points in the training help in modifying or tuning these networks, especially in the regions to which they belong. But at the same time, the global trends are respected. These ANNs make best efforts to correctly predict the outputs at the regions of known inputs as well as to correctly generalize the results over the entire input space to give good solutions in case of completely unknown inputs. This you may recall that this balance is maintained by the number of neurons. A very low number of neurons mean that we are allowing the ANN only to study the global trends and adjust itself accordingly. With a higher number of neurons, the ANN is able to adapt to the local regions and give correct output as per the demands of that region. This is because of the extra sensitivity added to the ANN by increasing the number of neurons. The number of neurons is a relative term that depends upon the problem and the behavior of the data.

Here, however, we assume that the inputs are complete and are the ones that affect the system performance, and also that the training had been judiciously done with the best possible architecture as regards generality. The generality of the ANN plays a big role in the multilevel generality of the system. The ANN is further able to learn as much as possible with the limited test cases.

19.3 OTHER ANN MODELS

Now that we have the background of the ANN with BPA and the various practical issues and concepts associated with it, let us turn our attention toward the other prominent models that we used in the text in connection with various real life problems. Most of the concepts are the same as discussed in ANN with BPA. Here we present some short notes on how the other models carry out the work of classification and function prediction. The input characteristics, output characteristics, and other parameters follow exactly the same guidelines presented in the preceding section.

19.3.1 Radial Basis Function Networks

The radial basis function network (RBFN) is the other ANN model that we discussed. We first discuss the roles of these networks in classificatory problems and then in functional prediction problems. Unlike ANN with BPA, which, as we discussed, is highly generalized, a different role is played by these ANNs, which are highly localized in nature.

The RBFNs select some training data sets that act as the data centers or guides for the rest of the training and testing data. The output to these guides can be assumed to be perfect. These are spread all across the input space. All these guides have parameters that are adjusted in the learning phase for optimal performance. Each guide has an effect on the entire input space. The effect decreases as we move further from the guide. The rate at which this decrease takes place is determined by the parameter of the guide. In order to know the output to an input, the influence of all the guides is taken into account and accordingly the final output is evaluated. Here the result is more dominated by the guides that were very close to the unknown inputs and very much less or negligibly by the ones that were far apart.

19.3.1.1 RBFNs as Classifiers

Dealing with localized networks, we expect better performance, as the classification problems are known to be highly localized in nature. In the training phase, if the input belongs to the same class, the guide tries to increase its output and vice versa. The output is mainly dominated by the guides that are very close to the input.

The localized nature of RBFNs works very well in the case of classificatory problems where every neuron covers a fixed set of regions that may even include the entire class. Each neuron has its effect reduced in the form of a radial basis function whose radius is a parameter that is adjusted at the training time. Ideally the function to be imitated is a binary function that has maxima at all regions where the class is found and minima at others. But in classificatory problems this is not a hard and fast rule. This function may even be replaced by a function whole value is above a threshold at regions where the class is found and below the threshold where the class is not found. Here we advocate that many times even a single neuron has the ability to represent the entire class region by using this property of the output function to be imitated. It will increase its radius to the order of the size of the class presence region. If not, then we can always put in more neurons to imitate the same functionality.

It must be noted that even the ANN with BPA actually is unable to imitate the binary function due to its characteristic nature where the outputs may not change at all over any change in inputs but then suddenly change. The same principle of a threshold applies there as well, except for the fact that it tries to predict the exact shape of this function as closely as possible using multiple neurons.

Now we know that the aim while training is to increase the final value of the output well above the threshold for the given input in case it lies at the region occupied by the class. Similarly, we need to decrease the output well below the threshold in case the input does not lie in the region occupied by the class. In this way, we intend to make the RBFN imitate the otherwise binary function. At every point, the contribution of all the data centers or guides is taken into account. We will see later that in order to achieve high generalization, the effect of every data center or guide must be the maximum at all other points in the input space. An RBFN in such a case tries to place the data centers such that they cover and dominate as wide a region as possible that denotes the presence of the class in the input space. Each guide dominates the region of its presence as well as having an influence at the other regions. The influence decreases as we move away. If the regions marking the presence and absence of the class intermingle with each other, a necessity arises for intelligent placement of these data centers or guides such that the net effect or output follows the threshold limits. One would observe that in the resultant system, the points in the input space well within the region marking the presence of the class have very large output values when compared to the threshold. At the same time the regions that lie at the class boundaries have an output close to the threshold.

In short, the RBFN works in a localized manner to predict a function that correctly classifies inputs into outputs according to the threshold values. The localized approach results in faster training.

19.3.1.2 RBFNs as Functional Predictors

For functional prediction problems, more or less the same concepts hold and work to make these great functional predictors as well. The RBFNs have a much localized approach where the output at a point is decided more by the nearby training data sets than the whole global set of inputs. This reduces the training time to some extent and may sometimes result in higher memory complexities. The result is that the output is largely dependent upon the outputs of the nearby inputs. The RBFN tries to find out the trends by looking at the nearby inputs, and, keeping all this in mind, it tries to calculate the output for unknown inputs. The nearer the data center or guide is to the unknown input, the greater is its role in deciding the final output. This influence decreases sharply as we move across or move farther in the input space. The training of the RBFN in case of functional prediction outputs involves the computation of the influence the various guides have at different parts of the input space.

Functional prediction problems usually involve smoothly changing functions as we move across the input space. These are usually marked by smooth surfaces that span the entire input space and whose behavior is more or less continuous in nature from which trends may be easily extracted and studied. The training incorporates the fixing of the contributions from the various data centers or guides such that the final outcome is as desired. Now there are two separate ways to study this. The first is what we have already discussed, to look at the local outputs that cast a deep impact on the output at any position. This presents a localized view of the RBFN, where it is possible to compute the output at a point depending upon the nearby outputs. This naturally is very easy for the smooth surfaces of the functional prediction problems. The second involves the globalized trends that need to be looked after. This is what happens when the distant guides make their contribution. They induce the effect of the global trends over these networks, and this makes them behave as per the demands as well as the requirements of the global map as well. Hence we are able to further predict the system output on the unknown inputs by all these kinds of inputs. They ensure that the continuity of the input surface is maintained and that the surface behaves as smoothly as possible considering the real life application.

19.3.1.3 Role of Radius of RBFNs in Generalization

These ANNs have a very important parameter that needs a special mention because of its functional characteristics and its effect on the resulting RBFN. This is the radius of the various radial basis functions, also termed spread. This parameter plays a very important role in controlling the localization as well as the globalization or the generalizing capability of the RBFN. This parameter hence has a big role to play and is usually set as a training parameter for the system's use in training itself. This parameter in the RBFN should be as large as possible, as it corresponds to the maximum degree of generalization. ANNs with less spread are very local in nature. Because of the same effect, they are more prone to noise and do not give very good output in the testing data set. The global trends that were a big source of global guidance before do not play a very important role here if the spread is less.

The spread looks after the level of influence that each of the data centers or guides may have over the other regions. If this is large, the guide has a very large area that it influences to varying degrees. This corresponds to the globalized nature of these networks, where the guide is trying to create a global impact. It is trying to say that there is a global trend that decreases in the manner depicted by the system. It tries to convey this global trend that helps the system in understanding and extracting the global patterns in the database. In this way, if all the guides start conveying information about the global trends and start creating a global impact, the generalization of the system may improve by a very large amount, especially for functional prediction problems.

We studied how the localized systems take the outputs in the close vicinity and on this basis decide the final system output. A system having a large contribution from far-away guides would naturally contradict this and would go a long way in reducing the localization. These guides help in the framing the globalized patterns and trends. On the other hand, if these guides have a very small coverage, the performance will be much localized, where the output mainly depends on a couple of surrounding guides. This would be more or less a weighted mean of a couple of values that becomes the final output of the system. The global trends would all be absent.

19.3.2 Self-Organizing Maps

Self-organizing maps (SOMs) constitute another implementation of the ANN. These employ an unsupervised training algorithm where the patterns are found in the inputs, which are also used to map the outputs to these inputs. We saw earlier how this property further enabled the use of these ANNs as natural classifiers. Here the given input data itself got classified into the feature map of the ANN to give clusters. For the classificatory problem, these clusters depict the various classes at various regions of the feature map. The SOMs are primarily used for the purpose of classification. They form one of the best techniques using ANNs as classifiers. Recall the working and the training of the SOMs that we discussed in Chapter 3. These try to find the outputs of the given input by calculating the distances or the norms between the unknown input and the neurons.

Given this characteristic, it is clear that SOM does not use global rules for calculating the outputs as is the case with ANN with BPA. On the contrary, these ANNs try to map the inputs on a feature map, and this is used and adjusted to calculate the outputs. This denotes quite a localized approach where the outputs are dependent on the nearby inputs to a very large degree. The inputs that lie very far away do not make that large a difference. The SOM uses the feature map to match the presented input to some output class, and this is returned by the system. The training involves the adjustment of this feature map, which carries a very high degree of importance for the SOM, as this decides the manner in which the classification is carried out and the natural clusters of the problem.

Such an approach serves as a very big boon for the classificatory problem where the output consists of classes that ideally have very small intraclass and very high interclass distances. The same can be easily mapped by the training algorithm over the feature map of the SOM. The feature map, in other words, will be easily able to depict in a two-dimensional structure the real life problem that spans numerous dimensions. The feature vector clearly depicts the various classes that reside at distinct locations of the feature map. This can even be generalized to the unknown inputs, which again will mostly lie very near to their classes. This ease of modeling along with the localized manner of action makes the SOMs very good classifiers that are extensively used for various real life applications. The results obtained by SOMs are usually better than those from many other approaches when applied to the classificatory problems. They hence form a natural choice for the system design.

Another characteristic of the SOM is its clustering nature. We saw how this clustering nature results in better performance in classificatory problems, but this is not the end. The clustering action of the SOM has numerous applications that further encourage their use in systems. They can be used in all situations and problems where clustering of the input data is needed. Here the SOM can examine the input data to distribute it into clusters. This property is again used in hybrid systems, where an SOM is used to cluster the input data among the various modules in a modular neural network (MNN). This forms a very fast technique of good clustering which leads to effective distribution of data among the modules of the MNN. It may be observed that as per characteristics of the classificatory problems, many times a clustering is all that needs to be done to solve the problem. The various clusters naturally distribute the various classes into their own clusters.

However, it may not always be assumed that the problem will have classes that support the classificatory codes of high interclass and low intraclass separation. They may fail to do so due to bad characteristics of the inputs selected, noise, and so forth. The SOMs are resistant to most of these issues and can perform even in the absence of these constraints to a large extent. This is again attributed to the training algorithm of the algorithm, where the negative effects of noisy data get canceled by the rest of the data. This makes the system perform well even in the presence of noise or bad training data. It may, however, be noted that any localized solution is adversely affected by the presence of noise. This is becausc the nearby inputs are consulted to calculate the final output of the system. In case the noise can damage one such input that is consulted by the system for the calculation of output to any system, the system may not give the correct results.

19.3.3 Learning Vector Quantization

The LVQs also represent a localized approach to problem solving by the use of ANNs. LVQs may be considered even more localized models as compared to the RBFNs due to their working. We say this because of their same fundamental mode of operation, where every neuron represents a point in the input space. These ANNs further work in a strict classificatory style where the closeness of the input vector is seen to the neurons, which makes them more localized in nature. Due to their localized nature, much of the discussion of RBFNs remains true of LVQs with respect to the training and the testing times and memory requirements. The major change between these and the other ANNs is that these networks do not try to predict the functions as the other two ANNs discussed did. This is chiefly because of the fact that these are by design classificatory models that try to return a single class as their output.

In these networks also, a single neuron may itself be able to predict a large number of training data sets by actually enclosing a large region in the input space. This is particularly the case when no other neuron lies close to it in the competitive layer. This may be very useful when the intraclass separation is high. The correct placement of these neurons may hence result in a big boost where each neuron identifies its region well and makes the correct prediction. For complex places, more neurons may be added as per the functional requirements.

For all these ANNs, we comment in general that they will be able to learn almost any training input data if we keep increasing the number of neurons. This might be very difficult but still not impossible for the ANNs with BPA. For the RBFN, SOM, and LVQ, this will happen when there are as many neurons as there are training data sets. Hence the resulting system will be more or less a lookup table with a highly localized extent. In RBFN, the radius needs to be kept low to avoid generalization. In ANN with BPA, the function that the ANN tries to predict keeps getting sensitive to inputs and changes rapidly on increasing the number of neurons. A time will come when it will have enough freedom to change and assume very complex shapes and hence classify the problem.

19.4 FUZZY INFERENCE SYSTEMS

Fuzzy inference systems (FISs) were another very active tool that we used for problem solving in this text. Although most of the time we used the method ANFIS for problem solving, yet the basis of ANFIS or the final system that the ANFIS results in after the training is the FIS. We saw the FIS as having large capabilities to work upon linguistic rules that drive the systems. In almost all systems, the basic inspiration was the belief that most natural systems run on rules that are known for their perfection. The motivation was to model the rules so that the resulting system is able to solve the real life application. In almost all these systems, we saw that we were able not only to solve the problems but also to receive very good results.

In the previous chapters, we also represented an analogy between the FIS and the ANNs. We recall the major points of this discussion here, as it will help us easily understand the FIS systems and the various problems associated with these systems by visualizing the effects of their ANN counterparts.

Both ANN and FIS are intelligent systems. They are used to give correct outputs for the inputs presented. The ANN learns from the historical database itself. The FIS, on the other hand, needs to be tuned manually so that it imitates the historical database, if available.

Knowledge exists in every intelligent system. Through this knowledge, they are able to map the inputs to the outputs or, in other words, to give the correct output for the inputs. This knowledge needs some kind of knowledge representation and usage. In case of ANN, this knowledge exists in form of weights between neurons. In case of fuzzy systems, the knowledge is in the form of rules that drive the system.

ANN training usually happens with a great ease when the rules that map the inputs to the outputs are simple enough. This is when the bulk of the data is in agreement with other data and covered by rules without great anomalies. In case the rules are not simple, the data training requires a lot of effort. It may be seen that not all kinds of data can be trained by ANN. We normally try to train the ANN with more neurons or for more time to expect decent performance. But this may not do well in the testing data due to loss of generalization as discussed in Chapter 2. In short, the ANN training is simple for the most simple and clearly defined problems.

The same is true for the FIS as well. If an FIS does not perform well, we may make necessary modifications. If we are unable to get high performance, we may add a lot of rules or MFs to expect high performance. This will not be required in most simple problems. These problems will give good performances with a few rules and MFs.

Hence, it may be seen that the neurons and connections in ANN are largely similar to MFs and rules in an FIS. The FIS, however, needs a fair idea of the initial rule and an understanding of the system. This helps in framing the correct rules, which are very necessary for the FIS. This is not the case with ANN, where rules are automatically extracted.

We now discuss the various issues of the FIS. The other issues may be easily tracked back to their ANN counterparts.

19.4.1 Number of Rules

In the preceding section, we studied the fact that the number of rules was analogous to the number of neurons in the ANN. This means that adding rules in an FIS has more or less the same effects in general as adding the neurons in an ANN.

Adding a rule in FIS makes the system perform better for some class of inputs. A rule gives us the autonomy to adjust the system output as per the demands of the system. The resulting system after the addition of a rule is able to perform much better as per the user instructions for some class of inputs. Now two things need to be understood here. The first is that the rules might many times be added in groups. This means that if a rule is being added for some membership function (MF) of the system, then it is likely that equivalent rules are added for all the MFs of the concerned input. This is usual given how people tend to design the FIS. This is to a very large extent good, as adding such rules is informative for the system. It helps us specify the relations while at the same time keeping the generality of the system high. We are able to control the outputs by varying the MFs and/or rules and at the same time avoiding the loss of generality.

The second important point here is that the rules may usually be added in the system when the designer realizes and confirms through testing that the system is not performing as per the requirements in certain conditions. Here we try to convert this condition and the associated output as a rule and add it as a rule base. This is more or less a fix that we have applied artificially to the system.

It may be seen that the system should perform well for this situation now. If it doesn't perform, then the designer will naturally be capable enough of modifying the rule set to make it perform. But this fix does not guarantee that the system performs well for all other cases as well. It doesn't even guarantee that the system performs well in similar situations at all times. Now we will be required to check the system for each and every type of input, which may not be possible considering the many-dimensional nature of the problem. Also the large number of rules may become bulky and messy to handle in the long run, further adding to the problem. Many times, addition of some rule, apart from making the system more localized in nature, may result in the system growing unstable for other inputs.

All in all, we may easily conclude that the addition of rules must only be done if we are not at all able to solve the problem using a lesser number of rules. Every problem requires rules, which depend upon the problem characteristics and the relations between the inputs and the outputs. The number of rules must be restricted to avoid the system growing complicated and more localized in nature.

19.4.2 Number of Membership Functions

The number of MFs in FIS plays more or less the same role as the number of layers in the ANNs. We definitely do not mean the same in the absolute sense. We do not mean that every FIS should be restricted to one or two MFs—this would make it impossible for most FISs to perform well. We also do not mean that adding an MF leads to the same massive increase in the complexity in the FIS as that of a hidden layer in an ANN. But the increase of MFs in general has the same effect as that of added layers in an ANN.

The MFs are usually added in case the system is unable to perform with a lower number of MFs. The addition of MFs gives a greater flexibility to the designer to map the inputs to the outputs. He can make more rules that are specialized over an input range. Hence as a result of this flexibility, the designer can make the system behave as per the requirements and adjust the system to the known cases. Every addition of an MF makes it possible to add many more rules to the rule base of the FIS. These rules go a long way in fixing the system, especially as per the known cases. But the addition of MFs has the same ill effects that we discussed for the number of rules earlier. It makes the system more localized. More MFs make the system more difficult to comprehend and understand. Also the addition of MFs inspired by the failure of the system to perform in specific conditions may result in neglecting numerous conditions. This again will have an ill effect on the system functioning in the real world.

Again most real life problems are quite complex in nature. They mostly involve the output changing in a characteristic manner with respect to the inputs. This will require some MFs to be used in the system below which the system might not perform well. The number again depends upon the nature of outputs and the manner in which they change with the inputs at the various regions of the input space. It is still always advisable to start with the least number of MFs per input and output. Then only, if very necessary, increase the number of MFs for any input or output. One must be aware of the cost of adding the MF.

19.4.3 Weights of Rules

The FIS gives an additional feature, that of weighing of rules. This is not necessarily the same as the role of weights in the ANN. The ANN has weights as a must for the system to perform, and these are the only parameter that the training algorithm may change to get the desired system in accordance with the training data set. Many times, in FIS the weights are kept at their defaults (or 1) without any change. The reason is that the MFs and rules give enough flexibility to adjust the system to meet the requirements as stated by the training data set. This hence suggests that weights are not major criteria of adjustment. In fact many times people prefer working over the rules and MFs more than the weights when the system does not function as per requirements.

Still, weights happen to be very strong agents to adjust the system as per requirements. The use of weights is even more important at design time when the initial rules are framed. At this time, we may think of some rules that are not that decisive or necessary for the system in general but may play a role in deciding the final output. Consider a system where rest of the rules are not able to decide a single class as the final output, as they all happen to cancel each other in favoring a class. Here we may require rules that act as tie breakers. Naturally, these rules, if given a greater weight, will influence the decision even in absence of ties. Hence one may consider keeping them in the system along with lower weights. Another common situation is when the system is supposed to adjust the final output by small amounts before it is returned to the intelligent system. The use of such adjustments is carried out many times in fuzzy control applications. Here also rules with reduced weight may be applied. Another common situation is where a group of rules start dominating the system. Here again we may consider a judicious distribution of weights among these rules.

Looking at all these situations, it is clear that rules may be used to improve the system in terms of design, logical understanding, or practical performance. Inspired by the fundamentals of rules, many times people try to reduce the weights of rules that may unfortunately be driving the system in the wrong direction when presented with some inputs. However, this again is more or less a random fix that may affect the system. One must avoid such things unless it is possible to appreciably test all the possible input types. Modification of weights is best if it is supported by a common sense understanding or an understanding of the linguistic representation of the system. In other cases, one may find the system fail to perform in many cases as a result of the weight modifications. Many times, adding rules may be better than trying to modify weights to overcome wrong outputs.

19.4.4 Input/Output Distribution Between MFs

The distribution of input and output ranges among the MFs is one of the most interesting things to observe in the FIS. Many times, people add more and more MFs even in cases where these many MFs may not actually be needed. It may be possible for the system to perform well with many fewer MFs. This is mainly because people make the mistake of imagining the distribution to be equitable between the MFs. They assume that if the input has three MFs, say '*high*', '*medium*', and '*low*' then the peak of these MFs will divide the input into equal parts.

It may be noted here that there is no clean distribution of the ranges among the MFs. Rather the distribution is fuzzy, starting from a membership value of 0, attaining its peak at a membership value of 1, and then again dissolving to a membership value of 0. In this section and the sections to follow, whenever we use the word distribution, we are referring to a fuzzy distribution. Similarly the terms range, size, and radius have a fuzzy meaning. These terms take a range of values with varying certainties rather than crisp values.

But this may not be true in general. It may be possible that the '*high*' and '*medium*' MFs together occupy only 15 percent of the entire input range and dissolve to a membership value of 0 after this, whereas the MF '*low*' covers the rest of the region. In any case, this distribution is worth experimenting with before we finally conclude that the system is unable to perform with the present architecture.

Now the major point of concern is how to decide the distribution among these MFs for any input or output. It would in any case be very tedious to try out all the combinations of distribution with varying ranges. Hence we need to be able to decide how to learn the optimal coverage of each MF used in the input or the output.

Ideally, any MF should effectively cover the range where the behavior of the system is the same or follows similar patterns. The various MFs basically group these ranges, and then rules are made based on these groups. The similarity in the output following the input needs to be studied whenever we decide how the MFs are to cover the entire input range.

In many problems, a large portion of the extreme values may be passive. Here the output may not change at all whenever there is a change in inputs. This mostly happens when these denote very exceptional or extreme behaviors of the system. We may consider keeping this entire range in one pool or MF. The MF may gradually decrease its membership value to zero as we march away from these regions. Very active regions where the output shows large changes may be given more than one MF to cover the active region. Here we require more MFs, so that we can carefully adapt the system output as per the requirements. Keeping more MFs in the very active regions gives us the flexibility to model the system and make it behave exactly as required. The more MFs we insert, the more flexible is the design. However, we need to take care of the ill effects of the addition of MFs.

The other regions may be given MFs as per the requirements. This will involve few MFs over large ranges if the output shows a gradual change in these regions. The radius or the sphere of influence of the MFs may again be adjusted, while we keep an eye on the system performance. If the system does not perform well for some input, we may consider rechecking the MF ranges. We may note the input values for which the system does not perform well and reconsider the MF that effectively covers this input as well as whether this MF is placed well enough to cover the input as per the practical inputs. It may happen that we fix MF *'high'* to a set of values at the extreme and practically no input may have these values. The larger inputs that the system produces may be spread over a much larger area.

This question of the judicious placement of the MFs and deciding their ranges again calls for an artful approach where we need to realize the problem requirements, analyze the possible solutions, and accordingly place the MFs. We then need to validate the assumptions against the practical outputs and inputs. If the system does not perform, there must be a mismatch between our expectations and the system inputs and outputs that needs to be corrected.

The MFs are usually placed at points around which the corresponding output is fairly well known. This helps in easy mapping of the input to the output by rules. Also usually an MF reaches its maxima (1) when the previous MF reaches its minima (0). At this point, there is only one active MF with a membership value of 1. This further allows us to easily map the inputs to outputs by rules.

It may again be recalled that we advocated modeling the problem in ANNs such that the inputs are equally distributed in the ANN. For this, we even recommended the use of a logarithmic or other suitable scale to model each of the inputs and the outputs. The preceding discussion speaks of the same issues where we adjust the parameters in place of adjusting the system inputs.

19.4.5 Shape of the MFs

Various functions are used in the design of MFs in FIS. All these model the MF in their own manner and denote the way in which the membership of the input or the output changes from a membership value of 1 to a membership value of 0. Much of the discussion on the shape of the MF follows the same guidelines that we discussed in addressing the distribution of the ranges among the MFs. In this section, we briefly try to relate the shape of the MF to the rules and the final output with respect to the same distribution of ranges that we talked about in the preceding section, thus continuing our previous discussion of ranges.

In order to decide the shape of the MF, we need to know how the applicability of some rules may change with respect to the change in the inputs and outputs. This may be taken to be a study of the manner in which the outputs change with respect to the inputs by the application of various rules. If a large part of the MF is passive where the output does not change under any circumstances, we may keep the membership value of 1 and constant for this passive region. However, this is only the case when the output does not change at all for a change in this input. Similarly if the output is likely to change uniformly if we uniformly change some input, we may assume a linear relationship and similar MFs may be chosen. Many times the output may suddenly change with a small change in input, and later the change may not be very high. These situations require sharper and smoother MF shapes. All MFs are largely the combination of these shapes.

Based on these discussions, let us analyze the MFs given in Figures 19.14a through 19.14d. These correspond to triangular, Gaussian, sigmoidal, and Z-shaped MFs.

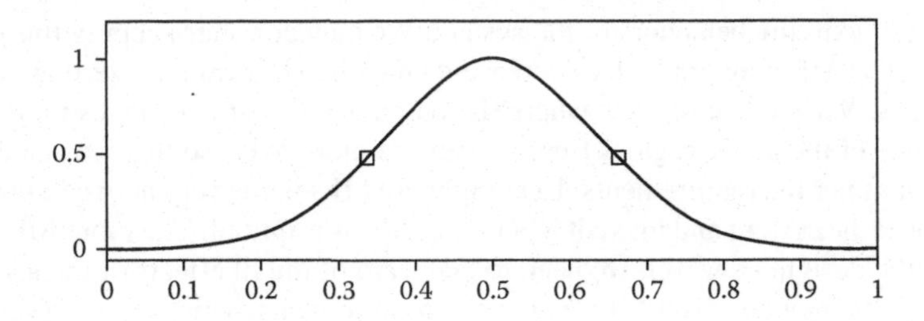

FIGURE 19.14a Common membership function curves: Gaussian.

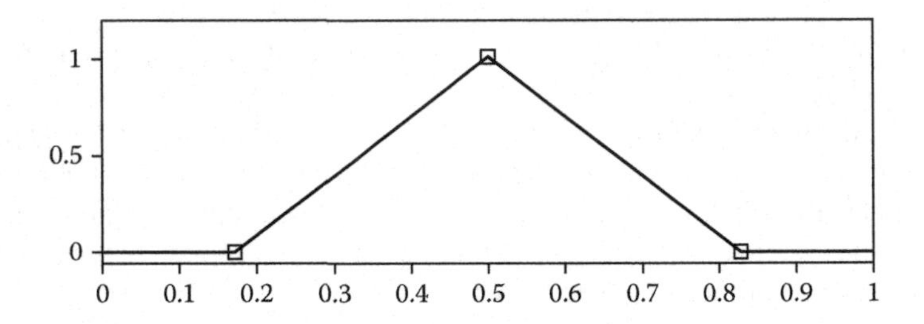

FIGURE 19.14b Common membership function curves: triangular.

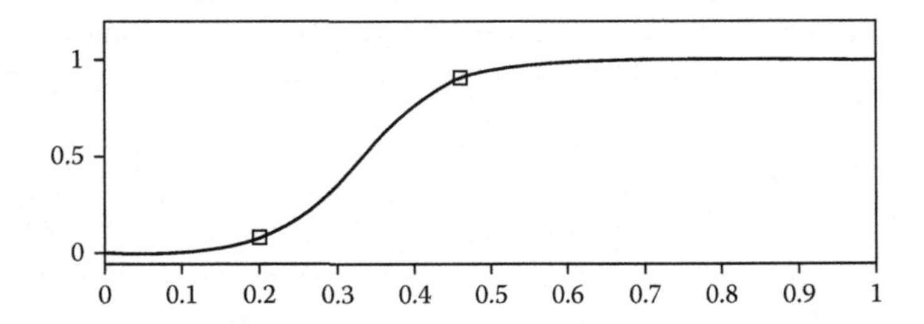

FIGURE 19.14c Common membership function curves: sigmoidal.

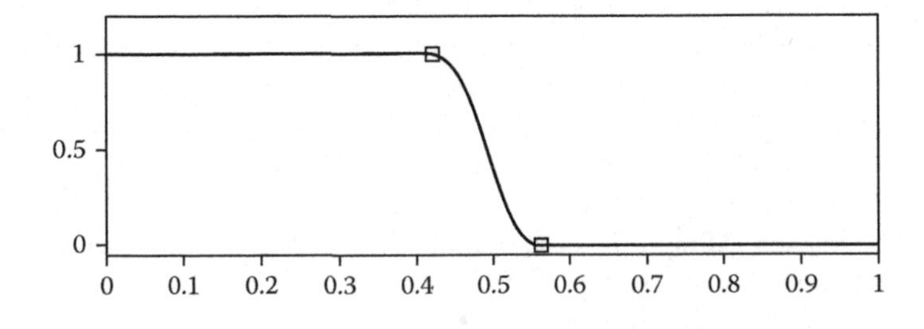

FIGURE 19.14d Common membership function curves: Z-shaped.

The Gaussian MF is used in situations that mark a rise in membership value to a maximum and then a drop again to a minimum. The range of the Gaussian MF can be increased or decreased as per demands that also control the rate at which the membership value grows. In any case, the membership value grows very slowly at the start. As we proceed, this rate gets higher. The value suddenly grows at the center of the curve. The other part of the curve is symmetric, where the membership value sharply decreases at the start but toward the extreme the decrease is slow. This function forms the basis of modeling many situations that are believed to be Gaussian in nature. Some common examples include the application of brakes in car, where the force on brakes usually goes up as we go closer to the destination. Also recall our discussion on RBFNs, which used a similar radial basis function for functioning. Here the effect of a class varied in a Gaussian manner as we moved away from it.

The triangular MF, on the other hand, represents a gradual increase and decrease in the membership value. This is again the case where the output may linearly vary with the change in the inputs. This MF forms the basis of design in many simple applications and functions mostly depicting a linear change of outputs with inputs.

The sigmoidal increases from maxima to minima gradually and uniformly. This is a much more flexible system where we can adjust the variation that governs the change from the least to the greatest membership value. This rate is important to be studied for any MF with respect to the problem demands discussed earlier. A similar MF is the Z-shaped function. Here there is a sharp region where the change from maxima to minima occurs. This region may be controlled by parameters.

19.4.6 Manual Modify and Test

We discussed this process of manually modifying the FIS and testing it again and again in the previous chapters. Naturally we did not apply this much over the real life problems apart from some control applications. This was because we used ANFIS, which not only automated this task but also resulted in automated framing of the initial model of the FIS. This was very simple to use. We repeat our discussion now that we are in much better position to understand the concepts presented there. This also completes the chapter objectives.

The problem solving in FIS is an iterative process where we make a model and keep modifying the model until satisfactory results are achieved. First we need to study the problem and decide the inputs and outputs. Once the inputs and the outputs are known, the next step is to decide the membership functions. Initially it would be preferable to go with a limited number of MFs rather than crowding the model with MFs. Another important aspect would be the placement of the MFs. These are normally placed by developing an insight into the kind of input the system would give or the kind of output that is expected out of the system.

Next, the rules are framed using common sense. Once this model is ready, it is tested by the simulation engine or known inputs or common sense. The discrepancies and errors are noted. Now we need to modify the input accordingly. Many times the wrong outputs may be due to the fact that did not consider many cases and hence did not frame rules for these cases. In many other cases, the errors may be due to the wrong placement of the MFs. These may be fine-tuned according to the requirements. If the problem is not due to these reasons, we may consider adding up MFs and replacing MFs in regions where the output was wrong. The modified FIS is again tested in a similar manner, and this process continues until we get the desired output for all inputs.

19.4.7 Generalization

We talked about generalization in all respects when we discussed the various models of ANN. We even discussed generalization under various heads of FIS. In this section, we consolidate the preceding discussions and have a look at the generalization that the FIS offers.

The FIS, being driven by rules, offers much less generalization than the ANN with BPA. However, the generalization is better than many other models of ANN that we discussed earlier. The ANN with BPA viewed the entire problem as a function that needed to be imitated as a whole. Even here we look at the problem as a whole and try to form rules that collectively map the inputs to the outputs. But every rule is mainly fired for some class of inputs. This presents a greatly localized representation of the problem. Here we are mainly trying to segment the inputs into parts and then map them in separate combinations to give the final output. The rules of the FIS effectively give most of their outputs for only some particular range of inputs or some inputs in the input space. For the others, they are passive with no output.

FIS is not completely a localized system. Here the rules are meant to depict the system in linguistic terms. These rules collectively represent the system and give the outputs as per the demands. The FIS has privileges for rules to be fired for most of the inputs that would be called the global rules and for other rules that fire for some very specific inputs that would be called the local rules. Hence the rules in FIS use a very flexible terminology that on the whole may be viewed as a generalized concept to give generalized systems. But the practical implementation is much localized in nature, where the input depends upon only some rules effectively rather than all the rules. These rules are the result of the neighboring inputs and outputs that were known to the designer.

The level of generalization largely depends upon the number of rules and MFs per input and output. The larger the number of rules or MFs, the lower will be the generalization of the resulting FIS. This may again be attributed to the range of inputs covered in case we increase the number of rules or MFs.

19.4.8 Can FIS Solve Every Problem?

We studied earlier that ANNs cannot solve all kinds of problems. The FIS forms a different type of solution, one that is driven by rules. Since the FIS follows patterns similar to ANN in the basic ideology of problem solving, it would be natural to expect that the FIS cannot solve all kinds of problems. These systems run by rules, and it is natural that rules have exceptions that cannot be modeled by the FIS. The only way to solve the problem if exceptions occur is to apply fixes in the form of MFs, and rules as we've seen is not a good technique for problem solving. Naturally we can solve any problem if we are allowed infinite rules and MFs, and if we know all the inputs that the system may face along with the outputs. The resulting solution would be much localized and naturally unlikely to be called an effective real life solution. Hence we may conclude by saying that FIS may not be able to effectively solve all problems. However, most real life problems with genuine inputs and outputs based on soft-computing principles may be solved by FIS. This is what we observed in numerous problems in which ANFIS was able to give good outputs to the presented problems.

Our discussion of FIS in this section was restricted in many ways to the general manner of problem solving in FIS. Here we discussed many issues that people face. The subsequent section on ANFIS will further extend this discussion. Here the whole orientation will be toward the manner in which the ANFIS depicts and solves the problem.

19.5 EVOLUTIONARY ALGORITHMS

The evolutionary algorithms (EAs) or in particular the genetic algorithms (GAs) were another widely used technique for solving real life problems in this text. Here we mainly used these algorithms for the purpose of optimization. In all the cases presented, the GA was able to effectively optimize the resulting objective function. This enabled the effective use of the GA in real life applications. The GA as well must be used with care. The GA involves numerous parameters, tips, and techniques that many times people may get wrong, and the resulting solution may be suboptimal in nature. Sometimes the solution may not be good at all. Here we discuss the major issues around the use of the GA.

19.5.1 Crossover Rate

The crossover is the operation where we try to mix two individuals in order to generate a new individual in the next population. This individual takes characteristics of both the parents. The motivation is that the new individual may take the better characteristics from the two parents and may result in good optimization. The crossover rate denotes the ratio of total population that goes on to the next generation by the crossover operation. The key point here is fixing up of the value of the crossover rate.

A very high crossover rate results in very fast convergence of the various solutions to a point. This can be verified by the general working of the algorithm. This mostly means that the GA gets stuck at some local minima. The various species will not at all be allowed to explore the configuration space and hunt for the possibility of the global minima. This is not a desirable characteristic of the system. Also the system will optimize very early, in a few generations. This is undesirable because we are getting suboptimal solutions even when we have time available to invest. This calls for a strategy that can take time but return a more globally optimal solution.

On the other hand, a very low crossover may lead to the algorithm behaving very randomly. The GA is used to enable generation of the best results in the finite period of time. It is natural that the randomness is not a good characteristic of the system, as no convergence to the global or local minima is carried out. Random approaches seldom result in good systems. Hence a very low crossover rate must be avoided.

In general a crossover rate of about 0.7 is justifiable. This may be increased or decreased, considering the convergence of the system as well as the total time taken for the convergence.

In order to study the effect of crossover in convergence, we solved the problem of optimization of the Rastrigin's function for two inputs, each with a range of –5 to 5. The graphs for the crossover rates of 0, 0.7, and 1 are given in Figures 19.15a through 19.15c, respectively. It can easily be observed that as we increase the crossover, the convergence happens early. Also as we lower the crossover, the mutation becomes effective and the randomness increases.

19.5.2 Mutation Rate

Mutation is the exact opposite of crossover. Crossover leads to the convergence of the various individuals, which ultimately results in the contraction of the search space. Mutation, on the other hand, tries to throw the individuals far apart and results in the expansion of the search space. This many

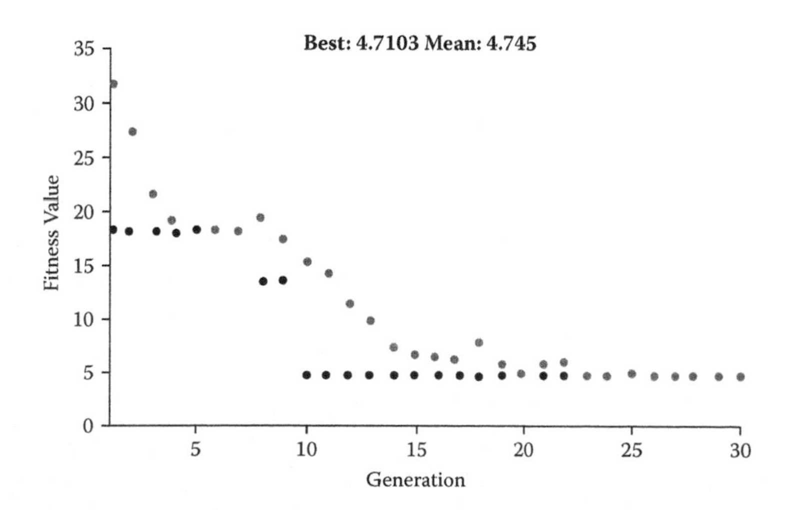

FIGURE 19.15a The convergence of the genetic algorithm for crossover of 0.

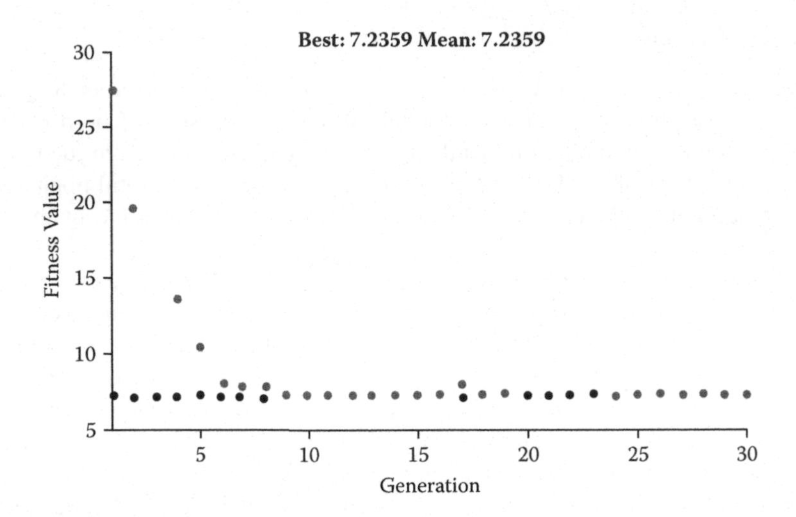

FIGURE 19.15b The convergence of the genetic algorithm for crossover of 0.7.

times leads to discovery of new areas. If these are interesting or suggest the existence of global minima, new individuals may join the search in the next generations. Else this solution might get eliminated or killed in subsequent generations. The mutation rate covers the percentage by which the individual may be thrown or displaced in the search space.

If the mutation rate is low, that means the newer individuals would be in almost the same position with almost the same fitness as the parent. This means that the mutation did not affect the system much and the system will continue to converge early as per the crossover operation. This is undesirable for the system for the same reasons that we presented in the case of early convergence with large crossover rates.

If the mutation operator is very large, the new individual lands at some unknown place in the search space. This means that the new individual will have any random fitness, depending upon its location. We may not be able to predict the fitness value at all. Since the GA individuals usually occupy positions that are better than the normal positions, it may be guessed that the new position

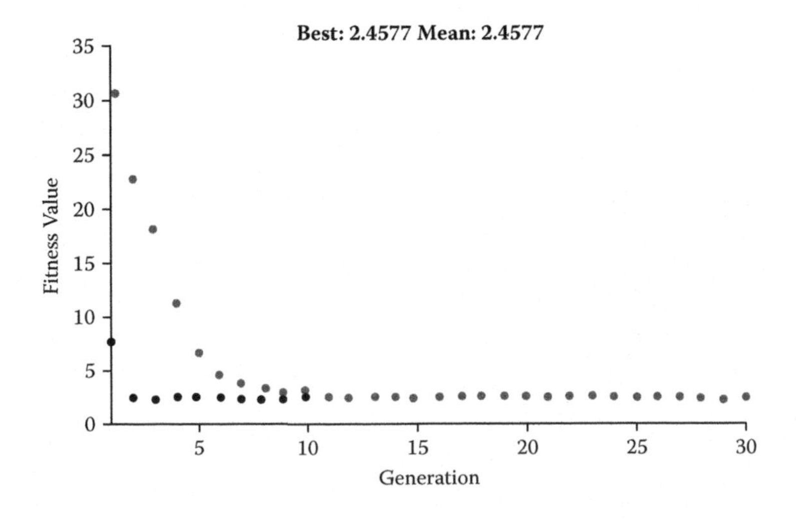

FIGURE 19.15c The convergence of the genetic algorithm for crossover of 1.

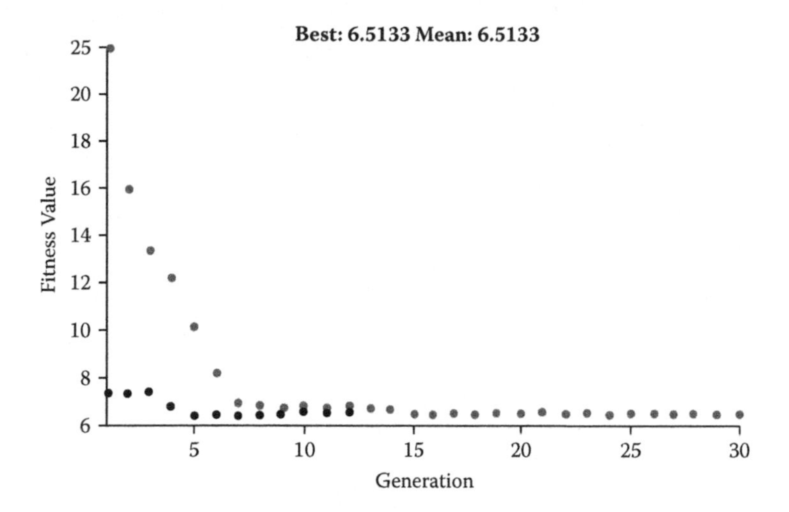

FIGURE 19.16a Randomness and convergence in the case of a mutation of 0.

may have a poor fitness value as compared to the earlier position. This, however, is not certain. This would affect the system convergence very badly as the search space expands randomly at each iteration. This affects the exploration of the minima or convergence to the minima that is an operation carried down by the crossover operator, where it tries to hunt the surroundings to converge to minima.

By contrast with the case of crossover, the mutation must be kept reasonably low to allow the system to converge. The mutation may be increased in case the system shows signs of fast convergence.

We again use Rastrigin's function to study the effects of mutation. Figures 19.16a through 19.16c show the system convergence when the mutation was fixed to values of 0, 0.1, and 1, respectively. The effect of the randomness may be easily observed by looking at the graphs. As we increase the mutation, the randomness increases and the convergence decreases.

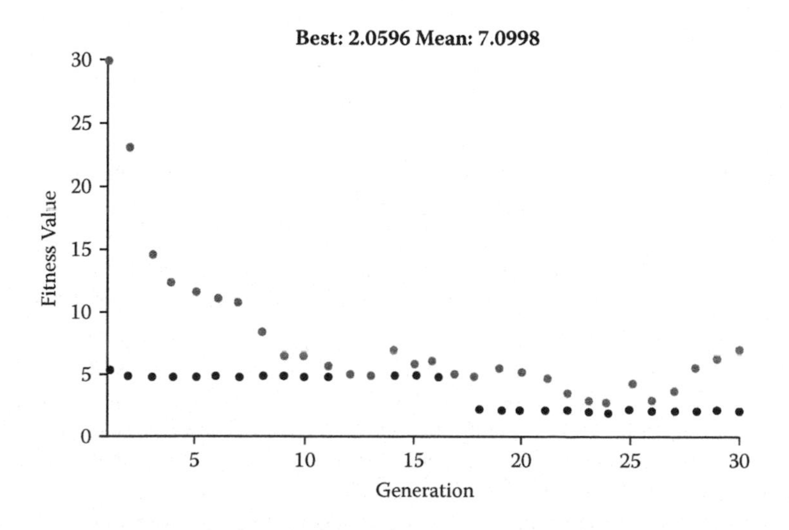

FIGURE 19.16b Randomness and convergence in the case of a mutation of 0.1.

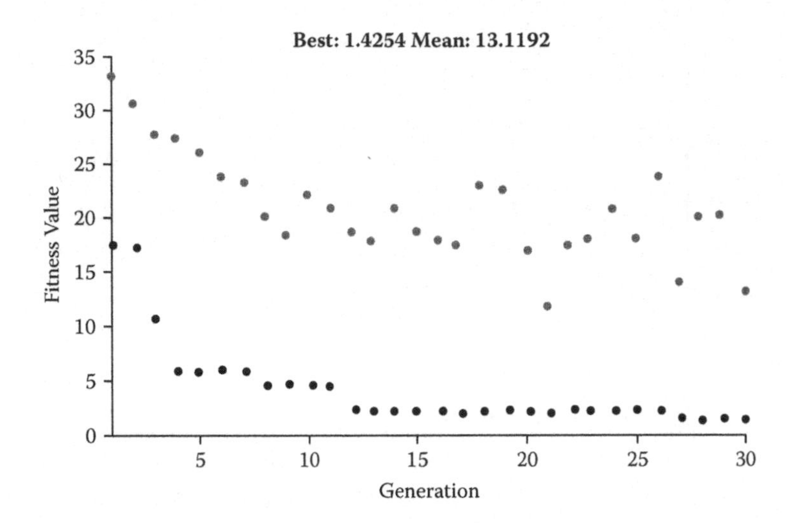

FIGURE 19.16c Randomness and convergence in the case of a mutation of 1.

It may hence be observed that we should be able to control the mutation and crossover in such a way that the system is able to search for the optimal solution in the finite time available. It must be able to converge to minima and ensure that there has been enough exploration to search for the global minima in the search space.

19.5.3 Individual Representation

This is one of the most artistic tasks of the GA. The whole working and performance of the GA depends upon the manner by which its individual has been represented. The individual representation is not a parameter that may be varied to control the system performance. This leads many people to overlook the mechanism of individual representation, which badly affects system performance. There are numerous issues that must be studied for the design of a good individual representation.

One of the major issues associated with the individual representation is that of feasibility. It must be ensured that the solution is feasible even if generated randomly. This adds a lot to the algorithmic performance, as infeasible solutions have a very bad impact on the performance of the system. The presence of a large number of infeasible solutions might result in the system not being optimized at all. Under all circumstances, the individual must be in a position to represent feasible solutions.

The same thing holds true for the generic operators as well. All the genetic operators, especially the mutation and crossover, must ensure that they result in feasible solutions. The individual representation must be such that it caters to the generation of feasible solutions to the largest extent. Also the crossover of two good solutions must usually mean a good solution. In general, the crossover operation must lead to a fair possibility of a good fitness solution being generated from the combination of characteristics.

One of the other major criteria is the search space. The algorithmic performance and optimality of the final result depends largely upon the search space that the algorithm is shown. A very large search space may result in the system being unable to converge or find a good solution. The search space must be as small as possible to fully enable the GA to return the best results. This again depends upon the correct representation of the individual.

A complete discussion of the numerous ways and techniques for representing the solutions and their pros and cons is beyond the scope of this text, and we go no further here.

19.5.4 Infeasible Solutions

The aim of the GA is to minimize the number of infeasible solutions at every step. The infeasible solutions restrict the performance of the optimization by wasting individuals that happen to lie in the infeasible areas. Many times too very good solutions after crossover result in an infeasible solution. This is undesirable, as our understanding of the GA does not expect this to happen.

There are two major ways of handling infeasible solutions. The first way is to assign the infeasible solutions a very low fitness value. As the GA believes in survival of the fittest, it may be seen that these infeasible solutions will be deleted from the population in some generations. This will solve our problem of handling infeasible solutions.

Here the infeasible solutions mark the sudden increase in the fitness value from moderate or sometimes even good fitness values. This makes the system unpredictable and difficult to optimize. Again as has already been discussed, if the number of infeasible solutions are very large, the system might not optimize at all. This is because the individuals in the infeasible areas, even if selected for crossover, would not make good offspring. Because of these individuals, the crossover would be carried out between the rest of the feasible solutions, which means a compromise with the number of individuals, as some individuals are completely wasted because of their infeasibility.

The second way is that we employ a repair operation in GA. The repair would result in correcting the inappropriate values and assigning feasible values as per the logic. This converts an infeasible solution into a feasible solution and hence solves the problem.

However, this method also has problems. The repair operation does correct the individuals and make them fit. However, the modified fitness value may not follow the trends of its parents. This value may be more or less random in nature, depending upon the repair operation used. This further affects the optimization of the GA and many times may result in an increase in the randomness of the algorithm. The randomness will be large, especially in cases where the total number of infeasible solutions is large. The problems of wasting of individuals and difficulty in the convergence of GA also hold for this method. If the numbers are high, it might be possible that the GA does not optimize at all.

The choice of method between the two is largely dependent on the problem. If the problem has a very high occurrence of infeasible solutions, the repair operation may be preferred. Also if the crossover of good, feasible solutions can give an infeasible solution, the repair method may be better. This is because we cannot expect a very large number of solutions to be deleted, as is the case with fitness method. Also if again and again infeasible solutions keep creeping up in large numbers, they will affect the total populations and hence cannot be entertained.

If, however, infeasible solutions are few in number and have a limited occurrence, the fitness function may be used. This method will result in deletion of the infeasible solutions as the generations increase.

19.5.5 Local and Global Minima

The local and global minima are always a point of concern in the design of any system. We saw this in the ANN with BPA, where we discussed the role of reduced learning rate and momentum for the convergence to the minima, as well as the problem of pulling the system out of the local minima in search of the global minima. We discussed that the ANN with BPA is prone to get struck at local minima due to its convergence and limited exploratory nature.

The GA is better when it comes to escaping from the local minima and trying to find the global minima in the entire search space. The various individuals of the GA may be regarded as the agents that try to pick up a location and search for the possibility of global minima there. The agent that has the highest fitness value is likely to be near global minima and hence attracts other agents. The diversity preservation technique facilitates the saving of features that further help in the search for global minima.

Hence it may easily be seen that there are numerous individuals, all trying to search for the minima in their own way. If any one of them lands at a point that is likely to represent global minima, it starts attracting the other individuals toward itself. In this manner more individuals are searching for the minima in the likely areas. If some individual does not enjoy good fitness value, it most probably is far away from the minima and hence joins some other individual in its search. This process goes on and on.

The convergence in this sense is when the system tries to give up further exploration and tries to concentrate more upon the searched areas. It may be viewed as a phenomenon where the system gives up its search for the global minima and decides to find the global minima in areas where so far the least fitness value is found. Mutation or the application of randomness, on the other hand, encourages the system to find the global minima and treats all the present minima as local minima.

The use of correct crossover and mutation rates may hence be appreciated in lieu of global and local minima as well. There is no doubt that these rates decide the convergence to local or global minima.

19.5.6 Optimization Time

In the ANN with BPA, we commented that it would not make much sense to leave the ANN to be trained for very long times. We supported our statements with the notions of overgeneralization, steepest descent approach, and low probability of the error reducing much. Here we more or less support the same views. It may sometimes look surprising to comment that a system should be made to work for a limited time span. This is because the use of GA in this problem and BPA in the case of ANN meant that it was not possible to try out all the combinations in the time available. Hence we needed a method that could return a solution in a finite amount of time. In other words, shortage of time was the basic reason for the use of GA in this problem and BPA in the case of ANN. We disregard the same motive when we say that the system must be left for a limited amount of time. This is, however, a basic difference between the iterative approach and the sequential approach. The beauty of the GA or BPA is that they convert the infinite possibilities to a framework that uses an iterative approach to finally come to an answer that maximizes the optimization with respect to time. One would realize that as per the same fundamentals of iterative working, further time for optimization would not result in lowering the fitness value remarkably. This, however, cannot be ascertained.

Unlike the ANN, the GA has no ill effects of long optimization time except that the optimization time may not result in improved performance, resulting in wastage of time. The hope or the probability of generating a more optimal solution is very much reduced after the system has converged. The mutation continues to keep some diversity and separation between the various individuals in the population. But all the individuals lie more or less in the same region. This kills the chance of better individuals to a very large extent. Hence it may not be profitable to keep the system running long after convergence. This was what we concluded in the case of ANNs as well. It may still be profitable if the entire population pool is spread far apart and the search goes on. This, however, is seldom carried out, as it is believed that the GA must have searched for most of the interesting areas and must have tried to converge if there was a possibility of global minima. The ANN counterpart of this would be to train the system again, but this is not usually carried out in training by BPA.

Hence we may generally accept the final optimized solution as the global minima. For most real-life applications, this indeed is thc global minima.

19.6 HYBRID ALGORITHMS

The hybrid algorithms are a mixture of the simple algorithms and make use of these algorithms in characteristic architectures to solve the problem given. These algorithms obey the laws of the constituent algorithms to a very large extent. Hence we may generalize the guidelines of the constituent systems to these systems as well. This means that if the ANFIS is being used, it will follow the rules and guidelines of both the ANNs and the FIS. We need to take care of the points presented in both these systems in the use of ANFIS. This makes these systems even more critical to use at many times, even though the motivation to combine systems and use them in a hybrid manner may be reasonable.

While most of the issues have been discussed that play an important role in the design of hybrid systems, we pay special attention to their use in classificatory problems. This is because classification is the major problem that is discussed in the text. The issue of classification has also been presented in earlier chapters. We summarize it in this section for the sake of completeness of the chapter.

19.6.1 ANFIS

ANFIS is a hybrid system where a fuzzy model is trained by use of ANN training. If we overlook the ANN that is used for the training purposes, the ANFIS happens to be an FIS. The FIS has a great capability to map inputs to outputs according to a set of rules that are cautiously framed considering the system behavior. These systems denote the natural way in which things work out. Here the output at any point depends upon the rules that drive the system and are made up from study of the system. In theory, FIS is able to form all kinds of rules, ranging from local to global. There are possibilities of making rules at all these levels whose combination gives the correct output or predicts the correct system behavior.

However, study of the generated fuzzy rules of any ANFIS will clearly point out the fact of the localized nature of these rules. Here all the problem inputs and outputs are broken down into membership functions (MFs), which lay consecutively one after the other, dividing the input and output axes in the input space. The whole problem may be supposed to be the calculation of the output with a set of inputs with reference to the points where the MF is centered. Hence the first task is that the output must be correctly calculated by the system at all combinations of inputs the MF is centered around. Say if an MF is centered at input I_1 and is called *low* and similarly the other sets of MFs called low at other inputs are centered about points I_i, we must be able to get the correct output when the input $I <I_1, I_2, \ldots I_n>$ is given to the system. It is not necessary that a training data set element be found near this input. If by any chance we are convinced that the output is correct for this input, then the task is relatively easy. The MF has a sudden or gradual decrease in its effect as we move away from its centers. This helps in calculating the output for any input in the input space. The natural systems use much the same analogy where the effect of MFs that ultimately drive the rules decreases sharply or gradually across these points where the MFs are centered and the output is assumed to be well defined.

Now we discuss the working of the fuzzy systems with relevance to the classificatory problems. We consider the problem of finding out the presence of disease, where the task is to classify the input in one of the two classes, depending on whether disease exists or not. Other types of classificatory problems may be broken down to this type by considering each output as to whether this class is found or not. This is the same modeling of the output that we considered in case of the ANNs for the classificatory problems.

In case of the classificatory or the binary classificatory problem, the FIS must be able to imitate the binary function that gives maxima at regions where the class is found and minima at the regions where the class is absent. Here we mainly study the rules, as they are the ones that drive the system outputs. The rules consist of the antecedents and consequents. The antecedents contain all combinations of the various MFs of the various inputs. Imagine only the antecedents in the input space. These cover the entire input space, with the membership degree varying at different parts of the input space. At regions where there is some combination of points where each input has some MF of degree 1, the membership value happens to be 1. As we move away from this point on any of the input axes, the MF value keeps dropping. It may be noted that we assume the AND operator is used to join the conditions. Now first we make a big assumption here that may not always be true. This is that the outputs at these points with membership degrees are well known. This means that here the output will be minimal or maximal, depending upon the presence or absence of a class. This also means that the whole input space has fairly separated regions with known or assured outputs. It may easily be seen that now the MFs may be easily adjusted to accommodate or not to accommodate the known outputs. The training algorithm so adjusts the graph of the concerned MF to easily accommodate the known input or to completely reject it. However, for the evaluation purposes, a value above a threshold is considered to belong to a class and vice versa.

Here readers may sometimes argue that because of this property, it may not be possible to adjust the functions according to the problem demands. It is as if you were asked to fix the width and height of an image, at the same time keeping the aspect ratio constant. We know that a change in width would naturally mean a change in height and vice versa. But remember that globalized networks aim at extracting the global rules, which are believed and experimentally confirmed to hold good. There is a lot of hope that these MFs follow globalized patterns. In other words, we would be able to adjust these functions such that the outputs are above and below a threshold wherever required. This is possible and is achieved by the training algorithm.

But it may even be possible that it is not at all possible to adjust the MFs such that this behavior is achieved. Then, clearly, we are trying to achieve more generalization than may be possible. Here we need to localize the FIS more by the addition of MFs or rules. This divides the axis on whose input the MF is added into more parts and naturally contributes toward the localization of the FIS.

But our entire discussion so far has been centered on the presumption that the outputs are clearly known at the points where the membership value is 1 or the input of which corresponds to all points where MFs are centered. This would very rarely be the case in real life applications. Even in case of RBFNs, we stated that the inputs are distant from each other and not uniformly distributed. Even though we are dividing the input axis into fuzzy regions, it may be recalled that this division is not uniform. The characteristic of a good division technique or what was previously referred to as a good MF selection criterion is that the MF width or coverage is as per the trends of the output. If the output changes by large magnitude over small regions, the width of MFs may be kept small in those areas. Similarly, if the output shows small and uniform change over large region, the MF width may be kept large.

To ensure or try to ensure that the outputs are correctly predicted at these points, the system uses the training algorithm. Here it tries various alterations to maximize the final output or performance of the system. This ensures that the overall performance is optimal, which would only be possible in the case of the correct mapping of the outputs corresponding to the discussed set of points.

The inputs form only one part of the rule; the other part or the consequent is composed of the output MFs. It may be naturally visualized that these function in the same way or rather support the work of the inputs. The clear mapping of the inputs to the outputs may be accomplished by placing the MFs of the outputs at desired locations. Again these MFs may be added to make the system more localized. The role of the output part is to map the previously discussed membership value considering both input and output into the corresponding hard-coded outputs. This transformation is done using the fuzzy arithmetic that we discussed earlier. It may be verified that our arguments hold whether we study the variations in the MFs of inputs or the final outputs.

The FIS is trained by use of ANNs. We have already seen the working of ANNs as classifiers and their learning in previous chapters.

19.6.2 Ensemble

The ensemble is a modular neural network (MNN) that forms a very promising solution to most of the problems presented. Here we divide the whole problem into modules. Each module represents an ANN that is trained using its own means independent of all other ANNs or modules. An integrator collects the results per ANN and uses them to calculate the final system output. Each time an input is given to the system, it is distributed among all modules or ANNs. Each decides its own output.

Then polling is done between ANNs by the integrator. The class that gets the maximum votes is declared the winner. In this section, we try to appreciate the novelty of the systems and the problem that they solve that was prevailing in the ANNs.

As discussed, the ANNs are always likely to be stuck at some local minima. The reason behind this may be the nature of the BPA that tries to calculate the minima from the present location of the network. Even though the momentum tries to pull the ANN out of being trapped by the local minima, yet chances exist that the ANN will again gets trapped at some minima or fail to get out at all. In any case, the global minima may lie at some very different place in the configuration space of the ANN. This would be a location that the network never even got close to.

The characteristic of the local minima, especially in the case of classificatory problems, is that even though they give correct output for most problems, yet there may be some inputs where they fail to give correct outputs. Since the total error is quite small. It may be assumed that the answer is correct for most problems. This is the reason we achieved good efficiencies in our systems using ANN with BPA. Hence even though most inputs do not create a problem, some inputs do lead to wrong results.

It may again be noted that the entire configuration space may be filled with numerous local minima. Each will be at some distinct region of the configuration space. Hence the ANN while training may get trapped in any one of them. This even depends on the initial weights and biases that are kept random. Hence the algorithm starts from a random location in the configuration space and may decide the local minima in which it may possibly get trapped. The momentum and epochs naturally play another important role. Of course if we add or delete some neurons, the configuration space changes with the decrease or addition of dimensions, but the issues remain the same.

Now if we train an ANN numerous numbers of times, it may again and again get trapped at different minima. Each minimum will give the correct results to most of the inputs in the input space and wrong answers to some of them. In other words this ANN can see a large part of the complete picture clearly, but another part is incorrectly visible to it. At the same time if it were at some different minima, another part would have been clearly visible to it with some other part unclear. This forms the basis of the ensemble.

Here we try to form a globalized view by taking or considering the localized view of the various ANNs or modules. Whenever an input is given, all the ANNs depict it using their own means. They all correctly or incorrectly calculate the final output and return the class. Now since most of the ANNs had a correct view of a large part of the picture or gave correct output to most of the inputs, the output returned by most of the ANNs is correct. This outperforms the wrong answers being returned by the ANNs that were incorrectly giving outputs to those inputs. The number of correct votes is large enough to make the genuine class win the poll.

Now it may sometimes be argued, why don't these ANNs return a 100 percent correct output? The reason is that many times the training data have noise due to which the inputs are present at the wrong locations. This makes the systems impossible to train unless they are localized to a very small level. Also many times the outputs have characteristic changes such that it is very difficult to make out the class to which they may belong. This happens due to the small interclass separations for some inputs that makes these systems very difficult to realize. All this poses a limitation on accuracy. Many times all the ANNs or modules give the wrong prediction of a class.

Hence it is clear that these approaches are primarily for classificatory problems where they try to use the localized or suboptimal performances of the individual ANNs to form a system with optimal performance. This may have a very deep and positive impact on the working of the algorithm.

19.6.3 Evolutionary ANN

These ANNs evolve with the help of evolutionary algorithms (EAs) or genetic algorithms (GAs). Here we can keep the entire architecture variable. Also these methods solve many problems that we face in the training of the ANNs where they get trapped in local minima. Hence these methods present promising solutions. The GA not only decides the correct values of the weights and the biases, but so does the entire ANN architecture. As the generations increase, the system keeps getting optimal and the fitness value of the performance improves.

The evolutionary ANNs are primarily ANNs that are formed and trained using the GA. We have already discussed the role and functionality of the ANNs numerous times. We saw how they imitated the behavior of the function and how the training over them was performed. We even discussed the problems that are associated with these ANNs and the cause of these problems.

Here we mainly study and visualize how the training is done in the ANNs. Before we move on with the discussion, let us clarify two very distinct words that we have been using throughout the text, although we may use these words together in this section. These are input space and configuration space. The *input space* is a high-dimensional space with each axis representing an input. The output is usually visualized as the vertically top axis. The *configuration space* is the space where each axis denotes some weight or bias that is used in the ANN. This is highly dimensional in nature and multiplies its dimensionality on the addition of every neuron to the system. The error (or negative performance) of the system for the training or the validation data set is usually marked as the vertical axis and plays the same role in our discussions as the output in the input space. In fixed architecture–type systems, this space is constant. On the other hand, when the architecture of the ANN changes, this space also changes.

Now let us see what happens at the time of training of the ANN. Here there is a transformation of the status or the position of the system in the configuration space. In any training algorithm, the intention is to reach a region in this configuration space that gives us the least error or the lowermost point in the output axis. This makes the training algorithm walk, wander, or jump in searching for such points. We discussed in Chapter 2 how the BPA uses the steepest descend approach in this search. We further studied, in considering the GA, how the GA invests various individuals in the configuration space and makes them move and jump in search of the point with the minimum error.

Now let us move from the configuration space to the input space. Here a position in the configuration space corresponds to an entire function being imitated in the input space. The point corresponding to the least error in the configuration space means the most ideal imitation of the function serving as a solution to the problem in the input space. As we move about the configuration space, the function changes its shape in the input space.

Now we have basically two phenomena that the evolutionary GA incorporates. First is the change of configuration with same architecture, and the other is the change in the architecture itself. We know the kind of results the former phenomenon produces in both the configuration and input spaces.

The change in the architecture is a complexity or the sensitivity control of the ANN. Architectures with limited connectivity and a limited number of weights prohibit the ANN from imitating complex functions that are prone to too many changes and sudden changes. The configuration space changes completely if we change the ANN architecture. Hence a jump from one particular architecture to another, or in other words the change in the configuration space, makes the resulting function being imitated simpler or more complex in the input space. A simpler structure due to limited dimensionality of configuration space is much easier to train and find the optimal solution, if the optimal error is within acceptable limits.

Now let us combine the two and see the working of the GA. As the configuration space is changing, the discussion requires a much broader insight into the functioning of the ANNs. Of course, the actual space that incorporates the GA encoding of the chromosome is much easier to discuss, as it obeys all laws of GA and is equally difficult to visualize and comprehend, especially with respect to the ultimate ANN and its performance.

The GA at any time consists of numerous individuals placed at multiple configuration spaces and multiple points at each configuration space. Here the convergence may be considered at two levels. One is that the GA at some configuration space tries to pull as many individuals as possible toward it that belong to the same configuration space. This is much like the trivial way of functioning of the GA. The other possibility is that the solution at any configuration space with decent performance tries to pull individuals from the other configuration spaces. These may land in a configuration space that is in between the two spaces. Hence there is jumping of configuration spaces in order to reach the configuration space where the optimal solution resides. At the same time, all the individuals of a configuration space try to converge to a point. At any time during the journey of any individual at any generation, a more nearly optimal solution is likely to be found. The necessary changes happen the next generation onward that attract the individuals toward this new optimal solution. This was the crossover operation. Mutation plays a similar role of adding to randomness.

Now we present a brief note about the classificatory problems where we tried to classify inputs or imitate a binary function. Imagine the condition of the input space for all the individuals as the GA runs across generations. Initially there are random functions ranging in complexity that claim to be imitating the ideal function. A convergence here toward the minima in the same configuration space would mean that the corresponding function in the input space approaches the ideal one. This means that the function that had a wider range of areas covered gets its area reduced and vice versa. Further, the jumping of neurons from a higher configuration space toward a lower configuration space means that the undue number of turns and sharp turns all get filtered out and the resulting function is retained. A transformation to the higher means that the function now adds more turns and makes more turns to be capable of imitating the actual function. Readers may visualize that as we proceed with the algorithm; the individual representing the best solution keeps adding/losing complexity and keeps getting more realistic closer to the ideal curve.

Appendices

Appendix A: MATLAB® GUIs for Soft Computing

A.1 INTRODUCTION

So far, we have seen all the ways to solve problems with the help of soft computing techniques. We made systems that could very well serve the purpose and attain high efficiencies. All the systems were based upon state-of-the-art technologies. Further we tested these systems by varying the underlying architecture and the various parameters in all the ways that we could. We studied various real life soft computing systems and effectively made and tested them. One might wonder if it is as simple to build and deploy soft computing systems as it looks. In fact it is. All this is made possible by the various GUIs and toolkits of MATLAB® that facilitate easy construction, training, and testing of soft computing systems. All this is done through a good GUI guide that makes the complete system design a task of a few mouse clicks, and then we are ready to anything with the system. Even the most complicated system involves just a little bit of setting of parameters and sometimes a few lines of code. This makes it possible for most novice users as well to build and test soft computing systems with great ease. Now one need not be a pro in order to use a soft computing system. This further underscores our comment that modern soft computing engineers and primarily designers will find in MATLAB an ideal design toolkit that gives a language to their designs and interprets and converts their designs into a real, working system. The engineer is relieved from all sorts of task and can continue experimenting with his designs.

The MATLAB GUIs to a very large extent make possible the contributions by millions of contributors toward this field. People just need to know the soft computing basics and can straightaway try their hands at the various GUIs that have been customized for the requirements in ways that ease use. It is now simple and easy to work with soft computing techniques. Further all the graphs, figures, and designs get generated by these GUIs only. This is of a lot of help in the system study. All this has in a large way contributed toward the intensive research in all the domains of soft computing and their related applications.

An extensive documentation of these GUIs and toolkits is already available in literature. The MATLAB guide and its help itself provide well-composed documentation to the use of these tools. Here they discuss the settings of various options, explain various features, and show how to use the system.

In this appendix, we briefly introduce this package to help readers get started in the use of these toolboxes and GUIs. This appendix is devoted to the use of various GUIs provided in MATLAB for soft computing techniques. The next appendix extends this by presenting the various codes for these systems.

Here, we study the GUIs of artificial neural networks (ANNs), genetic algorithms (GAs), fuzzy inference systems (FISs), and the adaptive neuro-fuzzy inference system (ANFIS).

A.2 ARTIFICIAL NEURAL NETWORKS

The ANN toolbox and GUI of MATLAB provides an easy means to design, develop, and test the ANNs. This GUI can be invoked by typing the command '*nntool*' or equivalently from the *start -> toolboxex -> neural network -> nntool* menu. This toolbox is represented in Figure A.1. This contains all the features to make a network, load data, train the network, test the network, and export the network.

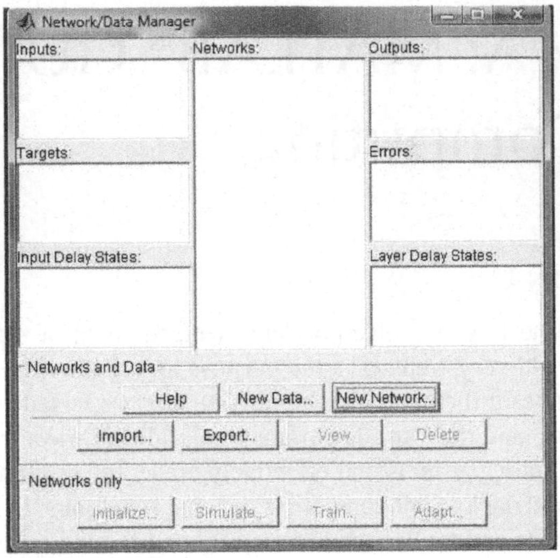

FIGURE A.1 The ANN GUI in MATLAB.

Before you begin, you may be interested in getting the data to be trained in the MATLAB environment. MATLAB provides a data import feature that is able to import data in most recognized formats. This includes the Excel xls format as well as the plain text format, which are widely used for import purposes. The data may be collectively imported as a collection of inputs and outputs and later on split between inputs and outputs using MATLAB statements. Alternatively, the inputs and outputs may be imported separately. The ANN follows a column major representation for the data, unlike the row major form that we are usually used to. Here the rows correspond to the attributes and the columns correspond to the data entries. The two forms are transposed from each other.

Once the data are in the MATLAB environment, it may be imported in the nntool using the import feature. This is shown in Figure A.2. Using this import, we can import any data, output, input, or even the entire network. This is the opposite of the export feature, where we may export

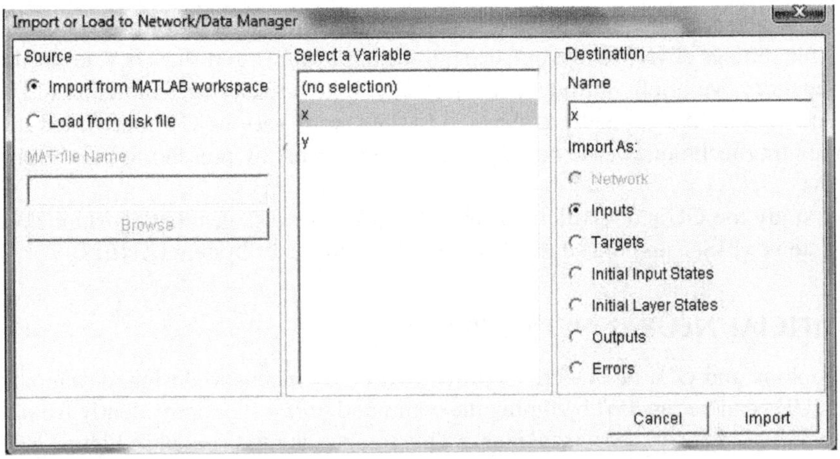

FIGURE A.2 Import network/data manager.

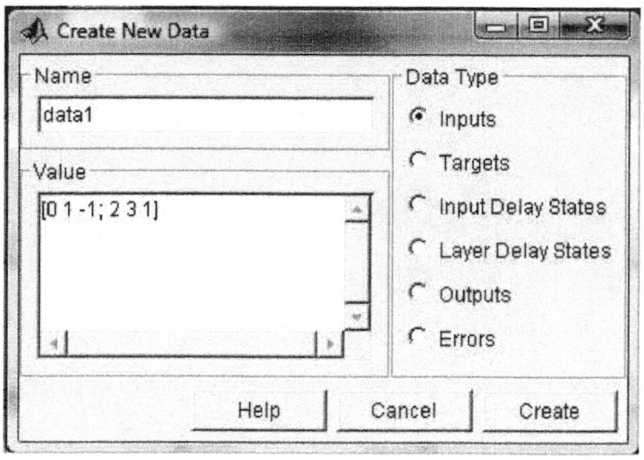

FIGURE A.3 Create new data.

these to the MATLAB environment. Alternatively, we may create the data manually using the new data feature of the nntool. This is shown in Figure A.3.

We then need to make an ANN to get started. This is done using the New Network feature. This provides us with a simple interface where we can create any type of ANN. All the ANN parameters are defined in this interface only. Here we can fix the ANN model that is to be used, the training function, the number of layers, and the number of neurons, as well as the activation function of all layers. When we click on the Create Network button, the network gets created. This may be viewed to give the general structure. The New Network feature is shown in Figure A.4, and the View Network feature is given in Figure A.5. The network so obtained may optionally be initialized. This GUI is shown in Figure A.6.

The next major task as per the ANN usage methodology is the training. MATLAB again provides a simple interface where we can specify the training parameters and carry on the training. Figure A.7a shows the interface where we specify the training inputs and targets. This is the data we imported or created earlier. We can specify the variables in nntool that store the final results and errors for future use in the toolkit. Figure A.7b shows the interface where we specify all the training

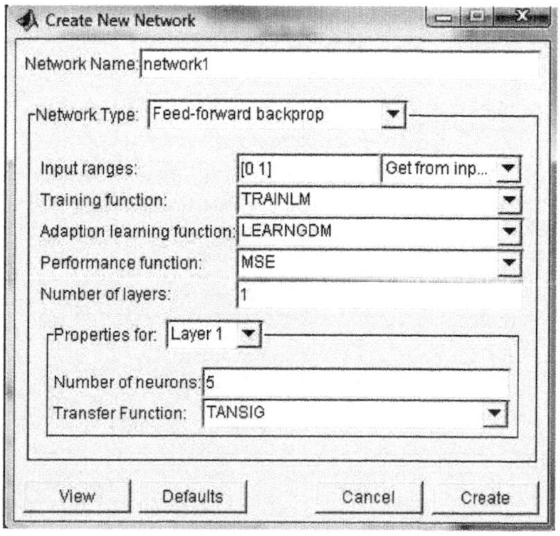

FIGURE A.4 Create new network.

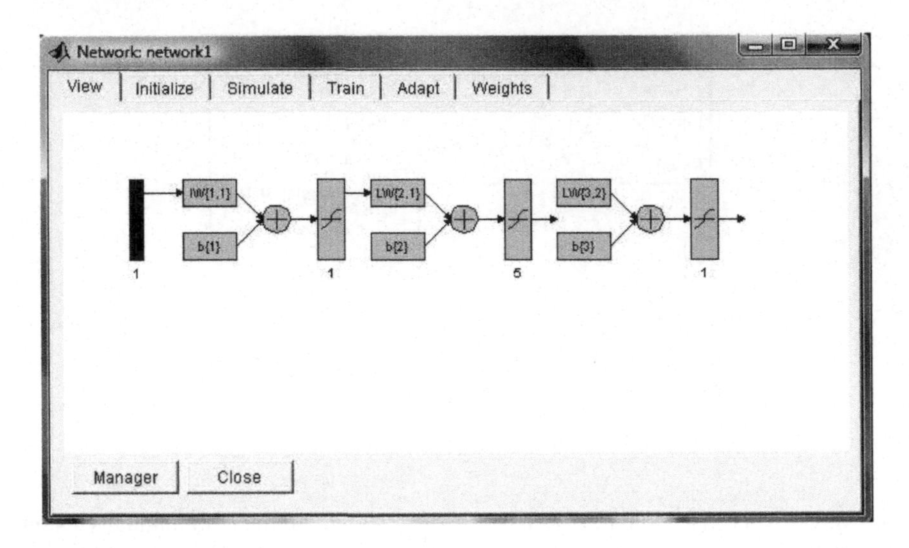

FIGURE A.5 View network.

parameters as per the requirement. Additional validation and checking data may also be given to the system. This is done using the interface of Figure A.7c.

The nntool also provides the features to carry on adaptive training. This interface is given in Figure A.8a, where we can specify the data to carry on the adaptive training. Figure A.8b shows the interface where we may specify the parameters of this adaptive training. When we click on Train, the training starts and the training curve is shown according to the parameters specified.

After training, we may wish to optionally see the weights of the ANN. For this also, an interface is provided that allows us to view the weights between the selected layers and the bias information for the various layers. This interface is shown in Figure A.9.

The last part left now is the system simulation. Here we are interested in checking the system performance for any of the data that we may specify. This is done using the simulation interface, where we may select any of the data as input for the simulation. If the output for these data is

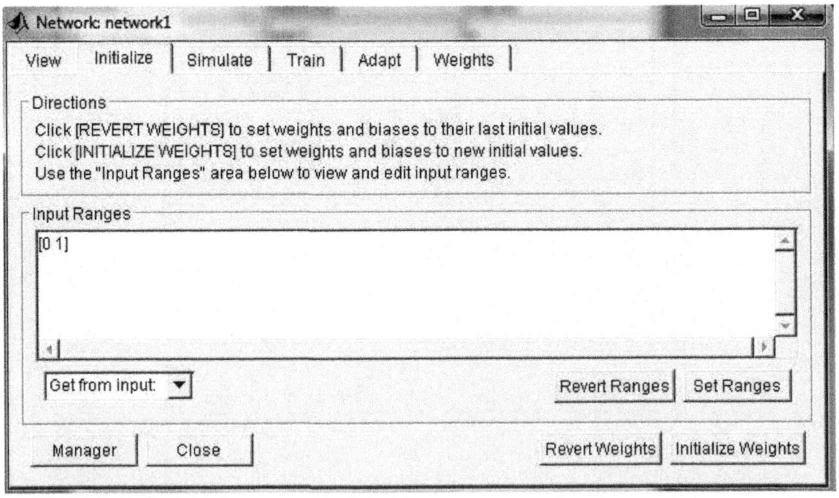

Network: network1
View
Initialize
Simulate
Train
Adapt
Weights
Directions
Click [REVERT WEIGHTS] to set weights and biases to their last initial values.
Click [INITIALIZE WEIGHTS] to set weights and biases to new initial values.
Use the "Input Ranges" area below to view and edit input ranges.
Input Ranges
[0 1]
Get from input:
Revert Ranges
Set Ranges
Manager
Close
Revert Weights
Initialize Weights

FIGURE A.6 Initialize network.

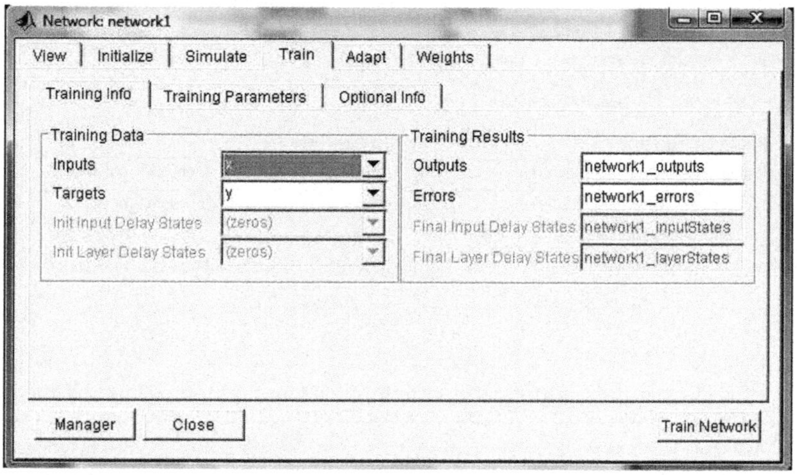

FIGURE A.7a Training data of network.

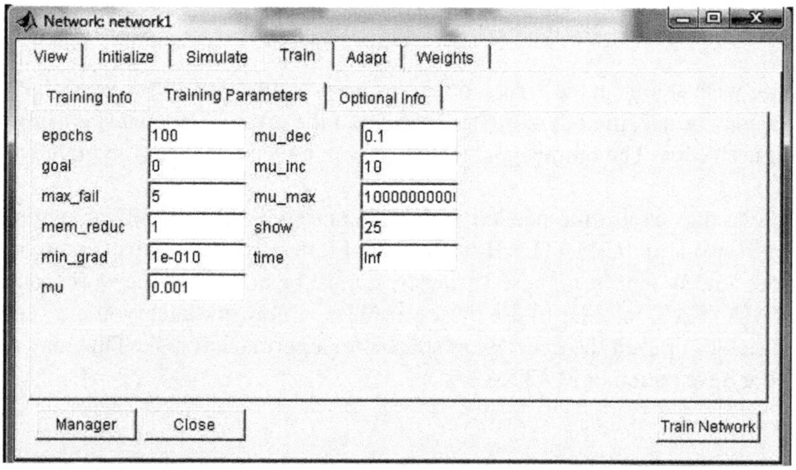

FIGURE A.7b Training parameters.

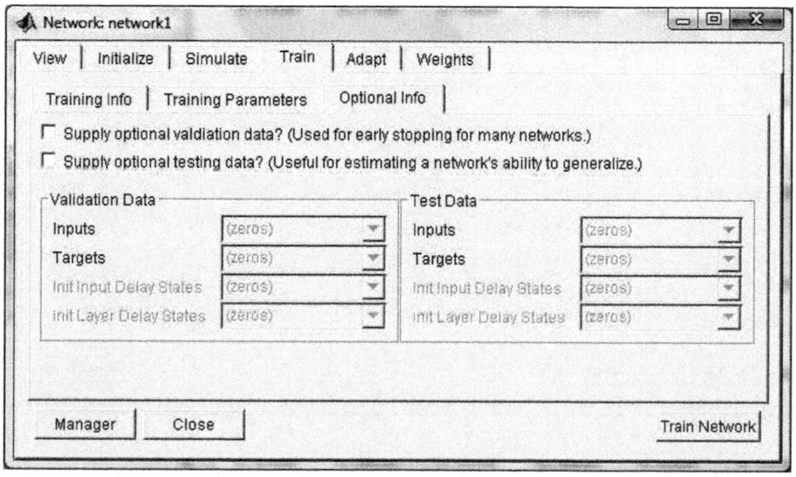

FIGURE A.7c Optional training information.

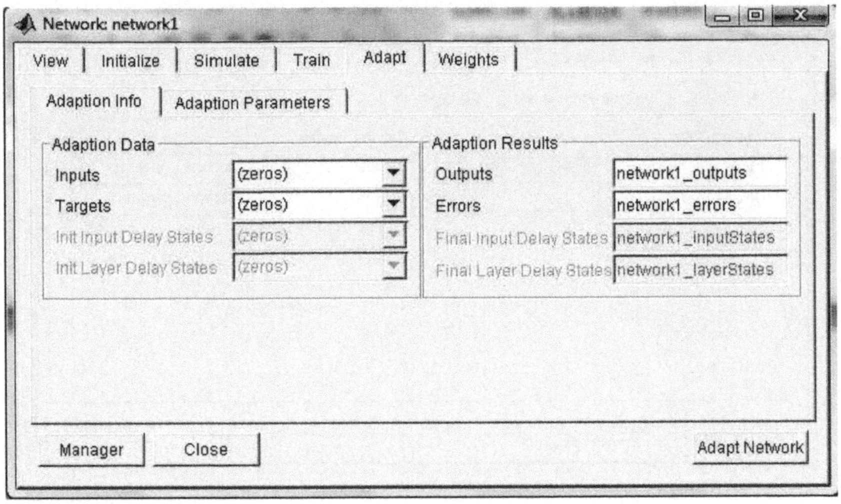

FIGURE A.8a Adaptive training.

available and we wish to look at the errors in the system, we may specify the target data as well. The variable names may be specified to store the output and the errors. This interface of the simulation is shown in Figure A.10a. The outputs and the errors can be viewed from the main window, shown in Figure A.10b.

At the end, one may export the needed variables consisting of data, inputs, outputs, targets, or even the entire network to the MATLAB workspace. From here we may use commands or m files to further process the system from these variables. It must be noted that the workspace gets cleared when you exit MATLAB. In order to use these variables or network later, you may consider saving the selected variables or even the entire workspace into a permanent disk. This may be done using the Export or the Save option of MATLAB.

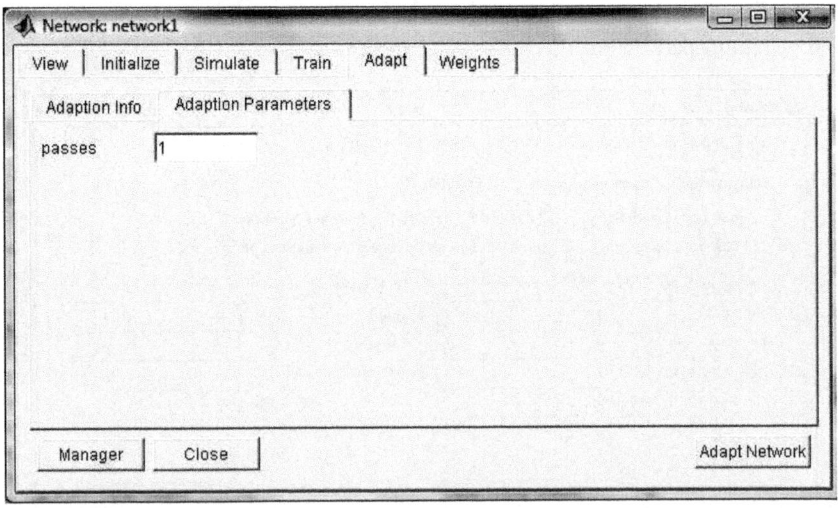

FIGURE A.8b Adaptive training parameters.

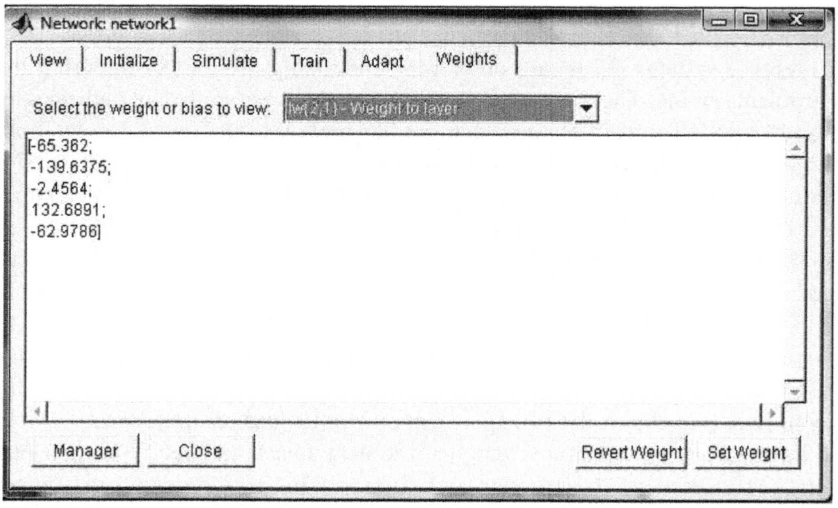

Network: network1
View | Initialize | Simulate | Train | Adapt | Weights
Select the weight or bias to view: lw{2,1} - Weight to layer
[-65.362;
-139.6375;
-2.4564;
132.6891;
-62.9786]
Manager | Close | Revert Weight | Set Weight

FIGURE A.9 Network weights.

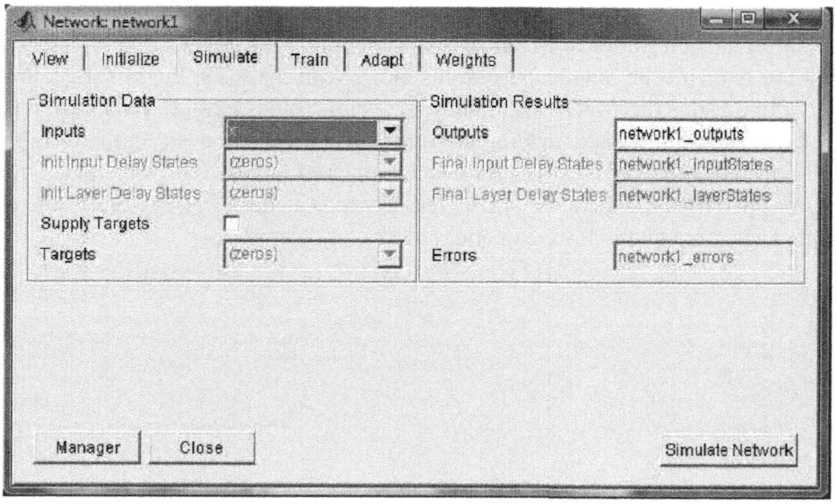

Network: network1
View | Initialize | Simulate | Train | Adapt | Weights
Simulation Data
Inputs
Init Input Delay States (zeros)
Init Layer Delay States (zeros)
Supply Targets
Targets (zeros)
Simulation Results
Outputs network1_outputs
Final Input Delay States network1_inputStates
Final Layer Delay States network1_layerStates
Errors network1_errors
Manager | Close | Simulate Network

FIGURE A.10a System simulation.

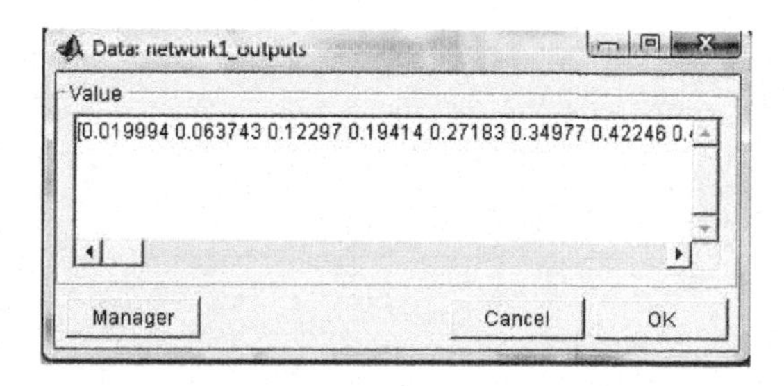

FIGURE A.10b System simulation output.

A.3 FUZZY INFERENCE SYSTEM

The fuzzy inference systems (FISs) are other very commonly used soft computing tools for the purpose of problem solving. These come in MATLAB with a very graphical and easy-to-use interface where we can use all sorts of features that the FIS provides. The Fuzzy Logic Toolbox can be invoked from the MATLAB command line by the command '*fuzzy*'. One may alternatively use the *Start -> Toolboxes -> Fuzzy Logic -> FIS Editor Viewer* menu. This opens the central window, which is the FIS editor of MATLAB. This editor is shown in Figure A.11.

The FIS editor enables you to select, add, or delete the various inputs and outputs used by the FIS. One may select each input or output and specify its name using the provided text box. The left part of the FIS editor shows the various available fuzzy operators that one may specify. All the standard fuzzy operators are already built in the Fuzzy Logic Toolbox. One may additionally specify custom or self-made fuzzy operators as well, if needed by the problem. This editor further provides functionality to import or export the FIS apart from the other general functions.

Once we have added the inputs and outputs, the next step is to specify the various membership functions (MFs) used by these inputs and outputs. This is done using the interface of the Membership Function Editor. Here we first select the variable at the right to view its present membership functions. We can add, delete, or modify these MFs. The shape can be selected from the drop-down menu provided at the right. The parameters of the selected MF may be modified by moving the MF over the range along with the modifiable points, or they may be manually specified in the text box. The left part of the interface lets one select the range of the selected input or output. The Membership Function Editor of MATLAB is shown in Figure A.12.

Once we have fixed all the inputs and outputs along with the MFs, the next task is to form the rules. This is done using the MATLAB interface for the Rules Editor. This editor is shown in Figure A.13. From here we can add, delete, or modify a rule. We can even set the weight associated with the rule. We select the various MFs for every input and output that participates in a rule. If a variable does not participate, then none is selected. There is a check box to specify the negation for every variable. Using this method, we can add, modify, or delete rules.

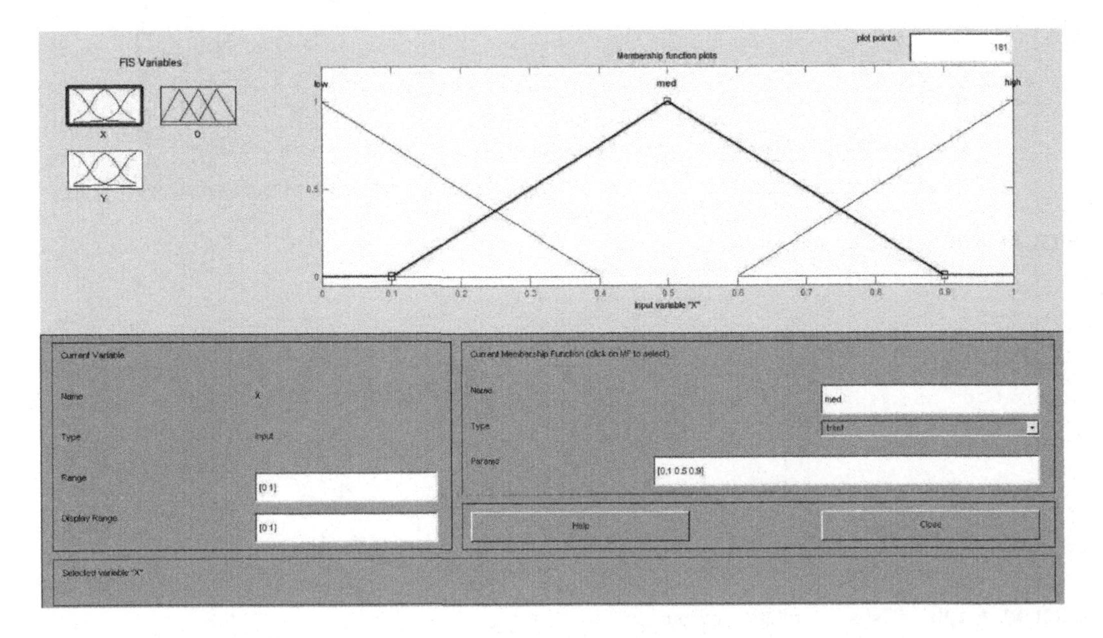

FIGURE A.11 The FIS editor.

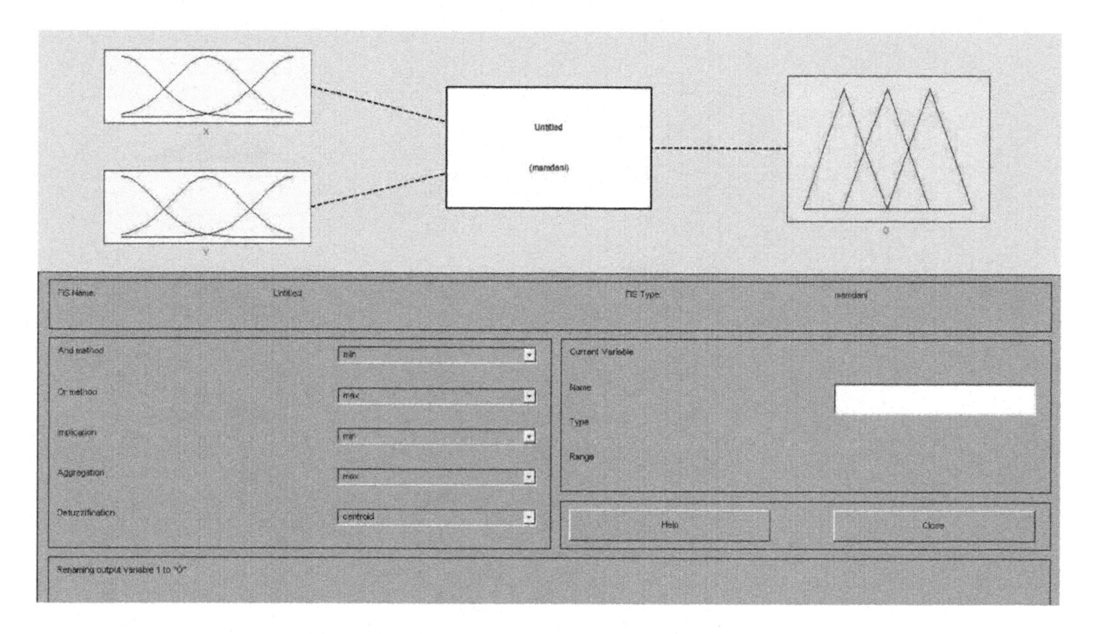

FIGURE A.12 The Membership Function Editor.

Now we are ready to test the system and give it the inputs and check its outputs. This is done using the Rule Viewer. The MATLAB interface for the Rule Viewer is shown in Figure A.14. This graphically depicts how the input is processed to give the output by the application of fuzzy rules and the fuzzy operator. Here we may move the vertical bars for all inputs to change the inputs and get the outputs. The inputs may also be changed by manually entering them in the text box provided.

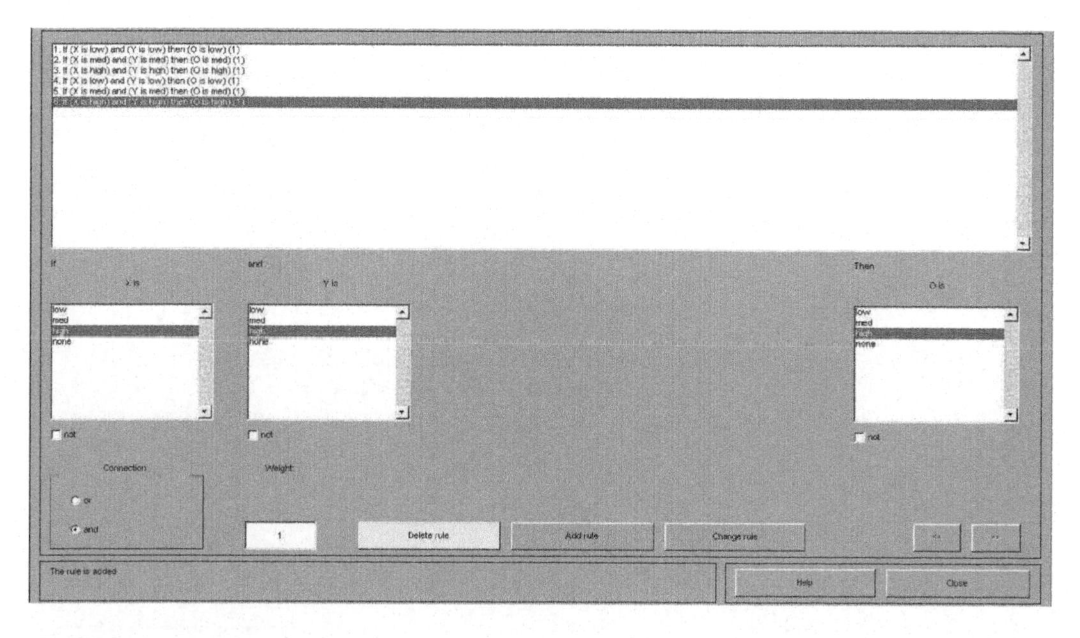

FIGURE A.13 The Rules Editor.

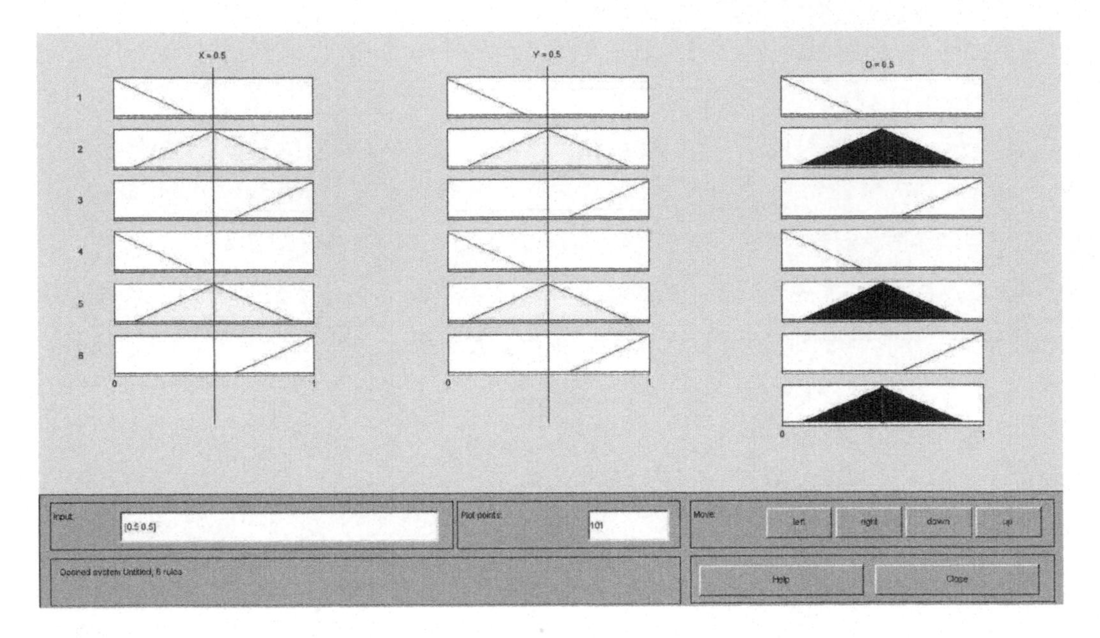

FIGURE A.14 The Rule Viewer.

The last task that one may be interested in is to analyze how the inputs connect to the outputs. For this, MATLAB provides the Surface Viewer. The Surface Viewer may be used to study how the variation of any two inputs affects the outputs. The other inputs if present in the system need to be kept constant. This is because of the fact that graphs with a maximum of three dimensions can be depicted to the human eye. One dimension is fixed for the output. We may hence modify the use of the other two remaining dimensions. We may specify the different values of the other inputs in the text box provided. The Surface Viewer is shown in Figure A.15.

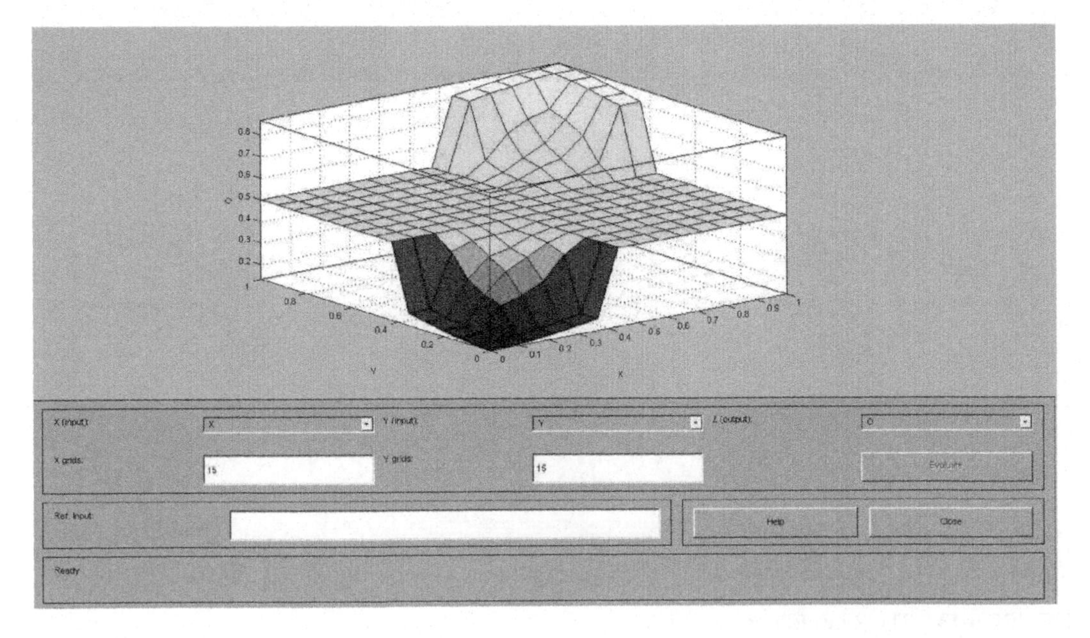

FIGURE A.15 The Surface Viewer.

A.4 GENETIC ALGORITHMS

The genetic algorithms (GA) has been used extensively for solving optimization problems. They form a very good means for optimization of functions by varying the parameters. The GA Toolbox is a very simple toolbox that lets you specify all the parameters and gives you the final result. The toolbox may be invoked by the command *gatool.* The various graphs are maps that may be generated to aid in better understanding of system functioning. The interface of this toolbox is given in Figure A.16.

First we specify the objective function that is to be optimized. This is a standard MATLAB function that returns the fitness value based on the entered parameters. The various parameters are given to the objective function as a vector. The objective function is specified in the GA toolbox prefixed with the '@' symbol. Then we specify the number of variables that the GA needs to optimize. The GA gives options to make numerous graphs that depict the solution evolving over time. This gives a multidimensional view of the problem and solution. We have been using the graph for the best fitness thoroughout the text. The needed graphs may be specified. These are made and updated as the GA runs over generations. The toolbox has its own console that shows the output of any operation being performed. If the objective function meets with an error or reports a warning, that may be visible on the GA console.

When the whole solution has been computed, it is visible in the output section. Here we may export the problem, solution, or individuals from the solution to the workspace. This would facilitate the further processing of the problem and the solutions. This may be re-imported into the GA toolbox when further optimization is needed or the solution needs to be computed again.

All the genetic settings may be specified at the various tabs shown in the middle. These have already been discussed in the Chapter 5. Here we may specify our own genetic operators in place of the built-in operators. This is done by making a custom selection in the menu of the genetic operator that we wish to use. This gives us a text box to specify the function name. The function name corresponds to the name of the function that replaces the built-in function preceded with the '@' sign. The modified function must use the same protocol as desired by the

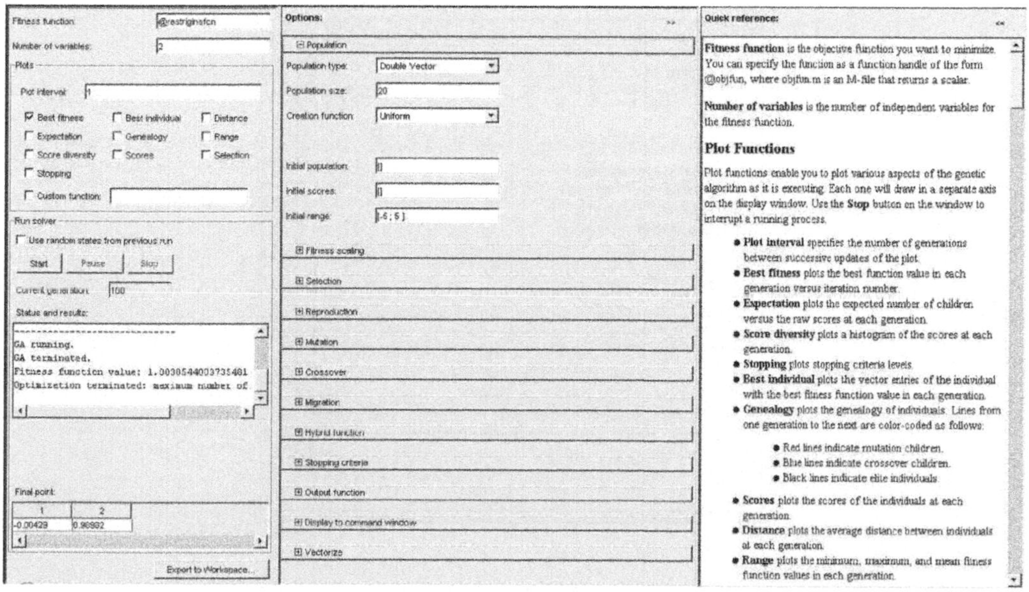

FIGURE A.16 The genetic algorithm GUI.

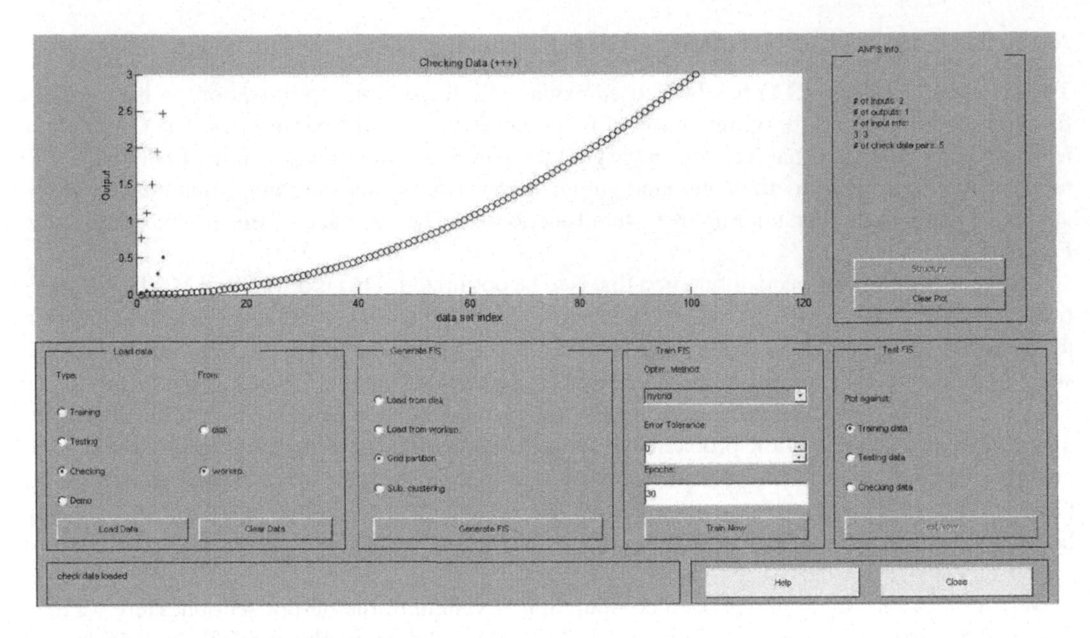

FIGURE A.17 The ANFIS editor GUI.

GA. A simple way to accomplish this is to copy any standard function of the desired generic operator and then change its code and function name. Many times the mutation and crossover need to be modified a little. This is again done by copying the original function and modifying the needed code. The function is renamed, and the GA is pointed to call this modified function.

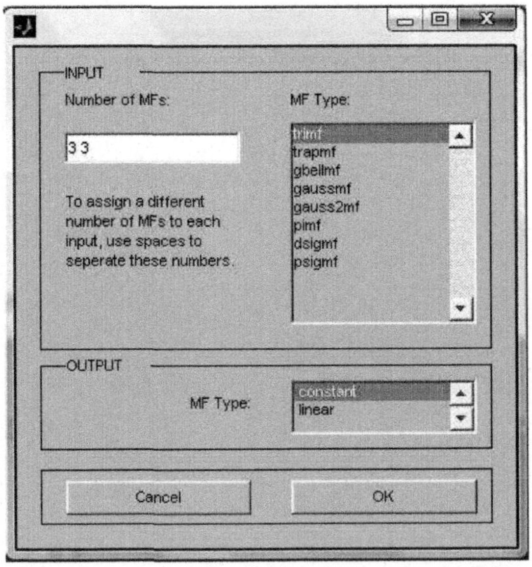

FIGURE A.18 The FIS generator.

At the extreme right-hand side the help manual is available. This is a good source of information that is readily available and is easy to use. One may refer to the help at any time to understand the meanings of the various settings and terms.

A.5 ADAPTIVE NEURO-FUZZY INFERENCE SYSTEMS

The last GUI that we study in this appendix is that for the adaptive neuro-fuzzy inference system (ANFIS). This is a customized GUI that implements the ANFIS training and testing. The system may be invoked by the MATLAB command *anfisedit*. The interface of this ANFIS editor is shown in Figure A.17. Here we first load the three types of data. These are the training data, the testing data, and the validation data. These data may be loaded from the workspace variable or from a file. The data need to be in a row major format, unlike the ANN. Here the rows represent input cases and the columns represent the attributes. These data get plotted at the ANFIS graph with different legends as shown in Figure A.17.

Once the data are ready, we now need to generate an initial FIS. This may be loaded from a file or created using the Generate FIS functionality of the ANFIS. The various methods of generation are listed, and any one of them may be chosen. In this FIS, we are free to specify as many MFs per input and of any shape as desired. This option is invoked at the next screen of the Generate FIS sequence. Here we specify the number of MFs per input and their types. This is shown in Figure A.18. The output may only be linear or constant.

Once the FIS is generated, we are ready to carry on training. The training parameters may be specified at the main interface shown in Figure A.17. Once the training starts, the ANFIS generates the training curve that shows the system error and the goal desired. This training was presented for numerous problems in the previous chapters.

At the end we test the system. This may be done using the rule interface of the ANFIS, which is exactly the same as that of the FIS. We have an additional functionality here at the ANFIS. Here

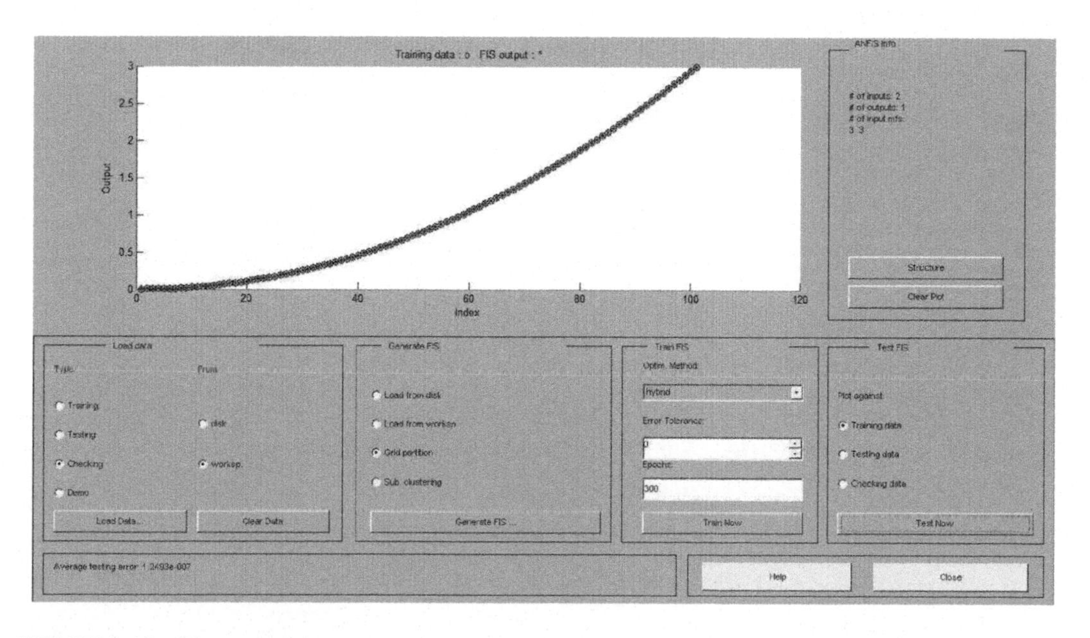

FIGURE A.19 Plot against data.

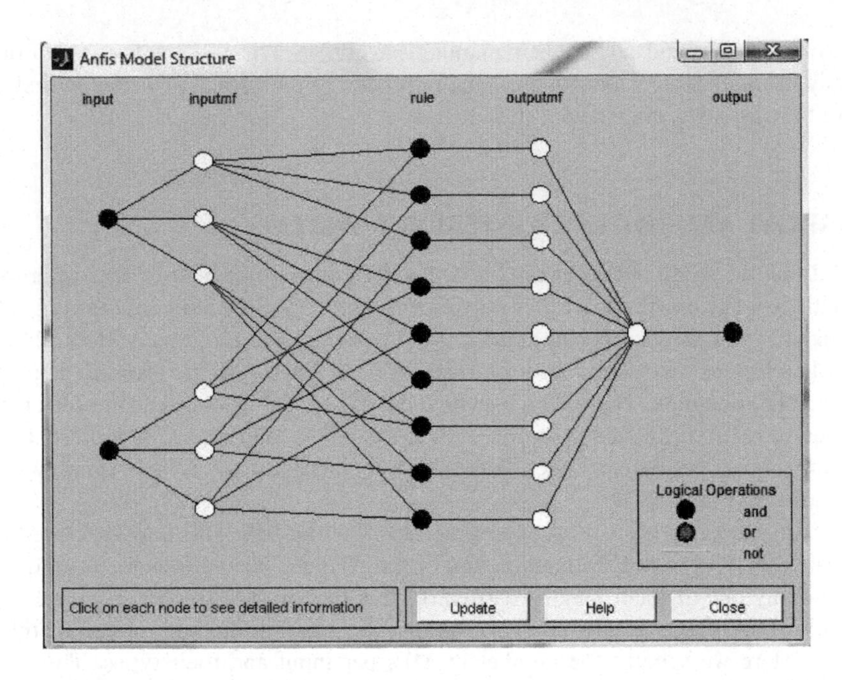

FIGURE A.20 Generated FIS architecture.

we can plot the training, testing, or checking data and compare their actual values and the values returned by the system. This is given in Figure A.19. Here we had taken a very simple synthetic problem. Hence the match between the received outputs and the targets is perfect.

We may also view the basic structure of the ANFIS constructed by the toolbox. This is shown in Figure A.20. This helps in attaining a logical understanding of the system.

Appendix B: MATLAB® Source Codes for Soft Computing

B.1 INTRODUCTION

In the Appendix A, we presented the GUI tools for the various applications that were discussed in this book. We saw that it was very simple to design and use these systems. This made the work very simple and made it possible to make real life applications with a very little effort. The simplicity of the GUIs has indeed resulted in making many people contribute toward this domain of soft computing. One may not be a pro or may not necessarily require professional training to master the simple yet powerful GUI tools that are present in MATLAB®.

But the source codes have their own weight and importance. They give a lot of flexibility and power to designers to use the system as per their own demands. The designer may integrate systems, implement them in hierarchies, or do any such task that may not be very simple to carry out if one is restricted to the GUI alone. Hence it is necessary to know the use of these systems as source codes as well that can be executed from a command line or as a batch file whenever required.

With the same objectives, this chapter aims at the study of the source codes of the various tools studied. The complete details and description of these source codes is outside the scope of the book. We present the source codes for the various systems to make the users aware of the way in which these systems can be used as MATLAB sources. The complete details of the source codes for every system along with the general help on MATLAB commands and statements can be found in the MATLAB guide itself. We assume here that the readers are aware of the general terminologies, commands, and statements used in MATLAB. We further assume that the readers have experience of programming in MATLAB.

Here we would present the source codes for numerous models of artificial neural networks, including back propagation algorithm, self-organizing maps, radial basis function networks, and more. We also discuss the source code of fuzzy inference systems and genetic algorithms. Here we would see how these systems can be worked with from the command line of a programmed m file.

B.2 ARTIFICIAL NEURAL NETWORK

The first system whose code we present is the artificial neural network (ANN). For the understanding and presentation of the source code, we will use the same character recognition problem that we presented in Chapter 3. For training, the ANN requires a matrix consisting of all the inputs and another matrix consisting of all the outputs to various cases in the training database. Every case here has multiple input or output attributes. The rows correspond to the attributes, and the columns correspond to the input test cases.

The first task before any model can be applied is to get an input vector per character. The character in the system was initially assumed to be made up of binary digits, with 0 representing the absence of ink and 1 representing the presence of ink at the block. This was assumed to be available in the system in the form of a two-dimensional array per character. In order to convert this into a valid ANN input, we are first supposed to convert all the two-dimensional arrays into one-dimensional arrays and then to append all the so-formed one-dimensional arrays one after the other as rows of the final input vector.

Box B.1 shows one of the letters of the system. Box B.2 is the code to convert this letter into a one-dimensional array. Later in Box B.2 we also join the various letters one after the other as columns of the input vector.

```
a=[
  0 0 0 1 0 0
  0 0 1 0 1 0
  0 1 0 0 0 1
  0 1 1 1 1 1
  0 1 0 0 0 1
  1 0 0 0 0 1
  ];
```

BOX B.1 The input letter 'A'.

The next task is to compute the output vector. Here we take as many output neurons as the number of classes (10 in this case). Each output neuron is the probability of occurrence of the class on a scale of 0 to 1. It may be observed that if we calculate the output for the training input using the order as given by the input, the final output matrix would come out as a unit matrix. Here the first output neuron corresponds to *'a'*, the second to *'b'*, the third to *'c'*, and so on. The output vector is simply the unit matrix given by Box B.3.

A similar procedure may be applied for testing inputs as well. Now since we have the inputs and outputs ready for training, the next step is the construction of ANN. This depends upon the model opted for. We hence discuss the various models one by one.

B.2.1 ANN with BPA

The first model is the ANN implementation with back propagation algorithm (BPA) as the training algorithm. The code of this network initialization along with the training parameters is given in Box B.4.

The min max function returns the minimum and maximum of the arguments, and this is used as the default range by the ANN. Although we write min max function here, when implementing an ANN in such problems it would be advisable to use the known range (0 to 1) for every input. This is true of all models of ANN we discuss in this appendix. The min max has been written to make the codes look more like the commonly written codes. It may be observed that according to this code we have two layers (denoted in the code in line 1 by [40,10]). The first is the hidden layer that has 40 neurons. The second layer is the hidden layer that has 10 output neurons as discussed. The activation functions of the two layers are tansig and purelin. The training function used is traingd. Lines 2 to 5 of the ANN set the training parameters of the ANN. The network is trained in line 6. Here 'le' is the input and 't' is the output matrix.

```
a1=[]
for i=1:6
   for j=1:6
      a1=[a1;a(i,j)];
   end
end

le=[a1 b1 c1 d1 e1 f1 g1 h1 i1 j1];
```

BOX B.2 Conversion of letters to input vector.

```
t=eye(10);
```

BOX B.3 The output vector.

B.2.2 Radial Basis Function Network

The radial basis function network (RBFN) is the other ANN model implemented. The code for this network is presented in Box B.5.

This network uses the newrb function of MATLAB to initialize itself. This code specifies le as input, t as output, 0.1 as the goal, and 15 as the spread. The spread is the radius of the individual radial basis function that must be large enough for generalization purposes. These networks do not require training separately. The newrbe is another function that may be used for the network initialization and training in place of newrb. The newrb function adds neurons until the goal is met. On the other hand, the newrbe function places as many neurons as the number of training inputs.

B.2.3 Learning Vector Quantization

The learning vector quantization (LVQ) forms a classificatory style of ANN model. Hence here the output representation is different from the ones discussed earlier. The code for this network is given in Box B.6.

Here the output is initially specified as the class labels. These are usually chosen such that they start with the label 1 and go on with each class being associated with a label. This can be seen in the code in line 1, which is the output vector for the 10 inputs. Line 2 makes an LVQ ANN. The first parameter is the input ranges. The next is the number of neurons. The last parameter comprises the class percentages of the various classes. These all are written as rations and add up together to give a total of 1. The output is a little specialized here. We need to first convert the class labels or indices into vectors and then convert the vectors into the full output matrix. This is the general way of solving problems using these ANNs. It may be noted that this gives the same output vector as we used for the other algorithms. These operations are done on lines 3 and 4. Lines 5 to 7 set the training parameters, and line 8 trains the system.

B.2.4 Self-Organizing Map

The self-organizing maps (SOMs) undergo an unsupervised training. These networks train by the inputs only. The outputs do not have a role to play in the training. Box B.7 shows the code for the SOM.

The network is created using the newsom function used in line 1. The first parameter is the range, and the second parameter is the size of the feature vector map. Line 2 specifies the training epochs,

```
Line 1: net=newff(minmax(le),[40,10],{'tansig','purelin'},'traingd');
Line 2: net.trainParam.epochs = 1000;
Line 3: net.trainParam.goal=le-2;
Line 4: net.trainParam.lr=0.01;
Line 5: net.trainParam.mc=0.7
Line 6: net = train(net,le,t);
```

BOX B.4 Artificial neural network with back propagation algorithm code.

```
Line 1: net = newrb(le,t,0.1,15);
```

BOX B.5 The radial basis function network code.

and line 3 trains the SOM. Here it may be seen that the outputs are not specified for the system training.

B.2.5 Recurrent Neural Network

The recurrent neural networks are implemented using the elman model. The code for this network is given in Box B.8.

Here we first create an elman network in line 1. The parameters are similar to that used in BPA. The inputs and outputs for these networks need to be in the form of sequences. This conversion is made using the con2seq function in lines 2 and 3. Lines 4 to 6 specify the training parameters, and line 7 trains the system.

The next part after the training is over is the testing. This may need to be done over the training data set, the testing data set, or the validation data set. We now look at the method by which we would simulate the system to get the outputs for the presented inputs. Here for every input we would be supplied by 10 outputs, each corresponding to letters from 'a' to 'j'. We need to select the maximum probability and return the same as output. This code is given in Box B.9.

In this problem 'le' represents the training, testing, or validation data set on which the simulation is to be performed. The corresponding outputs are assumed to be stored in the array t. Here we first use the network with the input to get the output matrix. This is done in line 1. The rest of the codes iterate for every input in the data set to calculate its output and compare it with the standard output. Line 14 gives the final efficiency of the system. The data set is supposed to contain n elements.

The system simulation is little different in the LVQ due to its classificatory nature, which gives the final class as its output. The code for LVQ is given in Box B.10. Hence we need not calculate the maximum of the probabilities to get the final output class. Here we need to convert the vectors back into indices that are easily comparable. It may be seen that the procedure is opposite to the one used in training. Again the whole method is similar to what we have been using, but this is the general manner of problem solving in LVQ.

The SOM simulation is similar to the LVQ simulation. Here the major point is that the training is in unsupervised mode. The SOM simulation returns vectors that may be converted to indices. Since the training is in unsupervised mode, the class labels are fixed by the system only. The system returns the closest mapping feature grid to every input. We may find the class it corresponds to by a lookup over the training data.

In case of a recurrent or elman neural network, the simulation performs an extra task of conversion of a sequence to concurrent vectors, which is then converted to a vector. The rest of the procedure is same as that in the ANN with BPA. The difference of code in the two models is given in Box B.11.

```
Line 1: t=[1 2 3 4 5 6 7 8 9 10];
Line 2: net=newlvq(minmax(le),15, [0.1 0.1 0.1 0.1 0.1 0.1 0.1 0.1 0.1 0.1])
Line 3: T=ind2vec(t);
Line 4: T=full(T);
Line 5: net.trainParam.epochs = 1500;
Line 6: net.trainParam.goal=1e-5;
Line 7: net.trainParam.lr=0.01;
Line 8: net = train(net,le,T);
```

BOX B.6 The learning vector quantization code.

```
Line 1: net = newsom(minmax(le) , [3 4]);
Line 2: net.trainParam.epochs = 1500;
Line 3: net = train(net,le);
```

BOX B.7 The self-organizing map code.

```
Line 1: net = newelm(minmax(le),[25 10],{'tansig','logsig'});
Line 2: Pseq = con2seq(le);
Line 3: Tseq = con2seq(t);
Line 4: net.trainParam.epochs = 1500;
Line 5: net.trainParam.goal=1e-2;
Line 6: net.trainParam.lr=0.1;
Line 7: net = train(net,Pseq,Tseq);
```

BOX B.8 The recurrent neural network code.

```
Line 1: Y = sim(net,le);
Line 2: cor=0;
Line 3: for i=1:n
Line 4:    max=1;
Line 5:    for j=1:10
Line 6:       if Y(j,i)>Y(max,i)
Line 7:          max=i;
Line 8:       end
Line 9:    end
Line 10:      if max==t(i)
Line 11:         cor=cor+1;
Line 12:   end
Line 13: end
Line 14: cor/n*100
```

BOX B.9 The simulation code for artificial neural networks.

```
Line 1: Y = sim(net,le);
Line 1: Y=vec2ind(Y)
Line 2: cor=0;
Line 3: for i=1:n
Line 4:    if y(i)==t(i)
Line 5:       cor=cor+1;
Line 6:    end
Line 7: end
Line 8: cor/n*100
```

BOX B.10 The simulation code for learning vector quantization.

```
Line 1: Y = sim(net,Pseq);
Line 2: Y=seq2con(Y);
Line 3: Y=Y{1}
```

BOX B.11 The code for recurrent neural networks.

```
a=newfis('fuz');
a.type='mmamdani';
a.andMethod='min';
a.orMethod='max';
a.defuzzMethod='centroid';
a.impMethod='min';
a.aggMethod='max';
```

BOX B.12 The fuzzy inference system creation code.

```
a.input(1).name='inp1';
a.input(1).range=[0 1];

a.input(1).mf(1).name='low';
a.input(1).mf(1).type='gaussmf';
a.input(1).mf(1).params=[0.1699 0];

a.input(1).mf(2).name='med';
a.input(1).mf(2).type='gaussmf';
a.input(1).mf(2).params=[0.1699 0.5];
…………..

a.input(2).name='inp2';
a.input(2).range=[0 1];

a.input(2).mf(1).name='low';
a.input(2).mf(1).type='gaussmf';
a.input(2).mf(1).params=[0.1699 0];
…………

a.output(1).name='out';
a.output(1).range=[0 1];

a.output(1).mf(1).name='cheap'
a.output(1).mf(1).type='trimf';
a.output(1).mf(1).params=[0 0.5 1];
…………..
```

BOX B.13 The membership, input, and output creation code.

```
a.rule(1).antecedent=[1 1];
a.rule(1).consequent=[1];
a.rule(1).weight=1;
a.rule(1).connection=2;
a.rule(2).antecedent=[2 0];
a.rule(2).consequent=[2];
a.rule(2).weight=1;
```

BOX B.14 The rule creation code.

```
evalfis([0.5 0.2], a)
```

BOX B.15 The fuzzy inference system evaluation code.

B.3 FUZZY INFERENCE SYSTEMS

In this section, we present a technique for working with FIS with the help of m file codes. This gives us the flexibility to specify all the FIS properties, MFs, and rules with the help of simple statements. The whole program is simple to frame and understand. This forms a very interesting use of the MATLAB Fuzzy Logic Toolbox by commands. The whole code may be divided into four parts. The first is to create an FIS and define the initial properties. The second part is to define the inputs and outputs and their MFs. The third part is to define the rules. The last part would be to run the FIS, give the inputs, and evaluate outputs. These are given in Boxes B.12, B.13, B.14, and B.15 respectively.

The code in Box B.12 only creates a new FIS and specifies its properties. The code in Box B.13 creates the inputs and outputs and the MFs for all these inputs and outputs. Here we specify a name and a range for all the input and output variables. We then specify all the MFs for the entire variables one after the other. Each MF consists of a name, type, and parameters as per the type selected.

Box B.13 represents a sample rule. The antecedents are the input variables one after the other that participate in a rule. The consequents are the outputs. The numbers denote the membership function used for every variable. Here, 0 represents the nonparticipation of a variable. Negated numbers mean the participation of the MF corresponding to the positive part of the number with a prefixed 'not' operator. The connection may be 1 or 2 representing 'and' or 'or'. The evaluation simply gives the specified input as an array to the mentioned FIS variable. This is given in Box B.15.

While working with any structure or variable, one may use a tab as an automatic code completion facility of MATLAB. Also the code may only be written for the nondefault settings.

B.4 GENETIC ALGORITHM

The genetic algorithm (GA) is another widely used technique for the optimization purposes. The code to carry this is given in Box B.16. We first make an option variable that holds all the options. The bulk of the code does exactly this. At the end we run the solver, which takes as argument the objective function, the number of variables, and the GA options. This returns a lot of information about the solution and the final result. It returns the reason why the algorithm terminated, the final output with details of the performance at every generation, the final population, and the final scores. These all may again be required to resume the solver or make it run from the same position again.

```
options = gaoptimset;
options.PopulationType='doubleVector';
options.PopInitRange=[-5;5];
options.PopulationSize=30;
options.EliteCount=2;
options.CrossoverFraction=0.7;
options.Generations=50;
options.TimeLimit=Inf;
options.FitnessLimit=-Inf;
options.StallGenLimit=Inf;
options.StallTimeLimit=Inf;
options.CreationFcn=@gacreationuniform;
options.FitnessScalingFcn=@fitscalingrank;
options.SelectionFcn=@selectionstochunif;
options.CrossoverFcn=@crossoverscattered;
options.MutationFcn={@Mutationuniform,0.02};
options.PlotFcns=[@gaplotbestf];

[x fval reason output population scores] = ga(@rastriginsfcn, 2, options);
```

BOX B.16 The genetic algorithm command line solver.

While working with any structure or variable or while specifying any function, one may use tab as an automatic code completion facility of MATLAB. Also the code may only be written for the nondefault settings.

The GUIs present an excellent means for easy use of the soft computing tools, yet the power and flexibility of command lines necessitate their use in many applications. MATLAB has good command line statements and functions that cater to this need to run tools using commands. We may solve soft computing problems using these means as well. It provides an easy way to take inputs from some systems and give the outputs to the other systems. The power of the command line may make many things possible that the GUI may not be able to.

Appendix C: Book Website

The authors in this book presented to best of their capabilities the foundations and real life applications of soft computing. Learning is a very dynamic concept that calls for dynamic changes and modifications. This is especially true in the domain of soft computing, which is developing at a rapid pace.

For these reasons the book has a separate website of its own. The website can be accessed at http://publication.iiitm.ac.in/softcomputing/.

This website contains presentations to the various chapters that may be used by the instructors in their courses. Presentations are a good means of learning. They enable us to convey ideas in an easy way that may not always be possible in case of text.

The first edition of this book does not have a separate instructor's manual so far, apart from this presentation. Instructors may require clarification or further explanation for some of the topics presented. They may also require hints or solutions to the unsolved exercises. These may be requested from the authors online at the book website itself.

Now that the readers have matured and become experienced soft computing professionals, the authors would like some genuine feedback on the book. Feedback may be filled in at the book website. This would help us further improve the book. Feedback is also a good source of active interaction between the authors and readers. This interaction would surely help us understand the readers' expectations and problems.

It is possible that the book may contains some errors or mistakes. All errors, if any, reported after book publication will be available at the book website. Readers may also submit bugs, if they happen to encounter any, on the website.

We strongly recommend that readers visit the book website, http://softcomputing.iiitm.ac.in/.

References

References

Abbass, H. 2003. Speeding Up Back-Propagation Using Multi-Objective Evolutionary Algorithms, *Neural Comput.*, vol. 15, no. 11, pp. 2705–2726.

Abonyi, J., Nagy, L., and Szeifert, F. 1999. Adaptive Sugeno Fuzzy Control: A case study. *Advances in Soft Computing: Engineering Design and Manufacturing,* R. Roy, T. Furuhashi, and P.K. Chawdhry, eds. London, UK: Springer-Verlag, pp. 135–146.

Abraham, A. and Nath, B. 2001. A Neuro-Fuzzy Approach for Modelling Electricity Demand in Victoria, *Appl. Soft Comput.,* vol. 1, no. 2, pp. 127–138.

Abraham, A. 2001. Neuro-Fuzzy Systems: State-of-the-Art Modeling Techniques. In J. Mira, A. Prieto eds. *Connectionist Models of Neurons, Learning Processes and Artificial Intelligence. Springer Lecture notes in computer science*, vol. 2084, pp. 269–276.

Accardi, A.J. and Cox, R.V. 1999. A Modular Approach to Speech Enhancement with an Application to Speech Coding, *Proc. IEEE Int. Conf. Acoustics, Speech, Signal Processing,* Phoenix, AZ, pp. 201–204.

Acierno, Antonio D. 2000. Back-Propagation Learning Algorithm and Parallel Computers: The CLEPSYDRA Mapping Scheme, *Neurocomputing,* vol. 31, no. 1–4, pp. 67–85.

Adamidis, P. and Petridis, V. 1996. Co-operating Populations with Different Evolution Behaviours, *Proc. 3rd IEEE Conf. Evolutionary Comput.,* New York: IEEE Press, pp. 188–191.

Adamidis, P. 1994. Review of Parallel Genetic Algorithms Bibliography, Aristotle Univ. Thessaloniki, Greece, Tech. Rep.

Adams, A. and Sasse, M.A. 1999. Users Are Not the Enemy, *Commun. ACM,* vol. 42, no. 12, pp. 41–46.

Adams, N.H., Bartsch, M.A., Shifrin, J., and Wakefield, G.H. 2004. Time Series Alignment for Music Information Retrieval, *Proc. of the Int. Symp. on Music Information Retrieval (ISMIR),* pp. 303–310, Barcelona, Spain.

Agakov, F., Bonilla, E., Cavazos, J., Franke, B., Fursin G., O'Boyle, M.F.P., Thomson, M.J., Toussaint, M., and Williams, C. 2006. Using Machine Learning to Focus Iterative Optimization, Proceedings of the International Symposium on Code Generation and Optimization, IEEE Computer Society Washington, DC, USA, pp. 295–305.

Agam, G. and Suresh, S. 2007. Warping Based Offline Signature Recognition, *Information Forensics and Security,* vol. 2, no. 3, pp. 430–437.

Agarwal, M. 1995. Combining Neural and Conventional Paradigms for Modeling, Prediction, and Control, *Proc. 4th IEEE Conf. Control Applications,* Albany, NY, pp. 566–571.

Agarwal, P., Agarwal, M., Shukla, A., Tiwari, R. 2009. Multilingual Speaker Recognition Using Artificial Neural Networks, accepted for publication in Springer's book series, *Advances in Intelligent and Soft Computing,* Proceedings Second International Workshop on *Advanced Computational Intelligence (IWACI2009)* to be held in Mexico City during June 22–23, 2009.

Agrawal, R. and Ramakrishnan, S. 1990. Fast Algorithms for Mining Association Rules, *Proc. of the 20th Int'l Conf. on Very Large Databases,* Santiago, Chile.

Agrawal, R., Imielinaki, T., and Swami, A. 1993. Mining Association Rules Between Sets of Items in Large Databases, *Proc. of the ACM SIGMOD Conference on Management of Data,* Washington, DC.

Aiyer, S.V.B., Niranjan, M., and Fallside, F. 1990. A Theoretical Investigation into the Performance of The Hopfield Model, *IEEE Trans. on Neural Networks,* vol. 1, no. 2, pp. 204–215.

Akers, S. 1978. Binary Decision Diagrams. *IEEE Transactions on Computers*, vol. 27, no. 6, pp. 509–516.

Akhmetov, D.F., Dote, Y., and Ovaska, S.J. 2001. Fuzzy Neural Network with General Parameter Adaptation for Modeling of Nonlinear Time-Series, *IEEE Trans. Neural Networks*, vol. 12, pp. 148–152.

Albayrak, S. and Amasyali, F. 2003. Fuzzy C-Means Clustering on Medical Diagnostic Systems, *Proc. International XII. Turkish Symposium on Artificial Intelligence and Neural Networks,* Antalya, Turkey.

Alex, G., Santiago, F., Marcus, L., Horst, B., and Jurgen, S. 2008. Unconstrained Online Handwriting Recognition with Recurrent Neural Networks, In J. Platt, D. Koller, Y. Singer, and S. Roweis, eds, *Advances in Neural Information Processing Systems,* vol. 20, pp. 577–584, MIT Press, Cambridge, MA, 2008.

Ali, M.M. and Nguyen, H.T. 1989. A Neural Network Implementation of an Input Access Scheme in a High-Speed Packet Switch, *Proc. IEEE Global Telecomm. Conf. Expo.,* pp. 1192–1196, New York, USA.

Aliev, R.A. and Aliev, R.R. 2001. *Soft Computing and Its Applications.* World Scientific Publishing Co. Pvt. Ltd., Singapore.

Alpaydin, E. 2004. *Introduction to Machine Learning.* Cambridge, MA: MIT Press.

Alves, R.Lde.S., De Melo, J.D., Neto, A.D.D., and Albuquerque, A.C.M.L. 2002. New Parallel Algorithms for Back-Propagation Learning, *Proceedings of the 2002 International Joint Conference on Neural Networks,* 2002. IJCNN 02, pp. 2686–2691, Honolulu, HI.

Amar, C. Ben and Jemai, O. 2005. Wavelet Networks Approach for Image Compression, *ICGST International Journal on Graphics, Vision and Image Processing,* vol. SI1, pp. 37–45.

Ambler, W.S. 2005. *Introduction to security Threat Modeling. Agile Modeling.* Available at http://www.agilemodeling.com/artifacts/securityThreatModel.htm.

Amin, Md. Faijul and Murase, K. 2008. Single-Layered Complex-Valued Neural Network for Real-Valued Classification Problems, *Neurocomputing,* doi:10.1016/j.neucom. 2008.04.006.

Aminzadeh, F. and Jamshidi, M. 1994. *Soft Computing, Fuzzy Logic, Neural Networks and Distributed Artificial Intelligence.* New Jersey: Prentice Hall.

Amjady, N. 2001. Short-Term Hourly Load Forecasting Using Time- Series Modeling with Peak Load Estimation Capability, *IEEE Transactions on Power Systems.* vol. 16, pp. 798–805.

An, P.E., Brown, B., and Harris, C.J. 1997. On the Convergence Rate Performance of the Normalized Least-Mean-Square Adaptation, *IEEE Trans. Neural Networks*, vol. 8, pp. 1211–1214.

Anderson, James A. 1995. *An Introduction to Neural Networks,* Cambridge, Massachusetts, London, UK: MIT Press.

Andrew Teoh, B.J. and David Ngo, C.L. 2005. Cancelable Biometrics Featuring with Tokenised Random Number. Pattern Recognition Letter, *Elsevier Science,* vol. 26, no. 10, pp. 1454–1460.

Andrew Teoh, B.J., Ong, T.S., and David, N.C.L. 2003. Automatic Fingerprint Center Point Determination. *Lecture Notes of Artificial Intelligent (LNAI),* 2903, T.D. Gedeon, Lance Chun Che Fung, eds., Springer-Verlag, pp. 633–640.

Andrews, G. 2000. *Foundations of Multithreaded, Parallel, and Distributed Programming.* Reading, MA: Addison Wesley.

Andrews, R., Diederich, J., and Tickle, A.B. 1995. Survey and Critique of Techniques for Extracting Rules from Trained Artificial Neural Networks, *Knowledge-Based Syst.,* vol. 8, no. 6, pp. 373–389.

Andrzejak, R.G., Mormann, F., Kreutz, T., Rieke, C., Kraskov, A., Elger, C.E., and Lehnertz, K. 2003. Testing the Null Hypothesis of the Nonexistence of a Preseizure State, *Phys Rev. E,* vol. 67, no. 1, 010901.

Andrzejak, R.G., Widman, G., Lehnertz, K., Rieke, C., David .P., and Elger, C.E. 2001. The Epileptic Process as Nonlinear Deterministic Dynamics in a Stochastic Environment: An Evaluation on Mesial Temporal Lobe Epilepsy, *Epilepsy Res.,* vol. 44, nos. 2–3, pp. 129–140.

Ang, J.H. et al. 2008. Training Neural Networks for Classification Using Growth Probability-Based Evolution, *Neurocomputing,* doi:10.1016/j.neucom.2007.10.011.

Ang, K.K. and Quek, C. 2005. RSPOP: Rough set-Based pseudo-Outerproduct Fuzzy Rule Identification Algorithm, *Neural Comput.,* vol. 17, no. 1, pp. 205–243.

Angel de la Terra et al. 2005. Histogram Equalization of Speech Representation for Robust Speech Recognition. *IEEE Transactions on Speech and Audio Processing*, vol. 40, no. 13, pp. 355–366.

Angulo, V. de and Torras, C. 1995. On-Line Learning with Minimal Degradation in Feedforward Networks, *IEEE Trans. Neural Netw.,* vol. 6, no. 3, pp. 657– 668.

Anile, A.M., Deodato, S., and Privitera, G. 1995. Implementing Fuzzy Arithmetic, *Fuzzy Sets Syst.,* vol. 72, pp. 239–250.

Antoine, A., Vandromme, J., et al. 2008. A Survey among Breast Cancer Survivors: Treatment of the Climacteric after Breast Cancer. *J. Climacteric,* vol. 11, no. 4, pp. 322–328.

Antsaklis, P.J., Lemon, M., and Sliver, I.A. 1996. Leaning to Be Autonomous Intelligent Supervisory Conrral, in M.M.M. Gupta, N.K. Sinha, eds. *Intelligent Control Systems,* IEEE Press, pp. 28–62, Piscataway, NJ.

Arakawa, K. and Nomoto, K. 2005. A System for Beautifying Face Images Using Interactive Evolutionary Computing, *Proc. IEEE ISPACS 2005,* pp. 9–12, Hong Kong.

Araokar S. Visual Character Recognition Using Artificial Neural Networks B.Sc thesis proposal, MGM college of Engineering & Technology, University of Mumbai, India.

Arbib, M.A., ed. 2003. *Handbook of Brain Theory and Neural Networks.* 2nd ed. Cambridge, MA: MIT Press.

Verma, A. 2005. *The Indian Police: A Critical Evaluation*, New Delhi: Regency Publications.

Asada, M. and Kitano, H., eds. 1999. RoboCup-1998: Robot Soccer World Cup II, *Lecture Notes in Computer Science,* vol. 1604, New York: Springer-Verlag.

Ashbourn, J. 2000. *Biometrics: Advanced Identity Verification.* New York: Springer-Verlag.

Ashbourn, J. 2003. *Practical Biometrics From Aspiration to Implementation.* Springer, ISBN 1852337745.

Atkinson, D.C., and Griswold, W.G. 2005. Effective Pattern Matching of Source Code Using Abstract Syntax Patterns, *SP&E,* pp 413–447, (www.interscience.wiley.com). DOI: 10.1002/spe.704.

Avineri, A., Prashker, J., and Ceder, A. 2000. Transportation Projects Selection Process Using Fuzzy Sets Theory, *Fuzzy Sets Syst.,* vol. 116, pp. 35–47.

Ayala, D.E., Hermida, R.C., Mojon, A., Fernandez, J.R., Alonso, I., Silva, I., Ucieda, R., and Iglesias, M. 2002. Blood Pressure and Heart Rate Variability During Gestation in Healthy and Hypertensive Pregnant Women, Engineering in Medicine and Biology Society, *Ann. Int. Conf. IEEE,* vol. 1, pp. 367–370.

Ayala, D.E., Hermida, R.C., Mojon, A., and Iglesias, M. 2002. Circadian Blood Pressure Variability in Healthy Human Pregnancy. Comparison with Gestational Hypertension, Engineering in Medicine and Biology Society, *IEEE Annual Conference,* vol. 1, pp. 639–640.

Ayala-Ramirez, V., Perez-Garcia, A., Montecillo-Puente, E.J., and Sanchez-Yanez, R.E. 2004. Path Planning Using Genetic Algorithms for Mini-Robotic Tasks, *IEEE International Conference on Systems, Man and Cybernetics,* vol. 4, pp. 3746–3750, The Hague, Netherlands.

Azari, N.G. and Lee, S.Y. 1991. Hybrid Partitioning for Particle-in-Cell Simulation on shared Memory Systems, *Proc. of 11th International Conference on Distributed Computing Systems,* pp. 526–533. Arlington, TX.

Baars, B.J. 1988. A Cognitive Theory of Consciousness. Cambridge: Cambridge University Press.

Baba, N. 1989. A New Approach for Finding the Global Minimum of Error Function of Neural Networks, *Neural Networks,* vol. 2, no. 3, pp. 367–373.

Babii, S., Cretu, V., and Petriu, E.M. 2007. Performance Evaluation of Two Distributed Back Propagation Implementations, *Proceedings of International Joint Conference on Neural Networks,* Orlando, Florida, USA.

Bacca, M. and Rabuzin, K. 2005. Biometrics in Network Security, *The Proceedings of the XXVIII International Convention MIPRO,* Rijeka, Croatia.

Back, T., Beielstein, T., Naujoks, B., and Heistermann, J. 1995. Evolutionary Algorithms for the Optimization of Simulation Models Using PVM, *Proc. EuroPVM'95: 2nd Euro. PVM User's Group Meeting,* J. Dongarra, M. Gengler, B. Tourancheau, and X. Vigouroux, eds. Paris, France: Hermes, pp. 277–282.

Back, T., Fogel, D.B., and Michalewicz, Z. eds. 1997. *Handbook of Evolutionary Computation.* Bristol, U.K.: Inst. of Physics and New York: Oxford Univ. Press.

Back, T., Hammel, U., and Schwefel, H.P. 1997. Evolutionary computation: Comments on the history and current state, *IEEE Trans. Evol. Comput.,* vol. 1, pp. 3–17. Journal Article; NA.

Baker, J.E. 1985. Adaptive Selection Methods for Genetic Algorithms, *Proc. 1st Int. Conf. Genetic Algorithms Appl.,* J.J. Grefenstette, ed. Hillsdale, NJ: Lawrence Erlbaum, pp. 101–111.

Baker, O.F., Abdul Kareem, and Kareem, S. 2004. Assessment of the Use of Soft Computing Models into Survival Analysis, *International Medical Journal (IMJ),* vol. 3, no. 2. Article F1.

Bakker, H.H.C. and Flemmer, R.C. 2009. Data Mining for Generalized Object Recognition, *Proceedings of the International Conference on Autonomous Robots and Agents (ICARA),* Wellington, New Zealand, *in press.*

Balakrishnan, S.N. and Biega, V. 1996. Adaptive-Critic-Based Neural Networks for Aircraft Optimal Control, *J. Guid. Control Dyn.,* vol. 19, no. 4, pp. 893–898.

Baldisserra, D., Franco, A., Maio, D., and Maltoni, D. 2001. Fake Fingerprint Detection by Odor Analysis, *Lecture Notes in Computer Science,* vol. 2013, pp. 369–376, *Proceedings of the International Conference on Biometric Authentication (ICBA06),* Hong Kong, Springer-Verlag.

Baldwin, J.F. 1987. Evidential Support Logic Programming, *Fuzzy Sets and Systems,* vol. 24, pp. 1–26.

Ballard, L., Monrose, F., and Lopresti, D. 2006. Biometric Authentication Revisited: Understanding the Impact of Wolves in Sheep's Clothing, *Proc. 15th Annu. USENIX Security Symp.,* pp. 29–41. Vancouver, BC, Canada.

Bandemer, H. and Nather, W. 1992. *Fuzzy Data Analysis, Theory and Decision Library, Series B: Mathematical and Statistical Methods.* Dordrecht: Kluwer Acadiemic Publishers.

Bandyopadhyay, S. 2004. An Automatic Shape Independent Clustering Technique, *Pattern Recog.,* vol. 37, no. 1, pp. 33–45.

Bandyopadhyay, S. and Saha, S. 2007. GAPS: A Clustering Method Using a New Point Symmetry Based Distance Measure, *Pattern Recog.* URL: http://dx.doi.org/10.1016/j.patcog.2007.03.026.

Banfield, J. and Raftery, A. 1993. Model-Based Gaussian and Non-Gaussian Clustering, *Biometrics,* vol. 49, pp. 803–821.

Banzhaf, W., Nordin, P., Keller, R.E., and Francone, F.D. 1998. *Genetic Programming: An Introduction on the Automatic Evolution of Computer Programs and Its Applications.* San Francisco, CA: Morgan Kaufmann.

Baraldi, A. and Blonda, P. 1999. A Survey of Fuzzy Clustering Algorithms for Pattern Recognition—Parts I and II, *IEEE Trans. Syst., Man, Cybern. B,* vol. 29, pp. 778–801.

Barbará, D. and Chen, P. 2003. Using Self-Similarity to Cluster Large Data Sets, *Data Mining Knowl. Discov.,* vol. 7, no. 2, pp. 123–152.

Barnsley, M. and Sloan, A.D. 1988. A Better Way to Compress Images, *BYTE* magazine, pp. 215–223, January 1988.

Barraquand, J. and Latombe, J.C. 1993. Nonholonomic Multibody Mobile Robots: Controllability and Motion Planning in the Presence of Obstacles, *Algorithmica,* vol. 10, nos. 2–4, pp. 121–155.

Barraquand, J., Langlois, B., and Latombe, J.C. 1992. Numerical Potential Field Techniques for Robot Path Planning. *IEEE Trans. Syst., Man, Cybern,* vol. 22, no. 2, pp. 224–241.

Barto, G., Sutton, R.S., and Anderson, C.W. 1983. Neuron-Like Adaptive Elements That Can Solve Difficult Learning Control Problems, *IEEE Trans. Syst., Man, Cybern.,* vol. SMC-13, no. 5, pp. 834–846.

Baruah, S., Howell, R., and Rosier, L. 1990. Algorithms and Complexity Concerning the Preemptive Scheduling of Periodic Real-Time Tasks on One Processor, *Real-Time Systems,* vol. 2, pp. 301–324.

Basheer, I.A. and Hajmeer, M. 2000. Artificial Neural Networks: Fundamentals, Computing, Design, and Application. *J. Microbiol. Methods,* vol. 43, pp. 3–31.

Basse, S. 1991. *Computer Algorithms: Introduction to Design and Analysis.* Reading MA: Addison-Wesley.

Bazin, A.I. and Nixon, M.S. 2004. Facial verification Using Probabilistic Methods. *Proc. British Machine Vision Association Workshop on Biometrics*, London, UK.

Behnke, S. 2003. Local Multiresolution Path Planning, preliminary version in *Proc. of 7th RoboCup Int. Symposium,* Padua, Italy.

Belacel, N., Hansen, P., and Mladenovic, N. 2002. Fuzzy J-Means: a New Heuristic for Fuzzy Clustering, *Pattern Recog.,* vol. 35, pp. 2193–2200.

Belding, T.C. 1995. The Distributed Genetic Algorithm Revisited, *Proc. 6th Int. Conf. Genetic Algorithms,* L. Eshelman, ed. San Francisco, CA: Morgan Kaufmann, pp. 114–121.

Bellman, R.E. and Zadeh, L.A. 1970. Decision-Making in a Fuzzy Environment, *Manag. Sci.*, vol. 17, no. 4, pp. B-141–B-164.

Bellucci, E. and Zeleznikow, J. 1999. AI Techniques for Modelling Legal Negotiation, *Proceedings of the Seventh International conference on Artificial Intelligence and Law,* New York, ACM, pp. 108–116.

Ben-Hur and Guyon, I. 2003. Detecting Stable Clusters Using Principal Component Analysis in Methods, *Molecular Biology*, M. Brownstein and A. Kohodursky, eds. Totowa, NJ: Humana Press.

Berghe, C.V., Riordan, J., and Piessens, F. 2005. A Vulnerability Taxonomy Methodology Applied to Web Services, *Proc. the 10th Nordic Workshop on Secure IT-systems (NORDSEC 2005),* Tartu, Estonia, pp. 20–21.

Bergstra, J. et al. 2006. Aggregate Features and ADABOOST for Music Classification, *Machine Learning,* vol. 65, nos. 2–3, pp. 473–484.

Bezdek, J. 1981. *Pattern Recognition with Fuzzy Objective Function Algorithms,* Norwell: Plenum Press.

Bhandarkar, S.M. and Zhang, H. 1999. Image Segmentation Using Evolutionary Computation, *IEEE Transactions on Evol. Comp.,* vol. 3, no. 1, pp. 1–21.

Bianchini, M., Frasconi, P., and Gori, M. 1995. Learning Without Local Minima in Radial Basis Function Networks, *IEEE Trans. Neural Netw.*, vol. 6, no. 3, pp. 749–756.

Biggs, J. 1999. *Teaching for Quality Learning at University.* Buckingham, UK: Open University Press.

Bilgic, T. and Turksen, I. 1997. Measurement of Membership Functions: Theoretical and Empirical Work, *Handbook of Fuzzy Sets and Systems,* Boston, MA: Kluwer, vol. 1, pp. 195–228.

Bimbard, F., George, L., and Cotsaftis, M. 2005. Design of Autonomous Modules for Self-Reconfigurable Robots, *3rd Intern. Conference on Computing, Communications and Control Technologies,* pp. 154–158.

Bishop, M. 1999, Vulnerabilities Analysis, *Proc. The 2nd International Workshop on Recent Advances in Intrusion Detection (RAID'99),* West Lafayette, Indiana, USA, pp. 7–9.

Blas, A.D., Jagota, A., and Hughey, R. 2005. Optimizing Neural Networks on SIMD Parallel Computers, *Parallel Computing*, vol. 31, pp. 97–115.

Bleha, S., Slivinsky, C., and Hussien, B. 1990. Computer-Access Security Systems Using Keystroke Dynamics, *IEEE Trans. Pattern Anal. Mach. Intell.*, vol. 12, no. 12, pp. 1217–1222.

Bobrow, J.E., Dubowsky, S., and Gibson, J.S. 1985. Time-Optimal Control of Robotic Manipulators along Specified Paths, *Int. J. Robot. Res.,* vol. 4, no. 3, pp. 3–17.

Boles, W.W. and Boashah, B. 1998. A Human Identification Technique Using Images of the Iris and Wavelet Transform. *IEEE Transaction on Signal Processing,* vol. 46, pp. 1185–1188.

Boll, S.F. 1979. Suppression of Acoustic Noise in Speech Using Spectral Subtraction, *IEEE Trans. Acoust., Speech, Signal Processing,* vol. 27, pp. 113–120.

Bolle, R.M., Connel, J.H., and Ratha, N.K. 2002. Biometrics Perils and Patches, *Pattern Recognition,* vol. 35, pp. 2727–2738.

Bologna, G. 2000. A Study on Rule Extraction from Neural Networks Applied to Medical Databases, *Proceedings of the 4th European Principles and Practice of Knowledge Discovery in Databases,* Lyon, France.

Bonissone, P.P., Chen, Y.T., Goebel, K., and Khedkar, P.S. 1999. Hybrid Soft Computing Systems: Industrial and Commercial Applications, *Proceedings of the IEEE,* vol. 87, no. 9, pp. 1641–1667.

Bonissone, P.P., Chen, Y.T., Goebel, K., and Khedkar, P.S. 1999. Hybrid Soft Computing Systems: Industrial and Commercial Applications, *Proc. IEEE,* vol. 87, pp. 1641–1667.

Bonissone, P.P. 1997. Soft Computing: The Convergence of Emerging Reasoning Techniques, *Soft Computing A Fusion of Foundations, Methodologies and Applications,* vol. 1, no. 1, pp. 6–18.

Borenstain, J., Everett, H.R., and Feng, L. 1996. *Navigating Mobile Robots: Systems and Techniques.* Wellesley: A.K. Peters.

Borg, I. and Groenen, P. 1997. *Modern Multidimensional Scaling—Theory and Applications.* Berlin, Germany: Springer-Verlag.

Borodin, A., Rosenthal, J.S., Roberts, G.O., and Tsaparas, P. 2005. Link Analysis Ranking: Algorithms, Theory and Experiments. *ACM Transactions on Internet Technologies,* vol. 5, no. 1, pp. 231–297.

Boser, E., Guyon, I.M., and Vapnik, V.N. 1992. A Training Algorithm for Optimal Margin Classifiers, *Proc. 5th Annu. ACM Workshop Comput. Learn. Theory,* D. Haussler, ed. Pittsburgh, PA, pp. 144–152.

Bossert, W. 1967. Mathematical Optimization: Are There Abstract Limits on Natural Selection? P.S. Moorehead and M.M. Kaplan, eds. Philadelphia, PA: Wistar Press, pp. 35–46.

Boualleg, A.H., Bencheriet, Ch., and Tebbikh, H. 2006. Automatic Face Recognition Using Neural Network-PCA, *Information and Communication Technologies,* ICTTA 06. 2nd vol. 1.

Bouqata, B., Bensaid, A., Palliam, R., and Gomez Skarmeta, A.F. 2000. *Proceedings of the International Conference on Computational Intelligence for Financial Engineering,* pp. 170–173, New York, USA.

Bouzerdoum and Pattison, T.R. 1993. Neural Network for Quadratic Optimization with Bound Constraints, *IEEE Trans. Neural Networks,* vol. 4, pp. 293–303.

Bowling, M., Browning, B., Chang, A., and Veloso, M. 2003. Plays as Team Plans for Coordination and Adaptation, Presented at the *IJCAI Workshop Issues Design. Phys. Agents Dyn. Real-Time Environ.: World Modeling, Plan., Learn., Commun.,* Acapulco, Mexico.

Boyer, R.S. and Moore, J.S. 1977. A Fast String Searching Algorithm, *Communications of ACM,* vol. 20, no. 10, pp. 762–772.

Bradley, B.S., Fayyad, U., and Reina, C. 1998. Scaling Clustering Algorithms to large Databases, *Proc. 4th ACM SIGKDD Int. Conf. Knowl. Discov. Data Mining,* pp. 9–15. New York, USA.

Brandt, R.D., Wang, Y., Laub, A.J., and Mitra, S.K. 1988. Alternative Networks for Solving the Traveling Salesman Problem and the List-Matching Problem, *Proc. IEEE Int. Conf. Neural Networks,* vol. II, San Diego, CA, pp. 333–340.

Branke, J. and Deb, K. 2004. Integrating User Preferences into Evolutionary Multi-Objective Optimization. In Y. Jin, ed., *Knowledge Incorporation in Evolutionary Computation.* Hiedelberg, Germany: Springer, pp. 461–477.

Branke, J., Kaubler, T., and Schmeck, H. 2001. Guidance in Evolutionary Multi-Objective Optimization. *Advances in Engineering Software,* vol. 32, pp. 499–507.

Fuzzy Expert System in the Prediction of Neonatal Resuscitation, *Braz. J. Med. Biol. Res.* May 2004, vol. 37, no. 5, pp. 755–764.

Breast cancer [internet]. Wikipedia Available via http://en.wikipedia.org/wiki/Breast_cancer, accessed Feb 2008.

Breast Cancer: Statistics on Incidence, Survival, and Screening [internet]. Available via http://imaginis.com/breasthealth/statistics.asp, accessed January 2008.

Bremermann, H.J., Rogson, M., and Salaff, S. 1966. Global Properties of Evolution Processes, Natural Automata and Useful Simulations, H.H. Pattee, E.A. Edelsack, L. Fein, and A.B. Callahan, eds. Washington, DC: Spartan, pp. 3–41.

Brockett, R. 1993. Hybrid Models for Motion Control Systems, *Perspectives in Control,* H. Trentelman and J. Willems, eds. Boston, MA: Birkhauser, pp. 29–54.

Brooks, R. 1986. A Robust Layered Control System for Mobile Robot, *IEEE J. Robot. Automat.*, vol. 2, pp. 14–23.

Brooks, R.A. 2003. *A Robot That Walks: Emergent Behaviours from a Carefully Evolved Network in Cambrian Intelligence: The Early History of the New AI.* Cambridge, UK: Bradford Books, pp. 27–36.

Brown, M., and Harris, C.J. 1995. A Perspective and Critique of Adaptive Neurofuzzy Systems Used for Modelling and Control Applications, *International Journal of Neural Systems,* vol. 6, no. 2, pp. 197–220.

Brunette, E.S., Flemmer, R.C., and Flemmer, C.L. 2009. A Review of Artificial Intelligence, *Proceedings of the International Conference on Autonomous Robots and Agents (ICARA),* Wellington, New Zealand, *in press.*

Brusco, M. and Nazeran, H. 2004. Digital Phonocardiography: A PDA-Based Approach, *Proceedings of the 26th Annual International Conference of the IEEE EMBS,* San Francisco, California, vol. 1, pp. 2299–2302.

Brussels 2002. International Diabetes Federation's World Diabetes Day 2002 Focuses on Eye Complications. WDD 2002. Available via http://www.idf.org/home/index.cfm?node=535.

Buckley, J. and Tucker, D. 1989. Second Generation Fuzzy Expert System, *Fuzzy Sets Syst.,* vol. 31, pp. 271–284.

Buckley, J.J. 1993. Sugeno Type Controller Are Universal Controllers, Fuzzy Sets Syst., vol. 52, pp. 299–303.

Buckley, J., Siler, W., and Tucker, D. 1986. Fuzzy Expert System, *Fuzzy Sets Syst.,* vol. 20, no. 1, pp. 1–16.

Buckner, G.D. 2002. Intelligent Bounds on Modeling Uncertainties: Applications to Sliding Mode Control, *IEEE Trans. Syst., Man, Cybern. C: Appl. Rev.,* vol. 32, no. 2, pp. 113–124.

Bum, D. 1988. Experiments on Neural Net Recognition of Spoken and Written Text. *IEEE Trans., on Acoust Speech and Signal Proc.,* vol. ASSP-36, no. 7, pp. 1162–1168.

Bunke, H. and Kandel, A. 2002. *Hybrid Methods in Pattern Recognition*, World Scientific Publishing. Singapore.

Burge, M. and Burger, W. 2000. Ear Biometrics in Computer Vision, *15th International Conference of Pattern Recognition,* pp. 822–826.

Burnham, K. and Anderson, D. 2002. *Model Selection and Multimodel Inference.* New York: Springer-Verlag.

Burt, P.J. and Adelson, E.H. 1983. The Laplacian Pyramid as a Compact Image Code, *IEEE Trans. Commun.* vol. COM-31, no. 4, pp. 532–540.

Bustard, D.W., Liu, W., and Stemtt, R. eds. 2002. *Computing in an Imperfect World, First International Conference, Soft-Ware 2002,* LNCS 231 1. Springer-Verlag, Belfard, Northern Ireland.

Cai, L., and Hofmann, T. 2003. Text Categorization by Boosting Automatically Extracted Concepts, *Proceedings of the 26th Annual International ACM SIGIR Conference on Research and Development in Information Retrieval*, pp 182–189, Toronto Canada.

Caikou, C., Jingyu, Y., and Jian, Y. 2004. Combined Subspace Based Optimal Feature Extraction and Face Recognition, *Signal Processing,* vol. 20 no. 6, pp. 609–612.

Calado, P., Cristo, M., Moura, E.S., Ziviani, N., Ribeiro-Neto, B.A., and Goncalves, M.A. 2003. Combining Link-Based and Content-Based Methods for Web Document Classification, *Proceedings of Conference on Information and Knowledge Management 2003,* pp. 394–401.

Calvo, R.A., Lee, J., and Li, X. 2004. Managing Content with Automatic Document Classification. *Journal of Digital Information*, vol. 5, no. 2.

Caminhas, Walmir M., Vieira, Douglas A.G., Vasconcelos, Joao A. 2003. Parallel Layer Perceptron, *Neurocomputing,* vol. 55, nos. 3–4.

Campbell, Joseph P., Jr. 1997. Speaker Recognition: A Tutorial. *Proceedings of the IEEE,* vol. 85, No. 9, pp. 1437–1462.

Campobello, G., Patane, G., and Russo, M. 2008. An Efficient Algorithm for Parallel Distributed Unsupervised Learning, *Neurocomputing,* vol. 71, nos. 13–15, pp. 2914–2928.

Cancer Research UK. [Internet]. Summary of Specific Cancers. Available via http://www.cancerresearchuk.org, accessed December 2007.

Canny, J. 1986. A Computational Approach to Edge Detection. *IEEE Transaction on Pattern Analysis and Machine Intelligence*, vol. 8, pp. 679–714.

Cano, A. and Moral, S. 1996. A Genetic Algorithm to Approximate Convex Sets of Probabilities, *Proc. 6th Int. Conf. Information Processing and Management of Uncertainty in Knowledge-Based Systems,* pp. 847–852.

Cao, L.J. 2003. Support Vector Machines Experts in Time Series Forecasting, *Neurocomputing,* vol. 51, pp. 321–339.

Casteele, S.V. 2005. Threat Modeling for Web Application Using STRIDE Model.

Castillo, E., Gutiérrez, J.M., and Hadi, A.S. 1997. Expert Systems and Probabilistic Network Models, *Monographs in Computer Science,* New York: Springer-Verlag.

Castillo, O. and Melin, P. 2004. Hybrid Intelligent System Using Fuzzy Logic, Neural Networks and Genetic Algorithms, *Nonlinear Studies*, vol. 11, no. 1, pp. 1–3.

Cavazos, J. and Moss, J.E.B. 2004. Inducing Heuristics to Decide Whether to Schedule, *SIGPLAN Not.,* vol. 39, no. 6, pp. 183–194.

Chakrabarti, S., Dom, B., and Indyk, P. 1998. Enhanced Hypertext Categorization Using Hyperlinks, *Proceedings of the ACM SIGMOD International Conference on Management of Data,* pages 307–318.

Chalup, S.K. and Murch, C.L. 2004. Machine Learning in the Four-Legged League, presented at the *3rd IFAC Symp. Mechatron. Syst.,* Sydney, Australia, Sep. 6–8.

Chalup, Stephan K. 2002. Incremental Learning in Biological and Machine Learning Systems, *International Journal of Neural Systems,* vol. 12, no. 6, pp. 447–465.

Chan, W.L., So, A.T.P., and Lai, L.L. 2000. Harmonics Load Signature Recognition by Wavelet Transform, *International Conference on Electric Utility Deregulation and Restructuring and Power Technologies 2000,* City University, London, UK. pp. 666–671.

Chandra and Yao, X. 2006. Evolving Hybrid Ensembles of Learning Machines for Better Election, *Neurocomputing,* vol. 69, pp. 686–700.

Ramchandran, K. and Vetterli, M. 1993. Best Wavelet packet Bases in a Rate-Distortion Sense, *IEEE Trans. on Image Processing*, vol. 2, no. 2, pp. 160–175.

Chang, Li-Chiu and Chang, Fi-John. 2002. An Efficient Parallel Algorithm for LISSOM Neural Network, *Parallel Computing*, vol. 28, pp. 1611–1633.

Chang, P.T. and Hung, K.C. 2005. Applying the Fuzzy-Weighted-Average Approach to Evaluate Network Security Systems, *Comput. Math. Appl.,* vol. 49, pp. 1797–1814.

Charniak, E. and McDermott, E. 1985. *Introduction to Artificial Intelligence.* Reading, MA: Addison-Wesley.

Chartrand, G. and Lesniak, L. 1979. *Graphs and Digraphs,* 2nd ed. Pacific Grove, California: The Wadsworth and Brooks/Cole Mathematics Series.

Chaudhuri, B.B and Bhattacharya U. 2000. Efficient Training and Improved Performance of Multilayer Perceptron in Pattern Classification, *Neurocomputing*, vol. 34, pp. 11–27.

Chen et al. 2000. A New LDA-Based Face Recognition System Which Can Solve the Small Sample Size Problem, *Pattern Recognition,* vol. 32, pp. 317–324.

Chen, A.C.J and Miikkulainen, R. 2001. Creating Melodies with Evolving Recurrent Neural Network, *Proceedings of the International Joint Conference on Neural Networks,* IJCNN'01, pp. 2241–2246.

Chen, C.H. and Chu, C.T. 2004. An High Efficiency Feature Extraction Based on Wavelet Transform for Speaker Recognition, *International Computer Symposium (ICS2004),* Taipei, pp. 93–98.

Chen, C.H. and Chu, C.T. 2004. Combining Multiple Features for High Performance Face Recognition System, *International Computer Symposium (ICS2004)* Taipei, pp. 387–392.

Chen, D.S. and Jain, R.C. 1994. A Robust Back-Propagation Learning Algorithm for Function Approximation, IEEE Trans. *Neural Networks,* vol. 5, no. 3, pp. 467–479.

Chen, J., Phua, K., Song, Y. and Shue, L. 2006. A Portable Phonocardiographic Feral Heart Rate Monitor, *Proceedings of IEEE International Conference ISCAS 2006,* Kos, Greece, pp. 1–4.

Chen, L.H. and Lieh, J.R. 1990. Handwritten Character Recognition Using a 2-Layer Random Graph Model by Relaxation Matching, *Pattern Recog.,* vol. 23, no. 11, pp. 1189–1205.

Chen, S., Cowan, C.F.N. and Grant, P.M. 1991. Orthogonal Least Squares Learning Algorithm for Radial Basis Function Networks, *IEEE Transactions on Neural Networks*, vol. 2, no. 2, pp. 302–309.

Chen, T.L., Cheng, C.H., and Teoh, H.J. 2008. High-Order Fuzzy Time-Series Based on Multi-Period Adaptation Model for Forecasting Stock Markets, *ScienceDirect Journal,* vol. 387, pp. 876–888.

Chen, Y. and Wang, J.Z. 2003. Support Vector Learning for Fuzzy Rule-Based Classification Systems, *IEEE Transactions on Fuzzy Systems*, vol. 11, no. 6, pp. 716–728.

Chen, Y., Zhang, Y., and Ji, X. 2005. Size Regularized Cut for Data Clustering, *Proc. NIPS,* pp. 211–218.

Cheng, L. and Chen, H. 1989. Fuzzy Reasoning Is the Inverse of Fuzzy Implication, *Proc. 19th ISMVL,* pp. 252–254.

Cheriet, Mohamed, Kharma, Nawwaf, Liu, Cheng-Lin, and Suen, Ching. 2007. Character Recognition Systems, Wiley Interscience, Wiley Interscience, Hoboken, New Jersey.

Cherkauer, K.J. 1996. Human Expert Level Performance on a Scientific Image Analysis Task by a System Using Combined Artificial Neural Networks, *Proc. 13th AAAI Workshop on Integrating Multiple Learned Models for Improving and Scaling Machine Learning Algorithms,* Portland, OR, pp. 15–21.

Chi, Chand, S., Moore, D., and Chaudhary, A. 1991. Fuzzy Logic for Control of Roll and Moment for a Flexible Wing Aircraft, *IEEE Contr. Syst. Mag.,* vol. 11, no. 4, pp. 42–48.

Chirillo, J. and Blaul, S. 2003. *Implementing Biometric Security,* John Wiley and Sons, Hungry Minds, Incorporated, 2003.
Chiu, S. 1994. Fuzzy Model Identification Based on Cluster Estimation, *Journal of Intelligent and Fuzzy Systems,* vol. 2, no. 3, pp. 267–278.
Cho, J. and Kim, J.H. 2001. Bayesian Network Modeling of Strokes and Their Relationships for On-Line Handwriting Recognition, *Proc. Sixth Int'l Conf. Document Analysis and Recognition,* pp. 86–90.
Chomsky, N. 2000. *Language and Thought,* Fifth Printing. Wakefield: RI Moyer Bell, p. 86.
Chopra, S., Mitra, R., and Kumar, V. 2004. Identification of Rules Using Substracting Clustering with Application to Fuzzy Controllers, *Proceeding of 2004 International Conference on Machine Learning and Cybernetics.* New York, IEEE, pp. 4125–4130.
Choras, M. 2005. Ear Biometrics Based on Geometrical Feature Extraction, Electronic Letters on Computer Vision and Image Analysis, vol. 5, no. 3, pp. 84–95.
Chou, H., Su, M.C., and Lai, E. 2002. Symmetry as a New Measure for Cluster Validity, *2nd WSEAS Int. Conf. on Scientific Computation and Soft Computing,* Crete, Greece, pp. 209–213.
Corke, P., Peterson, R., and Rus, D. 2005. Networked Robots: Flying Robot Navigation Using a Sensor Net. *Springer Tracts in Advanced Robotics*, vol. 15, pp. 234–243.
Choy, M.C., Srinivasan, D., and Cheu, R.L. 2003. Cooperative, Hybrid Agent Architecture for Real-Time Traffic Signal Control, *IEEE Transactions on Systems, Man, and Cybernetics, Part A: Systems and Humans,* vol. 33, no. 5, pp. 597–607.
Christmas, W.J., Kittler, J., and Petrou, M. 1995. Structural Matching in Computer Vision Using Probabilistic Relaxation, *IEEE Trans. Pattern Analysis and Machine Intelligence,* vol. 17, no. 8, pp. 749–764.
Christopher, J.C. Burges 1998. A Tutorial on Support Vector Machines for Pattern. Recognition, *Data Mining and Knowledge Discovery,* vol. 2, pp. 121–167.
Chuan Chen, T., and Liang Chung, K. 2001. An Efficient Randomized Algorithm for Detecting Circles, *Computer Vision and Image Understanding,* vol. 83, pp. 172–191.
Church. 1962. Logic, Arithmetic, and Automata, *Proc. Int. Congr. Mathematicians,* pp. 23–35.
Cincotti, G., Loi, G., and Pappalardo, D. 2001. Frequency Decomposition and Compounding of Ultrasound Medical Images with Wavelet Packet., *IEEE Trans. on medical image,* vol. 20, no. 8, pp. 764–771.
Clark, M.C. and Hall, L.O. 1995. MRI Segmentation Using Fuzzy Clustering Techniques: Integrating Knowledge, available at http://www.csee.usf.edu/.
Coates, Thomas D., Jr. 2000. Control and Monitoring of a Parallel Processed Neural Network via the World Wide Web, *Neurocomputing*, vols. 32–33, pp. 1021–1026.
Coello, C.A.C. and Christiansen, A.D. 1999. MOSES: A Multiobjective Optimization Tool for Engineering Design, *Eng. Opt.,* vol. 31, no. 3, pp. 337–368.
Coello, C.A.C., Hernandez, F.S., and Farrera, F.A. 1997. Optimal Design of Reinforced Concrete Beams Using Genetic Algorithms, *Int. J. Exp. Syst. Appl.,* vol. 12, no. 1, pp. 101–108.
Coello, C.A.C., Pulido, G.T., and Lechuga, M.S. 2004. Handling Multiple Objectives with Particle Swarm Optimization, *IEEE Trans. Evol. Comput.,* vol. 8, no. 3, pp. 256–279.
Cohen, M.E. and Hudson, D.L. 2002. Meta Neural Networks as Intelligent Agents for Diagnosis, *Neural Networks,* vol. 1, pp. 233–238.
Cohn, A. and Hofmann, T. 2000. The Missing Link: A Probabilistic Model of Document Content and Hypertext Connectivity, *Proceedings of Neural Information Processing Systems 2000,* December, pp. 430–436.
Cohoon, J.P., Martin, W.N., and Richards, D.S. 1990. Genetic Algorithms and Punctuated Equilibria in VLSI, *Parallel Problem Solving from Nature 1,* H.-P. Schwefel and R. Männer, eds. Berlin, Germany: Springer-Verlag, pp. 134–144.
Collins, R.J. 1992. *Studies in Artificial Evolution,* doctoral dissertation, Univ. California, Los Angeles, CA.
Connell, J. and Mahadevan S., eds. 1993. *Robot Learning.* Norwell, MA: Kluwer.
Cooper, A.S. 1995. Higher Order Neural Networks—Can They Help Us Optimize? *Proc. 6th Australian Conf. Neural Networks,* pp. 29–32. Sydney, Australia.
Cootes, T., Edwards, G., and Taylor, C. 2001. Active Appearance Models, *IEEE Trans. Pattern Analysis and Machine Intelligence,* vol. 23, no. 6, pp. 643–660.
Coppin, G. and Skrzyniarz, A. 2003. Human-Centered Processes : Individual and Distributed Decision Support, *IEEE Intelligent Systems,* vol. 18, no. 4, pp. 27–33.
Corke, P., Peterson, R., and Rus, D. 2003. Networked Robots: Flying Robot Navigation Using a Sensor Net, *Proc. of 11th International Symp. Of Robotics Research (ISRR),* pp. 234–243, Siena, Italy.
Cornelis, H. and Schutter, E.D. 2007. Neurospaces: Towards Automated Model Partitioning for Parallel Computers, *Neurocomputing*, vol. 70, nos. 10–12, pp. 2117–2121.
Cortes, C. and Vapnik, V. 1995. Support Vector Networks, *Mach. Learn.,* vol. 20, pp. 273–297.

Cotter, D. and Cotter, S.C. 1993. Algorithm for Binary Word Recognition Suited to Ultrafast Nonlinear Optics, *Electron. Lett.,* vol. 29, no. 11, pp. 945–947.

Cova, M., Felmetsger, V., and Vigna, G. 2007. Vulnerability Analysis of Web-Based Applications, *Test and Analysis of Web Services,* Baresi, L. and Nitto, E.D., eds. Berlin and Heidelberg, Germany: Springer, ch. IV: Reliability, Security, and Trust, pp. 363–394.

Cox, E. 1994. *The Fuzzy Systems Handbook: A Practitioner's Guide to Building, Using, and Maintaining Fuzzy Systems.* Academic Press, Inc., Boston

Craiger, J.P., Goodman, D.F., Weiss, R.J., and Butler, A. 1996. Modeling Organizational Behavior with Fuzzy Cognitive Maps, *Int. J. Computational Intelligence and Organisations,* vol. 1, pp. 120–123.

Craswell, N., Hawking, D., and Robertson, S. 2001. Effective Site Finding Using Link Anchor Information, *Proceedings of 24th SIGIR,* pp. 250–257, New Orleans, Louisiana, US.

Cristianini, N. and Shawe-Taylor, J. 2000. *An Introduction to Support Vector Machines and Other Kernel Based Learning Methods.* Cambridge University Press, Cambridge, UK.

Cross, J.M. 2008. Thyroid Disorders Linked to Glaucoma. *Medical News Today.* Available via http://www.medicalnewstoday.com/articles/126418.php.

Crowston, K. and H. Kwasnik, B. 2004. A FRAMEWORK for Creating a Faceted Classification for Genres: Addressing Issues of Multidimensionality, *Proceedings of the 37th Hawaii International Conference on System Sciences.*

Crowston, K. and Williams, M. 1997. Reproduced and Emergent Genres of Communication on the World-Wide Web, *Proceedings of the 30th Hawaii International Conference on System Sciences,* p. 30. IEEE Computer Society, pp. 1–9, Hawaii.

Crowston, K. and Williams, M. 1999. The Effects of Linking on Genres of Web Documents, *Proceedings of the 32nd Hawaii International Conference on System Sciences,* pp. 2–12, Hawaii.

Cunningham, Carney P.J. and Jacob, S. 2000. Stability Problems with Artificial Neural Networks and the Ensemble Solution, *Artif. Intell. Med.,* vol. 20, no. 3, pp. 217–225.

Cunningham, S.J., Littin, J., and Witten, I.H. 1997. Applications of Machine Learning in Information Retrieval. *Technical Report 97/6,* University of Waikato, New Zealand.

Da Ruan, ed. 1997. *Intelligent Hybrid Systems, Fuzzy Logic, Neural Networks and Genetic Algorithms.* Kluwer Academic Publishers, Norwell, MA.

Daming, Shi, Wenhao, Shu, and Haitao, Liu. 1998. Feature Selection for Handwritten Chinese Character Recognition Based on Genetic Algorithms, vol. 5, pp. 4201–4206.

Daniel, G. 2007. *Principles of Artificial Neural Networks,* 2nd ed, USA World Scientific.

Danielsson, P.E. and Ye, Q.Z. 1988. Rotation-Invariant Operators Applied to Enhancement of Fingerprints, *Proc. 8th ICPR,* Rome, pp. 329–333.

Daugman, J. 2002, How Iris Recognition Works, *Proceedings of International Conference on Image Processing,* vol. 1, pp. 33–36.

Daugman, J. How Iris Recognition Works, available at http://www.ncits.org/tc_home/m1htm/docs/m1020 044.pdf.

Dave, R.N. and Krishnapuram, R. 1997. Robust Clustering Method: A Unified View, *IEEE Trans. Fuzzy Systems,* vol. 5, pp. 270–293.

David, V. and Sanchez, A. 1995. Robustization of Learning Method for RBF networks, *Neurocomputing,* vol. 9, pp. 85–94.

Davida, G., Frankel, Y., and Matt, B.J. 1998. On Enabling Secure Applications Through Off-Line Biometrics Identification, *Proceeding Symposium on Privacy and Security,* pp. 148–157.

Davidor, Y. 1991. A Naturally Occurring Niche Species Phenomenon: The Model and First Results, *Proc. 4th Int. Conf. Genetic Algorithms,* R. Belew and L.B. Booker, eds. San Mateo, CA: Morgan Kaufmann, pp. 257–263.

Davies, A.R. 2005. *Machine Vision,* 3rd ed. Elsevier, Morgan Kaufmann, San Francisco.

Davis, M., Logemann, G., and Loveland, D. 1962. A Machine Program for Theorem-Proving. *Communications of the ACM,* vol. 5, no. 7, pp. 394–397.

Davis, N., Howard, M., Humphrey, W., McGraw, G., Redwine, S.T., Jr., Zibulski, G., and Graettinger, C. 2004. *Processes to Produce Secure Software: Towards More Secure Software. A Report at National Cyber Security Summit, vol. 1.* Available at http://www.cigital.com/papers/download/secure_software_process.pdf.

De Cock, D., Wouters, K., Schellekens, D., Singelee, D., and Preneel, B. 2005. Threat Modeling for Security Tokens in Web Applications, *Proceedings of the IFIP TC6/TC11 International Conference on Communications and Multimedia Security (CMS '04).* September 2004, pp. 183–193.

De Jong, K.A. and Spears, W.M. 1992. A Formal Analysis of the Role of Multipoint Crossover in Genetic Algorithms, *Annals Math. Artif. Intell.,* vol. 5, no. 1, pp. 1–26.

Deb, K. and Agrawal, S. 1995. Simulated Binary Crossover for Continuous Search Space, *Complex Systems,* vol. 9, no. 2, pp. 115–148.

Deb, K. and Goyal, M. 1996. A Combined Genetic Adaptive Search (Geneas) for Engineering Design, *Computer Science and Informatics,* vol. 26, no. 4, pp. 30–45.

Deb, K. 2002. *Multiobjective Optimization Using Evolutionary Algorithms.* New York: Wiley.

Deiri, Z. and Botros, N. 1990. LPC-Based Neural Network for Automatic Speech Recognition, *Proc. IEEE Engineering in Medicine and Biology Soc.,* IEEE Cat. no. 90 CH2936-3, pp. 1429–1430.

Dellar, J.R., Hansen, H.L., and Proakis, J.G. 2000. *Discrete-Time Processing of Speech Signals,* 2nd ed. New York: IEEE Press.

Demiris, J. and Birk, A., eds. 2000. *Interdisciplinary Approaches to Robot Learning.* Singapore: World Scientific.

Deng, Y., Huang, T. and Xu, B. 2000. Towards High Performance Continuous Mandarin Digit String Recognition, *Proc. Sixth International Conference on Spoken Language Processing,* Beijing, China, pp. 642–645.

Dhananjaya, N. 2006. Correlation-Based Similarity Between Signals for Speaker Verification with Limited Amount of Speech Data, Multimedia Content Representation, Classification and Security. *Proceedings of the Multimedia Content Representation, Classification and Security Conference, Springer Lecture Notes in Computer Science,* pp. 17–25, Istanbul, Turkey.

Dibike, Y.B., Velickov, S. and Solomantine, D. 2000. Support Vector Machines: Review and Applications in Civil Engineering, *AI Methods in Civil Engineering Applications,* pp. 45–58, *Proceedings of 2nd Joint Conference on application of AI in Civil Engineering,* Cottbus, Germany.

Doheny, K. 2008. Too Few Understand Diabetes Dangers, *HealthDay Reporter.* Available via http://health.usnews.com/articles/health/healthday/2008/10/28/ too-few-understand-diabetes-dangers.html.

Donald, A. and Xavier, P. 1995. Provably Good Approximation Algorithms for Optimal Kinodynamic Planning: Robots with Decoupled Dynamics Bounds, *Algorithmica,* vol. 14, no. 6, pp. 443–479.

Dong, H., Gao, J., and Wang, R. 2004. Fusion of Multiple Classifiers for Face Recognition and Person Authentication (in Chinese), *Journal of System Simulation,* vol. 16, no. 8, pp. 1849–1853.

Dong, Y., Zhang, Y., and Chang, C. 2004. Multistage Random Sampling Genetic-Algorithm-Based Fuzzy C-Means Clustering Algorithm, *Proceedings of the Third International Conference on Machine Learning and Cybernetics,* Shanghai, 26–29, pp. 2069–2073.

Dong, Y., Zhang, Y., and Chang, C. 2005. An Improved Hybrid Cluster Algorithm, *Fuzzy Systems and Mathematics,* vol. 19, no. 2, pp. 128–133.

Dorigo, M. 1996. Introduction to the Special Issue on Learning Autonomous Robots, *IEEE Trans. Syt., Man, Cybern.,* vol. 26, no. 3, pp. 361–364.

Doyle, J., Francis, B., and Tannenbaum, A. 1992. *Feedback Control Theory.* New York: Macmillan.

Doyle, R.S. and Harris, C.J. 1996. Multi-Sensor Data Fusion for Helicopter Guidance Using Neuro-Fuzzy Estimation Algorithms, *The Royal Aeronautical Society Journal,* vol. 100, pp. 43–58, pp. 241–251.

Draghici, S. 1997. A Neural Network Based Artificial Vision System for License Plate Recognition, *International Journal of Network Security, International Journal of Neural Systems,* vol. 8, no. 1.

Driankov, D., Hellendoorn, H., and Reinfrank, M. 1993. *An Introduction to Fuzzy Control.* New York: Springer-Verlag.

Drucker, H., Schapire, R., and Simard, P. 1993. Improving Performance in Neural Networks Using a Boosting Algorithm, *Advances in Neural Information Processing Systems,* S.J. Hanson, J.D. Cowan, and C.L. Giles, eds. San Mateo, CA: Morgan Kaufmann, vol. 5, pp. 42–49.

Dua, T., De Boer, Hanneke M., Prilipko, Leonid L., and Saxena, S. 2006. Epilepsy Care in the World: Results of an ILAE/IBE/WHO Global Campaign Against Epilepsy Survey. *Epilepsia,* ISSN 0013-9580, CODEN EPILAK, vol. 47, no. 7, pp. 1225–1231.

Dubois, D. and Prade, H. 1980. *Fuzzy Sets and Systems: Theory and Applications.* New York: Academic.

Dubois, D. and Prade, H. 1998. Soft Computing, Fuzzy Logic, and Artificial Intelligence, *Soft Computing: A Fusion of Foundations, Methodologies and Applications,* vol. 2, no. 1, pp. 7–11.

Duch, W., Adamczak, R., Grabczewski, K., Zal, G., and Hayashi, Y. 2000. Fuzzy and Crisp Logical Rule Extraction Methods in Application to Medical Data, *Computational Intelligence and Applications,* P.S. Szczepaniak, ed. Berlin, Germany: Springer-Verlag, vol. 23, pp. 593–616.

Duda, R.O. and Hart, P.E. 1973. *Pattern Classification and Scene Analysis.* New York: Wiley.

Dudgeon, D.E. and Mersereau, R.M. 1984. *Multidimensional Digital Signal Processing.* Prentice-Hall, Inc. Englewoods Cliffs, NJ.

Dutta, S. 1997. Strategies for Implementing Knowledge Based Systems, 20132, *IEEE Trans. Engineering Management,* pp. 79–90.

Dybowski, R. and Gant, V. 2001. *Clinical Applications of Artificial Neural Networks*. Cambridge University Press. Cambridge, UK.
Eck, A. and Schmidhuber, J. 2002. A First Look at Music Composition using LSTM Recurrent Neural Networks, *Technical Report: IDSIA-07-02.*
Eggermont, J., Eiben, A.E., and Van Hemert, J.I. 1999. A Comparison of Genetic Programming Variants for Data Classification, *Proc. 3rdSymp. IDA, Lecture Notes in Computer Science,* vol. 1642, pp. 281–290.
Eiben, A.E. and Van Kemenade, C.H.M. 1997. Diagonal Crossover in Genetic Algorithms for Numerical Optimization. *Journal of Control and Cybernetics,* vol. 26, no. 3, pp. 447–465.
Eiben, A.E., Van Kemenade, C.H.M., and Kok, J.N. 1995. Orgy in the Computer: Multi-Parent Reproduction in Genetic Algorithms, *Proceedings of the Third European Conference on Artificial Life,* vol. 929, pp. 934–945.
Eiben, A.E. 2000. Multiparent Recombination, *Evolutionary Computation 1: Basic Algorithms and Operators*. Institute of Physics Publishing, pp. 289–307, Ottawa, Canada.
Eiben, A.E. 2002. Multiparent Recombination in evolutionary computing. In *Advances in Evolutionary Computing*. Springer, pp. 175–192, New York.
Eldredge, N. and Gould, S.J. 1972. Punctuated Equilibria: An Alternative to Phyletic Gradualism, *Models of Paleobiology,* T.J.M. Schopf, ed. San Francisco, CA: Freeman, Cooper, pp. 82–115.
El-Essawi, D., Musial, J.L., Hammad, A., and Lim, H.W. 2007. A Survey of Skin Disease and Skin-Related Issues in Arab Americans, *J. American Academy of Dermatology,* vol. 56, pp. 933–938.
Elger, C.E. and Lehnertz, K. 2001. Can Chaos Analyses Be a Tool to Predict Epileptic Seizures, *Epilepsia,* Blackwell Science, Inc., Malden, vol. 42 suppl. 7, p. 3.
Ella, B. 2001. Reinforcement Learning in Neurofuzzy Traffic Signal Control, *European Journal of Operational Research,* vol. 161, pp. 232–241.
Epilepsy [internet] wikipedia. Available from http://en.wikipedia.org/wiki/Epilepsy, accessed Feb 2008.
Er, M.J., Wu, S., Lu, J., and Toh, H.L. 2002. Face Recognition with Radial Basis Function (RBF) Neural Networks, *IEEE Trans. on Neural Networks,* vol. 13, no. 3, pp. 697–710.
Er, M.J. and Zhou, Y. 2008. A Novel Framework for Automatic Generation of Fuzzy Neural Networks, *Elsevier Journal of Neurocomputing* vol. 71, pp. 584–591.
Ersson, T. and Hu, X. 2001. Path Planning and Navigation of Mobile Robots in Unknown Environments. Maui, USA.
Ertoz, L., Steinbach, M., and Kumar, V. 2002. A New Shared Nearest Neighbor Clustering Algorithm and Its Applications, *Proc. Workshop Clustering High Dimensional Data Appl.,* pp. 105–115.
Eshelman, L.J. and Schaffer, J.D. 1991. Preventing Premature Convergence in Genetic Algorithms by Preventing Incest, *Proc. 4th Int. Conf. Genetic Algorithms,* R. Belew and L.B. Booker, eds. San Mateo, CA: Morgan Kaufmann, pp. 115–122.
Eshelman, L.J. 1991. The CHC Adaptive Search Algorithm: How to Have Safe Search When Engaging in Nontraditional Genetic Recombination, *Foundations of Genetic Algorithms 1,* G.J.E. Rawlin, ed. San Mateo, CA: Morgan Kaufmann, pp. 265–283.
Eshelman, L.J., Mathias, K.E., and Schaffer, J.D. 1997. Convergence Controlled Variation, *Foundations of Genetic Algorithms 4,* R. Belew and M.Vose, eds. San Mateo, CA: Morgan Kaufmann, pp. 203–224.
Espinosa-Duro, V. and Faundez-Zanuy, M. 1999. Face Identification by Means of a Neural Net Classifier, *Proceedings of IEEE 33rd Annual 1999 International Carnahan Conf. on Security Technology,* pp. 182–186.
Essa and Pentland, A.P. 1997. Coding analysis, Interpretation, and Recognition of Facial Expressions,. *IEEE Tr. On Pattern Analysis Machine Intelligence,* vol. 19, no. 7.
Ester, M., Kriegel, H.P., Sander, J., and Xu, X. 1996. A Density-Based Algorithm for Discovering Clusters in Large Spatial Databases with Noise, *Proc. 2th ACM SIGKDD Int. Conf. Knowl. Discov. Data Mining,* pp. 226–231.
Estevez, Pablo A., Paugam-Moisy, H., Puzenat, D., and Ugarte, M. 2002. A Scalable Parallel Algorithm for Training a Hierarchical Mixture of Neural Experts, *Parallel Computing,* vol. 28, pp. 861–891.
Estlick, M., Leeser, M., Theiler, J., and Szymanski, J.J. 2001. Algorithmic Transformations in the Implementation of K-Means Clustering on Reconfigurable Hardware, *International Symposium on Field Programmable Gate Arrays,* pp. 103–110.
Eyal, Z. and Thomas, H. 1994. A Planning Approach for Robot-Assisted Multiple-Bent Profile Handling, *Robotics and Computer Integrated Manufacturing,* vol. 11, no. 1, pp. 35–40.
Fairhurst, M.C. 1997. Signature Verification Revisited: Promoting practical Exploitation of biometric Technology, *Electron. Commun. Eng. J.,* vol. 9, no. 6, pp. 273–280.

Faltlhauser, R. and Ruske G. 2002. Robust Speaker Clustering in Eigenspace, *Inst. for Human-Machine-Communication,* Technische Universitat Munchen, Munich, Germany, IEEE ASRU'01, pp. 57–60.

Farina, M. and Amato, P. 2004. A Fuzzy Definition of Optimality for Many Criteria Optimization Problems, *IEEE Trans. Syst., Man, Cybern. A Syst., Humans,* vol. 34, no. 3, pp. 315–326.

Fasal, B. 2002. Robust Face Analysis Using Convolutional Neural Network, *Proc. 7th Int.Conf. Pattern Recognition* 2, pp. 11–15.

Faundez-Zanuy, M. 2005. Biometric Verification of Humans by Means of Hand Geometry, *Proc. 39th Annu. Int. Carnahan Conf. Security Technology,* pp. 61–67.

Faundez-Zanuy, M. 2005. Data Fusion in Biometrics, *Aerospace and Electronic Systems Magazine, IEEE* vol. 20, no. 1, pp. 34–38.

Favata, J.T. and Srikantan, G. 1996. A Multiple Feature/Resolution Approach to Handprinted Digit and Character Recognition, *Int'l J. Imaging Systems and Technology,* vol. 7, pp. 304–311.

Feng, Z., Zhou, B., and Shen, J. 2007. A Parallel Hierarchical Clustering Algorithm for PCs Cluster System, *Neurocomputing,* vol. 70, pp. 809–818.

Ferrari, S., Maggioni, M., and Borghese, N.A. 2004. Multi-Scale Approximation with Hierarchical Radial Basis Functions Networks, *IEEE Transactions on Neural Networks,* vol. 15, no. 1, pp. 178–188.

Fernando, G.L., Claudio, F.L., and Zbigniew, M. 2007. *Parameter Setting in Evolutionary Algorithms,* Springer. Berlin, Heilderberg.

Fierrez-Aguilar, A., Ortega-Garcia, J., Garcia-Romero, D., and Gonzalez-Rodriguez, J. 2003. A Comparative Evaluation of Fusion Strategies for Multimodal Biometric Verification. Springer LNCS-2688. *4th Int'l. Conf. Audio- and Video-Based Biometric Person Authentication (AVBPA 2003).* Guildford. pp. 830–837.

Figueiredo, M. and Gomile, F. 1999. Design of Fuzzy System Using Neuro-Fuzzy Networks, *IEEE Trans. Neural Networks,* vol. 10, pp. 815–827.

Fikes, R. and Nilsson, N. 1971. STRIPS: A New Appraoch to the Application of Theorem Proving to Problem Solving, *Artificial Intelligence,* pp. 274–279.

Fink, G.A., and Thomas, P. 2006. Unsupervised Estimation of Writing Style Models for Improved Unconstrained Off-line Handwriting Recognition. *Proceedings of the Tenth International Workshop on Frontiers in Handwriting Recognition,* La Baule, France, pp. 429–434.

Fisher, M. and Everson, R. 2003. When Are Links Useful? Experiments in Text Classification, *Twenty-fifth European conference on IR Research,* pp. 41–56.

Fisher, Y. 1995. *Fractal Image Compression: Theory and Application.* New York: Springer-Verlag.

Frezza-Buet, H. and Alexandre, F. 1999. Modeling Prefrontal Functions for Robot Navigation. *Proceeding of IEEE International Joint Conference on Neural Networks,* 1999. IJCNN '99, vol. 1, pp. 252–257, Washington DC, USA.

Flanagan, J.L. 1985. Computer-Steered Microphone Arrays for Sound Transduction in Large Rooms, *J. Acoust. Soc. Amer.,* vol. 78, no. 11, pp. 1508–1518.

Flom, L. and Safir, A. 1987. *Iris Recognition System.* U.S. Patent No. 4641394.

Floreano, D. and Mattiussi, C. 2008. *Bio-Inspired Artificial Intelligence Theories, Methods, and Technologies,* MIT Press, USA.

Fogel, D.B. 1999. *Evolutionary Computation: Toward a New Philosophy of Machine Intelligence.* Piscataway, NJ: IEEE Press.

Fogel, L. J., Owens, A.J., and Walsh, M.J. 1998. Artificial Intelligence Through a Simulation of Evolution, *Evolutionary Computation: The Fossil Record,* D.B. Fogel, ed. Piscataway, NJ: IEEE Press, pp. 230–254.

Foggia, P., Sansone, C., Tortorella, F., and Vento, M. 1999. Combining Statistical and Structural Approaches for Handwritten Character Description, *Image and Vision Computing,* vol. 17, no. 9, pp. 701–711.

Fonseca, C.M. and Fleming, P.J. 1995. An Overview of Evolutionary Algorithms in Multiobjective Optimization, *Evol. Comput.,* vol. 3, no. 1, pp. 1–16.

Forsyth, D. A. and Ponce, J. 2003. *Computer Vision: A Modern Approach.* Prentice Hall, Upper Saddle River, NJ.

Fox, G., Chaza., C., and Reilly, R.B. 2003. Person Identification Using Automatic Integration of Speech, Lip and Face Experts, *WBMA '03.* Berkeley, California, USA.

Fox, N. and Reilly, R.B. 2003. Audio-Visual Speaker Identification Based on the Use of Dynamic Audio and Visual Features, *Proc. 4th International Conference on audio and video based biometric person authentication,* pp. 743–751, Guildford, UK.

Franklin, J.A. 2006. Recurrent Neural Networks for Music Computation, *Informs Journal on Computing,* vol. 18, no. 3, pp. 321–338.

Freitas, Alex A. 2003. A Survey of Evolutionary Algorithms for Data Mining and Knowledge Discovery, Postgraduate Program in Computer Science, Brazil, pp. 819–845.
Friinti, Kivijirvi, J. and Nevalainen, O. 1998. Tabu Search Algorithm for Codebook Generation in VQ, *Pattern Recogniiion.* vol. 31, no. 8, pp. 1139–1148.
Frischholz, R.W. and Dieckmann, U. 2000. BioID: A Multimodal Biometric Identification System, *IEEE Computer,* vol. 33, no. 2, pp. 64–68.
Frischholz, R.W. and Dieckmann, U. 2000. Bioid: A Multimodal Biometric Identification System, *IEEE Computer,* vol. 33, pp. 64–68.
Fujita, M. and Kitano, H. 1998. Development of an Autonomous Quadruped Robot for Robot Entertainment, *Autonom. Robots,* vol. 5, no. 1, pp. 7–18.
Fukase, T., Kobayashi, Y., Ueda, R., Kawabe, T., and Arai, T. 2002. Real-Time Decision Making under Uncertainty of Self-Localisation Results, *Lecture Notes in Computer Science,* vol. 2752, G.A. Kaminka, P.U. Lima, and R. Rojas, eds. NewYork: Springer-Verlag, vol. 2752, pp. 375–383.
Fukunaga, K. 1990. Introduction to Statistical Pattern Recognition, 2nd ed. San Diego, CA: Academic.
Fuller, R. and Carlsson, C. 1996. Fuzzy Multiple Criteria Decision Making: Recent Developments, *Fuzzy Set Syst.,* vol. 78, pp. 139–153.
Funda, Meric, Bernstam, Elmer V. et al. 2002. Breast Cancer on the World Wide Web: Cross Sectional Survey of Quality of Information and Popularity of Websites, BMJ, vol. 324, pp. 577–581.
Furuhashi, T. 2001. Fusion of Fuzzy/Neuro/Evolutionary Computing for Knowledge Acquisition, *Proc. IEEE,* vol. 89, no. 9, pp. 1266–1274.
Furundzic, D., Djordjevic, M., and Bekic, A.J. 1998. Neural Networks Approach to Early Breast Cancer Detection. *Systems Architecture,* vol. 44, no. 8, pp. 617–633.
Fuzzy Logic Toolbox, For Use with Matlab, User's Guide, Version 2, 1995.
Galindo, U.J. and Piattini, M. 2006. *Fuzzy Databases: Modeling, Design and Implementation.* Hershey, PA: Idea Group Publishing.
Garbi, Giuliani Paulaineli, Rosado, Victor Orlando Gamarra, and Grandinetti, Francisco Jose. 2007. Multivalued Adaptive Neuro-Fuzzy Controller for Robot Vehicle.
Garcia, C. and Delakis, M. 2002. A Neural Architecture for Fast and Robust Face Detection, *Proc. 16th Int. Conf. Pattern Recognition (ICPR'02)* 2. Quebec, Canada. pp. 20044–20048.
Gaves, A. and Schmidhuber, J. 2005. Framewise Phoneme Classification with Bidirectional LSTM and other Neural Network Architectures, *Neural Networks,* vol. 18, nos. 5–6, pp. 602–610.
Gelernter, H.L. and Rochester, N. 1958. Intelligent Behaviour in Problem-Solving Machines, *IBM Journal of Reasearch and Development*, vol. 2, p. 336.
Gelsema, E.S. 1995. Abductive Reasoning in Bayesian Belief Networks Using a Genetic Algorithm, *Pattern Recognit. Lett.,* vol. 16, no. 8, pp. 865–871.
Gelsema, E.S. 1996. Diagnostic Reasoning Based on a Genetic Algorithm Operating in a Bayesian Belief Network, *Pattern Recognit. Lett.,* vol. 17, pp. 1047–1055.
Genetic Algorithm and Direct Search, Toolbox™ 2, User's Guide.
Gennery, D.B. 1999. *Traversability Analysis and Path Planning for Planetary Rovers, Autonomous Robots,* vol. 6. Kluwer: Academic Publishers, pp. 131–146.
George, J.K. and Yuan, B. 1995. *Fuzzy Sets and Fuzzy logic,* Prentice Hall of India, pp. 280–300.
Gernot, A. Fink, and Thomas, Plotz. 2006. Unsupervised Estimation of Writing Style Models for Improved Unconstrained Off-line Handwriting Recognition. In *Proc. 10th Int. Workshop on Frontiers in Handwriting Recognition.*
Gesu, G., V. Di, Maccarone, M., and Tripiciano, M. 1993. Mathematical Morphology based on Fuzzy Operators, *Fuzzy Logic,* R. Lowen and M. Roubens, eds. Kluwer Academic Publishers, MA, USA.
Ghosh, A., Biehl, M., Freking, A., and Reents, G. 2004. A Theoretical Framework for Analyzing the Dynamics of LVQ: A Statistical Physics Approach. *Technical Report 2004-9-02,* Mathematics and Computing Science, University Groningen, P.O. Box 800, 9700 AV Groningen, Netherlands. Available via www.cs.rug.nl/biehl.
Girosi, F., Jones, M., and Poggio, T. 1995. Regularization Theory and Neural Network Architectures, *Neural Comput.,* vol. 7, pp. 219–269.
Glower, J.S. and Munighan, J. 1997. Designing Fuzzy Controllers from a Variable Structures Standpoint, *IEEE Trans. Fuzzy Syst.,* vol. 5, pp. 138–144.
Godinez, M., Jimenez, A., Ortiz, R., and Peria, M. 2003. On-line Fetal Heart Rate Monitor by Phonocardiography, *Proceedings of 25th Annual International Conference – IEEE EMBS*, Cancun, Mexico, pp. 3141–3144.

Goldberg, D.E. 1989. Genetic Algorithms in Search, Optimization, and Machine Learning. New York: *Addison-Wesley*.

Goldberg, D.E. 2002. A Meditation on the Application of Soft Computing and Its Future, *Soft Computing and Industry: Recent Applications,* R. Roy, M. K¨ooppen, S. Ovaska, T. Furuhashi, and F. Hoffmann, eds. London, UK: Springer-Verlag, pp. XV–XVIII.

Goldberg, D.E., Kargupta, H., Horn, J., and Cantu-Paz, E. 1995. Critical Deme Size for Serial and Parallel Genetic Algorithms, Univ. Illinois at Urbana–Champaign, Illinois Genetic Algorithms Laboratory, IlliGAL Rep. 95002.

Golubovic, D. and Hu, H. 2002. A Hybrid Evolutionary Algorithm for Gait Generation of Sony Legged Robots, *Proc. 28th Annu. Conf. IEEE Ind. Electron. Soc.,* Sevilla, Spain, Nov. 5–8, pp. 2593–2598.

Gonzalez, R.C. and Woods, R.E. 2002. *Digital Image Processing,* 2nd ed. Pearson Education, India.

Gonzalez, S., Traviesoa, C., Alonso, J., and Ferrer, M. 2003. Automatic Biometric Identification System by Hand Geometry, *Proc. IEEE 37th Int. Carnahan Conf. Security technology,* pp. 281–284.

Gordon, A.D. 1987. A Review of Hierarchical Classification, *Journal of the Royal Statistical Society,* vol. 150, no. 2, pp. 119–137.

Gordon, V.S. and Whitley, D. 1993. Serial and Parallel Genetic Algorithms as Function Optimizers, *Proc. 5th Int. Conf. Genetic Algorithms,* S. Forrest, ed. San Mateo, CA: Morgan Kaufmann, pp. 177–183.

Gordon, V.S., Whitley, D., and Böhm, A.P.W. 1992. Dataflow Parallelism in Genetic Algorithms, *Parallel Problem Solving from Nature 2,* R. Männer and B. Manderick, eds. Amsterdam, The Netherlands: Elsevier, pp. 533–542.

Gorinevsky, D., Kapitanovsky, A., and Goldenberg, A. 1996. Neural Network Architecture for Trajectory Generation and Control of Automated Car Parking, *IEEE Trans. Control Syst. Technol.*, vol. 4, pp. 50–56.

Gorman, L.O. and Nickerson, J.V. 1991. An Approach to Fingerprint Filter Design, *Pattern Recognition,* vol. 22, No. 1, pp. 29–38.

Goswami, A., Jin, R., and Agrawal, G. 2004. Fast and Exact Out-of-Core K-Means Clustering, *IEEE International Conference on Data Mining,* pp. 83–90.

Gray, H.F., Maxwell, R.J., Martinez-Perez, I., Arus, C., and Cerdan, S. 1996. Genetic Programming for Classification of Brain Tumours from Nuclear Magnetic Resonance Biopsy Spectra, *Proc. 1st Annu. Conf. Genetic Program.,* J.R. Koza, D.E. Goldberg, D.B. Fogel, and R.L. Riolo, eds., Jul. 28–31, p. 424.

Griewangk, A.O. 1981. Generalized Descent of Global Optimization, *J. Optim. Theory Appl.,* vol. 34, pp. 11–39.

Griggs, I. 2004. Browser Threat Model. Retrieved from http://iang.org/ssl/browser threat model.html.

Grigore, O. and Gavat, I. 1998. Neuro-Fuzzy Models for Speech Pattern Recognition in Romanian Language, *CONTI98 Proceedings,* Timisoara, Romania: CONTI, pp. 165–172.

Grossman, J. 2007. *WhiteHat Website Security Statistics Report,* WhiteHat Security.

Groumpos, P.P. and Stylios, C.D. 1999, Modeling Supervisory Control Systems Using Fuzzy Cognitive Maps, Chaos, Solitons and Fractals, vol. 11, nos. 1–3, pp. 303–308.

Gu, D., Hu, H., Reynolds, J., and Tsang, E. 2003. GA-Based Learning in Behavior Based Robotics, Proc. IEEE Int. Symp. Comput. Intell. Robot. Autom., vol. 3, Kobe, Japan, Jul. 16–20, pp. 1521–1526.

Guan, L. and Ward, R.K. 1989. Restoration of Randomly Blurred Images by the Wiener Filter, *IEEE Transactions on ASSP*, vol. 37, no. 4, pp. 589–592.

Guenoche, A. 2004. Clustering by Vertex Density in a Graph, Proceedings of IFCS Congress Classification, pp. 15–24.

Guest, R.M. 2004. The Repeatability of Signatures, *Proc. 9th Int. Workshop Frontiers Handwriting Recog.*, pp. 492–497.

Guh, Y.Y., Hon, C.C., and Lee, E.S. 2001. Fuzzy Weighted Average: The Linear Programming Approach via Charnes and Cooper's rule, *Fuzzy Sets Syst.,* vol. 117, pp. 157–160.

Gupta, P., Mehrotra, H., Rattani, A., Chatterjee, A., and Kaushik, A.K. 2006. Iris Recognition Using Corner Detection, *Proc. 23rd Int'l Biometric Conference.*

Gutta, S. and Wechsler, H. 1996. Face Recognition Using Hybrid Classifier Systems, *Proc. IEEE Int. Conf. Neural Networks*, pp. 1017–1022.

Halkidi, M., Batistakis, Y., and Vazirgiannis, M. 2002. Cluster Validity Methods: Part I, *SIGMOD Rec.,* vol. 31, no. 2, pp. 40–45.

Hammer, B. and Villmann, T. 2002. Generalized Relevance Learning Vector Quantization, *Neural Networks,* vol. 15, pp. 1059–1068.

Hanzalek, Zdenek, 1998. A Parallel Algorithm for Gradient Training of Feed forward Neural Networks, Pattern Computing, Elsevier Science Publishers B. V., vol. 24, nos. 5–6, pp. 823–839.

Handa, I. and Baba, M. 2002. A Novel Hybrid Framework of Coevolutionary GA and Machine Learning, *International Journal of Computational Intelligence and Applications,* vol. 2, no. 1, pp. 33–52.

Hansen, L. and Salammon, P. 1990. Neural network ensembles, *IEEE Trans. Pattern Anal. Mach. Intell.*, vol. 12, no. 10, pp. 993–1101.

Hansen, L.K., Liisberg, L., and Salamon, P. 1992. Ensemble Methods for Handwritten Digit Recognition, *Proc. IEEE Workshop Neural Networks Signal Processing,* pp. 333–342.

Hanzalek Z. 1998. A Parallel Algorithm for Gradient Training of Feedforward Neural Networks, *Pattern Computing,* vol. 24, pp. 823–839.

Hao, G., ZongYing, O., and Yang, H. 2003. Automatic Fingerprint Classification Based on Embedded Hidden Markov Models, *International Conference on Machine Learning and Cybernetics,* vol. 5, pp. 3033–3038.

Haralick, R., Sternberg, S., and Zhuang, X. 1987. Image Analysis Using Mathematical Morphology, *IEEE Transactaons on Pattern Analyszs and Machane Intellagence,* vol. 9, pp. 532–550.

Hargreaves, D.J. and North, A.C. 2005. eds. *The Social Psychology of Music.* Oxford University Press, Oxford, UK.

Harmon, L.D., Khan, M.K., Lasch, R., and Ramig, P.F. 1981. Machine Identification of Human Faces, *Pattern Recognit.,* vol. 13, pp. 97–110.

Harris, C. and Hong, X. 2001. Neuro-Fuzzy Mixture of Experts Network Parallel Learning and Model Construction Algorithms, *IEEE Proc. Control Theory Appl.*, vol. 148, no. 6, pp. 456–465.

Hartigan, J. 1975. *Clustering Algorithms.* New York: Wiley.

Hashemi, J. and Fatemizadeh, E. 2005. Biometric Identification Through Hand Geometry, *Proc. Int. Conf. Comput. Tool.* vol. 2, pp. 1011–1014. Haykin, S. 1999. *Neural Networks—A Comprehensive Foundation,* 2nd ed, New Jersey, USA Prentice Hall.

Hawking, D. and Craswell, N. 2002. Overview of the TREC 2001 Web Track, Voorhees, E., Harman, D. eds., *Proceedings of the Tenth Text Retrieval Conference,* NIST Special Publication, pp. 25–31, Gaithersburg, Maryland.

Hawkins, D.M. 1980. *Identification of Outliers.* Chapman and Hall, London.

Hayashi, Umano, M., Maeda, T., Bastian, A., and Jain, L.C. 1998. Acquisition of Fuzzy Knowledge by NN and GA—A Survey of the Fusion and Union Methods Proposed in Japan, *Proc. 2nd Int. Conf. Knowledge-Based Intelligent Electronic Systems*, Adelaide, Australia, pp. 69–78.

Hayashi, Y. 1992. A Neural Expert System Using Fuzzy Teaching Input, *Pwc. IEEE Conference on Fuzzy Systems*, San Diego, pp. 485–491.

Haykin, S. 1994. *Neural Networks: A Comprehensive Foundation.* Upper Saddle River, NJ: Prentice-Hall, Inc.

He, Y., Liu, G., Rees, D., and Wu, M. 2007. Stability Analysis for Neural Networks with Time-Varying Interval Delay, *IEEE Trans. Neural Netw.,* vol. 18, no. 6, pp. 1850–1854.

Heart Diseases [internet]. http://www.drugs.com/condition/heart-disease.html

Hebb, O.L. 1949. *The organisation of behaviour.* New York: Wiley.

Hecht-Nielsen, R. 1989. Theory of the Backpropagation Neural Network, *Proceedings of the International Joint Conference on Neural Networks*, vol. 1, pp. 593–606. Helsinki. Neural Networks Research Centre. Bibliography on the self-organizing maps (SOM) and learning vector quantization (LVQ), Otaniemi: Helsinki Univ. of Technology. Available via http://liinwww.ira.uka.de/bibliography/Neural/SOM.LVQ.html.

Henrion, M. 1988. Propagating Uncertainty in Bayesian Networks by Probabilistic Logic Sampling, *Uncertainty in Artificial Intelligence 2,* J. Lemmer and L. Kanal, eds. Amsterdam, The Netherlands: North-Holland, pp. 149–263.

Hernandez, E. and Weiss, G. 1996. A First Course on Wavelets. Boca Raton, FL: CRC Press. With a foreword by Yves Meyer.

Herrera, F. and Lozano, M. 1996. Adaptation of Genetic Algorithm Parameters Based on Fuzzy Logic Controllers, *Genetic Algorithms and Soft Computing,* F. Herrera and J.L. Verdegay, eds. Berlin, Germany: Physica-Verlag, vol. 8, Studies in Fuzziness and Soft Computing, pp. 95–125.

Herrera, F. and Lozano, M. 1997. Heterogeneous Distributed Genetic Algorithms Based on the Crossover Operator, *2nd IEE/IEEE Int. Conf. Genetic Algorithms Eng. Syst.: Innovations Appl.*, pp. 203–208.

Herrera, F., Lozano, M., and Verdegay, J.L. 1996. Dynamic and Heuristic Fuzzy Connectives Based Crossover Operators for Controlling the Diversity and Convergence of Real-Coded Genetic Algorithms, *Int. J. Intell. Syst.,* vol. 11, pp. 1013–1041.

Herrera, F., Lozano, M., and Verdegay, J.L. 1997. Fuzzy Connectives Based Crossover Operators to Model Genetic Algorithms Population Diversity, *Fuzzy Sets Syst.,* vol. 92, no. 1, pp. 21–30.

Herrera, F., Lozano, M., and Verdegay, J.L. 1998. Tackling Real-Coded Genetic Algorithms: Operators and Tools for Behavioral Analysis, *Artif. Intell. Rev.*, vol. 12, no. 4, pp. 265–319.

Heskes, T. and Wiegerinck, W. 1996. A Theoretical Comparison of Batch-Mode, On-Line, Cyclic, and Almost-Cyclic Learning, *IEEE Trans. Neural Netw.*, vol. 7, no. 4, pp. 919–925.

Hewavitharana, S., Fernando, H.C., and Kodikara, N.D. 2002. Off-line Sinhala Handwriting Recognition using Hidden Markov Models, *In Proceedings of the 3rd Indian Conference on Computer Vision, Graphics and Image Processing (ICVGIP'02),* pp. 266–269, Ahmedabad, India.

Hintz, T. and Wu, Q. 2003. Image Compression on Spiral Architecture, *Proceedings of The International Conference on Imaging Science, Systems and Technology,* pp. 201–204.

Hochreiter, S. and Schmidhuber, J., Long Short-Term Memory, *Neural Computation,* vol. 9, no. 8, pp. 1735–1780.

Hochreiter, S., Bengio, Y., Frasconi, P., and Schmidhuber, J. 2001. Gradient Flow in Recurrent Nets: The Difficulty of Learning longterm Dependencies, *A Field Guide to Dynamical Recurrent Networks.* New York: IEEE Press.

Hoffmann, F., Koppen, M., Klawonn, F., and Roy, R. 2005. *Soft Computing: Methodologies and Applications.* Birkhäuser Publications. ISBN 3540257268, 9783540257264.

Hoglund, G. and McGraw, G. 2004. *Exploiting Software: How to Break Code.* Addison-Wesley Professional, Canada.

Holland, H. 1992. Adaptation in Natural and Artificial Systems. Ann Arbor, MI: Univ. Michigan Press, pp. 211.

Hong, L., Jain, A.K., and Pankanti, S. 1999. Can Multibiometrics Improve Performance?, *Proc. AutoID '99,* pp. 59–64.

Hong, T. and Lee, C. 1996. Induction of Fuzzy Rules and Membership Functions from Training Examples. *Fuzzy Sets and Systems,* vol. 84, pp. 33–47.

Hooper, K., Meikle, S.R., Eberl, S., and Fulham, M.J. 1996. Validation of Post Injection Transmission Measurements for Attenuation Correction in Neurologic FDG PET Studies, *J. Nucl. Med.,* vol. 37, pp. 128–136.

Hopfield, J.J. and Tank, D.W. 1985. Neural Computation of Decisions in Optimization Problems, *Biol. Cybern.,* vol. 52, pp. 141–152.

Hopfield, J.J. 1982. Neural Network and Physical Systems with Emergent Collective Computational Abilities, *Proc. Nat. Acad. Sci. USA,* vol. 79, pp. 2554–2558.

Hoppner, F., Klawonn, F., Kruse, R., and Runkler, T. 1999. *Fuzzy Cluster Analysis.* New York: Wiley Press.

Hornik, K., Stinchcombe, M., and White, H. 1989. Multilayer Feedforward Networks are Universal Approximators, *Neural Networks,* vol. 2, pp. 359–366.

Howarth, J.W., Bakker, H.H.C., and Flemmer, R.C. 2009. Feature-Based Object Recognition, *Proceedings of the International Conference on Autonomous Robots and Agents (ICARA),* Wellington, New Zealand, *in press.*

Huan-cheng, Zhang, and Miao-liang, Zhu. 2005. Self-Organized Architecture for Outdoor Mobile Robot Navigation, Journal of Zhejiang University Science, vol. 6, no. 6, pp. 583–590.

Huang, F.J., Zhou, Z.H., Zhang, H.J., and Chen, T.H. 2000. Pose Invariant Face Recognition, *Proc. 4th IEEE Int. Conf. Automatic Face Gesture Recognition,* pp. 245–250.

Huang, G.B., Zhu, Q.Y., and Siew, C.K. 2006. Real-Time Learning Capability of Neural Networks, *IEEE Trans. Neural Netw.*, vol. 17, no. 4, pp. 863–878.

Huang, S.J. and Lee, J.S. 2000. A Stable Self-Organizing Fuzzy Controller for Robotic Motion control, *IEEE Trans. Ind. Electron.,* vol. 47, pp. 421–428.

Hughes, J. 2001. The Future of Death: Cryonics and the Telos of Liberal Individualis, *Journal of Evolution and Technology,* vol. 6.

Humphrey, W.S. 1995. *A Discipline for Software Engineering.* Reading, MA: Addison-Wesley.

Husbands, P., Holland, O., and Wheeler, M. 2008. *The Mechanical Mind in History.* Cambridge, London, UK: Bradford Books.

Hussain, M.A. 1999. Review of the Applications of Neural Networks in Chemical Process Control—Simulation and Online Implementation, *Artif. Intell. In Eng.,* vol. 13, pp. 55–68.

Hutchinson, S.A. and Kak, A.C. 1989. Planning Sensing Strategies in a Robot Work Cell with Multi-Sensor Capabilities, *IEEE Trans. On Robotics and Automation,* vol. 5, no. 6.

Hwang, G.C. and Lin, S.C. 1992. A Stability Approach to Fuzzy Control Design for Nonlinear Systems, Fuzzy Sets Syst., vol. 48, no. 3, pp. 279–287.

Hwang, Gwo-Haur, Chen, Jun-Ming, Hwang, Gwo-Jen, and Chu, Hui-Chun 2006 . A Time Scale-Oriented Approach for Building Medical Expert Systems, *Expert Systems with Applications,* vol. 31, pp. 299–308.

Hwang, J.N., Lay, S.R., Maechler, M., Martin, R.D., and Schimert, J. 1994. Regression Modeling in Back-Propagation and Projection Pursuit Learning, *IEEE Trans. Neural Netw.*, vol. 5, no. 3, pp. 342–353.

Ichihashi, H., Honda, K., Notsu, A., and Eri Miyamoto 2008. FCM Classifier for High-Dimensional Data, *IEEE World Congress on Computational Intelligence,* pp. 200–206.

Ignizio, James P. 1991. *Introduction to Expert Systems: The Development and Implementation of Rule Based Expert Systems.* New York: MGH Inc.

Ihlstrom, C. and Akesson, M. 2004. Genre Characteristics: a Front Page Analysis of 85 Swedish Online Newspapers, *Proceedings of the Thirty-Seventh Annual Hawaii International Conference on System Sciences,* vol. 2.

Ihlstrom, C. and Eriksen, Lars B. 2000. Evolution of the Web News Genre: the Slow Move Beyond the Print Metaphor. *Proceedings of the 33rd Hawaii International Conference on System Sciences.*

Inoue, T. and Abe, S. 2001. Fuzzy Support Vector Machines for Pattern Classification, *Proc. International Joint Conference on Neural Networks (IJCNN '01),* vol. 2, pp. 1449–1454.

Inza, Larrañaga, P. and Sierra, B. 2001. Feature Subset Selection by Bayesian networks: a Comparison with Genetic and Sequential Algorithms, *Artificial Intelligence in Medicine, Elsevier,* vol. 27, no. 2, pp. 143–164.

Ishibuchi, H. and Yamamoto, T. 2003. Evolutionary Multi-Objective Optimization for Generating an Ensemble of Fuzzy Rule-Based Classifiers, *Proc. Genetic Evol. Comput. Conf. Lecture Notes in Computer Science,* vol. 2723, pp. 1077–1088.

Ishibuchi, H., Murata, T., and Turksen, I. 1997. Single-Objective and Two Objective Genetic Algorithms for Selecting Linguistic Rules for Pattern Recognition, *Fuzzy Sets Syst.,* vol. 89, pp. 135–150.

Jain and Ross, A. 2002. Learning User-Specific Parameters in Multibiometric System., *Proc. Int'l Conf. of Image Processing (ICIP 2002),* New York, pp. 57–70.

Jain and Dubes, R. 1998. *Algorithms for Clustering Data.* Englewood Cliffs, NJ: Prentice-Hall.

Jain, A.K., Griess, F.D., and Connell, S.D. 2002. On-Line Signature Verification, *Pattern Recognit.,* vol. 35, no. 12, pp. 2963–2972.

Jain, A.K. and Duta, N. 1999. Deformable Matching of Hand Shapes for Verfication, *IEEE International Conference on Image Processing,* pp. 857–861.

Jain, A.K., Ross, A., and Pankanti, S. 1999. A Prototype Hand Geometry-Based Verfication System, *2nd International Conference on Audio-and Video-based Biometric Person Authentication,* pp. 166–171.

Jain, A.K., Ross, A., and Prabhakar, S. 2004. An Introduction to Biometric Recognition, *IEEE Transactions on Circuits and Systems for Video Tech–nology, Special Issue on Image- and Video-Based Biometrics,* vol. 14, no. 1, pp. 4–20.

Jain, A., Bolle, R., and Pankanti, S. 1999. *Biometrics: Personal Identification in a Networked Society.* Norwell: Kluwer.

Jain, A.K. and Dubes, R.C. 1988. *Algorithms for Clustering Data.* Englewood Cliffs, NJ: Prentice-Hall.

Jain, A.K. 1981. Image Data Compression: A Review, *Proc. IEEE,* vol. 69, pp. 349–389.

Jain, K., Murthy, M., and Flynn, P. 1999. Data Clustering: a Review, *ACM Computing Reviews.*

Jain, B.R. and Pankanti, S. 1999. *Biometrics: Personal Identification in Networked Society.* 2nd printing. Kluwer Academic Publishers, Massachusetts, USA.

Jain, L.C. and Martin, N.M. 1998. *Fusion of Neural Networks, Fuzzy Systems and Genetic Algorithms: Industrial Applications,* CRC Press, Boca Raton, FL.

Jang, J.R. 1992. Self-Learning Fuzzy Controllers Based on Temporal Back Propagation, *IEEE Trans. Neural Networks,* vol. 3, pp. 714–723.

Jang, J.S.R. 1993. ANFIS: Adaptive Network Based Fuzzy Inference System, *IEEE Trans. Syst.*, Man, Cybern., vol. 23, no. 3, vol. 23, no. 3, pp. 665–685.

Jang, J.S.R. and Sun, C.T. 1998. *Neuro-Fuzzy and Soft Computing—ANFIS (Adaptive Neuro-Fuzzy Inference Systems).* Englewood Cliffs, NJ: Prentice-Hall, ch. 12, Englewood Cliffs, NJ.

Jang, J.S.R., Sun, C.T., and Mizutani, E. 1997. *Neuro-Fuzzy and Soft Computing: A Computational Approach to Learning and Machine Intelligence.* Upper Saddle River, NJ: *Prentice-Hall.*

Jang, J.S.R. 1993. ANFIS: Adaptive-Network-Based Fuzzy Inference System. *IEEE Trans. Systems, Man, and Cybernetics,* pp. 665–685.

Jang, J.S.R. and Sun, C.T. 1995. Neuro-fuzzy Modeling and Control, *Proceedings of the IEEE,* vol. 83, no. 3, pp. 378–406.

Jang, R., Sun, C., and Mizutani, E. 1997. *Neuro-fuzzy and Soft Computing.* Prentice-Hall, pp. 353–360.

Janikow, C.Z. 1998. Fuzzy decision trees: Issues and methods. *IEEE Trans. Syst. Man Cybern.* vol. 28, no. 1, pp. 114.

Jankowski, N. 1999. Approximation with RBF-type Neural Networks Using Flexible Local and Semi-Local Transfer Functions. *4th Conference on Neural Networks and Their Applications*, pp. 77–82.

Jarvis, R. and Patrick, E. 1973. Clustering Using a Similarity Measure Based on Shared Near Neighbors," *IEEE Trans. Comput.*, vol. C-22, no. 11, pp. 1025–1034.

Jayadeva, Khemchandani, R., and Chandra, S. 2004. Fast and Robust Learning Through Fuzzy Linear Proximal Support Vector Machines, *Neurocomputing,* vol. 61, pp. 401–411.

Jayasekra, B., Jayasiri, A., and Udawatta, L. 2006. An Evolving Signature Recognition System, *Proceedings of first ICIIS 2006,* Srilanka, pp. 529–534.

Jensen, F.V. 1996. *An Introduction to Bayesian Networks.* London, UK: UCL Press.

Jensen, M.T. 2003. Reducing the Run-Time Complexity of Multiobjective EAs: The NSGA-II and other Algorithms, *IEEE Trans. Evol. Comput.,* vol. 7, no. 5, pp. 503–515.

Jia, Z., Balasuriya, A., and Challa, S. 2008. Sensor Fusion-Based Visual Target Tracking for Autonomous Vehicles with the Out-Of-Sequence Measurements Solution, *Robotics and Autonomous Systems,* vol. 56, no. 2, pp. 157–176.

Jiang, Ming-Yan and Chen, Zhi-Jian 1997. Diabetes Expert System, *ICIPS 97,* vol. 2, pp. 1076–1077.

Jianguang, G., Weihua, Z., Zhongwei, W., and Hongyu, X. 2007. Method of Machine Learning Based on Partitioned Clustering and Fuzzy Neural Network, *Journal of System Simulation,* vol. 19, no. 23, pp. 5581–5586.

Jiasheng, H., Chuangxin, G., and Yijia, C. 2004. A Hybrid Particle Swarm Optimization Method for Unit Commitment Problem, *Proceedings of the CSEE,* vol. 24, no. 4, pp. 24–28.

Jim, W., Gleb, B., and Berend, V.Z. 2000. Fuzzy Logic in Clinical Practice Decision Support Systems, Proceedings of the 33rd Hawaii IEEE International Conference on System Sciences, pp. 1–10.

Jimenez, A., Ortiz, R., Peria, M., Charleston, S., Aljama, A.T., and Gonzalez, R. 2000. Performance of Two Adaptive Sub Band Filtering Scheme for Processing Fetal Phonocardiogram: Influence of the Wavelet and the level Decomposition, *Proceedings of IEEE International Conference – Computers in Cardiology 2000,* Cambridge, USA, pp. 427–430.

Jin, Y. 2000. Fuzzy Modeling of High-Dimensional Systems: Complexity Reduction and interpretability Improvement, *IEEE Trans. Fuzzy Syst.,* vol. 8, no. 2, pp. 212–221.

Jin, Y. 2003. *Advanced Fuzzy Systems Design and Applications.* Heidelberg, Germany: Physica Verlag.

Jin, Y., ed. 2006. *Multi-Objective Machine Learning.* New York: Springer-Verlag.

Jin, Y. and Sendhoff, B. 2006. Alleviating Catastrophic Forgetting via Multiobjective Learning, *Proc. Int. Joint Conf. Neural Netw.,* pp. 6367–6374.

Jin, Y., Okabe, T., and Sendhoff, B. 2004. Evolutionary Multi-Objective Approach to Constructing Neural Network Ensembles for Regression, *Applications of Evolutionary Multi-Objective Optimization,* C. Coello Coello, ed. Singapore: World Scientific, pp. 653–672.

Jin, Y., Wen, R., and Sendhoff, B. 2007. Evolutionary Multi-Objective Optimization of Spiking Neural Networks, *Proc. Int. Conf. Artif. Neural Netw. (ICANN) Part I, Lecture Notes in Computer Science,* vol. 4668. New York: Springer-Verlag, pp. 370–379.

Jinshan, C. and Gang, W. 2002. A Hybrid Clustering Algorithm Incorporating Fuzzy C-Means into Canonical Genetic Algorithm, *Journal of Electronics and Information Technology,* vol. 24, no. 2, pp. 210–215.

Jobling, R. 2007. A Patient's Journey: Psoriasis. *J Br Med,* vol. 334, pp. 953– 954.

John, R. and Birkenhead, J.R., eds. 2002. Developments in Soft Computing, Advances in Soft Computing. Springer-Verlag, Heilderberg.

Johnston, D., Millett, D.T. and Ayoub, A.F. 2003. Are Facial Expressions Reproducible?, *Cleft Palate-Craniofacial Journal,* vol. 40, no. 3, pp. 291– 296.

Jolliffe, I. 2002. *Principal Component Analysis,* 2nd ed. Berlin, Germany: Springer-Verlag.

Jolliffe, T. 1986. *Principal Component Analysis.* NewYork: Springer-Verlag.

Jonson L.L. 1988. *Expert System Architectures.* Kopan Page Limited, London, UK.

Jonyer, I., Cook, D.J., and Holder, L.B. 2001. Graph-Based Hierarchical Conceptual Clustering, *J. Mach. Learn. Res.,* vol. 2, no. 2, pp. 19–43.

Jordan, A. and Neri, F. 1995. Search-Intensive Concept Induction, *Evolutionary Computation,* vol. 3, no. 4, pp. 375–416.

Juang, C.F. and Lin, C.T. 1998. An On-Line Self-Constructing Neural Fuzzy Inference Network and Its Application, *IEEE Trans. Fuzzy Syst.,* vol. 6, pp. 12–32.

Juang, J.G., Chin, K.C., and Chio, J.Z. 2004. Intelligent Automatic Landing System Using Fuzzy Neural Networks and Genetic Algorithm, *Proc. Amer. Control Conf.,* pp. 5790–5795.

Jun-jie, Yang et al. 2004. A Hybrid Intelligent Genetic Algorithm for Large-Scale Unit Commitment, *Power System Technology,* vol. 28, no. 19, pp. 47–50.

Junqua, J.C. and Haton, J.P. 1996. *Robustness in Automatic Speech Recognition.* Norwell, MA: Kluwer.

Kak, Subhash C. 1998. On Generalization by Neural Networks, *Elsevier Journal of Information Sciences*, vol. 111, pp. 293–302.

Kala, R., Shukla, A., and Tiwari, R. 2009a. Fuzzy Neuro Systems for Machine Learning for Large Data Sets, *Proceedings of the IEEE International Advance Computing Conference, ieeexplore,* pp. 223–227, Patiala, India.

Kala, R., Shukla, A., and Tiwari, R. 2009b. A Novel Approach to Classificatory Problem Using Neuro-Fuzzy Architecture, *International Journal of Systems, Control and Communications (IJSCC),* pp. 367–373.

Kala, R., Shukla, A., Tiwari, R., Roongta, Sourabh, and Janghel, R.R. 2009c. Mobile Robot Navigation Control in Moving Obstacle Environment using Genetic Algorithm, Artificial Neural Networks and A* Algorithm, *Proceedings of the IEEE World Congress on Computer Science and Information Engineering (CSIE 2009), ieeexplore,* DOI 10.1109/CSIE.2009.854, pp. 705–713, Los Angeles/Anaheim, USA.

Kala, R., Shukla, A., and Tiwari, R. 2009d. Fast Learning Neural Network Using Modified Corners Algorithm, *Proceedings of the IEEE Global Congress on Intelligent* System, ieeexplore, May 2009, Xiamen, China.

Kala, R., Shukla, A., and Tiwari, R. 2009e. Optimized Graph Search Using Multi Level Graph Clustering, *Proceedings of the International Conference on Contemporary Computing (IC3),* Springer Communications in Computer and Information Science ISSN: 1865-0929.

Kala, R., Shukla, A., and Tiwari, R. 2009f. Self Adaptive Parallel Processing Power Using Hierarchical Positioning, *J. Neural Network World,* vol. 19, no. 6, pp. 657–680.

Kalitzin, S.N., Parra, J., Velis, D.N., and Lopes da Silva, F.H. 2002. Enhancement of Phase Clustering in the EEG/MEG Gamma Frequency Band Anticipates Transitions to Paroxysmal Epileptiform Activity in Epileptic Patients with Known Visual Sensitivity, *IEEE Trans Biomed Eng,* vol. 49, pp. 1279–1286.

Kals, S., Kirda, E., Kruegel, C., and Jovanovic, N. 2006. SecuBat: A Web Vulnerability Scanner, *Proc. The 15th International Conference on World Wide Web (WWW 2006),* Edinburgh, Scotland, 23–26, pp. 247–256.

Kam, A. and Cohen, A. 2000. Separation of twins Fetal ECG by Means of Blind Source Separation (BSS), *Electrical and Electronic Engineers in Israel. The 21st IEEE,* pp. 342–345.

Kam, M., Fielding, G., and Conn, R. 1997. Writer Identification by Professional Document Examiners, *J. Forensic Sci.,* vol. 42, no. 5, pp. 778–785.

Kam, M., Gummadidala, K., and Conn, R. 2001. Signature Authentication by Forensic Document Examiners, *J. Forensic Sci.,* vol. 46, no. 4, pp. 884–888.

Kamal et al. 2007. Text Independent Time Domain Speech Features Extraction Using AANN and Speaker Identification Using PNN, *MLMTA'07- International Conference on Machine learning Models, Technology and Applications,* Monte Carlo Resort, Las Vegas, USA.

Kamruzzaman, S. and Islam, M. 2005. Extraction of Symbolic Rules from Artificial Neural Networks, *Trans. Eng., Comput. Technol.,* vol. 10, pp. 271–277.

Kandel, B., and Cao, S.Q. 1987. Some Fuzzy Implication Operators in Fuzzy Inference. *ISPK,* pp. 78–88.

Kanungo, T., Mount, D.M., Netanyahu, N.S., Piatko, C.D., Silverman, R., and Yu, A.Y. 2002. An Efficient K-Means Clustering Algorithm: Analysis and Implementation, *IEEE Trans. On Pattern Analysis and Machine Intelligence,* vol. 24, no. 7, pp. 881–892.

Kaotian, J. et al. 2000. Designing Automatic Neuro Fuzzy Controller, *Ladkrabang Engineering Journal,* vol. 17, no. 1, pp. 79–84.

Karayiamis, N.B. et al. 1996. Fuzzy Algorithm for Learning Vector Quantization. *IEEE Trans NN,* vol. 7, pp. 1196–1121.

Karnin, B.D. 1990. A Simple Procedure for Pruning Back-Propagation Trained Neural Networks, *IEEE Transactions on Neural Networks,* vol. 1, no. 2, pp. 239–242.

Karu, K. and Jain, A.K. 1996. Fingerprint Classification, *Pattern Recognition,* vol. 29, no. 3, pp. 389–404.

Karydis, Y., Nanopoulos, A., Papadopoulos, A.N., and Manolopoulos, Y. 2005. Evaluation of Similarity Searching Methods for Music Data in Peer-to-Peer Networks, *International Journal of Business Intelligence and Data Mining,* vol. 1, pp. 210–228.

Karypis, G. and Kumar, V. 1999. Multilevel K-Way Hypergraph Partitioning, *Proc. DAC,* pp. 343–348.

Kashinath P., Threats to India's Stability by Koshur Samachar.

Kass, L. 2001. Preventing a Brave New World, *The New Republic.*

Kaufman, L. and Rousseeuw, P.J. 1987. *Clustering by Means of Medoids.* Y. Dodge, ed. Amsterdam: North Holland Elsevier, pp. 405–416.

Keith, J.L., Mathias, E., and Whitley, L.D. 1994. Changing Representations During Search: A Comparative Study of Delta Coding, *Evolutionary Comput.,* vol. 2, no. 3, pp. 249–278.

Kelkboom, E.J.C. et al. 2007. 3D Face: Biometric Template Protection for 3D Face Recognition, *Proceedings of the Second International Conference on Biometrics, Springer's Lecture Notes in Computer Science,* pp. 566–573, Seoul, Korea.

Keller, J.M., Yager, R.R., and Tahani, H. 1992. Neural network implementation of fuzzy logic, *Fuzzy Sets Syst.*, vol. 45, no. 1, pp. 1–12.

Kelly, S.D. and Murray, R.M. 1995. Geometric Phases and Robotic Locomotion, *J. Robot. Syst.,* vol. 12, no. 6, pp. 417–431.

Kennedy, M.P. and Chua, L.O. 1988. Neural Networks for Nonlinear Programming, *IEEE Trans. Circuits Syst.*, vol. 35, pp. 554–562.

Khan, F.S., Maqbool, F., Razzaq, S., Kashif, I., and Zia, T. 2008. The Role of Medical Expert Systems in Pakistan, *Proceedings of World Academy of Science, Engineering and Technology,* vol. 27, pp. 296–298.

Kiguchi, K. and Fukuda, T. 1999. Fuzzy Selection of Fuzzy-Neuro Robot Force Controllers in an Unknown Environment, *IEEE Int. Conf. on Robotics and Automation*, pp. 1182–1187.

Kim D.J. and Bien, Z. 2004. A Novel Feature Selection for Fuzzy Neural Networks for Personalized Facial Expression Recognition,. *IEICE Tran. on Fundamentals of Electronics, Communications and Computer Science,* vol. E87-A, no. 6, pp. 1386–1392.

Kim, D.J., Lee, S.W., and Bien, Z. 2005. Facial Emotional Expression Recognition with soft Computing Techniques, *Proc. IEEE Int. Conf. Fuzzy Syst.*, pp. 661–666.

Kim, H.R. and Lee, H.S. 1991. Postprocessor Using Fuzzy Vector Quantiser in HMM-Based Speech Recognition, *Electron. Lett.,* vol. 17, no. 22, pp. 1998–2000.

Kim, H.Y. and Kim, J.H. 2001. Hierarchical Random Graph Representation of Handwritten Characters and Its Application to Hangul Recognition, *Pattern Recognition,* vol. 34, no. 2, pp. 187–201.

Kim, I.J. and Kim, J.H. 2000. Statistical Utilization of Structural Neighborhood Information for Oriental Character Recognition, *Proc. Fourth Int'l Workshop Document Analysis Systems,* pp. 303–312.

Kim, Y., Street, W., and Menczer, F. 2002. Evolutionary Model Election in Unsupervised Learning, *Intell. Data Anal.,* vol. 6, pp. 531–556.

Sharon, K. 2005. Growing Awareness of Skin Disease Starts Flurry of Initiatives. *Bulletin of the World Health Organization,* vol. 83, no. 12. London, UK: Genebra.

Kiss, L., Annamaria, R., and Varkonyi-Koczy. 2003. A Universal Vision-Based Navigation System for Autonomous Indoor Robots. *In Proc. of the 1st Slovakian-Hungarian Joint Symposium on Applied Machine Intelligence, SAMI 2003,* Herlany, Slovakia, pp. 183–196.

Kiszka, J.B. et al. 1985. The Inference of Some Fuzzy Implication Operators on the Accuracy of a Fuzzy Model Part I and Part 11, *Fuzzy Sets and Systems,* vol. 15, pp. 111–128 and 223–240.

Kita, H., Ono, I., and Kobayashi, S. 1999. Multi-Parental Extension of the Unimodal Normal Distribution Crossover for Real-Coded Genetic Algorithms, *Proceedings of the Congress on Evolutionary Computation,* vol. 2, pp. 1581–1588. IEEE Press, Washington DC.

Kittler, A., Messer, K., and Czyz, J. 2002. Fusion of Intramodal and Multimodal Experts in Personal Identity Authentication Systems, *Proc. Cost 275 Workshop,* Rome, pp. 17–24.

Klein, S.A. 2005. Position paper on voting system threat modeling. Available at http://vote.nist.gov/threats/papers/threat-modelling.pdf.

Kleinberg, J.M. 1999. Authoritative Sources in a Hyperlinked Environment, *Journal of the ACM,* vol. 46, no. 5, pp. 604–632.

Kleiner, Dietl M. and Nebel, B. 2002. Towards a Life-Long Learning Soccer Agent, *Lecture Notes in Computer Science,* vol. 2752, G.A. Kaminka, P.U. Lima, and R. Rojas, eds. New York: Springer-Verlag.

Klir, G.J. and Yuan, B. 1995. *Fuzzy Sets and Fuzzy Logic: Theory and Applications.* Englewood Cliffs, NJ: Prentice Hall.

Klir, G.J. and Yuan, B. 1995. *Fuzzy Sets and Fuzzy Logic: Theory and Applications.* Englewood Cliffs, NJ: Prentice Hall.

Knuth, D.E. and Moore, R.W. 1975. An Analysis of Alpha-Beta Pruning, *Artificial Intelligence*, vol. 6, no. 4, pp. 293–326.

Kohl, N. and Stone, P. 2004. Machine Learning for Fast Quadrupedal Locomotion, *Proc. 19th Nat. Conf. Artif. Intell.,* pp. 611–616.

Kohonen, T. 1990. The Self-Organizing Map, *Proc. IEEE,* vol. 78, no. 9, pp. 1464–1480.

Kohonen, T., Kangas, J.A., and Laaksonen, J.T. 1990. Variants of Self-Organizing Maps, *IEEE Transactions on Neural Networks,* vol. 1, pp. 93–99.

Kohonen, T., Oja, E., Simula, O., Visa, A., and Kangas, J. 1996. Engineering Applications of the Self-Organizing Map, *Proc. IEEE*, vol. 84, pp. 1358–1384.

Kohonen, T. 1997. Self-Organizing Maps, *Springer Series in Information Science,* Berlin.

Koller, D. and Sahami, M. 1996. Toward Optimal Feature Selection, *Proceedings of the Thirteenth International Conference on Machine Learning,* Morgan-Kaufman Publishers, pp. 284–292, Bari, Italy.

Kolousek, G. 1997. *The System Architecture of an Integrated Medical Consultation System and Its Implementation Based on Fuzzy Technology.* Doctoral thesis, Technical University of Vienna, Austria.

Konar, A. 2000. *Artificial intelligence and soft computing: behavioral and cognitive modeling of the human,* CRC Press, Inc. Boca Raton, FL, USA.

Konar, A. and Pal, S. 1999. *Modeling Cognition with Fuzzy Neural Nets in Fuzzy Systems Theory: Techniques and Applications.* Leondes, C.T., ed., New York: Academic Press.

Kong, Hui, Li, Xuchun, Wang, Lei, Teoh, Earn Khwang, Wang, Jian-Gang, and Venkateswarlu, R. 2005. Generalized 2D Principal Component Analysis. Neural Networks 2005. IJCNN 05. *Proceedings, 2005 IEEE International Joint Conference*, vol. 1.

Kononenko. 2001. Machine Learning for Medical Diagnosis: History, State of the Art and Perspective, *Artif. Intell. Med.,* vol. 23, no. 1, pp. 89–109.

Kosko, B. 1992. *Neural Networks and Fuzzy Systems: A Dynamical Systems Approach to Machine Intelligence.* Prentice-Hall International Editions, Englewood Cliffs.

Kosko, B. 1996. *Fuzzy Engineering.* Upper Saddle River, NJ: Prentice-Hall.

Kosko, B. 1992. Neural Networks and Fuzzy Systems. Englewood Cliffs, NJ: Prentice-Hall.

Koutri, M., Daskalaki, S., and Avouris, N. 2002. Adaptive Interaction with Web Sites: an Overview of Methods and Techniques, *4th International Workshop on Computer Science and Information Technologies CSIT'2002*, Patras, Greece.

Kovacs, F. and Torok, M. 1998. An Improved Phonocardiographic Method for Fetal Heart Rate Monitoring, *Proceedings of the 20th Annual International Conference of the IEEE Engineering in Medicine and Biology Society, Hong Kong Sar*, vol. 20, no. 4, pp. 1719–1722.

Kovacs, F., Torok, M., and Habermajer, I. 2000. A Rule Based Phonocardiographic Method for Long Term Fetal Heart Rate Monitoring, *IEEE Transactions on Biomedical Engineering,* vol. 47, no. 1, pp. 124–130.

Kovalenko, O., Liu, D.R., and Javaherian, H. 2004. Neural Network Modeling and Adaptive Critic Control of Automotive Fuel-Injection Systems, *Proceedings of the 2004 IEEE International Symposium on Intelligent Control,* Taipei, Taiwan, pp. 368–373.

Koza, J.R. 1990. The Genetic Programming Paradigm: Genetically Breeding Populations of Computer Programs to Solve Problems, *Dynamic, Genetic, and Chaotic Programming,* B. Soucek and the IRIS Group, eds. New York: John Wiley, pp. 203–321.

Koza, J.R. 1992. *Genetic Programming: On the Programming of Computers by Means of Natural Selection.* Cambridge, MA: MIT Press.

Krause, P.J. 1998. Learning Probabilistic Networks, *Knowledge Eng. Rev.,* vol. 13, no. 4, pp. 321–351.

Kribeche, A., Benderbous, S., Tranquart, F., Kouame, D., and Pourcelot, L. 2004. Detection and Analysis of Fetal Movements by Ultrasonic Multi-Sensor Doppler (ACTIFOETUS), *Ultrasonics Symposium, 2004 IEEE,* vol. 2, pp. 1457–1460.

Krishna, K. and Narasimha Murty, M. 1999. Genetic K-Means Algorithm, *IEEE Trans. Syst., Man, Cybern. B, Cybern.,* vol. 29, no. 3, pp. 433–439.

Kroger, B., Schwenderling, P., and Vornberger, O. 1993. Parallel Genetic Packing on Transputers, *Parallel Genetic Algorithms: Theory and Applications: Theory Applications,* J. Stender, ed. Amsterdam, The Netherlands: IOS.

Matjaz K., Ciril, G. et.al. 1997. An Application of Machine Learning in the Diagnosis of Ischaemic Heart Disease, *AIME,* France: 461–464.

Kukolj, A. 2002. Design of Adaptive Takagi–Sugeno–Kang Fuzzy Models, *Appl. Soft Comput,* vol. 2, pp. 89–103.

Kumar, A. and Shukla, A. 2008. Improving Document Classification Using Anchor Text, *IKE,* pp. 90–95.

Chellapilla, Kumar, Larson, Kevin, Simard, Patrice, and Czerwinski, Mary. 2005. Computers Beat Humans at Single Character Recognition in Reading Based Human Interaction Proofs (HIPs), Proceedings of the Second Conference on Email and Anti-Spam (July 21–22).

Kumar, P. and Srivastava V.J. 2002. Selecting the Right Interestingness Measure for Association Patterns, *Proc. of SIGKDD.* pp. 32–41.

Kumar, S.C. and Li, H. 2004. Language Identification System for Multilingual Speech Recognition Systems, *Proceedings of the 9th International Conference Speech and Computer (SPECOM 2004),* St. Petersburg, Russia.

Kumar, S., Dharmapurikar, S., Crowley, P., Turner, J., and Yu, F. 2006. Algorithms to Accelerate Multiple Regular Expression Matching for Deep Packet Inspection, *SIGCOMM 2006,* Pisa, Italy.

Kumar, Santhosh C., Mohandas, V.P., and Haizhou, Li. 2005. Multilingual Speech Recognition: A Unified Approach, *Proceedings of Interspeech 2005, (Eurospeech: 9th European Conference on Speech Communication and Technology),* Lisboa.

Kumar, U.P. and Desai, U.B. 1996. Image Interpretation Using Bayesian Networks, *IEEE Trans. Pattern Anal. Machine Intell.,* vol. 18, pp. 74–78.

Kumari, V., Sundara Rao, C., Madhusudhana Rao, G., and Reddy, G.K.V.V.S. 2002. Expert System on Neonatal Birth Injuries, vol. 6, 2002-08-06, pp. 1999–2000.

Kuncheva, L. and Whitaker, C. 2003. Measures of Diversity in Classifier Ensembles and Their Relationship with the Ensemble Accuracy, *Mach. Learn.,* vol. 51, no. 2, pp. 181–207.

Kuncheva, L.I., Vezdek, J.C., and Duin, R.P.W. 2001. Decision Templates for Multiple Classifier Fusion: An Experimental Comparison, *Pattern Recognition,* vol. 34, pp. 299–314.

Kuo, T. and Hwang, S.Y. 1996. A Genetic Algorithm with Disruptive Selection, *IEEE Trans. Syst., Man, Cybern.,* vol. 26, no. 2, pp. 299–307.

Kwok, D.P., and Sheng, F. 1994. Genetic Algorithm and Simulated Annealing for Optimal Robot Arm PID Control, *Proc. IEEE Cond. Evol. Comp.,* 1994, vol. 2, pp. 707–713.

Kwok, T., Smith, R, Lozano, S., and Taniar, D. 2002. Parallel Fuzzy C-Means Clustering for Large Data Sets, in Burkhard Monien and Rainer Feldmann, eds., *EUROPARO2,* vol. 2400 of LNCS, pp. 365–374.

Kwok, T.Y. and Yeung, D.Y. 1996. Use of Bias Term in Projection Pursuit Learning Improves Approximation and Convergence Properties, *IEEE Trans. Neural Netw.*, vol. 7, no. 5, pp. 1168–1183.

Kwok, T.Y. and Yeung, D.Y. 1997. Objective Functions for Training New Hidden units in Constructive Neural Networks, *IEEE Trans. Neural Netw.*, vol. 8, no. 5, pp. 1131–1148.

Ky Van H. 1998. Hierarchical Radial Basis Function Networks, *Neural Networks Proceedings,* vol. 3, pp, 1893–1898.

Laarhoven, P.J.M.V. and Aarts, E.H.L. 1988. *Simulated Annealing.* Amsterdam, The Netherlands: Reidel.

Laden, A. and Keefe, D.H. 1989. The Representation of Pitch in a Neural Net Model for Chord Classification, *Computer Music Journal,* vol. 13, no. 4.

Lakhotia, A. and Gravely, J.M. 1995. Toward Experimental Evaluation of Subsystem Classification Recovery Techniques, *Proceedings of the Second Working Conference on Reverse Engineering,* pp. 262–269, Toronto.

Lakov, D.V. and.Vassileva, M.V. 2005. Decision Making Soft Computing Agent, *International Journal of Systems Science*, vol. 36, no. 14, pp. 921–930.

Larranaga, P., Kuijpers, C.M., Poza, M., and Murga, R.H. 1997. Decomposing Bayesian Networks: Triangulation of the Moral Graph with Genetic Algorithms, *Statistics Comput.,* vol. 7, no. 1, pp. 19–34.

Latombe, J.C. 1991. *Robot Motion Planning.* Boston: Kluwer Academic Publishers.

Laumanns, M., Thiele, L., and Zitzler, E. 2004. Running Time Analysis of Multi Objective Evolutionary Algorithms on Pseudo-Boolean Functions, *IEEE Trans. Evol. Comput.,* vol. 8, no. 2, pp. 170–182.

Laumond, J.P., Jacobs, P., Taix, M., and Murray, R.M. 1994. A Motion Planner for Nonholonomic Mobile Robots, *IEEE Trans. Robot. Automat.,* vol. 10, pp. 577–593.

Law, M.H.C., Topchy, A.P., and Jain, A.K. 2004. Multiobjective Data Clustering, *Proc. IEEE Conf. Comput. Vis. Pattern Recognit.*, Piscataway, NJ: IEEE Press, vol. 2, pp. 424–430.

Lawrence, R.R. 1989. *A Tutorial on Hidden Markov Models and Selected Applications in Speech Recognition,* Proceeding of The IEEE, vol. 77, no. 2, pp. 257– 285.

Lazkano, E. and Sierra, B. 2003, BAYES-NEAREST: a new Hybrid Classifier Combining Bayesian Network and Distance Based Algorithms, *Progress in Artificial Intelligence,* Springer, LNCS vol. 2902, pp. 171–183.

Le, H.T. and Loh, P.K.K. 2007. Unified Approach to Vulnerability Analysis of Web Applications, presented at *The International e-Conference on Computer Science (IeCCS 2007),* pp. 14–23 *in press.*

Leclerc, F. and Plamondon, R. 1994. Automatic Signature Verification: The State of the Art 1989–1993, *Int. J. Pattern Recogn. Artif. Intell.,* vol. 8, no. 3, pp. 643–660.

Lee, C.C. 1990. Fuzzy Logic in Control Systems: Fuzzy Logic Controller—Part I, *IEEE Trans. Syst., Man, Cybern.,* vol. 20, pp. 404–418.

Lee, C.C., Chung, P.C., Tsai, J.R., and Chang, C.I. 1999. Robust Radial Basis Function Neural Networks, *IEEE Trans. on Systems. Man, and Cybernetics Part B: Cybernetics,* vol. 29, no. 6, pp. 674–685.

Lee, C.H. and Huo, Q. 2000. On Adaptive Decision Rules and Decision Parameter Adaptation for Automatic Speech Recognition. *Proc. IEEE,* vol. 88, pp. 1241–1268.

Lee, Henry C. and Gaensslen, R.E., eds. *Advances in Fingerprint Technology.* New York: Elsevier.

Lee, L., Berger, T., and Aviczer, E. 1996. Reliable On-Line Human Signature Verification Systems, *IEEE Trans. Pattern Anal. Mach. Intell.,* vol. 18, no. 6, pp. 643–647.

Lee, M.R. 1998. Generating Fuzzy Rules by Genetic Method and Its Application. *International Journal on Artificial Intelligence Tools,* vol. 7, no. 4, pp. 399–413.

Lee, S.W., Kim, D.J., Park, K.H., and Bien, Z. 2004. Gabor Wavelet Neural Network-Based Facial Expression Recognition, presented at the World Multi-Conf. Syst. Cybern. Inf., Orlando, FL.

Lee, Sang G. et al. 1999. A Neuro-Fuzzy Classifier for Land Cover Classification, *1999 IEEE International Fuzzy Systems Conference Proceedings,* Seoul, Korea.

Lee, T.W. 1998. *Independent Component Analysis: Theory and Applications.* Dordrecht, The Netherlands: Kluwer Academic.

Lee, Y.M. and Moghavvemi, M. 2002. Remote Heart Rate Monitoring System Based on Phonocardiography, *Proceedings of Student Conference on Research and Development – IEEE,* Shah Alam Malaysia, pp. 27–30.

Leggett, J.J. and Williams, G. 1988. Verifying Identity via Keystroke Characteristics, *Int. J. Man-Mach. Stud.,* vol. 28, no. 1, pp. 67–76.

Lendasse, A., De Bodt, E., Wertz, V., and Verleysen, M. 2000. Non-Linear Financial Time Series Forecasting – Application to the Bel 20 Stock Market Index, *European Journal of Economic and Social Systems,* vol. 14, pp. 81–91.

Leszek, R. 2004. *Flexible Neuro-fuzzy Systems: Structures, Learning and Performance Evaluation,* Kluwer Academic publishers, New York.

Levy, D.N.L. and Newborn, M. 1982. *All about Chess and Computers.* Computer Science Press.

Lewis, A.D. 1999. When Is a Mechanical Control System Kinematic? *IEEE Conf. Decision Contr.,* Phoenix, AZ, pp. 1162–1167.

Li, A. and Liao, X. 2006. Robust Stability and Robust Periodicity of Delayed Recurrent Neural Networks with Noise Disturbance, *IEEE Trans. Circuits Syst. I, Reg. Papers,* vol. 53, no. 10, pp. 2265–2273.

Li, T.H., Lucasius, C.B., and Kateman, G. 1992. Optimization of Calibration Data with the Dynamic Genetic Algorithm, *Analytica Chimica Acta,* vol. 2768, pp. 123–134.

Li, W., Ng, W.K., Liu, Y., and Ong, K.L. 2007. Enhancing the Effectiveness of Clustering with Spectra Analysis, *IEEE Trans. Knowl. Data Eng.,* vol. 19, no. 7, pp. 887–902.

Li, Y.F. and Lan, C.C. 1989. Development of Fuzzy Algorithms for Servo Systems, *IEEE Control Syst. Mag.,* pp. 65–72.

Li, Yuelong, Li, Jinping, and Li Meng. 2006. Character Recognition Based on Hierarchical RBF Neural Networks, Proceedings of the Sixth International Conference on Intelligent Systems Design and Applications (ISDA06), vol. 01, pp. 127–132.

Li, Z.G., Soh, Y.C., and Wen, C.Y. 2002. Robust Stability of a Class of Hybrid Nonlinear Systems, *IEEE Trans. Autom. Control,* vol. 46, no. 6, pp. 897–903.

Likas, A., Vlassis, N., and Verbeek, J.J. 2003. The Global K-Means Clustering Algorithm, *Pattern Recognition,* vol. 36, no. 2, pp. 451–461.

Likas, A., Vlassis, N., and Verbeek, Jakob J. 2003. The Global K-Means Clustering Algorithm, *Pattern Recognition,* vol. 36, pp. 451–461.

Lim, S., Lee, K., Byeon, O., and Kim, T. 2001. Efficient Iris Recognition Through Improvement of Feature Vector and Classifier, *ETRI J,* vol. 23, no. 2, pp. 1–70.

LiMin F. 1994. Rule Generation from Neural Network, *IEEE Bans. on SMC,* vol. 24, no. 8, pp. 1114–1124.

Lin, C. 1994. *Neural Fuzzy Control Systems with Structure and Parameter Learning.* Singapore: World Scientific.

Lin, C.T. and Lee, C.S.G. 1996. *Neural Fuzzy Systems: A Neuro-Fuzzy Synergism to Intelligent Systems.* Englewood Cliffs, NJ: Prentice-Hall.

Lin, C.T., Juang, C.F., and Huang, J.C. 1999. Temperature Control of Rapid Thermal Processing System Using Adaptive Fuzzy Network, *Fuzzy Sets Syst.,* vol. 103, pp. 49–65.

Lin, C., Wu, H., Liu, T., Lee, M., Kuo, T., and Young, S. 1997. A Portable Monitor for Fetal Heart Rate and Uterine Contraction, *IEEE Engineering in Medicine and Biology,* vol. 16, no. 6, pp. 80–84.

Lin, C.F. and Wang, S.D. 2002. Fuzzy Support Vector Machines, *IEEE Trans. Neural Networks,* vol. 13, pp. 464–471.

Lin, C.T. and Lee, C.S.G. 1994. Reinforcement Structure/Parameter Learning for Neuro-network-based Fuzzy Logic Control Systems *IEEE Tran. on Fuzzy Systems,* vol. 2, no. 1, pp. 46–63.

Lin, Cheng-Jian, and Hong, Shang-Jin. 2007. The Design of Neuro-Fuzzy Networks Using Particle Swarm Optimization and Recursive Singular Value Decomposition, *Elsevier Journal of Neurocomputing*, vol. 71, pp. 297–310.

Lin, Cheng-Jian, Chung, I-Fang, and Chen, Cheng-Hung. 2007. An Entropy-Based Quantum Neuro-Fuzzy System for Classification Applications, *Neurocomputing,* vol. 70, pp. 2502–2516.

Lin, D.T. 1997. Computer-Access Authentication with Neural Network Based Keystroke Identity Verification, *Proc. IEEE Int. Conf. Neural Netw.*, vol. 1, pp. 174–178.

Lin, J.Y., Ke, H.R., Chien, B.C., and Yang, W.P. 2007. Designing a Classifier by a Layered Multi-Population Genetic Programming Approach, *Pattern Recognition, Elsevier*, vol. 40, no. 8, pp. 2211–2225.

Lin, S.C., Punch III, W.F., and Goodman, E.D. 1994. Coarse-Grain Genetic Algorithms: Categorization and new Approach, *Proc. 6th IEEE Parallel Distributed Processing,* New York: IEEE Press, pp. 28–37.

Lipmann, Richard P. 1989. Review of Neural Networks for Speech Recognition, *Neural Computation,* vol. 1, pp. 1–38, Massachusets Institute of Technology.

Lipton, R.J., DeMillo, R.A., and Sayward, F.G. 1978. Hints on Test Data Selection: Help for the Practicing Programmer, *IEEE Computer,* vol. 11, no. 4, pp. 34–41.

Lis, J. and Eiben, A.E. 1996. A Multi-Sexual Genetic Algorithm for Multiobjective Optimization, *Proc. IEEE 1996 Int. Conf. Evol. Comput.,* Nagoya, Japan, pp. 59–64.

Lisboa, J.G., Ifeachor, E.C., and Szczepaniak, P.S., eds. 2000. *Artificial Neural Networks in Biomedicine.* London, UK: Springer-Verlag.

Lisboa, P.J.G., Vellido, A., and Wong, H. 2000. Outstanding Issues for Clinical Decision Support with Neural Networks, *Proceedings of the International Conference on Artificial Neural Networks in Medicine and Biology,* Springer, pp. 63–71, Göteborg, Sweden.

Litt, A. and Echauz, R. 2002. Prediction of epileptic seizures. *Lancet,* vol. 1, pp. 22–30.

Liu, H. et al. 2005. Evolving Feature Selection, *IEEE Intelligent Systems*, pp. 64–76.

Liu, Y. and Yao, X. 1999. Ensemble Learning via Negative Correlation, *Neural Netw.,* vol. 12, pp. 1399–1404.

Liu, C.L., Kim, I..J., and Kim, J.H. 2001. Model-Based Stroke Extraction and Matching for Handwritten Chinese Character Recognition, *Pattern Recognition,* vol. 34, no. 12, pp. 2339–2352.

Liu, F., Quek, C., and Ng, G.S. 2007. A Novel Generic Hebbian Ordering Based Fuzzy Rule Base Reduction Approach to Mamdani Neuro-Fuzzy System, *Neural Comput.,* vol. 19, no. 6, pp. 1656–1680.

Liu, G., Li, T., Peng, Y., and Hou, X. 2005. The Ant Algorithm for Solving Robot Path Planning Problem, *Proceedings of 3rd International Conf on Information Technology and Applications,* vol. 2, IEEE Press, pp. 25–27.

Liu, J. and Shiffman, R. 1997., Operationalization of Clinical Practice Guidelines Using Fuzzy Logic, *Proceedings of AMIA Ann. Fall Symp.,* pp. 283–287.

Liu, J.G. 2000. Smoothing Filter-Based Intensity Modulation: A Spectral Preserve Image Fusion Technique for Improving Spatial Details, *Int. J. Remote Sens.,* vol. 21, no. 18, pp. 3461–3472.

Liu, J., Bai, Y., and Li, B. 2007. A New Approach to Forecast Crude Oil Price Based on Fuzzy Neural Network, *Fourth International Conference on Fuzzy Systems and Knowledge Discovery,* vol. 3, pp. 273–277.

Liu, Wei, W. Yunhong, Li, Stan Z., and Tan, T. 2004. Null Space Approach of Fisher Discriminant Analysis for Face Recognition, *ECCV workshop on Biometric Authentication,* pp. 32–44, Prague.

Liwicki, M. and Bunke, H. 2005. Handwriting Recognition of Whiteboard Notes, *ieeexplore* 29 Aug.–1 Sept. 2005.

Lofholm, P.W. 2000. The Psoriasis Treatment Ladder: A Clinical Overview for Pharmacists. *US Pharm,* vol. 25, no. 5, pp. 26–47.

Logan, B. 2000. Mel Frequency Cepstral Coefficients for Music Modeling, Proceedings of the First International Symposium on Music Information Retrieval, Plymouth, Massachusetts.

Loia, V. and Sessa, S. 2001. Soft Computing Agents: New Trends for Designing Autonomous Systems, *Studies in Fuzziness and Soft Computing,* Physica-Verlag, Springer, vol. 75, pp. 301–320.

Long, C.J. and Datta S. 1996. Wavelet Based Feature Extraction for Phoneme Recognition, *Proc. of 4th Int. Conf. of Spoken Language Processing,* pp 264–267.

Looney, Carl G. 1997. *Pattern Recognition Using Neural Networks.* New York: Oxford University Press.

Lopes da Silva, F.H., Blanes, W., Kalitzin, S., Parra, J., P, Gomez, S., and Velis, D.N. 2003. Dynamical Diseases of Brain Systems: Different Routes to Epileptic Seizures. *IEEE Trans Biomed Eng.* vol. 50, no. 5, pp. 540–548.

Lopes da Silva, F.H., Pijn, J.P., and Wadman, W.J. 1994. Dynamics of Local Neuronal Networks: Control Parameters and State Bifurcations in Epileptogenesis, *Prog. Brain Res.,* vol. 102, pp. 359–370.

Louis, S. and Rawlins, G. 1993. Pareto Optimality, GA-Easiness and Deception, *Proc. Int. Conf. Genetic Algorithms,* pp. 118–123.

Love, C. and Kinsne, W. 1991. *A Speech Recognition System Using a Neural Network Model for Vocal Shaping.* Department of Electrical and Computer Engineering, University of Manitoba Winnipeg, Manitoba, Canada R3T-2N2, p. 198.

Lu, Juwei, Plataniotis, K.N., and Venetsanopoulos, A.N. 2003. Face Recognition Using LDA Based Algorithms, *IEEE Transactions on Neural Networks,* vol. 14, no. 1, pp. 195–200.

Lu, Q. and Getoor, L. 2003. Link-Based Classification Using Labeled and Unlabeled Data, *Proceedings of the International Conference on Machine Learning.*

Lu, Q. and Getoor, L. 2003. Link-Based Classification, *Proceedings of The Twentieth International Conference on Machine Learning,* pp. 496–503, Washington DC, USA.

Lu, X., Wang, Y., and Jain, A.K. 2003. Combining Classifier for Face Recognition, *Proc. of IEEE 2003 Intern. Conf. on Multimedia and Expo.*, vol. 3. pp. 13–16.

Lu, Y., Sundararajan, N., and Saratchandran, P. 1998. Performance Evaluation of a Sequential Minimal Radial Basis Function (RBF) Neural Network Learning Algorithm, *IEEE Transctions on Neural Networks,* vol. 9, no. 2, pp. 308–318.

Lu, Juwei, Plataniotis, K.N., and Venetsanopoulos, A.N. 2005. Regularization Studies of Linear Discriminant Analysis in Small Sample Size Scenarios with Application to Face Recognition, *Pattern Recognition Letter,* vol. 26, no. 2, pp. 181–191.

Luersen, M.A., Riche, R.L., and Guyon, F. 2004. A Constrained, Globalized, and Bounded Neldermead Method for Engineering Optimization, Structural and Multidisciplinary Optimization, vol. 27, no. 1–2, pp. 43–54.

Lungarella, M., Lida, F., Bongard, J., and Pfeifer, R. 2007. *50 Years of artificial Intelligence, Essays Dedicated to the 50th Anniversary of Artificial Intelligence.* New York: Springer.

Luo, F., Khan, L., Bastani, F., Yen, I-Ling, and Zhou, J. 2004. A Dynamical Growing Self-Organizing Tree (DGSOT) for Hierarchical Clustering Gene Expression Profiles, *The Bioinformatics Journal,* UK: Oxford University Press, pp. 2605–2617.

Luo, F., Tang, K., and Khan, L. 2003. Hierarchical Clustering of Gene Microarray Expression Data, *Proc. of Workshop on Clustering High Dimensional Data and its Applications in Conjunction with the Third SIAM International Conference on Data Mining (SDM 2003),* pp. 6–17, San Francisco, CA.

Lynch, K.M., Shiroma, N., Arai, H., and Tanie, K. 2000. Collision-Free Trajectory Planning for a 3-DOF Robot with a Passive Joint, *Int. J. Robot. Res.,* vol. 19, no. 12, pp. 1171–1184.

Ma, Bin, Guan, Cuntai, Li, Haizhou, and Lee, Chin-Hui. 2002. Multilingual Speech Recognition with Language Identification, *International Conference on Spoken Language Processing (ICSLP),* Denver, Colorado.

Ma, L. 2003. Personal Identification Based on Iris Recognition, Ph.D dissertation, Inst. Automation, Chinese Academy of Sciences, Beijing, China.

Ma, L., Tan, T., Wang, Y., and Zhang, D. 2003. The Personal Identification Based on Iris Texture Analysis, *IEEE Transactions on pattern analysis and machine intelligence,* vol. 25, no. 12.

Maaly, I.A. and El-Obaid, M. 2006. Speech Recognition Using Artificial Neural Networks, *Proc. Second Information and Communication Technologies (ICTTA '06), Damascus, Syria,* pp. 1246–1247.

Mahfoud, S. 1995. Niching Methods for genetic algorithms, Ph.D. dissertation, Univ. of Illinois at Urbana-Champaign, Urbana, IL.

Mahmoudi, S.E., Ali, A.B., Forouzandeh, B., and Marandi, A.R. 2004. A New Genetic Method for Mobile Robot Navigation, *10th IEEE International Conference on Methods and Models in Automation and Robotics*, Miedzyzdroje, Poland.

Mak, M.W., Allen, W.G., and Sexton, G.G. 1993. Speaker Identification Using Radial Basis Functions, IEEE 3rd international conference on neural networks, ieeexplore, pp. 138–142, San Francisco, California.

Makrehchi M., Basir O.A., Kamel M. 2003. Generation of Fuzzy Membership Function Using Information Theory Measures and Genetic Algorithm. *Proceedings of the Fuzzy Sets and Systems—IFSA 2003,* Springer Lecture Notes in Computer Science, T. Bilgiç, B. D. Baets, and O. Kaynak, eds., vol. 2715. Springer, 2003, pp. 603–610, Istanbul, Turkey.

Makrehchi, O.B.M. and Kamel, M. 2003. Generation of Fuzzy Membership Function Using Information Theory Measures and Genetic Algorithm, *IEEE Transactions on Fuzzy Systems,* pp. 603–610.

Mallat, S.G. 1989. A Theory for Multiresolution Signal Decomposition: The Wavelet Representation, *IEEE Trans. Pattern Anal. Machine Intell.,* vol. PAMI-11, pp. 674–693.

Mallat, Stephane G. 1989. A Theory for Multiresolution Signal Decomposition: The Wavelet Representation, *674 IEEE Transactions On Pattern Analysis and Machine Intelligence.* Vol. II, No. 7.

Maltoni, A., Maio, D., Jain, Anil K., and Prabhakar, Salil 2003. *Handbook of Fingerprint Recognition,* New York: Springer-Verlag.

Manikas, Theodore W., Ashenayi, K., and Wainwright, Roger L. 2007. Genetic Algorithms for Autonomous Robot Navigation, *IEEE Instrumentation and Measurement Magazine.*

Mansfield, A.J. and Wayman, J.L. 2002. *Best Practices in Testing and Reporting Performance of Biometric Devices.* Centre Math.Sci. Comput., Nat. Phys. Laboratory, UK, Tech. Rep. NPL CMSC 14/02.

Manyika, J. and Durrant-Whyte, H. 1994. *Data Fusion and Sensor Management: A Decentralized Information-Theoretic Approach.* New York: Ellis Horwood.

Marill, T. and Green, D. 1963. On the Effectiveness of Receptors in Recognition Systems, *IEEE Trans. Inf. Theory,* vol. IT-9, no. 1, pp. 1–17.

Marolt, M. 2004. A Connectionist Approach to Automatic Transcription of Polyphonic Piano Music, *IEEE Trans on Multimedia,* vol. 6, pp. 439–449.

Martin and Przybocki, M. 2001. Speaker Recognition in Multi-Speaker Environment, *Proc. 7th Euro. Conf. Speech Communication and Technology (Eurospeech 2001),* Aalborg, Denmark, pp. 780–790.

Martin, R. and Cox, R.V. 1999. New Speech Enhancement Techniques for Low Bit Rate Speech Coding, *Proc. 1999 IEEE Workshop on Speech Coding,* Porvoo, Finland, pp. 165–167.

Martinez R., J.C., Lopez V., J.de.J., and Luna Rosas, F.J. 1999. A Low Cost System for Signature Recognition, *IEEE,* pp. 101–104.

Martinez, A.M. and Kak, A.C. 2001. PCA versus LDA, *IEEE Trans. on pattern Analysis and Machine Intelligence,* vol. 23, no. 2, pp. 228–233.

Maruyama, T. 2006. Real-Time K-Means Clustering for Color Images on Reconfigurable Hardware, *International Conference on Pattern Recognition,* vol. 2, pp. 816–819, Hong Kong.

Masek, L. 2003. *Recognition of Human Iris Patterns for Biometric Identification.* Bachelor of Engineering thesis, School of Computer Science and Software Engineering, The University of Western Australia.

Mataric, M. 1998. Coordination and Learning in Multirobot Systems, *IEEE Intelligent Systems,* vol. 13, no. 2, pp. 6–8.

Matlin, Margaret W. 1996. *Cognition.* Hault Sounders, printed and circulated by Prism books, India.

Matsugu, M., Mori, K., Mitari, Y., and Keneda, Y., 2003. Facial Expression Recognition Combined with Robust Face Detection in A Convolutional Neural Network, *Proc. of IJCNN'03,* pp. 2243–2246.

Matsuyama, Y. 1996. Harmonic Competition: A Self-Organizing Multiple Criteria Optimization, *IEEE Trans. Neural Netw.,* vol. 7, no. 3, pp. 652–668.

Matyas, J. 1965. Random optimization, *Automat. Rem. Control,* vol. 26, pp. 243–253.

Matyas, S.M. and Stapleton, J. 2000. A Biometric Standard for Information Management and Security. Computers & Security, vol. 19, no. 2, pp. 428–441.

Maulik, U. and Bandyopadhyay, S. 2003. Fuzzy Partitioning Using A Realcoded Variable-Length Genetic Algorithm for Pixel Classification, *IEEE Transactions Geoscience and Remote Sensing,* vol. 41, no. 5, pp. 1075–1081.

McClelland, J.L., Rumelhart, D.E., and Hinton, G.E. 1986. The Appeal of Parallel Distributed Processing, *Parallel Distributed Processing: Explorations in the Microstructure of Cognition, vol. 1: Foundations.* Cambridge: MIT Press.

McClintock, S., Lunney, T., and Hashim, A. 1998. A Genetic Algorithm Environment for Star Pattern Recognition, *J. Intell. Fuzzy Syst.,* vol. 6, pp. 3–16.

Tamara ML. 2008 .Women wrong on heart disease: survey. Available via http://news.smh.com.au/national/women-wrong-on-heart -disease-survey-20080613-2pqt.html.

Meesad, P. and Yen, G.G. 2003. Combined Numerical and Linguistic Knowledge Representation and Its Application to Medical Diagnosis, *IEEE Transactions on Systems, Man, and Cybernatics* vol. 33, no. 2, pp. 1083–4427.

Melanie, M. 1998. *An Introduction to Genetic Algorithms,* the MIT Press.

Melin, P. and Castillo, O. 2005. *Hybrid Intelligent Systems for Pattern Recognition Using Soft Computing,* Springer.

Mendis, D.S.K., Karunananda, A.S., Samarathunga, U., and Ratnayaka, U. 2007. Tacit Knowledge Modeling in Intelligent Hybrid Systems, *Second International Conference on Industrial and Information Systems,* Peradeniya 9-11 Aug. 2007 pp. 279–284.

Mendis, D.S.K., Karunananda, A.S., and Samarathunga, U. 2004. Multi-Techniques Integrated Tacit Knowledge Election System, *International Journal of Information Technology,* vol. 9, pp. 265–271.

Meng, A. and Xu, C. 2006. Iris Recognition Algorithms Based on Gabor Wavelet Transforms, *Proceedings of the IEEE. Proceedings of the 2006 IEEE International Conference on Mechatronics and Automation,* pp. 1785–1789, Luoyang, Henan.

Merck and Co. 2008. Survey Helps Identify Strategies to Address Diabetes Care and Global Economic and Social Impact of Type 2 Diabetes, *Medical News Today.* Available via http://www.medicalnewstoday.com/articles/120658.php. Meric, F., Elmer, V.B., et.al. 2002. Breast cancer on the World Wide Web: cross sectional survey of quality of information and popularity of websites. *BMJ,* vol. 324, pp. 577–581.

Michel, N. 1999. Recent Trends in the Stability Analysis of Hybrid Dynamical Systems, *IEEE Trans. Circuits Syst. I, Fundam. Theory Appl.,* vol. 46, no. 1, pp. 120–134.

Miine, R. Fault Diagnosis and Expert Systems, *The 6th International Workshop on Expert Systems and Their Applications,* Avington, France.

Minsky, M. 1988. *The Society of Mind.* New York: Simon and Schuster.

Miosso, C.J. and Bauchspiess, A. 2001. Fuzzy Inference System Applied to Edge Detection in Digital Images, *Proceedings of the V Brazilian Conf. on Neural Networks,* Rio de Janeiro, Brazil, pp. 481–486.

Mitchell, M. 1996. *An Introduction to Genetic Algorithms.* Cambridge, MA: MIT Press.

Mitchell, T. 1997. *Machine Learning.* New York: McGraw Hill.

Mitra, A.K., Shukla, A., and Zadgaonkar, A.S. 2007. System Simulation and Comparative Analysis of Foetal Heart Sound De-Noising Techniques for Advanced Phonocardiography, *Int. J. Biomedical Engineering and Technology,* vol. 1, no. 1, pp. 73–85.

Mitra, S. and Hayashi, Y. 2000. Neuro-Fuzzy Rule Generation: Survey in Soft Computing Framework, *IEEE Trans.NeuralNetw.*, vol. 11, no. 3, pp. 748–768.

Mitra, S. and Hayashi, Y. 2006. Bioinformatics with Soft Computing, *IEEE Systems, Man and Cybernetics – Part C: Applications and Reviews,* vol. 36, no. 5, pp. 616–635.

Mitra, S. and Pal, S.K. 1995. Fuzzy Multi-Layer Perceptron, Inferencing and Rule Generation, *IEEE Trans. on NN,* vol. 6, no. 1, pp. 51–62.

Mitra, S., Pal, S.K., and Mitra, P. 2002. Data Mining in Soft Computing Framework: A Survey, *IEEE Transactions on Neural Networks*, vol. 13, no. 1, pp. 3–14.

Mittra, A.K. 2004. Functional analysis of sensors used in cardiotocograph: a trans abdominal fetal heart rate and uterine contraction monitoring machine, *Proceedings of National Conference – NCST – 2004,* Sensor Technologies Gwalior, India, pp. 28–31.

Mittra, A.K., Shukla, A., and Zadgaonkar, A.S. 2005. Improvisation in Technique for Trans Abdominal Monitoring of Fetal Heart Rate and Uterus Contraction, *Proceedings of National Conference – BIOCON – 2005,* Bharati Vidyapeeth Deemed University, Pune, India, pp. 25–28.

Mittra, A.K., Shukla, A., and Zadgaonkar, A.S. 2006. Development of Non-Invasive Portable Fetal Heart Sound Monitoring Machine: an Experimental Approach, *The Journal of Lab Experiments,* vol. 6, No. 2, pp. 104–110.

Mittra, A.K., Shukla, A., and Zadgaonkar, A.S. 2006b. Design and Development of PC-Based Fetal Heart Sound Monitoring System, *The Indian Journal of Information Science and Technology,* vol. 1, no. 2, pp. 1–8.

Mizumoto, M. and Shi, Y. 1997. A New Approach of Neurofuzzy Learning Algorithm. In Ruan, D., ed., *Intelligent Hybrid Systems, Fuzzy Logic, Neural Networks, Genetic Algorithms,* Kluwer Academic Publishers, pp. 10–129.

Mizumoto, M. 1989. Pictorial Representations of Fuzzy Connectives—Part I: Cases of T-Norms, T-Conorms and Averaging Operators, *Fuzzy Sets Syst.,* vol. 31, pp. 217–242. US.

Mohanty, S. and Bhattacharya, S. 2008. Recognition of Voice Signals for Oriya Language Using Wavelet Neural Network, *ACM International Journal of Expert Systems with Applications,* vol. 34, no. 3, pp. 2130–2147.

Momoh, J.A., Dias, L.G., and Laird, D.N. 1997. An Implementation of a Hybrid Intelligent Tool for Distribution System Fault Diagnosis, *IEEE Trans. Power Delivery*, vol. 12, pp. 1035–1040.

Monrose, F. and Rubin, A. 2000. Keystroke Dynamics as a Biometric for Authentication, *Future Generation Comput. Syst. (FGCS),* vol. 16, no. 4, pp. 351–359.

Mozaffari, S., Faez, K., and Faradji, F. 2006. One Dimensional Fractal Coder for Online Signature Recognition, *Proceedings of the 18th International Conference on Pattern Recognition,* vol. 2, pp. 857–860, Hong Kong.

Mozer, M.C. 1994. Neural Network Music Composition by Prediction: Exploring the Benefits of Psychoacoustic Constraints and Multiscale Processing. *Connection Science,* vol. 6, no. 2–3, pp. 247–280.

Mrusco, A. and Chaurasia, V. 2004. Fetal heart rate detection and monitoring techniques: a comparative analysis and literature review, *Proceedings of National Conference – NCST – 2004,* Sensor Technologies Gwalior, India, pp. 75–84.

Mu, Zhichun et al. 2004. Shape and Structural Feature Based Ear Recognition, *Proceedings of International Conference on Advances in Biometric Person Authentication, Springer Lecture Notes in Computer Science,* Guangzhou, pp. 663–670.

Mukherjee, S., Osuna, E., and Girosi, F. 1997. Nonlinear Prediction of Chaotic Time Series Using Support Vector Machines, *NNSP97: Neural Networks for Signal Processing VII: Proc. of the IEEE Signal Processing Society Workshop,* pp. 511–520.

Murakami, M. and Honda, N. 2007. A Study on the Modeling Ability of the IDS Method: A Soft Computing Technique Using Pattern-Based Information Processing, *Int. J. Approx. Reason.*, vol. 45, no. 3, pp. 470–487.

Murray, J.J., Cox, C.J., Lendaris, G.G., and Saeks, R. 2002. Adaptive Dynamic Programming, *IEEE Trans. Syst., Man, Cybern. C: Appl. Rev.,* vol. 32, no. 2, pp. 140–153.

Murtagh, F. 2000. Clustering Massive Data Sets, *Handbook of Massive Data Sets*. Norwell, MA: Kluwer.
Na, M.G. 1999. Application of a Genetic Neuro-Fuzzy Logic to Departure from Nucleate Boiling Protection Limit Estimation, *Nucl. Tech.,* vol. 128, pp. 327–340.
Na, M.G. 1998. Design of a Genetic Fuzzy Controller for the Nuclear Steam Generator Water Level Control, *IEEE Trans. Nucl. Sci.,* vol. 45, no. 4, pp. 2261–2271.
Na, Yong-Kyun and Oh, Se-Young 2003. *Hybrid Control for Autonomous Mobile Robot Navigation Using Neural Network Based Behavior Modules and Environment Classification.* The Netherlands: Kluwer Academic Publishers.
Nadler, M. and Smith, E.P. 1993. *Pattern Recognition Engineering*. New York: Wiley.
Neapolitan, R.E. 1990. *Probabilistic Reasoning in Expert Systems: Theory and Algorithms*. New York: Wiley.
Nefian, A.V. and Hayes III, M.H. 1999. Face Recognition Using an Embedded-HMM, *Proceedings of IEEE International Conference on Audio and Video-based Biometric Person Authentication,* Washington, DC, USA, pp. 19–21.
Negnevitsky, M. 2002. *Artificial Intelligence:A Guide to Intelligent Systems,* 1st ed. England: Pearson Education, pp. 266–271.
Nehmzow, U. and Owen, C. 2000. Robot Navigation in the Real World: Experiments with Manchester's Forty Two in Unmodified Large Environments, *Robotics and Autonomous Systems,* vol. 33, Elsevier Science, pp. 223–242.
Nelson, W. and Kishon, E. 1991. Use of Dynamic Features for Signature Verification, *Proc. IEEE Int. Conf. Syst. Man, Cybern.,* vol. 1, pp. 201–205, Charlottesville.
Newth, D., Kirley, M., and Green, D.G. 2000. An Investigation of the Use of Local Search in NP-Hard Problems, *Proc. 26th Ann. Conf. IEEE Industrial Electronics Soc.*, vol. 4, Nagoya, Japan, pp. 2710–2715.
Ng, Kim C., and Trivedi, Mohan M. 1998. A Neuro-Fuzzy Controller for Mobile Robot Navigation and Multirobot Convoying. vol. 28, no. 6, pp. 829–840.
Nguyen, D.H. and Widrow, B. 1990. Neural Networks for Self-Learning Control Systems, *IEEE Control Syst. Mag.,* pp. 18–23.
Nigam, V. and Priemer, R. 2006a. Fuzzy Logic Based Variable Step Size for Blind Delayed Source Separation, *Journal of Fuzzy Sets and Systems*, vol. 157, no. 13, pp. 1851–1863.
Nigam, V. and Priemer, R. 2006b. A Dynamic Method to Estimate the Time Split Between the A2 and P2 Components of the S2 Heart Sound, *Journal of Physiological Measurement*, vol. 27, no. 1, pp. 553–567.
Nikola, K.K. 1998, *Foundations of Neural Networks, Fuzzy Systems, and Knowledge Engineering,* The MIT Press, Cambridge, Massachusetts, London, UK.
Nishi, T., Inoue, T., and Hattori, Y. 2000. Development of a Decentralized Supply Chain Optimization System, *Proc. Int. Symp. Design, Operation and Control of Next Generation Chemical Plants,* Kyoto, Japan, pp. 141–151.
Noak, V.V. 2000. *Discovering the World with Fuzzy Logic.* Springer-Verlag, pp. 3–50, Heidelberg, Germany.
Noraini, A.J., Raveendran, P., and Selvanathan, N. 2005. A Comparative Analysis of Feature Extraction Methods for Face Recognition System Sensors, *The International Conference on new Techniques in Pharmaceutical and Biomedical Research,* Asian Conference on 2005.
Norman, D.A., Ortony, A., and Russell, D.M. 2003. Affect and Machine Design: Lessons for the Development of Autonomous Machines, *IBM Systems Journal*, vol. 42, no. 1, pp. 38–44.
Novovicova, J., Pudil, P., and Kitttler, J. 1996. Divergence Based Feature Selection for Multimodal Class Densities, *IEEE Trans. Pattern Anal. Mach. Intell.*, vol. 18, no. 2, pp. 218–223.
Nussey, S. and Whitehead, S. 2001. *The Thyroid Gland in Endocrinology: An Integrated Approach.* BIOS Scientific Publishers Ltd, Oxford, UK.
O'Dunlaing, A. 1987. Motion Planning with Inertial Constraints, *Algorithmica,* vol. 2, no. 4, pp. 431–475.
Obaidat, M. and Sadoun, B. 1997. Verification of Computer Users Using Keystroke Dynamics, *IEEE Trans. Syst., Man Cybern.—Part B: Cybern.,* vol. 27, no. 2, pp. 261–269.
Odusanya, A.A., Odetayo, M.O., Petrovic, D., and Naguib, R.N.G. 2001. The Use of Evolutionary and Fuzzy Models in Oncological Prognosis. In: John, R., and Birkenhead, R. eds. *Developments in Soft Computing.* Springer-Verlag, pp. 207–215.
Oh, Byung-Joo 2005. Face Recognition by Using Neural Network Classifiers Based on PCA and LDA. *Systems, man & Cybernetics,* IEEE international conference.
Oh, K.W. and Bandler, W. 1987. Properties of Fuzzy Implication Operators. *Internat. J. Approximate Reasoning,* vol. 1, no. 3, pp. 273–286.
Oliveira, L., Sabourin, R., Bortolozzi, F., and Suen, C. 2003. Feature Election for Ensembles: A Hierarchical Multi-Objective Genetic Algorithm Approach, *Proc. 7th Int. Conf. Anal. Recognit.*, pp. 676–680.

Oosterom, Babuska, R., and Verbruggen, H.B. 2002. Soft Computing Applications in Aircraft Sensor Management and Flight Control Law Reconfiguration, *IEEE Trans. Syst., Man, Cybern. C: Appl. Rev.,* vol. 32, no. 2, pp. 125–139.

Opitz, D.W. and Shavlik, J.W. 1996. Actively Searching for an Effective Neural Network Ensemble, *Connection Sci.,* vol. 8, no. 3–4, pp. 337–353.

Orr, M.J.L. 1995. Regularisation in the Selection of Radial Basis Function Centres. *Neural Computation,* vol. 7, no. 3, pp. 606–623.

Otsu, N. 1978. A Threshold Selection Method from Gray-Scale Histogram, *IEEE Transaction Syst., Man, Cybern.,* vol. 8, pp. 62–66.

Oudeyer, P.Y. and Kaplan, F. 2004. Intelligent Adaptive Curiosity: A Source of Self-Development, presented at the *4th Epigenetic Robot. Workshop,* Genoa, Italy.

Ovaska, S.J., Dote, Y., Furuhashi, T., Kamiya, A., and VanLandingham, H.F. 1999. Fusion of Soft Computing and Hard Computing Techniques: A Review of Applications, *h c. JEEE Int. Conf. System, Man and Cybernetics,* vol. 1, pp. 370–375.

Ozcan, N. and Arik, S. 2006. Global Robust Stability Analysis of Neural Networks with Multiple Time Delays, *IEEE Trans. Circuits Syst. I, Reg. Papers,* vol. 35, no. 1, pp. 166–176.

Ozdzynski, P., Lin, A., Liljeholm, M., and Beatty, J. 2002. A Parallel General Implementation of Kohonens Self-Organizing Map Algorithm: Performance and Scalability, *Neurocomputing,* vols. 44–46, pp. 567–571.

Pagac, D., Nebot, E.M. and Durrant, W.H. 1998. An Evidential Approach to Map Building for Autonomous Robots, *IEEE Trans. On Robotics and Automation,* vol. 14, no. 2, pp. 623–629. PAHO/WHO (Pan American Health Organization/World Health Organization) [internet] Diabetes Increasing along U.S.-Mexico Border. 2007. Available via http://www.fep.paho.org/eng/TechnicalCooperation/ Diabetes/ SurveyResults/tabid/318/language/en-US/Default.aspx

Paiva, R.P. and Dourado, A. 2004. Interpretability and Learning in Neuro-Fuzzy Systems, *Fuzzy Sets and Systems,* vol. 147, pp. 17–38.

Pal, S.K. and Dutta Majumder, D.K. 1986. *Fuzzy Mathematical Approaches to Pattern Recognition.* New York: John Wiley (Halstead).

Pan, J., DeSouza, G.N., and Kak, A.C. 1998. FuzzyShell: A Large-Scale Expert System Shell Using Fuzzy Logic for Uncertainty Reasoning, *IEEE Transactions on Fuzzy Systems,* vol. 6, no. 4, pp. 563–581.

Pandey, A., Shukla, A., and Ray, L. 2009. Uptake and Recovery of Lead by Agarose Gel Polymers, *American Journal of Biochemistry and Biotechnology,* vol. 5, no. 1, pp. 14–20.

Papageorgiou, E.I. and Groumpos, P.P. 2005. A *New Hybrid Learning Algorithm for Fuzzy Cognitive Maps Learning.* Elsevier: Applied Soft Computing.

Papageorgiou, E.I., Stylios, C.D., and Groumpos, P.P., 2004. Active Hebbian Learning Algorithm to Train Fuzzy Cognitive Maps, *Int. J. Approx. Reason.,* vol. 37, no. 3, pp. 219–245.

Papakostas, G.A., Karras, D.A., Mertzios, B.G., and Boutalis, Y.S. 2007. An Efficient Feature Extraction Methodology for Computer Vision Applications Using Wavelet Compressed Zernike Moments, *ACM International Journal of Information Sciences,* vol. 177, no. 13.

Papaloukas, A., Fotiadis, D.I., Likas, A., and Michalis, L.K. 2003. Automated Methods for Ischemia Detection in Long Duration ECGs, *Cardiovascular Reviews and Reports,* vol. 24, no. 6, pp. 313–320.

Papez, J.W. 1995. A Proposed Mechanism of Emotion, *Journal of Neuropsychiatry and Clinical Neurosciences,* vol. 7, no. 1, pp. 103–112.

Paplinski, P. 1999. Basic Structures and Properties of Artificial Neural Networks, Lecture notes: CSE5312, Monash University, Australia.

Park, G.T. and Bien, Z. 2000. Neural network-Based Fuzzy Observer with Application to Facial Analysis, *Pattern Recognition Letters,* vol. 21, pp. 93–105.

Park, H.S. and Lee, S.W. 1996. Off-Line Recognition of Large-Set Handwritten Characters with Multiple Hidden Markov Models, *Pattern Recognition,* vol. 29, no. 2, pp. 231–244.

Park, J., Govindaraju, V., and Srihari, S.N. 2000. OCR in a Hierarchical Feature Space, *IEEE Trans. Pattern Analysis and Machine Intelligence,* vol. 22, no. 4, pp. 400–407.

Patnaik, S., Jain, L.C., Tzafestas, S.G., Resconi, G., and Konar, A. 2005. *Innovations in Robot Mobility and Control,* Springer.

Peacock, A.M., Renshaw, D., and Hannah, J. 1999. Fuzzy Data Fusion Approach for Image Processing, *Electron. Lett.,* vol. 35, no. 18, pp. 1527–1529.

Peacock, X. Ke and Wilkerson, M. 2004. Typing Patterns: A Key to User Identification, *IEEE Secur. Privacy Mag.,* vol. 2, no. 5, pp. 40–47.

Pearl, J. 1988. *Probabilistic Reasoning in Intelligent Systems—Networks of Plausible Inference.* Morgan Kaufmann.

Pearlmutter, A.A. 1995. Gradient Calculations for Dynamic Recurrent Neural Networks: A Survey. *IEEE Transactions on Neural Network,* vol. 6, no. 5, pp. 1212–1228.

Pedro, M., Aguiar, Q., Jasinschi, R., M.F. Moura, Jose, and Pluempitiwiriyawej, C. 2004. Content-Based Image Sequence Representation, Todd Reed, ed, *Digital Image Sequence Processing: Compression and Analysis,* Boca Raton, FL: CRC Press Handbook, Chapter 2, pp. 5–72.

Pei, M., Goodman, E.D., and Punch, W.F. 1997. Pattern Discovery from Data Using Genetic Algorithms, *Proc. 1st Pacific-Asia Conf. Knowledge Discovery and Data Mining (PAKDD-97).* Singapore.

Pena-Reyes, C.A. and Sipper, M. 1999. Designing Breast Cancer Diagnostic System via a Hybrid Fuzzy-Genetic Methodology. *IEEE International Fuzzy Systems Conference Proceedings,* IEEE Press, vol. 1, pp. 135–139, Piscataway, NJ.

Peng, Y. and Reggia, J.A. 1987. A Probabilistic Causal Model for Diagnostic Problem Solving, Part One, *IEEE Trans. Syst., Man, Cybern.,* vol. SMC-17, pp. 146–162.

Pennebaker, W.B. and Mitchell, J.L. 1993. *JPEG Still Image Data Compression Standard.* New York: Van Nostrand Reinhold.

Pérez-Ilzarbe, M.J. 1998. Convergence Analysis of a Discrete-Time Recurrent Neural Network to Perform Quadratic Real Optimization with Bound Constraints, *IEEE Trans. Neural Networks,* vol. 9, pp. 1344–1351.

Périaux, J., Sefrioui, M., Stoufflet, B., Mantel, B., and Laporte, E. 1995. Robust Genetic Algorithms for Optimization Problems in Aerodynamic Design, *Genetic Algorithms in Engineering and Computer Science,* G.Winter, J. Périaux, M. Galán, and P. Cuesta, eds. Chichester, England: Wiley, pp. 371–396.

Perrone, M.P. and Cooper, L.N. 1993. When Networks Disagree: Ensemble Method for Neural Networks, *Artificial Neural Networks for Speech and Vision,* R.J. Mammone, ed. New York: Chapman and Hall.

Pfeiffer, F. and Johanni, R. 1987. A Concept for Manipulator Trajectory Planning, *IEEE Trans. Robot. Automat.,* vol. 3, pp. 115–123.

Phatak, D.S. and Koren, I. 1995. Complete and Partial Fault Tolerance of Feedforward Neural Nets, *IEEE Trans. Neural Netw.*, vol. 6, no. 2, pp. 446–456.

Phelps, S. and Koksalan, M. 2003. An Interactive Evolutionary Metaheuristic for Multiobjective Combinatorial Optimization, *Management Science,* vol. 49, no. 12, pp. 1726–1738.

Phillips, P.J., Flynn, P.J., Scruggs, T., Bowyer, K.W., Chang, J., Hoffman, K., Marques, J., Min, J., and Worek, W. 2005. Overview of the Face Recognition Grand Challenge, *IEEE CVPR,* IEEE Computer Society Press, Los Alamitos, vol. 2, pp. 454–461.

Pinero, P. et al. 2004. Sleep Stage Classification Using Fuzzy Sets and Machine Learning Techniques, *Neurocomputing,* vol. 58–60, pp. 1137–1143.

Plamondon, R. and Srihari, S.N. 2000. On-Line and Off-Line Handwriting Recognition: A Comprehensive Survey, *IEEE Trans. Pattern Analysis and Machine Intelligence,* vol. 22, no. 1, pp. 63–84.

Plamondon, R., ed. 1994. *Progress in Automatic Signature Verification.* Singapore: World Scientific.

Plataniotis, K.N., Androutsos, D., and Venetsanopoulos, A.N. 1999. Adaptive Fuzzy Systems for Multichannel Signal Processing, *Proc. IEEE,* vol. 87, pp. 1601–1622.

Poggio, T. and Girosi, F. 1990. Networks for Approximation and Learning, *Proc. IEEE*, vol. 78, pp. 1481–1497.

Poli, R. and Cagnoni, S. 1997. Genetic Programming with User-Driven Selection: Experiments on the Evolution of Algorithms for Image Enhancement, *2nd Annual Conference on Genetic Programming,* pp. 269–277.

Poli, R. 1996. Genetic Programming for Image Analysis, *Proc. 1st Annu. Conf. Genetic Program.,* J.R. Kosa, D.E. Goldberg, D.B. Fogel, and R.L. Riolo, eds. pp. 363–368.

Polvere, M. and Nappi, M. 2000. Speed-Up in Fractal Image Coding: Comparison of Methods, *IEEE Trans. Image Process.,* vol. 9, pp. 1002–1009.

Poole, D., Macworth, A., and Goebel, R. 1998. *Computational Intelligence: a Logical Approach.* Oxford, UK: Oxford Press.

Porwik, P. 2007. The Compact Three Stage Method of the Signature Recognition, *Proceeding on 6th International conference on computer Information system and Industrial Management applications.*

Potts, J.C., Giddens, T.D., and Yadav, S.B. 1994. The Development and Evaluation of an Improved Genetic Algorithm Based on Migration and Artificial Selection, *IEEE Trans. Syst., Man, Cybern.,* vol. 24, pp. 73–86.

Pouget and Snyder, L. 2000. Computational Approaches to Sensorimotor Transformations, *Nature Neuroscience supplement,* vol. 3, pp. 1192–1198.

Powell, J.D. 1987. *Radial Basis Function Approximations to Polynomials, Numerical analysis.* White Plains, NY: Longman Publishing Group.

Pradhan, S.K., Parhi, D.R., and Panda, A.K. 2006. Navigation of Multiple Mobile Robots Using Rule-Based Neuro-Fuzzy Technique, *Proceedings of International Journal of Computational Intelligence,* vol. 3, no. 2, pp. 142–151.

Price, R.C., Willmore, P., Robert, W.J., and Zyga, K.J. 2000. Genetically Optimized Feedforward Neural Networks for Speaker Identification, *Fourth International Conference on Knowledge-Based Engineering System & Allied. Technologies,* Brighton, UK.

Psaltis, D., Sideris A., and Yamamura, A. 1988. A Multilayered Neural Network Controller, *IEEE Control Syst. Mag.,* vol. 8, pp. 17–21.

Purwin, O., D'Andrea, R., and Lee, J.-W. 2008. Theory and Implementation of Path Planning by Negotiation for Decentralized Agents, *Robotics and Autonomous Systems,* vol. 56, no. 5, pp. 422–436.

Qingshan, L., Hanqing, L., and Songde, M. 2003. A Survey: Subspace Analysis for Face Recognition, *ACTA Automatica Sinica,* vol. 29, no. 6, pp. 900–911.

Qiping, Y. and Wude, X. 2000. The Application of Artificial Intelligence (AI) in Power Transformer Fault Diagnosis, *CICED,* Shanghai, China, pp. 173–175.

Quammen, Cory. 2001. Evolutionary learning in mobile robot navigation, The ACM Student Magazine, vol. 8, no. 2, pp. 10–14.

Quinlan, J.R. 1979. *Discovering Rules from Large Collections of Examples: A Case Study in Expert Systems in the Microelectronic Age.* D. Michie, ed. Edinburgh, Scotland: University Press.

Rabiner, L. and Juang, Biing-Hwang. 1993. *Fundamentals of Speech Recognition.* Prentice Hall.

Rabunal, J.R. and Dorado, J. 2005. *Artificial Neural Networks in Real-Life Applications,* Idea Group Reference.

Rabuzin, Kornelije, Miroslav, Bace, and Sajko, Mario. 2006. E-Learning: Biometrics as a Security Factor, Proceedings of the International Multi-Conference on Computing in the Global Information Technology (ICCGI'06), IEEE, pp. 646–654.

Raidl, G.R. 1998. An Improved Genetic Algorithm for the Multiconstrained 0-1 Knapsack Problem, *Proc. IEEE Int. Conf. Evolutionary Computation*, Anchorage, AK, pp. 207–211.

Rajagopal, P. 2003. *The Basic Kak Neural Network with Complex Inputs.* Master of Science in Electrical Engineering Thesis, University of Madras.

Rajasekaran, J.S. and Vijalakshmi Pai, G.A. 2006. *Neural Networks, Fuzzy Logic, and Genetic Algorithms.* New Delhi: PHI.

Ram Chandran, K. and Vetterli, M. 1993. Best Wavelet Bases in a Rate-Distortion Sense, *IEEE Trans. On Image Processing,* vol. 2, no. 2, pp. 160–175.

Ramesh, V.E. and Murty, M.N. 1999. Offline Signature Verification Using Genetically Optimized Weighted Features, *Pattern Recognition,* vol. 32, no. 2, pp. 217–233.

Ramos, C. 2007. Ambient Intelligence—a State of the Art from Artificial Intelligence Perspective, *Proc. 13th Portuguese Conf. Artificial Intelligence Workshops,* LNAI 4874, Springer, pp. 285–295.

Rao, P. Shankar, and Aditya, J., Handwriting Recognition: Offline Approach.

Ratha, N., Karu, K., Chen, S., and Jain, A.K. 1996. A Real-Time Matching System for Large Fingerprint Databases, *IEEE Transactions on Pattern Analysis and Machine Intelligence,* vol. 18, no. 8, pp. 799–813.

Ratha, N.K. 2003. *Guide to Biometrics.* Springer, ISBN 0387400893.

Reynolds, D.A. 2002. An Overview of Automatic Speaker Recognition Technology. *Proc. Int. Conf. on Acoustics, Speech and Signal Processing (ICASSP 2002),* Orlando FL, pp. 4072–4075. Rheumatic heart disease – Demonstration projects [internet]. WHF (World Heart Federation). Available via HYPERLINK "http://www.world-heart-federation.org/what-we-do/demonstration%20-pr%20ojects/rheumatic-heart-disease-demonstration-projects/"http://www.world-heart-federation.org/what-we-do/demonstration-projects/rheumatic-heart-disease-demonstration-projects/

Rich, E. and Knight, K. 1991. *Artificial Intelligence.* New York: McGraw-Hill, pp. 29–98.

Richards, D. and Bush, P. 2003. Measuring, Formalizing and Modeling Tacit Knowledge, *IEEE/Web Intelligence Conference (WI-2003),* Bejing.

Richards, G., Rayward-Smith, V.J., Sönksen, P.H., Carey, S., and Weng, C. 2000. Data Mining for Indicators of Early Mortality in a Database of Clinical Records, *Artif. Intell. Med.,* vol. 22, no. 3, pp. 215–231.

Richardt, J., Karl, F., and Muller, C. 1998. Connections Between Fuzzy Theory, Simulated Annealing, and Convex Duality, Fuzzy Sets and Systems, vol. 96, pp. 307–334.

Robert, P. and Anthony, B. Predicting Credit Ratings with a GA-MLP Hybrid, *Artificial Neural Networks in Real Life Applications*, Idea Group Publishing, pp. 202–237.

Bertolami, Roman, Zimmermann, Matthias, and Bunke, Horst, 2006. Rejection Strategies for Offline Handwritten Text Line Recognition, ACM Portal, vol. 27, no. 16. pp. 2005–2012.

Rosen, B. 1996. Ensemble Learning Using Decorrelated Neural Networks, *Connection Sci.,* vol. 8, pp. 373–384.
Rosenblum, M., Yacoob, Y., and Davis, L.S. 1996. Human Expression Recognition from Motion Using a Radial Basis Function Network Architecture, *IEEE Transactions on Neural Network,* vol. 7, no. 5, pp. 1121–1138.
Rosenfeld, A. and Kak, A.C. 1982. *Digital Picture Processing.* San Diego: Academic Press.
Ross, A., Jain, A.K., and Qian, J.Z. 2001. Information Fusion in Biometrics, *Proc. 3rd International Conference on Audio- and Video-Based Biometric Person Authentication,* Halmstad, Sweden, vol. 2091, pp. 354–359.
Ross, A. and Jain, A.K. 2004. Multimodal biometrics: an Overview, *Proceedings of 12th European Signal Processing Conference,* pp. 1121–1224.
Ross, Arun and Jain, Anil 2003. Information Fusion in Biometrics, *Pattern Recognition Letters,* vol. 24, pp. 2115–2125.
Ross, A. and Jain, A. 2004. Multimodal Biometrics: An Overview, *Proc. Of 12th European Processing Conference,* pp. 1221–1224.
Rukhin, Andrew L. and Malioutov, I. 2001. Fusion of Biometric Algorithm in the Recognition Problem, *Pattern Recogition Letters,* pp. 299–314.
Rumelhart, D., Hinton, G., and Williams, R. 1986. Learning Representations by Backpropagating Errors, *Nature,* vol. 323, no. 9, pp. 318–362.
Russell, S. and Norvig, P. 2003. *Artificial Intelligence: A Modern Approach,* 2nd ed. Upper Saddle River, NJ: Pearson Education.
Rusu, P., Petriu, E.M., Whalen, T.E., Cornell, A., and Spoelder, H.J.W. 2003. Behavior-Based Neurofuzzy Controller for Mobile Robot Navigation, *IEEE Trans. Instrum. Measur,* vol. 52, no. 4, pp. 1335–1340.
Rutkowski, L. 2004. *Flexible Neuro-Fuzzy Systems. Structures, Learning and Performance Evaluation.* The Springer International Series in Engineering and Computer Science. (771). ISBN: 978-1-4020-8042-5.
Rungta, S., Tripathi, N., Verma, A.K., and Shukla, A. 2009. Designing and Optimizing of Codec H-263 for Mobile Applications, *International Journal of Computer Science and Network Security.*
Ryan, C. 1995. Niche and Species Formation in Genetic Algorithms, *Practical Handbook of Genetic Algorithms: Applications,* L. Chambers, ed. Boca Raton, FL: CRC Press, pp. 57–74.
Ryan, C., Collins, J.J., and Neill, M O. 1998. Grammatical Evolution: Evolving Programs for an Arbitrary Language, *Proceedings of the First European Workshop on Genetic Programming,* W. Banzhaf, R. Poli, M. Schoenauer, and T.C. Fogarty, eds. vol. 1391. Paris: Springer-Verlag, pp. 83–95.
S. Hayashi. 1991. Auto-Tuning Fuzzy PI Controller, *Proc. IFSA'91,* pp. 41–44.
Saastamoinen, K. and Luukka, P. 2003. Testing Continuous T-Norm Called Łukasiewicz Algebra with Different Means in Classification, *Proceedings of the FUZZ-IEEE Conference,* St Louis, USA.
Saastamoinen, K. 2004. On the Use of Generalized Mean with T-Norms and T-Conorms, *Proceedings of the IEEE Conference on Cybernetic and Intelligent Systems,* Singapore.
Saastamoinen, K., Kononen, V., and Luukka, P. 2002. A Classifier Based on the Fuzzy Similarity in the Łukasiewicz-Structure with Different Metrics, *Proceedings of the FUZZ-IEEE Conference,* Hawaii, USA.
Sachin, A. 2008. *Automatic Person Identification and Verification Using Online Handwriting.* Thesis International Institute of Information Technology, Hyderabad, India.
Saffiotti, A., Konolige, K., and Ruspini, H. 1995. A Multivalued Logic Approach to Integrating Planning and Control, *Artificial Intelligence,* vol. 76, pp. 481–526.
Saha, R.K. 1996. Track-to-Track Fusion with Dissimilar Sensors, *IEEE Transactions on Aerospace and Electronic Systems,* vol. 32, no. 3, pp. 1021–1029.
Saha, R.K. and Chang, K.C. 1998. An Efficient Algorithm for Ultisensory Track Fusion, *IEEE Transactions on Aerospace and Electronic Systems,* vol. 34, no. 1, pp. 200–210.
Saha, S. and Bandyopadhyay, S. 2008. Application of a New Symmetry Based Cluster Validity Index for Satellite Image Segmentation, *IEEE Geoscience and Remote Sensing Letters,* vol. 5, no. 2, pp. 166–170.
Sakoe and Chiba, S. 1978. Dynamic Programming Algorithm Optimization for Spoken Word Recognition, *IEEE Trans. Acoustics, Speech, and Signal Processing.*
Samuel, A.L. 1959. Some Studies in Machine Learning Using the Game of Checkers, IBM Journal, vol. 3, no. 3, pp. 210–299.
Sanchez-Reillo, R. 2000. Hand Geometry Pattern Recognition Through Gaussian Mixture Modeling, *15th International Conference on Pattern Recognition,* vol. 2, pp. 937–940.

Sanchez-Reillo, R., Sanchez-Avila, C., and Gonzalez-Marcos, A. 2000. Biometric Identification Through Hand Geometry Measurements, *IEEE Transactions on Pattern Analysis and Machine Intelligence,* vol. 22, no. 10, pp. 1168–1171.

Sanderson, C. and Paliwal, K.K. 2000. Adaptive Multimodal Person Verification System, *Proc. 1st IEEE Pacific-Rim Conf. On Multimedia,* Sydney, pp. 210–213.

Sanderson, C. and Paliwal, K.K. 2002. Information Fusion and Person Verification Using Speech & Face Information, *IDIAP, Martigny, Research Report,* pp. 02–33.

Sanderson, C. 2002. Information Fusion and Person Verification Using Speech and Face Information, *IDIAP Research report*, pp. 02–33.

Sandhu, P.S., Salaria, D.S., and Singh, H. 2008. A Comparative Analysis of Fuzzy, Neuro-Fuzzy and Fuzzy-GA Based Approaches for Software Reusability Evaluation, *Proceedings of World Academy of Science, Engineering and Technology*, vol. 29, ISSN 1307-6884.

Savasere, A., Omiecinski, and E., Navathe. 1998. Mining for Strong Negative Associations in a Large Database of Customer Transactions, *Proc. of ICDE*, pp. 494–502.

Savvides, M., Vijaya Kumar, B.V.K., and Khosla, P.K. 2004. Cancelable Biometrics Filters for Face Recognition, *Int. Conf. of Pattern Recognition,* vol. 3, pp. 922–925.

Sbarciog, M., Wyns, B., Ionescu, C., De Keyser, R., and Boullart, L. 2004. Neural Network Model for a Solar Plant, *Proceedings (413) Neural Networks and Computational Intelligence.*

Schaal, S., Ijspeert, A., and Billard, A. 2004. Computational Approaches to Motor Learning by Imitation, *The Neuroscience of Social Interaction,* vol. 1431, C.D. Frith and D. Wolpert, eds. Oxford University Press, pp. 199–218.

Schaffer, J.D. 1985. Multiple Objective Optimization with Vector Evaluated Genetic Algorithms, *Proc. Int. Conf. Genetic Algorithms and Their Applications,* Pittsburgh, PA, pp. 93–100.

Schapire, R.E. 1990. The Strength of Weak Learnability, *Machine Learning,* vol. 5, no. 2, pp. 197–227.

Schlierkamp-Voosen, D. and Mühlenbein, H. 1994. Strategy Adaptation by Competing Subpopulations, *Parallel Problem Solving from Nature 3,* Y. Davidor, H.-P. Schwefel, and R. Männer, eds. Berlin, Germany: Springer-Verlag, pp. 199–208.

Schlkopf, B. and Smola, A.J. 2001. *Learning with Kernels: Support Vector Machines Regularization, Optimization, and Beyond.* Cambridge, MA: MIT Press.

Schwartz, R.A., Janusz, C.A., and Janniger, C.K. 2006. Seborrheic Dermatitis: an Overview. *Am Fam Physician* vol. 74, no. 1, pp. 125–130.

Schwefel, H.P. 1981. *Numerical Optimization of Computer Models.* Chichester, England: Wiley.

Schwefel, H.P. 1995. *Evolution and Optimum Seeking.* John Wiley & Sons, Inc.

Sebastiani, F. 2002. Machine Learning in Automated Text Categorization, ACM Computing Surveys (CSUR), vol. 34, no. 1, pp. 1–47.

Seiffert, Udo 2004. Artificial neural networks on massively parallel computer hardware, *Neurocomputing*, vol. 57, pp. 135–150.

Selekwa, M.F., Dunlap, D.D., and Collins, E.G. 2005. Implementation of Multi-Valued Fuzzy Behavior Control for Robot Navigation in Cluttered Environments, Proceedings of IEEE International Conference on Robotics and Automation, Barcelona, Spain, pp. 3699–3706.

Selekwa, M.F. and Collins, E.G., Jr. 2003. A Centralized Fuzzy Behavior Control for Robot Navigation, *Proceedings of IEEE International Symposium on Intelligent Control,* Houston, Texas, pp. 602–607.

Senior, A. 2001. A Combination Fingerprint Classifier, *IEEE Transactions on Pattern Analysis and Machine Intelligence,* vol. 10, no. 23, pp. 1165–1174.

Seno, Sadakane, S., Baba, T., Shikama, Y., Koui, T., and Nakaya, Y. 2003. A Network Authentication System with Multi-Biometrics, *IEEE,* vol. 3, pp. 914–918.

Sequin, C.H. and Clay, R.D. 1990. Fault Tolerance in Artificial Neural Networks, *in Proc. Int. Joint Conf. Neural Netw. (IJCNN'90),* San Diego, CA, pp. 703–708.

Sethuram, J. 2005. Soft Computing Approach for Bond Rating Prediction, *Artificial Nural Networks in Real Life Applications,* Idea Group Publishing, pp. 202–219.

Setiono, R. 1996. Extracting Rules from Pruned Neural Networks for Breast Cancer Diagnosis, *Artif. Intell. Med.,* vol. 8, no. 1, pp. 37–51.

Setiono, R. 2000. Generating Concise and Accurate Classification Rules for Breast Cancer Diagnosis, *Artificial Intelligence in Medicine*, pp. 205–219. Setiono, R. 2000. Generating concise and accurate classification rules for breast cancer diagnosis. Artificial Intelligence in Medicine, vol. 18, no. 3, 205–219.

Shadbolt, N. 2003. Ambient Intelligence, *IEEE Intelligent Systems*, vol. 18, no. 4, pp. 2–3.

Shapiro, Linda G., and Stockman, George C. 2001. *Computer Vision.* Prentice Hall, Upper Saddle River, NJ.

Sharkey, D., ed. 1999. *Combining Artificial Neural Nets: Ensemble and Modular Multi-Net Systems.* London, UK: Springer-Verlag.

Sharma, J.R. and Mittra, A.K. 2005. Cardiotocograph Fetal Heart Rate and Uterine Contraction Monitoring Machine: a TQM Approach to Optimization, *Proceedings of International Conference – EISCO – 2005,* Kumar Guru College of Technology, Coimbatore, India, pp. 251–257.

SHARP (Safety & Health Assessment & Research for Prevention) [internet] A report on Occupational Skin Disorders, Washington State Department of Labor and Industries. 1998. Available via http://www.wa.gov/lni/sharp.

Sheik, S.S., Aggarwal, S.K., and Poddar, A. 2004. A Fast Pattern Matching Algorithm, *Journal of Chemical Information and Computer Sciences,* vol. 44, pp. 1251–1256.

Shields, Mike W. and Casey, Matthew C. 2008. A Theoretical Framework for Multiple Neural Network Systems, *Neurocomputing*, vol. 71, pp. 1462–1476.

Shiller, Z. and Lu, H.H. 1992. Computation of Path Constrained Time Optimal Motions with Dynamic Singularities, *ASME J. Dynamic Syst., Measurement, Contr.,* vol. 114, pp. 34–40.

Shiu, S. C.K., Li, Y., and Wang, X.Z. 2001. Using Fuzzy Integral to Model Case-Base Competence, *Proceedings of Soft Computing in Case-Based Reasoning Workshop in Conjunction with the ICCBR-02,* Vancouver, Canada, pp. 206–212.

Shi-zhuo, L. and Man-fu, Y. 2007. Machine Learning Problems Based on Data. *Journal of Tangshan Teachers College,* vol. 9, pp. 66–67.

Shukla, A., Tiwari, R., Ranjan, A., and Kala, R. 2009a. Multi Lingual Character Recognition Using Hierarchical Rule Based Classification and Artificial Neural Network, *Proceedings of the Sixth International Symposium on Neural Networks,* part II, Springer Verlag Lecture Notes in Computer Science, vol. 5552, pp. 821–830, Wukan, China.

Shukla, A., Tiwari, R., Meena, H.K., and Kala, R. 2009b. Speaker Identification using Wavelet Analysis and Artificial Neural Networks, *Journal of Acoustic Society of India.*

Shukla, A., Tiwari, R., and Kaur, P. 2009c. Diagnosis of Epilepsy Disorders Using Artificial Neural Networks, *Proceedings of the Sixth International Symposium on Neural Networks*, book chapter in Springer-Verlag book series on *Advances in Intelligent and Soft Computing,* ISNN 1615-3871 (Print) 1860-0794 (Online), vol. 56/2009, ISBN 978-3-642-01215-0, pp. 807–815.

Shukla, A., Tiwari, R., and Kaur, P. 2009d. Knowledge Based Approach for Diagnosis of Breast Cancer, *Proceedings IEEE International Advance Computing Conference (IACC),* March 6–7, 2009, Patiala, India, pp. 6–12.

Shukla, A., Tiwari, R., and Rungta, S. 2009e. A New Heuristic Channel Assignment in Cellular Networks, *IEEE World Congress on Computer Science and Information Engineering (CSIE)*, ieeexplore, March 31–April 2, Los Angeles/Anaheim, USA, pp. 473–478.

Shukla, A. et al. 2009f. Face Recognition Based on PCA, R-LDA and Supervised Neural Networks, accepted for *International Journal of Engineering Research and Applications (IJERIA).*

Shukla, A. et al. 2009g. Intelligent Biometric System Using PCA and R-LDA, *IEEE Global Congress on Intelligent Systems (GCIS)*, ieeexplore, Xiamen, China.

Shukla, A. et al. 2009h. Face Recognition Using Morphological Method, *IEEE International Advance Computing Conference (IACC),* Patiala, India.

Shukla, A., Tiwari, R., Janghel, R.R., and Kaur, P. 2009i. Diagnosis of Thyroid Disorders Using Artificial Neural Networks, *IEEE International Advanced Computing Conference,* Patiala, India, pp. 1016–1020.

Shukla, A. and Tiwari, R. 2009j. Book Chapter on Intelligent Biometric System Using Soft Computing Tools, *Breakthrough Discoveries in Information Technology Research: Advancing Trends (Advances in Information Technology Research series),* IGI Global.

Shukla, A. and Tiwari, R. 2009k. Book Chapter on "Robot Navigation with Speech Commands." accepted in the *INTECH Publications book on "Advanced Technologies"*. ISBN 978-953-7619-X-X.

Shukla, A., Tiwari, R., Janghel, R.R., and Tiwari, P. 2009l. Clinical Decision Support System for Fetal Delivery Using Artificial Neural Network, *Proceedings of the NISS-2009,* Beijing, China.

Shukla, A. et al. 2009m. An Efficient Mode Selection Algorithm for H.264 Encoder for Application in Low Computational Power Devices, *International Conference on Digital Image Processing (ICDIP 2009).*

Shukla, A., Tiwari, R., and Kaur, P. 2009n. Intelligent System for the Diagnosis of Epilepsy, *IEEE World Congress on Computer Science and Information Engineering (CSIE),* ieeexplore, Los Angeles/Anaheim, USA. IEEE DOI 10.1109/CSIE.2009.652, pp. 755–758.

Shukla, A. and Kala, R. 2008a. Multi Neuron Heuristic Search, *International Journal of Computer Science and Network Security*, vol. 8, no. 6, pp. 344–350.

Shukla, A. and Kala, R. 2008b. Predictive Sort, *International Journal of Computer Science and Network Security*, vol. 8, no. 6, pp. 314–320.

Shukla, A. and Tiwari, R. 2008c. A Novel Approach of Speaker Authentication by Fusion of Speech and Image Features using ANN, *International Journal of Information and Communication Technology (IJICT).* Inderscience Publishers, vol. 1, no. 2, pp. 159–170.

Shukla, A., Tiwari, R., and Kala, R. 2008d. Mobile Robot Navigation Control in Moving Obstacle Environment using A* Algorithm, Intelligent Systems Engineering Systems Through Artificial Neural Networks, ASME Publications, vol. 18, pp. 113–120; Also: *Proceedings of the International conference on Artificial Neural Networks in Engineering (ANNIE)*, ASME Publications.

Shukla, A., Tiwari, R., Meena, H.K., and Kala, R. 2008e. Speaker Identification Using Wavelet Analysis and Artificial Neural Networks, *Proceedings of the National Symposium on Acoustics (NSA).*

Shukla, A., Tiwari, R. 2008f. Intelligent Biometric System: A Case Study. *International Journal of Information Technology and Research (JITR)*, vol. 1, no. 3, pp. 41–56.

Shukla, A., Tiwati, R., Goyal, M., and Gutpa, S. 2008g. Speaker Independent Sentence Reconstruction Using Artificial Neural Networks, *Proc. National Symposium on Acoustics,* ASI, p. 119.

Shukla, A. and Tiwari, R. 2008h. Intelligent Biometric System: A Case Study, *International Journal of Information Technology and Research (JITR),* vol. 1, no. 3, pp. 41–56.

Shukla, A. and Tiwari, R. 2006. Speaker Authentication Using Speech Synthesis and Analysis by Artificial Neural Network, *Proceedings of FRSM-2006, Frontiers of Research on Speech and Music,* Lucknow, 9–10th January, pp. 126–131.

Shukla, A. and Tiwari R. 2008i. Speech Synthesis, Analysis and Its Application for Automatic Gender Identification of Speakers Using Artificial Neural Network, *Research Journal by CSVTU (Chhattisgarh Swami Vivekanand Technical University),* Bhilai, 2008.

Shukla, A. and Tiwari, R. 2008j. Diabetes Diagnosis Through the Means of Radial Basis Function Network, *National Confernce IST_ID 2008, RCET BHILAI (C.G).*

Shukla, A. and Tiwari, R. 2008k. Artificial Neural Networks: An AI Tool for Biometrics, *National Conference IST_ID 2008,* RCET Bhilai (C.G).

Shukla, A. and Sharma, P. 2008l. Hand Geometry Based Biometric Recognititon System, *Proceedings of the 2008 IEEE Conference TENCON 2008,* Hyderabad, India.

Shukla, A. and Tiwari, R. 2008m. Speech Synthesis, Analysis and its Application for Automatic Gender Identification of Speakers Using Artificial Neural Network, *Research Journal by CSVTU (Chhattisgarh Swami Vivekanand Technical University),* Bhilai, India.

Shukla, A. and Tiwari, R. 2008n. Diabetes Diagnosis Through the Means of Radial Basis Function Network, *National Conference IST_ID, RCET* Bhilai (C.G), India.

Shukla, A. and Tiwari, R. 2008o. Artificial Neural Networks: An AI Tool for Biometrics, *National Conference, National Confernce IST_ID, RCET* Bhilai (C.G), India.

Shukla, A. and Tiwari, R. 2007a. Fusion of Face and Speech Features with Artificial Neural Network for Speaker Authentication, *IETE Technical Review,* vol. 24, no. 5, pp. 359–368.

Shukla, A. and Tiwari, R. 2007b. Recent Trends in Nano Technology and Its Application in Biomedical Engineering, *National Conference on Nanotechnology*, RCET Bhilai (C.G).

Shukla, A. and Tiwari, R. 2007c. An Improved Method of Long-Term Fatal Heart Sound Monitoring in High-Risk Pregnancies, *Journal of Electronics and Telecommunication Engineers (IETE).*

Shuzhi, S.G. and Frank, L.L. 2006. *Autonomous Mobile Robots Sensing, Control, Decision Making and Applications,* CRC Press.

Siarry, P. and Guely, F. 1998. A Genetic Algorithm for Optimizing Takagi-Sugeno Fuzzy Rule Bases. *Fuzzy Sets and Systems,* vol. 99, no. 1, pp. 37–47.

Siddall, James N., 1990. *Expert System for Engineers.* New York, NY: Marcel Dekker Inc. Signal Processing Toolbox™ 6, User's Guide, Matlab.

Simon, G. and Horst, B. 2005. Off-Line Cursive Handwriting Recognition: On the Influence of Training Set and Vocabulary Size in Multiple Classifier Systems, *Elsevier*, vol. 43, no. 3–5, pp. 437–454.

Sinha, D. and Dougherty, E. 1975. Fuzzification of Set Inclusion: Theory and Applications, *Fuzzy Sets and Systems,* vol. 55, pp. 15–42.

Sipser, M. 1997. *Introduction to the Theory of Computation.* Boston: PWS Publishing, pp. 209–210.

Sirohey, S., Rosenfeld, A., and Duric, Z. 1999. *Eye Tracking. Technical Report CAR-TR-922,* Center for Automation Research, University of Maryland, College Park.

Slotine, J.J.E. and Yang, H.S. 1989. Improving the Efficiency of Time-Optimal Path-Following Algorithms, *IEEE Trans. Robot. Automat.,* vol. 5, pp. 118–124.

Smets, P. and Magrez, P. 1987. Implication in Fuzzy Logic, *Int. J. Approx. Reasoning*, vol. 1, no. 4, pp. 327–347.
Smola, J. 1996. *Regression Estimation with Support Vector Learning Machines.* Master's Thesis, Technische Universitt Mnchen.
Snelick, R., Indovina, M., Yen, J., and Alan M. 2003. Multimodal Biometrics: Issues in Design and Testing, *ICMI'03,* Canada, pp. 68–72.
Sodiya, A.S., Onashoga, S.A., and Oladunjoye, B.A. 2007. Threat Modeling Using Fuzzy Logic Paradigm, *Issues in Informing Science and Information Technology,* vol. 4.
Sodiya, A.S, Onashoga, S.A, and Ajayi, O.B. 2006. Toward Building Secure Software Products, *Journal of Issues in Informing Science and Information Technology,* vol. 3, pp. 635–646. Available at http://informingscience.org/proceedings/InSITE2006/IISITSodi143.pdf.
Som, T. and Saha, S. Handwritten Character Recognition by Using Neural-Network and Euclidean Distance Metric, Social Science Research Network.
Soong, F.K., Rosenberg, A., Rabiner, L., and Juang, B. 1985. A Vector Quantization Approach to Speaker Recognition, *Proc. ICASSP,* pp. 387–390.
Sorin, D. 1997. A Neural Network Based Artificial Vision System for licence Plate Recognition, *International Journal of Network Security, International Journal of Neural Systems,* vol. 8, no. 1.
Soutar, C., Roberge, D., Stoianov, A.R., Gilroy, and Vijaya Kumar, B.V.K. 1999. Biometrics Encryption. *ICSA Guide to Cryptography,* R.K. Nichols, ed. New York: McGraw-Hill, pp. 649–675.
Spillane, R. 1975. Keyboard Apparatus for Personal Identification, *IBM Tech. Disclosure Bulletin,* vol. 17, no. 3346.
Sporns, O. and Tononi, G. 2002. Classes of Network Connectivity and Dynamics, *Complexity,* vol. 7, pp. 28–38.
Spyridonos, P., Cavouras, D., Ravazoula, P., and Nikiforidis, G. 2002. Neural Network Based Segmentation and Classification System for the Automatic Grading of Histological Sections of Urinary Bladder Carcinoma, *Anal. Quantit. Cytol. Histol.,* vol. 24, no. 6, pp. 317–324.
Spyridonos, P., Glotsos, D., Papageorgiou, E.I., Stylios, C.D., Ravazoula, P., Groumpos, P.P., and Nikiforidis, G.N. 2005. Fuzzy Cognitive Map-Based Methodology for Grading Brain Tumors, Proc. 3rd European Medical and Biological Engineering Conference, EMBEC'05.
Spyridonos, P., Papageorgiou, E.I., Groumpos, P.P., and Nikiforidis, G. 2006. Integration of Expert Systems with Image Analysis Techniques for Medical Diagnosis, *Proc ICIAR, Lecture Notes in Computer Science,* vol. 4142, pp. 110–121.
Srinivas, M. and Patnaik, L. 1994. Adaptive Probabilities of Crossover and Mutation in Genetic Algorithms, *IEEE Transactions on Systems, Manand Cybernatics,* vol. 24, no. 4, pp. 656–667.
Srinivas, N. and Deb, K. 1994. Multiobjective Function Optimization Using Nondominated Sorting Genetic Algorithms, *Evol. Comput.,* vol. 2, no. 3, pp. 221–248.
Srinivasan, D., Chang, C.S., and Liew, A.C. 1994. Multiobjective Generation Schedule Using Fuzzy Optimal Search Technique, *Proc. Inst. Elect. Eng. Gen. Trans. Distrib.,* vol. 141, no. 3, pp. 231–241.
Stamp, M. 2006. *Information Security: Principles and Practice.* John Wiley and Sons.
Stanley, K.O. and Miikkulainen, R. 2002. *Evolving Neural Nets Through Augmenting Topologies, Evolutionary Computation.* MIT Press, vol. 10, no. 2, pp. 99–127.
Steinbach, M., Karypis, G., and Kumar, V. 2000. A Comparison of Document Clustering Techniques, *Proc. Workshop Text Mining, 6th ACMSIGKDD Int. Conf. Knowl. Discov. Data Mining,* pp. 20–23.
Stemtt, R. and Bustard, D.W., Fusing Hard and Soft Computing for Fault Management in Telecommunications Systems, *IEEE Trans. Systems Man and Cybernetics part C*, vol. 32, no. 2.
Reinberg S. 2008. Thyroid Problems Boost Glaucoma Risk, *Health Day Reporter*. Available via http://www.medicinenet.com/script/main/art.asp?articlekey=93453.
Stone, P. and Veloso, M. 2000. Multiagent Systems: A Survey from a Machine Learning Perspective, *Auton. Robots,* vol. 8, no. 3, pp. 345–383.
Stone, P., Sutton, R.S., and Kuhlmann, G. 2005. Reinforcement Learning for RoboCup-Soccer Keepaway, *Adapt. Behav.,* vol. 13, no. 3, pp. 165–188.
Storer, J.A. 1992. *Image and Text Compression.* Norwell, MA: Kluwer Academic Publisher.
Stork, D.G. and Yom-Tov, E. 2002.*Classification Toolbox for Use with MATLAB: User's Guide in Computer Manual in MATLAB to Accompany Pattern Classification.* New York: Wiley.
Storn, R. and Price, K. 1995. Differential Evolution—A Simple and Efficient Adaptive Scheme for Global Optimization over Continuous Spaces, *Int. Comput. Sci. Inst.,* Berkeley, CA, Tech. Rep. TR-95-012.
Stylios, Chrysostomos D. and Groumpos, Peter P. 2000. Fuzzy Cognitive Maps in Modeling Supervisory Control Systems, *Journal of Intelligent and Fuzzy Systems: Applications in Engineering and Technology,* vol. 8, no. 2, pp. 83–98.

Su, M.C. and Chou C.H. 2001. A Modified Version of the K-Means Algorithm with a Distance Based On Cluster Symmetry, *IEEE Transactions Pattern Analysis and Machine Intelligence,* vol. 23, no. 6, pp. 674–680.

Su, M.-C., Chou, C.-H., Lai, E., and Lee, J. 2006. A New Approach to Fuzzy Classifier Systems and Its Application in Self-Generating Neuro-Fuzzy Systems, *Neurocomputing,* vol. 69, pp. 586–614.

Suckling, J., Sigmundsson, T., Greenwood, K., and Bullmore, E. 1999. A Modified Fuzzy Clustering Algorithm for Operator Independent Brain Tissue Classification of Dual Echo MR Images, *Magnetic Resonance Imaging,* vol. 17, pp. 1065–1076.

Sudharsanan, S.I. and Sundareshan, M.K. 1991. Exponential Stability and a Systematic Synthesis of a Neural Network for Quadratic Minimization, *Neural Networks,* vol. 4, pp. 599–613.

Sugeno, M. and Nishida, M. 1985. Fuzzy Control of Model Car, *Fuzzy Sets Syst.,* vol. 16, pp. 103–113.

Sugeno, M. and Yasukawa, T. 1993. A Fuzzy-Logic-Based Approach to Qualitative Modeling, *IEEE Trans. Fuzzy Syst.,* vol. 1, pp. 7–31.

Sugeno, M. 1985. An Introduction Survey of Fuzzy Control, *Information Sciences*, vol 36, pp. 59–83.

Suhua, L., Xinmei, Y., and Xinyin, X. 2005. Unit Commitment Using Improved Discrete Particle Swarm Optimization Algorithm, *Proceedings of the CSEE,* vol. 25, no. 8, pp. 30–35.

Sun, B., Liu, W., and Zhong, Q. 2003. Hierarchical Speaker Identification Using Speaker Clustering, *Proc. Int. Conf. Natural Language Processing and Knowledge Engineering.* Beijing. China. pp. 299–304.

Sun, Hanwu, Ma, Bin, and Li, Haizhou. 2008. An Efficient Feature Selection Method for Speaker Recognition, International Symposium on Chinese Spoken Language Processing, China, pp. 1–4.

Sun, Y., Karray, F., and Al-Sharhan, S. 2002. Hybrid Soft Computing Techniques for Heterogeneous Data Classification, *Proceedings of the 2002 IEEE International Conference on Fuzzy Systems*, vol. 2, pp. 1511–1516.

Suresh, S., Omkar, S.N., and Mani, V. 2005. Parallel Implementation of Back-Propagation Algorithm in Networks of Workstations, *IEEE Transactions on Parallel and Distributed Systems,* vol. 16, no. 1, pp. 24–34.

Sutton, R. and Barto, A. 1998. *Reinforcement Learning: An Introduction.* Cambridge, MA: MIT Press.

Suykens, J.A.K., Brabanter, J.D., Lukas, L., and Vandewalle, J. 2002. Weighted Least Squares Support Vector Machines: Robustness and Sparse Approximation, *Neurocomputing,* vol. 48, no. 1–4, pp. 85–105.

Suzuki, M., Uchida, S., and Nomura, A. 2005. A Ground-Truthed Mathematical Character and Symbol Image Database, *Proc. Of ICDAR 2005 – 8th Int'l Conf. on Doc. Anal. And Recog.*, pp. 675–679.

Suzuki, Y., Ovaska, S., Furuhashi, T., Roy, R. and Dote, Y. eds. 2000. *Soft Computing in Industrial Applications.* London, UK: Springer-Verlag.

Syswerda, G. 1989. Uniform Crossover in Genetic Algorithms, *Proceedings of the 3rd International Conference on Genetic Algorithms,* pp. 2–9. Morgan Kaufmann.

Szczerbicki, Edward. 2002. Advances in Soft Modeling Techniques and Decision Support, *Special Issue on Soft Computing and Intelligent Systems for Industry,* vol. 33, no. 4, pp. 293–296.

Tachibana, K. and Furuhashi, T. 2001. A Structure Identification Method of Sub Models for Hierarchical Fuzzy Modeling Using the Multiple Objective Genetic Algorithm, *Int. J. Intell. Syst.,* vol. 17, pp. 495–513.

Takagi, H. and Hayashi, I. 1991. NN-Driven Fuzzy Reasoning, *Int. J. Approx. Reason.,* vol. 5, pp. 191–212.

Takagi, H. 1997. R&D in Intelligent Technologies: Fusion of NN, FS, GA, Chaos, and Human, *IEEE Int. Conf. Systems, Man, and Cybernetics*, Orlando, FL, half-day tutorial/workshop.

Takagi, H., Suzuki, N., Koda, T., and Kojima, Y. 1992. Neural Networks Designed on Approximate Reasoning Architecture and Their Applications, *IEEE Trans. Neural Netw.,* vol. 3, no. 5, pp. 752–760.

Takagi, T. and Sugeno, M. 1985. Fuzzy Identification of Systems and Its Applications to Modeling and Control, *IEEE Trans. Systems, Man Cybern.*, vol. SMC-15, pp. 116–132.

Takahashi, O. and Schilling, R.J. 1989. Motion Planning in a Plane Using Generalized Voronoi Diagrams, *IEEE Trans. on Robotics and Automation,* vol. 5, no. 2.

Takefuji, Y. 1992. *Neural Network Parallel Computing.* Norwell, MA: Kluwer.

Takenaka, Y., Lee, K.C., and Aiso, H. 1992. An ×work: A ×ing the st×ate of the syste×m in a s×ain, *Biol. Cybern.,* vol. 67, pp. 243–251.

Takiguchi, Okada, M. and Miyake, Y. 2005. A Fundamental Study of Output Translation from Layout Recognition and Semantic Understanding System for Mathematical Formulae, *Proc. Of ICDAR 2005 – 8th Int'l Conf. on Doc. Anal. and Recog.,* pp. 745–749.

Tan, P.N., Steinbach, M., and Kumar, V. 2005. *Introduction to Data Mining.* Reading, MA: Addison-Wesley.

Tan, Y. and Nanya, T. 1993. Fault-Tolerant Back-Propagation Model and Its Generalization Ability, *Proc. Int. Joint Conf. Neural Netw. (IJCNN'93),* Nagoya, Japan, pp. 2516–2519.

Tang, Y.-L. and Hung, C.-J. 2005. Recoverable Authentication of Wavelet-Transformed Images, *ICGST International Journal on Graphics, Vision and Image Processing,* vol. S11, pp. 61–66.

Taniguchi, S., Dote, Y., and Ovaska, S.J. 2000. Control of Intelligent Agent Systems (Robots) Using Extended Soft Computing, *Proc. IEEE Int. Conference on Systems, Man & Cybernetics,* pp. 3568–3572.

Tank, D.W. and Hopfield, J.J. 1986. Simple 'Neural' Optimization Networks: An A/D Converter, Signal Decision Circuit, and a Linear Programming Circuit, *IEEE Trans. Circuits Syst.,* vol. CS-33, pp. 533–541.

Tanomaru, J. and Omatu, S. 1992. Process Control by On-Line Trained Neural Controller, *IEEE Trans. Ind. Electron.,* vol. 39, pp. 511–521.

Tao, C.W. and Taur, J.S. 1999. Design of Fuzzy Controllers with Adaptive Rule Insertion, *IEEE Trans. System, Man, Cybern.,* vol. SMC-29, no. 3, pp. 389–397.

Taur, J.S. and Tao, C.W. 2000. A New Neuro-Fuzzy Classifier with Application to On-Line Face Detection and Recognition, *Journal of VLSI Signal Processing*, vol 26, pp. 397–409.

Teddy, D., Lai, E., and Quek, C. 2007. Hierarchically Clustered Adaptive Quantization CMAC and Its Learning Convergence, *IEEE Trans. Neural Netw.,* vol. 18, no. 6, pp. 1658–1682.

Teixeira, R. de A., Braga, A., Takahashi, R., and Saldanha, R. 2000. Improving Generalization of MLP with Multi-Objective Optimization, *Neurocomputing,* vol. 35, pp. 189–194.

Teng, W., Hsieh, M., and Chen, M. 2002. On the Mining of Substitution Rules for Statistically Dependent Items, *Proc. of ICDM.* pp. 442–449.

Thakur, J.S., Negi, P.C., Ahluwalia, S.K., and Vaidya, N.K. 1996. Epidemiological Survey of Rheumatic Heart Disease among School Children in the Shimla Hills of Northern India: Prevalence and Risk Factors, *J Epidemiol Community Health*, vol 50, no. 1, pp. 62–67.

Thakur, S., Sing, J.K., Basu, D.K., Nasipuri, M., and Kundu, M. 2008. Face Recognition Using Principal Component Analysis and RBF Neural Networks, *Proceedings of the First International Conference on Emerging Trends in Engineering and Technology,* ICETET '08. pp. 695–700, Nagpur, India.

Thorndike, E.L. 1932. *Fundamentals of Learning.* Columbia University Teacher College, New York.

Thorpe, J., van Oorschot, P.C., and Somayaji, A. 2005. Pass-Thoughts: Authenticating with Our Minds, *Proceedings of New Security Paradigns Workshop.* Lake Arrowhead: ACM Press, pp. 45–56.

Thrun, S. 1994. Extracting Rules from Artificial Neural Networks with Distributed Representation, *Proc. Adv. Neural Inf. Process. Syst. (NIPS),* pp. 505–512. Thyroid [internet] Wikipedia. Available from http://en.wikipedia.org/wiki/Thyroid

Tian, Y., Kanade, T., and Cohn, J.F. 2000. Recognizing Upper Face Action Units for Facial Expression Analysis, *Proceedings of CVPR'00.*

Tian, Y.L., Kanade, T., and Cohn, J. 2001. Recognizing Action Units for Facial Expression Analysis, *IEEE Trans. Pattern Anal. Mach. Intell.*, vol. 23, no. 2, pp. 97–115.

Tickle, B. and Andrews, R. 1998. The truth Will Come to Light: Directions and Challenges in Extracting the Knowledge Embedded within Trained Artificial Neural Networks, *IEEE Trans. Neural Networks*, vol. 9, pp. 1057–1068.

Ting, C.K. 2005. On the Convergence of Multi-Parent Genetic Algorithms, *Proceedings of the 2005 Congress on Evolutionary Computation,* vol. 1, pp. 396–403. *IEEE Press.*

Todd, P.M.A. 1997. Connectionist Approach to Algorithmic Composition. *Computer Music Journal*, vol. 13, no. 4, 1989.

Toh, K.A. and Mao, K.Z. 2002. A Global Transformation Approach to RBF Neural Network Learning, *ICPR02,* vol. 2, pp. 96–99. Toh, K.A., and Mao, K.Z. 2002. A global transformation approach to RBF neural network learning, *ICPR,* vol 2, pp. 96–99.

Tom, Ritchey. 2002. General Mophological Analysis, Swedish Morphological Society.

Tong, R.M., Beck, M.B., and Latten, A. 1980. Fuzzy Control of the Activated Sludge Wastewater Treatment Process, *Automatica,* vol. 16, no. 6, pp. 695–701.

Toorn, A. and Antanas, Z. 1989. *Global Optimization.* Berlin, Germany: Springer, vol. 350, Lecture Notes in Comput. Sci.

Torrence, C. and Compo, Gilbert P. 1998. A Practical Guide to Wavelet Analysis, *Bulletin of the American Meteorological Society,* vol. 79, pp. 61–78.

Torresen, J., Shin-ichiro, M., Nakashima, H., Tomita, S., and Landsverk, O. 1994. Parallel Back Propagation Training Algorithm for MIMD Computer with 2D-Torus Network, *The Third Parallel Computing Workshop (PCW94),* Kawasaki, Japan.

Trentin, E. and Gori, M. 2001. A Survey of Hybrid ANN/HMM Models for Automatic Speech Recognition, *Neurocomputing,* vol. 37, pp. 91–126.

Tripathi, N., Shukla, A., and Zadgaonkar, A. 2006. Under Mining Hidden Emotion from Speech Parameter, *Journal of Acoustic Society of India.*

Tripathi, N., Shukla, A., and Zadgaonkar, A. 2006. Emotional Status Recognition in Mobile Communication, *The Indian Journal of Telecommunication*, vol. 56, no. 1, pp. 66–70.

Tripathi, S.N. 1978. Clinical Diagnosis, Science and Philosophy of Indian medicine. Shree Beldyanath Ayurved Bhawan Ltd.

Tsai, W.H. and Fu, K.S. 1980. Attributed Grammar: A Tool for Combining Syntactic and Statistical Approaches to Pattern Recognition, *IEEE Trans. Systems, Man and Cybernetics,* vol. 10, no. 12, pp. 873–885.

Tsuji, T., Ito, K., and Morasso, P.G. 1996. Neural Network Learning of Robot Arm Impedance in Operational Space, IEEE Trans. on Systems, Man, and Cybernetics – Part B: Cybernetics, vol. 26, no. 2, pp. 290–298.

Tsutsui, S. and Jain, L.C. 1998. On the Effect on Multi-Parent Recombination in Real Coded Genetic Algorithms, *Proceedings of the 2nd International Conference on Knowledge-based Intelligent electronic systems,* pp. 155–160.

Tsutsui, S., Yamamura, M., and Higuchi, T. 1999. Multi-Parent Recombination with Simplex Crossover in Real Coded Genetic Algorithms, *Proc. of the Genetic and Evolutionary Computation Conference (GECCO-99),* pp. 657–664.

Tung, W.L. and Quek, C. 2002. GenSoFNN: A Generic Self-Organizing Fuzzy Neural Network, *IEEE Trans. Neural Netw.,* vol. 13, no. 5, pp. 1075–1086.

Tunstel, E.W., Jr. 1996. Mobile Robot Autonomy via Hierarchical Fuzzy Behavior Control, *Proceedings of the 6th International Symposium on Robotics and Manufacturing,* Montpellier, France, pp. 837–842.

Turing, M. 1950. Computing Machinery and Intelligence, *Mind,* vol. 59, no. 236, pp. 433–460.

Turk, M. and Pentland, A. 1991. Eigenfaces for Recognition, *Cognitive Neuroscience,* vol. 3, no. 1, pp. 71–86.

Turk, M.A. and Pentland, A.P. 1991. Face Recognition Using Eigenfaces, *IEEE Conf. on Computer Vision and Pattern Recognition,* pp. 586–591.

Twaakyondo, H.M. and Okamoto, M. 1995. Structure Analysis and Recognition of Mathematical Expression, *Proc. of ICDAR 1995 – 3rd Int'l Conf. on Doc. Anal. and Recog.,* pp. 430–437.

Tzanetakis, George, Essl, Georg, and Cook, Perry 2001. Audio Analysis Using the Discrete Wavelet Transform, *Proc. WSES Int. Conf. Acoustics and Music: Theory and Applications (AMTA 2001),* Skiathos, Greece.

Uludag, U., Pankanti, S., Prabhakar, S., and Jain, A.K. 2004. Biometric Cryptosystems: Issues and Challenges. *Proceedings of the IEEE,* vol. 92, No. 6, pp. 948–960.

Umbaugh, S.E. and Moss, R.H. 1993. Automatic Color Segmentation Algorithms, with Application to Skin Tumor Feature Identification. *Engineering in Medicine and Biology Magazine, IEEE,* vol. 12, no. 3, pp. 75–82.

Upstill, T., Craswell, N., and Hawking, D. 2003. Query Independent Evidence in Home Page Finding. *ACM Transactions on Information Systems (TOIS),* pp. 286–313.

Vaidya, B. et al. 2007. Radioiodine Treatment for Benign Thyroid Disorders: Results of a Nationwide Survey of UK Endocrinologists, *The Clinical Journal of the Society for Endocrinology,* vol. 68, no. 5, pp. 814–820. Vaidya, B., and Williams, G.R., et.al. 2008. Radioiodine treatment for benign thyroid disorders: results of a nationwide survey of UK endocrinologists. *Journal of Clinical Endocrinology,* vol. 68, no. 5, pp. 814–820.

Valova, I., Szer, D., Gueorguieva, N., and Buer, A. 2005. A Parallel Growing Architecture for Self-Organizing Maps with Unsupervised Learning, *Neurocomputing,* vol. 68, pp. 177–195.

Van den Bout, A.E. and Miller, T.K. 1989. Improving the Performance of the Hopfield-Tank Neural Network through Normalization and Annealing, *Biol. Cybern.,* vol. 62, pp. 129–139.

VanDoren, V. ed., 2003. *Techniques for Adaptive Control.* Elsevier, Burlington, MA.

Vapnik, V. 1995. *The Nature of Statistical Learning Theory.* Springer-Verlag, New York.

Vapnik, V. 1998. *Statistical Learning Theory.* Wiley, New York.

Varady, P. 2001. Wavelet-Based Adaptive De-Noising of Phonocardiographic Records, *Proceedings of IEEE-23rd Annual EMBS International Conference*, Istanbul, Turkey, vol. 2, pp. 1846–1849.

Varma, V.K. 1993. Testing Speech Coders for Usage in Wireless Communications Systems, *Proc. IEEE Workshop on Speech Coding for Telecommunications,* QC, Canada, pp. 93–94.

Veera Ragavan, S. and Ganapathy, V. 2007. A Unified Framework for a Robust Conflict-Free Robot Navigation, *Proceedings of World Academy of Science, Engineering and Technology,* vol. 21, ISSN 1307-6884.

Veldhuizen, D.A.V. and Lamont, G.B. 2000. Multiobjective Evolutionary Algorithms: Analyzing the State-of-the-Art, *Evol. Comput.,* vol. 8, no. 2, pp. 125–147.

Vieira, Armando and Barradas, Nuno. 2001. A Training Algorithm for Classification of High-Dimensional Data, Neurocomputing 50 (2003), pp. 461–472.

Vielhauer, C. and Steinmetz, R. 2004. Handwriting: Feature Correlation Analysis for Biometric Hashes, *EURASIP J. Appl. Signal Process.,* vol. 4, pp. 542–558.

Villasenor, J.D. 1993. Alternatives to the Discrete Cosine Transform for Irreversible Tomographic Image Compression, *IEEE Trans. Med. Imag.*, vol. 12, pp. 803–811.

Voigt, H.M., Muhlenbein, H., and Cvetkovicc, D. 1995. Fuzzy Recombination for the Breeder Genetic Algorithm, in *Proc. 6th Int. Conf. Genetic Algorithms,* L. Eshelman, ed. San Mateo, CA: Morgan Kaufmann, pp. 104–111.

Von Neumann, J. 1958. *The Computer and the Brain.* New Haven, CT: Yale Univ. Press, republished in 2000, p. 82.

Von Neumann, J. 1966. *Theory of Self-Reproducing Automata.* Fifth Lecture edited and completed by A.W. Burks, Urbana: University of Illinois Press, p. 77.

Voss, A. , Baumert, M., Baier, V., Stepan, H., Walther, T., and Faber, R. 2003. Analysis of Interactions Between Heart Rate and Blood Pressure in Chronic Hypertensive Pregnancy, *Engineering in Medicine and Biology, Ann.Conf. of Biomedical Engineering Society,* vol. 2, pp. 1615–1616.

Wald, L., Ranchin, T., and Mangolini, M. 1997. Fusion of Satellite Images of Different Spatial Resolutions: Assessing the Quality of Resulting Images, *Photogramm. Eng. Remote Sens.,* vol. 63, no. 6, pp. 691–699.

Walter, J.A. and Schulten, K. 1993. Implementation of Self-Organizing Neural Networks for Visuo-Motor Control of an Industrial Robot, *IEEE Transactions on Neural Networks,* vol. 4, no. 1, pp. 86–95.

Wan, S. and Banta, L. 2006. Parameter Incremental Learning Algorithm for Neural Networks, *IEEE Trans. Neural Netw.,* vol. 17, no. 6, pp. 1424–1438.

Wang, C., Zhang, X., Liu, Y., and Vasana, S. 2006. Adaptive Noise Canceller for Cardiac Sounds of Fetus, *The IASTED Conference on Applied Simulation and Modelling,* Rhodes, Greece, pp. 275–282.

Wang, H., Kwong, S., Jin, Y., Wei W., and Man, K. 2005. Multi-Objective Hierarchical Genetic Algorithm for interpretable Fuzzy Rule Based Knowledge Extraction, *Fuzzy Sets Syst.,* vol. 149, no. 1, pp. 149–186.

Wang, H., Wang, M., Hintz, T., Wu, Q., and He, X. 2005. VSA Based Fractal Image Compression, *Journal of WSCG,* vol. 13, Nos. 1–3, pp. 89–96.

Wang, J. 1996. Recurrent Neural Networks for Optimization, *Fuzzy Logic and Neural Network Handbook,* C.H. Chen, ed. New York: McGraw-Hill, pp. 4.1–4.35.

Wang, J.-S. and George Lee, C.S. 2002. Self-Adaptive Neuro-Fuzzy Inference Systems for Classification Applications, *IEEE Transactions on Fuzzy Systems,* vol. 10, no. 6, p. 790.

Wang, L. and Langari, R. 1992. Building Sugeno-Type Models Using Fuzzy Discretization and Orthogonal Parameter Estimation Techniques, *IEEE Trans. Fuzzy Syst.,* vol. 3, pp. 724–740.

Wang, L.X. 1992. Fuzzy Systems Are Universal Approximators, *Proc. IEEE Int. Conf. Fuzzy Systems,* pp. 1163–1169.

Wang, L.X. and Mendel, J.M. 1992. Generating Fuzzy Rules by Learning from Examples, *IEEE Transactions on Systems, Man, and Cybernetics,* vol. 22, no. 6, pp. 1414–1427.

Wang, X.Z. and Hong, J.R. 1999. Learning Optimization in Simplifying Fuzzy Rules, *Fuzzy Sets and Systems,* vol. 106, pp. 349–356.

Wang, Y., Tan, T., and Jain, A.K. 2003. Combining Face and Iris Biometrics for Identity Verification, *Proceedings of 4th International Conference on AVBPA,* Guildford, UK, pp. 805–813.

Wassermann, Philip P. 1989. *Neural Computing: Theory and Practice.* New York: VNR.

Wassner, H. and Chollet, G. New Sepstral Representation Using Wavelet Analysis and Spectral Transformation for Robust Speech Recognition.

Waterman, D.A., Paul, J., and Peterson, M. 1986. Expert Systems for Legal Decision Making, *Expert Systems,* vol. 3, pp. 212–226.

Waterman, D.A. and Peterson, M. 1980. Rule-Based Models of Legal Expertise, *Proceedings of the First National Conference on Artificial Intelligence*, Stanford University: AAAI: 272-275.

Waterman, Donald, A. 1985. *A Guide to Expert Systems.* Reading, MA: Addison-Wesley.

Wei, Li, Chenyu, Maa and Wahlb, F. M., 1997. Neuro-Fuzzy Systems for Intelligent Robot Navigation and Control Under Uncertainty, pp. 133–140.

Weisstein, Eric W. *Wavelet Transform.* From MathWorld: A Wolfram Web Resource. http://mathworld.wolfram.com/WaveletTransform.html.

Weitzenfeld, A. 2008. From schemas to neural networks: A multi-level modelling approach to biologically-inspired autonomous robotic systems. *Robotics and Autonomous Systems*, vol. 56, pp. 177–197.

Weizenbaum, J. 1966. ELIZA: A Computer Program for the Study of Natural Language Communication Between Man and Machine, *Communications of the Association for Computing Machinary (ACM),* vol. 9, pp. 36–45.

Weng, Y.C., Chang, N.B., and Lee, T.Y. 2008. Nonlinear Time Series Analysis of Ground-Level Ozone Dynamics in Southern Taiwan, *Journal of Environmental Management,* vol. 87, pp. 405–414.

Whalen, T. 2003. Parameterized R-Implications, *Fuzzy Sets and Systems,* vol. 134, no. 2, pp. 231–281.

Whitley, D., Beveridge, R., Graves, C., and Mathias, K. 1995. Test Driving Three 1995 Genetic Algorithms: New Test Functions and Geometric Matching, *J. Heuristics,* vol. 1, pp. 77–104.

Whitley, D., Rana, S., Dzubera J., and Mathias, E. 1996. Evaluating Evolutionary Algorithms, *Artif. Intell.,* vol. 85, pp. 245–276.

WHO Epilepsy [internet]. Available from http://www.who.int/mediacentre/factsheets/fs166/en/

Widrow, A. and Lehr, M.A. 1990. 30 Years of Adaptive Neural Networks: Perceptron, Madaline and Backpropagation. *Proc. IEEE,* vol. 78, no. 9, pp. 1415–1441. Widrow B. and Lehr M.A. 1990. 30 Years of Adaptive Neural Networks: Perceptron, Madeline and Back propagation. *Proc. IEEE,* vol. 78, no. 9, 1415–1441.

Widrow, B. and Stearns, S.D. 1985. *Adaptive Signal Processing.* Englewood Cliffs, NJ: Prentice-Hall.

Wierzbicki, A.P. 1980. The Use of Reference Objectives in Multiobjective Optimization. In G. Fandel and T. Gal, eds., *Multiple Criteria Decision Making Theory and Applications.* Berlin, Germany: Springer-Verlag, pp. 468–486.

Wildes, R., Asmuth, J., Green, G., Hsu, S., Kolczynski, R., Matey, J., and McBride, S. 1996. A Machine-Vision System for Iris Recognition, *Machine Vision and Applications,* vol. 9.

Wildes, Richard P. 1997. Iris Recognition: An Emerging Biometric Technology, *Proceedings of the IEEE,* vol. 85, no. 9, pp. 1348–1363.

Wilson, G.V. and Pawley, G.S. 1988. On the Stability of the Traveling Salesman Problem Algorithm of Hopfield and Tank, *Biol. Cybern.,* vol. 58, pp. 63–70.

Winer, B.J., Brown, D.R. and Michels, K.M. 1991. *Statistical Principles in Experimental Design,* 3rd ed. *McGraw-Hill Series in Psychology.*

Wong, Alexandra L.N. and Shi, P. 2002. Peg-Free Hand Geometry Recognition Using Hierarchical Geometry and Shape Matching, *IAPR Workshop on Machine Vision Applications,* Nara, Japan, pp. 281–284.

Wong, K.P., Feng, D., Meikle, S.R., and Fulham, M.J. 2001. Segmentation of Dynamic PET Images Using Cluster Analysis, *IEEE Trans. Nucl. Sci.* (*in press*).

Woodward, J.D., Orlans N.M., and Higgins P.T. 2002. *Biometrics.* McGraw- Hill/Osborne Media, ISBN 0072222271.

World Top Ten countries by highest Death Rate from Heart Disease [internet]. Maps of world available via http://www.mapsofworld.com/world-top-ten/countries-by-highest-death-rate-from-heart-disease.html.

Wright, S. 1934. The Roles of Mutation, Inbreeding, Crossbreeding and Selection in Evolution, *Proc. 6th Int. Congr. Genetics,* vol. 1, pp. 356–366.

WSEC (Western Sydney Endocrine Centre) [internet]. 2004. Available via http://designchip.net/endocrine/index.php?option=com_ content task=view&id=6&Itemid=29.

Wu, Jian-Da, Hsu, Chuang-Chin and Chen, Hui-Chu 2009. An Expert System of Price Forecasting for Used Cars Using Adaptive Neuro-Fuzzy Inference, Expert Systems with Applications, vol. 36, issue 4, pp. 7809–7817.

Wu, J. and Zhou, Z. 2003. Efficient Face Candidates Selector for Face Detection, *The journal of the Pattern Recognition Society.*

Wu, X., Zhang, C., and Zhang, S. 2002. Mining Both Positive and Negative Association Rules, Proc. of ICML, pp. 658–665.

Xia, Y. and Wang, J. 1998. A General Methodology for Designing Globally Convergent Optimization Neural Networks, IEEE Trans. Neural Networks, vol. 9, no. 6, pp. 1331–1343.

Xiang, B. and Berger, T. 2003. Efficient Text-Independent Speaker Verification with Structural Gaussian Mixture Models and Neural Network, *IEEE Trans. Speech Audio Process,* vol. 11, no. 5, pp. 447–456.

Zeng, X., Zhang, Z., and Zhang, Z. 2005. Heuristic Skill of Computer Virus Analysis Based on Virtual Machine, *Computer Applications and Software,* vol. 22, no. 9, pp. 125–126.

Xiao, J., Michalewicz, Z., Zhang, L., and Trojanowski, K. 1997. Adaptive Evolutionary Planner/Navigator for Robots. *IEEE Trans. on Evolutionary Computation.*

Xie, X.L. and Beni, G. 1991. A Validity Measure for Fuzzy Clustering, *IEEE Transactions on Pattern Analysis and Machine Intelligence,* vol. 13, pp. 841–847.

Xin, Du, Hua-hua, Chen, and Wei-kang, Gu. 2004. Neural Network and Genetic Algorithm Based Global Path Planning in a static Environment, Journal of Zhejiang University SCIENCE.vol. 6, pp. 554–549.

Xiong, H., Wu, J., and Chen, J. 2006. K-Means Clustering Versus Validation Measures: A Data Distribution Perspective, *Proc. 12th ACM SIGKDD Int. Conf. Knowl. Discov. Data Mining,* pp. 779–784.

Xu, Z.B., Hu, G.Q., and Kwong, C.P. 1996. Asymmetric Hopfield-Type Networks: Theory and Applications, *Neural Networks,* vol. 9, pp. 483–501.

Xudong, Jiang, Mandal, B., and Kot, A. 2008. Eigenfeature Regularization and Extraction in Face Recognition, IEEE Transactions on Pattern Analysis and Machine Intelligence, vol. 30, no. 3, pp. 383–394. Jiang M.Y. and Chen Z.J. 1997. Diabetes expert system, *ICIPS*; vol. 2, 1076–1077.

Yahja, G.A., Singh, S., and Stentz, A. 2000. An Efficient On-Line Path Planner for Outdoor Mobile Robots. *Robotics and Autonomous Systems,* vol. 32, Elsevier Science, pp. 129–143.

Yamamoto, Y. and Nikiforuk, P.N. 2000. A New Supervised Learning Algorithm for Multilayered and Interconnected Neural Networks, *IEEE Trans. Neural Networks,* pp. 36–46.

Yan, B. 2005. Empirical Evaluation of Ear Biometric, *The Biometric Consortium Conference,* Arlington USA.

Yan, Ping and Bowyer K.W. 2007. Biometric Recognition Using 3D Ear Shape, IEEE Trans. On Pattern Analysis and Machine Intelligence, vol. 29, no. 8, pp. 1297–1308.

Yang, J. and Honavar, V. 1998. Feature Subset Selection Using a Genetic Algorithm, IEEE Intelligent Systems, vol. 13, no. 2, pp. 44–49.

Yang, J. and Yang, J.Y. 2001. Optimal FLD Algorithm for Facial Feature Extraction, *SPIE Processing of the Intelligent Robots and Computer Vision: Algorithms, Techniques, and Active Vision,* vol. 4572, pp. 438–444.

Yang, L., Yen, J., Rajesh, A., and Kihm, K.D. 1999. A Supervisory Architecture and Hybrid GA for the Identifications of Complex Systems, *Proc. Cong. Evolutionary Computation,* Washington, DC, pp. 862–869.

Yao, X. and Liu, Y. 1997. Fast Evolutionary Strategies, *Evolutionary Programming VI,* P.J. Angeline, R.G. Reynolds, J.R. McDonnell, and R. Eberhart, eds. Berlin, Germany: Springer, pp. 151–161.

Yao, X. 1999. Evolving Artificial Neural Networks, *Proc. IEEE,* vol. 87, no. 9, Sep, pp. 1423–1447.

Yao, X., Liu, Y., and Lin, G. 1999. Evolutionary Programming Made Faster, *IEEE Transactions on Evolutionary Computation,* vol. 3, no. 2, pp. 82–102.

Yasunaga, M., Yoshida, E., and Yoshihara, I. 1999. Parallel Back-Propagation Using Genetic Algorithm: Real-Time BP Learning on the Massively Parallel Computer CP-PACS, *Proc. the IEEE and INNS Intl. Joint Conf. on Neural Networks.*

Yen, J., Liao, J.C., Lee, B., and Randolph, D. 1998. A Hybrid Approach to Modeling Metabolic Systems Using a Genetic Algorithm and Simplex Method, *IEEE Transactions on Systems, Man, and Cybernetics – PART B: Cybernetics,* vol. 28, no. 2, p. 173.

Yokoyama, R., Bae, S.H., Morita, T., and Sasaki, H. 1988. Multiobjective Generation Dispatch Based on Probability Security Criteria, *IEEE Trans. Power Syst.,* vol. 3, no. 1, pp. 317–324.

Yoshino, K., Watanabe, Y., and Kakeshita, T. 1994. Hopfield Neural Network Using Oscillatory Units with Sigmoidal Input-Average Out Characteristics, *IEICE Trans.,* vol. J77-DII, pp. 219–227.

Yu, Wen and Li, Xiaoou. 2004. Fuzzy Identification Using Fuzzy Neural with Stable Learning Algorithms. IEEE Transactions on Fuzzy System, vol. 12, no. 3, pp. 411–420.

Yu, Hua and Yang, Jie. 2001. A Direct LDA Algorithm for High Dimensional Data with Application to Face Recognition, Pattern Recognition, vol. 34, no. 11, pp. 2067–2070.

Yuan, L., Chun Mu, Z., Zhang, Y., and Liu, K. 2006. Ear Recognition Using Improved Non-Negative Matrix Factorization, *The 18th International Conference on Pattern Recognition,* Hong Kong, August.

Yuan, Y. and Suarga, S. 1995. On the Integration of Neural Networks and Fuzzy Logic Systems, *IEEE International Conference on Systems, Man and Cybernetics,* pp. 452–457.

Yunhe, H. et al. 2004. Enhanced Particle Swarm Optimization Algorithm and Its Application on Economic Dispatch of Power Systems, *Proceedings of the CSEE,* vol. 24, No. 7, pp. 95–100.

Z. Selim, Skokri and Ismail, M.A. 1984. Soft Clustering of Multidimensional Data: A Semi-Fuzzy Approach, *Pattern Recognition,* vol. 17, No. 5, pp 559–568. Printed in Great Britain.

Zadeh, L.A. 1965, Fuzzy Sets, *Inform. Contr.*, vol. 8, pp. 338–353.

Zadeh, L.A. 1991. A Definition of Soft Computing in Greater Detail. Available at http://http.cs.berkeley.edu/~mazlack/BISC/BISC-DBM-soft.html.

Zadeh, L.A. 1992. *Calculus of Fuzzy Restrictions, Fuzzy Sets and Their Application to Conitive and Decision Processes 683-697.* Academic Press, pp. 1–39.

Zadeh, L.A. 1994. Soft Computing and Fuzzy Logic, *IEEE Softw.*, vol. 11, no. 6, pp. 48–56.

Zadeh, L.A. 1994. Fuzzy Logic and Soft Computing: Issues, Contentions and Perspectives, *Proc. IIZUKA,* pp. 1–2.

Zaheeruddin, G. A Neuro-Fuzzy Approach for Prediction of Human Work Efficiency in Noisy Environment, *Applied Soft Computing*, vol. 6, no. 3, March 2006, pp. 283–294.

Zanuy, M.F. 2005. Signature Recognition State of the Art, *IEEE A&E System Magazine,* pp. 28–32.

Zebdum, R.S., Guedes, K., Vellasco, M., and Pacheco, M.A. 1995. Short Term Load Forecasting Using Neural Nets, *Proceedings of the International Workshop on Artificial Neural Networks, LNCS No. 930.* Torremohos, Spain: Springer-Verlag.

Zeleznikow, J. and Bellucci, E. 2003. Family-Winner: Integrating Game Theory and Heuristics to Provide Negotiation Support, *Proceedings of Sixteenth International Conference on Legal Knowledge Based Systems*. Amsterdam, Netherlands: IOS Publications, pp. 21–30.

Zeng, X.C. and Martinez, T. 1999. A New Relaxation Procedure in the Hopfield Neural Networks for Solving Optimization Problems, *Neuron Processing Lett.,* vol. 10, pp. 211–222.

Zeng, Z., Wang, J., and Liao, X. 2003. Global Exponential Stability of a General Class of Recurrent Neural Networks with Time-Varying Delays, *IEEE Trans. Circuits Syst. I, Fundam. Theory Appl.,* vol. 50, no. 10, pp. 1353–1358.

Zengbo, A. and Yan, Z. 2007. The Application Study of Machine Learning, *Journal of Changzhi University,* vol. 24, no. 2, pp. 21–24.

Zhang, H., Basin, M., and Skliar, M. 2007. Itô–Volterra Optimal State Estimation with Continuous, Multirate, Randomly Sampled, and Delayed Measurements, *IEEE Trans. Autom. Control,* vol. 52, no. 3, pp. 401–416.

Zhang, J., Li, G., and Freeman, W.J. 2006. Application of Novel Chaotic Neural Networks to Mandarin Digital Speech Recognition, *Proc. Of International Joint Conference on Neural Networks 2006 (IJCNN'06),* Vancouver, BC, Canada, pp. 653–658.

Zhang, J.S. and Leung, Y.W. 2003. Robust Clustering by Pruning Outliers, *IEEE Trans. Syst., Man, Cybern. B, Cybern.,* vol. 33, no. 6, pp. 983–998.

Zhang, Q., Wei, X., and Xu, J. 2007. Delay-Dependent Global Stability Results for Delayed Hopfield Neural Networks, *Chaos, Solitons & Fractals,* vol. 34, pp. 662–668.

Zhang, X., Durand, L., Senhadji, L., Lee, H., and Coatrieux, J. 1998. Time-Frequency Scaling Transformation of the Phonocardiogram Based of the Matching Pursuit Method, *IEEE Transactions on Biomedical Engineering,* vol. 45, no. 8, pp. 972–979.

Zhang, Y. and Rockett, P.I. 2005. Evolving Optimal Feature Extraction Using Multi-Objective Genetic Programming: A Methodology and Preliminary Study on Edge Detection, *Proc. Genetic Evol. Comput. Conf.*, pp. 795–802.

Zhao, W., Chellappa, R., Rosenfeld, A., and Phillips, P.J. 2000. Face Recognition: A Literature Survey, *UMD CfAR Technical Report CAR-TR-948.*

Zhao, Y. and Karypis, G. 2004. Criterion Functions for Document Clustering: Experiments and Analysis, *Mach. Learn.,* vol. 55, no. 3, pp. 311–331.

Zhao, Y., Edwards, R.M., and Lee, K.Y. 1997. Hybrid Feedforward and Feedback Controller Design for Nuclear Steam Generators over Wide Range Operation Using Genetic Algorithm, *IEEE Trans. Energy Convers.*, vol. 12, no. 1, pp. 100–105.

Zhao, Z.Y., Tomizuka, M., and Isaka, S. 1993. Fuzzy Gain Scheduling of PID Controllers, *IEEE Trans. Syst., Man, Cybern.,* vol. 23, pp. 1392–1398.

Zheng, J. and Zhu .G. 2006. On-Line Handwriting Recognition Based on Wavelet Energy Feature Matching, *Proceedings of the 6th World Congress on Intelligent Control and Automation,* pp. 9885–9888.

Zhou, Z.H., Wu, J., and Tang, W. 2002. Ensembling Neural Networks: Many Could Be Better Than All, *Artif. Intell.,* vol. 137, nos. 1–2, pp. 239–263.

Zhu, Y., Tan, T., and Wang, Y. 2000. Biometric Personal Identification Based on Iris Patterns. *Proceedings of ICPR, International Conference on Pattern Recognition,* vol. II, pp. 805–808.

Zhu, Y., Tan, T., and Wang, Y. 1999. Biometric Personal Identification System Based on Iris Pattern, Chinese Patent Application. No. 9911025.6.

Zitzler, E. and Thiele, L. 1998. An Evolutionary Algorithm for Multiobjective Optimization: The Strength Pareto Approach, *TIK-Rep.,* vol. 43.

Zitzler, E. and Thiele, L. 1998. Multiobjective Optimization Using Evolutionary Algorithms—A Comparative Case Study, *Parallel Problem Solving from Nature.* Amsterdam, Netherlands: Springer-Verlag, pp. 292–301.

Ziv, J. and Lempel, A. 1978, Compression of Individual Sequence via Variable-Rate Coding, *IEEE Trans. On Information Theory,* vol. 24, no. 5, pp. 530–536.

Zoltan, Z., Kalmar, Z., and Szepesvari, C. 1998. Multi-Criteria Reinforcement Learning, *Proc. Int. Conf. Mach. Learn.,* pp. 197–205.

Zuckerwar, A.J., Pretlow, R.A., Stoughton, J.W., and Baker, D.A. 1993. Development of a Piezopolymer Pressure Sensor for Portable Fetal Rate Monitor, *IEEE Transactions on Biomedical Engineering,* vol. 40, no. 9, pp. 963–969.

Standard Data Sets Used

Quarterly S&P 500 index, 1900–1996. Source: Makridakis, Wheelwright, and Hyndman (1998). Available at http://robjhyndman.com/TSDL/data/9-17b.dat.

Box and Jenkins furnace data. Source: G.E.P. Box and G.M. Jenkins, Time Series Analysis, Forecasting and Control, San Francisco, Holden Day, 1970, pp. 532–533. Available at: http://neural.cs.nthu.edu.tw/jang/dataset/box-jenkins.dat.

RCSB protein data bank, available at http://www.rcsb.org/pdb.

Wolberg, William H., Mangasarian, Olvi L., and Aha, David W. 1992. UCI Machine Learning Repository [http://www.ics.uci.edu/~mlearn/MLRepository.html], University of Wisconsin Hospitals. Available at http://archive.ics.uci.edu/ml/datasets/Breast+Cancer+Wisconsin+(Original).

Coomans, D. and Aeberhard, S. 1992. UCI Machine Learning Repository [http://www.ics.uci.edu/~mlcarn/MLRepository.html], James Cook University, Available at http://archive.ics.uci.edu/ml/machine-learning-databases/thyroid-disease/new-thyroid.names.

Ilter, N. and Guvenir, H. Altay, 1998. UCI Machine Learning Repository [http://www.ics.uci.edu/~mlearn/MLRepository.html], Gazi University and Bilkent University. Available at http://archive.ics.uci.edu/ml/datasets/Dermatology.

Janosi, A., Steinbrunn, W., Pfisterer, M., and Detrano, R. 1988. UCI Machine Learning Repository [http://www.ics.uci.edu/~mlearn/MLRepository.html]. Available at http://archive.ics.uci.edu/ml/datasets/Heart%2BDisease.

Sigillito, V. 1990. UCI Machine Learning Repository [http://www.ics.uci.edu/~mlearn/MLRepository.html], The Johns Hopkins University. Available at http://archive.ics.uci.edu/datasets/Pima+Indians+Diabetes.

Fetal Delivery Database, Pt. jawahar lal Nehru Government Medical College, Raipur, Chhattisgarh, India.

Legal Threats Database, Citizen Media Law Project, Berkman Center for Internet and Society, Available at http://www.citmedialaw.org/database.

Spacek, L., Grimace Face database. Available at http://cswww.essex.ac.uk/mv/allfaces/grimace.html.

Registered Trademarks

MATLAB is the registered trademark of The MathWorks, Inc.
Microsoft, Vista, and Tablet PC are the registered trademarks of Microsoft Corporation
Adobe and Photoshop are the registered trademarks of Adobe Systems
GoldWave is the registered trademark of GoldWave Inc.
Java is the registered trademark of Sun Microsystems, Inc.

Index

[NOTE: Page numbers followed by *t* and *f* indicate tables and figures, respectively.]

A

B

C

D

E

F

G

H

I

Q

R

S

T

U

V

W

Z